AF179750

Bernhard A. Koch (Ed.)
Damage Caused by Genetically Modified Organisms

Tort and Insurance Law
TIL 27

Tort and Insurance Law

Vol. 27

Edited by the
European Centre of Tort
and Insurance Law

together with the

Institute for European Tort Law
of the Austrian Academy of Sciences

De Gruyter

Bernhard A. Koch (ed.)

Damage Caused by Genetically Modified Organisms

Comparative Survey of Redress Options for Harm to Persons, Property or the Environment

With Contributions by

Bjarte Askeland
Ewa Bagińska
Eugène Buttigieg
Eugenia Dacoronia
A. Bryan Endres
Thomas K. Epprecht
Matthias Erhardt
Michael Faure
Jörg Fedtke
Federico Fusco
Märten Geiger
Patrick Goergen
Ingrid Greveling
Jiří Hrádek
Thomas Kadner Graziano
Bernhard A. Koch
Irene Kull
Rok Lampe
David Langlet
Mark Lunney
Miquel Martín-Casals
Jane Matthews Glenn
Attila Menyhárd
Philip Mielnicki
Alberto Monti
Ana Paula Myszczuk
Ken Oliphant
Roseli Rocha dos Santos
Albert Ruda
Björn Sandvik
Simon Taylor
Vibe Ulfbeck
Willem H. van Boom
Clarissa Bueno Wandscheer
Manuela Weissenbacher
Andri Wibisana
Vanessa Wilcox

De Gruyter

European Centre of Tort and Insurance Law
Reichsratsstraße 17/2
A-1010 Vienna
Tel.: +43 1 4277-29650
Fax: +43 1 4277-29670
E-Mail: ectil@ectil.org

Austrian Academy of Sciences
Institute for European Tort Law
Reichsratsstraße 17/2
A-1010 Vienna
Tel.: +43 1 4277-29651
Fax: +43 1 4277-29670
E-Mail: etl@oeaw.ac.at

This publication contains the results of a study on "Liability Issues and Redress Mechanisms in the Food and Feed Supply Chains" submitted as a deliverable to and financed by Co-Extra, which is an integrated project under the Sixth Framework Programme for Research and Technological Development of the EU.

The conclusions, recommendations and opinions presented in this study reflect those of the authors and do not necessarily reflect the opinion of the sponsors.

ISBN 978-3-89949-811-0
e-ISBN 978-3-89949-812-7
ISSN 1616-8623

Bibliografische Information der Deutschen Nationalbibliothek

Die Deutsche Nationalbibliothek verzeichnet diese Publikation in der Deutschen Nationalbibliografie; detaillierte bibliografische Daten sind im Internet über http://dnb.d-nb.de abrufbar.

© 2010 Walter de Gruyter GmbH & Co. KG, Berlin/New York

Druck: Hubert & Co. GmbH & Co. KG, Göttingen

∞ Gedruckt auf säurefreiem Papier

Printed in Germany

www.degruyter.com

Preface

Farming with genetically modified organisms is a highly sensitive issue, not only in Europe, driven to a large extent by deep concerns of the public that transgenic products may be harmful for the people, for animals or other living beings, and for the environment at large. Whether or not those fears are substantiated will of course not be addressed here. This book starts from the theoretical assumption that there may be damage caused by GMOs in the future and examines to what extent current legal systems are prepared to respond to such potential losses by offering indemnification to those suffering harm. The European Centre of Tort and Insurance Law already participated in an earlier research project dealing with one specific aspect of this question. This previous study was published two years ago as volume 24 of this book series under the title "Economic Losses Caused by Genetically Modified Organisms. Liability and Redress for the Adventitious Presence for GMOs in Non-GM Crops".

The present book aims at addressing the remaining varieties of possible damage apart from the purely economic losses already covered. In the following, the focus will therefore be on damage to persons or property and harm to the environment. National reporters from twenty European and four non-European jurisdictions analyze on the basis of a uniform questionnaire how such losses would be categorized in their respective legal system, what options the law of delict provides in order to remedy such harm, and what requirements have to be met by potential victims in order to successfully claim compensation. Alternative redress schemes are equally addressed to the extent already available or at least foreseeable. Furthermore, a special report looks at developments in the field of international environmental law by presenting work done on the basis of the Cartagena Protocol on Biosafety in the field of liability. Two authors outline EU norms governing jurisdiction and choice of law applicable to cross-border disputes. Two further special reports present an insurers' perspective on biotechnology and provide an economic analysis of liability to third parties for damage caused by GMOs. A comparative report tries to highlight some key aspects of the afore-mentioned contributions to this volume in an overview.

The work was conducted under the auspices of and financed by an integrated project under the 6th Framework Programme of the European Union called "Co-Extra", which stands for "GM and non-GM supply chains: their CO-EXistence and TRAceability". This large-scale multidisciplinary network of researchers around the world brought together 57 partner institutions to deal with scientific, technical, economic, legal and other aspects of how GM and non-GM farming and the distribution of its produce can coexist along the food and feed supply chain, including both practical and economic aspects of identifying transgenic products along the way, in order to guarantee freedom of choice for all stakeholders. More details can be found on the project's website at www.coextra.eu. ECTIL was invited to join Co-Extra at the beginning of 2008 in order to contribute a study on delictual liability for GMOs in the food and feed supply chain. This book presents the outcome of this study.

At the occasion of concluding this research endeavour with the publication of its outcome, I would like to thank all authors for participating and for allowing us to conclude the project exactly on time in September 2009. Several individuals have provided assistance at various stages of this project, including in particular Claudia Hirsch, Susanne Knasmüller, Sara Löw, and Stuart Wallace. Edina Toth has produced the index to this book. Fiona Salter-Townshend, Lisa Stadler and Donna Stockenhuber have taken care of proof-reading the contributions as perfectly as always. Kathrin Karner-Strobach has checked the citation and style uniformity. I would like to express my deep gratitude to all of them for their support in bringing about this volume.

Bernhard A. Koch
Innsbruck, June 2010

Table of Contents

Introduction

Bernhard A. Koch

Co-existence of modern and more traditional farming techniques is not only a matter of agricultural law and practice. Biotechnology in particular triggers concerns and even fears whether its use in the food and feed supply chain may cause losses to farmers or seed producers, to distributors or processors, to consumers, or to the environment. Consequently, the question who should compensate such losses and how is an important item on the agenda of those regulating the market. No matter how likely or improbable such losses are, the mere existence of such worries calls for at least clarifications as to what the remedies could be and who should provide them, and these answers should ideally be given upfront in order to enable the peaceful coexistence of conventional, organic and GM farmers. 1

The present study is the product of a follow-up project to a previous endeavour by almost the same team of authors some three years ago dealing with a more limited question of losses caused by genetically modified organisms, i.e. the economic loss resulting from the adventitious presence of GMOs to neighbouring (conventional or organic) farmers. The initiative at the time came from the European Commission directly, whereas the current work is financed indirectly with EU monies through an FP6 integrated project called "Co-Extra"[1]. The outcome of the earlier research was published in 2008[2]. To the extent that the issues overlap the authors in this current volume will obviously rely on cross-referencing to their earlier contributions rather than duplicating the texts. In both studies, the views presented are those of the authors and do not necessarily reflect the opinion of the sponsors. 2

The following contributions aim at tackling aspects of delictual liability not yet covered by their predecessors. While the original task was to concentrate on the (pure) economic loss of adjacent farmers only, here we are trying to address further possible losses in the entire food and feed supply chain, including (but not limited to) losses of individuals as well as harm 3

1 http://www.coextra.eu.

2 B.A. Koch (ed), Economic Loss Caused by Genetically Modified Organisms. Liability and Redress for the Adventitious Presence of GMOs in Non-GM Crops (2008).

to the environment. However, it is impossible to provide a comparative study of all imaginable losses in all jurisdictions covered, analyzing all aspects of potential claims. Instead, the focus will primarily be on deviations from standard tort law as well as on alternative routes to compensation. Claims based on contract law will be mostly disregarded, since these problems will be covered by yet another study within the Co-Extra framework. However, some pointers to contractual obligations will be inevitable, particularly in those jurisdictions where the borderlines between tortious and contractual liability are less strict.

4 Unlike the previous study, the current survey includes non-European jurisdictions for the sake of comparison, selected in light of their market importance and particular experience with GMOs. Not all European countries could be covered, of course, but the twenty jurisdictions chosen from all legal families should constitute a representative sample.

5 In order to asses the state of the law in all these legal systems, a standardized questionnaire was sent out to all reporters, to which they provided replies to the extent that the questions were relevant for their respective jurisdiction. In light of the harmonization in the field of conflict of laws and jurisdiction within the EU, a separate report was commissioned presenting the (at least in part just recently) harmonized aspects of cross-border claims in the Member States. A further paper analyzes international environmental law, in particular with an eye to ongoing developments initiated by the Cartagena Protocol on Biosafety. Specific problems that the insurance industry faces are presented by a report drafted by a senior risk expert working for a major reinsurer. An economic analysis gives insight into the legal problems from a different, but equally important perspective. The comparative report at the end does not try to summarize the state of the law in all jurisdictions considered – this would by far exceed the scope of this volume and the task pursued. Instead, it is meant to serve as a pointer to more detailed information throughout this volume, but also to highlight the key areas of concern where problems may lie when aiming at co-existence despite potential risks of admixture.

6 When considering the various types of losses imaginable, the assessment starts from the assumption that such damage was indeed caused, without questioning the scientific likelihood that such risks may indeed materialize. This is the task of scientists, not of lawyers. Tort law in particular steps in once something has gone wrong and proceeds from there. Furthermore, as in the previous volume, even the adventitious presence of GMOs in non-GM crops will occasionally be referred to as "contamination". The person that thereby (directly or indirectly) suffers harm will be

called the "victim", who seeks indemnification in tort law from a "tortfeasor". The use of such words is purely technical and is not meant to have some pejorative undertone.

Questionnaire

I. General overview

1. Are losses caused by GMOs covered by any specific liability or other compensation/redress regime in your jurisdiction, or are these losses at least addressed expressly by a special regime with an otherwise wider scope of application? If so, please present these in overview and describe the extent to which they overlap/interact with more general tort law regimes, and where they differ.
2. Can the state be held liable for any such losses under certain conditions, or do general or special rules of sovereign immunity apply? To what extent (if at all) does state liability deviate from the standard tort law regimes? If you mentioned a special regime under 1, to what extent is the state involved therein?

II. Damage

1. Which losses caused by GMOs are recoverable in your jurisdiction in general?
2. Is pure economic loss recoverable, and if so, under what conditions?
3. Is mere fear of a loss triggered by GMOs recoverable (e.g. non-pecuniary loss due to the fear of developing an illness in the future)? If applicable, please mention how the fear of mobile phone radiation or of adverse consequences of other modern technology is considered in your jurisdiction.
4. What is the standard of proof with respect to establishing losses envisaged by this study, particularly with an eye to future losses?
5. Are nominal/symbolic losses recognised in your jurisdiction?
6. Are there any special rules for mass losses (apart from caps – e.g. special rules on establishing/proving loss)?

III. Causation

1. How does your jurisdiction react to the uncertainty of merely potential causes (as in cases of alternative causation, for example), particularly in light of causes possibly lying within the victim's own sphere? Possible scenarios relevant for this study include cases where it is unclear who violated segregation rules/good farming practice, for example.
2. Do special rules apply in case of more complex cases such as scenarios comparable to the DES cases (e.g. where a food producer has bought maize from several producers in differing quantities, and it remains unclear which batch was contaminated)?
3. What is the impact of force majeure when it comes to establishing causation?
4. Threshold to prove causation
5. Are there any special rules on causation (including those developed in court practice) that may apply to GMO cases?

IV. Types of liability

1. Fault liability
   (a) If fault rules are applied in the cases envisaged here, are they handled in any special way (e.g. by reversing the burden of proving fault, or by altering the standard of care)?
   (b) Does it make any difference if specific (statutory or customary) rules governing GM or non-GM farming (e.g. good farming practice, segregation rules, food safety or hygiene provisions, etc.) have been violated?
2. Product liability
   (a) Did your jurisdiction incorporate the development risk defence when implementing the EC Product Liability Directive (85/374/EC)? Whether or not this is the case, how is this aspect handled in practice?
   (b) Apart from the regime implementing Directive 85/374/EC, does your legal system have any alternative route for compensating losses caused by (agricultural) products? What are the conditions thereof, and do you think it will survive the scrutiny of the ECJ (which in

recent rulings has repeatedly insisted on the Directive's regime as being not only a minimal, but an exclusive standard)?

(c) How does it affect liability if it is proven that the defendant has obeyed all rules and regulations governing his production process?

3. Environmental liability

(a) How did your country implement the Environmental Liability Directive (2004/35/EC)? Do you foresee financial guarantees?

(b) Is there any liability regime covering environmental harm in your jurisdiction which exceeds the scope of Directive 2004/35/EC?

(c) Who can recover what losses in the case of harm to biodiversity or other harm to the environment as such?

(d) Is there any special liability regime covering losses sustained by individuals in the course of damage to the environment as such (e.g. nuisance or neighbourhood regimes)?

(e) What is the position of your jurisdiction with respect to international agreements such as the Cartagena Protocol?

4. Other strict liability regimes

Are there any (other) strict liability regimes applicable in cases of harm caused by GMOs (apart from product or environmental liability as just mentioned)?

V. Vicarious liability

1. Please mention very briefly the scope of vicarious liability in your jurisdiction, in particular whether someone can be held liable for an independent contractor hired by him/herself.
2. If you consider the feed and food production chain, who would be liable for people further up (e.g. the crop retailer for the crop wholesaler, or for the farmer, or the latter for the seed producer etc.)? Please disregard contractual liability for warranties relating to the products.
3. Can someone further down the feed or food chain include someone further up it in a trial against him/herself? Can, e.g., the farmer bring his seed producer into a lawsuit filed against him by a customer (or is he even under an obligation to include him)? Can he thereby escape liability (by shifting the loss onto the seed producer)?

VI. Multiple tortfeasors

Please describe very briefly whether your jurisdiction applies joint and several liability and/or proportional liability in cases of multiple tortfeasors.

VII. Defences

Which of the following defences may be raised against liability for GMO related losses?

1. Does the licence/permission to grow GM material (if applicable) serve as a defence against liability triggered by its use? Does it make any difference if the GMOs are approved for testing only?
2. What would the impact of consent/assumption of risk be on liability, e.g. if the victim had knowingly consumed GM products?
3. To what extent will third-party behaviour (such as sabotage or improper conduct by a neighbouring farmer) be considered?
4. Which statutes of limitation apply to the cases envisaged by this study? Please mention specifically if these deviate from standard tort cases.
5. Are there other (e.g. specific) defences that might be relevant in a GMO liability case?

VIII. Remedies

1. Pecuniary compensation
   (a) Are there any special rules on compensating bodily harm caused by GMOs, or do the regular remedies apply? (such as caps)
   (b) Are there any special rules on compensating property losses caused by GMOs, or do the regular remedies apply? (such as caps)
   (c) Are there any special rules on compensating economic losses caused by GMOs, or do the regular remedies apply? Can an organic farmer, for example, recover the full costs of restoring a field that was contaminated with GMOs to meet organic standards, or would deductions apply for the seasons that he could have grown conventional crops there?
   (d) If animals are harmed by contaminated feed, what can their owner recover? How is the value of an animal calculated, and does it in-

clude, e.g., a cow's potential for producing milk or meat? Does the mere fact that an animal eats GM contaminated feed constitute harm to the animal and therefore, recoverable damage for its owner?

(e) Are the costs of disposing of contaminated production/animals fed with GM feed recoverable?

2. Non-compensatory damages

Does your jurisdiction recognise punitive, exemplary or any other forms of non-compensatory damages?

3. Other remedies

Does your legal system foresee other remedies such as reparation in kind, substitution in kind, or further types (e.g. community contributions)?

4. Costs of pursuing a claim

(a) What is the general rule on the recoverability of costs in a civil trial (e.g. the "loser pays" principle)?

(b) What are the conditions for recovering the costs of establishing causation (e.g. in case a food producer was held liable – can he recover the costs of detecting GMOs at farming level)?

5. Advance cover

Do any of the liability or other redress schemes foresee financial guarantees such as mandatory insurance cover by, e.g., the GMO producers?

IX. Cross-border issues – Conflict of laws

1. If the Rome II Regulation comes into force in your jurisdiction (which will be dealt with by a separate report), please briefly describe the torts conflicts regime that it will replace, focusing on the main differences in a nutshell. Otherwise, please present the key aspects of the conflicts rules applicable to the cases under survey here.
2. Is there any special regime that applies to resolve bilateral or multilateral cross-border claims for compensation, either in tort or under any compensation scheme (since, e.g., a GMO fund regime may foresee provisions for cross-border claims)?

X. Cases

1. Due to the adventitious presence of GMOs in a field, maize which is normally sold as conventional contains GMOs beyond the legal labelling threshold. This is not discovered before the final stage of the food production chain by the producer of taco chips. The whole production is lost since the supermarket chains refuse to accept delivery from the producer.
   (a) Please solve the case by cross-referring to your general statements above. Who can sue along the chain of distribution – the taco producer, the wholesaler of the maize, etc.?
   (b) Would the case be solved differently if the GMO content was below the labelling threshold?
   (c) Would the case be solved differently if the admixture was not adventitious, but occurred due to the disregard of segregation rules, for example?
   (d) Would the case be solved differently if the GMO found was not admitted for production in your jurisdiction?
   (e) If the admixture had occurred on a non-GM field and it transpires that the GM seeds were blown from: (i) neighbouring fields; or (ii) a truck passing by, would the farmer of the affected field be liable for all or part of the loss caused further down the distribution chain?
2. Twenty years after the sale of GM maize used for food products, it turns out that it has certain disadvantageous health effects for humans.
   (a) Can the producers be held liable at this point for risks unknown at the time of growing the maize? Who would be liable – the seed producer/farmer/food producer/distributor/etc?
   (b) Can compensation already be claimed at a point when the negative health effects have not yet materialised, but are to expected to according to scientific expertise/mere rumours?
   (c) Would it make any difference if the GM maize had only been in use for feed, causing harm to the animals, which may or may not cause harm to humans consuming the meat as well?
3. The driver of a food logistics company discovers that a farmer, from where he regularly picks up agricultural products fails to obey mandatory segregation rules or food or feed hygiene standards, which may

lead to the admixture of GM and non-GM produce sold and packaged separately by that farmer. Does he or his employer have a duty to warn, i.e. notify the recipient of the allegedly "non-GM" produce?

European Reports

Damage Caused by GMOs under Austrian Law

Manuela Weissenbacher

I. General overview

1. Special liability or redress scheme for GMOs

1 Under Austrian law there is a special liability regime for losses caused by GMOs, which is laid down in the Gene Technology Act (*Gentechnikgesetz*, GTG).[1] The GTG came into force in 1994. It deals with installations for GMOs, the use, release and putting into circulation of GMOs, the labelling of products containing GMOs and the genetic analysis and therapy of human beings (§ 2 par. 1 GTG). Yet it did not provide for any special civil liability law provisions at that time.[2] It was the 1997 amendment of the GTG[3] which introduced a specific new liability regime following the gene technology referendum in 1997. §§ 79a–j GTG provide for a strict liability of the operator covering the risks of the production, use, increase, storage, destruction or disposal of genetically modified organisms, as well as their intentional or unintentional release (§ 4 Z 4 and § 20 GTG). According to § 79a par. 1 GTG, the operator is liable regardless of fault for harm suffered by a human being (death, personal injury or damage to health) and for damage to property and the environment. The cause of the damage must lie in the specific properties of the organism, derived from the genetic modification, or in the combination of these properties with other dangerous properties of the organism. The operator is liable irrespective of the fact that the use of GMOs or their release has already been

1 Bundesgesetzblatt (BGBl) 1994/510 as amended by BGBl I 2006/13.

2 Liability issues were to be covered by a general environmental strict liability law which was expected to be enacted soon. In fact, it was 15 years later that the Federal Environmental Liability Act (*Bundes-Umwelthaftungsgesetz*,B-UHG, BGBl I 2009/55) came into force. It transposes the Environmental Liability Directive (2004/35/EC) and still does not entitle the individual injured party to a damages claim regardless of fault under civil law. For further details with respect to the scope of the B-UHG see IV.3.

3 BGBl I 1998/73.

authorised. His liability is not limited to damage emerging from an accident or a malfunction either.[4] Liability is imposed on the operator also for damage which arises under normal conditions and during normal operation.

Whenever products, which consist of or contain GMOs, are put into circulation and cause damage, §§ 79a ff. GTG are not applicable.[5] These damage claims are to be asserted under the Product Liability Act (*Produkthaftungsgesetz*, PHG),[6] which also provides for no-fault liability. Although both liability regimes are no-fault based, they follow different approaches: According to the GTG, the operator can only be held liable if the specific danger of the genetic modification has become effective, while liability according to the PHG arises from the risk of defective products for the private consumer. Hence it follows that as soon as GMOs are put into circulation, considerable loopholes concerning liability emerge. They ought to be closed by tightening the rules of the PHG; although this has been planned since 1998, it has not been realised yet.[7] **2**

In 2004 another important amendment of the GTG came into force, which covers a special strict liability regime limited to farming and neighbouring parcels (§§ 79k–m GTG).[8] According to § 79k par. 1 GTG, the owner of land in agricultural use, or the holder of another property right, is entitled to an injunction against emissions from neighbouring land provided that the neighbour cultivates products that consist of or contain GMOs. An injunction according to this provision will be granted in case of contamination by GMOs originating from agriculturally used land which was either caused directly (e.g. by sowing or planting) or by indirect effects during the growth phase, the harvest or even later. Furthermore, the interference must exceed a certain tolerance threshold and cause a substantial impairment of the use of the affected farmland. If the above-described interference results in damage to person and/or property, the neighbour who caused the interference is liable regardless of fault vis-à-vis the affected landowner or holder of the property right (§ 79k par. 2 GTG). As compared to § 79a GTG, the strict liability regime as laid down in § 79k par. 2 GTG also covers damage which occurs after the products **3**

4 ME 214 BlgNR 20. GP 17.

5 *F. Kerschner* in: F. Kerschner/C. Lang/G. Satzinger/E. Wagner (eds.), Kommentar zum Gentechnikgesetz (2007) § 79a no. 4 f.

6 BGBl 1988/95 as amended by BGBl I 2001/98.

7 *J. Stabentheiner*, Die neue Gentechnikhaftung, Österreichische Juristenzeitung (ÖJZ) 1998, 521, 535 ff.; *Kerschner* (fn. 5) § 79a no. 1, 5, 19; *F. Kerschner/E. Wagner*, Koexistenz zwischen Gentechnik, Landwirtschaft und Natur – Rechtliche Rahmenbedingungen (2003) 145 ff.

8 BGBl I 2004/126.

consisting of or containing GMOs were put into circulation, although it is restricted to agricultural uses and neighbouring territories.[9]

4 Liability for GMOs is not governed solely by the federal liability regime as laid down in the GTG – several provinces have also enacted their own Genetic Engineering Precautionary Measures Acts (*Gentechnikvorsorgegesetze*).[10] They provide for civil law regulations with respect to liability for GMOs which at least partly overlap with § 79a GTG. As far as liability according to the state regulations and liability under federal law (GTG) coexist, they will be competing claims.[11]

2. State liability

5 State liability is governed by the rules of the Public Liability Act (*Amtshaftungsgesetz*, AHG).[12] It provides for a fault based liability of a public body as specified in § 1 par. 1 AHG (state, provinces, districts, municipalities, other bodies of public law and the institutions of social insurance) under the provisions of civil law. State liability covers any damage to a person or property, which is caused by unlawful acts of persons at fault when implementing the law on behalf of such public bodies. The persons implementing the law, however, are not liable themselves vis-à-vis the persons who have suffered the loss. Damages are to be paid in terms of money only.[13]

6 State liability as laid down in the AHG is considered to be conclusive.[14] Nevertheless, it is generally accepted by the prevailing opinion of scholars and legal practice that acts of persons implementing the law may be actionable under other general no-fault based liability regimes as well.[15] Thus, the EKHG with its strict liability rules is applicable in addition to the AHG.[16] The same holds true for the regulations of the Aviation Act (*Luftfahrtgesetz*,

9 *Kerschner* (fn. 5) § 79a no. 20.

10 E.g. Salzburger Gentechnik-Vorsorgegesetz, Landesgesetzblatt (LGBl) 2004/75; Kärntner Gentechnikvorsorgegesetz LGBl 2005/5; Burgendländisches Gentechnikvorsorgegesetz LGBl 2005/94.

11 *Kerschner* (fn. 5) § 79a no. 22.

12 BGBl 1949/20 as amended by BGBl I 1999/194.

13 See § 1 par. 1 AHG.

14 *P. Mader* in: M. Schwimann (ed.), Praxiskommentar zum ABGB VII (3rd ed. 2005) Vor § 1 AHG no. 4.

15 OGH 22.08.1996, 1 Ob 2047/96b, Entscheidungen des Österreichischen Obersten Gerichtshofes in Zivil- und Justizverwaltungssachen (SZ) 69/188; *W. Schragel*, Kommentar zum Amtshaftungsgesetz (3rd ed. 2003) no. 252; *A. Zechner*, Amtshaftung – Recht im Wandel, Zeitschrift für Verwaltung (ZfV) 1985, 591f.

16 *Schragel* (fn. 15) no. 252; OGH 29.10.1964, 2 Ob 328/64, SZ 37/158; 05.11.1965, 2 Ob 239/65, SZ 38/183; 15.01.1970, 2 Ob 396, 397/69, SZ 43/10; 19.02.1975, 1 Ob 187/74,

LuftFG)[17] as well as for the Nuclear Liability Act (*Atomhaftungsgesetz*, AtomHG 1999)[18] and §§ 1319 and 1319a General Civil Code (*Allgemeines bürgerliches Gesetzbuch*, ABGB).[19] With respect to the GTG, which pursuant to § 79i GTG does not affect the applicability of other provisions, the AHG does not preclude its additional applicability to acts of State, either.[20]

Whenever the State or its public bodies cause damage in the performance of private sector administration, the AHG will not be applicable. Thus, it follows that the State and respectively the public bodies will be held liable under general tort law as well as under the above cited special strict liability regimes. 7

II. Damage

1. Recoverable losses

Liability under § 79a par. 1 GTG, emerging from the use or release of GMOs, covers damage to person and property, including loss of profit as well as economic loss arising from the damage.[21] The same holds true for § 79k par. 2 GTG if the damage results from interference as described under par. 1 leg cit. In this case, the affected landowner is entitled, for instance, to claim the loss of profits he suffers because he has to destroy the crop or as a consequence of a failure of crop, because he obtains a lower price for the crop or because the value of his real property has decreased.[22] Whenever the damage to property at the same time presents a significant impairment to the environment, the damage claim also covers the costs of measures of reinstatement as provided by § 79b GTG. These remediation costs are actionable even if they exceed the market value of the impaired good. The plaintiff is entitled to ask for advance payment. If, however, he does not restore the damaged good to its original condition within reasonable time, he will be obliged to refund the amount exceeding the market value of the impaired good. 8

SZ 48/17; 23.03.1977, 8 Ob 218/76, SZ 50/45; 21.09.1983, 1 Ob 27/83, SZ 56/133; 22.11.1995, 1 Ob 49, 54/95, SZ 68/220; 25.06.2002, 1 Ob 129/02f, SZ 2002/87 etc.

17 §§ 146 ff. LuftFG; OGH 25.08.1992, 1 Ob 23/92, SZ 65/111.

18 §§ 3 ff. and § 16 AtomHG 1999.

19 OGH 25.06.2002, 1 Ob 129/02f, SZ 2002/87; 05.12.1995, 1 Ob 29/95, Zeitschrift für Verkehrsrecht (ZVR) 1996/109.

20 ME 214 BlgNR 20. GP 42.

21 *Kerschner* (fn. 5) § 79a no. 10 f.

22 *Kerschner* (fn. 5) § 79k no. 36.

2. Pure economic loss

9 The extent and manner of compensation for damage as described above is governed by the civil law rules of the General Civil Code (§§ 1323 ff. ABGB). Austrian tort law is very reluctant when it comes to compensation of pure economic loss. § 1295 par. 1 ABGB basically states that everybody who suffers damage, be it within or beyond a contractual relationship, is entitled to claim damages from the tortfeasor who was at fault. Although this provision does not express any restraints, the OGH (*Oberster Gerichtshof* – Supreme Court) consistently holds the opinion that beyond contractual relationships pure economic loss is only recoverable upon violation of a protective law (*Schutzgesetz*) or if the tortfeasor has acted contra bonos mores.[23] In fact, the ABGB recognizes compensation for pure economic loss if the tortfeasor consciously gives wrong advice (§ 1300 ABGB) and in case of conscious fraudulent misrepresentation (§ 874 ABGB). Apart from that there are further examples, approved by legal doctrine and legal practice, which clearly show that there are cases which justify the compensation of pure economic loss beyond the OGH's narrow interpretation.[24] Despite all this it is common opinion that, apart from contractual relationships, the compensation of pure economic loss needs to be granted only in justified cases. That is why pure economic loss caused by the use or release of GMOs will not be recoverable.

3. Mere fear of a loss

10 An action according to § 79a par. 1 GTG (damage claim), requires an actual loss with respect to persons or property. Therefore, mere fear of a loss triggered by GMOs will not be recoverable under this provision. Neither will it be recoverable under § 79k par. 2 GTG, which entitles the plaintiff to damages only if an actual interference in the sense of § 79k par. 1 GTG can be established. Remediation costs pursuant to § 79b GTG can only be claimed if the impairment of the environment can also be qualified as damage to the plaintiff's property. Again, this does not cover mere fear of a loss triggered by GMOs. Even under the general tort rules of the General Civil Code (§§ 1293 ff. ABGB), the subsidiary application of which is regulated in § 79h par. 1 GTG, any compensation claim requires the establish-

23 OGH 12.06.1979, 2 Ob 193/78, SZ 52/93; 21.12.1983, 8 Ob 78/83, SZ 56/199; 26.06.1986, 7 Ob 598/86, Juristische Blätter (JBl) 1986, 650; 20.12.1988, 2 Ob 151/88, SZ 61/279; 29.01.1994, 2 Ob 557/93, SZ 67/17 = JBl 1994, 687.

24 *H. Koziol*, Schadenersatz für reine Vermögensschäden, JBl 2004, 273.

ment of an actual loss to persons or property. Consequently, the mere fear of adverse consequences usually does not constitute compensable damage.

4. Standard of proof

In general, the burden of proof for all requirements of the damage lies with the injured person. Strict liability regimes usually provide for some alleviation of the burden of proof (e.g. presumption of causation, unlawfulness of the act and fault of the tortfeasor need not be established). The presence of a loss, however, always needs to be established by the plaintiff, be it under fault-based liability or strict liability. In the event that the plaintiff is not able to prove the exact quantum of damage or if the proof thereof is unreasonably difficult, which will particularly be the case with respect to future losses, the court is allowed to determine the respective amount at its absolute discretion pursuant to § 273 Civil Procedure Act (*Zivilprozessordnung*, ZPO). The judicial estimation of the loss requires that the loss is determined on its merits.[25] **11**

5. Nominal losses

Nominal losses are not recognized. As mentioned above (supra no. 11), the courts are allowed to estimate the quantum of damage at their absolute discretion pursuant to § 273 ZPO, if such cannot be established by the plaintiff or if the proof would be unreasonably difficult. **12**

6. Mass losses

There are no special provisions governing mass losses, neither under the GTG nor under the General Civil Code. **13**

III. Causation

1. Uncertainty of merely potential causes

The GTG does not provide for any special rules governing these problems. Thus, in cases of alternative or overtaking (intervening) causation, the pro- **14**

---

25 *M. Gimpel-Hinteregger*, Grundfragen der Umwelthaftung (1994) 139.

visions of general tort law have to be applied. If two or more potential tortfeasors might have contributed to the damage (alternative causation), all actors are jointly and severally liable pursuant to the analogous application of § 1302 ABGB.[26] In the case of overtaking (intervening) causation, courts usually hold the opinion that the person who caused the damage first is wholly liable.[27] In personal injury cases, however, where the victim would have contracted a disease anyway due to a personal disposition, the defendant, whose action merely caused the outbreak of the disease earlier, will not be fully liable. His liability is limited to the point in time in which the victim would have contracted the disease anyway.[28] There are several authors, however, who suggest a different solution for all cases of intervening causation: They would like to consider the liability of the first tortfeasor with due regard to the action of the second one.[29] As a result, the first injurer may be totally or partly exculpated, or, under certain conditions, jointly and severally liable.

15 If the injured person has actually contributed to the damage him/herself, he/she will not be awarded full compensation. The plaintiff has to bear part of the damage, depending on the proportion of his/her fault compared to the fault of the tortfeasor (contributory negligence, § 1304 ABGB).[30] If both of them are equally at fault or if the proportions cannot be determined, the plaintiff and the tortfeasor will be liable in equal shares. In strict liability cases, which establish a liability of the tortfeasor regardless of fault, the victim's share of responsibility with respect to the damage sustained is also taken into due consideration.[31] The extent of the hazard will be weighed against the contributory negligence of the injured person.[32] If it is not clear whether the injured person has contrib-

26 *H. Koziol/R. Welser*, Grundriss des bürgerlichen Rechts. Schuldrecht Allgemeiner Teil, Schuldrecht Besonderer Teil, Erbrecht II (13th ed. 2007) 334.

27 See OGH 08.04.1959, 2 Ob 166, 167, Evidenzblatt der Rechtsmittelentscheidungen (EvBl) 1959/244; 19.10.1966, 2 Ob 216/66, SZ 39/172; 15.12.1992, 1 Ob 642/92, JBl 1993, 663.

28 OGH 09.05.1973, 1 Ob 65/73, JBl 1974, 318; 14.09.1977, 8 Ob 116/77, ZVR 1978/165; 03.09.1996, 10 Ob 2350/96t, SZ 69/199; 05.05.1998, 4 Ob 23/98f, JBl 1999, 246.

29 See *H. Koziol*, Österreichisches Haftpflichtrecht. Allgemeiner Teil I (3rd ed. 1997) no. 3/58 ff.

30 See in detail *M. Hinteregger*, Contributory Negligence under Austrian Law, in: U. Magnus/M. Martin-Casals (eds.), Unification of Tort law: Contributory Negligence (2004) 9 ff.

31 See § 79h par. 3 GTG, which refers to § 1304 ABGB.

32 E.g. OGH 28.03.1990, 2 Ob 164/89, ZVR 1991/52; *Koziol/Welser* (fn. 26) 328.

uted to the damage or not, § 1304 ABGB is applied and the loss will be allocated equally among the actors.[33]

2. Complex causation scenarios

There are no special rules for more complex cases under Austrian civil law. **16**

3. Force majeure

The presence of force majeure excludes the liability of the tortfeasor. This limitation of liability is expressly regulated in § 79c Z 1 GTG. Force majeure in the sense of this provision is described as a natural phenomenon, which needs to be extraordinary,[34] inevitable and not reasonably preventable[35] as far as its consequences are concerned. **17**

4. Threshold to prove causation

Pursuant to the general liability rules, the burden of establishing the causal link between the damage and the tortious act is with the plaintiff. In general, in order to convince judges of causation, the plaintiff must show to a very high level of probability close to certainty (to the satisfaction of the court) that the defendant caused the injury.[36] There are a few more recent decisions of the OGH, however, according to which a high degree of likelihood which does not need to come close to certainty shall be sufficient, when causality is to be established.[37] According to these rulings, the very high level of probability close to certainty shall only be necessary in **18**

33 See *F. Bydlinski*, Haftung bei alternativer Kausalität. Zur Frage der ungeklärten Verursachung, besonders nach österreichischem Zivilrecht, JBl 1959, 1, 13; *F. Bydlinski*, Probleme der Schadensverursachung nach deutschem und österreichischem Recht (1964) 77; *F. Bydlinski*, Aktuelle Streitfragen um die alternative Kausalität, in: FS (Festschrift) Beitzke (1979) 26, 30.

34 This will be judged by the frequency and probability of occurrence of the natural phenomenon. See *Kerschner* (fn. 5) 79c no. 4.

35 In case of an extraordinary and inevitable natural phenomenon the operator will still be liable if he could have reasonably avoided its consequences (e.g. by taking preventive measures against floodwaters). See *Kerschner* (fn. 5) § 79c no. 5.

36 E.g. OGH 22.10.2002, 10 Ob 98/02p, ecolex 2003/46; *H. Fasching*, Lehrbuch des österreichischen Zivilprozessrechts (2nd ed. 1990) no. 815; *W. Rechberger/D. Simotta*, Grundriss des österreichischen Zivilprozessrechts (6th ed. 2003) no. 580.

37 See OGH 24.09.1998, 2 Ob 185/98i; 17.11.2004, 7 Ob 260/04t, EvBl 2005/77 = JBl 2005, 464. See also *Kerschner* (fn. 5) § 79d no. 3.

cases which require an elevated (general) standard of evidence (*Regelbeweismaß*). In 7 Ob 260/04t, the OGH required a high degree of likelihood and qualified the standard of evidence as flexible, depending on the objective circumstances of the single case as well as on the subjective evaluation of the judge.

19 Under certain circumstances, the burden of proof may be eased for the plaintiff. This is the case whenever causation can be inferred against the defendant from a typical course of events.[38] The application of prima facie evidence is especially justified if the defendant has violated a protective law, that is a law that was designed to protect persons like the plaintiff from the sort of damage that occurred.[39]

5. Special rules on causation

20 The GTG provides for special rules of conduct and further notification and documentation duties with respect to GMOs. They can be considered protective laws and therefore allow that the burden of proof be eased for the plaintiff. Provided that the defendant has violated his duties according to the GTG, the plaintiff will be able to establish the causal link between the damage and the tortious act by prima facie evidence.

21 Apart from that, the GTG explicitly provides for another alleviation in favour of the plaintiff when it comes to establishing the proof of causality. § 79d GTG and § 79k par. 4 GTG ease the burden of proof for the plaintiff by way of a presumption of causation. § 79d GTG applies when the GMO which was the object of an activity as prescribed under § 79a GTG, is prone to cause the damage under the particular circumstances of the case. A certain level of probability is not required. If the injured person can show the required specific ability of the GMO to cause the damage inflicted on him/her, it will be presumed that the damage was caused by the GMO as a consequence of its special nature as caused by its genetic modification. The presumption is rebutted if the operator can show that it is probable that the damage was not caused by its specific nature, but by another cause. In this case, the burden of proof is with the plaintiff again. Pursuant to § 79k par. 4 GTG, the owner of the affected land can rely on a presumption of causation if he/she can plausibly show that, considering the circumstances of the particular case, a certain act or omission of the neighbour was prone

38 *Gimpel-Hinteregger* (fn. 25) 136 ff.

39 See M. *Gimpel-Hinteregger*, Umwelthaftung in Österreich, Produkthaftpflicht International (Phi) 1996, 202, 204, 207.

to cause the interference in the sense of par. 1 leg cit. In this case it is presumed that the interference was caused by the act or omission. Again, the presumption is rebutted if the neighbour can show that it is probable that the interference was not caused by his act or omission. Then, it is up to the injured landowner to establish the causal link between the damage and the tortious act.

IV. Types of liability

1. Fault liability

(a) Special rules governing fault

N/A.

(b) Impact of specific rules of conduct

N/A.

2. Product liability

(a) Development risk defence

The development risk defence was implemented into § 8 Z 2 PHG. Thus, 22
liability of the producer is excluded if he/she is able to show that the properties of the product could not be identified as defective according to the state of science and technology. The relevant moment is the point in time when the product was put into circulation by the defendant. It is settled case law of the OGH to clearly distinguish the development risk from damage caused by so-called *Ausreißerschäden*.[40] These only affect single products and can be defined as exceptional defects. Therefore, the impossibility of discovering or avoiding the exceptional defect of a single product does not exclude the liability of the producer under the PHG. The defendant will be able to successfully assert the development risk defence only if he can show that the potential hazard of the product could not be recog-

40 See OGH 08.04.1997, 4 Ob 87/97s, ecolex 1997, 749 = ZVR 1998/19; 22.10.2002, 10 Ob 98/02p, ecolex 2003/46; *H. Fitz/M. Purtscheller*, Produkthaftung (2nd ed. 2004) § 8 no. 11; *R. Welser/Ch. Rabl*, Produkthaftungsgesetz. Kommentar (2nd ed. 2004) § 8 no. 13.

nized according to the state of science and technology when it was put into circulation. The producer of a full-automatic coffee machine, for instance, which caught fire when in the standby mode due to a technical defect and caused a house fire, was denied the development risk defence pursuant to § 8 Z 2 PGH and held liable under the PHG by the OGH,[41] as well as the producer of a mineral water bottle, which shot away the sealing cap like a sparkling wine cork on opening.[42]

(b) Alternative routes

23 Whenever products which are already in circulation cause a loss, apart from the PHG only general tort law is applicable. It is a fact that this general fault-based liability regime is not at all suitable to fully compensate the injured person: (1) the general burden of proof for all requirements of the damage lies with the plaintiff (§ 1296 ABGB); (2) the producer is not at fault if the loss was caused by a single deficiently manufactured product, for which he is not liable (*Ausreißerschaden*); (3) vicarious liability is limited to § 1315 ABGB.[43]

(c) Impact of compliance with rules and regulations

24 The fact that the defendant has complied with all rules and regulations applicable to his business does not save him from liability or lessen his liability. Product liability as regulated in the PHG is a strict liability regime, which aims at compensating the victim regardless of fault.

3. Environmental liability

(a) Implementation of the Environmental Liability Directive

25 First of all, it ought to be mentioned that Austria implemented the Environmental Liability Directive after much delay.[44] The new Environmental Liability Act (*Bundes-Umwelthaftungsgesetz*, B-UHG) was adopted by the National Assembly in March 2009 and came into force on 20 June 2009.

---

41 OGH 22.10.2002, 10 Ob 98/02p, ecolex 2003/46.
42 OGH 08.04.1997, 4 Ob 87/97s, ecolex 1997, 749 = ZVR 1998/19.
43 For more details with respect to vicarious liability and § 1315 ABGB see below no. 36 ff.
44 The implementation of the Directive should have been completed by 30 April 2007.

Being a federal law, the B-UHG only covers water and land damage; the Austrian federation is not competent to enact regulations with respect to damage to protected species and natural habitats. Nature conservation is subject to the legislative competence of the provinces (Art. 15 par. 1 *Bundes-Verfassungsgesetz*, B-VG[45]). This was not the only reason why the Austrian legislator had difficulties with the implementation of the Environmental Liability Directive. Being an instrument of administrative law, the directive overlaps with other Austrian laws, like the Water Protection Act (*Wasserrechtsgesetz*, WRG), the Industrial Code (*Gewerbeordnung*, GewO), the Waste Management Act (*Abfallwirtschaftsgesetz*) and the Nature Conservation Laws of the provinces, all of which at least partly deal with the protection of land and water as well.

The B-UHG, which consists of 20 provisions and three annexes, remains 26
close to the Directive as regards content. The Austrian legislator, however, decided not to implement a permit defence or a state of the art defence. According to § 11 par. 1 B-UHG, any natural or legal person whose rights might be infringed as defined under par. 2 leg cit[46] due to actual environmental damage, is entitled to call on the administrative body of the district to act in accordance with § 6 (remedial action) and § 7 par. 2 (determination of remedial measures) B-UHG upon a written complaint. The environmental counsellor (*Umweltanwalt*) and some accredited environmental institutions are also entitled to file an environmental complaint (*Umweltbeschwerde*). Besides the operator, all of them have legal standing in the proceedings pursuant to § 6 and § 7 par. 2 B-UHG.[47] The parties are entitled to appeal against notifications issued upon application of the B-UHG to the respective UVS (*unabhängiger Verwaltungssenat*, independent administrative chamber) of the province (§ 13 B-UHG).

The B-UHG does not foresee any financial guarantees, nor does it provide 27
for a system of mandatory financial security; it does not even mention how the operator should be prepared to meet his financial obligations arising from § 8 B-UHG (prevention and remediation costs). This is an interesting point, since the government bill to the B-UHG of February 2007 still obliged the operator to provide for a reasonable risk management in its

45 BGBl 1930/1 as amended by BGBl I 2009/127.

46 These rights are: 1. the protection of life and health of human beings; 2. with respect to water: established rights in the sense of § 12 par. 2 WRG 1959 and 3. with respect to land: property or other real rights pertaining to the affected land, but not the possibility of a mere depreciation of the market value.

47 § 12 B-UHG.

§ 13.[48] The explanatory remarks to the government bill point out that Austria (initially) intended to evaluate the effectiveness of the financial securities until 30 April 2010 as provided by Art. 14 par. 2 of Directive 2004/35/EC.[49]

(b) Environmental liability regime beyond the scope of the Directive

28 There is no liability regime which goes beyond the scope of Directive 2004/35/EC in Austria.[50] Although the GTG (§ 79b) as well as the Nuclear Liability Act 1999 (§ 11 par. 2) explicitly refer to the impairment of the environment when they regulate the subject matter of compensation and although both of them allow for the compensation of remediation costs even if these costs exceed the market value of the impaired good, impairment of the environment is only recoverable if it qualifies as damage to the property of the injured person at the same time. Apart from the liability regime of the B-UHG, which is subject to public law, compensation for mere environmental harm which does not become manifest in a private legal sphere is not covered by any liability regime under civil law.

(c) Claimants in cases of environmental harm

29 The liability regimes in force do not foresee a special claim covering the impairment of biodiversity or mere environmental losses. The legislator of the B-UHG did not make use of the possibility as laid down in recital 29 of Directive 2004/35/EC to adopt more stringent and more comprehensive rules in relation to the prevention and remedying of environmental damage; that is to say it did not enact a comprehensive strict liability regime for environmental damage which would entitle individuals to file claims for compensation under civil law.

30 As indicated above, environmental harm is only recoverable if it can be qualified as damage to the plaintiff's property pursuant to general tort law as well as special strict liability schemes (GTG, PHG, law of neighbours, AtomHG 1999). Environmental harm is mostly reflected as damage which

48 RV 95 BlgNR 23. GP.

49 EB RV 95 BlgNR 23. GP 16.

50 For a comprehensive overview regarding compensation for environmental harm on a European level as well as in 14 European countries see M. *Hinteregger* (ed.), Environmental Liability and Ecological Damage in European Law (2008).

is not only within the individual sphere of a legal person but also impairs the ecological dimension of the respective environmental good.[51] Losses to biodiversity rarely represent damage to property, which is why they usually will not be recoverable at all. The same is true for bodies of flowing water, which are not exploited economically.[52] However, if the plaintiff is able to establish damage to his property, he can first of all claim actual damages, which are limited in amount to the market value of the impaired good. In addition, compensation for loss of profits is generally accepted by legal and court practice with respect to strict liability regimes,[53] whereas fault-based liability requires the tortfeasor to have acted in a grossly negligent manner.[54] Pursuant to § 364a ABGB, the damage claim also covers cost of repairs and other remedial work and diminution of value of property. Under the Nuclear Liability Act 1999, compensation for property damage shall also include decontamination costs (§ 11 par. 1 AtomHG 1999). Further, the person who has suffered the loss or damage is entitled to claim damages for consequent economic loss, which is recognized within the scope of the AtomHG 1999 and GTG.[55] Finally, if the damage to property represents significant impairment to the environment, the injured person may also claim remediation costs according to § 11 par. 2 AtomHG 1999 and § 79b GTG, which are not limited to the market value of the impaired good. If the plaintiff asks for advance payment regarding remediation costs, he is obliged to either perform restoration of the environment within reasonable time or refund the amount which exceeds the market value of the damaged good.

(d) Special liability regime for losses sustained by individuals

The law of neighbours in the General Civil Code provides for an injunction 31 against emissions (§ 364 par. 2 ABGB) and a damage claim if the impairment is caused by an activity covered by a licence (§ 364a ABGB). The owner of land whose property is affected by wastewater, smoke, gas, heat, smells, noises, vibrations and similar interferences may claim an injunction against his neighbour if the following prerequisites are met: The interference must exceed a certain tolerance threshold (this is the level customary under local conditions), it must result in a substantial impairment of the enjoyment of

51 See in detail *Gimpel-Hinteregger* (fn. 25) 216 ff.
52 *M. Hinteregger/S. Kissich*, Atomhaftungsgesetz 1999 (2004) § 11 AtomHG no. 14.
53 *Koziol* (fn. 29) no. 10/9f; *Stabentheiner*, ÖJZ 1998, 527 f.
54 See § 1324 ABGB.
55 *Kerschner* (fn. 5) § 79a GTG no. 10.

land and the interference must be attributable to human behaviour. The injunction claim, however, does not cover the compensation for personal injury caused by the above-described interferences. In order to be compensated, the victim has to rely on the fault-based rules of general tort law.[56]

32 Whereas § 364 par. 2 ABGB only allows for injunctive relief and refers the injured person to fault-based general tort law in order to be compensated, § 364a ABGB entitles the owner of land to a compensation claim if the impairment is caused by an activity covered by a licence. In contrast to § 364 par. 2 ABGB, injunctive relief may not be obtained under this provision. The claim for damages is regardless of fault and requires the plaintiff to satisfy the prerequisite conditions for an injunction provided by § 364 par. 2 ABGB. The burden of proof that the interference is not beyond the tolerance threshold is with the defendant.[57] It is not required that the interfering and affected estates be contiguous. § 364a ABGB covers damage to real estate, such as the cost of repairs and other remedial work, diminution of value of property and loss of profits.[58] Loss of life, however, and personal injury are not recoverable under this provision.[59]

33 Court practice has enlarged the scope of application of § 364a ABGB insofar as under certain conditions the owner of land may resort to § 364a ABGB even if the harmful activity of the neighbour is unlicensed. The OGH applies § 364a ABGB analogously to cases where the injured owner of land was legally or factually deprived of preventing the damage by an injunction,[60] or where the operation of a plant or engagement in an activity exposes his/her neighbours to imminent offensive effects.[61] Consequently, the liberal application of § 364a ABGB by the OGH has led to a wide comprehension of this provision in the sense of a general strict liability rule for environmental damage that covers all types of real property damage caused by activities dangerous or offensive to the environment.

---

56 *H. Koziol*/*R. Welser*, Grundriss des bürgerlichen Rechts. Allgemeiner Teil, Sachenrecht, Familienrecht I (13th ed. 2006) 283 ff.

57 OGH 11.10.1995, 3 Ob 508/93, JBl 1996, 446 (Sandstrahl-decision); *P. Oberhammer* in: M. Schwimann (ed.), Praxiskommentar zum ABGB II (3rd ed. 2005) § 364 no. 22.

58 OGH 31.03.1925, Ob III 234/25, SZ 7/115; 01.12.1965, 7 Ob 298/65, JBl 1966, 319; *K. Spielbüchler* in: P. Rummel (ed.), Kommentar zum Allgemeinen bürgerlichen Gesetzbuch I (3rd ed. 2002) § 364a no. 9; *Oberhammer* (fn. 57) § 364a no. 10.

59 Compare the critical remarks of *Gimpel-Hinteregger* (fn. 25) 322 ff.

60 OGH 15.10.1992, 7 Ob 601/92, JBl 1993, 387 (annotated by *F. Kerschner*); 17.11.1993, 1 Ob 19/93, Recht der Umwelt (RdU) 1994/9 (annotated by *F. Kerschner*); 19.12.1995, 1 Ob 31/95, RdU 1996/122 (annotated by *F. Kerschner*).

61 OGH 24.10.1990, 1 Ob 21/90, JBl 1991, 110 (detergent manufacturing plant); 16.01.1991, 1 Ob 39/90, JBl 1991, 580 (illegal disposal of industrial waste); 17.11.1993, 1 Ob 19/93, RdU 1994/9 (excessive manuring by a farmer); 26.01.1999, 5 Ob 3/99y, JBl 1999, 520 (wood logging).

This development, however, has been subject to heavy criticism by some legal scholars.[62]

(e) Cartagena Protocol

The Cartagena Protocol on Biosafety was adopted by the Conference of the Parties to the Convention on Biological Diversity as supplementary agreement to the Convention on 29 January 2000. The Protocol seeks to protect biological diversity from the potential risks posed by living modified organisms resulting from modern biotechnology. Austria was among the first countries who signed the Protocol on 24 May 2000. After its ratification on 27 August 2002, it came into force on 2 February 2006.[63] The Federal Environmental Agency (*Umweltbundesamt*), which is the expert authority of the federal government in Austria for environmental protection and environmental control and the only expert institution in Austria that deals with all areas of environmental protection, acts as a connecting link (National Focal Point) as regards the Secretariat and the parties to the Protocol.[64] It is also in charge of the so-called "Biosafety Clearing House", which can be described as a specific platform aimed at the exchange of information. The fact that Austria was among the first countries to sign the Protocol already clearly shows that Austria is very much concerned about the integrity and preservation of our environment and biodiversity and that it is prepared to join the efforts in developing adequate safety measures for the environment and human health on an international level. **34**

4. Other strict liability regimes

There only remains the law of neighbours in its direct or analogous application (§ 364a ABGB) as described above.[65] **35**

---

62 *F. Kerschner*, Kausalitätshafung im Nachbarrecht? RdU 1998, 10; *P. Rummel*, annotation to OGH 26.01.1999, 5 Ob 3/99y, JBl 1999, 523.
63 BGBl III 2003/94.
64 See http://www.umweltbundesamt.at.
65 See in detail *I. Bernert*, Haftung für Genmais (Teil I), Juristische Ausbildung und Praxis (JAP) 2004/05/29.

V. Vicarious liability

1. Scope of vicarious liability

36 In general, every individual person can only be held liable for his/her own conduct. Vicarious liability is to be regarded as an exception to the rule. It is a very important concept, because contractors very often employ third parties in the performance of their obligation. The chances of the injured person to enforce a damage claim against the principal, who usually is economically well-situated, are much better than against the agent (*Gehilfe*). The scope of vicarious liability of the principal differs, depending on the legal qualification of the agent as *Erfüllungsgehilfe* in accordance with § 1313a ABGB or as *Besorgungsgehilfe* in the sense of § 1315 ABGB.

37 Pursuant to § 1313a ABGB, the *Erfüllungsgehilfe* is an agent who is employed by the principal in the performance of a valid contractual obligation. The definition also comprises independent contractors hired by him/her.[66] It is disputed, however, whether the principal must be entitled to exercise discretionary power.[67] It is not necessary that there be a special legal relationship (e.g. employer/employee relationship) between the principal and the agent. According to recent legal and court practice, the principal is also considered to be liable for agents who are only seemingly in the position of *Erfüllungsgehilfe*.[68] This appearance, however, must have been provoked by the conduct of the principal. The vicarious liability of the principal is limited to harmful activities of the agent which have an inner link to the performance of the obligation (violation of principal or ancillary obligations, violation of duty of care).

38 The principal who employs an incapable person or one known to be dangerous (as a *Besorgungsgehilfe*) in the performance of his obligations is liable for any loss the agent inflicts on a third party in this capacity (§ 1315 ABGB). The *Besorgungsgehilfe* is an auxiliary person who deals with an actual or legal affair of the principal while supervised and instructed by him. Contrary to § 1313a ABGB, the principal's liability under § 1315

---

66 OGH 14.11.1990, 1 Ob 711/89, SZ 63/201; 05.02.1992, 2 Ob 593/91, SZ 65/16; 19.12.2000, 1 Ob 178/00h, Recht der Wirtschaft (RdW) 2001/366; 07.02.2000, 8 Ob 315/99b, Österreichisches Bank-Archiv (ÖBA) 2001, 328 (annotated by P. *Apathy*); 17.10.2001, 7 Ob 237/01f, EvBl 2002/50.

67 E. *Karner* in: H. Koziol/P. Bydlinski/R. Bollenberger (eds.), Kurzkommentar zum ABGB (2nd ed. 2007) § 1313a no. 4 (with further references); *Koziol/Welser* (fn. 26) 356.

68 OGH 09.04.1996, 10 Ob 528/94, ecolex 1997, 151 (annotated by G. *Wilhelm*); 24.02.1998, 4 Ob 365/97y, ÖBA 1998, 556 (annotated by G. *Iro*); 28.06.2005, 10 Ob 17/04d, RdW 2005/757; A. *Kletecka*, Der Anscheinserfüllungsgehilfe, JBl 1996, 84.

ABGB does not require the existence of a contractual obligation between the principal and the injured party. The principal is only liable if he has hired an incapable person or one known to be dangerous (*culpa in eligendo*). There are some court decisions which qualify the independent contractor as *Besorgungsgehilfen* in the sense of § 1315 ABGB,[69] whereas the prevailing doctrine emphasizes that an independent contractor who is not bound by instructions cannot be considered a *Besorgungsgehilfe* pursuant to § 1315 ABGB.[70] The strict prerequisites for vicarious liability under § 1315 ABGB, however, have been penetrated in many cases by other legal regulations. Apart from § 1319a ABGB, according to which the holder of a path is liable for any grossly negligent behaviour of his agents vis-à-vis people using it, recent strict liability regimes provide for an extended vicarious liability: Pursuant to § 19 par. 2 EKHG, § 17 AtomHG 1999, § 56 *Forstgesetz* (Forestry Act) 1975 and § 10 *Rohrleitungsgesetz* (Pipelines Act) 1975, the holder of a dangerous good is liable for any negligent behaviour of his agents. By analogous application of these provisions, the holder of a dangerous good is generally held liable at least for the grossly negligent behaviour of agents.[71]

Vicarious liability of the principal as regulated in §§ 1313a and 1315 ABGB **39**
does not exclude liability of the agent himself vis-à-vis the injured person. In any case, however, the agent will only be liable for tortious acts, since there is no contractual obligation between the agent and the creditor of the principal. Even in cases where § 1313a ABGB is applicable, the agent will only be liable for the loss inflicted on the plaintiff by him if his behaviour can be qualified as an unlawful act irrespective of the duties emerging from the contractual obligation (e.g. violation of an absolute right of the injured person like health or property). Thus, as a general rule, the principal is liable for the violation of contractual obligations and the agent only for tortious acts. If the principal has compensated the loss, he is able to seek regress from the agent, provided that the agent has culpably violated his contractual duties vis-à-vis the principal by the harmful act (§ 1313 ABGB). There are special provisions governing regress pursuant to the Employee Liability Act (*Dienstnehmerhaftpflichtgesetz*, DHG),[72] in the case that there is an employer/employee relationship between the principal and the agent.

---

69 OGH 08.02.1968, 2 Ob 164, 165/67, JBl 1968, 473; 18.12.1973, 3 Ob 167/73, EvBl 1974/109; 28.10.1975, 3 Ob 217/75, JBl 1978, 91 (annotated by *H. Koziol*).

70 *Karner* (fn. 67) § 1315 no. 2.

71 *F. Harrer* in: M. Schwimann (ed.), Praxiskommentar zum ABGB VI (3rd ed. 2006) § 1315 no. 15 ff; *Karner* (fn. 67) § 1315 no. 6.

72 BGBl 1965/80 as amended by BGBl 1983/169.

2. Liability for people further up the food or feed production chain

40 Independent contractors can be qualified as *Erfüllungsgehilfe* in accordance with § 1313a ABGB, whereas it is not clear whether the existence of discretionary power exercised by the principal is essential. In fact, the crucial point is the interpretation of the contractual agreement between the principal and the third party and the determination of their respective duties (obligation to perform and duty of care), the performance of which shall be effected with the help of the independent contractor.[73] The applicability of § 1313a ABGB depends on whether the independent contractor was hired in the performance of the principal's contractual obligation. That is why the producer cannot be considered an *Erfüllungsgehilfe* of the seller, because he is not obliged to produce.[74] Similarly, the supplier of raw material or components is regularly not qualified as *Erfüllungsgehilfe* of the producer.[75] There is no hard and fast rule which makes the distributor or farmer automatically liable for tortious acts of anyone further up in the feed or food production chain. Rather, each case requires separate consideration of whether the respective independent contractor actually was called in by the principal in the performance of his contractual duties.

41 With respect to § 1315 ABGB, the situation is similar: It is also disputed whether an independent contractor, who is not bound by the instructions of the principal, qualifies as *Besorgungsgehilfe* in the sense of § 1315 ABGB or not. Within the feed and food production chain, there will hardly ever be a relationship which allows the independent contractor below to exercise discretionary power with regard to the independent contractor further up. Further, lacking a contractual relationship between the principal and the injured third party, the principal is only liable for culpa in eligendo, the proof of which will be very hard to establish for the injured party.

42 Thus, the scope of vicarious liability within the feed and food production chain, which typically consists of independent contractors, is limited. Nevertheless, the GTG provides for an extended vicarious liability within the range of fault liability:[76] Under § 79h par. 2 GTG, the holder is liable –

73 *Karner* (fn. 67) § 1313a no. 4.

74 OGH 08.05.1979, 2 Ob 514/79, SZ 52/74; 27.08.1981, 6 Ob 521/81, SZ 54/116; *H. Koziol*, Österreichisches Haftpflichtrecht. Besonderer Teil II (2nd ed. 1984) 340 f.

75 OGH 04.02.1988, 7 Ob 516/88, JBl 1988, 650; *F. Bydlinski*, Zur Haftung für Erfüllungsgehilfen im Vorbereitungsstadium, JBl 1995, 477, 480 ff.

76 According to § 79h par. 1 GTG, the rules of the General Civil Code are applicable to claims provided by the GTG unless otherwise regulated.

beyond § 1315 ABGB – for the behaviour of each person who with his approval contributed to an activity pursuant to § 79a GTG. The holder's vicarious liability also covers independent contractors irrespective of the existence of discretionary power. It is sufficient that the participation of the agent is approved by the holder's consensus.[77] With respect to defective products, § 12 par. 1 PHG provides for a right of recourse, which entitles the person liable under the PHG to take recourse against the producer, provided that he or his agents (in the sense of §§ 1313a, 1315 ABGB) did not cause the defect of the product. If the producer cannot be determined, recourse can be taken further up in the supply chain against the entrepreneur who put the product into circulation. The entrepreneur can escape recourse if he names the producer or the person who delivered the product to him (§ 12 par. 3 PHG).

3. Can someone further down the feed or food chain include someone further up it in a trial against him/herself?

Someone further down the feed or food chain is not able to bring someone further up into a lawsuit filed against him, nor is he obliged to include him, nor can he escape liability by shifting the loss. Provided that vicarious liability is applicable, the principal who has compensated the plaintiff is able to take recourse against the agent (§ 1313 ABGB). The right to take recourse requires that the agent has culpably violated his contractual duties vis-à-vis the principal. The principal is more likely to be taken to trial, since there is no contractual relationship between the agent and the injured party, neither under § 1313a ABGB nor under § 1315 ABGB, from which it follows that the agent is only liable for tortious acts (violations of an absolute right). In this case, the burden of proof for all requirements of the damage claim is with the injured party, whereas a claim against the principal pursuant to § 1313a ABGB leads to a reversal of the burden of proof concerning fault, because the principal is liable for the violation of a contractual obligation (§ 1298 ABGB). Under § 1315 ABGB, the principal is liable for culpa in eligendo, which in general is difficult to establish for the injured party. 43

If there is an employer/employee relationship between the principal and the agent in accordance with the DHG, there are special rules governing the recourse: There is no recourse if the employee is only liable for a very slight degree of negligence (*entschuldbare Fehlleistung*; § 4 par. 3 DHG). 44

77 *Kerschner* (fn. 5) § 79h no. 3.

The recourse is subject to reduction in accordance with § 2 par. 2 DHG in case of slight and gross negligence, which may result in the elimination of liability as far as slight negligence is concerned (§ 4 par. 2 DHG). If the injured party claims compensation from the employer, he is obliged to give notice of the litigation to the employee. If the employee himself has compensated the plaintiff, § 3 DHG provides for a corresponding right of recourse against the employer.

VI. Multiple tortfeasors

45 If several tortfeasors have caused damage, they are jointly and severally liable, unless the damage can be apportioned among them (§ 1302 ABGB). That is to say that the injured person is entitled to claim full damages from each defendant if the court is not able to estimate each defendant's share. According to § 1302 ABGB, joint and several liability of the tortfeasors will also be applied if the damage was caused with intent. The tortfeasor who has paid compensation to the injured person has a right of recourse against the other tortfeasors pursuant to § 896 ABGB. The internal allocation of the damages among the tortfeasors primarily depends on the degree of fault of each single tortfeasor.[78]

VII. Defences

1. Licence/permission to grow GM material

46 Liability under the GTG is regardless of fault, unlimited in amount and does not allow for a defence excluding liability for the approved use of GMOs. The operator is also liable for damage caused by GMOs which occur under normal conditions in the course of their approved use (e.g. in the laboratory) or their approved release.[79]

2. Consent/assumption of risk

47 As long as GMOs and GM products are considered to have no adverse effects on human health, the knowingly consumption of these products by the victim will have no impact on the tortfeasor's liability. From today's

78 *Karner* (fn. 67) § 1302 no. 14.
79 *Kerschner* (fn. 5) § 79a no. 3.

perspective, a person who knowingly consumes GM products neither exposes himself to a known or at least perceivable risk, nor does he consent to his own impairment, which – under certain circumstances – would justify the harmful act of the tortfeasor or result in contributory negligence.[80]

3. Third-party influence

§ 79c Z 2 GTG allows for a defence excluding liability of the operator if the damage was caused by a wilful act of a third party. This defence shall mainly cover terrorist acts.[81] The operator is entitled to invoke this defence only if the third party did not participate in the laboratory activities or the release of the GMOs. Liability will not be excluded if the operator violated his duties under §§ 10 and 45 GTG, which provide for specific safety measures. **48**

4. Prescription

Since the GTG does not provide for any specific statutes of limitation, the general rules of the General Civil Code apply.[82] In general, the limitation period with respect to damage claims arising from contractual obligations or from tortious acts is three years pursuant to § 1489 ABGB. It commences as soon as the injured person knows about the damage and the tortfeasor. The limitation period amounts to 30 years if the injured person does not acquire knowledge of either the damage or the tortfeasor, or if the damage results from a criminal offence, which can only be committed with intent and which is punishable by imprisonment exceeding one year. **49**

5. Other defences

§ 79c GTG provides for a set of defences which entirely exclude liability of the operator. According to Z 1 leg cit, this will be the case if the damage was caused by an act of war or similar conflicts (e.g. hostilities, civil war or insurrection) or by a natural phenomenon, which needs to be extraordinary, inevitable and not reasonably preventable as far as its conse- **50**

80 *Koziol* (fn. 29) no. 4/90 ff.
81 ME 214 BlgNR 20. GP 27.
82 See § 79h par. 1 GTG.

quences are concerned. Further, the operator will not be liable for harm caused by GMOs, if a third party caused them with intent (Z 2). Finally, the operator may invoke a defence under Z 3 if the damage was caused upon compliance with a legal rule, a specific official order or a subpoena. According to this provision, the operator will not be liable if he was obliged to use GMOs by the legislator or public authority. At present, such compulsory rules of conduct do not exist with respect to the use or release of GMOs.[83]

VIII. Remedies

1. Pecuniary compensation

(a) Bodily harm

51 The GTG does not foresee any special remedies regarding the compensation of bodily harm caused by GMOs. § 79a par. 1 GTG states the general rule that the operator is liable for the death of a person, personal injury and damage to health if the damage was caused by the specific properties of GMOs which result from their genetic modification. According to § 79h par. 1 GTG, the extent and manner of the compensation are based on the rules of the General Civil Code. Thus, a compensation claim for bodily harm primarily covers healing costs, loss of earnings and damages for pain and suffering (§ 1325 ABGB). In the event that the injured person is permanently deformed, which prevents him/her from better advancement regarding his/her professional and private career, he/she is additionally entitled to claim compensation for economic loss (§ 1326 ABGB). If the injured person dies, the tortfeasor has to compensate the surviving dependants for the treatment costs as well as for all costs connected with the death of the injured person (e.g. funeral costs) and for their lost alimony (§ 1327 ABGB).

(b) Property losses

52 Under § 79a par. 1 GTG, the operator is liable for property damage caused by the specific properties of GMOs resulting from their genetic modification. Again, the GTG does not provide for any special rules on compensat-

83 *Kerschner* (fn. 5) § 79c no. 8.

ing property losses. Consequently, the regular remedies pursuant to the General Civil Code apply (see also § 79h par. 1 GTG). The tortfeasor is primarily liable for restitution in kind as far as it is feasible and possible (§ 1323 ABGB). According to prevailing opinion however, the plaintiff is entitled to choose pecuniary compensation instead without being obliged to restore the damaged object to its original condition.[84] The measure of indemnity depends on the tortfeasor's degree of fault. Pursuant to §§ 1323 f. ABGB, only the actual damage is recoverable if the defendant is found to have acted with slight negligence, whereas loss of profits can be claimed in case of gross negligence. Clear distinction between these two types of losses is very difficult. Court rulings usually consider profit as actual damage if it is highly probable (close to certainty) that the profit would actually have been gained.[85]

§ 79b GTG provides for a remedy which deviates from the above-described general rules on compensation for property damage. If the damage to property constitutes a significant impairment to the environment, the damage claim also covers the costs of measures of reinstatement. First of all, the plaintiff may claim restitution in kind from the operator – this is possible even if these costs far exceed the market value of the damaged good.[86] If restitution in kind by the operator is not feasible or the operator is not willing to undertake it, the plaintiff is entitled to claim the costs of such measures. They cover expenses for the assessment, reduction and elimination of the environmental harm, restoration costs as well as expenses spent in order to prevent further impairment.[87] Under general tort law, restoration costs can be claimed only if they do not considerably (by about 10%) exceed the market value of the good.[88] Pursuant to § 79b GTG, all essential and necessary restoration costs are recoverable (usually unlimited in amount). The market value of the good does not serve as a limit. Under § 79b GTG, the plaintiff is also entitled to advance payment of the restoration costs, but has to refund the amount exceeding the market value of the impaired good if he does not restore the damaged good to its original condition within reasonable time. **53**

The same principles apply for damage suffered under § 79k par. 2 GTG. **54**

84 *R. Reischauer* in: P. Rummel (ed.), ABGB II/2b (3rd ed. 2004) § 1323 no. 7 ff; *Kerschner* (fn. 5) § 79a GTG no. 14.

85 OGH 24.06.1992, 1 Ob 15/92, JBl 1993, 399; 17.10.1995, 1 Ob 20/94, ÖBA 1996, 549; 07.12.1995, 2 Ob 72/94, ecolex 1996, 357; 24.03.1998, 1 Ob 315/97y, ecolex 1998, 392.

86 *Kerschner* (fn. 5) § 79b GTG no. 3.

87 Ibid., no. 5.

88 *Reischauer* (fn. 84) § 1323 no. 8 with further references.

(c) Economic losses

55 § 79a GTG as well as § 79k par. 2 GTG provide for compensation for personal injury and damage to property. Again, the GTG does not provide for special rules with respect to the compensation of economic loss. According to general tort law, pure economic loss is defined as any loss which is not based on an infringement of an absolutely protected legal interest (personal rights, property, etc.). Pure economic loss is only recoverable within the scope of contractual liability, in case of damage infliction contra bonos mores, and in case of the violation of a protective law, provided that the violated protective law is designed, inter alia, to protect from pure economic loss.[89]

56 The organic farmer involved in the above-cited example would not be able to recover the full costs of restituting his contaminated field if he remained inactive until he was able to sell his agricultural products as organic products again instead of growing conventional crops there during the qualifying period. According to § 1304 ABGB, the plaintiff is obliged to keep the pending loss as low as possible. This obligation, which is not a genuine legal duty, is called "duty to avert, minimise or mitigate loss" (*Schadensminderungspflicht*). The violation of this obligation results in a reduction of compensation.[90] The courts usually allocate the augmented damage to the plaintiff.[91]

(d) Harm to animals

57 In 1988 the Federal Law on the Legal Status of Animals[92] came into force, which introduced a new provision, § 285a ABGB, into the General Civil Code. This provision expressly declares that animals are not things in the sense of property law and are protected by specific laws. The same provision, however, determines that, in the absence of specific regulations, the rules applicable to things also apply to animals. Apart from § 1332a ABGB,[93] the ABGB does not provide for a specific compensation scheme governing animals. Consequently, if an animal is hurt or killed by a third

89 *Koziol/Welser* (fn. 26) 314.
90 *Koziol/Welser* (fn. 26) 329; *Karner* (fn. 67) § 1304 no. 9 f.
91 OGH 18.10.1966, 8 Ob 227/66, SZ 39/170; 24.05.1967, 2 Ob 155/67, ZVR 1968/123; 07.04.1970, 8 Ob 67/70, JBl 1970, 621; 21.11.1972, 8 Ob 231/72, ZVR 1973, 196; 17.01.1991, 8 Ob 38/90, ÖBA 1991, 535 = RdW 1991, 261.
92 BG über die Rechtsstellung von Tieren BGBl 1988/179.
93 See below.

party, the owner of the animal is entitled to claim compensation according to the general rules governing property damage. If the animal was killed, the owner is entitled to claim compensation according to its replacement value.[94] Pursuant to § 1331 ABGB, the compensation claim also covers the immaterial damage (*Affektionsinteresse*) sustained due to the loss of the animal if the animal was injured by an activity punishable under criminal law or out of mischief and malicious pleasure. A cow's potential for producing milk or meat would be recoverable insofar as the market value of the cow decreases. The owner of the cow would be entitled to claim the balance. The Federal Law on the Legal Status of Animals, however, introduced another provision into the ABGB, which was designed to highlight the special legal status of animals under tort law. According to § 1332a ABGB, the owner of an injured animal is entitled to treatment costs which exceed the value of the animal if a rational animal owner in the situation of the injured party would also have incurred these expenses. Under general tort law, reparation costs are limited by the market value of the damaged good.[95] Furthermore, the injured party may recover not only the successful treatment costs but also the treatment costs which failed in the end.[96] Treatment costs that have never been incurred are not recoverable at all.[97]

(e) Costs of disposal

It is generally recognized that in warranty cases the plaintiff who is **58**
entitled to a redhibitory action can demand the disposal of the defective good by the defendant regardless of fault.[98] Apart from contractual liability for warranties, the costs of disposing of the harmed goods or animals are recoverable under general tort law as well as under the strict liability regime of the GTG from the tortfeasor, who is liable for the contamination of the products or animals with GMOs pursuant to § 79a or § 79k par. 2 GTG. There is definitely a causal link between the damage (disposal costs) and the wrongful behaviour of the tortfeasor.[99]

94 M. *Gimpel-Hinteregger*, Das Tier als Sache und Ersatz der Heilungskosten für ein verletztes Tier, ÖJZ 1989, 65.
95 *Harrer* (fn. 71) § 1323 no. 42 ff.
96 *Gimpel-Hinteregger*, ÖJZ 1989, 65.
97 Ibid.
98 I. *Kurschel*, Die Gewährleistung beim Werkvertrag (1989) 49 f.
99 See also *Reischauer* (fn. 84) § 1293 no. 10.

2. Non-compensatory damages

59 The primary function of tort law is to compensate the injured party for the sustained loss.[100] There is no intention of the legislator to punish the tortfeasor, which is why non-compensatory damages are not recognized under the Austrian law of damages.

3. Other remedies

60 § 1323 ABGB expressly states the principle that any damage should primarily be compensated by way of reparation in kind. If reparation in kind is not feasible, the tortfeasor has to pay pecuniary damages. This, for instance, will be the case if the restoration proves disproportionally expensive.[101] Although § 1323 ABGB does not entitle the injured party to choose between reparation in kind effected by the tortfeasor and pecuniary damages, which enable the injured party to restore the damage him/herself or have the damage restored by a third party, the prevailing view actually recognizes the injured party's right to choose. That is why compensation in money has meanwhile become the most common way of compensation.[102]

4. Costs of pursuing a claim

(a) General cost rule

61 According to § 41 par. 2 of the Code of Civil Procedure (*Zivilprozessordnung*, ZPO),[103] the party who loses the lawsuit is obliged to pay all legal expenses which were necessary and adequately caused by the lawsuit ("loser pays" principle). If one party only partially prevails over the other party, the successful party can recover the legal costs proportionally. Thus, the general rule on the recoverability of costs in a civil trial is that the parties shall bear the costs of these proceedings in proportion to their respective failure to prevail.[104]

---

100 *Koziol/Welser* (fn. 26) 301.

101 OGH 29.05.1995, 1 Ob 620/94, SZ 68/101; 21.12.1999, 4 Ob 343/99s, EvBl 2000/104.

102 *Harrer* (fn. 71) § 1323 no. 1, 11.

103 Reichsgesetzblatt (RGBl) 1895/113 as amended by BGBl I 2009/137.

104 For further principles (principle of fault: *Verschuldensprinzip*, principle of causation: *Verursachungsprinzip*) governing legal costs see *M. Bydlinski* in: H. Fasching (ed.), Kommentar zu den Zivilprozessgesetzen II/1 (2nd ed. 2002) Vor §§ 40 ff. ZPO no. 5.

(b) Costs of establishing causation

Costs which are necessarily incurred during the collection of evidence before the procedures are opened (pre-procedural expenses) or outside the court (extra-judicial costs) are generally considered legal costs in the sense of § 41 ZPO and follow the "loser pays" principle. The courts, however, are strict upon determination of whether the pre-procedural or extra-judicial costs were necessary or not.[105] A private opinion, for instance, will often be essential for the plaintiff in order to establish causation in cases involving GMOs. It is considered to be necessary only if the current state of a person or thing requires an immediate expert opinion, which could not be achieved by way of judicial proceedings for the preservation of evidence, or if the private opinion results in the reduction of the costs of litigation.[106] If the proceedings regarding the main claim do not take place because the main claim, for instance, was settled before the proceedings started, the pre-procedural or extra-judicial costs can only be claimed in accordance with the rules of substantive law, namely tort law or contract.[107] **62**

5. Advance cover

§ 79j GTG obliges the operator who uses or releases GMOs according to **63** § 79a par. 1 GTG to provide for advance cover, be it by contracting third party liability insurance or by other appropriate means (e.g. bank guarantee, guarantee of the parent company, indemnification by governmental units).[108] This provision also covers farmers, who are strictly liable under § 79k par. 2 GTG, provided that they use GM seeds and plants. There is no duty to provide for advance cover if the operator is the state or a federal province (§ 79j par. 2 GTG). The extent and manner of advance cover are not specified by the law; they just need to be suitable for advance cover as well as consistent with the fair course of business. Third party liability insurance is only mandatory in two cases: a) use of GMOs at security level 3 on a large scale or release of GMOs on a low scale (minimum insured sum € 712,200 for each liability case) and b) use of GMOs at security level 4 or

105 *R. Fucik* in: W. Rechberger (ed.), Kommentar zur ZPO (3rd ed. 2006) Vor § 40 no. 5 with further references.
106 *Fucik* (fn. 105) Vor § 40 no. 5 with further references.
107 Ibid., Vor § 40 no. 5; *Bydlinski* (fn. 104) Vor § 40 ff. ZPO no. 4.
108 *Kerschner* (fn. 5) § 79j no. 7.

release of GMOs on a large scale (minimum insured sum € 4,069,700 for each liability case). The violation of the duty to provide third party liability insurance is sanctioned by an administrative penalty (up to € 21,800). Apart from that the GTG does not foresee any sanctions if the duty to provide for advance cover pursuant to § 79j GTG is violated.

IX. Cross-border issues – Conflict of laws

1. Conflicts rules applicable before (or instead of) Rome II

64 The Rome II Regulation entered into force on 11 January 2009. According to Art. 4 par. 1 of the Rome II Regulation, an international claim for damages based on general tort law is to be treated pursuant to the right of the state where the damage occurred (*lex loci damni*). Under the Austrian conflict of laws rules (*Internationales Privatrechts-Gesetz*, IPRG),[109] the law of the state where the conduct that caused the damage was carried out had to be applied (§ 48 IPRG).

2. Special regime for cross-border claims

65 § 79 l GTG provides for a special rule applicable to non-contractual claims for damages on the basis of interferences according to § 79k GTG. According to this provision, the law of the state where the damage occurred has to be applied. This was designed to ensure the application of the strict rules as provided by §§ 79k–m GTG in the case that the damage was caused on agriculturally used land situated in Austria.[110]

109 BGBl 1978/304 as amended by BGBl I 2009/135.
110 *Kerschner* (fn. 5) § 79l no. 1.

X. Cases

1. Due to the adventitious presence of GMOs in a field, maize which is normally sold as conventional contains GMOs beyond the legal labelling threshold. This is not discovered before the final stage of the food production chain by the producer of taco chips. The whole production is lost since the supermarket chains refuse to accept delivery from the producer.

(a) Who can sue along the chain of distribution?

A compensation claim requires the presence of actual damage. Therefore, only the party who actually suffered a loss will be able to sue. In the case above, this will only be the taco producer along the chain of distribution, because he will not receive any money for his production from the supermarket chains that refuse to accept delivery of the contaminated taco chips. The taco producer cannot sue the supermarket chains for violation of contractual duties since he is not able to deliver taco chips produced with conventional maize according to his contractual obligation and the supermarket chains consequently are not obliged to accept a defective or wrong delivery. Neither can he sue the wholesaler of the maize under tort law, who would only be liable if his fault with respect to the contamination of the maize with GMOs could be established, which will not be the case. The taco producer would have to find out who originally caused the contamination of the maize with GMOs and he would have to prove this tortfeasor's fault, which might not be possible to achieve in the case of adventitious presence of GMOs in a field. Strict liability in accordance with the GTG will not be applicable either, since the damage was not caused by the specific properties of the GM maize as required by § 79a GTG. Moreover, liability under § 79k par. 2 GTG only entitles the owner of the contaminated field to sue. As a result, the taco producer will not receive any compensation for his loss at all. 66

(b) Would the case be solved differently if the GMO content was below the labelling threshold?

If the GMO content is below the labelling threshold, the maize will be treated as conventional and the supermarket chains would violate their contractual obligations if they refused delivery of the taco chips. In this case, nobody along the chain of distribution would suffer actual damage and be entitled to sue under fault-based tort law. 67

68 For the affected farmer, who intends to grow his maize free from genetic engineering, even the fact that his maize contains GMOs below the labelling threshold might prove fatal. He must expect to have difficulties selling the maize or at least to be forced to sell it at a lower price. Consequently, even the contamination of an agricultural product below the labelling threshold might already qualify as a substantial impairment of the use of the affected farmland as required for the injunction under § 79k par. 1 GTG as well as respectively for the compensation claim in accordance with § 79k par. 2 GTG.[111]

(c) Would the case be solved differently if the admixture was not adventitious, but occurred due to the disregard of segregation rules, for example?

69 Yes, the disregard of segregation rules would fundamentally change the situation: Fault-based liability would be applicable and the taco producer would be able to sue the person who was liable for the disregarding of segregation rules under general tort law, provided that he succeeds in establishing the tortfeasor's fault.

(d) Would the case be solved differently if the GMO found was not admitted for production in your jurisdiction?

70 For the taco producer the fact that the GMO was not admitted for production would not change much: The supermarket chains would all the more refuse delivery of his production and the taco producer would be obliged to destroy it. The strict liability regime of the GTG would still not be applicable and compensation under the fault-based tort law regime would still depend on the taco producer's ability to establish the fault of the person who caused the contamination of the maize with the GMO.

111 F. *Kerschner*, Neue Gentechnikhaftung in der Landwirtschaft (§§ 79k–79m GTG), RdU 2005, 112, 115; *Kerschner* (fn. 5) § 79k no. 20.

(e) If the admixture had occurred on a non-GM field and it transpires that the GM seeds were blown from: (i) neighbouring fields; or (ii) a truck passing by, would the farmer of the affected field be liable for all or part of the loss caused further down the distribution chain?

If the contamination of the farmer's field was caused by a truck passing by, the farmer would definitely not be liable, neither under the GTG (since it was not him who deliberately used GM seed) nor under general tort law (since no fault can be established). If the neighbouring land is cultivated with GM seed, the farmer of the affected land will not be liable either. He has no way of preventing GM seeds from getting onto his soil, nor is he obliged to warn or inform third parties that his maize might contain GMOs. 71

2. Twenty years after the sale of GM maize used for food products, it turns out that it has certain disadvantageous health effects for humans.

(a) Can the producers be held liable at this point for risks unknown at the time of growing the maize? Who would be liable?

Nobody along the chain of distribution would be liable, because there is no basis for a claim – be it based on fault, be it regardless of fault – for the following reasons: First of all the applicability of general tort law under the General Civil Code (§§ 1293 ff. ABGB) would fail. Provided that the risks of the GM maize were detected only years after its sale, the plaintiffs would not be able to establish a fault on the part of anybody belonging to the chain of distribution. Moreover, the GTG would not be applicable either. § 79a GTG only provides for a liability of the operator who uses or releases GMOs in accordance with § 4 GTG, and therefore only concerns the period in time before the GM product was put into circulation. Further, the special strict liability regime of § 79k par. 2 GTG only covers the agricultural use of GM seed and plants and is consequently not applicable to the facts of the above case. Finally, there would be no liability of the producers under the PHG: On the one hand, the period of limitation ends ten years after the product has been put into circulation (§ 13 PHG), and on the other hand the development risk defence pursuant to § 8 Z 2 PHG would be applicable. 72

(b) Can compensation already be claimed at a point when the negative health effects have not yet materialised, but are to expected to according to scientific expertise/mere rumours?

73 No, a claim for compensation requires the presence of actual damage. The mere fear of negative health effects as well as the mere expectation of damage to health by experts does not entitle the affected person to file a compensation claim. Some authors, however, are of the opinion that even the detected presence of a harmful contaminant in the body of a person may already entitle the affected person to claim for compensation, even if the negative health effects have not yet materialized.[112]

(c) Would it make any difference if the GM maize had only been in use for feed, causing harm to the animals, which may or may not cause harm to humans consuming the meat as well?

74 As already stated above, the affected person is entitled to a claim for compensation only if he/she can prove that he/she suffered damage to health.

3. The driver of a food logistics company discovers that a farmer, from where he regularly picks up agricultural products fails to obey mandatory segregation rules or food or feed hygiene standards, which may lead to the admixture of GM and non-GM produce sold and packaged separately by that farmer. Does he or his employer have a duty to warn, i.e. warn the recipient of the allegedly "non-GM" produce?

75 Under Austrian civil law, an omission is unlawful only if the defendant is obliged to actively ward off the damage. Such an obligation to act may result from (pre-) contractual duties, whereas under general tort law there is no general duty to ward off damage. Under general tort law, an omission will only be considered to be unlawful for specific reasons, e.g. if a duty to take care (*Verkehrssicherungspflicht*) was violated.[113] Therefore, neither the driver of the food logistics company nor his employer has a duty to warn under general tort law. A duty to warn may only arise from a contractual obligation, the presence of which cannot be established from the facts of the case.

---

112 *Gimpel-Hinteregger* (fn. 25) 246 ff.

113 OGH 18.10.1966, 8 Ob 227/66, SZ 39/170; 15.01.1986, 1 Ob 713/85, SZ 59/7; 14.12.1993, 4 Ob 170/93, Wirtschaftsrechtliche Blätter (wbl) 1994, 210; *Koziol* (fn. 29) no. 4/60.

Damage Caused by GMOs under Czech Law

Jiří Hrádek

I. General overview

1. Special liability or redress scheme for GMOs

The Czech system of regulating genetically modified products is basically based on two groups of legislative measures: (i) on Act No. 78/2004 Coll., on the Use of Genetically Modified Organisms and Genetic Products ("Act on GM")[1] (substantially amended by Act No. 346/2005) and on a statutory instrument providing the Act on GM, Decree No. 209/2004 Coll., on Detailed Conditions for the Use of Genetically Modified Organisms and Genetic Products,[2] and further (ii) on Act No. 257/1997 Coll., on Agriculture, as amended ("Act on Agriculture"),[3] and on a statutory instrument providing the Act on Agriculture, Decree No. 89/2006 Coll., on Detailed Conditions for the Production of Genetically Modified Strains.[4] 1

The Czech legislation on GMOs does not stipulate any specific rules for liability. The potential private law liability for losses caused by GMOs is covered by the provisions of Act No. 40/1964 Coll., the Civil Code.[5] 2

---

1 Zákon č. 78/2004 Sb., o nakládání s geneticky modifikovanými organismy a genetickými produkty.

2 Vyhláška č. 209/2004 Sb., o bližších podmínkách nakládání s geneticky modifikovanými organismy a genetickými produkty.

3 Zákon č. 257/1997 Sb., o zemědělství.

4 Vyhláška č. 89/2006 Sb., o bližších podmínkách pěstování geneticky modifikované odrůdy.

5 The regulation of liability in non-labour relations can basically be divided into two branches of legislation: business and civil law legislation. The regulation of liability for damage in business relations is based on the speciality of the regulation stipulated in the Commercial Code with respect to the Civil Code. In that respect, the Commercial Code contains its specific regulation of the liability issue in sec. 373 ff., and this regulation is regarded as "comprehensive", i.e. when the relationship qualifies as a business relationship, the provisions of the Commercial Code shall apply in full regardless of regulations in the Civil Code. However, no provision of the Commercial Code enables the application of the regulation of the liability issue based on the provisions of sec. 373 ff.

3 The general provisions relating to liability in the Civil Code are based on sec. 420, and the regulation includes the general clause defining the conditions for liability of legal and natural persons in delict. Sec. 420 of the Civil Code provides that *every person is liable for damage which he causes by breaching a legal obligation.* This means that under this condition, the distinction between liability based on breach of contract, and liability based on delicts cannot be determined.

4 The civil law theory requires the following elements for liability to be established: (i) breach of a legal duty or an event qualified by the law, (ii) damage and (iii) causation between the breach and the harm consequently inflicted. In most cases of liability, fault is required, either in the form of negligence or intention. The first three elements of the liability relationship are regarded as objective; fault, on the other hand, is a subjective criterion of the liability relationship, i.e. connected with the particular person.

5 The provisions relating to strict liability are located in sec. 420a–437 of the Civil Code (with the exception of sec. 422–424). These are cases which do not need fault to be established. For fulfilment of the facts of a particular case just three conditions must be met: a legally specified event causing damage, damage and the causation between the incident, and the harm incurred.

6 The wrongful and qualified event that results in the harm presents a sufficient reason for liability and therefore no fault of the liable person is required. As no fault shall be required, the wrongdoer cannot be availed of the right of exoneration, as opposed to a comparable situation where liability is based on fault. In some cases, however, the legislator allows for the wrongdoer to release himself from liability if specific legal conditions are met.[6]

7 It has long been a subject of discussion in Czech legal theory whether the Civil Code contains a general provision for strict liability in sec. 420a of the Civil Code, which should have a subsidiary effect on all cases regulated by Czech law, i.e. for provisions of the Civil Code as well as other statutes. Experts maintain both views. According to the majority opinion, however, there is no general clause for strict liability, as opposed to liability based on fault. The provision in sec. 420a of the Civil Code merely presents a case of strict liability without being a general provision.[7]

---

to cases of damage caused by the dangerous nature of a product or organism unless such damage results from a contractual relationship. For these cases, the general provisions of the Civil Code apply.

6 Sec. 420a, 421, 427 ff., 432 ff. of the Civil Code.

7 M. *Pokorný*/J. *Salač* in: O. Jehlička/J. Švestka/M. Škárová et al. (eds.), Občanský zákoník – komentář (Civil Code – Commentary) (8th ed. 2003) 501; other view M. *Škárová* in:

Sec. 420a of the Civil Code provides for the regulation of the liability of operational activity: 8

> "Any person shall be liable for damage which he causes to another person while operating a business (sec. 420a (1)). Damage is considered to have been caused while operating a business if it was caused: (a) by an activity performed in the operation of a business or by an item used in that activity, (b) by the physical, chemical or biological impacts of the operation on its surroundings, (c) by the lawful performance or by making arrangements for such performance of those kinds of work which cause damage to someone else's immovable or which substantially impede or make impossible to use someone else's immovable (sec. 420a (2)). A person shall only exempt himself from liability for damage caused upon proving that such damage was caused either by an unavoidable event not arising from the operation of a certain business or by the conduct of the injured party (sec. 420a (3))."[8]

These provisions of the Civil Code should be especially relevant for the purpose of this study. 9

2. State liability

The liability of the state can be divided into: (i) liability of the state when exercising state power (public law relationship) and (ii) liability of the state as a legal entity if the state acts as a legal entity under sec. 420 of the Civil Code (private law relationship). 10

The state liability as the holder of the public authority is regulated by Act No. 82/1998 Coll., on Liability for Damage Based Either on Maladministration or on Illegal Decisions and on changes in Act No. 358/1992 Coll., on Notaries and their Activity (Notary Order) ("State Liability Act").[9] This special statute in relation to the general provision of the Civil Code establishes the liability of the state and limits its extent to harm caused by state bodies, bodies of self-government, or by individuals or legal persons carrying out public government as a result of a transferred power. 11

J. Švestka/J. Spáčil/M. Škárová/M. Hulmák et al. (eds.), Občanský zákoník – komentář (Civil Code – Commentary) (1st ed. 2008) 1077.

8 Translation: *TradeLinks, s.r.o.*, Civil Code – Občanský zákoník (2005).

9 Zákon č. 82/1998 Sb., o odpovědnosti za škodu způsobenou při výkonu veřejné moci rozhodnutím nebo nesprávným úředním postupem a o změně zákona České národní rady č. 358/1992 Sb., o notářích a jejich činnosti (notářský řád).

12 Damage recoverable under the State Liability Act must be caused either by an illegal decision or by maladministration. The liability of the state based on the State Liability Act shall be deemed strict and the Civil Code has a subsidiary role to the State Liability Act (see sec. 26 of the State Liability Act).

13 If damage is caused by an illegal decision or maladministration, the causal relationship is based on the *inter partes* administration act and only the party to the decision or other administration act can claim damages. Thus, it is impossible for third parties to claim damages from the state, because under Czech case law, direct causation is missing.[10]

14 The issue of the state's liability for legislative activity has not been covered expressly by the State Liability Act. The case law of the Supreme Court plays a crucial role.[11]

15 According to the current opinion of the Supreme Court, not every activity of the public authorities shall be considered an administrative procedure and a possible source of maladministration or illegal decision. This term should only cover activities in which the administrative body issued a decision as an individual legal act, i.e. a result of the decision-making activity in specific cases. Based on this conclusion, the legislative activity of any state or public body, i.e. the Government or the Parliament, cannot become subject to liability provisions stipulated in the State Liability Act because it does not meet the conditions for qualification as maladministration or illegal decision.

16 Provided that the state acts as a normal legal entity, the general provisions on liability for damage shall apply in full. No sovereign immunity applies to such cases.

---

10 Czech case law denies this possibility because of the absence of causality between the damage and the damage sustained by a third party (Supreme Court, published in the Collection of Decisions and Opinions of the Supreme Court under R 7/1979, R 21/1992, 25 Cdo 1354/2005, 25 Cdo 1355/2005). Pursuant to the Supreme Court's case law, the result must be in direct connection with the cause. It may be the case that the result arises in consequence of another circumstance that was caused by a cause that can be attributed to a wrongdoer, assuming that this consequential damage was foreseeable and therefore attributable to the wrongdoer. Causality as the necessary condition of liability must therefore also be concluded in the case when the relation between the cause and the result is indirect; however, this result is the consequence of the cause.

11 For example Supreme Court, 25 Cdo 2064/2005.

II. Damage

1. Recoverable losses

As a general rule, the Czech law of compensation for damage is based on the principle that full compensation for damage sustained should be granted to the injured party. This principle is expressed especially in sec. 442 of the Civil Code under which the injured party shall be compensated for both actual damage and lost profit. Pecuniary damage in Czech civil law is divided into two categories:[12] actual damage (*damnum emergens*) and lost profit (*lucrum cessans*); in accordance with the literature and standard judicial interpretation, damage means any loss of property which can be objectively calculated in an equivalent value, i.e. a monetary value. 17

Actual damage can be defined as damage caused to property which can be assessed by calculating the reduction or devaluation of the existing property of the injured party. Lost profit, on the other hand, is a loss sustained as the result of a wrongful event which caused the property of the injured party not to increase even though such an increase could have been expected in the usual circumstances. However, the inevitable nature of such damage is the uncertain probability of the increase of the value, amount or other aspects of property. Therefore, if the damages were not proved as the probable consequence of the defendant's wrongful act, the Czech courts would deny compensation. 18

The concept of damage is not defined by the current legislation in the Czech Republic. However, the case law in connection with the doctrine generally defines damage as "any loss of property which can be objectively calculated in an equivalent value, i.e. a monetary value."[13] Pursuant to sec. 442 of the Civil Code, damage is always recoverable either in money or in restitution in kind if the injured party so requires and if it is possible and expedient. 19

Alternatively, Czech law also acknowledges damage to the intangible sphere of the injured party for which the term "*(nehmotná) újma*" – (non-pecuniary) harm is usually used. However, in contrast to damage as described above, such harm can only be recoverable in damages (compensation, satisfaction) in certain cases stipulated by law. 20

---

12 Recently for instance Supreme Court, 25 Cdo 1307/2003.

13 Supreme Court, R 55/1971 in *M. Škárová/M. Pokorný/J. Salač* in: O. Jehlička/J. Švestka/M. Škárová, Civil Code – Commentary (10th ed. 2006) 771; includes *damnum emergens* and *lucrum cessans*.

2. Pure economic loss

21 Concerning pure economic loss, Czech law does not explicitly acknowledge this kind of damage. However, pure economic loss could fall under the category of actual damage or the lost profit category if it is proved that such loss fulfils the conditions set out by the case law and doctrine. It cannot be said in general what kind of damage pure economic loss is and whether it is recoverable as this issue is very complex and concerns the issue of causation.

3. Mere fear of a loss

22 As regards the compensation for pecuniary damage suffered, the liability of the farmer using the GMOs should be considered pursuant to sec. 420a of the Civil Code, i.e. as a case of strict liability regulated in the Civil Code. An indispensable condition for liability is the proof of the existence of the wrongful event qualified by law, the establishment of damage and causality between the operational activity and the damage sustained, in particular, the physical, chemical or biological impacts on the surroundings. Such a fact can be proved only by proving the contamination.

23 In our opinion, proof of the existence of consumer fear of crop contamination on own products would not give rise to a claim for damages, as this is not a sufficient cause for the lost profit suffered, as required under sec. 420a of the Civil Code. Thus, pecuniary damage would not be recovered under this provision.

24 The Civil Code has no clause for compensation for non-pecuniary damage which could qualify as general. Nevertheless, the rules for compensation for non-pecuniary damage to personality rights of an individual are set forth in particular in sec. 11 ff. of the Civil Code. Under this provision, any individual shall be entitled to protect his person, especially his life and health, civic honour and human dignity, as well as his privacy, his name and expressions of a personal nature. Infringement shall be compensated either by cessation of and desistance from the unlawful infringement, rectification of the consequences of such infringement or by appropriate satisfaction.

25 The protection of the individual's personality rights shall be granted only against an infringement of such personality rights which can be qualified

as unlawful.[14] However, even if the farmer using GMOs objectively caused the grounds for fear suffered by third parties, he has not breached any legal duty which could qualify his behaviour as unlawful.

Therefore, in addition to the refusal to grant damages for pecuniary damage, no satisfaction for fear or directly connected non-pecuniary loss due to the fear of developing an illness in the future would be granted. 26

4. Standard of proof

In sec. 420 (3), the Czech Civil Code regulates fault as a presumed fact and the defendant-wrongdoer has to prove that he did not act with fault. However, the theory concludes that in such a case only unconscious negligence could be presumed. In fact, this rule presents a reversal of the burden of proof for the benefit of the injured party. 27

With respect to the subject matter of this study, however, no fault shall be required in order to establish liability. 28

Any other preconditions of liability shall be proven by the injured party. 29

5. Nominal losses

Czech law does not recognise such a loss. 30

6. Mass losses

Czech law does not establish any special rules for mass losses. 31

III. Causation

1. Uncertainty of merely potential causes

Czech legal theory acknowledges that causality is based on the existence of cause and result in such a manner that without the cause no result would have occurred. The result must be in direct connection with the cause. 32

14 *J. Švestka* in: J. Švestka/J. Spáčil/M. Škárová/M. Hulmák et al. (eds.), Občanský zákoník – komentář (1st ed. 2008) 96.

33 It may be the case that the result arises as a consequence of another circumstance which was caused by something that can be attributed to a wrongdoer, assuming that this consequential damage was foreseeable and therefore attributable to the wrongdoer. Causality as an indispensable condition of liability can therefore also be concluded in the case when the relation between the cause and the result is indirect; however, this result is the consequence of the cause. Nevertheless, this conclusion is not always accepted by case law and causality is rejected in many cases.[15]

34 Contemporary Czech legal theory acknowledges two basic theories concerning the examination of causality which are presented by the authors of a textbook on civil law:[16] the theory of equivalency or *conditio sine qua non* (*teorie ekvivalence*) and the theory of adequacy (*teorie adekvátnosti, teorie adekvátní příčinnosti*).

35 The theory of adequacy, which has been predominantly applied in private law unlike the theory of adequacy uses the following criterion: the damage is regarded as a result of the wrongful activity if, besides being the condition of the damage, the wrongful act or wrongful event is due to the general nature, or, in the usual course of events and experience, a common result of the damage. For the theory of adequacy, therefore, a cause of a wrongful result can only be constituted by such a wrongful act or event as would foreseeably cause such result in a manner objectively foreseeable to any average person, i.e. also to the person to whom the relevant cause is attributable. The theory of adequacy is used predominantly in civil law which in addition to subjective fault also uses an objective examination of cause and result.[17]

36 The above-mentioned approach is influenced by a 20-year-old decision of the Supreme Court of the Czech Republic, published under R 7/1979. However, this approach was changed, or a change has been commenced, by a decision of the Supreme Court dated 24 May 2001[18] in which it concluded that even in a circumstance where the defendant is at the same time responsible for the damage to the plaintiff's item, this does not

15 Supreme Court, R 7/1979 – "The health of the plaintiff was damaged as a consequence of the reaction to the death of her child. The alleged cause therefore consists of the fact which alone is the result for which the defendant is held liable. [...] Therefore, the causality as the legal condition of the liability is missing. The direct result of the breach of the legal duty of the defendant was the death of the plaintiff's child and not the damage to the plaintiff's health."

16 *J. Švestka* in: M. Knappová/J. Švestka et al. (eds.), Občanské právo hmotné, vol. II (Substantive Civil Law) (3rd ed. 2002) 457 ff.

17 *Švestka* (fn. 16) 458 f.

18 Supreme Court, 25 Cdo 1946/2000.

exclude the causal connection between the breach of duty resulting in the damage and the damage which the plaintiff incurred in the form of lost profit.[19] Unfortunately, the Supreme Court turned again to applying the previous conclusion in its later decisions.[20]

It can be accepted that in terms of causation, the factual relationship is always the deciding factor and not the chronological relation between the cause and the wrongful result. In this respect, the chronological relationship only helps to clarify the factual causation. 37

It is the duty of the injured party to prove the relevant circumstances. The only exemption from the duty to allege and prove the relevant circumstances of the case is fault that has been presumed under the current Civil Code. 38

Pursuant to consistent case law, the case will only be further examined with respect to the fault provided the causality between the wrongful event or breach of duty and the wrongful result is proved (Supreme Court R 47/84).[21] However, it is not only the duty to prove the relevant elements of the claim which creates the duty to present the relevant facts. The injured party must present and give evidence of everything that could be relevant for the assessment of the case. In other words, it is the duty of the injured party to claim and prove all facts (burden of allegation and burden of proof). 39

The facts must always be proven without doubt. Czech tort law does not explicitly establish the categories of multiple causes. Therefore, in each particular case all of the causes leading to the wrongful result must be examined and each cause must be analysed with regard to its relation to the wrongful result. The result of this examination must be the discovery of the relevant cause, i.e. the cause which inflicted the damage in question.[22] Especially in cases covered by this study, the wrongful event qualified by law is important due to the application of sec. 420a of the Civil Code. 40

19 The Supreme Court argued in its reasoning that in the present case the logical chain of causes and results was not interrupted because the direct cause of the lost profit was a direct result of the damage to an item caused by the wrongdoer. No new fact had therefore entered into the chain of causes and results, but only a fact which had already been foreseeable for the wrongdoer before he caused the damage in question. The chronological point of view for the establishment of damage is not conclusive because it cannot be required that harm should arise immediately after the wrongdoer's action.

20 Supreme Court, 25 Cdo 1354/2005; 25 Cdo 1355/2005; 25 Cdo 1961/2005-2.

21 *Švestka* (fn. 16) 454.

22 *Švestka* (fn. 16) 457.

41 The answer to this question must be that the Czech civil law does not include any special rules on alternative or potential causation. The particular adjudication of that issue depends on the circumstances of the individual case, the allegation and the proof of the parties, and finally on the free considerations of the judge. Czech law offers only a general clause stating that anybody is liable for damage caused by the breach of a legal obligation. This provision means that a particular liable person must always be found. If there is no liable third party proven under the above-stated conditions, Czech law concludes that the damage in question must be borne solely by the injured party. This corresponds with the traditional principle *casus sentit dominus*.

2. Complex causation scenarios

42 Czech law does not establish any special rules for more complex cases.

3. Force majeure

43 Czech law does not explicitly establish the term of *force majeure*. Nevertheless, the existence of *vis maior* as a case of a qualified coincidence in the case of liability based on fault means that fault as the substantial element does not exist between the act or omission and the result.

44 However, in case of liability for GMOs, liability shall qualify as strict. Despite this qualification, legal theory acknowledges that there are some grounds which exclude the tortfeasor even from liability which is regarded as strict.

45 Pursuant to sec. 420a (3) of the Civil Code, a person shall exempt himself from liability for damage if he proves that the damage was caused either by an unavoidable event not arising from the operation of a certain business or by the conduct of the injured party.

46 Thus, in case of strict liability, causation can be established between the wrongful event and the damage because no fault is required. However, the law enables the liable party to exempt himself from liability.

4. Threshold to prove causation

47 To establish the existence of causation between damage and breach of duties or a legally qualified wrongful event, causation must always be

proved with respect to the particular case. In this respect, only the probability or expectation that a similar breach leads "beyond a doubt" to damage is not sufficient. The same applies to cases of strict liability.[23] This approach may, of course, lead to the impossibility of proving the causality between the damage and the breach of duties or the legally qualified event.

This strict approach is supported by the legal theory. The courts which **48**
decide on the particular case must consider all facts and allegations individually, and the court is entitled to evaluate all evidence brought freely and under its own consideration. As a result, each case might be assessed in a different way.

5. Special rules on causation

There are no special rules on causation that may apply to GMO cases. **49**

IV. Types of liability

1. Fault liability

(a) Special rules governing fault

The liability of the farmer using the GMO should be considered pursuant **50**
to sec. 420a of the Civil Code, i.e. as a case of strict liability regulated in the Civil Code. Indispensable conditions for liability are the proof of the existence of the event qualified by law, the establishment of damage and causality between the operational activity and the damage sustained, in particular, the results of physical, chemical or biological impacts on the surroundings. However, no fault is required in such a case.

(b) Impact of specific rules of conduct

For the establishment of strict liability based on sec. 420a of the Civil **51**
Code, it is decisive that an event qualified by law occurs. It therefore does not make any difference whether specific (statutory or customary) rules

23 *M. Holub/J. Bičovský/M. Pokorný/J. Hochman/I. Kobliha/R. Ondruš*, Odpovědnost v občanském, obchodním, pracovním a správním právu (Liability for Damage in Civil, Commercial, Labour and Administrative Law) (2003) 18.

governing GM or non-GM farming (e.g. good farming practice, segregation rules, food safety or hygiene provisions, etc.) have been violated.

52 If liability based on fault was involved, the existence of specific rules would play an important rule.

53 When considering the liability of the farmer in a similar case based on the omission to carry out various duties, the provisions of sec. 415[24] of the Civil Code (applicable also to the business relationships) must be taken into account. This provision sets out the duty to behave in such a manner that no damage to health, property or other value is inflicted. The breach of this legal duty consequentlylves the liability of the wrongdoer for damage inflicted by breach of duties pursuant to sec. 420[25] of the Civil Code or sec. 373 of the Commercial Code. Thus, since the Civil Code does not specify the applicable rules more closely, any standard rules shall be taken into account.

2. Product liability

(a) Development risk defence

54 Council Directive 85/374/EEC (Liability for Defective Products) was implemented into Czech law by Act No. 59/1998 Coll., on liability for damage caused by defective products *Způsobe vadou výrobku.*

55 Under the Product Liability Act, a product shall be deemed as any movable, including primary agricultural products and game, even if incorporated into another movable or into an immovable.

56 The Czech Republic did not implement any special provision based on sec. 7e) of the Directive. Thus, the producer may exempt himself from liability if he proves that the state of scientific and technical knowledge at the time when he put the product into circulation was not such as to enable the existence of a defect to be discovered.

57 The damages based on the Product Liability Act are governed by the same principles as other cases of compensation for damage, namely by the general provisions of the Civil Code, because sec. 10 of the Product Liability Act sets out that unless stated otherwise, the provisions of the Civil Code

24 Sec. 415: Everybody is obliged to behave in such a way that no damage to health, property, nature and the environment occurs.

25 Sec. 420 (1): Every person is liable for damage which he/she caused by breaching a legal obligation.

shall be applied hereto. However, the injured party is entitled to request damages either in accordance with the general provisions on compensation for damage or under the provisions of the Product Liability Act. In spite of this concurrence of remedies, there is no conflict with the previously existing above-mentioned categories.

(b) Alternative routes

There is no alternative system of compensating losses caused by agricultural products. **58**

(c) Impact of compliance with rules and regulations

The producer can exempt himself from liability only if he proves that he (i) **59** did not put the product into circulation, (ii) with regard to the circumstances, it is probable that the defect which caused the damage did not exist at the time when the product was put into circulation by him or that this defect came into being afterwards, (iii) the product was neither manufactured by him for sale or any form of distribution for economic purpose nor manufactured or distributed by him in the course of his business, (iv) the defect is due to compliance of the product with mandatory regulations issued by the public authorities, (v) the state of scientific and technical knowledge at the time when he put the product into circulation was not such as to enable the existence of the defect to be discovered and (vi) in the case of a manufacturer of a component, that the defect is attributable to the design of the product in which the component has been fitted or to the instructions given by the manufacturer of the product.

Under sec. 5 (3) of the Product Liability Act, the producer may also exempt **60** himself from liability if he proves that the occurrence of damage was caused by the injured party by its act or omission or other party for which the injured party is liable.

Thus, under the condition that the farmer is able to prove that he obeyed **61** all rules and regulations governing his production process, such facts should present legal reasons for exemption from liability if the legally defined conditions are met.

3. Environmental liability

(a) Implementation of the Environmental Liability Directive

62 In 2008, the Czech Parliament approved Act No. 167/2008 Coll., on Prevention of Ecological Harm and its Remedying ("Act on Prevention") which implements into Czech legislation Directive 2004/35/CE of the European Parliament and of the Council of 21 April 2004 on environmental liability with regard to the prevention and remedying of environmental damage.

63 The Act on Prevention also fully applies to GMO matter.

64 The Act on Prevention sets forth that an operator who carries out an operative activity specified in Annex 1 to the Act on Prevention is obliged to provide the financial instruments to cover his responsibilities under this Act on Prevention. These instruments shall not be provided by an operator whose operative activity cannot cause ecological harm with costs of compensation higher than CZK 20 million or ecological harm where the costs of compensation are higher than CZK 20 million, but the operator is registered in the EMAS-Programme or he has received the certificate under the ČSN EN ISO 14000 standard.

65 In this regard, the Czech Republic has not yet developed any programme of financial security instruments. The Act on Prevention stipulates in sec. 14 (5) that the Government shall specify the manners of risk assessment, the criteria for assessing adequate financial security for operators as well as conditions for implementing financial security for the preventive and corrective measures. This provision shall only become effective as of April 2011 and the duty to arrange for financial instruments from 2013.

(b) Environmental liability regime beyond the scope of the Directive

66 Conditions for public law liability for ecological harm are also stipulated in the Act on the Environment (Act No. 17/1992 Coll.).[26] The Act on the Environment is uncertain regarding ecological harm. It can therefore happen that to a certain extent it may be applicable besides the new law.

26 Zákon č. 17/1992 Sb., o životním prostředí.

The Act on the Environment sets forth a definition of ecological harm in sec. 10. Pursuant to this provision, ecological harm is a loss or impairment of the natural function of ecosystems, caused by damaging their components or disturbing their internal relations and processes as a result of human activity. 67

Pursuant to sec. 27 of the Act on the Environment, anyone who by damaging the environment or by other criminal activity causes ecological harm shall restore the natural functions of the damaged ecosystem or of its parts. If this is not possible or if there is a serious reason which makes it undesirable, he shall compensate the ecological harm in a different manner. If this is not possible, he shall compensate the damage in money. A public authority decides on the imposition of the obligation. 68

(c) Claimants in cases of environmental harm

The individual cannot receive any compensation based on the ecological harm because the state shall receive all funds from the liable party. 69

The competent authority shall recover from the operator who has caused the damage or the imminent threat of damage the costs it has incurred in relation to the preventive or remedial actions that it has taken. If the ecological harm was caused by multiple operators, they shall bear the costs jointly and severally. 70

(d) Special liability regime for losses sustained by individuals

Czech law provides for special rules applicable to cases of causing a nuisance or similar neighbourhood issues. These cases are ranged under the Second Part of the Civil Code called Rights to Real Property, and the Civil Code provides for regulation of the ownership rights to an item. To the same extent as the owner of an item, the holder of that item is also protected. 71

The provisions of sec. 127 of the Civil Code provide for special regulation of the interference with neighbours' rights. This section states that the owner of an item must abstain from anything that would cause an unreasonable amount of annoyance to another person or seriously endanger the latter's ability to exercise his rights. The owner may not endanger his neighbour's buildings or plot of land by making alternations to his own plot of land or to any building erected on such land without having taken adequate mea- 72 sures in respect of proper reinforcement of his building or other appropriate measures in respect of his plot of land. He may not vex his neighbours to an unreasonable extent by noise, dust, ash, smoke, gases, fumes, odours, solid or liquid waste, light, shadows and vibrations.[27]

73 Another crucial regulation is established in sec. 127 (3) of the Civil Code which constitutes the right of a neighbour with plots of land to get access to their plots of land, to the buildings standing upon them to the extent necessary, and for the necessary period for the required maintenance and management of the neighbour's plot of land and buildings. Where damage to a plot of land or building occurs, the person who caused the damage is obliged to compensate it. The person responsible cannot release himself from this liability.

74 Sec. 127 (3) of the Civil Code further provides that liability arising in connection with the entry to the neighbouring land cannot be excluded. That means that such liability is a special case of strict liability which does not allow application of the general provision of sec. 420 of the Civil Code. Other cases of damage caused to ownership rights are subject to the general provisions of sec. 420 ff.[28]

75 The protection of the ownership right cannot be statute-barred. However as the right to compensation is a monetary receivable, which must be regarded as a property right, it must also be subject to the statutory limitation period pursuant to sec. 106 of the Civil Code. Under this provision, the right to damages becomes statute-barred two years after the day on which the injured party became aware of the damage and of the identity of the liable party. The right to damages becomes statute-barred after three years at the latest. If the damage was caused intentionally, it is ten years from the day on which the event resulting in the damage occurred.[29]

76 Pursuant to sec. 9 of the Act on Prevention, the employees of the operator or of the competent public authority shall also be entitled to access the land affected by the ecological harm.

---

27 Translation: *TradeLinks, s.r.o.*, Civil Code – Občanský zákoník (2005).

28 O. *Jehlička* in: O. Jehlička/J. Švestka/M. Škárová et al., Občanský zákoník – komentář (8th ed. 2003) 385.

29 Pursuant to sec. 398 in connection with sec. 397 of the Commercial Code, the subjective term amounts for business relations to four years and the objective to ten years.

(e) Cartagena Protocol

The Czech Republic ratified the Cartagena Protocol on 8 October 2001. It became valid on 11 September 2003 as international treaty No. 89/2005 Coll. 77

An international treaty becomes directly applicable in the Czech Republic and it precedes standard laws. Pursuant to sec. 10 of the Constitution, promulgated international agreements, the ratification of which have been approved by the Parliament and which are binding on the Czech Republic, shall constitute a part of the legal order. Should an international agreement make a provision contrary to standard national law, the international agreement shall prevail. 78

V. Vicarious liability

1. Scope of vicarious liability

Liability for third parties shall be considered pursuant to sec. 420 (2) of the Civil Code. In accordance with this provision, damage is deemed to have been caused by a legal entity or an individual if it was caused within the scope of performance of activity by persons whom the legal entity or individual assigned to perform such activity or certain tasks within the scope of such activity. These persons shall be liable only to the extent provided by the provisions of labour law set out in the Labour Code and the extent is very limited.[30] 79

The Civil Code uses the term "employee" as an expression for the relationship between the legal entity and the subordinate person. Even though this term tends towards the premise that the relationship between the employer and the employee should be based on the provisions of the Labour Code, the scope of employment relationship must be understood in a wider manner and shall include all dependent working relationships. 80

The Commercial Code also contains relevant provisions for third-party liability. Pursuant to sec. 375 of the Commercial Code, if a breach of duty arising from a contractual relationship was caused by a third party to whom the liable party entrusted performance of its obligation, the liabil- 81

30 The maximum amount may not exceed an amount equal to four and a half times the average monthly earnings of the employee, unless the damage is caused by inebriation or while under the influence of other addictive substances. In that event, this limit shall not be applied and the member must compensate the actual damage in full.

ity of the liable party is excluded only if it is excluded under sec. 374 and when, under the same section, the third party's liability would also be excluded if the third party was bound to perform directly to the aggrieved party instead of the liable party.

82 However, the application of that provision to cases of damage caused by GMOs is disputable and unclear and should be rejected. The reason is that liability shall be based on sec. 420a of the Civil Code and not sec. 373 of the Commercial Code which would automatically provide for application of sec. 375 of the Commercial Code.[31]

2. Liability for people further up the food or feed production chain

83 There is a typical case of the co-existence of a liability relationship based on the legal guarantee regime and liability for damage. Thus, the seller shall always be liable for the defects of the sold product towards his buyer just as the wrongdoer shall be liable towards the injured party if damage occurs.

84 The distinction between general liability and liability for defects is very important because under Czech law, claims which can be asserted based on liability for defects (guarantee or warranty claims) cannot be asserted under the general compensation scheme.

85 Since sec. 420a of the Civil Code presents a case of strict liability, only a causal link must exist between the damage and the legally qualified event. The knowledge of the wrongdoer is not decisive since it presents a subjective element.

86 As the causal link must always be direct, the primarily liable party would be a person who has a relationship to the legally qualified event, even without having knowledge thereof. In the particular case, such an event would be the physical, chemical or biological impact of working with GMOs.

---

31 The case law of the Supreme Court is unclear. The Court adjudicated many times that a relationship between businessmen, which arose based on a provision outside the Commercial Code, shall be considered a relationship pursuant to Civil Code. An example would be the relationship upon the lease of non-residential premises, which is subject to specific law outside the scope of the Commercial Code (Supreme Court, 28 Cdo 2181/2007).

In our opinion, if an individual suffers damage due to consumption of products containing GMOs, the liable party should be not only the producer of such food (baker, etc.) but also the producer of the grains. The same rule applies for other cases of liability, e.g. if a person suffers damage consisting of diminishment of property or lost future profit. The multiplicity of liable parties is caused by the close relationship of all parties to the physical, chemical or biological impact. **87**

Pursuant to the case law of the Supreme Court, if conditions for the liability of more parties are met, the injured party shall be entitled to choose which relationship he will base his claim on. The liability of one party does not exclude the liability of another party.[32] **88**

3. Can someone further down the feed or food chain include someone further up it in a trial against him/herself?

The title to bring a suit against another based on applicable law denies such a possibility. It is always the claimant who determines the defendant. If the defendant brings the other party into a lawsuit filed against him, this fact will establish a new case. **89**

VI. Multiple tortfeasors

The joint liability of multiple tortfeasors is currently regulated in sec. 438–441[33] of the Civil Code. Pursuant to sec. 438, if damage is caused by multiple tortfeasors, they shall be held jointly and severally liable. This provision covers the following situations: (i) where damage was caused by contributory fault, i.e. where each wrongdoer has a psychological relationship not only to his own act or omission but also to the activities of other persons, or (ii) in a case of concurrent contribution, i.e. a case when only **90**

32 *Škárová* (fn. 7) 1068, e.g. Supreme Court published in the Collection of Decisions and Opinions of the Supreme Court under R 3/1969.

33 Sec. 438: (i) If damage is caused by two or more persons, they shall be liable for it jointly and severally. (ii) In warranted cases, a court may rule that those who caused damage shall be held liable for it to the extent of their proportionate share of the damage.
Sec. 439: Any person who is jointly and severally liable with others for damage shall settle with these persons in proportion to their share of the damage that occurred.
Sec. 440: Whoever is liable for damage caused by another person has a right of recourse against such person.
Sec. 441: If the damage caused was also the fault of the injured person, he bears corresponding liability for the damage; if the damage was exclusively his own fault, he alone bears the liability.

damage based on independent acts of wrongdoers occurs.[34] The contributory fault refers, however, not only to the concurrence of cases of liability based on fault, but also to cases of the concurrence of liability based on fault and strict liability or to cases of strict liability.

91 The primary type of joint liability is, in accordance with sec. 438 (1) of the Civil Code, joint and several liability, i.e. liability of one wrongdoer for an activity of other wrongdoers and all wrongdoers for the activity of each of them,[35] whereas each of them is entitled to recourse if he compensates more than his share of the damage. The exception to this principle is several liability, i.e. liability of the wrongdoer for a certain part of the damage which he individually caused. The application of this exception is not obvious, however, and must always be sufficiently reasoned in respect to the particulars of the case (R 80/1985).[36]

VII. Defences

1. Licence/permission to grow GM material

92 In general, the following defences based on justification are acknowledged both by legal theory and case law:

- Fulfilment of legal obligations
- Exercise of a subjective right (*neminem laedit qui iure suo utitur*). The exercise thereof must not interfere with the rights of third parties without a legal reason and must not be in contradiction to "proper morals".
- Self-help
- Self-defence
- Necessity
- Approval of the injured party

93 Based on the above conditions, the existence of a licence can generally be understood as a possibility to exercise a subjective right. Thus, the lawful exercise of such rights can be considered a defence.

94 The limitation of the licence to testing cannot be considered the lawful exercise of a subjective right if the farmer breaches the conditions stipu-

34 *Holub/Bičovský/Pokorný/Hochman/Kobliha/Ondruš* (fn. 23) 85.
35 *Pokorný/Salač* (fn. 7) 538.
36 Supreme Court, R 80/1985 in *Pokorný/Salač* (fn. 7) 538.

lated in the licence. Consequently, the farmer must be held liable provided that he cannot prove that that such damage was caused either by an unavoidable event not arising from the operation of a certain business or by the injured person's own conduct (see sec. 420a of the Civil Code).

2. Consent/assumption of risk

One of the cases of justification is the approval of the injured party. Moreover, the contributory conduct of the injured party is covered by sec. 420a (3), when the conduct of the injured party may present a reason for release from liability arising under the conditions stipulated in sec. 420a of the Civil Code. However, the operator of the operational activity bears the burden of proof. 95

3. Third-party influence

The contributory conduct of the injured party is covered by sec. 420a (3) of the Civil Code. Pursuant to this provision, the conduct of the injured party may present a reason for exemption from liability arising under the conditions of sec. 420a of the Civil Code. However, the operator of the operational activity bears the burden of proof. 96

The Commercial Code also contains a provision on contributory fault of the injured party. Pursuant to sec. 382, the injured party has no right to compensation for that part of the damage which was caused by non-fulfilment (non-performance) of its own duty as defined by the statutory provisions which were issued for the purpose of preventing the occurrence of such damage or limiting its extent. 97

Whether this specific provision for a business relationship shall also apply to cases of liability pursuant to sec. 420a of the Civil Code is unclear und should be rejected. 98

4. Prescription

Save for the exceptions mentioned (e.g. ownership rights), pursuant to sec. 106 of the Civil Code, a claim for compensation for pecuniary damage becomes statute-barred two years from the date when the injured party realised for the first time that he had suffered damage and determined who was liable for the damage. Such a claim generally runs out three years 99 after the wrongful event regardless of when the knowledge was gained; however, case law also considers the occurrence of the damage for the commencement of such a period[37] as decisive. Furthermore, should such damage have been caused intentionally, the right only becomes statute-barred ten years after the objective occurrence.

100 This differentiation means that two kinds of limitation periods are laid down by law in the Czech Republic; firstly, a subjective two-year limitation period, and secondly, a three-year or ten-year objective period. The subjective period commences with knowledge of the damage, the objective with the wrongful event (occurrence of the damage).

101 There is an exception regarding damage to the health of the injured person; only a subjective limitation period is allowed for this damage and all components of the damage to health, for instance, the loss of earnings or payments for compensation of physical injury,[38] which are understood as independent claims to compensation.

102 A claim for compensation of non-material harm pursuant to sec. 11 ff. of the Civil Code is not subject to any limitation period. However, monetary compensation for the interference with personality rights is, in accordance with the opinion expressed in the case law and legal writing, subject to the general three-year limitation period pursuant to sec. 101 of the Civil Code.[39]

103 The Commercial Code also contains a specific limitation period for business relationships. Pursuant to sec. 397, the period of the statute of limitations is four years. Under sec. 398 of the Commercial Code, in the case of the right to damages, the period of the statute of limitations begins to run as of the day when the aggrieved party learned, or could have learned, of the damage and of the identity of the party liable for its compensation; however, it shall expire no later than ten years from the day when such a breach of duty occurred.

104 Since the liability for GMOs shall be based on sec. 420a of the Civil Code, the application of the statute of limitations stipulated in the Commercial Code is unclear. On the other hand, the regulation of statutes of limitations is rather autonomous. We are of the opinion that unlike in the case of contributory fault or liability for third parties, which are directly con-

---

37 Supreme Court, 1 Cz 29/1990 in *Švestka* (fn. 14) 551.

38 *Pokorný/Salač* (fn. 7) sec. 106.

39 Higher Court of Olomouc, 1 Co 63/2003 (published in the Collection of Decisions and Opinions of the Supreme Court under R 4/2008).

nected with the regulation of compensation for damage pursuant to sec. 373 ff. of the Commercial Code, the statute of limitations presents comprehensive and autonomous legislation and should also apply to cases of sec. 420a of the Civil Code.

5. Other defences

Sec. 420a provides for exoneration reasons in the case of damage caused under the conditions of that section if the wrongdoer proves that such damage was caused either by an unavoidable event not arising from the operation of a certain business, or by the injured person's own conduct. **105**

In that respect, the interpretation and application of the term "unavoidable event" is extremely important with regard to the subject of this study. Doctrine defines an unavoidable event as an event which could not have been stopped despite exercising all possible care. These circumstances must be considered with respect to the particular conditions of the case; however, the objective point of view must also be taken into account.[40] **106**

VIII. Remedies

1. Pecuniary compensation

(a) Bodily harm

There is no special rule with regard to the liability for GMOs. **107**

As regards the scope of damages, the Civil Code includes a reduction clause in sec. 450 which sets forth the discretionary power of a judge to reduce damages in favour of the wrongdoer. Under this rule, the judge shall consider the property owned by both parties to find out if reasons meriting special consideration exist. When such a situation allows a reduction in favour of the defendant-wrongdoer, the judge shall reduce damages. **108**

Certain limits apply also to the amount of damages granted, especially in cases of non-pecuniary injuries. Under Decree No. 440/2001 Coll. on Compensation for Pain Suffering and for Aggravation of Social Position, both categories (i.e. compensation for pain suffering and for aggravation of social position) are compensated by a lump sum and the amount is deter- **109**

40 *Švestka* (fn. 16) 518.

mined by the court pursuant to a point scale set by the Decree. The judge shall apply this schedule to the particular case (the value is determined by a physician). Even in exceptional cases, the special circumstances of the particular case can be taken into account. Hereafter the judge may use his discretionary power to increase the amount of compensation payable.[41] The Decree allows "reasonable" variation from the set amount and the judge must always give the reason for his decision.

110 If an injury is fatal, a cash annuity pursuant to sec. 448 of the Civil Code shall cover the cost of maintenance of the surviving dependents whose support was provided by, or was the responsibility of, the deceased. Compensation for the cost of maintenance shall be due to the surviving dependents, unless such cost is covered by pension benefits paid for this reason. The calculation of the compensation is based on the average earnings of the deceased, although compensation for the cost of maintenance for all the surviving dependents may not in aggregate exceed the amount which the deceased could have claimed as compensation for loss of earnings under sec. 447 (2) of the Civil Code, i.e. compensation which, together with the earnings of the injured person and any disability pension does not exceed the amount determined by labour law provisions on compensation for injury resulting from accidents at work and occupational disease (sec. 448 (2) and sec. 447 (2) of the Civil Code). This limit is set at the maximum of 50% or 80% of the employee's average salary.

111 Unlike the Civil Code, the Commercial Code does not allow any reduction in the compensation for damage.

(b) Property losses

112 There is no special rule with regard to the liability for GMOs.

113 The Czech law of compensation for damage is based on the principle that full compensation for sustained damage should be granted to the injured party. This principle is expressed in sec. 442 of the Civil Code, under which both actual damage and lost profit shall be compensated to the injured party.

114 In the case of damage to a thing, actual damage shall be compensated. However, if the injured party had no possibility to buy a new thing (as is

41 Sec. 7 subs. 3 of Decree No. 440/2001 Coll.

probable in the case of older things), then actual damage also covers the costs of procurement or manufacturing a new thing.[42]

This conclusion is then decisive for cases when the damaged item is not new and its value has therefore been influenced by its previous use. In such a case the objective value must be assessed based on the market value at the moment of damage occurrence. The Supreme Court has stated this in several cases, particularly in a case published in the Collection of decisions and opinions under R 27/1977[43] and R 54/2003.[44] In the latter case the Supreme Court ruled that when assessing the amount of damage occurred by damaging a used or partly worn-out thing, the usual (market) value of the thing at the moment of its damage as well as the scope of the damage shall be taken into account. Then, from the amount representing the costs for reparation of the thing, the amount representing the appreciation against the previous state shall be deducted. **115**

On the other hand, the Czech Civil Code does not allow compensation for *pretium affectionis*. **116**

(c) Economic losses

In accordance with the decision-making of the Supreme Court, the actual damage is harm caused to property, which consists of destruction, loss, reduction or other devaluation of the existing property of the injured party.[45] **117**

With respect to the extent of the compensation, the depreciation of the products would be compensable as this alone represents the actual damage. Such damage is typically subject to an expert's appraisal and valuation; however, if the injured party is able to calculate the damage suffered precisely enough, such evidence could be found sufficient. **118**

The actual damage consisting in the diminishment of the value of the land is represented by the costs actually expended by the injured party. Case law establishes that actual damage includes expenditure which is to be expended in the future in order to restore the previous state or to mitigate all disadvantages resulting from the fact that restitution in kind was not **119**

42 Supreme Court, R 5/1978 in *Škárová/Pokorný/Salač* (fn. 13) 772.
43 *P. Vojtek*, Přehled judikatury ve věcech náhrady škody (2006), case 101.
44 Ibid., case 102.
45 *Švestka* in: O. Jehlička/J. Švestka/M. Škárová, Civil Code – Commentary (10th ed. 2006) 447.

provided (R 25/90).[46] The injured party must always prove the actual damage. In exceptional cases, however, when the damage could be determined only against many obstacles or the determination of the damage is absolutely impossible, the relevant court may use its discretion and determine the damage pursuant to its free consideration (sec. 136 of the act No. 99/1963 Coll., Civil Procedure Code).

120 In the case 25 Cdo, the Supreme Court ruled that if the damage consists of the devaluation of land due to pollution coming from a neighbour's land, it is evident that – if proved and correct – the value of the land and the property owned by the landowner diminishes, regardless of whether the landowner has already expended costs on limiting or mitigating the source of the contamination or on restoring the land to its previous condition.[47]

121 The Supreme Court has commented on another case (S IV, p. 628)[48] the merits of which are very similar to the subject of this study. The Supreme Court included this case in its commentary on damage cases and it confirmed the following qualification: *a change of the value of a vineyard caused by the fertilisation of neighbouring lands which resulted in the low productivity of the vineyard, presents actual damage.*

122 As regards lost profit, hypothetical profit is not sufficient: the lost profit must have been reasonably expected under the usual circumstances. Therefore, in every case the court must consider all the circumstances at hand to finally decide on the nature of the damage.

123 The Commercial Code provides in sec. 381 that instead of the profit actually lost, the injured party may demand compensation based on the profit attained as a rule in fair business dealings in the injured party's line of business, under the conditions similar to those in the breached contract.

(d) Harm to animals

124 Since a change of the value of a vineyard caused by the fertilisation of neighbouring lands which resulted in the low productivity of the vineyard presents actual damage, the same should apply to animals harmed by contaminated feed.

---

46 Ibid., 447.
47 *J. Hrádek* in: H. Koziol/B.C. Steininger (eds.), European Tort Law 2004 (2005) 212.
48 Supreme Court, 1 Cz 137/80 in *Škárová* (fn. 7) 1141.

The actual damage is represented by the actually expended costs of the injured party. Case law acknowledges, however, that actual damage also includes expenditure which is to be incurred in the future to restore the previous state or to mitigate all disadvantages resulting from the fact that restitution in kind was not provided (R 25/90).[49] **125**

This damage must also include lost milk production potential or the lost profit from the meat sold. **126**

(e) Costs of disposal

For relationships among business entities, damage shall be assessed and calculated pursuant to the provisions of the Commercial Code. In the Commercial Code, damage is divided into the same categories; however, there are two important differences: (i) the regulation of the Commercial Code is not of a mandatory nature and can be altered by the consent of the parties (i.e. some part of the damage can be excluded for compensation or the principles of the compensation can be changed); (ii) pursuant to sec. 380, costs incurred by the injured party in connection with breach of duties by the liable party also qualify as damage. **127**

The costs of disposing of contaminated production/animals fed with GM feed should therefore be recoverable pursuant to sec. 380 of the Commercial Code. **128**

As regards the provisions of the Civil Code, the same principles apply from sec. 420 or sec. 420a of the Civil Code. Provided that all elements for establishing liability are fulfilled, the costs incurred by the injured party in connection with the breach of duties by the liable party shall also represent damage. **129**

2. Non-compensatory damages

As regards non-compensatory damage, no *pretium affectionis* shall be compensated. Only the real value of a thing at the time of damaging shall be compensated. **130**

Another category of loss is damage to value acquired by illegal activity. Since the activity whereby the value was acquired is considered illegal (e.g. contrary to good morals), there is no interest in protecting the owner **131**

49 Supreme Court, R 25/1990 in *Pokorný/Salač* (fn. 7) 546.

or possessor from damage. This conclusion can be drawn from a decision of the Supreme Court 25 Cdo 1702/2002 in which the court refused compensation for lost profit gained as a result of activity in contravention of a non-competition clause.[50]

3. Other remedies

132 The injured party could also claim compensation for infringement of his personality rights pursuant to sec. 11 ff. of the Civil Code. This provision also applies to legal entities, even though for infringement of the name of a legal entity and good reputation the Civil Code contains a specific provision in sec. 19b.

133 Under this provision, any individual shall be entitled to protect his person, especially his life and health, civic honour and human dignity, as well as his privacy, his name and expressions of a personal nature. Infringement shall be compensated either by cessation of and desistance from the unlawful infringement, rectification of the consequences of such infringement or by appropriate satisfaction.

134 The scope of the protection is very wide and the protection is not subject to any limitation period. In accordance with established case law and legal doctrine, however, monetary compensation for infringement of personality rights must be subject to a three-year limitation period pursuant to sec. 101 of the Civil Code.[51]

4. Costs of pursuing a claim

(a) General cost rule

135 The basic rule for recoverability of costs in a civil trial is the principle "the loser pays". Pursuant to sec. 142 (1) of Act No. 99/1963 Coll., Civil Procedure Act ("CPA"), the court shall award a participant who was fully successful in a case reimbursement of the costs necessary for a useful exercise or defence of a right against the participant who lost the case.

136 The CPA contains certain rules under which exceptional procedures can be acceptable. Pursuant to sec. 142 (2) of the CPA, if the participant only had

50 *Hrádek* (fn. 47) 206.

51 *Švestka* (fn. 14) 102; Higher Court of Olomouc 1 Co 63/2003 (R 4/2008); for another view see Supreme Court, 30 Cdo 1542/2003 (minority opinion).

partial success in the case, the court shall divide the reimbursement of the costs proportionately; accordingly, the court may decide that none of the participants has a right to the reimbursement of costs. Pursuant to sec. 142 (3) of the CPA, even if the participant was only partially successful in the case, the court may award him a full reimbursement of the costs of the proceedings if the participant failed only in an inappreciable part or if the decision on the amount of the performance depended on an expert's report or on the court's consideration.

Moreover, pursuant to sec. 143 of the CPA, a defendant who lost the case **137** shall have the right to reimbursement of the costs of the proceedings against the plaintiff if the defendant's behaviour was not the cause of the submission of the petition for commencement of the proceedings.

(b) Costs of establishing causation

The costs of proceedings are in particular as follows: cash expenses of the **138** participants and their representatives including the court fee, lost earnings of the participants and their legal representatives, costs of evidence, remuneration of a notary for acts of the judicial commissioner and his cash expenses, remuneration of inheritance trustee and his cash expenses, interpreter's remuneration and remuneration for representation. The remuneration for representation shall be considered costs of proceedings only if the participant is represented by an attorney or by a notary.

Since the costs of testing can be considered as actual damage, the food pro- **139** ducer could claim compensation for such costs.

5. Advance cover

Czech law does not provide for advance cover. **140**

IX. Cross-border issues – Conflict of laws

1. Conflicts rules applicable before (or instead of) Rome II

There are no special jurisdictional or conflict of laws rules that apply to **141** harm caused by GMOs.

Consequently, the general provisions of Act No. 97/1963 Coll. on private **142** international law and procedure law (IPL) apply. Pursuant to sec. 15 of the IPL, "*claims for damages which do not result from breach of legal duties based on contracts or other legal actions shall be governed by the laws of the place where the damage occurred or of the place where a fact establishing the claim for damages came from.*"

143 This provision covers cases of damage caused either by breach of a legal duty resulting from generally binding legal provisions or damage arising from strict liability.[52]

144 The Czech regulation of the determination of a legal order applicable to the extra-contractual relationship (*lex loci delicti*) is based on an alternative application: either the laws of the place of the damage's occurrence or the place where a fact establishing the claim for damages came from. Unlike certain other legal orders, when it comes to the application of the particular legal order, the relevant court shall decide, not the parties to the extra-contractual relationship. In other words, Czech law does not allow for the choice of law in extra-contractual relations.

145 Pursuant to legal theory,[53] a rule for the determination states that the court should select the most important legal order for the particular relationship. In other words, the relationship established by the delictually caused damage should become the subject of the legal order it relates to in the closest way. The Czech Supreme Court decided in a recent case on cross-border relations that in cases of damage to health and the consequential claim of the Health Insurance Company against the wrongdoer, the law of the place where the damage occurred must govern such a relationship.[54]

2. Special regime for cross-border claims

146 Czechoslovakia concluded many bilateral treaties with other socialist countries[55] which are still valid and effective and which, pursuant to sec. 2 of the IPL, shall take precedence over the general rules of the IPL,

52 *Z. Kučera*, Mezinárodní právo soukromé (Conflict of Laws) (5th ed. 2001) 307. Damage caused in a road accident is not subject to this provision, as these cases are subject to international regulation based on international treaties, neither does it cover damage within a labour law relationship, which is subject to the special regulation of labour relations within the IPL. However, both these areas are not very relevant for the topic of this study.

53 *Kučera* (fn. 52) 308.

54 Supreme Court, 25 Cdo 2881/2004.

55 Albania, Bulgaria, Hungary, Cuba, North Korea, Poland, Yugoslavia, Soviet Union, etc.

i.e. also the provisions of sec. 15 of the IPL. These bilateral treaties also include the specific rules for the liability issue.

There is no particular bilateral or multilateral rule for GMO cross-border cases. However, the treaties with Bulgaria, Mongolia, Poland and the former Soviet Union contain a provision based on which the laws of such country shall apply where the wrongful act or other event occurred.[56] **147**

X. Cases

1. Due to the adventitious presence of GMOs in a field, maize which is normally sold as conventional contains GMOs beyond the legal labelling threshold. This is not discovered before the final stage of the food production chain by the producer of taco chips. The whole production is lost since the supermarket chains refuse to accept delivery from the producer.

(a) Please solve the case by cross-referring to your general statements above. Who can sue along the chain of distribution – the taco producer, the wholesaler of the maize, etc.?

As regards compensation for damage, liability would be based on sec. 420a of the Civil Code, which establishes a case of strict liability. The liability relationship shall arise only between the injured party and a party who is responsible for (has a relationship to) the event qualified as the reason for the damage. **148**

Since the causal link must always be direct, the primarily liable party would be a person who has a relationship to the legally qualified event. In the particular case, such an event would be the physical, chemical or biological impact of the dealings with the GMO. **149**

In our opinion, if the taco producer suffers damage due to contamination of products by GMOs, the liable party should be, on one hand, his distributor (wholesaler) but, on the other hand, also the producer of the grain. Such a multiplicity of liable parties is possible because pursuant to the case law of the Supreme Court, once the conditions for the liability of multiple parties are met, the injured party shall be entitled to choose which **150**

56 *Kučera* (fn. 52) 308.

relationship he will use as the basis for his claim for damages. The liability of one party does not exclude the liability of another party.[57]

151 The liable party may then take recourse against another liable party if this party is not involved in the compensation relationship.

(b) Would the case be solved differently if the GMO content was below the labelling threshold?

152 The liability based on sec. 420a of the Civil Code shall apply regardless of the set labelling threshold because sec. 420a establishes a case of strict liability for damage caused by the physical, chemical or biological impact of the dealings with GMOs.

153 Even though the labelling threshold was not reached, the damage suffered through the refusal by the supermarket chains can be qualified as being in a direct causal link with the qualified event. The threshold for labelling only sets out a rule on how to deal with GMOs or products and does not lead to any different qualification of the GM product.

(c) Would the case be solved differently if the admixture was not adventitious, but occurred due to the disregard of segregation rules, for example?

154 Whether the admixture was adventitious or as a result of breach of a legal duty plays no role. The liability based on sec. 420a of the Civil Code is considered objective and thus only limited reasons for release from liability shall apply.

(d) Would the case be solved differently if the GMO found was not admitted for production in your jurisdiction?

155 The results would be the same as under no. 54. However, there are administrative penalties which can be imposed in case of breach of rules for dealing with GMOs.

---

57 *Škárová* (fn. 7) 1068, e.g. Supreme Court published in the Collection of Decisions and Opinions of the Supreme Court under R 3/1969.

(e) If the admixture had occurred on a non-GM field and it transpires that the GM seeds were blown from: (i) neighbouring fields; or (ii) a truck passing by, would the farmer of the affected field be liable for all or part of the loss caused further down the distribution chain?

The farmer would not need to bear the whole liability since the law allows for release from liability. The farmer may prove that the damage was caused either by an unavoidable event not arising from the operation of a certain business (wind which brings the seeds unexpectedly far from the field) or by the conduct of the injured party (the truck driven by the aggrieved farmer). **156**

On the other hand, if the admixture was caused by a car passing along the field and driven by a third party, such an event would hardly present a reason for release from liability. The operational activity under sec. 420a of the Civil Code includes all its typical side-effects, including the transmission of genetic material from neighbouring fields by winds that are common to the area. Thus, since a car passing by is an equally common event, it would equally not cause the release from liability. **157**

2. Twenty years after the sale of GM maize used for food products, it turns out that it has certain disadvantageous health effects for humans.

(a) Can the producers be held liable at this point for risks unknown at the time of growing the maize? Who would be liable?

Since sec. 420a of the Civil Code presents a case of strict liability, the only causal link must exist between the damage sustained and the legally qualified event. The liable party's knowledge is not decisive since it presents a subjective element. **158**

As the causal link must always be direct, the primarily liable party would be a person who has a relationship to the legally qualified event, i.e. the physical, chemical or biological impact of the dealings with GMOs. In our opinion, if an individual suffers damage due to consumption of products containing a GMO, the liable parties would be both the producer of such food (baker, etc.) and also the producer of the grains. **159**

160 The limitation period of the statute of limitations in both the Civil and Commercial Codes differentiates between subjective and objective terms. The subjective term begins to run as of the day when the aggrieved party learned, or could have learned, of the damage and of the identity of the party liable for its compensation; however, it shall expire no later than three or ten years from the day when such a breach of duty occurred. Based on the case law, damage must also exist for the beginning of the objective period. A breach of duty or a certain event alone does not suffice.

(b) Can compensation already be claimed at a point when the negative health effects have not yet materialised, but are expected to according to scientific expertise/mere rumours?

161 Damage can be claimed only when the aggrieved party learned, or could have learned, of the damage and of the identity of the party liable for its compensation. As to the exact time of the appearance of damage pursuant to sec. 154 of the CPA, damage must exist (have materialised) at the latest when the court decides on the matter.

(c) Would it make any difference if the GM maize had only been in use for feed, causing harm to the animals, which may or may not cause harm to humans consuming the meat as well?

162 The difference would be only as to the direct link possible. Based on the decision of the Supreme Court (R 7/1979), if the alleged cause consists of a fact which alone is the result for which the defendant is held liable, causality as the legal condition of the liability is missing.

163 In this case, the direct result of the wrongful event attributable to the defendant was the harm to the animals and not the damage to the plaintiff's health. Thus, if a consumer directly buys food from a farmer and he suffers such damage, the causal link between the damage and the physical, chemical or biological impact of the dealings with GMOs may be (at least based on the case law of the Supreme Court) questionable.

164 On the other hand, if another party who uses the meat within its operation for further products, e.g. a butcher, and the biological or chemical impact can also have resulted from its activity enters into the chain, it can also be held liable.

3. The driver of a food logistics company discovers that a farmer, from where he regularly picks up agricultural products fails to obey mandatory segregation rules or food or feed hygiene standards, which may lead to the admixture of GM and non-GM produce sold and packaged separately by that farmer. Does he or his employer have a duty to warn, i.e. warn the recipient of the allegedly "non-GM" produce?

Neither the Civil Code nor the Commercial Code contains any general obligation to act if a third party suffers damages. Sec. 415[58] of the Civil Code (applicable also to business relationships) sets out the duty to behave in such a manner that no damage to health, property or other value is inflicted. The breach of this legal duty involves, consequently, liability of the wrongdoer for damage sustained as a result of the breach of duties pursuant to sec. 420[59] of the Civil Code. However, this obligation does not qualify as duty to protect a third party's interest. **165**

Neither the driver nor his employer has a duty to warn third parties. However, the employer may be obliged to undertake actions which ensure that his behaviour and products comply with the rule stipulated in sec. 415 of the Civil Code. **166**

As regards the employee (the driver), specific information duties and duties to act on his part towards the employer may be derived from Act No. 262/2006 Coll. of the Labour Code. **167**

58 Sec. 415: Everybody is obliged to behave in such a way that no damage to health, property, nature and the environment occurs.

59 Sec. 420 (1): Every person is liable for damage which he/she caused by breaching a legal obligation.

Damage Caused by GMOs under Danish Law

Vibe Ulfbeck

I. General overview

1. Special liability or redress scheme for GMOs

1 As of now, cultivation by means of GMOs requires permission from the Plant Directorate[1] and cultivation has not yet been practised on a large scale in Denmark.[2] Consequently, there is no case law that can illustrate the liability issues.

2 Losses caused by GMOs are covered by a special compensation regime to some extent.[3] The regime was introduced by the Act on *cultivation of genetically modified crops*[4] (the Co-existence Act). In addition, an executive order on *compensation for loss due to the presence of genetically modified material* was issued (Executive Order on Compensation).[5] According to § 1 of the Co-existence Act, it is applicable to commercial cultivation, handling, sale and transport of genetically modified crops. The system is not a liability regime. It is meant to work by way of a compensation fund. The compensation fund is financed by the state and the GMO cultivators. The system covers economic loss resulting from actual GMO presence in non-GM crops. The person suffering damage is entitled to compensation if he can prove the existence of a loss caused under the specific circumstances described in the Co-existence Act and in the Executive Order on Compensation. Compensation

1 See Executive Order no. 220 of 31 March 2005 on Cultivation of Genetically Modified Crops (Executive order on Cultivation of GM crops), § 1, sec. 1.
2 According to the Plant Directorate, only one permit had been granted by June 2006.
3 For a detailed description of this, see *V. Ulfbeck* in: B.A. Koch, Economic Loss Caused by Genetically Modified Organisms (2008) 145 ff.
4 Act no. 436 of 9 June 2004. The Act entered into force on 9 April and 17 December 2005, see Executive Order no. 224 of 31 March 2005 and Executive Order no. 1178 of 17 December 2005.
5 Executive Order no. 1170 of 7 December 2005. The Executive Order entered into force on 17 December 2005.

will be paid by the Plant Directorate (the state), provided the injured party fulfils the requirements. If the special compensatory regime does not apply, the person suffering damage may have a cause of action under ordinary tort law rules. The ordinary tort law rules have a broader area of applicability than the special compensatory regime. For instance, any type of damage can be recovered. If ordinary tort law rules are to apply, the plaintiff must prove negligence on the part of the GMO cultivator.

2. State liability

There is no principle of sovereign immunity, so in principle the state can 3
be held liable for losses caused by GMOs if there is a basis for liability (negligence). As a starting point, state liability does not deviate from ordinary tort law. However, state liability is sometimes taken to be a milder form of liability than that following from ordinary tort law. As mentioned above, the state partly finances the special compensation scheme which is in force.[6]

II. Damage

1. Recoverable losses

Under the compensation scheme, a farmer can obtain compensation if his 4
crops have been contaminated with GMOs. The farmer can claim compensation for three different types of losses: the reduction in the sales price, expenses in relation to sample taking and compensation in relation to the re-establishment of organic areas and animals. A detailed description of this has been given elsewhere.[7] For losses not covered by the compensation scheme, such as personal injury, the injured party can rely on ordinary tort law rules.

2. Pure economic loss

In principle, pure economic loss is recoverable under the same conditions 5
as other types of losses.[8]

6 See *Ulfbeck* (fn. 3) 151, para. 21.
7 See ibid., 157, para. 45.
8 See ibid., 158, para. 45.

3. Mere fear of a loss

6 These issues have been dealt with elsewhere.[9] There are no general rules on this question in Danish law. However, case law shows that in the context of neighbourhood conflicts, a diminution in value of the neighbour's property caused by the mere risk that the neighbour will suffer some kind of inconvenience is sufficient for the claiming of compensation.[10] Thus, in U 1998.1515 H, fear of a health risk was accepted as a basis for awarding damages for diminution in value. The case concerned a house which had been bought by A. Subsequently, the municipality (M) placed a high tension line near the house. At the time there was a debate as to whether high tension lines could cause health problems such as cancer. The fear that this might be so caused the value of A's house to drop. A sought damages for the lost value from M. The Supreme Court found that A was entitled to damages for the lost value. It was argued that compensation should only be payable to the extent that the inconveniences exceeded the level of what had to be tolerated. However, the Supreme Court disregarded this argument and awarded full compensation.[11] This case can be compared to the GMO situation described above, where the value of a crop drops because of fear that it contains GMO admixture. Although U 1998.1515 H concerned a different kind of harm (possible health risk), it must be assumed that a diminution in value caused by mere fear will also be sufficient for awarding damages in other cases.[12] Normally, however, damages will only be awarded to the extent that the inconveniences exceed the level of what should be tolerated – taking into account ordinary developments in society. On the basis of this decision, it must be assumed that it would also be possible to claim damages from a GMO farmer in the case that there is fear that the GMO has spread and such fear has led to a decrease in the value of the crops belonging to the conventional farmer.

4. Standard of proof

7 The compensation scheme contains no special rules on the standard of proof. In general, it is difficult to say what the standard of proof would

9 See ibid., 158, para. 46.

10 *B. von Eyben/P. Mortensen/P. Pagh*, Fast ejendom (1999) 147.

11 The fact that damages were not reduced in U 1998.1515 H was probably because the case concerned fear of *health* risk, see *Lene Pagter Kristensen*, Ugeskrift for Retsvæsen (UfR, Weekly Law Report) 2000B.403, at 412–413.

12 *Von Eyben/Mortensen/Pagh* (fn. 10) 215.

be in this type of case since none has been decided yet. However, it is often said that a mere balance of probabilities likelihood is not sufficient for something to be regarded as proved. On the other hand, there is no requirement of "certainty".

5. Nominal losses

Nominal losses are not recognized. 8

6. Mass losses

There are no special rules for mass losses. 9

III. Causation

1. Uncertainty of merely potential causes

Under the compensation scheme, compensation for loss due to the pre- 10
sence of GMOs is awarded if (1) in the same cultivation season within a certain area, a genetically modified crop of the same kind or a kind which is next of kin has been cultivated and it can be crossed with the crops of the injured party, and 2) genetically modified crops above a certain threshold level can be identified among the crops belonging to the injured party. Thus, there is no specific requirement regarding proof of causation.[13] Therefore, in the example above, it would be possible for the injured farmer to claim compensation under this Act if he fulfils the above requirements, regardless of the fact that it is unclear who violated segregation rules or good farming practices.

For losses not covered by the compensation scheme, ordinary tort law 11
rules apply. In the example above, as a starting point both potential tortfeasors would be acquitted of liability if it is not possible to prove who actually violated segregation rules/good farming practice. However, if it is considerably more likely that the one farmer rather than the other person involved violated the rules, this farmer will most likely be held responsible and vice versa.[14]

---

13 For a more detailed description see *Ulfbeck* (fn. 3) 146, para. 2.
14 *B. von Eyben/H. Isager*, Lærebog i erstatningsret (6th. ed. 2007) 278.

2. Complex causation scenarios

12 As a starting point, no special rules apply. In particular, Danish case law has not recognised the concept of proportionate liability.

3. Force majeure

13 Force majeure can have different impacts when it comes to establishing causation. A force majeure event can influence a chain of events so that the outcome becomes unforeseeable. In this case, the potential tortfeasor (T) is relieved of liability. Force majeure events can also complicate questions of (natural) causation. This is the case, for instance, if an act by T would normally lead to damage but before this happens, a force majeure event leads to the same damage. In this case it would be doubtful whether T would be held liable since this conclusion would not follow from the sine qua non rule which is the general rule on causation in Danish law.

4. Threshold to prove causation

14 See supra no. 7 (II 4).

5. Special rules on causation

15 See supra III.1. It can be added that as regards ecologically cultivated crops the Co-existence Act contains a special provision in § 9, subsec. 4, making it even easier to obtain compensation under the Act.

IV. Types of liability

1. Fault liability

(a) Special rules governing fault

16 Fault rules are applied.[15] In the case of fault-based liability, the main parameter for determining fault is the question of whether the tortfeasor has acted in a way that differs from recognised standards of behaviour in the

15 See *Ulfbeck* (fn. 3) 155, para. 40.

specific context.[16] Thus, the focus is not so much on the psychological experience of the tortfeasor as on objective standards.

(b) Impact of specific rules of conduct

If the pertinent area of the law is regulated by statutory rules defining the **17**
required conduct, these rules may be decisive for determining the question of fault. As a general rule, the burden of proof is on the injured party. He must prove that the tortfeasor has acted negligently. However, if statutory rules lay down rules of required conduct and these rules have been violated, the burden of proof will often be reversed. In these cases the tortfeasor will be liable unless he can prove that in spite of the violation of the statutory rules, he has not acted negligently.[17] As to GMO cultivation, the Government Notice on cultivation of GMO crops[18] contains several formal rules that must be observed by the GMO cultivator.[19] If these rules are violated, it is not unlikely that the courts will find that there is a presumption of fault. In that case, the GMO cultivator will be regarded as having acted negligently unless he can prove otherwise.[20]

2. Product liability

(a) Development risk defence

The development risk defence was incorporated into the Danish Product **18**
Liability Act.[21] However, there have not yet been any cases dealing with this rule under the Product Liability Act under Danish law. In contrast, one case, U 1997.203 V, has dealt with the problem outside the scope of the Act. This case concerned a glue product which had been put on the market and caused damage. The court held the manufacturer liable and applied the same strict standard of risk development defence as applied in the Directive. In general, therefore, the extent to which a manufacturer can claim protection under the development risk defence seems to be limited under Danish law.

---

16 *Von Eyben/Isager* (fn. 14) 62.
17 Ibid., 87.
18 Executive Order no. 220 of 31 March 2005.
19 For instance, appendix 1 contains rules as to the required distance between fields where GMO crops are grown and fields with conventional or ecological crops.
20 See in general, *Ulfbeck* (fn. 3) 155, para. 40.
21 Act no. 261 of 20 March 2007.

(b) Alternative routes

19 Under Danish law, the special compensation scheme described above is in force. This is not in conflict with the Product Liability Directive since the scheme is not a liability regime but a compensatory regime. Secondly, it does not apply to the types of losses covered by the Product Liability Directive (personal injuries and consumer injuries) but to financial loss suffered by a farmer.

(c) Impact of compliance with rules and regulations

20 As a general rule, there will be a presumption of no liability if the producer has obeyed all rules and regulations governing his production process.

3. Environmental liability

(a) Implementation of the Environmental Liability Directive

21 The Directive is implemented primarily by Act no. 466 of 17 June 2008 (the Environmental Damage Act, EDA). In addition, changes were made in other pieces of legislation dealing with environmental issues. Financial guarantees are not mentioned.

(b) Environmental liability regime beyond the scope of the Directive

22 According to EDA § 49, the EDA does not limit the right of the authorities to claim compensation according to ordinary tort law rules.

(c) Claimants in cases of environmental harm

23 According to EDA § 23 ff., the Minister for the Environment can demand that the person liable repairs environmental damage. If the person liable does not carry out the required repairs, the Minister has the power to order the repair work to be done and require payment from the person liable, cf. EDA § 33, sec. 2.

(d) Special liability regime for losses sustained by individuals

Act no. 225 of 6 April 1994 covers losses sustained by individuals due to damage done to the environment. The Act applies to personal injuries, property damage and certain other types of losses, cf. § 2. Liability is strict but the Act only applies to certain, narrowly defined activities described in an appendix to the Act. In addition, ordinary liability rules on conflicts between neighbours apply to cases concerning environmental damage.[22] **24**

(e) Cartagena Protocol

Denmark has signed and ratified numerous international agreements. The Cartagena Protocol was signed and was ratified on 27 August 2002. It entered into force on 11 September 2003. **25**

4. Other strict liability regimes

N/A. **26**

V. Vicarious liability

1. Scope of vicarious liability

Under Danish law the employer is liable if his employee has caused damage to a third party in the course of his work and in so doing the employee has acted negligently. In contrast, as a starting point, someone who hires an independent contractor is not liable in the event that the independent contractor causes damage to a third party. **27**

2. Liability for people further up the food or feed production chain

This depends on the type of injury suffered and the basis of liability. If there is personal injury, it may be possible to invoke product liability rules, provided the product sold can be considered "defective" because of GMO presence. In this case, the person injured would be able to sue the **28**

22 See *Ulfbeck* (fn. 3) 156, para. 42 ff.

producer of the food product (the farmer) and the producer would be liable, for instance, for the seed producer. In contrast, as a general rule, it would not be possible to sue intermediaries under the product liability rules. If there is no personal injury but only economic loss (the product can not be sold because of GMO content), product liability rules would most likely not apply since this would hardly be considered a case in which one product has caused “damage” to another. Consequently, the plaintiff would have to rely on the ordinary rules on sales of goods. In this case, as a general rule, the crop retailer would be liable for the wholesaler on the basis of the crop retailer’s contract with the plaintiff, etc.

3. Can someone further down the feed or food chain include someone further up it in a trial against him/herself?

29 Yes, under procedural law it is possible to include someone further down the chain in a trial against oneself by suing that party. In this way it may be possible to shift the loss onto someone else.

VI. Multiple tortfeasors

30 In case of multiple tortfeasors, the general rule in Danish law is joint liability. This is so even if one of the tortfeasors has only contributed on a minor scale. The loss is distributed between the tortfeasors according to the principles stated in § 25 of the Danish Liability Act.[23]

VII. Defences

1. Licence/permission to grow GM material

31 There are no cases addressing these questions.

2. Consent/assumption of risk

32 As a general rule, consent/assumption of risk does play a role in tort law but there are no cases dealing with the specific question of the impact of such rules in a GMO case. Most likely, the outcome would depend on the extent

23 Act no. 885 of 20 September 2005.

of the injured party's knowledge of any risks related to the consumption of GM products. The mere fact that the victim was aware that the product was a GM product would not necessarily have any impact at all.

3. Third-party influence

If third-party behaviour contributes to the causation of the loss, the tortfeasor and the third party will be jointly liable, cf. supra VI. Third- party interference in the cause of events may render the loss suffered unforeseeable and thereby relieve the tortfeasor of liability. **33**

4. Prescription

If a claim is brought under the special compensation scheme, the limitation rules of this scheme apply. Thus, according to § 10, sec. 2 in the Co-existence Act, the right to receive compensation will be forfeited if the claim has not been filed by 1 August in the first calendar year after harvesting the crop. Outside the scope of the compensation scheme, applicable limitation rules may be those of the Danish Product Liability Act (3, 5 or 10 years), the ordinary tort law limitation rules (3 years/20 years) or the limitation rules of the Sales of Goods Act[24] (2 years). **34**

5. Other defences

Ordinary defences would apply. In addition, a special defence applies under the compensation scheme: according to § 11, sec. 2 of the Co-existence Act, the Danish Plant Directorate may entirely or partially refuse to pay compensation if the loss is covered by an insurance benefit or by other benefits in the nature of compensation for damage. **35**

VIII. Remedies

1. Pecuniary compensation

(a) Bodily harm

The regular rules apply. **36**

24 Act no. 261 of 28 March 2003.

(b) Property losses

37 The regular rules apply.

(c) Economic losses

38 Special rules apply under the compensation scheme. According to § 9, sec. 3 of the Co-existence Act, the amount a farmer is entitled to in compensation can not exceed 1) the reduction in the sales price of the crop caused by the admixture of genetically modified material, 2) the costs for sampling and analysis and 3) any losses sustained as a consequence of requirements for the conversion of organic areas or animals subsequent to the admixture of genetically modified material.[25] According to § 9, sec. 4, the loss an organic farmer incurs is to be calculated in the same fashion.

(d) Harm to animals

39 Harm caused to animals might be compensable under the special compensation scheme, cf. § 3, sec. 3. Moreover, it must be assumed that the mere fact that an animal has eaten GM contaminated feed would entitle its owner to compensation. However, under this compensation scheme, it must also be assumed that it would not be possible to take into consideration the animal's potential for producing milk. According to ordinary tort rules, damage to animals would be considered property damage and compensable in principle. When assessing the loss, it must be assumed under normal tort rules that it would be possible to take into consideration the animal's potential for producing milk or meat. There are no cases as to the specific question on the impact of GM intake.

(e) Costs of disposal

40 According to ordinary tort law rules, the injured party has a right to obtain full compensation. As a starting point, therefore, the costs mentioned would be recoverable. However, ordinary rules on foreseeability may limit the amount of damages recoverable.

---

25 The rules have been described in more detail elsewhere, see *Ulfbeck* (fn. 3) 148, para. 8 ff.

2. Non-compensatory damages

Punitive damages are unknown under Danish law. 41

3. Other remedies

Apart from damages, a buyer would be able to claim a reduction of the price if he has bought a product which does not conform to specifications due to presence of GMO. The reduction in price can be claimed regardless of whether the seller has acted negligently or not. 42

4. Costs of pursuing a claim

(a) General cost rule

The general rule is that the court distributes the costs between the parties taking into consideration all the circumstances of the case. Often the loser will pay the greater part of the costs. 43

(b) Costs of establishing causation

There are no cases as to this question. 44

5. Advance cover

There are no rules allowing for advance cover under Danish law. 45

IX. Cross-border issues – Conflict of laws

1. Conflict rules applicable before (or instead of) Rome II

Because of Denmark's reservation to the EU treaty, Rome II will not come into force in Denmark. Instead, national rules on conflict of laws will apply. In a GMO case, one basic question would be how to deal with multi-state torts, i.e. situations in which the component factors of a case are spread over several countries. In a GMO case, this would apply if GMO crops are grown illegally in one country and spread by the wind to another country, where a non-GM farmer's production is contaminated. 46

Under Danish law, the main rule is the lex loci delicti. Thus, as a starting point, the applicable law would be the law of the country in which the tort has been committed.[26] Sometimes however, the courts have not applied the lex loci delicti rule but instead "the centre of gravity" rule.[27] It is not quite clear when the one rule is preferred to the other.

2. Special regime for cross-border claims

47 There are no special rules as to these issues.

X. Cases

1. Due to the adventitious presence of GMOs in a field, maize which is normally sold as conventional contains GMOs beyond the legal labelling threshold. This is not discovered before the final stage of the food production chain by the producer of taco chips. The whole production is lost since the supermarket chains refuse to accept delivery from the producer.

(a) Who can sue along the chain of distribution?

48 The taco producer would be able to sue the wholesaler of the maize on the basis of sales law. The wholesaler may in turn sue his contractual partner, etc., see supra no. 28, provided there is a breach of contract. Under sales law, a variety of different remedies are available to the buyer in case of breach of contract. However, in order to be able to claim damages there must be a basis of liability. According to the Danish Sales of Goods Act § 24, liability is nearly strict but is in practice interpreted as being somewhat milder. Thus, it seems doubtful whether the sellers in the chain would be liable to pay damages if there is no negligence on their part. Regardless of this, the buyers would be able to claim a reduction in price. Most likely, the producer of taco chips would not be able to base a claim on product liability rules. Firstly, it seems doubtful whether the maize containing GMOs would be regarded as "defective". Secondly, since the maize is an integrated

26 *J. Lookofsky/K. Hertz*, EU-PIL, European Union Private International law in Contract and Tort (2008) 118.
27 Ibid., 122.

part of the taco chips, it would be hard to argue that the contaminated maize has caused damage to property other than the product itself.

(b) Would the case be solved differently if the GMO content was below the labelling threshold?

Since it must be assumed that this would be a sales law case, the answer to this would depend on the agreements made between the parties. If no particular agreement as to the GMO aspect has been concluded, the question of whether the maize product should be regarded as failing to conform would have to be decided according to the principles of the Danish Sales of Goods Act § 76. **49**

(c) Would the case be solved differently if the admixture was not adventitious, but occurred due to the disregard of segregation rules, for example?

As a starting point, the same rules would apply. Thus, in general, a claim would have to be based on sales law. However, assuming that it is the maize producer who has violated segregation rules, this may have the effect of establishing a basis of liability in tort, thus enabling the producer of taco chips to make a direct claim against the maize producer. However, it is uncertain to what extent such direct cause of action would exist. Most likely, it would at least be a requirement that the violation of segregation rules could be qualified as gross negligence. **50**

(d) Would the case be solved differently if the GMO found was not admitted for production in your jurisdiction?

Under Danish law, GMOs are only permitted for production to a very limited extent.[28] If the GMO found were not permitted at all, the taco producer would probably have a stronger case in arguing non-conformity of the goods in the action against his contractual (see above under a and c) partner and also a stronger case arguing that the farmer has acted with gross negligence (see above under a and c). **51**

---

28 See *Ulfbeck* (fn. 3) 145.

(e) If the admixture had occurred on a non-GM field and it transpires that the GM seeds were blown from: (i) neighbouring fields; or (ii) a truck passing by, would the farmer of the affected field be liable for all or part of the loss caused further down the distribution chain?

52 If there is no negligence on the part of the farmer, it would not be possible for buyers further down the distribution chain to make a claim in tort law against said farmer. As to the farmer's buyer, it seems doubtful whether there would be a basis for claiming damages according to § 24 of the Danish Sales of Goods Act, since this rule does not establish entirely strict liability for the seller, see supra X.1.a.

2. Twenty years after the sale of GM maize used for food products, it turns out that it has certain disadvantageous health effects for humans.

(a) Can the producers be held liable at this point for risks unknown at the time of growing the maize? Who would be liable?

53 If the health effects are of a nature that must be qualified as personal injury, there is a prima facie case of product liability for the producers under the Danish Product Liability Act. According to this act, which implements the EU Directive on Product Liability, the liability of the producer is strict. The seed producer, the farmer and the food producer would all qualify as producers under these rules. However, the development risk defence is incorporated into the Danish Product Liability Act, see supra para. 18. Consequently, depending on the circumstances, the producers might be relieved of liability. The distributor would not qualify as a producer but would be subject to liability based on negligence with a reversed burden of proof according to § 10b in the Danish Product Liability Act. However, the distributor would also be able to rely on the development risk defence.

(b) Can compensation already be claimed at a point when the negative health effects have not yet materialised, but are expected to according to scientific expertise/mere rumours?

54 As a starting point, compensation can only be claimed if negative health effects have materialised. However, sometimes special "ad hoc" compen-

sation funds are created by the legislator. This has not yet been the case in relation to GMOs.

(c) Would it make any difference if the GM maize had only been in use for feed, causing harm to the animals, which may or may not cause harm to humans consuming the meat as well?

As a starting point, the product liability rules would also be applicable in **55**
this situation. The harm caused to the animals would be considered property damage. The damage would not be recoverable under the Product Liability Act since there is no damage to consumer goods. Thus, ordinary (judge-made) product liability rules would apply (i.e. liability based on negligence). Although there is a risk that the products would also cause damage to humans, the risk in itself would not be sufficient to establish a product liability claim.

3. The driver of a food logistics company discovers that a farmer, from where he regularly picks up agricultural products fails to obey mandatory segregation rules or food or feed hygiene standards, which may lead to the admixture of GM and non-GM produce sold and packaged separately by that farmer. Does he or his employer have a duty to warn, i.e. warn the recipient of the allegedly "non-GM" produce?

There are no cases on this. In general, however, it is the case that there is **56**
no liability for omissions in Danish law. It seems unlikely that there would be a duty to warn in the above-described situation.

Damage Caused by GMOs under English Law

Ken Oliphant

I. General overview

1. Special liability or redress scheme for GMOs

1 There is currently no civil liability or other compensation regime applying specifically to liability for GMOs (cf. the administrative liability scheme described below). The general law of tort applies.[1] In 2006, the Government conducted a public consultation about the introduction of a statutory redress scheme in respect of economic damage resulting from GMO presence in non-GM crops, but it is not yet clear whether this will result in concrete proposals.[2] There are no plans to introduce new statutory liability or compensation provisions for other damage caused by GMOs, though liability may arise in some cases under existing legal principles. The proposed scheme relates to England only, it being for the devolved authorities elsewhere in the United Kingdom to develop their own policy in the area. In June 2009, the Welsh Assembly Government launched a public consultation on co-existence – including civil liability issues – in Wales.[3]

---

1 My account draws intermittently on *K. Oliphant*, Economic Loss Caused by GMOs in the United Kingdom: England and Wales, in: B.A. Koch (ed.), Economic Loss Caused by Genetically Modified Organisms Liability and Redress for the Adventitious Presence of GMOs in Non-GM Crops (2008). See further *M. Lee/R. Burrell*, Liability for the Escape of GM Seeds: Pursuing the "Victim" (2002) 65 MLR 517 and *C. Rodgers*, Liability for the Release of GMOs into the Environment: Exploring the Boundaries of Nuisance, [2003] CLJ 371.

2 *Department for Environment, Food, & Rural Affairs DEFRA*, Consultation on proposals for managing the coexistence of GM, conventional and organic crops, June 2006. See further *id.*, Summary of responses to Defra consultation paper on proposals for managing the coexistence of GM, conventional and organic crops, November 2007.

3 *Welsh Assembly Government*, Consultation on proposals for managing the co-existence of genetically modified, conventional and organic crops in Wales, June 2009.

As noted above, an *administrative* liability scheme already applies to damage to the environment arising from the escape or release from human control of GMOs. But the scheme is of very limited scope. Under Part VI of the Environmental Protection Act (EPA) 1990, a person who contravenes the duties that the Act imposes in connection with (*inter alia*) the release or marketing of GMOs – for example, failure to comply with risk assessment requirements[4] or releasing GMOs when there is a risk of damage to the environment as a consequence, despite the precautions that can be taken[5] – may be convicted of an offence[6] and required to take such steps as the court deems appropriate to remedy matters.[7] The Act also provides for the Secretary of State to arrange for reasonable steps to be taken towards remedying harm caused by the offence and to recover the cost from any person convicted of it.[8] Only a person convicted of one of the specified offences can be made to remedy, or bear the cost of remedying, the harm caused by the GMOs. Proceedings are by way of criminal prosecution initiated by the state, not civil action initiated by an individual suffering loss. Because of this, and the regime's limited scope, I shall not consider it further in this report. **2**

2. State liability

In general, public bodies and public officials are subject to the same tortious liabilities as private persons (e.g. negligence, breach of statutory duty, nuisance and trespass). By way of exception, there is a specific tort – misfeasance in public office – that is exclusively targeted at public officials, while the statutory liability under the Human Rights Act (HRA) 1998 applies only to a public authority. The former is wider than the general liability for negligence in some respects (e.g. regarding the recovery of pure economic loss), but is limited by the requirement that the defendant have either an intention to injure the claimant or knowledge that the act in question is unlawful and will probably injure the claimant.[9] The latter liability arises where a public authority breaches its duty under the Act not to act incompatibly with a person's Convention rights,[10] i.e. the substan- **3**

4 EPA 1990, sec. 108.
5 EPA 1990, sec. 109(4).
6 EPA 1990, sec. 118.
7 EPA 1990, sec. 120.
8 EPA 1990, sec. 121.
9 *Three Rivers District Council v Governor and Company of the Bank of England* (No. 3) [2003] 2 AC 1.
10 HRA 1998, sec. 6(1).

tive rights arising under the ECHR and designated protocol rights (including protocol 1, art. 1: protection of property). Damages are to be awarded for breach of this duty only if necessary to afford just satisfaction to the person in whose favour it is made.[11]

4 I think it is unlikely that either of these specific liabilities will play much of a role in respect of losses caused by GMOs.

II. Damage

1. Recoverable losses

5 What constitutes actionable damage differs according to the tort in question. The principal torts that could be relied on in this context are negligence, private nuisance, and the rule in *Rylands v Fletcher*. The last is regarded as a subset of private nuisance[12] but has its own distinct requirements (see no. 35 ff., below).

6 In negligence, the claimant must normally establish physical injury to his person or property. Should an unwanted GM presence in the claimant's crop be so treated? It must first be admitted that not every physical change in the claimant's property warrants the conclusion that it has been damaged. In a recent case,[13] the House of Lords considered analogous issues in the context of personal injury, ruling that physical changes in the claimant's body (the development of pleural plaques) did not on the facts satisfy the damage requirement of the tort of negligence because they were asymptomatic, not visible or disfiguring, and not the first stage of any other, significantly debilitating condition. There was some difference of opinion in the Lords on whether this was an application of the maxim *de minimis non curat lex*,[14] but the stronger view is that it was.[15] Although this was a case of (alleged) personal injury, the same principles undoubtedly apply to property damage. A crucial question for the court would be whether a GM presence in non-GM crops as small as 0.1%, but resulting in loss of organic certification, could be considered "more than

11 Human Rights Act 1998, sec. 8(3).

12 *Cambridge Water Co Ltd* v *Eastern Counties Leather plc* [1994] 2 AC 264.

13 *Rothwell v Chemical & Insulating Co Ltd, Re Pleural Plaques Litigation* [2007] UKHL 39, [2008] 1 AC 281.

14 See *K. Oliphant*, England and Wales, in: H. Koziol/B.C. Steininger (eds), European Tort Law 2007 (2008) no. 5 ff.

15 This is the interpretation advanced by, e.g., *D. Leczykiewicz* (2008) 124 LQR 548 and *M. Jones* (2008) 24 PN 13.

minimal". The stronger view, implicit in the court's approach in the pleural plaques case, is that regard should be had to the consequences that flow from the physical change, and that a loss of organic certification resulting from GM presence in the crop should be regarded as sufficiently significant to warrant the conclusion that it has been damaged.[16] Nevertheless, what constitutes damage in the tort of negligence remains to be fully explored by the courts.

A different approach prevails in the tort of private nuisance. A private nuisance is the unreasonable interference with the claimant's use or enjoyment of land. It is not necessary to show that the claimant suffered physical damage: interference with his amenity interests will suffice. But the interference must be substantial. In a well-known dictum,[17] James LJ stated that the damage must be "visible" and that purely "scientific evidence, such as the microscope of the naturalist, or the tests of the chemist", would not suffice to establish it: "The damage must be such as can be shewn by a plain witness to a plain common juryman." It has been doubted whether adventitious GM presence would be visible in this way.[18] However, a lack of visible damage does not preclude liability in private nuisance in other contexts,[19] and it is submitted that visibility is not required in a literal sense, but only insofar as the alleged damage must manifest itself in some way that would be appreciable to an ordinary, informed person. 7

2. Pure economic loss

In English law, there is generally no duty to take reasonable care to avoid causing purely economic loss to another,[20] and the only exceptions (e.g. voluntary assumption of responsibility) would appear to be of limited application in the GM context. So a claim for (e.g.) a farmer's loss when a feared GM presence leads a retailer to cancel its repeat order would be 8

16 *Contra, M. Lee/R. Burrell* (supra fn. 1) 530 and, in Canada, *Hoffmann v Monsanto Canada Inc*, 2005 Saskatchewan Court of Queen's Bench (SKQB) 225, [2005] 7 Western Weekly Reports (WWR) 665.

17 *Salvin v North Brancepeth Coal Co* [1873] LR 9 Ch App 705, 709. Cf. Mellish LJ at 713: the damage must be such that "every fairly instructed eye can really and clearly see it." See further Rodgers (fn. 1), 382–7.

18 *Agriculture and Environment Biotechnology Commission*, GM Crops? Coexistence and Liability (2003), para. 270.

19 E.g. water pollution: see *Cambridge Water Co Ltd* v *Eastern Counties Leather plc* [1994] 2 AC 264.

20 This general exclusionary rule was recently affirmed by the House of Lords in *Customs and Excise Commissioners v Barclays Bank plc* [2006] UK HL 28, [2007] 1 AC 181.

very unlikely to succeed on the basis of negligence. There is a stronger possibility of success in the law of nuisance as there is no requirement that the claimant prove physical damage, and the recovery of pure economic loss raises no special problems.[21] But there might be issues whether the loss was properly attributed to the defendant or only to the retailer.

3. Mere fear of a loss

9 In *Rothwell v Chemical & Insulating Co Ltd*,[22] in the context of occupational exposure to asbestos, the House of Lords confirmed that neither the risk of future harm, nor anxiety at that risk, nor risk combined with anxiety was actionable damage in the tort of negligence. Their Lordships also ruled out liability for a recognised psychiatric condition caused by such anxiety, even if combined with the reasonably foreseeable risk of the physical harm in future, unless some psychiatric illness had been the reasonably foreseeable consequence of the defendant's negligence.

10 In private nuisance, the claimant's well-founded fear of injury may serve to establish an unreasonable interference with his or her use or enjoyment of land, and so give rise to liability, but the claimant will have to prove on the balance of probabilities the reality of the danger that has given rise to the fear.[23] As private nuisance is a tort to land, no liability can arise in respect of the fear itself.

4. Standard of proof

11 The claimant must prove the elements of the tort, including the damage necessary to establish a cause of action, on the balance of probabilities. Once proven, each of these elements is treated as a certainty. However, the balance of probabilities test does not apply in quantifying the damages to be paid in respect of a proven tort. The court must simply make an estimate of the chance that a given loss will occur, or that its occurrence was attributable to the tort, and adjust the damages to reflect the relevant probability.[24]

---

21 For examples of recovery of pure economic loss, see *Rose v Miles* (1815) 4 M&S 101 (public nuisance); *Andreae v Selfridge & Co Ltd* [1938] Ch 1 (private nuisance).

22 [2008] 1 AC 281.

23 *Birmingham Development Co Ltd v Tyler* [2008] EWCA Civ 859, [2008] BLR 445.

24 *Mallett v McMonagle* [1970] AC 166.

5. Nominal losses

It is generally accepted in English law that a de minimis loss is not actionable damage. But a few exceptional torts, including trespass to land, are actionable per se, meaning that liability arises for the mere interference with the claimant's protected interests (e.g. the mere invasion of the claimant's land), and there is no need to prove any consequential harm. In such a case, it is normal to award the successful claimant nominal damages. However, the "invasion" of the claimant's land by GMOs is unlikely to be regarded as a trespass unless the GMOs were deliberately released onto the land,[25] and the more likely causes of action are in private nuisance or (if physical damage results) negligence. **12**

6. Mass losses

There are no special rules of substantive English tort law relating to mass losses. **13**

III. Causation

1. Uncertainty of merely potential causes

The GMO must be shown to have caused or at least to have made a material contribution to the damage,[26] or – conceivably – to have made a material contribution *to the risk* that the damage might occur.[27] The claimant must normally show that the damage would not have occurred *but for* the conduct alleged to be tortious, and that it was of a type that was reasonably foreseeable.[28] **14**

According to normal principles of tort law, where two sources of a harmful thing combine to cause the claimant injury, and the defendant is responsible for one of the sources, he can be held liable on the basis of his material contribution to the claimant's injury, without having to show that the **15**

25 See *Hoffman v Monsanto* [2005] 7 WWR 665 at [133].

26 *Bonnington Castings Ltd v Wardlaw* [1956] AC 613.

27 *McGhee v National Coal Board* [1973] 1 WLR 1; *Fairchild v Glenhaven Funeral Services Ltd* [2003] 1 AC 32.

28 *Overseas Tankship (UK) Ltd v Morts Dock & Engineering Co Ltd, The Wagon Mound* [1961] AC 388. This rule applies even under the strict liability rule in *Rylands v Fletcher*: *Cambridge Water Co Ltd* v *Eastern Counties Leather plc* [1994] 2 AC 264.

injury would not have occurred but for his contribution.[29] Damages are then awarded on a proportionate basis, according to the extent of the defendant's contribution to the injury insofar as this can be assessed; if necessary, a surrogate criterion may be employed (e.g. the length of time the claimant was exposed to each source).[30]

16 Where there are two or more potential causes of the damage, and it is disputed whether or not the damage resulted (wholly or partially) from the potential cause for which the defendant is responsible, the claimant is normally required to prove the defendant's contribution to the damage on the balance of probabilities.[31] However, by way of exception to the general approach, it is sufficient in some cases that the defendant materially increased the risk of the claimant's injury, even if it cannot be shown on the balance of probabilities that he actually contributed to the injury.[32] The exception is of uncertain scope, and it cannot be lightly assumed that it would be applied to liability for GMOs. But the House of Lords has recently affirmed that the principle applies not just where all sources of the risk were tortious,[33] or – if not all tortious – were at least all within the defendant's control,[34] but also where there were a number of quite independent tortious and non-tortious exposures to the same risk, and even if part of the exposure was the claimant's own fault.[35] Quite how far the courts are willing to take the exception is at present rather unclear, but there is certainly a possibility that it might be applied in a case where the damage might have been caused by any one of several GM farmers in the claimant's vicinity, but it cannot be established against any of them individually that they were more likely than not to have contributed.[36]

17 It should be noted that liability under the exception is attributed on a proportionate basis, relative to the extent of the defendant's contribution to the risk.[37]

29 *Bonnington Castings Ltd v Wardlaw* [1956] AC 613; *Bailey v Ministry of Defence* [2008] EWCA Civ 883; [2009] 1 WLR 1052.

30 *Holtby v Brigham & Cowan (Hull) Ltd* [2000] 3 All ER 421.

31 *Wilsher v Essex Area Health Authority* [1988] AC 1074.

32 *McGhee v National Coal Board* [1973] 1 WLR 1; *Fairchild v Glenhaven Funeral Services Ltd* [2003] 1 AC 32.

33 *Fairchild v Glenhaven Funeral Services Ltd* [2003] 1 AC 32.

34 *McGhee v National Coal Board* [1973] 1 WLR 1.

35 *Barker v Corus UK Ltd* [2006] UKHL 20, [2006] 2 AC 572.

36 Cf. *Fairchild v Glenhaven Funeral Services Ltd* [2003] 1 AC 32.

37 *Barker v Corus UK Ltd* [2006] UKHL 20, [2006] 2 AC 572.

2. Complex causation scenarios

There are no special rules for complex cases other than those explained above. Market-share liability has not yet been recognised in English law, but neither has it been ruled out.[38] **18**

3. Force majeure

There is no tortious doctrine of force majeure as such, but *vis major* is exceptionally recognised as a defence to the strict liability arising under *Rylands v Fletcher* (considered below).[39] Further, an act or event that intervenes in point of time between the defendant's act or omission and the claimant's injury may have the effect of "breaking the chain of causation" and so relieve the defendant of liability even if the claimant's injury would not have occurred but for the act or omission in question. Broadly speaking, the intervening occurrence must be independent of the defendant's conduct, and neither reasonably foreseeable nor reasonably preventable by him.[40] **19**

4. Threshold to prove causation

More likely than not. **20**

5. Special rules on causation

N/A. **21**

IV. Types of liability

1. Fault liability

(a) Special rules governing fault

There are no special rules here. As in the general law, fault is established by reference to the standards of the reasonable person, balancing the prob- **22**

38 *Fairchild v Glenhaven Funeral Services Ltd* [2002] UKHL 22, [2003] 2 AC 32, para. 74, per Lord Hoffmann.

39 See, e.g., *Nichols v Marsland* (1876) 2 Ex D 1, and compare *Greenock Corporation v Caledonian Railway* [1917] AC 556.

40 See, e.g., *Carslogie Steamship Co v Royal Norwegian Government* [1952] AC 292 (heavy storm).

ability of harm resulting from the activity in question, and the likely gravity of that harm if it should result, against the cost to the defendant of taking precautions against the risk, and activity's social utility. The burden of proving fault always rests on the claimant at common law, and it has been argued that discharging the burden in the area with which we are concerned will be "difficult" as, for example, "it is in the nature of 'reasonable' GM farming techniques that they will bring about cross-pollination."[41]

(b) Impact of specific rules of conduct

23 In English law, violation of a statutory safety rule or norm of customary practice may be treated as rebuttable evidence of fault. Breach of a statute may also give rise to liability in the tort of breach of statutory duty, but this is limited by the requirement that Parliament must have intended such breach to be actionable in damages, rather than merely resulting in (say) the application of a criminal or administrative sanction.

2. Product liability

(a) Development risk defence

24 The development risk defence is incorporated in English law in the Consumer Protection Act (CPA) 1987, sec. 4(1)(e). The defence applies where

> "the state of scientific and technical knowledge at the [time the product was supplied by the producer] was not such that a producer of products of the same description as the product in question might be expected to have discovered the defect if it had existed in his products while they were under his control..."

25 This formulation could be seen as more generous than that in art. 7 of the Directive ("the state of scientific and technical knowledge at the time when he put the product into circulation was not such as to enable the existence of the defect of the defect to be discovered") as its apparent focus is on what might reasonably be expected of producers rather than what was possible. However, in infringement proceedings brought against the United Kingdom by the European Commission, the European Court of Justice ruled that there was no clear conflict between the Act and the

41 *M. Lee/R. Burrell* (fn. 1) 530.

Directive.[42] The UK courts could be relied upon to interpret s. 4(1)(e) so as to achieve the result which the Directive had in view, in accordance with their general interpretative obligation under s. 1(1). The s. 4(1)(e) defence did not arise simply because a producer in the sector in question could not reasonably have been expected to know of the risk: the question was whether any scientific and technical data that was "accessible" at the time the product was put into circulation was such as to enable the existence of the defect to be discovered. The English courts have subsequently clarified that, where the existence of a defect is known to be theoretically possible, there is no development risk defence even if there is no known way of preventing its occurrence.[43]

(b) Alternative routes

There is no alternative regime of this nature. **26**

(c) Impact of compliance with rules and regulations

Doing what is required by statute cannot amount to negligence, but if it is **27** reasonable to expect the statutory requirements to be fulfilled without causing the claimant damage, yet the defendant chooses to fulfil them in such a way that damage results, that prima facie constitutes fault.[44] A similar analysis could doubtless be adopted in considering the question of defect in a product liability context.

3. Environmental liability

(a) Implementation of the Environmental Liability Directive

The Directive is implemented by the Environmental Damage (Prevention **28** and Remediation) Regulations 2009 (SI 2009/153).[45] The Regulations apply to environmental damage (i.e. damage to protected species or natural habi-

---

42 *European Commission* v *United Kingdom* [1997] ECR I-2649.

43 *A v National Blood Authority* [2001] 3 All ER 289. The analysis in the section draws upon *A. Mullis/K. Oliphant*, Torts (3rd ed. 2003) ch. 4.7.

44 *Geddis v Proprietors of Bann Reservoir* (1878) 3 App Cas 430.

45 See further *DEFRA*, The Environmental Damage (Prevention and Remediation) Regulations 2009: Guidance for England and Wales, May 2009. Available online at http://www.defra.gov.uk/environment/policy/liability/pdf/indepth-guide-regs09.pdf.

tats, or a site of special scientific interest, or surface water or groundwater, or land)[46] that is caused by designated activities. In the case of damage to protected species, natural habitats or a site of special scientific interest, the application of the Regulations is extended to other activities where the operator intended to cause environmental damage or was negligent as to whether environmental damage would be caused.[47] The designated activities to which the Regulations apply include waste management operations, water abstraction and impoundment, the manufacture, use, storage, processing, filling, release into the environment and on-site transport of dangerous substances, plant protection products and biocidal products, and the transport of dangerous or polluting goods.[48] Of note for present purposes is that the designated activities also include:

> "(1) Any contained use, including transport, involving genetically modified organisms (including genetically modified micro-organisms as defined by Council Directive 90/219/EEC on the contained use of genetically modified micro-organisms)."

29 And:

> "(2) Any deliberate release into the environment, transport and placing on the market of genetically modified organisms as defined by Directive 2001/18/EC of the European Parliament and of the Council on the deliberate release into the environment of genetically modified organisms."[49]

30 The regulations provide that the operator of an activity that has caused environmental damage, or an imminent threat of environmental damage, must immediately take all practicable steps to prevent the damage, or further damage, and notify the appropriate enforcing authority.[50] In an emergency, or if the operator cannot be ascertained, or if the operator fails to comply, the enforcing authority may carry out these duties, and any other duty of the operator, instead.[51] In such event, the enforcing authority is entitled to recover the reasonable costs of its reasonable action from the operator.[52]

46 Reg. 4.
47 Reg. 5.
48 Schedule 2.
49 Schedule 2, para. 9.
50 Reg. 13–14.
51 Reg. 15.
52 Reg. 24.

Similar provisions apply to the remediation of environmental damage.[53] Remediation entails the removal of any significant risk to human health, and, in the case of natural resources other than land, the reinstatement of the same level of resource or service as would have existed if the damage had not occurred (primary remediation), or the provision of substitute resources (complementary remediation), and the provision of (non-financial) compensation for interim losses pending such reinstatement (compensatory remediation).[54] Remediation does not extend to financial compensation for losses suffered in consequence of the environmental damage (e.g. loss of income from tourism). **31**

The Regulations make no provision in respect of financial guarantees. However, many of the activities designated in them are subject to permits or authorisations, and it is certainly possible that these could be conditional on appropriate financial guarantees.[55] In the present context, it is material to note that the Secretary of State has an unfettered discretion, in granting a consent for the deliberate release of a GMO, to include in it such limitations and conditions as he may think fit.[56] The current general policy, however, is not to require insurance but to encourage operators to extend their insurance policies or to take out other financial guarantees to cover their potential liabilities for damage to the environment.[57] **32**

(b) Environmental liability regime beyond the scope of the Directive

A number of specific liability regimes cover environmental harm,[58] and may exceed the scope of Directive 2004/35/EC in particular respects. **33**

53 Reg. 17–23 and Schedule 4.

54 Reg. 18(c) and Schedule 4.

55 The financial guarantees required by the Mining Waste Directive 2006/21/EC, for example, will be introduced by way of the Environmental Agency's granting of permits: see *DEFRA*, Environmental Permitting: Environmental Permitting Guidance: The Mining Waste Directive: For the Environmental Permitting (England and Wales) Regulations 2007: Draft guidance for Consultation, May 2009, para. 3.7 and 3.65 ff.

56 EPA 1990, sec. 112(1).

57 *DEFRA*, Environmental Damage (Prevention and Remediation) Regulations 2009: Frequently Asked Questions, URL: http://www.defra.gov.uk/environment/policy/liability/pdf/faq.pdf, 6.

58 E.g. under the Water Resources Act 1991, the Wildlife and Countryside Act 1981, and the Control of Major Accident Hazards Regulations 1999.

(c) Claimants in cases of environmental harm

34 The Regulations provide for recovery by an enforcing authority of its own costs incurred in preventing or remediating environmental damage. Other affected parties have no right to compensation in respect of their losses. In general tort law, harm to biodiversity and harm to the environment as such – as distinct from harm to the proprietary rights of particular persons – cannot give rise to liability.

(d) Special liability regime for losses sustained by individuals

35 The tort of private nuisance is committed where the defendant unreasonably interferes with the claimant's use or enjoyment of land. It is "a tort to land" and therefore actionable only by a person with an interest in the land affected.[59] In determining the reasonableness of any interference, the courts may take into account a wide variety of circumstances, including the duration and timing of the interference, its severity, the locality in which it occurs,[60] the defendant's motive, and whether or not the claimant was being unduly sensitive.[61] Regulatory consents, even if they do not *per se* give rise to a defence of statutory authorisation, may serve to "crystallise" what is a reasonable land-use in the area in question.[62] As noted above, it is not necessary to show that the claimant suffered physical damage: interference with his amenity interests will suffice provided it is substantial. If the interference is found to have been unreasonable, it is no defence that the defendant took all reasonable steps to reduce its effects. Private nuisances normally involve an element of continuity or repetition, but even a single occurrence can give rise to liability in appropriate circumstances (though the liability here is often indistinguishable from that in negligence). The successful claimant is prima facie entitled to an injunction to prevent the continuation or repetition of the nuisance in the future, as well as damages for harm already suffered.

59 *Hunter v Canary Wharf Ltd* [1997] AC 655.

60 *C. Rodgers* (fn. 1) 381 plausibly suggests that GM crop farming is more likely to give rise to liability in private nuisance if the area is one which has declared itself "GM free" via collective land-use decisions made within the community.

61 See further *Network Rail Infrastructure Ltd v Morris (t/a Soundstar Studio)* [2004] EWCA Civ 172, [2004] Env LR 41.

62 *Gillingham Borough Council v Chatham and Medway Dock Co Ltd* [1993] QB 343. Cf. *Wheeler v Saunders* [1996] Ch 19. *C. Rodgers* (fn. 1) 395 argues that the licensed planting of GM crops does not change the character of the area as such, or what it is reasonable land-use in it, but effects merely a subtle change in the nature of local agricultural production.

(e) Cartagena Protocol

The United Kingdom signed the Cartagena Protocol on Biosafety on 24 May 2000 and ratified it on 19 November 2003. It entered into force on 17 February 2004.[63] 36

4. Other strict liability regimes

The common law rule in *Rylands v Fletcher*[64] should be mentioned here. In the classic formulation of Blackburn J, "the person who for his own purposes brings on his lands and collects and keeps there anything likely to do mischief if it escapes, must keep it in at his peril, and, if he does not do so, is prima facie answerable for all the damage which is the natural consequence of its escape."[65] He went on to say that the defendant could excuse himself by showing that the escape was owing to the claimant's default, *vis major*, or an act of God. The rule is further subject to a number of important preconditions which have been restrictively construed by the courts in the intervening years. First, the rule applies only to dangerous things, whose presence on the defendant's land creates "an exceptionally high risk of danger or mischief if there should be an escape, however unlikely an escape may have been thought to be."[66] Secondly, the damage must be attributable to the thing's escape from the land, not merely from the defendant's control. The rule does not apply to damage suffered on rather than outside the land.[67] Thirdly, the defendant must have been engaging in a non-natural use of land. This means not only that the thing which escapes was not naturally on his land, but was brought onto it by the defendant, or accumulated there by virtue of his activities, but also that the defendant's use of land was "extraordinary and unusual".[68] According to the classic test, it must be a "special use… and must not merely be the ordinary use of the land or such a use as is proper for the general benefit of the community."[69] The exception relating to public benefit was doubted by the House of Lords in the *Cambridge Water* case,[70] 37

63 http://www.cbd.int/biosafety/signinglist.shtml.
64 *Fletcher v Rylands* (1866) LR 1 Ex 265; *Rylands v Fletcher* (1868) LR 3 HL 330.
65 *Fletcher v Rylands* (1866) LR 1 Ex 265, 279–280 approved by the House of Lords in *Rylands v Fletcher* (1868) LR 3 HL 330.
66 *Transco plc v Stockport MBC* [2004] 2 AC 1 at [10] per Lord Bingham.
67 *Read v J Lyons & Co* [1947] AC 156.
68 *Transco plc v Stockport MBC* [2004] 2 AC 1 at [11] per Lord Bingham.
69 *Rickards v Lothian* [1913] AC 263, 280 per Lord Moulton.
70 *Cambridge Water Co Ltd v Eastern Counties Leather plc* [1994] 2 AC 264, 308–9 per Lord Goff.

but the expression of the test in terms of "ordinary" rather than "natural" use has since then received the Law Lords' approval.[71]

38 The application of the rule to GMOs may well prove problematic: for example, it is far from obvious that GM farming would be regarded as a "non-natural" use of land.[72]

39 Although it was for a long time thought to represent an independent category of liability, the House of Lords ruled in the *Cambridge Water* case that liability under *Rylands v Fletcher* was a species of liability in private nuisance, albeit dealing with isolated escapes from land rather than continuous or repetitive interference. It therefore appears to be subject to the same limitations as follow from the recognition of private nuisance as a tort to land, namely, that it provides no remedy for personal injury as such, and that it is actionable only by a person with an interest in the land affected.

V. Vicarious liability

1. Scope of vicarious liability

40 Under English law, vicarious liability arises where the defendant's employee commits a tort in the scope of (or closely connected with) his employment. It does not generally arise in respect of independent contractors. Vicarious liability for independent contractors may, however, arise in the case of statutory non-delegable duties and certain common law liabilities, for example, for the undertaking of inherently hazardous activities[73] or work involving a special danger of a private nuisance.[74]

2. Liability for people further up the food or feed production chain

41 It is hard to see how a party further down the production chain could be tortiously liable for someone higher up. Even the supplier of a defective prod-

71 *Transco plc v Stockport MBC* [2004] 2 AC 1 at [11] per Lord Bingham.

72 See further *M. Lee/R. Burrell* (fn. 1) 532–3 and *C. Rodgers* (fn. 1) 377 ("improbable" that growing GM crops would be seen as a non-natural use).

73 *Honeywill and Stein Ltd v Larkin Bros ltd* [11934] 1 KB 191.

74 *Matania v National Provincial Bank* [1936] 2 All ER 633. But see the criticisms of *P. Giliker*, Vicarious Liability, in: K. Oliphant (ed.), The Law of Tort (2nd edn. 2007) para. 3.78. It is in any case clear that there is no general rule of vicarious liability for independent contractors in private nuisance: *Alcock v Wraith* (1991) 59 BLR 20.

uct's contingent tortious liability, which may be negated by identifying someone higher up the chain,[75] is not really a liability *for* anyone further up the chain, but rather a liability for his/her own supply of the product.

3. Can someone further down the feed or food chain include someone further up it in a trial against him/herself?

In accordance with general principles, any number of parties may be joined as defendants to a claim if this would assist in the resolution of the matters in dispute or if there is an issue between the new party and an existing party which is connected to the matters in dispute, and it is desirable to add the new party so that the court can resolve that issue. Once the claim form has been served, the court's permission is necessary to add a new party.[76] This would result in a "shifting" of the loss only where the existing and new defendants are liable in respect of the same damage and the former is able to claim contribution from the latter.[77] **42**

VI. Multiple tortfeasors

There are three ways in English law in which multiple tortfeasors may incur liability to a single claimant.[78] First, they may be "joint tortfeasors" who are responsible for the same damage because of their action in concert or in the course of certain relationships (e.g. the employment relationship). The liability of joint tortfeasors is joint and several. Secondly, they may be "several concurrent tortfeasors" who are responsible for the same damage because of their independent causal contributions to it. Their liability too is joint and several. Thirdly, they may be "several non-concurrent tortfeasors" who cause different damage and therefore are responsible for different damage. There is no joint and several liability in such a case. **43**

Joint tortfeasors are liable whether or not they contribute causally to the same damage. It follows that, where there is no action in concert, nor **44**

75 Consumer Protection Act 1987, sec. 2(3).
76 Civil Procedure Rules (CPR), Part 19.
77 Civil Liability (Contribution) Act 1978, sec.1 and CPR, rule 20.6.
78 See further *V. Harpwood/K. Oliphant*, Joint and Several Liability in Tort, in: K. Oliphant (ed.), The Law of Tort (2nd ed. 2007).

any other relationship importing liability as joint tortfeasors, it is crucial to ask whether the damage to which the tortfeasors have contributed causally is the same or different. In this context, a crucial distinction is drawn between divisible and indivisible injuries: in the case of the latter, there will be joint and several liability to the full extent of the injury; in the former, damages will be proportioned to the defendant's (partial) contribution to the total injury, i.e. the defendant is found responsible for only part of the loss.[79]

VII. Defences

1. Licence/permission to grow GM material

45 This is unlikely to be regarded as a defence, but may be relevant to the determination of fault or, in respect of private nuisance, unreasonableness.[80]

2. Consent/assumption of risk

46 The defence *volenti non fit iniuria* applies where the claimant freely assumes the risk of harm in full knowledge of the danger. It seems unlikely that, at the current time, the average consumer would be found to possess the necessary knowledge.

3. Third-party influence

47 "Act of stranger" is a defence to liability under *Rylands v Fletcher*. In other torts, such conduct is not strictly a defence, but may amount to a *novus actus interveniens* breaking the chain of causation, unless perhaps the defendant had special reason to foresee its occurrence and the capacity to prevent it.[81]

79 *Holtby v Brigham & Cowan Ltd* [2000] 3 All ER 421 (damages for asbestosis apportioned on a time-exposure basis).
80 Cf. *Gillingham Borough Council v Chatham and Medway Dock Co Ltd* [1993] QB 343; *Wheeler v Saunders* [1996] Ch 19 (both relating to planning permission).
81 Cf. *Smith v Littlewoods Organisation Ltd* [1987] AC 241.

4. Prescription

The ordinary limitation periods apply: three years for personal injury,[82] and six years for all other actions in tort (unless specifically provided).[83] Time starts to run when the injury, damage or loss occurs or was reasonably discoverable.[84] An overriding time limit of 15 years from the date of the negligence or the damage (whichever is the later) applies in respect of latent damage not involving personal injuries.[85] The court has a discretion to allow a claim to proceed out of time in the case of personal injuries and death,[86] but not otherwise. 48

5. Other defences

Other general defences like contributory negligence and illegality might conceivably be relevant. 49

VIII. Remedies

1. Pecuniary compensation

(a) Bodily harm

The ordinary law of damages applies. The injured party would be entitled to compensation for both pecuniary and non-pecuniary loss, if any, consequent on the injury. 50

(b) Property losses

The ordinary law of damages applies. The normal measure of damages in cases of property damage is the item's diminution in value. In the case of commodities, this is normally determined by reference to prices in the market in which the commodity is traded. If as a result of the tort the property is worth nothing, its previous value is in principle recoverable 51

82 Limitation Act 1980, sec. 11(4).
83 Limitation Act 1980, sec. 2.
84 Limitation Act 1980, sec. 11(4), 14 and 14A.
85 Limitation Act 1980, sec. 14B.
86 Limitation Act 1980, sec. 33.

in its entirety. In exceptional cases, the defendant may be entitled to damages on a "cost of cure" basis.

(c) Economic losses

52 As noted above, damages for property damage may be awarded on a "cost of cure" basis, and in principle would cover all reasonable costs incurred in restoring a contaminated field to meet organic standards. I assume that that objective would be considered reasonable.

(d) Harm to animals

53 In principle, the owner would be entitled to full replacement costs (calculated by reference to market values) and consequential losses such as loss of profit resulting from lost production. Whether the eating of GM feed is in itself enough to prove damage is debatable, but if the animal thereby loses value, it is arguable that it is indeed "damaged". See further no. 6, above.

(e) Costs of disposal

54 In principle, yes, but subject to the ordinary obligation to take reasonable steps to mitigate the loss.

2. Non-compensatory damages

55 Punitive or exemplary damages are recognised by English law but confined at common law to two established categories: oppressive, arbitrary or unconstitutional action by servants of the government, and wrongful conduct by the defendant which has been calculated by him to make a profit for himself which may well exceed the compensation payable to the claimant.[87] It remains unclear whether exemplary damages are avail-

87 *Rookes v Barnard* [1964] AC 1129. See further *V. Wilcox*, Punitive Damages in England, in: H. Koziol/V. Wilcox (eds), Punitive Damages: Common Law and Civil Law Perspectives (2009).

able for the main torts likely to be relied on in the present context, i.e. negligence and private nuisance.[88]

3. Other remedies

Injunctions may be awarded, at the court's discretion, to restrain a private nuisance – but not negligence. Reparation in kind and substitution are not generally available remedies in English law, though some specific statutory remediation provisions (e.g. the remediation provisions of the Environmental Damage (Prevention and Remediation) Regulations: considered above) may be considered to effect such outcomes. **56**

4. Costs of pursuing a claim

(a) General cost rule

English law applies the "loser pays" principle. **57**

(b) Costs of establishing causation

The party found liable must generally pay the successful party's reasonable costs in fighting the action, including the cost of expert witness reports, etc., though the award of costs is at the court's discretion.[89] The award of legal costs does not extend to the general costs of gathering relevant evidence but these may be awarded as damages, subject to the ordinary rules of the tort in question.[90] **58**

5. Advance cover

As mentioned above (fn. 55), financial guarantees are a condition for the grant of a permit in respect of certain activities covered by the Mining Waste Directive 2006/21/EC. **59**

---

88 See *K. Oliphant*, England, in: H. Koziol/B.C. Steininger (eds), European Tort Law 2001 (2002), no. 45, discussing *Kuddus v Chief Constable of Leicestershire* [2001] UKHL 29, [2002] 2 AC 122.

89 Civil Procedure Rules, Part 44.

90 See, e.g., *Jan De Nul (UK) Ltd v Axa Royale Belge SA* [2002] EWCA Civ 209, [2002] 1 Lloyd's Rep 583 (private nuisance; costs of investigating the damage).

IX. Cross-border issues – Conflict of laws

1. Conflicts rules applicable before (or instead of) Rome II

60 Under the Private International Law (Miscellaneous Provisions) Act 1995, which remains in force, the general rule is that the applicable law is the law of the country in which the events constituting the tort in question occur.[91] Where elements of those events occur in different countries, the applicable law under the general rule is, in a property damage claim, the law of the country where the property was when it was damaged,[92] and, in a claim for pure economic loss, the law of the country in which the most significant element or elements of those events occurred.[93] The general rule may be displaced if the significance of the factors which connect the tort with some other country make it substantially more appropriate to apply that country's law.[94]

61 The main change introduced by the Rome II Regulation is that the assessment of damages is now, contrary to the previous approach,[95] treated as a matter of substance rather than procedure, and so follows the generally applicable law,[96] not (as was previously the case) the law of the forum.

2. Special regime for cross-border claims

62 N/A.

91 Sec. 11(1).
92 Sec. 11(2)(b).
93 Sec. 11(2)(c).
94 Sec. 12.
95 See *Harding v Wealands* [2006] UKHL 32, [2007] 2 AC 1.
96 Art. 15(c) of the Regulation.

X. Cases

1. Due to the adventitious presence of GMOs in a field, maize which is normally sold as conventional contains GMOs beyond the legal labelling threshold. This is not discovered before the final stage of the food production chain by the producer of taco chips. The whole production is lost since the supermarket chains refuse to accept delivery from the producer.

(a) Who can sue along the chain of distribution?

The excess GM presence in the maize will probably be considered damage, **63**
though this is not free from doubt (see no. 6, above). However, this damage seems to be suffered by the maize producer, whereas the taco producer and others in the chain of distribution appear to suffer only pure economic loss which would not be recoverable in tort (though there could be claims in contract).

(b) Would the case be solved differently if the GMO content was below the labelling threshold?

No. **64**

(c) Would the case be solved differently if the admixture was not adventitious, but occurred due to the disregard of segregation rules, for example?

No. **65**

(d) Would the case be solved differently if the GMO found was not admitted for production in your jurisdiction?

No. **66**

(e) If the admixture had occurred on a non-GM field and it transpires that the GM seeds were blown from: (i) neighbouring fields; or (ii) a truck passing by, would the farmer of the affected field be liable for all or part of the loss caused further down the distribution chain?

67 No.

2. Twenty years after the sale of GM maize used for food products, it turns out that it has certain disadvantageous health effects for humans.

(a) Can the producers be held liable at this point for risks unknown at the time of growing the maize? Who would be liable?

68 Any claim would fall outside the 10-year longstop period for (statutory) product liability claims (from time of supply),[97] and (depending on when the damage occurred) may also fall outside the overriding 15-year time limit for negligence actions not involving personal injuries (see no. 48). Even assuming limitation is not otherwise a problem, it seems that no liability would arise because there is no fault and no foreseeability of injury.

(b) Can compensation already be claimed at a point when the negative health effects have not yet materialised, but are expected to according to scientific expertise/mere rumours?

69 English law does not allow a claim for mere exposure to risk (see no. 9). The risk must materialise.

97 Limitation Act 1980, sec. 11A. This would in any case appear to be a clear case for application of the development risk defence.

(c) Would it make any difference if the GM maize had only been in use for feed, causing harm to the animals, which may or may not cause harm to humans consuming the meat as well?

No. The farmer's possible claim for injury to the animals does not entitle persons at risk from having eaten the meat to sue before they have suffered any actual injury. **70**

3. The driver of a food logistics company discovers that a farmer, from where he regularly picks up agricultural products fails to obey mandatory segregation rules or food or feed hygiene standards, which may lead to the admixture of GM and non-GM produce sold and packaged separately by that farmer. Does he or his employer have a duty to warn, i.e. warn the recipient of the allegedly "non-GM" produce?

I know of no comparable case in English law but such a duty might arguably exist in respect of risks to person or property. But there could be no claim if the recipient suffers only pure economic loss. **71**

Damage Caused by GMOs under Estonian Law

Irene Kull*

I. General overview

1. Special liability or redress scheme for GMOs

1 Regulation regarding civil liability for GMOs can be found in § 32 of the Deliberate Release into the Environment of Genetically Modified Organisms Act (DREGMOA),[1] according to subsec. 1 of which, damage caused by the illegal deliberate release of GMOs into the environment or damage from the illegal marketing of GMOs or genetically modified products will be compensated for as provided by the Law of Obligations Act (LOA),[2] i.e. under the general rules of civil liability. There are no differences in liability for GMOs in comparison to the general tort liability rules. As there are no court cases, this answer is based only on the legal rules.

2 If the tortfeasor is a public authority, the State Liability Act[3] (SLA) will apply and general tort law rules from the LOA will be used only to fill the gaps in the SLA. See also no. 4 below.

3 The most probable legal ground for liability for GMOs would be liability for damage caused by a major source of danger. § 1056 LOA provides that if damage is caused by a major source of danger or an extremely dangerous activity, the person who manages the source of danger shall be liable for

* I am very grateful to my colleagues for writing parts of the report: in environmental law to H. Veinla, in international private law to M. Torga and in solving cases to V. Kõve.

1 Deliberate Release into the Environment of Genetically Modified Organisms Act (*geneetiliselt muundatud organismide keskkonda viimise seadus*). Passed 14 April 2004, Riigi Teataja (State Gazette, RT) I 2004, No. 30, Art. 209. All legal Acts are available in English: www.legaltext.ee.

2 Law of Obligations Act (*võlaõigusseadus*). Passed 28 September 2001, entered into force 1.07.2002. RT I 2001, No. 81, Art. 487.

3 State Liability Act (*riigivastutuse seadus*). Passed 2 May 2001, entered into force 1.01.2002. RT I 2001, No. 47, Art. 260.

the causing of damage regardless of the person's culpability. A thing or an activity is deemed to be a major source of danger if, due to its nature or to the substances or means used in connection with the thing or activity, major or frequent damage may arise even if it is handled or performed with due diligence by a specialist. It is obvious that fault-based liability can be enforced with difficulty in environmental cases – mainly due to problems of proof. In case of strict liability schemes, a presumption of causal link is applied – if a dangerous activity is a potential cause of damage, it shall be presumed that the damage is caused as a result of the particular danger arising from the activity. However, this presumption does not apply if the "operator" proves that the activity is operated in compliance with all requirements and if the normal operation of the activity has not been disturbed.

2. State liability

The State Liability Act of 2002 aimed at the protection and restoration of rights violated upon the exercise of powers of a public authority and performance of other public duties and compensation for damage caused (§ 1 (1) SLA). § 7 (3) SLA provides that the provisions of private law concerning compensation for damage also apply in addition to the SLA with respect to the compensation for damage caused in the exercise of public authority unless otherwise provided by law and provided this is not in conflict with the nature of public law relationships. 4

The general liability regime of public authorities is direct liability. The only exceptions are cases of loss of profit and non-pecuniary damage (§§ 13 (2) and 9 (1) SLA) where fault liability will be applied. However, in these cases a public authority shall be relieved of liability for damage caused in the course of performance of public duties if the damage could not have been prevented even if the diligence necessary for the performance of public duties had been fully observed (§ 13 (3) SLA). Differences between state liability and liability under general tort law rules are not remarkable, at least as to the main concepts. Important is that the differences are not connected with the losses caused by GMOs. 5

II. Damage

1. Recoverable losses

6 Estonian law does not provide for any restrictions concerning compensable damage caused by GMOs. Pecuniary and non-pecuniary damage, loss of profit and pure economic loss (§ 128 LOA) are recoverable. § 128 subsec. 3 LOA provides that direct pecuniary damage includes, primarily, the value of the lost or destroyed property or the decrease in the value of property due to deterioration even if such decrease occurs in the future, and reasonable expenses which have been incurred or will be incurred in the future due to the damage, including reasonable expenses relating to prevention or mitigation of damage and receipt of compensation, including expenses relating to establishment of the damage and submission of claims relating to compensation for the damage. Loss of profit according to § 128 subsec. 4 LOA is loss of the gain which a person would have been likely to receive in the circumstances, in particular as a result of the preparations made by the person, if the circumstances on which compensation for damage is based had not occurred. Loss of profit may also include the loss of an opportunity to receive a gain.

7 In the case of an obligation to compensate for the damage arising from the death of a person, according to § 129 of LOA, the obligated person shall compensate for the expenses arising from the death of the deceased person, in particular for reasonable funeral expenses, reasonable medical expenses relating to the health damage or bodily injury which caused the death of the person, and the damage arising from the aggrieved person's interim incapacity for work and maintenance costs for the dependants of the deceased.

8 In the case of an obligation to compensate for damage arising from health damage or bodily injury caused to a person, according to § 130 subsec. 1 LOA, the obligated person shall compensate the aggrieved person for expenses arising from such damage or injury, including expenses arising from the increased needs of the aggrieved person, and damage arising from total or partial incapacity to work, including damage arising from a decrease in income or deterioration of the future economic potential of the aggrieved person. The tortfeasor shall also pay to the aggrieved person a reasonable amount of money as compensation for non-pecuniary damage caused to the person by such damage or injury (§ 130 (2) LOA).

9 In the case of an obligation to compensate for damage arising from the death of a person or a serious bodily injury or health damage caused to

the person, the persons close to the deceased or the aggrieved person may also claim compensation for non-pecuniary damage if payment of such compensation is justified by exceptional circumstances (§ 134 (3) LOA).

2. Pure economic loss

Taking into account the existing court practice[4] pure economic loss was recoverable only in cases of contractual liability and in cases directly provided for in the law (e.g. state liability, see supra no. 4 f.). There is no notion of pure economic loss used or defined in legal acts but it is used in legal literature as a generally known legal term[5] and the concept of loss of profit (which is defined in the LOA) will cover some cases of pure economic loss. § 128 subsec. 2 LOA provides that recoverable pecuniary damage primarily includes direct pecuniary damage and loss of profit. Loss of profit is defined as loss of the gain which a person would have been likely to receive in the circumstances, in particular as a result of the preparations made by the person, if the circumstances on which compensation for damage is based had not occurred. Loss of profit may also include the loss of an opportunity to receive gain (§ 128 (4) LOA). **10**

In a decision from 2008[6] the Estonian Supreme Court opened the gate to the possibility that pure economic loss (loss of profit) might be recoverable under the tort liability rules, overruling the judgment of the district court and sending it back. The Supreme Court emphasized at the same time that the court should evaluate the purpose of the compensation of damage and regarding tort it was noted that pure economic loss is not recoverable under the delictual liability rules as it is not listed among the protected interests. There are some court decisions by the Supreme Court from 2009 where this previously formulated strong position has been softened and reconsidered at least for some cases of delictual liability. **11**

The SLA provides that pure economic damage may be compensated only on the basis of fault liability (§ 13 (2) SLA). According to law, there should be breach of obligations (protected interests), loss and a causal link between the act and consequences proved by the claimant. Losses can be proved on the basis of probability of future profit which the claimant **12**

4 *I. Nõmm*, Compensation for Pure Economic Loss under the General Conditions of Delictual Liability, Juridica 2 (2008) 84–92.

5 See *J. Lahe*, Fault in the Three-stage Structure of the General Elements of Tort, Juridica Inter national 1 (2007) 152–160.

6 Ruling of the Supreme Court in civil matters No. 3-2-1-75-08 (only in Estonian).

would have earned in the absence of the unlawful act of the tortfeasor. § 127 subsec. 4 LOA provides that a person shall compensate for damage only if the circumstances on which the liability of the person is based and the damage caused are related in such a manner that the damage is a consequence of the circumstances (causation *conditio sine qua non*). To establish causation, elimination and substitution methods are used. It must be observed together with the general purpose of compensation (§ 127 (2) of LOA) which limits the damage to be compensated only to the extent that prevention of damage was the purpose of the obligation or provision due to the non-performance of which the compensation obligation arose.

3. Mere fear of a loss

13 There is no court practice in compensating mere fear of loss triggered by GMOs, mobile phone radiation or other modern technology. Existing court practice concerning compensation for non-pecuniary losses is not consistent and is moving seemingly towards a more flexible application of law which does not provide very many possibilities to claim compensation for non-pecuniary losses. The Civil Chamber of the Estonian Supreme Court has always been more modest in providing compensation for non-pecuniary losses in comparison to the Administrative Chamber which, according to one of its judgments,[7] takes the position that if illegal acts and the causal link between such acts and consequent circumstances (fright, concern, anger, stress, etc.) as the content of the non-pecuniary damage are shown by the claimant, compensation should be awarded. As § 128 subsec. 5 LOA provides that non-pecuniary damage primarily involves the physical and emotional distress and suffering caused to the aggrieved person, claimants have to substantiate the distress or suffering in order to claim compensation (burden of proof).

4. Standard of proof

14 It is difficult to describe a certain standard of proof. The Supreme Court applies the concept of consequences, which means that if the claimant proves that somebody acted illegally in breach of his or her obligations and the tortfeasor cannot prove that there was no fault, the court will award compensation taking into account the probability of losses and the

7 Ruling of the Supreme Court in administrative matters No. 3-3-1-27-02 (only in Estonian).

degree of illegality of the act. According to § 230 subsec. 1 of the Code of Civil Procedure,[8] each party shall prove the facts on which the claims and objections of the party are based insofar as the law does not provide otherwise. Thus, if the aggrieved party demands compensation, he as the claimant has to present the circumstances that prove his claim; in case of a delictual claim he has to prove that the other party's actions were illegal and that these actions caused the damage.

5. Nominal losses

§ 127 subsec. 6 LOA provides that if damage is established but the exact extent of the damage cannot be established, also in the event of non-pecuniary damage or future damage, the amount of compensation shall be determined by the court. Amounts of damages are not restricted by law or other statute, and suggestions based on previous court practice do not exist. Questions of nominal damages will be solved on a case by case basis. **15**

6. Mass losses

There are no special rules for mass losses. **16**

III. Causation

1. Uncertainty of merely potential causes

It is provided by the LOA that there must be causation to receive compensation for the damage – according to § 127 subsec. 4 LOA, a person shall compensate for damage only if the circumstances on which the liability of the person is based and the damage caused are related in such a manner that the damage is a consequence of the circumstances (*conditio sine qua non*). Causation does not have to be a direct link between the actions of the person and the consequences (damage), i.e. the damage does not have to be the result of breaking the law, but it can occur due to a sequel of events that is started by the person's actions. To establish causation, elimination and substitution methods are used. With the elimination method, the damage is in **17**

8 Code of Civil Procedure. Passed 20 April 2005, entered into force 1 January 2006, RT I 2005, No. 26, Art. 197.

a causal relationship with the actions only when the person's actions were an unavoidable prerequisite for such damage, i.e. there would not have been any damage if these certain actions had not taken place. Thus, in order to make sure that there is causation we need to answer the question of whether there would have been damage if the defendant had not acted in the way he did. If the answer is no, then the defendant has to prove that there would have been damage even without his tort.

18 The LOA also contains a special rule concerning joint liability (§ 138 (1) LOA) which provides that several persons may be liable for damage caused and if it has been established that any of the persons could have caused the damage, compensation for the damage may be claimed from all such persons. In cases where there are two or more potential tortfeasors or causal links, the court may decide that they should be jointly liable. Questions of risk will be solved under the rules of contributory fault.[9]

2. Complex causation scenarios

19 There are no special rules.

3. Force majeure

20 In cases of establishing liability there should be a causal link between the damage and force majeure e.g. the damage should be the result of the force majeure circumstances. Therefore, if losses were the result of force majeure circumstances, the liability will be shifted.

4. Threshold to prove causation

21 It is difficult to describe or generalize. I would say that in most cases the court uses the words: "more likely than not". The word "certainly" is not often used in judgments.

9 LOA § 139 (1): If damage is caused in part by circumstances dependent on the aggrieved party or due to a risk borne by the aggrieved party, the amount of compensation for the damage shall be reduced to the extent that such circumstances or risk contributed to the damage.

5. Special rules on causation

There are no special rules. 22

IV. Types of liability

1. Fault liability

(a) Special rules governing fault

There are no special rules applicable in cases envisaged here concerning standard of fault. In case of unlawfully caused damage, fault-based liability is presumed if the law does not provide otherwise (§ 1043 LOA). According to § 1050 subsec. 1 LOA, a tortfeasor is not liable for the causing of damage if the tortfeasor proves that he is not guilty of causing the damage, unless otherwise provided by law. If the victim (injured person) claims compensation for the damage, he as the claimant bears the burden of proving the facts on which the claim is based; in case of delictual liability this includes the unlawful action of the tortfeasor, the damage sustained and the causality between the actions and damage. In case of fault-based liability, a tortfeasor must prove that he is not guilty of the damage in order to be free from liability. According to § 1050 subsec. 2 LOA, the situation, age, education, knowledge, abilities and other personal characteristics of a person shall be taken into consideration when assessing the guilt of the person (i.e. the tortfeasor's subjective characteristics shall be taken into account). 23

(b) Impact of specific rules of conduct

There will be no differences in applying fault liability (if it is applicable at all in cases of GMOs). In case of fault liability, the subjective standard of care will be evaluated also on the basis of the violation of any specific statutory provisions if there are any. General rules will apply to GM and non-GM farming. 24

2. Product liability

(a) Development risk defence

The development risk defence has been incorporated into Estonian law (§ 1064 (1) (5) LOA). There is no case law and also no discussions on the 25 handling of the development risk defence. There are opinions that producers are not entitled to rely on the development risk defence to escape liability for defective products which were put into circulation and were not recalled after defects in the product were discovered.

(b) Alternative routes

26 There are no other alternative routes for compensating losses caused by agricultural products in Estonian law.

(c) Impact of compliance with rules and regulations

27 It is not clear if the fact that defendant obeyed all rules and regulations governing production processes would be good ground for release from liability. Also the question is whether minimum or maximum standards in regulations should be followed by the producer in order to be released from liability. As there is no court practice yet on this, the answer to the question is partly speculative.

3. Environmental liability

(a) Implementation of the Environmental Liability Directive

28 The Environmental Liability Directive was transposed by the Environmental Liability Act[10] (ELA). The Act regulates the prevention and remediation of damage caused to the environment based on the principle that polluters should pay. The notion of "prevention of environmental damage" means implementation of measures to eliminate a threat of damage caused by an incident, an act or failure to act or to reduce the extent of possible environmental damage or to manage the contaminants or other damage factors, to prevent, eliminate or otherwise influence the spread in order to limit or to prevent further environmental damage and adverse effects on human health or further impairment of the quality of the benefit of the protected area or water. The concept of "remedying of environmental damage" means implementation of measures to restore, replace or compensate for natural resources or the benefits of it and to eliminate the significant risks dangerous to human health.

10 Environmental Liability Act. Passed 14 November 2007.

The Estonian method of transposition of the Directive could be characterised as minimum harmonisation, no attempts to go further and to be innovative were made. As an exception, the Estonian ELA is indeed stricter when it comes to the definition of operator. The Directive defines the "operator" in connection with occupational activities. The Estonian Act defines an operator (person who is liable for prevention or remediation) irrespective of the nature of its activity – a person is liable even when the activity is not carried out in the course of an economic activity, a business or an undertaking. Operations carried on in the public sphere are also covered. 29

Estonian law does not foresee obligatory schemes for financial guarantees. 30

(b) Environmental liability regime beyond the scope of the Directive

§ 133 LOA specifically regulates environmental liability. This article draws a distinction between two kinds of environment-related damage – personal damage and environmental damage per se. § 133 LOA prescribes that if damage is caused by environmentally hazardous activities, damage relating to a deterioration in environmental quality shall also be compensated for in addition to the damage caused to persons or the property thereof. Expenses relating to preventing an increase in the damage and to applying reasonable measures for mitigating the consequences of the damage, and the damage arising from the application of such measures should also be compensated. 31

Personal damage shall be compensated pursuant to the general law of torts and environmental damage pursuant to environmental law (ELA) concerning particular sectors of the environment and the new Environmental Liability Act which aims to transpose the Directive. See also no. 3. 32

The sphere of application of ELA is limited. It covers: 33

- substantial adverse effects on reaching or maintaining a favourable conservation status of a habitat or species;
- substantial adverse effects on a protected area, a special conservation area, a species protection site, a protected individual natural object;
- substantial adverse effects on the ecological or chemical status of a surface water body or coastal waters or on the ecological potential or on the chemical or quantitative status of a groundwater body;

- soil damage caused by the direct or indirect introduction of substances, preparations, organisms or micro-organisms into the environment which results in a significant risk of human health being adversely affected.

34 In addition to the ELA, remedying or compensation of environmental damage per se is covered also by the Waste Act[11] and Nature Conservation Act.[12] The Waste Act (§ 128) prescribes that damages related to the release of waste into the environment and to the pollution created by waste, including the costs related to the handling of waste and to the remedy of the effects of environmental pollution caused by waste, shall be covered by the person who released the waste into the environment. The Nature Conservation Act (§ 77) provides for fixed rates (up to EEK 1 million) to compensate the damage caused to the environment by destroying or damaging protected natural objects and specimens of protected species or other species of animals or birds.

(c) Claimants in cases of environmental harm

35 The Estonian public law environmental liability regime is in principle based not on monetary recovery of environmental losses but on the obligation to take appropriate measures to prevent or remedy environmental damage per se. As for persons who are liable for prevention or remediation – Estonian law defines them irrespective of the nature of their activity – a person is liable even where their activity is not carried out in the course of an economic activity, business or an undertaking. Operations carried out in the public sphere are also covered. See supra no. 28 f. as far as notions of "prevention" and "remediation" are concerned, and no. 31–34 as far as the sphere of application of the Environmental Liability Act is concerned.

(d) Special liability regime for losses sustained by individuals

36 Any such special regime is lacking in Estonian law; the general law of torts and property law are implemented in these cases.

---

11 Waste Act. Passed 28 January 2004, entered into force 1 May 2004, RT I 2004, No. 9, Art. 52.

12 Nature Conservation Act. Passed 21 April 2004, entered into force 1 May 2004, RT I 2004, No. 38, Art. 258.

Environmental liability in civil law covers foremost personal damage (including property damage), but at least in theory, certain aspects of environmental damage per se can be covered as well. 37

(e) Cartagena Protocol

Estonia ratified the Cartagena Protocol on Biosafety on 24 March 2004 and the Convention on Biodiversity on 11 May 1994. 38

4. Other strict liability regimes

N/A. 39

V. Vicarious liability

1. Scope of vicarious liability

There are two sets of rules regulating vicarious liability in the Estonian legal system. Vicarious liability for persons involved in the performance of contractual obligations is provided for in the GPCCA[13] (§ 132) and liability for violations committed by other persons in tort law is provided for in the LOA (§ 1054). In the tort liability regime there are three different situations where vicarious liability will be applied: (1) if one person engages another person in his or her economic or professional activities on a regular basis (employers' liability for employees); (2) if one person engages another person in the performance of his or her duties (to perform somebody's legal or general duties); (3) if a person performs an act at the request of another person and the person at whose request the act is performed has control over the behaviour of the person who causes the damage (independent subcontractors). 40

Persons who hired independent subcontractors can be held liable in torts for the damage caused in the course of the performance of obligations of the persons who hired them on contractual (§ 132 GPCCA) and extra-contractual bases (§ 1054 LOA). 41

---

13 General Part of Civil Code Act (*tsiviilseadustiku üldosa seadus*). RT I 2002, No. 35, Art. 216, adopted 27 March 2002.

2. Liability for people further up the food or feed production chain

42 The general rule is that one is not, in principle, liable for torts of independent contractors chosen in the chain of the food or feed production. There is also no liability recognised in the law for the negligent selection of independent contractors or for failure to control their performance. Liability for people further up the food or feed production chain can arise on the basis of the solidary liability regime if the claim was filed against one (or some) of the solidary liable persons in the chain.

3. Can someone further down the feed or food chain include someone further up it in a trial against him/herself?

43 Persons in the feed or food chain cannot include persons further down or further up in a trial by themselves. The only possibility to file a claim against other persons in the chain would be under the rules on multiple tortfeasors (see no. 44 f. below) or on some other legal ground (like recourse actions against the previous seller or producer). The fact that there are some other persons who might be liable for the damage does not excuse the person against whom the claim was filed.

VI. Multiple tortfeasors

44 Firstly, if several persons are liable, on the same or different grounds, to a third party for the same damage caused to the third party, they shall be solidarily liable for payment of compensation (§ 137 (1) LOA). If several persons may be liable for damage caused and it has been established that any of the persons could have caused the damage, compensation for the damage may be claimed from all such persons.

45 Secondly, if several persons may be liable for damage caused and it has been established that any of the persons could have caused the damage, compensation for the damage may be claimed from all such persons. In that case a person obligated to compensate for damage shall be released from liability if the person proves that the damage was not caused thereby. Compensation for damage may be claimed from each person to an extent in proportion to the probability that the damage was caused by the person concerned.

VII. Defences

1. Licence/permission to grow GM material

46 Taking into account the provisions of the DREGMOA[14] and court practice in delictual liability, there are no reasons to find the licence or permission to grow GM material a defence against the liability of the tortfeasor. There are no differences either in cases of GMOs approved only for testing. It can only be used as a defence against the amount of damages compensable to the victim (§ 139 LOA – contributory fault).

2. Consent/assumption of risk

47 Causing damage is not unlawful if the victim consents to the damage being caused, except in the case where giving consent is contrary to law or good morals (§ 1045 (2) (2) LOA). This rule is based on the German legal doctrine, and as there are no cases in Estonian court practice, the German interpretation can be taken into account as a basic model for the application of the rule.

3. Third-party influence

48 The liability of the producer shall not be reduced if the damage occurs due to both a deficiency of the product and the behaviour of a third party which means that third-party conduct will be taken into account only if there was no defect in the product (§ 1064 (3) LOA).

4. Prescription

49 The limitation period for a claim arising from unlawfully caused damage shall be three years as of the moment when the entitled person became or should have become aware of the damage and of the person obligated to compensate for the damage (§ 150 (1) GPCCA). Regardless of the general three years limitation period, a claim arising from unlawfully caused damage expires not later than ten years after performance of the act or

14 Deliberate Release into the Environment of Genetically Modified Organisms Act (*geneetiliselt muundatud organismide keskkonda viimise seadus*). Passed 14 April 2004. New version in force from 1 February 2009.

occurrence of the event which caused the damage. The limitation period for a claim arising from causing death, a bodily injury or damage to health expires not later than thirty years as of performance of the act or occurrence of the event which caused the damage. There are no special limitation statutes for claims based on GMO cases.

5. Other defences

50 In cases of GMOs, all general defences will apply with those special rules applicable to the special kind of liability (product liability, vicarious liability, fault liability, liability for damage caused by major source of danger, etc.).

VIII. Remedies

1. Pecuniary compensation

(a) Bodily harm

51 There are no special rules. The regular remedies apply.

(b) Property losses

52 There are no special rules. The regular remedies apply.

(c) Economic losses

53 There are no special rules. This is difficult to answer; the possibility of deduction will depend on the circumstances provided for in § 139 subsec. 2 LOA and it depends on the court which arguments will be taken into account: e.g. whether the aggrieved person failed to draw the attention of the person causing the damage to an unusually high risk of damage or to prevent the risk of damage or to perform any act which would have reduced the damage caused if the aggrieved person could have reasonably been expected to do so. Usually the court will use the test of reasonableness which means the application of not only ethical but also economical arguments.

(d) Harm to animals

If animals are harmed by contaminated feed, the rules on things will apply. The Supreme Court of Estonia has found that in cases of damage or destruction to a thing, economic damage usually cannot be compensated. The owner can recover the reasonable expenses incurred to acquire new animals of equal value. In determining the amount of compensation for the damage, the decrease in value of the damaged animals shall be taken into account in a reasonable manner if the value considerably decreased (§ 132 (1) LOA). If the replacement of damaged animals with animals of equal value is not possible, the value of the destroyed animals shall be compensated. The owner of the animals can also recover the reasonable costs of treatment of the animals and the potential decrease in their value. Only if treatment is unreasonably expensive in comparison to the value of the animals shall compensation not be paid. Compensation for other damage beside the costs concerning the restoration of a thing becomes relevant only if it was the purpose of the provision due to the non-performance of which the damage arose (see Supreme Court ruling from 13 May 2005 on civil matters No. 3-2-1-64-05). 54

The value of the animal will be calculated under the rules of the General Part of Civil Code Act § 65: the usual value of an object is deemed to be the value of the object unless otherwise prescribed by law or a transaction. The usual value of an object is its average local selling price (market price) and will also cover animals' productivity (cow's potential for producing milk or meat). 55

The mere fact that an animal eats GM contaminated feed can constitute harm to the animal and be illegal. Courts may establish that there will be future damage even if the extent of the damage cannot be established and determine the amount of the compensation later (§ 127 (7) LOA). If damage is established but the exact extent of the damage cannot be established, including in the event of non-pecuniary damage or future damage, the amount of compensation shall be determined by the court (§ 127 (6) LOA). 56

(e) Costs of disposal

§ 127 subsec. 3 of the LOA defines as direct pecuniary damage among other expenses also reasonable expenses which have been incurred or will be incurred in the future due to the damage, including reasonable expenses relating to prevention or mitigation of damage. The costs of dis- 57 posing of contaminated production can be qualified as expenses incurred in relation to prevention or mitigation of damage.

2. Non-compensatory damages

58 Non-compensatory damages are not recoverable in Estonian tort law.

3. Other remedies

59 In the cases provided by law or a contract and in other cases where this is reasonable under the circumstances, the aggrieved person may claim compensation for damage in a manner other than monetary compensation. For example repairs, taking over obligations of the aggrieved person and other measures for prevention and remedying of damage. The Environment Liability Act (passed 14 November 2007) provides for prevention and remedying of environmental damage.

4. Costs of pursuing a claim

(a) General cost rule

60 There is a general rule in the Code of Civil Procedure (CCP) that if the claimant wins the case, the expertise costs and other legal costs will be paid by the other party (§ 162 (1) CCP). If a party is in difficult economic straits, such may demand state legal aid to pay the costs, i.e. leave all the costs partly or fully to be paid by the state (§ 180 (1) (1) CCP). The general rule of "loser pays" will apply also in cases of state legal aid.

(b) Costs of establishing causation

61 Costs regarding establishing causation are connected to the definition of direct pecuniary damage as in § 128 subsec. 3 LOA and are thus compensable. Fees for experts shall be paid in advance by the party who submits the application from which such costs arise (§ 148 (3) CCP).

5. Advance cover

According to Estonian law, it is not necessary for people operating with GMOs to present liability insurance or some other sort of compensation system to cover potential losses and neither are farmers required to take out insurance which covers such losses. 62

IX. Cross-border issues – Conflict of laws

1. Conflicts rules applicable before (or instead of) Rome II

Currently, the questions of the law applicable to non-contractual obligations are regulated by the Estonian Private International Law Act (PIL) 2002. 63

According to Art. 50(1) of the PIL, the law applicable to tort claims is the law of the state where the act or event which forms the basis for causing the damage was performed or occurred. Thus, the current regulation differs from the Rome II Regulation Art. 4(1), which gives priority to the law of the country in which the damage occurred. It should be noted however, that in certain exceptional cases, the PIL does allow the law of the country in which the damage occurred to be the applicable law. According to Art. 50(2) of the PIL, this can happen if the consequences of the act do not become evident in the state where the act or event which formed the basis for causing the damage was performed or occurred and when the injured person requests the application of the law of the country in which the damage occurred. 64

Another difference to the current regime is that the Regulation gives guidelines on certain aspects of conflict of laws regulation of torts which have not yet been resolved by the Estonian legislator. For example, the instructions in Recital 30 and Art. 2(1) on *culpa in contrahendo* indisputably constitute an efficient improvement for Estonian private international law. 65

Like the Rome II Regulation Art. 4(3), the PIL also provides for an exception clause (Art. 53 of the PIL). 66

2. Special regime for cross-border claims

All tort claims are covered by PIL. 67

X. Cases

1. Due to the adventitious presence of GMOs in a field, maize which is normally sold as conventional contains GMOs beyond the legal labelling threshold. This is not discovered before the final stage of the food production chain by the producer of taco chips. The whole production is lost since the supermarket chains refuse to accept delivery from the producer.

(a) Who can sue along the chain of distribution?

68 This case would be solved under Estonian law as a case of contractual liability between sellers and buyers. If there is non-conformity of the goods, the seller as the last in the contractual chain will be liable to the buyer (the producer of the taco chips). The buyer can sue the seller if the preconditions of contractual liability are fulfilled. The first precondition for the use of remedies is the non-performance of the obligations. Non-performance is defined in § 100 LOA as a failure to perform or defective performance of the obligation, including a delay in performance. If the maize contained GMOs beyond the labelling threshold, there is non-conformity as described in § 217 (2) (1) LOA. In cases of non-conformity, the general rule on contractual liability does not apply (the debtor shall be liable for non-performance unless the non-performance is excused, § 103 LOA) and the seller will be liable for any lack of conformity of a thing which exists at the time when the risk of accidental loss or damage of the thing passes to the purchaser even if the lack of conformity becomes apparent after that time (§ 218 (1) LOA). In this case the producer can claim compensation for direct pecuniary damage and loss of profit.

69 § 217 (1) (2) LOA provides that goods do not conform to the contract in cases of lack of agreement concerning the characteristics of the thing if the thing is not fit for the particular purpose for which the purchaser needs it and of which the seller was or ought to have been aware at the time of entry into the contract. If the court concludes that the goods are not in conformity with the contract, contractual liability will apply and all remedies are open to the buyer (§ 101 (1) LOA).

70 In all other cases, the buyer can sue the persons in the chain under the rules on product liability or extra-contractual liability (tort) if the preconditions for liability are met. In cases of multiple tortfeasors, see supra. no. 33–36.

(b) Would the case be solved differently if the GMO content was below the labelling threshold?

If the GMO content was below the labelling threshold, the probable result would be that there is no non-conformity of the goods and the buyer would have no remedies under contractual liability. The same result will be applicable also in case of other kinds of liability. **71**

(c) Would the case be solved differently if the admixture was not adventitious, but occurred due to the disregard of segregation rules, for example?

No. **72**

(d) Would the case be solved differently if the GMO found was not admitted for production in your jurisdiction?

No. **73**

(e) If the admixture had occurred on a non-GM field and it transpires that the GM seeds were blown from: (i) neighbouring fields; or (ii) a truck passing by, would the farmer of the affected field be liable for all or part of the loss caused further down the distribution chain?

The farmer of the affected field would be liable for the non-conformity of the maize further down the distribution chain only under the contractual liability rules in both cases. Due to the special rule applicable to the contracts of sale which provides as a general rule that the seller is liable for all defects existing at the time of passing the risk to the buyer, the question of applicability of liability under the *force majeure* rule is not finally decided.[15] Damages can be claimed only if the party in breach is liable for the damage. In commentaries to the rules in LOA, liability for defects in the product should be based only on the fact that defects existed already at the moment when the risk passes. The Estonian Supreme Court men- **74**

15 *A. Värv/P. Karu*, The Seller's Liability in the Event of Lack of Conformity of Goods, Juridica International 2009, 85–93 (available at http://www.juridicainternational.eu/last-issue/2006-2/vol-xvi/the-sellers-liability-in-the-event-of-lack-of-conformity-of-goods).

tioned in one judgment that the general rule on warranty liability (§ 103 LOA) should also be applicable in cases of non-conformity. The amount of damages which can be claimed will be limited under the foreseeability rule which provides that if there was no intentional behaviour or negligence on the side of the farmer, compensation will be limited to the amount which the obligor foresaw or could have foreseen at the moment when the contract was concluded (§ 127 (3) LOA). See also supra no. 68.

2. Twenty years after the sale of GM maize used for food products, it turns out that it has certain disadvantageous health effects for humans.

(a) Can the producers be held liable at this point for risks unknown at the time of growing the maize? Who would be liable – the seed producer/farmer/food producer/distributor/ etc.?

75 According to § 1066 subsec. 2 LOA, claims which arise from product liability rules shall terminate after ten years have passed as of the date on which the product which causes damage is placed on the market, unless an action has been filed with a court by that time. The limitation period for a claim arising from causing death, a bodily injury or damage to health expires not later than thirty years as of performance of the act or occurrence of the event which caused the damage (§ 153 (2) GPCCA). As the seller is not liable for personal injuries under contractual liability, the producer/distributor will be liable under tort liability rules and rules on solidary liability may apply (§ 137 (1) LOA). As tort liability is a fault liability, the question of subjective fault will arise and in this case it is doubtful that the fault of the producer or distributor will be found.

(b) Can compensation already be claimed at a point when the negative health effects have not yet materialised, but are expected to according to scientific expertise/mere rumours?

76 Future damage is compensable under the rules of LOA. § 127 subsec. 7 provides that, if future damage is established but the extent of the damage cannot be established, the court may determine the amount of the compensation later. If there are mere rumours, the court may establish that there is no reasonable ground to provide compensation for future damage.

(c) Would it make any difference if the GM maize had only been in use for feed, causing harm to the animals, which may or may not cause harm to humans consuming the meat as well?

If the maize had only been in use for feed and the harm to humans is not proved, the persons in the chain will be liable only for the harm caused to animals. The court may establish that there is a causal link between the circumstances on which the liability of the person is based and the future damage caused to humans and leave the amount of the damage open, to be determined in the future. See also no. 17 ff. **77**

3. The driver of a food logistics company discovers that a farmer, from where he regularly picks up agricultural products fails to obey mandatory segregation rules or food or feed hygiene standards, which may lead to the admixture of GM and non-GM produce sold and packaged separately by that farmer. Does he or his employer have a duty to warn, i.e. warn the recipient of the allegedly "non-GM" produce?

There are no provisions in the law concerning the obligation to warn. The legal ground for such obligation may be found under the rules of principles of good faith and reasonableness (§ 23 (1) (4) LOA). It is doubtful in a case where the food logistics company is not a party to the distribution contract. Under the rules of tort liability, the only possibility seems to be liability on the ground that the behaviour of the food logistics company was intentional behaviour contrary to good morals (§ 1045 (1) (8) LOA). **78**

Damage Caused by GMOs under Finnish Law

Björn Sandvik

I. General overview*

1. Special liability or redress scheme for GMOs

1 In Finland, a special liability regime for damage caused by GMOs was established in 1995 by the Gene Technology Act (377/1995) (GTA). A liability provision is found in GTA sec. 36, which reads as follows:

> "*Liability for damage*. Compensation for damage in the environment caused by activities referred to in this Act is subject to the provisions of the Environmental Damage Compensation Act (737/1994).
>
> Compensation for personal injury or for damage to property intended for private use or consumption and used by the injured party mainly for such purpose are subject to the provisions of the Product Liability Act (694/1990).
>
> Compensation for other damage caused by activities referred to in this Act is subject to the provisions of the Tort Liability Act (412/1974). The operator is liable to compensate for such damage even if it was not caused wilfully or through negligence.
>
> The provisions of par. 1–3 shall not restrict the right of the injured party to compensation on the basis of an agreement or by virtue of other statutes than those referred to in par. 1–3."

2 According to the bill to the GTA, sec. 36(1) will lead to the application of the Environmental Damage Compensation Act (EDCA) in relation to damage caused by contained use of GMOs[1] (chap. 4 of the GTA) or by

* This report does not take into consideration materials published after November 18, 2008.

1 "Contained use" is defined in sec. 3 of the GTA as "any activity in which organisms are genetically modified or in which such organisms are cultured, stored, transported,

deliberate release of GMOs[2] into the environment for any other purpose than placing them on the market (chap. 5 of the GTA). The EDCA provides compensation for a wide range of damage caused in the environment. Compensation is payable for both individual losses under traditional heads of damage (sec. 5) and public (collective) losses in the form of damage to the environment per se (sec. 6). (See also no. 14 below.) In the bill to the GTA, it is envisaged that the EDCA will be the most important statute of the statutes referred to in sec. 36 of the GTA.[3]

The PLA will, of course, apply if a GM product that has been placed on the 3
market[4] (chap. 6 of the GTA) causes personal injury, or damage to property of the kind referred to in sec. 36(2) of the GTA ("consumer property"). This does not seem to alter the applicability of the PLA compared with the sphere of application that would follow in any case directly pursuant to the PLA (see sec. 1 and 2). The PLA implements Directive 85/374/EEC of the Council of 25 July 1985 on the approximation of the laws, regulations and administrative provisions of the Member States concerning liability for defective products.[5]

Under sec. 36(3) of the GTA, compensation for other damage caused by 4
activities referred to in this Act is subject to the general provisions of torts in the Tort Liability Act (TLA). This seems to include, for example, situations where a GM product that has been placed on the market causes damage to property which is not intended or used for private use or consumption ("non-consumer property").[6] The TLA itself is based on fault liability. However, sec. 36(3) of the GTA lays down a strict (no-fault) liability if the TLA is to be applied pursuant to sec. 36(3) of the GTA.

---

destroyed or disposed of or used in any other way, and for which specific containment measures are used to limit their contact with the general population and the environment".

2 "Deliberate release" is defined in sec. 3 of the GTA as "introduction into the environment of genetically modified organisms without using any specific containment measures to limit their contact with the general population and the environment or to provide a high level of safety for the general population and the environment".

3 See *Government Bill* 1994, 349 at 36.

4 "Placing on the market" is defined in sec. 3 of the GTA as "making a product available to third parties either in return for payment or free of charge; as placing on the market is not considered supplying genetically modified organisms for contained use or their deliberate release into the environment for any other purpose than for placing on the market".

5 Official Journal (OJ) L 210, 7.8.1985, 29–33.

6 See also, e.g. *B. Sandvik*, Economic Loss Caused by GMOs in Finland, in: B.A. Koch (ed.), Economic Loss Caused by Genetically Modified Organisms. Liability and Redress for the Adventitious Presence of GMOs in Non-GM Crops (2008) no. 55.

Thus, since also the EDCA and the PLA are based on strict liability, the GTA generally imposes a strict non-contractual liability upon activities under the GTA.[7]

5 Sec. 36(4) of the GTA makes it clear that the liability regime in the GTA does not restrict the right of the injured party to compensation on a contractual basis[8] or by virtue of statutes other than the EDCA, the PLA or the TLA. In this context, particular reference may be made to the Feed Act (86/2008) regarding other statutes on non-contractual liability. The Feed Act is also applicable to GM feed.[9] A liability provision is laid down in sec. 48 of the Act.[10] This provision reads as follows:

> "The feed producer, the person on behalf of whom the feed is produced and the feed importer are liable to pay compensation for damage caused to the buyer if, in professional use of the feed, the feed does not meet the requirements in the legislation of the European Community, or the requirements in this Act or in other statutes promulgated by virtue of this Act. Compensation shall be paid for such damage even if it was not caused wilfully or through negligence.
>
> Liability under par. 1 is excluded if the liable party shows with probability that the feed was not defective when it was placed on the market.
>
> If the feed causes personal injury or damage to property intended for private use or consumption and used by the injured party mainly for such purposes, the liability of the feed producer, the person on behalf of whom the feed is produced and the feed importer is subject to the provisions of the Product Liability Act (694/1990)."

6 Despite the rather general wording of sec. 48(1) – and although the bill to the Feed Act is rather confusing on this point – it should be clear that sec. 48 deals with non-contractual product liability only, without affecting

7 In *Government Bill* 1994:349 at 36 f. it is generally stressed that compensation for damage caused by the use of gene technology shall be based on strict liability, and that the provisions on liability in the GTA also aim at preventing damage. Cf. also *Government Bill* 1989:119 at 16 regarding the strict liability in the PLA, and *Government Bill* 1992:165 at 15 regarding the strict liability in the EDCA. But see also for criticism regarding the preventive role of tort law and strict liability, e.g. *B. Sandvik*, Skadeståndsinstitutets preventiva betydelse – särskilt om strikt ansvar och ekonomisk prevention [The Deterrent Role of Tort Law – In Particular on Strict Liability and Economic Deterrence], Retfærd 86 (1999) 22 ff.

8 The most important statutory rules on contractual liability with regard to the present study are found in the Sale of Goods Act (355/1987) and in the provisions on consumer sales in chap. 5 of the Consumer Protection Act (38/1978).

9 See sec. 3, 5 and 11.

10 Cf. also a somewhat similar provision in sec. 22 of the previous Feed Act (396/1998).

the contractual liability of the liable parties.[11] Cf. also the Seed Trade Act (728/2000) which is also applicable to GM seeds,[12] and which contains a provision on non-contractual liability in sec. 14(1):

> "The person who has produced or packed the seed, or imported the seed to Finland is liable to pay compensation for damage caused to the user of the seed if the seed does not correspond to given information on it, or if it is otherwise defective. Liability is excluded if the liable person shows with probability that the seed was not defective when it was placed on the market."

It may be noted that the Food Act (23/2006) is also applicable to GMOs.[13] **7**
But this Act lacks special provisions on liability for damage. Cf. also the Act on Trade with Forest Cultivation Products (241/2002).[14]

Further, at the time of writing, Finland is in the process of implementing **8**
Directive 2004/35/CE of the European Parliament and of the Council of 21 April 2004 on environmental liability with regard to prevention and remedying of environmental damage.[15] A draft government bill has recently been published.[16] In the draft, a new Act on Compensation for Certain Damage Caused to the Environment is proposed. This Act will serve as a framework piece of legislation to which references will be made in several existing environmental statutes regarding liability for such damage as is covered by the framework Act. According to the draft, a reference to the framework Act will be included also in sec. 23 of the GTA. However, no amendments are proposed to either the EDCA or sec. 36 of the GTA.

Furthermore, an umbrella law on the coexistence of GM and non-GM **9**
agricultural production is under preparation. A draft governmental bill has recently been published.[17] A liability regime is also proposed in the draft umbrella law. In simplified terms, compensation would be payable

---

11 It seems that the drafters of the Feed Act did not fully grasp the distinction between non-contractual product liability and contractual liability for non-conforming goods. See *Government Bill* 2007:27 at 42. Cf. also *T. Wilhelmsson/M. Rudanko*, Tuotevastuu [Product Liability] (2nd ed. 2004) 28 f. for some critical remarks on sec. 22 in the previous Feed Act of 1998 (see fn. 10).

12 See ibid., sec. 3 and 8.

13 See ibid., sec. 3.

14 See ibid., sec. 3 and 8 regarding the applicability of the Act to GM forest cultivation products. Note, however, that the previous Act (684/1979) contained (in sec. 14) a liability provision quite similar to sec. 14 of the present Seed Trade Act.

15 OJ L 143, 30.4.2004, 56–75.

16 Draft Government Bill of 2008 concerning legislation on liability for damage caused to the environment (*Environmental Liability Draft* 2008).

17 Draft Governmental Bill of 2008 on legislation concerning the coexistence of GM, traditional and organic agricultural production (*Coexistence Draft* 2008).

for certain economic losses suffered by producers of non-GM crops due to GMO admixture caused by production of GM crops, and for the costs of certain investigations and of restoring cultivations of non-GM crops that have been exposed to GMO admixture (sec. 27). Compensation would be payable by the state through the Finnish Food Safety Authority (sec. 26, 29). The state would have right of recourse against any producer of GM crops who has negligently or wilfully caused the admixture by breaching statutory requirements on GM cultivation laid down in the proposed coexistence law (sec. 30).

10 It should be noted in this context that a special liability regime is considered necessary since, in the draft bill to the umbrella law, existing liability rules are held to be insufficient. It is especially emphasised that the EDCA is inapplicable because the GTA does not deal with, e.g. cultivation of GM crops that have been approved for the market. Thus, it is held that the EDCA cannot be applied in coexistence situations by virtue of GTA sec. 36(1). In addition, it is held in rather categorical, short and one-sided terms that the EDCA is inapplicable to damage caused by GMO admixture outside the scope of the GTA.[18]

11 However, according to sec. 1 of the EDCA, compensation for damage caused in the environment by an activity in a specific area shall be payable. The damage shall be caused by "pollution of water, air or land, or noise, vibration, radiation, light, heating or smell, or other comparable disturbance". Even if GMO admixture could not be regarded as "pollution" under the EDCA, the majority view seems to be that GMOs can still be regarded as a "comparable disturbance".[19] And damage suffered by, for example, a

18 Ibid. at 19.

19 See e.g. *Sandvik* (fn. 6) no. 4 with further references. But see also *E. Hollo*, Perinteisen maanviljelyn ja geenimaatalouden tuotantomuotojen rinnakkaiselon järjestämisestä [The Arrangement of the Coexistence of Traditional Farming and GMO Agricultural Production Forms], in: E. Kolehmainen (ed.), Business Law Forum (2006) 205. Hollo seems to assert that it is unclear whether the EDCA provides compensation for damage caused by GMOs even under sec. 36(1) of the GTA. Hollo finds that the concept of environmental damage under sec. 1 of the EDCA is linked to pollution of the environment, and that it is a matter of interpretation if and when damage caused by GMOs is to be considered pollution of the environment. However, sec. 1 of the EDCA refers to "*pollution* of water, air or land, *or* ... [certain other disturbances]". See further *B. Sandvik*, Miljöskadeansvar [Environmental Impairment Liability] (2002) 128 ff., in particular 143 ff. on the interpretation of "other comparable disturbance". Cf. also the reference to microorganisms in the bill to the EDCA regarding "pollution of water, air, or land"; see *Government Bill* 1992:165 at 20. And even if GMO admixture cannot be regarded as such pollution, it is still difficult to see why GMOs could never be considered a comparable disturbance. In any case, the explicit reference to the EDCA in sec. 36(1) of the GTA should make it perfectly clear that the EDCA is applicable to GMOs irrespective of the enumeration of disturbances in sec. 1 of the EDCA.

farmer of non-GM crops as a result of GMO admixture from another farmer's cultivation of GM crops in the vicinity is clearly caused both "in the environment" and by an "activity in a specific area" within the meaning of sec. 1 of the EDCA.[20] (See also no. 14 below on the wide range of damage for which compensation is payable under the EDCA.) With this in mind, it has even been held that the reference to the EDCA in sec. 36(1) of the GTA is perhaps more informative than normative in nature[21] (in much the same way as the reference to the PLA in sec. 36(2) of the GTA and in sec. 48(3) of the Feed Act[22]).

In Finland, preparatory documents to legislation are important authorita- **12**
tive (albeit not binding) sources of law. Nevertheless, the draft bill on coexistence does not seem well-considered or particularly convincing in regard to the EDCA. (Cf. also no. 40 below as regards the draft bill and the PLA.) Several strong positive arguments clearly speak against the holdings in the draft bill. Therefore, the present country report starts from the assumption that the applicability of the EDCA to damage caused by GMOs cannot be excluded in coexistence situations. This is important because an application of the EDCA will lead to a wider scope of protection than the liability under the proposed coexistence legislation.

2. State liability

As seen above (no. 9), compensation would be payable by the state under the **13**
proposed umbrella law on coexistence. In other respects, it is typical for the liability regimes described above to be construed upon rules on strict (non-fault) liability that is channelled to certain liable parties such as the "operator", the "producer" or the "importer". It seems rather unlikely that the

20 But see also *Coexistence Draft* 2008 (fn. 17) at 19 holding that cultivation of a non-GM crop is not damaged as a result of GMO admixture, rather economic loss is suffered as a consequence of the changed market valuation after the admixture occurred. It is held that the product itself is still in faultless condition according to provisions on product safety and health. Therefore, it is concluded that compensation is not payable under the EDCA. However, this interpretation is extremely problematic for several reasons (see also fn. 29). For example, if this interpretation was correct, it would also mean that compensation could never be paid under the EDCA for pure economic loss suffered, for instance, by a restaurant owner who loses customers as a result of the totally harmless but disgusting smell from a factory or other sources in the vicinity. However, such an interpretation of the EDCA would clearly be incorrect.

21 See *Sandvik* (fn. 6) no. 4.

22 Cf. also *Government Bill* 2007:27 at 42, stating that a reference to the PLA is made in sec. 48(3) of the Feed Act "for the sake of clarity". Despite the good intentions, however, sec. 48(3) of the Feed Act is not unproblematic; see no. 42 below.

state would be directly engaged in such activities as would make the state a liable party under those rules. The only tort law rules on some kind of sovereign immunity are found in the TLA chap. 3, sec. 5[23] and 7,[24] but these rules seem to be of limited importance in the present context.

II. Damage

1. Recoverable losses

14 As indicated above and as will be demonstrated in greater detail below, compensation may be payable for a wide range of damage and losses under the various liability rules applicable to damage caused by GMOs. Of these rules, the EDCA has by far the widest scope of protection. Under sec. 5 and 6 of the EDCA, compensation may be payable for individual losses in the form of personal injury, property damage, pure economic loss, or certain non-pecuniary losses (e.g. discomfort due to smell) as well as for public (collective) losses in the form of damage to the environment per se.

2. Pure economic loss

15 In Finnish law, pure economic loss is defined in chap. 5, sec. 1 of the TLA as "economic loss that is not connected to personal injury or damage to property". According to chap. 5, sec. 1 of the TLA, compensation for such loss is payable only where the loss is caused (1) by a criminal act, (2) by a public body in the exercise of its authority, or (3) in other cases, where there are especially weighty reasons for compensating such loss. Case law on "especially weighty reasons" is casuistic and does not provide any clear general guidance, rather the interpretation by the Supreme Court (KKO)

23 Chap. 3, sec. 5 of the TLA read as follows: "No action for damages can be brought for injury or damage caused by a decision of the Government, a Ministry, the Cabinet Office, a court of law or a judge unless the decision has been amended or overturned or unless the person committing the error has been found guilty of misconduct or rendered personally liable in damages. Where a decision of an administrative authority has been appealed against in the Government or the Supreme Administrative Court, no action in damages can be brought for injury or damage caused by the decision insofar as it has been allowed to stand".

24 According to chap. 3, sec. 7 of the TLA, the "state and the municipalities shall not be liable in damages for injury or damage caused in maritime piloting". The ratio of this rule is the fairly high accident risk in shipping, which risk is often covered by insurance. According to chap. 7, sec. 1 of the Maritime Code, the shipowner is liable in damages for damage caused by the pilot, and the provisions in chap. 4 of the TLA on liability of employees and public officials are applicable to the pilot's personal liability.

seems to be somewhat restrictive.[25] In particular, it may be pointed out that the Supreme Court has held that the magnitude of the loss or the gravity of fault do not as such constitute especially weighty reasons under chap. 5, sec. 1 of the TLA.[26] However, the Supreme Court decision KKO 2005:105 now indicates that at least a wilfully caused loss may as such constitute an especially weighty reason.

16 In the draft government bill to the umbrella law on coexistence, it is emphasised that "it is unclear what could constitute a so-called particularly weighty reason in instances of damage caused by GMO admixture".[27] By stating this, the draft bill throughout seems to refer to situations where a non-GM producer suffers economic loss as a consequence of his production having been exposed to GMO admixture.[28] Therefore, it is proposed in sec. 27 of the draft coexistence law that compensation shall be payable for loss of income suffered by producers of non-GM cereals or non-GM livestock as a consequence of their products having been affected by GMO admixture to a degree exceeding threshold values established in EC legislation.

17 One could assert, however, that the situations referred to in the draft bill and in sec. 27 of the draft coexistence law do not fall under the definition of pure economic loss. Rather, there are good reasons to maintain that such situations should be regarded as consequential economic losses, i.e. economic losses that are connected to damage to property.[29]

18 In any case, the proposed sec. 27 of the draft coexistence law does not provide for compensation for economic losses other than those referred to

25 See further, e.g. *B. Sandvik*, Skada orsakad av otillbörligt förfarande i näringsverksamhet – HD:s praxis pekar på behovet av en särreglering [Damage Caused by Unjust Competition – Supreme Court Case Law Points to the Need for a Special Rule], Tidskrift utgiven av Juridiska Föreningen i Finland (JFT) 2006, 524 ff., especially 525 f.

26 This was held by the Supreme Court in its decision KKO 1991:66.

27 *Coexistence Draft* 2008 (fn. 17) 21.

28 See ibid., 17 ff.

29 See also *Sandvik* (fn. 6) no. 27. To some extent the position in the coexistence draft is understandable if linked to the interpretation in the draft bill according to which a non-GM product is not damaged as a result of GMO admixture (see fn. 20). Consequently, the resulting economic loss would not be connected to any property damage and therefore the loss would constitute a pure economic loss. But such an interpretation does not fit neatly into the concept of property damage. See further on the concept of property damage, e.g. *H. Saxén*, Skadeståndsrätt [Tort Law] (1975) 69 ff., *E. Routamo/ P. Ståhlberg/J. Karhu*, Suomen vahingonkorvausoikeus [The Finnish Law on Damages] (5th ed. 2006) 295 ff. (cf. also 29 ff. on the general notion of damage). See also *Sandvik* (fn. 19) 236 ff., and *B. Sandvik*, Tredjemansskada, sakskada eller ren förmögenhetsskada? Några kommentarer med anledning av HD 2003:124 [Third Party Damage, Property Damage or Pure Economic Loss? Some Comments Regarding KKO 2003:124], JFT 2004, 75 ff. on the borders between property damage, including consequential economic loss, and pure economic loss.

above. Thus, compensation could not be paid, for example, to a farmer of non-GM crops whose own production is not affected by GMO admixture but who nevertheless loses customers due to the fear that his products may also be affected as a result of a GMO admixture in the vicinity. However, compensation for such losses may, in principle, be payable under sec. 5(1) of the EDCA. (See also no. 20–24 below, especially no. 24.) Sec. 5(1) of the EDCA provides that compensation shall also be payable for pure economic losses with the exception of minor losses. The exception for minor loss does not apply if the loss is caused by criminal behaviour.[30]

19 Further, sec. 48(1) of the Feed Act and sec. 14(1) of the Seed Trade Act (see no. 5–6 above) do not distinguish between different kinds of damage or loss and the short explanations in the bills to the Acts do not shed any light on exactly what compensation can be paid.[31] However, since no clear limitations are made in this respect, one may assume that they cover both property damage (including consequential economic loss) and pure economic loss.[32] But this is not to say that pure economic loss should necessarily enjoy the same actual protection as property damage and consequential economic loss. (cf. also no. 24 below).[33]

3. Mere fear of a loss

20 Compensation is not payable for mere fear as such. However, it seems that the possibility that damages will be awarded cannot be ruled out if the claimant seeking damages succeeds in proving that the fear of GMOs has caused damage of a kind for which compensation is payable.

21 In the Supreme Court decision KKO 1999:102, for example, five police constables had been bitten and scratched by an HIV positive person. None of the constables became infected with HIV. Nevertheless, the Supreme Court held that four of them had suffered compensable damage in the form of mental distress comparable to personal injury and pain and suffering due to the fear of having become infected. In so holding, the Supreme Court stressed that the four constables had turned to a psychologist in order to control their fear. Further, the fear had lasted for a long

30 See further *Sandvik* (fn. 6) no. 22.

31 See *Government Bill* 2007:27 at 42 and *Government Bill* 2000:45 at 18.

32 Compensation for personal injury is, of course, subject to the provisions of the PLA; see also sec. 48(3) of the Feed Act.

33 Cf. also, e.g. the discussion by *L. Sisula-Tulokas*, Rena ekonomiska skador i skuggan av sakskador [Pure Economic Loss in the Shadow of Property Damage], JFT 2007, 415 ff., especially 425 ff.

time since the infection could not be reliably excluded by blood tests until six months after the incident. The fear had also affected the personal lives of the constables. One constable had no right to compensation because he had not even taken a blood test to determine whether he was infected. Therefore, that constable had not proven any damage comparable to pain and suffering. Cf. also KKO 2002:110 concerning the fear of both HIV and hepatitis C virus infection.

On the other hand, in the Supreme Court decision KKO 1998:80, a person **22**
actually became infected with the chronic hepatitis C virus. The infection as such was not regarded a personal injury for which compensation is payable, but the anxiety caused by the infection and its possible long term effects was compensable as damage comparable to personal injury and pain and suffering.[34]

Still, one could perhaps argue that the fear of GMOs is scientifically **23**
unfounded and therefore completely different to the fear of the real and severe danger associated with HIV and the hepatitis C virus. In this context, however, reference may also be made to the Supreme Court decision KKO 1999:61. In this case, it was held that alleged health risks of electromagnetic fields from power transmission lines are unproven except for cardiac patients with pacemakers. Yet, the Supreme Court found that compensation was payable for the diminished market values of real estate due to the fear of health risks from a high voltage power line on or in the immediate vicinity of the estates.[35]

Also in this context, it may be noted that the EDCA is probably not applic- **24**
able to the mere fear of GMO admixture (or of other disturbances; see no. 11 above on disturbances under the Act). The mere fear of GMO admixture is probably not a disturbance within the meaning of sec. 1 of the Act.[36] Thus, if no actual GMO admixture has occurred, compensation cannot be paid under the EDCA, for instance, to farmers of non-GM crops who suffer pure economic loss as a consequence of their customers' fear

34 One could argue that the general notion of damage suggests that a chronic virus infection – which may have long-range effects in the form of cancer or cirrhosis of the liver and which may require immediate medication – constitutes personal injury as such. Cf. also, e.g. *Routamo/Ståhlberg/Karhu* (fn. 29) 29 ff. on the general notion of damage. However, the Court held that not even a chronic virus infection as such is clinically an illness. Further, the Court stressed that the risk of developing an illness due to the infection was distant and uncertain. Therefore, the infection as such did not constitute personal injury.

35 Although the case concerns the right to compensation under the Act on Redemption (Expropriation) of Immovable Property and Special Rights (603/1977) it may perhaps be said to reflect a more general principle; see also *Sandvik* (fn. 19) 167 f.

36 See further *Sandvik* (fn. 6) no. 27.

that their products are no longer GMO-free because of the presence of GMO cultivation in the vicinity. But if an admixture of GMOs has actually occurred in the vicinity (i.e. in the environment), the EDCA is applicable and, in principle, compensation is probably payable under sec. 5(1) for the pure economic loss suffered by farmers of non-GM crops as a result of their customers' fear that their products may also have been affected by GMO admixture. As indicated by KKO 1999:61 above, this may possibly be the case even if the fear as such is scientifically unfounded. Needless to say, however, accepting the right to compensation for such losses poses extremely difficult causality problems. A non-restrictive attitude could cause large, complicated and unforeseeable compensation issues. However, in deciding where the line should be drawn between those claims which should be allowed and those which should be dismissed as too remote or indirect, each claim must obviously be considered on its own merits.[37] Further, even if compensable in principle, pure economic loss caused by fear should not necessarily enjoy the same actual protection as, e.g. health problems caused by fear (cf. also no. 19 above).

4. Standard of proof

25 As a rule, full proof is required, but the general principle of free judgement of proof applies. Further, chap. 17, sec. 6 of the Code of Judicial Procedure lays down a special rule in issues relating to the quantum of damages: If no evidence is available on the quantum of damages or if such evidence can only be presented with difficulty, the court shall have the power to assess the quantum, within reason.

26 Moreover, chap, 5, sec. 2a and 2b of the TLA lay down certain rules relating to future loss of income in personal injury cases. According to the general rule in sec. 2a, loss of income shall be determined on the basis of an assessment of the income the claimant would have had, had the damage not occurred. A reduction shall be made for the income the claimant had or could have had irrespective of the damage, or for the claimant's estimated future income taking into account the claimant's working capacity, education, previous activities, re-education possibilities, age, conditions of living, and other comparable circumstances. According to sec. 2b, the income the claimant would have had, had the damage not occurred shall be assessed according to standard incomes if the claimant was under the age of 18

37 See further ibid., no. 28 ff.

when the damage occurred and if nothing else is shown. Sec. 2b is based on median taxable incomes for different age groups.[38]

Furthermore, a special rule on future loss is laid down in sec. 9 of the EDCA. If the future environmental damage resulting from a disturbance (nuisance) can be assessed in advance, compensation for it shall, on demand, be pre-set either as a lump sum or as an annual payment. If there arises later an essential change in circumstances or the assessed loss is otherwise essentially different from the actual loss resulting from the disturbance, the compensation set in this manner may be adjusted to a reasonable extent considering the circumstances. According to the bill to the EDCA, this rule corresponds to advance assessment of compensation ex officio in permit matters under the Water Act (264/1991). Since compensation for damage caused by water pollution shall still be determined in proceedings pursuant to the Water Act, there was no need to establish essentially different proceedings under the EDCA for damage caused by disturbances other than water pollution.[39] 27

5. Nominal losses

In principle, it seems possible that damages for nominal/symbolic losses be awarded if claimed by the victim. But there is no tradition in this regard. 28

6. Mass losses

There are no specific tort rules for mass losses. However, from a procedural point of view, the new Act on Class Actions (444/2007) may be noted in this context. This Act is applicable in civil cases, within the limits of the competence of the Consumer Ombudsman, if several consumers have claims against the same business, based on the same or similar circumstances. The Consumer Ombudsman, as the plaintiff, has exclusive standing to bring a class action and to exercise the right of a party to the case to be heard in court. Class membership is based on an opt in-system. It is evident that this Act also opens the way for tort class actions concerning, for 29

38 The Government shall issue an annual decree on the amount of the median incomes referred to in sec. 2b. The latest decree is Government Decree 358/2008. Sec. 2a–b in chap. 5 of the TLA is based on a total amendment of sec. 2 by Law 509/2004.

39 See *Government Bill* 1992:165 at 29.

example, liability for personal injuries under the PLA.[40] It may also be noted that the power of courts to assess the quantum of damages to a reasonable amount under chap. 17, sec. 6 of the Code of Judicial Procedure (see no. 25 above) is believed to become of particular importance in class actions for damages.[41] There is so far no practical experience of class actions under the Act.

III. Causation

1. Uncertainty of merely potential causes

30 In addition to showing proof of the damage suffered, the claimant must show that a causal link exists between the alleged activity and the damage. It may, of course, often be difficult for the claimant to prove such a causal link. This is especially true for merely potential causes, including causes possibly lying within the victim's own sphere.[42] However, although full proof of causality is required as a rule, the principle of free judgement of proof also applies in this context (cf. no 25 above). As far as personal injury is concerned, the Supreme Court has held that medical proof is highly influential, but at the same time stressed that causality is ultimately a legal matter by nature.[43] In the Supreme Court decision KKO 1995:53, for instance, causality was held to be sufficiently proven although the causal link between the activity (mass vaccination against polio) and the injury (paralysis; the Guillain-Barré syndrome) was not medically proved.

31 The drafters of the EDCA, however, recognised that much of the practical significance of the liability rules could be lost if onerous demands were placed on the claimant in terms of the burden of proving the causal link. Therefore, the EDCA contains, in sec. 3, a special rule that the claimant seeking compensation has to prove that there exists "a probability" of a causal link between the activity and the alleged damage. Thus, full proof

40 See also *Government Bill* 2006:154 at 16 f. As a concrete example of a possible product liability class action under the Act, the bill refers to the situation that led to the Spanish legislation on class actions. This situation occurred in the 1980s when several hundred persons died and several thousand fell ill after having had poisonous cooking oil.

41 See ibid., 17.

42 This is illustrated, for example, by case law on liability for health problems alleged to have been caused by mildew and damp at the place of work; see the Supreme Court decisions KKO 2000:117, KKO 2000:118 and KKO 1998:22.

43 See, e.g., KKO 2005:99. Cf. also, e.g., KKO 2008:117.

of causality is not required under the EDCA. In judging probability, one shall take into account, among other things, the nature of the activity and the damage, as well as other possible causes of damage. According to the bill to the EDCA, however, "probability" means a rather high probability; in mathematical terms "clearly over 50 per cent".[44]

A somewhat similar rule on causality is proposed in sec. 26(2) of the draft law on coexistence (see also no. 9–12 above on this draft). According to the proposed rule, compensation would be payable if it is proven "with probability that the damage is caused by an activity referred to in sec. 2(1)–(3),[45] and provided the damage cannot be considered to have been caused by the victim's own activity". In the draft bill, it is only explained that the provision aims at lowering the victim's burden of proving causality.[46] However, the proposed provision may be understood as lowering the burden of proving the causal link combined with a rule on contributory negligence similar to the rule on contributory negligence in chap. 6, sec. 1 of the TLA.[47] 32

However, it has been called into question whether a rule such as sec. 3 of the EDCA really improves the victim's position when compared to the result that can already be achieved under the principle of free judgement of proof. With regard to the EDCA, it has been asked whether the victim's interest should have been protected more progressively by a rule reversing the burden of proof.[48] 33

2. Complex causation scenarios

There are no special rules on causation relevant to the context of the present study other than the rules referred to in III.1 above. 34

44 See *Government Bill* 1992:165 at 23. See also on sec. 3 of the EDCA *Sandvik* (fn. 6) no. 11.

45 That is, cultivation of GMOs, handling and storing of GMOs on a farm, and transportation of GMOs on a farm or between farms.

46 See *Coexistence Draft* 2008 (fn. 17) 41.

47 Under chap. 6, sec. 1 of the TLA, the damages may be adjusted (reduced) as reasonable if there has been a contribution to the injury or damage from the side of the person sustaining it.

48 See especially P. *Wetterstein*, The Finnish Environmental Damage Compensation Act – and Some Comparisons with the Norwegian and Swedish Law, in: Environmental Liability, vol. 3, issue 3 (1995) 45. In this context, Wetterstein refers to the solution in sec. 59 of the Norwegian Pollution Act. This provision alters the burden of proof in situations where it has emerged that pollution which could have caused the damage has occurred but it is unclear whether the damage may have some other cause(s).

3. Force majeure

35 In academic literature it has been argued that in Finnish law a general principle holds that a force majeure event has the effect of excluding strict liability.[49] One of the reasons for such a holding is that a force majeure event may break the causal link between the damage and the activity upon which the strict liability is imposed.[50] Thus, even if a force majeure defence is not explicitly provided for in, for example, the EDCA, a force majeure defence may still be available.[51] However, the notion of force majeure should be given a narrow interpretation. In any event, to suffice as an exclusion of liability, the force majeure event should cause the damage in its entirety and not merely contribute to it. On the other hand, if the event has merely contributed to the damage, the damages may be adjusted accordingly as is found reasonable (cf. the causality rule in chap. 6, sec. 1 of the TLA).[52]

4. Threshold to prove causation

36 See III.1 above.

5. Special rules on causation

37 See III.1 above.

---

49 See in particular *B. Sandvik*, Hur strikt är det strikta ansvaret enligt lagen om ersättning för miljöskador? [How Strict Is the Strict Liability Under the Environmental Damage Compensation Act?], JFT 1998, 544 ff. Cf. also *Sandvik* (fn. 6) no. 16.

50 See *Sandvik*, JFT 1998, 554 f. with further references. But see also, e.g. *M. Hemmo*, Vahingonkorvausoikeus [The Law on Damages] (2nd ed. 2005) 105 f. who finds it somewhat unclear whether a force majeure defence is available as a general principle. However, Hemmo also seems to recognise the importance of causality aspects when stressing that the operator should at least not escape liability for events resulting from the typical risk associated with the operator's activity.

51 Support for this interpretation is found also in the bill to the EDCA; see *Government Bill* 1992:165 at 27. (See further also no. 75 below.).

52 Chap. 6, sec. 1 of the TLA provides that damages may be adjusted (reduced) as is found reasonable if a circumstance other than the fault of the person liable contributed to the injury or damage. Although this rule may be said to reflect a more general principle of causality, it is as such entirely adapted to fault liability under the TLA. Therefore, the underlying principle should be applied with discretion in relation to strict liability. See also no. 77 below.

IV. Types of liability

1. Fault liability

All liability rules referred to above are based on strict liability and there seems to be no or only limited room for fault liability under the TLA. (Even so, aspects of fault may, of course, be relevant with regard to matters such as contributory negligence[53] or right to recourse.[54]) 38

2. Product liability

(a) Development risk defence

Finland has not incorporated the development risk defence into the PLA. 39 Therefore, it has even been held that the whole concept of "development defect", "development damage" or "development risk" is unnecessary with regard to the PLA.[55] The conditions for liability will not differ from the liability for defective products in general under the Act.[56] Thus, whether the GMO product in question was defective within the meaning of sec. 3 (cf. also Art. 6 of Dir. 85/374/EC) will undoubtedly be of utmost importance. There is no reported case law on the PLA.

However, in the draft bill to the proposed coexistence law (see no. 9 40 above), it is held that GMO admixture "does not make a product defective ... [as required by sec. 3 of the PLA] ... Thus, the PLA cannot be applied to the compensation issue in cases of damage caused by admixture".[57] But sec. 3 of the PLA has nothing to do with the applicability of the Act. And it is difficult to understand why every product containing GMOs should always be considered non-defective even after having caused damage. Again, the reasoning by the drafters of the proposed coexistence law does not seem particularly convincing with regard to all the assumed insufficiencies of other liability rules, other than those proposed in the draft coexistence law. (See also no. 10–12 above regarding the assumed inapplicability of the EDCA.) In this context, it may also be recalled that the Feed Act is also applicable to GM feed (see no.5 above). Further, in

53 See e.g. no. 32 above.
54 See e.g. no. 9 above.
55 See *Wilhelmsson/Rudanko* (fn. 11) 149 f.
56 See also *Government Bill* 1989:119 at 30.
57 *Coexistence Draft* (fn. 17) 20. Cf. also fn. 20 above regarding the holding in the draft bill according to which no damage has occurred as a result of GMO admixture.

sec. 48(3) of that Act, a reference to the PLA is expressly made solely "for the sake of clarity".[58]

(b) Alternative routes

41 According to sec. 2(2) of the EDCA, this Act also applies to environmental damage where compensation is due by virtue of the PLA. This provision was included by the Legal Affairs Committee of the Parliament during the parliamentary readings of the bill to the EDCA.[59] The Committee noted that the PLA had been amended in 1993 – i.e. after the bill to the EDCA was published in 1992 – to implement Dir. 85/374/EEC. As a consequence, several differences between issues covered by the PLA and the draft EDCA (such as proof of causality) had come to a head. Therefore, and in order to strengthen the position of damage victims, the Committee found that the victims should have a right to compensation by virtue of the EDCA irrespective of the applicability of the PLA. Unfortunately, however, not only is the precise meaning of EDCA sec. 2(2) somewhat blurry,[60] but for several reasons this provision is also extremely problematic in relation to Art. 13 of Dir. 85/374/EEC as interpreted by the European Court of Justice (ECJ).[61]

42 Further, it may be recalled that the strict liability under sec. 48 of the Feed Act (see no. 5 above) is channelled to "the feed producer, the person on behalf of whom the feed is produced and the feed importer" – also in

58 *Government Bill* 2007:27 at 42; see also fn. 22. (Cf. also fn. 21 regarding the reference to the EDCA in sec. 36(2) of the GTA.).

59 See *Report of the Legal Affairs Committee* 1994:10 at 4.

60 *E. Hollo/P. Vihervuori,* Ympäristövahinkolaki [The Environmental Damage Compensation Act] (1995) 94 ff. assume that sec. 2(2) of the EDCA will be of limited practical importance, since, according to them, the damage must be quite unusual from the perspective of both the EDCA and the PLA to constitute environmental damage and product damage simultaneously. However, this is not necessarily the case.

61 See e.g. ECJ C-183/00, *González Sánchez v. Medicina Asturiana SA* [2002] ECR I-3901, at par. 30: "... Article 13 of the Directive cannot be interpreted as giving the Member States the possibility of maintaining a general system of product liability different from that provided for in the Directive". The Court further held at par. 31: "The reference in Article 13 of the Directive to the rights which an injured person may rely on under the rules of law of contractual or non-contractual liability must be interpreted as meaning that the system of rules put in place in the Directive, which in Article 4 enables the victim to seek compensation where he proves the damage, the defect in the product and the causal link between that defect and the damage, does not preclude the application of other systems of contractual or non-contractual liability *based on other grounds, such as fault* [emphasis added] or warranty in respect of latent defects." Cf. also ECJ C-154/00, *Commission v. Greece* [2002] ECR I-3879; ECJ C-52/00, *Commission v. France* [2002] ECR I-3827 and ECJ C-402/03, *Skov v. Bilka* [2006] ECR I-199.

sec. 48(3) which refers to the PLA. (Cf. also e.g. sec. 40 of the Act on Fertilizer Products (539/2006).) However, such a channelling of the liability may be problematic in relation to Art. 3 and 13 of Dir. 85/374/EEC.[62]

It cannot be entirely excluded that Dir. 85/374/EEC as interpreted by the ECJ may also cause other problems in relation to Finnish law. (Cf. also, e.g. the discussion in no. 87 below regarding the PLA sec. 8(2) and Art. 9 of the Directive.) **43**

(c) Impact of compliance with rules and regulations

Since liability under the PLA is strict, and since the situation referred to in the present question is not mentioned as a defence, it will not affect liability. Under sec. 7(1) (2) of the PLA, liability shall be exempted if the defect in the product which caused the injury or damage is attribut able to compliance of the product with mandatory regulations issued by a public authority. **44**

3. Environmental liability

(a) Implementation of the Environmental Liability Directive

See no. 8 above on the proposed Act on Compensation for Certain Damage Caused to the Environment (the Environmental Liability Draft). Financial guarantees are not foreseen. **45**

In this context, it may especially be noted that the Environmental Liability Draft is seemingly not intended to impose any liability upon, for example, a farmer of GM crops by virtue of the proposed reference to it in sec. 23 of the GTA. As seen above (no. 10), it has been held that the GTA does not deal with, e.g. cultivation of GM crops that have been approved and released on the market. Consequently, a cultivator of such crops is not a liable operator under sec. 3 of the GTA. And in the draft bill to the Environmental Liability Draft,[63] it is emphasised that such an interpretation of the GTA is in line with Directive 2001/18/EC[64] (which is referred to in Annex III of Dir. 2004/35/CE). Con- **46**

62 Among other things, Art. 3 of the Directive does not impose liability upon a person on behalf of whom the product is produced (note especially that such a person does not "present himself as ... [the] producer" within the meaning of Art. 3.1). See in this respect also especially *Skov* [2006] ECR I-199.

63 See *Environmental Liability Draft* 2008 (fn. 16) 43 f.

64 Directive 2001/18/EC of the European Parliament and of the Council of 14 April 2001 on the deliberate release into the environment of genetically modified organisms, OJ L 106,

sequently, it is provided in quite general terms that "a farmer cannot become liable" for damage to such resources as are covered by the Environmental Liability Draft.[65] However, this interpretation in the draft bill does not seem to be in line with Art. 3(1)(b) of Dir. 2004/35/CE. In fact, it is even hard to see how this interpretation could fit into the general scope of activities covered by the strict liability pursuant to the Environmental Liability Draft (see further no. 47 below).

(b) Environmental liability regime beyond the scope of the Directive

47 The EDCA exceeds the scope of Dir. 2004/35/CE (see also no. 11 on the applicability of the EDCA). The civil liability under the EDCA also covers damage to the environment per se (see also no. 14 above and in more detail no. 52 below). But in contrast to the Directive (and the Environmental Liability Draft), the applicability of the EDCA is not restricted to significant damage to certain resources (protected species and natural habitats, water, and land[66]). Further, the strict liability under the EDCA is not limited to particular activities.[67] Note also that neither does the Environmental Liability Draft only impose strict liability upon certain activities; any operator having caused damage may be strictly liable under the various environmental statutes referring to the Environmental Liability Draft.

48 In addition, several existing provisions on public (administrative) liability for environmental harm also exceed the scope of Dir. 2004/35/CE. For example, sec. 84 of the Environmental Protection Act (86/2000) (EPA) on the obligation to restore damaged environment is – in much the same way as the EDCA – not limited to significant damage to particular resources and does not impose strict liability only upon certain activities.

17.4.2001, 1–39, as amended by Regulation (EC) No. 1830/2003 of the European Parliament and of the Council of 22 September 2003 concerning the traceability and labelling of genetically modified organisms and the traceability of food and feed products produced from genetically modified organisms and amending Directive 2001/18/EC, OJ L 268, 18.10.2003, 24–28.

65 See *Environmental Liability Draft* 2008 (fn. 16) 44.

66 See Art. 3.1 and 2.1 of the Directive.

67 Regarding the Directive; see Art. 3.1 and Annex III. In principle, the EDCA may apply to any activity meeting the criteria in sec. 1, and even if the activity is performed, e.g., by a private person outside business.

(c) Claimants in cases of environmental harm

According to the Environmental Liability Draft, the operator shall bear the costs of remedial (restoration) actions taken pursuant to the Draft. According to sec. 4 of the Draft, there are three categories of remedial actions; cf. Art. 7 and Annex II.1 of Dir. 2004/35/CE. "Primary remediation" aims at restoring baseline conditions, i.e. the situation that existed prior to the damage. Where baseline conditions cannot be fully restored, "complementary remediation" shall be undertaken on or outside the damaged site. The purpose of such remediation is to provide a similar level of resources and benefits as would have been provided if the damaged site had been restored to its baseline conditions. Finally, "compensatory remediation" shall be undertaken on or outside the damaged site to compensate for interim losses during the time of primary and/or complementary remediation. **49**

Regarding remediation of damage to land and groundwater, the Environmental Liability Draft refers, in sec. 2(3), to chap. 12 of the EPA which lays down provisions on clean-up of contaminated land and groundwater. In the draft bill to the Environmental Liability Draft, the existing provisions in chap. 12 of the EPA are held to be sufficient to meet the requirements of Dir. 2004/35/CE regarding damage to land and groundwater.[68] This may be true regarding land damage; see Annex II.2 of the Directive on remediation of land damage. But according to the Directive, damage to waters – seemingly including groundwater[69] – shall be remedied pursuant to the system in Annex II.1 (see no. 49 above on this system) which goes far beyond only clean-up. **50**

The Environmental Liability Draft does not contain any special provisions on preventive actions. Instead, the Draft refers (in sec. 3(3)) to several existing statutes on preventive actions which are considered to meet the requirements of such actions under the Directive. **51**

Further, there are also several existing provisions on the liability to restore damage to the environment per se, such as sec. 6(2) of the EDCA[70] and **52**

68 See *Environmental Liability Draft* 2008 (fn. 16) 29.

69 See Art. 2.1(b) of the Directive, according to which "environmental damage" includes "water damage, which is any damage that significantly adversely affects the ecological, chemical and/or quantitative status and/or ecological potential, as defined in Dir. 2000/60/EC, of the waters concerned, with the exception of adverse effects where Article 4(7) of that Directive applies". As regards groundwater; see Art. 4.1(b) of Dir. 2000/60/EC.

70 See also on this provision *Sandvik* (fn. 6) 25.

sec. 84 of the EPA. Although these provisions exceed the scope of the Environmental Liability Draft (see also no. 47 and 48 above), the concept of "restoration" contained therein is not as extensive as the concept of "remedial actions" under the Environmental Liability Draft and Annex II.1 of Dir. 2004/35/CE. It is true that a broad interpretation of restoration under, for example, sec. 6(2) of the EDCA has been advocated,[71] but it is at least unclear whether this provision (or other similar provisions) really can be interpreted to also include "complementary" restoration (remediation) not to mention "compensatory" restoration.

53 Insofar as nothing else follows from particular other statutes such as the GTA or the EPA (see no. 54 and 55 below), the regional environmental centre shall have the right to require that remedial measures pursuant to the Environmental Liability Draft are taken by the liable operator (sec. 3). The regional environmental centre itself may also take the necessary remedial measures if the matter is urgent or if the liable operator cannot be identified without difficulties (sec. 8(1)). But if later identified and if no defences are at hand, the operator shall, of course, bear the costs of the measures (sec. 8(2), 9 and 10).

54 According to the proposed reference to the Environmental Liability Draft in sec. 23 of the GTA (see also no. 8 above), the Board of Gene Technology shall order the operator to take remedial actions pursuant to the Environmental Liability Draft. In addition, it is proposed (in a new sec. 36c to be included in the GTA) that such actions may also be initiated in writing by a person whose rights or interests may be affected by the matter, or a registered association or foundation whose purpose is the promotion of nature conservation or environmental protection and whose sphere of operation relates to the environmental effects concerned.

55 Under the EPA, the regional environmental centre and the municipal environmental protection committee shall take legal actions in matters concerning clean-up of land and groundwater or restoration of the environment (sec. 22 and 92). But if legal action is not taken on their initiative, action may (according to sec. 92) be initiated in writing by, for example, a person whose rights or interests may be affected, any authority watching over the public interest in the matter, or a registered association or foundation whose purpose is the promotion of environmental protection, the protection of

71 See in particular *Sandvik* (fn. 19) 309 ff., especially 390 ff., 399 ff. Cf. also, e.g. *P. Wetterstein*, Ekonomiskt ansvar enligt EG:s miljöskadedirektiv [Economic Liability Under the EC Directive on Environmental Liability], JFT 2007, 459 ff. at 470.

health, nature conservation or pleasant living environment and whose sphere of operations relates to the environmental effects concerned.

Under sec. 6(2) of the EDCA, authorities have the right to claim reasonable costs from the operator liable for measures undertaken to restore damage to the environment per se. 56

(d) Special liability regime for losses sustained by individuals

Even the general rule of torts in the TLA may cover losses sustained by individuals in the course of damage to the environment per se, although compensation would not be payable for the damage caused to the environment per se.[72] There is really nothing inherently special in compensating, for example, property damage caused by emissions from a factory which at the same time may also have caused non-compensable damage to the environment as such. However, if the question is understood as referring to regimes covering both individual losses and damage to the environment per se, reference may be made to no. 14 above regarding the EDCA. Also, for instance, the provisions on oil pollution damage in chap. 10 of the Maritime Code (674/1994) – which are mainly based on international treaties – cover both individual losses in the form of personal injury, property damage and pure economic loss, and damage to the environment per se in the form of clean-up and restoration costs.[73] 57

72 However, a famous decision by the Swedish Supreme Court (NJA 1995, s. 249) illustrates that compensation for damage to the environment per se may be possible even under the general rules of torts. In this case, two wolverines had been unlawfully killed. The Swedish state claimed damages for the loss of the animals. The Court found that the state is under a public law duty to protect and conserve wolverines. The Court categorised the harm as a hybrid between economic and non-economic damage. However, since the wolverine is a protected animal it lacks an economic value. But as the state bears large expenses for the duty to conserve and protect the species, and those expenses have rendered partly useless due to the killing – which decreased the species' chances of propagation – the measure of compensation should be the breeding value of the wolverines. Reasonable compensation should therefore be awarded under the Swedish TLA with the conservation expenses as a starting point. The Court found that a value of SEK 20,000 for each wolverine was reasonable. See further on the case, e.g. *M.L. Larsson*, The Law of Environmental Damage (1999) 497 ff., and from a Finnish point of view, e.g. *B. Sandvik*, Ekonomisk eller icke ekonomisk skada? Allmänna läror i kritisk granskning [Economic or Non-Economic Loss? A Critical Review of General Principles], JFT 2007, 431 ff., 449 ff.

73 See also on chap. 10 of the Maritime Code and the international treaties *Sandvik* (fn. 6) at no. 30 ff.

(e) Cartagena Protocol

58 Finland ratified the Cartagena Protocol on 9 July 2004.[74]

4. Other strict liability regimes

59 There does not seem to be any strict liability regime of particular importance with regard to the present study other than those already referred to above.

V. Vicarious liability

1. Scope of vicarious liability

60 According to chap. 3, sec. 1 of the TLA, an employer shall be vicariously liable in damages for injury or damage negligently caused by an employee at work. Further, a person shall, according to the same provision, also be deemed to be an employer where he/she assigns work to an independent contractor who, in view of the permanent nature of the assignment, the nature of the work and other circumstances, parallels an employee.

61 Thus, the main rule under the TLA is that an employer is not liable for damage caused by an independent contractor, but special circumstances may lead to liability.[75] However, the TLA is based on fault liability and the mere fact that the damage is caused by an independent contractor will probably not exempt a strictly liable operator from liability.[76] In this context reference may be made, for example, to the Supreme Court decision KKO 2001:66 concerning the strict liability under the EDCA.[77] However, under some rules of strict liability, the operator will be liable for damage caused by an independent contractor only if the contractor has been negligent.[78]

---

74 See http://www.cbd.int/biosafety/parties/list.shtml (visited 18 November 2008).

75 As a comparison it may be noted that contract law follows an opposite rule.

76 See also, e.g. *Routamo/Ståhlberg/Karhu* (fn. 29) 221 f.

77 See further on the decision *Sandvik* (fn. 6) at no. 15. The decision may be compared, for example, to KKO 2004:53 concerning liability for an independent contractor under fault liability and chap. 3, sec. 1 of the TLA.

78 See further *Routamo/Ståhlberg/Karhu* (fn. 29) 222.

2. Liability for people further up the food or feed production chain

It could be argued that people further up in a production or distribution chain, such as retailers or wholesalers, do not "assign work" to suppliers further down in the chain, such as seed producers or farmers, in the meaning of chap. 3, sec. 1 of the TLA. Thus, people further up in the chain would have no liability for people further down in the chain. Some support for this interpretation can be found in case law. **62**

For example, in the well known Supreme Court decision KKO 1981 I 1, a metal piece in a sausage damaged the tooth of the person eating the sausage. The sausage packaging identified the producer, whose identity had thus been brought to the knowledge of the injured person who had bought the sausage. Since the employees of the seller had not acted negligently and since the seller could not be held liable for the injury solely on the ground that he had sold the sausage, the injured person's claim for damages against the seller was disallowed. See also, e.g. KKO 1984 II 225 concerning liability for health problems and other damage caused by defective construction material in a prefabricated house bought by a family. The Supreme Court held the producer of the construction material liable in damages. Cf. also an early case KKO 1955 II 31 holding the producer of defective curling-tongs liable in damages for injury caused to a hairdresser's customer when the hairdresser had not contributed to the injury. **63**

Still, it has been pointed out that the liability issue was not settled until the PLA entered into force on 1 September 1991.[79] See in particular PLA sec. 5 and 6; cf. Dir. 85/374/EC Art. 3.[80] However, the TLA is based on fault liability and it is an open question whether non-contractual strict liability outside the applicability of the PLA may also lead to liability for people further down in a production or distribution chain.[81] There is no guiding case law on the matter. As a comparison, it may be noted that in contract law, the seller's strict liability not only covers third persons whom the seller has engaged to perform the whole or part of the contract; the seller is also liable for general suppliers of, e.g. raw materials, and for any previous level in the chain of supply.[82] But it should also be recognised that **64**

---

79 See *O. Norros*, Vastuu sopimusketjussa [Liability in a Contract Chain] (2007) 51.

80 Note in this context also the case law of the ECJ above in fn. 61.

81 *H. Saxen*, Tillägg till skadeståndsrätt [Supplement to Tort Law] (1983) 423 stresses that in KKO 1981 I 1 (above), the seller was not held strictly liable for the damage.

82 See sec. 27(2) and 40(1) of the Sale of Goods Act (355/1987). This is said to reflect a general principle of strict liability in contract law; see, e.g. *B. Sandvik*, Säljarens kontrollansvar [Seller's Control Liability] (2004) 140 ff., in particular 156 ff. with further references.

all reasoning behind that solution (such as the doctrine of privity of contract[83]) is not relevant in non-contractual relations.

65 Moreover, the whole issue of non-contractual liability for people further down in a production or distribution chain seems to be of most practical importance in product liability cases. In regard to GMOs and non-contractual product liability outside the scope of the PLA, sec. 36(3) of the GTA (see no. 1 and 4 above) generally imposes strict liability upon the operator pursuant to the GTA, i.e. the producer (developer) of the GMOs. See also the channelling of the strict liability under the special rules in sec. 48(1) of the Feed Act (see no. 5 above[84]) and in sec. 14(1) of the Seed Trade Act (see no 6 above).

3. Can someone further down the feed or food chain include someone further up it in a trial against him/herself?

66 Probably not. If found liable, it seems that the loss can be shifted onto another person in the chain only through a separate trial.

VI. Multiple tortfeasors

67 Where the injury or damage has been caused by two or more persons, they shall be jointly and severally liable according to chap. 6, sec. 2 of the TLA. However, a person who has not been rendered liable in full damages shall be liable only to the amount of the award. The damages payable shall, according to chap. 6, sec. 3, be allocated to those liable as is deemed reasonable in view of the guilt apparent in each person liable, the possible benefit accruing from the event and other circumstances. Further, a person who has paid damages beyond his/her allocated share shall have the

---

It is unclear to what extent the seller's fault liability under sec. 27(4) and 40(3) of the Sale of Goods Act may extend also to general suppliers and previous levels in the chain of supply of the goods. However, it has been suggested that in this respect, the interpretation should be coordinated with sec. 27(2) and 40(1) of the Sale of Goods Act. See ibid., 159 ff.

83 See also ibid., 137 ff. with further references. See also KKO 2008:31 concerning an entrepreneur's possible liability for a construction fault in a building built by the entrepreneur. The Supreme Court found that liability of the entrepreneur vis-à-vis person B, who had bought the building from person A, could not be based on contractual principles or the TLA (which, according to sec. 1, does not apply to liability for damages under contract unless otherwise provided in this or another Act).

84 See also no. 42 above regarding sec. 48(3) of the Feed Act.

right to recover from the other persons liable the amount paid on behalf of such persons.

A somewhat similar rule is included in the EDCA sec. 8. The most notable difference compared to the TLA is that under the EDCA, liability for compensation shall not be imposed by judgment, in a degree exceeding the appropriate share, on a person whose share in inflicting the damage is manifestly minor.[85] 68

Regarding public (administrative) environmental liability and damage caused by more than one operator's activity, the legal picture is less clear.[86] But the primary rule seems to be several liability apportioned to each of the liable operators in accordance with their respective share in inflicting the damage or, if their shares cannot be estimated, equally apportioned to each operator. In a matter concerning the duty to clean-up polluted land, however, the Supreme Administrative Court (KHO) held two operators jointly liable when their respective share in inflicting the pollution could not be estimated; see KHO 1996 A 26 (leaving open how the clean-up costs should finally be allocated between the operators).[87] Yet, in the Environmental Liability Draft, an explicit provision in line with the general rule is proposed (sec. 13). The drafters emphasise that such a provision "would be analogous to the Supreme Administrative Court decision KHO 2005:11"[88] concerning the apportionment of the clean-up duty and the costs thereof when the pollution had been caused by the activities of two operators. (Note also Art. 9 of Dir. 2004/35/CE.) 69

VII. Defences

1. Licence/permission to grow GM material

Under the EDCA, an authorisation to grow GM material may be relevant as one factor among others in relation to the tolerance level pursuant to sec. 4 of the Act. See further no. 83 below. 70

85 See further *Sandvik* (fn. 6) no. 13 and 53.

86 See further, e.g. *J. Tuomainen*, Vastuu saastuneesta ympäristöstä [Liability for Polluted Environment] (2001) 150 ff.

87 *Id.*, 156 finds such a solution unacceptable as contrary to the polluter pays principle. *C. Herler*, Markföroreningsansvaret [Liability for Pollution of Land] (2008) 314, 326, on the other hand, seems to advocate that joint and several liability under the TLA should be analogously applicable to public environmental liability.

88 See *Environmental Liability Draft* 2008 (fn. 16) 33.

71 Moreover, according to sec. 12 of the Environmental Liability Draft, the authority who decides upon remedial actions pursuant to the proposed Act may allow the operator to avoid bearing the full cost of such actions provided that the operator demonstrates he was not at fault. Further conditions for such adjustment (reduction) of the liability would be that the environmental damage was caused by an emission or event in accordance with the conditions of an authorisation or other decision by an authority, or that the activity which caused the damage fully complied with statutory requirements for that activity.[89] In the draft bill to the proposed legislation, it is stressed that such a provision on adjustment of the liability for the cost of remedial actions is authorised by sec. 8(4)(a) of Dir. 2004/35/CE.[90] Still, the author of the present country report finds it doubtful whether the conditions under the proposed provision really are in keeping in every respect with the leeway left to the Member States in Art. 8(4)(a) of the Directive. But note also Art. 8(3) (b).

72 It may be further noted in this context that Finnish civil liability law recognises a general principle according to which damages may be adjusted (reduced) if the liability is deemed unreasonably onerous.[91] This principle is manifested in chap. 2, sec. 2 of the TLA. Under this provision, regard shall be had to the financial status of the tortfeasor and the victim, and the other circumstances when deciding whether the liability would be unreasonably onerous. If the injury or damage has been caused deliberately, full damages shall be awarded unless it is deemed that there are special reasons for a reduction in the damages. It cannot be precluded that authorisation to grow GM material could be taken into account as one "other circumstance" in much the same way as authorisation may justify adjustment under the proposed sec. 12 of the Environmental Liability Draft. However, it seems rather unlikely that authorisation alone would justify adjustment under chap. 2, sec. 2 of the TLA.

---

89 In the draft bill, it is emphasised that adjustment under the proposed sec. 12 would always be a matter of reasonableness, and that adjustment under the provision would be exceptional in practice. See ibid., 32 f.

90 Ibid., 32.

91 This principle applies to non-contractual as well as contractual liability. See, e.g. *Sandvik* (fn. 82) 277 ff., especially at 278 f., and M. *Hemmo*, Vahingonkorvauksen sovittely ja moderni korvausoikeus [The Adjustment of Damages and Contemporary Compensation Law] (1996) 319 ff. (discussing whether the principle of full compensation has in fact been substituted by a principle of reasonable compensation). See further also *Herler* (fn. 87) 311 ff. regarding public (administrative) environmental liability (arguing that the general civil liability principles on reasonableness and adjustment of damages are analogously applicable).

Finally, if, for example, GM materials that have been placed on the market have in fact been authorised for testing only, the authorisation will naturally not serve as any defence or as ground for any reduction of liability. **73**

2. Consent/assumption of risk

The mere fact that the victim has knowingly consumed GM products will likely not have any impact on liability.[92] It seems likely that the victim would be considered to have negligently contributed to the injury by knowingly consuming GM products only under very special conditions which would probably be quite rare in practice. **74**

3. Third-party influence

As seen above (III.3), a force majeure event may be considered to break the causal link between the damage and the activity upon which strict liability is imposed. Therefore, force majeure may have the effect of an exclusion of strict liability even if a force majeure defence is not explicitly provided for in, for example, the EDCA. Moreover, the bill to the EDCA supports such reasoning. According to the bill, if a third party has trespassed upon the area in which the activity is performed and caused an accident by mischief, the resulting damage is not caused by the operator's activity provided the operator has not contributed to the damage.[93] However, this example should be interpreted narrowly;[94] it is not necessarily an indisputable example of force majeure. **75**

Further, sec. 11 of the Environmental Liability Draft provides, inter alia, that an operator shall not be required to bear the costs of remedial actions when he can prove that the damage was caused by a third party and occurred despite the fact that appropriate safety measures were in place. Cf. Art. 8(3) (a) of Dir. 2004/35/CE which, however, explicitly covers both remedial and preventive actions. (See no. 51 above regarding the Environmental Liability Draft and preventive actions.) **76**

However, even if a certain event (such as sabotage or improper conduct by a neighbouring farmer) does not exclude liability, the event may still be relevant to the rule on contributing causes in chap. 6, sec. 1 of the TLA. **77**

---

92 Cf. also the reasoning by *Wilhelmsson/Rudanko* (fn. 11) 238 ff.

93 See *Government Bill* 1992:165 at 27.

94 See further *Sandvik*, JFT 1998, 553 ff.; cf. also *Sandvik* (fn. 6) no. 16.

This provision provides, among other things, that the damages may be adjusted (reduced) as is found reasonable if a circumstance, other than the fault of the liable person, contributed to the injury or damage. This causality rule may be said to reflect a general principle of tort law. But in cases of strict liability the expression "circumstance other than the fault of the liable person" should be interpreted as "circumstance unconnected with the activity of the operator" upon which strict liability is imposed.[95]

78 In this context, however, regard should be had also to Art. 8(1) of Dir. 85/374/EEC. According to this provision, which is without prejudice to the provisions of national law concerning the right to contribution or recourse, the liability of the producer shall not be reduced when the damage is caused both by a defect in the product and by the act or omission of a third party. Although the PLA lacks a corresponding provision, it should obviously be interpreted in the same way.

79 Finally, the author of the present report finds it unclear how Art. 8(3) (a) of Dir. 2004/35/CE (see no. 76 above) should be interpreted in this regard. Does this provision also allow a reduction of liability in instances when a third party has contributed to the damage? Or is the rule restricted to the exclusion of liability in cases where all of the damage is caused by a third party?

4. Prescription

80 According to sec. 4 and 7 of the Limitation Act (728/2003), actions for non-contractual damages are limited to three years after the victim became aware, or should have become aware, of the injury or damage and the liable person. However, the right to take action is extinguished upon the expiry of a period of ten years from the date of the occurrence of the event that caused the damage. However, the ten year rule does not apply to personal injury or environmental damage.[96] In these two respects, the three year rule applies exclusively. The Limitation Act applies in standard cases where there are no special provisions on limitation.

81 Special provisions on limitation are found in both the PLA and the Environmental Liability Draft. Sec. 9 of the PLA lays down a limitation period

95 See further *Sandvik*, JFT 1998, 563 ff.; cf. also *Sandvik* (fn. 6) no. 12.

96 See further, e.g. *Routamo/Ståhlberg/Karhu* (fn. 29) 459, noting that personal injury and environmental damage are special and need a separate rule since they may both become apparent only a long time after the occurrence of the event that caused the injury or damage.

in conformity with Art. 10 and 11 of Dir. 85/374/EEC. Sec. 8(2) of the Environmental Liability Draft lays down a limitation period of five years which is thought to be in line with Art. 10 of Dir. 2004/35/CE.[97] However, whereas Art. 10 of Dir. 2004/35/CE seems to cover the costs of both preventive and remedial actions, the Environmental Liability Draft as a whole covers only remedial actions and refers to other legislation regarding preventive actions (see also no. 51 above). This seems to lead to a deviation from the Directive in respect of preventive actions and limitation. Further, the Environmental Liability Draft lacks a 30 year rule corresponding to Art. 17 of the Directive.

5. Other defences

An obligation to tolerate disturbance is laid down in sec. 4 of the EDCA. **82** According to sec. 4(1), compensation for environmental damage is payable only if it is not reasonable to tolerate the disturbance taking into account factors such as the local circumstances, the situation as a whole that led to the disturbance, and how common the disturbance in question is in comparable circumstances. According to sec. 4(2), however, the obligation to tolerate disturbance does not apply to personal injury or non-minor property damage, and it does not affect damage caused by criminal or intentional behaviour.[98]

In this context, it may be significant that the obligation to tolerate distur- **83** bance is not expressly linked to licenses/permissions by authorities. Nevertheless, according to the bill to the EDCA, regard shall be had to the content of different environmental permits (nuisance thresholds, measures of health safeguard, etc.) as one factor among others in deciding whether a disturbance should be tolerated or not.[99]

For the sake of clarity, it may also be mentioned that the PLA and the **84** Environmental Liability Draft contain some exhaustive defences (not all of which have necessarily been addressed above) in conformity with Dir. 85/374/EEC and Dir. 2004/35/CE. It may be recalled in this context, however, that Finland chose not to incorporate into the PLA the "development risk defence" under Art. 7(e) of Dir. 85/374/EEC (see also no. 39 above).

97 See *Environmental Liability Draft* 2008 (fn. 16) 31.

98 See further on sec. 4 of the EDCA *Sandvik* (fn. 6) no. 17 referring to the Supreme Court decisions KKO 1999:124, KKO 2001:61 and KKO 2004:89.

99 See *Government Bill* 1992:165 at 15.

Further, neither is the corresponding defence under Art. 8(4) (b) of Dir. 2004/35/CE incorporated into the Environmental Liability Draft.

VIII. Remedies

1. Pecuniary compensation

(a) Bodily harm

85 Regular rules apply.

(b) Property losses

86 See no. 82 and 83 above regarding the tolerance level under sec. 4 of the EDCA.

87 Further, according to sec. 8(2) of the Act, liability under the PLA is, excluded if damage to "consumer property" does not exceed FIM 2,350. This threshold corresponds to approximately € 395 and is thus lower than the threshold of € (ECU) 500 under Art. 9 of Dir. 85/374/EEC. But it has been emphasised that sec. 8(2) of the PLA was enacted before the transition to Euro, that the threshold under sec. 8(2) was calculated in accordance with Art. 18 of the Directive and that subsequent change of currencies or variation in exchange rates do not require amendments of national rules once they have properly implemented a Directive.[100] Still, even if correct, it is obvious that this interpretation will not result in a complete harmonisation of the laws of the Member States with regard to the matters for which the Directive makes provisions as envisaged by the ECJ.[101] (See also no. 41–43 above.)

(c) Economic losses

88 See II.2 above regarding pure economic loss.

---

100 See *Wilhelmsson/Rudanko* (fn. 11) 234 f.

101 See in particular *Commission* [2002] ECR I-3827 and *Commission* [2002] ECR I-3879 as regards the threshold under Art. 9 of the Directive. Cf. also the other cases referred to above in fn. 61. *Wilhelmsson/Rudanko* (fn. 11) 234 f. do not address these cases when discussing (see the previous fn.) the consistency of sec. 8(2) of the PLA vis-à-vis Art. 9 of the Directive (although Wilhelmsson/Rudanko refer to the cases in an appendix on page 308).

An explicit rule on compensation for the costs of necessary measures to restore a GMO contaminated field to meet organic standards is included in sec. 27 of the proposed coexistence law (see also no. 9 above). The draft bill does not shed much light upon the proposed rule on restoration.[102] But it seems that no reduction is intended to apply for the seasons that the farmer could have grown conventional crops on the field concerned. Further, if such a reduction shall be made at all, it seems more logical to link it to possible compensation for loss of income (see also no. 16–18 above on compensation for loss of income under the proposed coexistence law). **89**

(d) Harm to animals

The mere fact that an animal has eaten GM contaminated feed does not seem to constitute any harm or damage. But damage has occurred if, for example, the animal needs curative treatment as a result of having eaten GM contaminated feed or if its owner suffers loss due to the fact that the meat or the milk of the animal can no longer be sold as organic. Such damage is compensable under sec. 48 of the Feed Act (see also no. 5 above). If the animal harmed constitutes "consumer property", however, the PLA will naturally apply (see also no. 42 above on the reference to the PLA in sec. 48(3) of the Feed Act). **90**

Neither the Feed Act nor the PLA lay down any rules on exactly how the compensation should be calculated. Thus, the standard tort rules apply. Under chap. 5, sec. 5 of the TLA, compensation for property damage shall cover reasonable costs of repair, other costs arising from the damage and diminution in value of the property, as well as loss of income and maintenance, that is, consequential economic loss. If repair is not feasible or reasonable, damages shall cover reduction in value and other costs arising from the damage, as well as consequential economic loss.[103] **91**

The principal rule is that costs of repair will be deemed unreasonable if the costs exceed the diminution of value and the other losses for which compensation is payable if repair is not undertaken.[104] This reflects the duty **92**

---

102 See *Coexistence Draft* 2008 (fn. 17) 42. In principle, it is only stressed that the proposed rule on restoration would also cover compensation for the costs of necessary rotation.

103 See further also *Sandvik* (fn. 6) at no. 34–37.

104 See also, e.g. the Supreme Court decisions KKO 1961 II 137 and KKO 1978 II 95 on the reasonableness of costs of repair in relation to diminution of value. As rightly stressed by *Routamo/Ståhlberg/Karhu* (fn. 29) 409, however, the reasonableness of costs of repair

to mitigate the damage. Vice versa, compensation will not be awarded for the diminution of value if that exceeds the costs of repair and other losses for which compensation is payable if repair is undertaken. At least regarding pets, however, it seems that expenditures incurred as a result of curative treatment should not be deemed unreasonable merely because the expenditures even significantly exceed the value of the injured pet (and other possible losses).[105]

93 In addition, the whole compensation issue is complicated by the fact that the TLA lacks rules on how the diminution of value should be calculated. Generally speaking, there are several conceivable valuation methods (such as sale value, use value, day value, replacement value, or yield value). But it is difficult to draw any general conclusions on the basis of case law since the choice of valuation method(s) is closely linked to the lawsuit, the presentation of evidence and other individual circumstances in case.[106] However, for example in KKO 1998:30 concerning damage to cattle, the claimant used the replacement value decreased by the sale value and the Supreme Court apparently did nothing to invalidate that method.[107] Further, compensation for consequential economic loss may naturally include, for instance, a cow's potential for producing milk or meat to the extent that has not already been reflected in the calculation of the cow's value. For example, if the sale value is used, this value can (depending on the actual situation) be seen as the value of the cow if sold for slaughter (cf. also KKO 1998:30) and include the value of both the meat and the skin of the cow.[108]

(e) Costs of disposal

94 It seems that the costs of such disposing measures would be recoverable.

---

should be related not only to diminution of value but also to other losses (e.g. loss of income) for which compensation would be payable if repair were not undertaken.

105 See further *Sandvik*, JFT 2007, 440 f. (referring, inter alia, also to two decisions by the Swedish Supreme Court; NJA 2001 s. 65 I and II).

106 See also, e.g. *Routamo/Ståhlberg/Karhu* (fn. 29) 405 ff., especially 407.

107 In other cases, however, it may not be clear at all which method was used to calculate the value; see, e.g. KKO 1992:138 and KKO 1991:162 regarding animals. However, KKO 1992:138 makes it clear the valuation may reflect that the animal was pregnant.

108 Cf. also M. *Radetzki*, Skadeståndsberäkning vid sakskada [The Assessment of Compensation for Property Damage] (2004) 102.

2. Non-compensatory damages

In Finland, all damages are compensatory. Punitive damages – which are non-compensatory by nature – cannot be awarded. **95**

3. Other remedies

Other remedies relevant in a GMO context may include injunctive relief,[109] threat of performance at the defaulter's expense and conditional fines,[110] and even criminal sanctions.[111] **96**

4. Costs of pursuing a claim

(a) General cost rule

Under chap. 21 of the Code of Judicial Procedure, the losing party generally bears the legal expenses in civil procedures where settlements are allowed. **97**

(b) Costs of establishing causation

To the extent that such costs are regarded as legal expenses, they are allocated in accordance with the rules in chap. 21 of the Code of Judicial Procedure; cf. the answer to the previous question. **98**

5. Advance cover

At present, neither liability nor first-party insurance products covering GMO risks seem to be available. In an earlier draft to the proposed umbrella law on the coexistence of GM and non-GM agricultural production, a compensation fund was proposed. However, during the subsequent drafting process, the financing of such a fund was considered problematic for several reasons. Therefore, under the present draft governmental bill to the proposed legislation, compensation would be payable by the state (see no. 9 above). **99**

---

109 Under, e.g. the GTA (see further *Sandvik* (fn. 6) no. 47) or the *Coexistence Draft* (sec. 23).

110 Under, e.g. the GTA or the *Coexistence Draft*.

111 See, e.g. sec. 42 of the GTA referring to several existing provisions in the Penal Code, and sec. 25 of the *Coexistence Draft* which introduces a new coexistence offence.

IX. Cross-border issues – Conflict of laws

1. Conflicts rules applicable before (or instead of) Rome II

100 In Finland, there are no generally applicable statutory provisions on the choice of law in cross-border cases involving tort liability. However, the principle of *lex loci delicti commissi* has been established. Further, it is widely (albeit not unanimously) held that in at least some circumstances, the claimant has the right to choose between the law of the state in which the event giving rise to damage occurred and the law of the state in which the damage occurred.[112] Still, the overall picture is rather unclear.

101 However, apart from clarifying the choice of law issue, the new Regulation (EC) No. 864/2007 of the European Parliament and of the Council of 11 July 2007 on the law applicable to non-contractual obligations (Rome II)[113] will bring some changes.[114] Most notably, the main rule under the Rome II Regulation is that "the law applicable to non-contractual obligations arising out of a tort/delict shall be the law of the country in which the damage occurs irrespective of the country in which the event giving rise to the damage occurred and irrespective of the country or countries in which the indirect consequences of that event occur" (Art. 4(1)). Special rules apply, for example, with respect to both product liability (Art. 5) and environmental damage (Art. 7). Regarding product liability, however, it should be noted that Finland is a party to the Hague Convention on the Law Applicable to Product Liability. And the Rome II Regulation does not prejudice the application of the Hague Convention (see Rome II, Art. 28).

2. Special regime for cross-border claims

102 There is no special regime.

---

112 See further, e.g. *H.T. Klami*/*E. Kuisma*, Suomen kansainvälinen yksityisoikeus [Finnish International Private Law] (3rd ed. 2000) 152 ff., especially 155. Cf. *Sandvik* (fn. 6) no. 42 (cf. also at no. 61 regarding jurisdiction).

113 OJ L 199, 31.7.2007, 40–49.

114 Cf. also *Sandvik* (fn. 6) no. 63.

X. Cases

1. Due to the adventitious presence of GMOs in a field, maize which is normally sold as conventional contains GMOs beyond the legal labelling threshold. This is not discovered before the final stage of the food production chain by the producer of taco chips. The whole production is lost since the supermarket chains refuse to accept delivery from the producer.

The PLA will of course not apply to this situation; there is no personal **103**
injury or damage to "consumer property" in the sense of sec. 1 of the PLA (cf. also Art. 9 of Dir. 85/374/EEC).[115] Outside the scope of the PLA, the borders between non-contractual product liability and contractual liability are rather esoteric from a Finnish point of view.[116] Nevertheless, it is beyond any doubt that the present situation relates to non-conformity in the form of defective ingredients and contractual liability under the Sale of Goods Act (355/1987).[117] Cf. also, e.g. the situation in the Supreme Court decision KKO 1997:61 concerning the liability under the Sale of Goods Act (the SGA) of an importer of defective components that had been used in goods manufactured and sold by another person.[118] And since the present study should deal with non-contractual liability, I will take the liberty of briefly commenting on a few aspects of that liability.

In most respects, lawyers acquainted with the 1980 UN Convention on **104**
Contracts for the International Sale of Goods (CISG) should find it relatively easy to orientate themselves with the Finnish SGA. The provisions on sales in Part III of the CISG were used as an important model by the drafters of the SGA. But there are also several fundamental differences between the CISG and the SGA. One of the most notable differences concerns liability in damages. Among other things, the SGA distinguishes between liability for direct and indirect losses; this is not paralleled in the CISG.

---

115 Note especially that for the PLA (and the Directive) to be applicable not only must the damaged property be intended for private use or consumption, but also the property must actually have been used by the injured party mainly for such purpose.

116 See also, e.g. *Sandvik* (fn. 82) 324 ff. (Cf. also on the difficulties fn. 11 above.).

117 See *Government Bill* 1986:93, 120 f. Cf. further also, e.g. *T. Wilhelmsson*, Köprätten och produktansvaret [The Sales Law and the Product Liability], JFT 1994, 627 ff.

118 See further on the decision *B. Sandvik*, Direct and Indirect Loss under "Catch-22" in the Nordic Law of Sales, 38 Scandinavian Studies in Law (Sc.St.) 1999, 25 ff., especially 39 ff., 44, 48.

105 Under the SGA, the seller's strict liability (see also no. 64 above on the strict liability under the SGA) covers only direct losses. The buyer is entitled to damages for both direct and indirect losses only if negligence is attributable to the seller or if the goods did not, at the time of the conclusion of the contract, conform to an express warranty of the seller (sec. 27 and 40). Indirect losses are defined in sec. 67(2).[119] The differentiation between the two categories of losses is as such difficult to comprehend or even confusing and it is further complicated by the fact that the buyer may convert indirect losses into direct ones. Under sec. 67(3), loss incurred by the buyer for mitigation of loss not covered by sec. 67(2) shall not be considered indirect loss. The whole differentiation in damages liability between direct and indirect losses has been heavily criticised.[120]

106 Now, looking at the position of the producer of taco chips, it could be argued that he has suffered an indirect loss under sec. 67(2) of the SGA in the form of "loss due to damage to property [the taco chips] other than the goods sold" to him (the maize). This would at least be consistent with the holding by the Supreme Court in its decision KKO 1997:61 (although the holding in that decision has been questioned). Thus, the wholesaler of the maize would be liable for the chip producer's loss only if negligence were attributable to the wholesaler, or if the maize did not, at the time the contract was concluded, conform to an express warranty. But as noted above, the differentiation between direct and indirect loss is extremely unclear and problematic. And when applied to the strict liability for non-conformity under sec. 40 of the SGA, the exemption pursuant to sec. 27 (cf. CISG Art. 79) can, from a practical point of view, hardly ever exempt a seller from his liability for direct losses.[121]

---

119 Indirect losses consists of the following: loss due to reduction or interruption in production or turnover; other loss arising because the goods cannot be used as intended; loss of profit arising because a contract with a third party has been lost or breached; loss due to damage to property other than the goods sold; and other similar loss that is difficult to foresee.

120 See especially *Sandvik* (fn. 82) 301 ff., in particular 349 ff., 362 ff. and *Sandvik*, 38 Sc.St. 1999, 25 ff., in particular 43 ff. with further references.

121 See further *Sandvik* (fn. 82) 203 ff. The exemption under sec. 27 of the SGA (cf. also CISG Art. 79) is elaborated with a view to typical situations of late performance or non-performance and in academic literature, it is generally recognised that it has very little bearing in relation to non-conformity. Indeed, the extremely limited practical importance of the exemption possibility in relation to non-conformity was one of the main reasons for introducing into the SGA the distinction between liability for direct and indirect losses.

2. Twenty years after the sale of GM maize used for food products, it turns out that it has certain disadvantageous health effects for humans.

(a) Can the producers be held liable at this point for risks unknown at the time of growing the maize? Who would be liable?

As seen above in VII.4, the standard limitation period in personal injury cases is three years after the victim became aware, or should have become aware, of the injury or damage and the liable person. Thus, actions could be taken even 20 years after the sale of the product under the standard rule. **107**

As noted above in VII.4, however, sec. 9 of the PLA lays down a limitation period in conformity with Art. 10 and 11 of Dir. 85/374/EEC. But it should be noted that the expiry period of 10 years under sec. 9(2) of the PLA and Art. 11 of the Directive is explicitly linked to the date on which the producer placed the actual damage-causing product on the market. Thus, if the damage is caused by use of the same GM product over many years (e.g. more than 20 years), the expiry period of 10 years would be calculated from the date the last item of the product which caused the damage was placed on the market.[122] **108**

Also in this context, it should be recalled that Finland has not incorporated the development risk defence under Art. 7(e) of Dir. 85/374/EEC into the PLA (see nos. 39 and 84 above). Thus, the person liable can be held liable at this point even for risks unknown at the time of growing the maize. **109**

As to who would be liable; see V.2 above, in particular no. 64 (cf. also no. 42) on the channelling of the strict liability under sec. 5 and 6 of the PLA and Art. 3 of Dir. 85/374/EEC. Where, as a result of the provisions of the PLA and the Directive, two or more persons are liable for the same damage, they shall be jointly and severally liable in accordance with chap. 6, sec. 2 and 3 of the TLA (see no. 67 above; see also Art. 5 of Dir. 86/374/EEC). **110**

---

122 See also *Wilhelmsson/Rudanko* (fn. 11) 260 f. It may be further noted that the 10 year rule under the Limitation Act (see no. 80 above) has also been interpreted in a corresponding fashion; see *Routamo/Ståhlberg/Karhu* (fn 29) 460.

(b) Can compensation already be claimed at a point when the negative health effects have not yet materialised, but are to expected to according to scientific expertise/mere rumours?

111 Compensation likely cannot be awarded for personal injury before the negative health effects have actually materialised, at least not if the health effects are distant and uncertain. See no. 22 above on the Supreme Court decision KKO 1998:80. But compensation may be payable if the fear of possible future health effects causes proven mental distress comparable to personal injury and pain and suffering, or other loss for which compensation is payable. See KKO 1998:80, no. 21 above on KKO 1999:102 and KKO 2002:110, and no. 23 above on KKO 1999:61. Further, as demonstrated by KKO 1999:61, the feared health effects do not even necessarily have to be scientifically founded to justify an award of damages.

(c) Would it make any difference if the GM maize had only been in use for feed, causing harm to the animals, which may or may not cause harm to humans consuming the meat as well?

112 Probably not.

3. The driver of a food logistics company discovers that a farmer, from where he regularly picks up agricultural products fails to obey mandatory segregation rules or food or feed hygiene standards, which may lead to the admixture of GM and non-GM produce sold and packaged separately by that farmer. Does he or his employer have a duty to warn, i.e. warn the recipient of the allegedly "non-GM" produce?

113 It seems that the driver or his employer would be under no statutory duty to warn about such "non-GM" procedures.

Damage Caused by GMOs under French Law

Simon Taylor

I. General overview

1. Special liability or redress scheme for GMOs

There is no general liability regime in France specifically for loss caused by GMOs. However, a 2008 Law on genetically modified organisms[1] imposes a new strict liability regime on GM farmers for pure economic loss incurred by non-GM farmers as a result of contamination of their crops by neighbouring GM crops.[2] General civil and administrative liability principles will nevertheless remain relevant, since the farmer whose crops are contaminated can choose to base his claim on these parallel rules, which will in any event apply to all other forms of personal loss not covered by the statutory scheme. With respect to environmental harm, the 2004 Environmental Liability Directive was implemented in France by the Law on Environmental Liability of 1 August 2008.[3] **1**

2. State liability

The state may be liable in administrative law. Where the state is itself the direct cause of the damage, which could, for example, be where a state **2**

1 Law no. 2008–595 of 25 June 2008. Art. 8, dealing with the liability regime, is inserted in the Rural Code as art. L.663-4 and L.663-5.

2 For details of the statutory scheme (based on the provisions of the government Bill, shortly before the final adoption of the 2008 GMO Act by the French Parliament) and analysis of its interaction with general tort liability rules, see my earlier report, *S. Taylor*, Economic Loss Caused by GMOs in France, in: B.A. Koch (ed.), Economic Loss Caused by Genetically Modified Organisms (2008) 203 ff. At the time of writing, in January 2009, regulations defining certain aspects of the new regime in more detail which are necessary for the regime's application are still to be published.

3 Law no. 2008–757 of 1 August 2008 *relative à la responsabilité environnementale.*

research institute carries out field tests on genetically modified plants which then contaminate neighbouring crops, then the administrative courts impose strict liability for nuisance on the state for damage which exceeds "the normal levels of disturbance to neighbours".[4]

3 The administrative authorities could potentially also be liable for fault in a number of other circumstances: for example, where there has been a failure to attach adequate conditions to an authorisation for dissemination, where the authorities have failed to use their powers to enforce regulations and conditions of exploitation,[5] or where they have inadequately enforced food safety regulations. In such cases, any action by a claimant would again be brought before the French administrative courts. Liability would be based on fault, applying administrative law rules, and may be based on proof of a "faute lourde"[6] or "faute simple" depending on the circumstances.[7]

4 Under the new liability regime established by the 2008 Law on GMOs, compensation will be paid by GM farmers, and this compensation will be financed by private liability insurance rather than the state.

II. Damage

1. Recoverable losses

5 French liability rules will allow the recovery of a broad range of losses caused by GMOs: pure economic loss, personal injury, damage to property, and moral damage. To date, cases that allow damages to be recovered for environmental damage have been rare, although there are signs that recovery of this form of loss could be more common in the future.[8]

---

4 *M. Prieur*, Droit de l'environnement (5th ed. 2005) 934. No compensation will be ordered, or at least its level will be reduced, when the nuisance, at its current level, already existed at the time that the claimant came to the land, CE, 19 March 1915, Rec., 84; *Prieur*, 935.

5 Liability of the administrative authorities, for example, for not having enforced an order for a company to eradicate solvents which had polluted soil, CAA Paris, 21.1.1997, Cne Saint-Chéron, req. no. 94PA00119. See the numerous examples cited by *S. Galand-Carval*, La causalité, Rapport français, in: B. Dubuisson/G. Viney (eds.), Les responsabilités environnementales dans l'espace européen. Point de vue franco-belge (2006) 71, at 76.

6 Meaning a particularly serious fault, caused by grave negligence or recklessness. CE, 24 January 1964, Archand, Actualité Juridique: Droit Administratif (AJDA) 1964, 187.

7 *Prieur* (fn. 4) 937. Where the administrative task is perceived as particularly complex, the courts are more likely to require a "faute lourde", *J.-F. Brisson/A. Rouyère*, Droit administratif (2004) 57.

8 See the later section on environmental liability, infra no. 43 ff.

2. Pure economic loss

Pure economic loss is recoverable in French tort law, provided that the general requirements for liability are fulfilled, namely that the damage is sufficiently certain and that a causal link between that loss and the act or omission generating liability is established. For the new strict liability regime for pure economic loss introduced by the 2008 Law, refer to my earlier report.[9] 6

3. Mere fear of a loss

The general rule in French law is that in order to be recoverable the damage must be certain. The mere fear of a future loss will not normally be considered as fulfilling this requirement. However, there are examples of cases where such loss has been deemed recoverable by the courts. Hence, in a decision of the Cour de cassation[10] of 19 December 2006,[11] the claimant learnt that his pacemaker was defective, and that there was a consequent risk of heart failure and death. The claimant underwent a second operation to have the pacemaker replaced, and was allowed to recover damages for the fear of future harm he had suffered which had been engendered by the faulty apparatus, classified by the judges as moral damage. 7

There would appear however to be a considerable difference between the fear of harm in this case and the much vaguer fears associated with mobile phone radiation, which, for the moment, poses a much less certain risk. Nevertheless, the Aix-en Provence Court of Appeal was prepared to base liability on such a fear in a decision in 2004.[12] In that case, a local council brought an action against a mobile telephone company to require a telephone aerial to be moved away from its position near a primary school. The application brought by the local authority was based on the possible health risk posed by radiation. The Court ordered the post to be moved on the grounds that the potential risk that it posed to residents' health constituted a common law nuisance (*trouble anormal de voisinage*), despite the fact that there was no conclusive scientific evidence of a risk to health. 8

9 *Taylor* (fn. 2) 206 ff.

10 The highest French civil court.

11 Cass. 1st Civ., 19.12.2006, Revue Trimestrielle de Droit Civil (RTDCiv) 2007, 352, observations *P. Jourdain*; Responsabilité civile et assurances (RCA) 2007, Comm. no. 64, observations *C. Radé*; La semaine juridique édition générale (JCP) 2007, II, 10052, note *S. Hocquet-Berg*; Receuil Dalloz (D.) 2007, 2897, note *P. Brun* et *P. Jourdain*.

12 CA Aix-en-Provence, 8.6.2004, D. 2004, 2678, note *M. Boutonnet*.

Doctrinal writers see this case as evidence of the influence of the precautionary principle on such questions.[13] The same approach has recently been followed by the Nanterre first instance court in another case relating to possible risks to health from a mobile telephone mast, and the court in this case actually awarded the claimants damages for the risk to health they had supposedly suffered.[14] Some reserve must be maintained as to whether such a generous approach to victims would be adopted by the Cour de cassation.

4. Standard of proof

9 The claimant must establish the loss suffered with certainty. The defendant will not be held liable for loss which is considered merely hypothetical or "possible".[15] However, the notion of "certainty" is often approached by the judges with a great deal of flexibility, and whether loss is merely hypothetical is clearly a question of degree.

10 With respect to future loss, the approach of the courts will vary depending on whether the loss has already occurred and it is merely the extent of the damage and its future evolution that is uncertain, or the occurrence of the damage itself is an uncertain future event. In the first type of situation, the court will willingly make an estimation of the future loss.[16] Hence, where a victim has been intoxicated by a GMO, but the evolution of his illness remains uncertain, he will be entitled to damages based on an estimation of his future loss. This will include all normally foreseeable future harm where its effects can already start to be perceived at the date of the evaluation of the loss.[17] The court may in such cases award provisional damages pending the stabilisation of the loss.[18] Likewise, a farmer whose crop has been contaminated by GMOs will be able to recover for his future loss of profits. French courts also frequently award damages for the loss of a

13 *P. Jourdain*, RTDCiv 2005, chroniques, 146.

14 TGI Nanterre, 18.9.2008, D. 2008, 2916, note *M. Boutonnet*. The claimants were awarded € 3,000 damages per couple (3 couples) for the health risk that they had been subjected to.

15 Cass. Com. 19.7.1971, D. 1972, abstract/résumé. 62; Cass. 2nd Civ. 3.11.1971, D. 1972, 666, note *Ch. Lapoyade-Deschamps*; *G. Viney*/*P. Jourdain*, Traité de Droit Civil. Les Conditions de la Responsabilité (3rd ed. 2006) 83 ff.

16 Sirey (S.) 1933, 1, 49, note *H. Mazeaud*; Cass. Civ., 13.3.1967, D. 1967, 591; *Viney*/*Jourdain* (fn. 15) 84 ff.

17 Cass. Crim. 2.6.1964, Bulletin des Arrets de la Chambre Criminelle (Bull.Crim.) 402; Cass. 2nd Civ. 4.1.1978, JCP 1978, IV, 79; Cass. 3rd Civ. 3.3.1993, RCA. 1993, chap. no. 186.

18 Cass. 2nd Civ. 19.4.1958, Bulletin des Arrets del la Chambre Civile (Bull.Civ.) II, no. 264.

chance of future gain or for the loss of a chance to avoid a future loss. Thus, non-GM farmers whose crops have been contaminated by GMOs could recover for the loss of a chance of obtaining a valuable contract for the sale of their crops, just as a consumer intoxicated by GM food could potentially claim for the loss of a chance of career success.[19]

Different considerations apply where the uncertainty applies to the occurrence of the damage itself, that is to say where there is a mere risk of future damage. Although the French courts do not specifically recognise compensation for the creation of a risk of future damage, such loss is in practice compensated by the courts.[20] This was the case, for example, where the installation of a fuel storage facility in a residential area created a potential fire hazard.[21] Likewise, a defendant was found liable for creating a risk of flooding due to the building of a dam which prevented the normal flow of water from a pond.[22] Since there are no strict requirements on the court to justify the level of damages awarded, it is often not clear to what extent the court has reduced the level of damages in such cases to reflect the fact that they are compensating for the mere risk of an occurrence rather than for the occurrence itself.[23] 11

However, despite the fact that compensation received by a claimant for the presence of a risk may in practice be used to pay for measures to combat this risk, French courts have often been unwilling to compensate for the cost of preventative measures taken before damage occurs. The Cour de cassation recently refused to compensate a patient fitted with a pacemaker for the costs and pain incurred in preventative surgery to remove the equipment when a design defect was found to affect the reliability of the machines,[24] despite the fact that they did award compensation to the same victim for his moral loss due to the fear of future harm during the time that he had the defective pacemaker. The same approach has, until recently, been applied generally, and therefore to measures taken to prevent future environmental damage.[25] However, a recent 2008 decision of 12

19 *Viney/Jourdain* (fn. 15) 92 ff.; Cass. 1st Civ. 19.12.1983, Bull.Civ. I, no. 304; Cass. 2nd Civ. 13.11.1985, Bull.Civ. II, no. 172.

20 *Viney/Jourdain* (fn. 15) 88 ff.

21 Cass. 2nd Civ. 16.7.1982.

22 Cass. 2nd Civ. 17.12.2002, Revue de droit immobilier (RD imm.) 2003, 322, obs. *F.-G. Trébulle*; JCP 2004, I, 101, no. 5, obs. *G. Viney*.

23 *Viney/Jourdain* (fn. 15) 89 ff.

24 Cf. fn. 11.

25 *P. Jourdain*, Le dommage écologique et sa réparation. Rapport français, in: B. Dubuisson/G. Viney (eds.), Les responsabilités environnementales dans l'espace européen. Point de vue franco-belge (2006) 175 ff.

the Cour de cassation goes against this trend by allowing a claimant to recover from the defendant the cost of work aimed at constructing a barrier on his land in order to protect it from the risk of erosion due to work carried on the defendant's neighbouring property,[26] and this follows the same approach taken by an earlier decision in 2007.[27] These cases perhaps mark a new tendency to increased flexibility with respect to the prevention of environmental damage. Recent proposals to redraft the obligations provisions of the French Civil Code propose to include liability for the reasonable costs incurred in preventing imminent damage.[28]

5. Nominal losses

13 Nominal/symbolic losses are recognised in French law.[29] Where associations for the defence of the environment bring a tort action for compensation for their "moral" damage, the courts often award a nominal or symbolic sum.[30]

6. Mass losses

14 French law states that in order for an individual to bring an action for damages he must have suffered the loss himself. An individual is not entitled to bring an action for the recovery of loss to other members of a group. In situations of mass loss, where the damage is extremely diffuse, an individual claimant may on occasions see his claim refused by the court due to the minimal nature of his personal loss.[31]

15 However, French law does provide for actions to be brought for collective loss. Hence, it is possible for a licensed association to bring an action on

26 Cass. 2nd Civ. 15.5.2008, JCP G 2008, I, 186, chron. *P. Stoffel-Munck*; D. 2008, 2900, chron. *P. Brun*.

27 Cass. 1st Civ. 28.11.2007, Bull.Civ. 2007, I, no. 372; JCP G 2008, I, 125, no. 7, note *Ph. Stoffel-Munck*.

28 Art. 1344 avant-projet de réforme du droit des obligations et du droit de la prescription. The reform proposals were presented to the Justice Ministry in September 2005. See http://lesrapports.ladocumentationfrancaise.fr/BRP/054000622/0000.pdf.

29 For example, symbolic awards have been made to associations for the defence of the environment for the death of a bird of prey (CA Pau, 17.3.2005, no. 00/400632) and for the death of a wolf (CA Aix-en Provence, 21.3.2005, no. 534/M/2005).

30 *F. Nei/D. Guihal*, Le nouveau dispositif de responsabilité environnementale et le droit commun, Droit de l'environnement 2007, 230 ff.

31 CA Paris, 14.10.1992, D. 1993, 111, conclusions *D. Fortin*; or alternatively on the basis that there is no direct causal link between the fault and the claimant's loss.

behalf of its members to recover damages for economic or moral harm.[32] In such cases, the courts will not calculate the loss of each particular member, and the loss will be evaluated at a global level and paid to the association. The association may make a claim with respect to harm which occurred before it was created, but the objects of the association must specifically allow it to act on behalf of its members in this way.[33] Rather paradoxically, the case law states that the individual members of the association can bring an action at the same time, creating the risk that the same loss will be compensated twice.[34]

A representative action ("action en représentation conjointe") is also available for associations for the defence of consumers and for the protection of the environment.[35] Damages in this case will be paid directly to the victims. The representative action must be brought on the basis of specific written mandates which cannot be solicited by the association. This form of action has been practically ignored by associations who perhaps fear facing heavy liability in cases where mistakes are made in the administration of the action.[36] **16**

Art. 1343 of the recent proposals for the redrafting of the obligations sections of the Civil Code provides for the compensation of collective interests through a form of class action brought by authorised consumer groups.[37] **17**

III. Causation

1. Uncertainty of merely potential causes

It will be for the victim to prove the causal link between the potential cause and his loss. The causal link must in principle be proved with certainty. However, French judges are often willing to accept the existence of a causal link where there are shown to be "serious, specific and concordant"[38] indications of such a link, in other words that causation is probable. **18**

---

32 Cass. Civ. 25.11.1929, Gazette du Palais (Gaz. Pal.) 1930, I, 29; Cass. Com. 17.7.1951, JCP 1952, II, 7128, note *D.B.*

33 *Viney/Jourdain* (fn. 15) 136.

34 Ibid., 137.

35 Art. L 422–1 ff. Consumer Code; Art. L 142–3 Environment Code.

36 *L. Boré*, L'action en représentation conjointe: class action française ou action mort-née? D. 1995, chron. 267 ff.

37 See fn. 28.

38 Art. 1353 Civil Code permits judges to rely on presumptions provided that they are: "graves, précises et concordantes" (serious, precise and concordant), Cass. 2nd Civ.

19 The courts sometimes proceed by elimination. The causal link is presumed by the fact that there is no other apparent cause of the harm. Using this approach, the courts have allowed claims where walls have collapsed or greenhouses smashed as a result of sonic booms from aeroplanes.[39] Examples of this approach can be found with respect to loss connected with environmental pollution. Hence, a first instance court was willing to accept the presumption that, in the absence of other available explanations, the higher than normal death rate of the claimant's bees was due to pollution on nearby land.[40] In another case, the loss of part of a herd of cows was judged to be due to pollution caused by the defendant, since, according to the veterinary expert "a link between the death of the cattle and the pollution could not be excluded", and no other explanation was forthcoming.[41] This approach has been used to establish the causal link between contamination by the hepatitis C virus and blood transfusions. Blood transfusion centres have been found liable on this basis where the victim shows that the contamination occurred consecutively with the blood transfusion and that he was not within a category of patients having a high risk of contamination.[42]

20 In three recent cases concerning the claims of victims of multiple sclerosis against the manufacturer of the hepatitis B vaccine, the Cour de cassation has drawn a clear distinction between the proof of legal and scientific causation. The Court ruled that the absence of clear scientific proof of a causal link between the vaccine and the illness should not prevent the first instance judges from concluding in favour of a causal link based again on the existence of "serious, precise and concordant" evidence provided by the circumstances of the case and the absence of alternative explanations.[43] Such evidence on the facts may, for example, include the short time that elapsed between receiving a vaccine and the occurrence of the disease. The highest French civil court thus authorises here the finding of a causal link despite the absence of a scientifically established risk. This goes even further than the position of the Conseil d'Etat[44] on the same question, which accepted the possibility of a causal link based on

14.12.1965, D. 1966, 453; Cass. 1st Civ. 24 January 2006, JCP G 1996 II, 10082, note *L. Grynbaum* (causal link between growth hormones and Creuzfeld-Jacob disease).

39 Cass. 2nd Civ. 29 April 1969, D. 1969, 534.

40 TGI Albertville, 26.8.1975, JCP G 1976, II, no. 18384, note *W. Rabinovitch*.

41 CA Caen, 13.1.2005.

42 Cass. 1st Civ. 10.6.2002; 2.7.2002; 10.7.2002.

43 Three decisions of the Cour de cassation: Cass. 1st Civ. 22.5.2008, D. 2008, 1544, note *I. Gallmeister*.

44 The highest French administrative court.

the circumstances "provided that scientific evidence has not excluded the existence of such a link".[45]

In certain cases the courts even impose liability on the basis of the creation of a risk of damage without strict proof of a causal link.[46] In such cases the courts have found the defendant liable on the basis that, voluntarily or by negligence, he has created a situation which is objectively dangerous and as a result the victim had suffered damage which is the foreseeable consequence of the creation of the risk.[47] The application of this principle tends to be restricted to cases where the creator of the risk has committed a fault.[48] **21**

Where there are a number of potential causes of the claimant's harm, then it will often be sufficient to show that the defendant's act or omission contributed to the harm. Hence, in 2003 the Cour de cassation stated that a factory owner who had contributed, amongst others, to the pollution of a stream was liable for the loss of the claimant's calves who had drunk from the water.[49] **22**

In a scenario where it is unclear who violated segregation rules or good farming practice, a neighbouring farmer whose crop is contaminated will often be able to rely on the new statutory liability regime in order to recover his pure economic loss caused by GMOs. Under this scheme, liability will be imposed on the GM farmer without proof of a causal link, provided that the non-GM farmer can establish that his contaminated crop comes from a field "near" a field where crops containing that GM organism were grown during the same season, and that his crops, which were intended to be sold as a product not subject to European Community GM labelling requirements, now have to be labelled due to the contamination.[50] **23**

For other claims, the flexibility shown by the French courts with respect to establishing a causal link could enable causation to be established where the circumstances indicate that a causal link is probable. Where the claimant himself has contributed to the contamination by not respecting segregation **24**

45 CE 9.3.2007, JCP 2007, II,10142, note *A. Laude*.
46 *Viney/Jourdain* (fn. 15) 224 ff.
47 Ibid., 226.
48 Cass. 2nd Civ. 11.3.1976, JCP 1976, IV, 157; Cass. 3rd Civ. 2.12.1980, JCP 1981, IV, 69; Cass. 1st Civ. 5.5.2005, Bull.Civ., I, no. 173; D. 2005, 2256, note *A. Gorny*; JCP 2005, I, 149 no. 7–9, observations *G. Viney*; RTDCiv 2005, 607, observations *P. Jourdain*; Cass. 1st Civ. 24.1.2006, Bull.Civ. I, no. 34; JCP 2006, II, 10082, note *L. Grynbaum*.
49 Cass. 2nd Civ. 23.1.2003.
50 *Taylor* (fn. 2) 206 f.

rules or good farming practice, then his damages will be reduced to reflect his contributory negligence, in proportion to the seriousness of the fault.[51]

2. Uncertainty of merely potential causes

25 This type of situation has caused considerable discussion amongst doctrinal writers in France. Where it is unclear which of a number of defendants has actually caused the victim's loss, the courts have found a number of solutions to enable the victim to overcome causal difficulties. In the classic case of a victim injured by a shot fired by an unidentified member of a group of hunters, the courts have imposed joint and several liability on the entire group, basing their reasoning on a "collective fault", or even on the "collective control" of the bullets fired.[52] Many doctrinal writers have identified as a common feature here that the defendants were acting collectively as a group. This would seem to prevent this technique from being used in the case cited in the present questionnaire, where damage could have been caused by any one of the various batches of maize coming from various suppliers and where there is no evidence that the different suppliers were in any way acting as a coordinated group. However, it has been argued that the hunting decisions should be interpreted more broadly as a technique used by the courts to facilitate the victim's task of establishing a causal link where it is impossible to identify the defendant at the origin of the damage.[53] According to this view, the decisions should be seen as establishing a presumption of a causal link once the claimant has shown the existence of damage and that this loss has been caused by a tortious act. According to this reading of the case law, where the victim produces evidence that one of a group of defendants has caused the loss, it is for the members of the group to then prove that they individually did not contribute. If these decisions can be interpreted in these broad terms, this approach could be used to establish joint liability of GM farmers for the contamination of neighbouring fields, and of seed or maize suppliers for damage caused by a single unidentified batch. It would then be for the various defendants to prove that their product did not in fact cause the victim's loss, either in the initial action, or in a recourse action against another defendant. This is indeed the solution put forward by the proposals for the reform of the French law of obligations, which

51 *Viney/Jourdain* (fn. 15) 328.
52 CA Angers, 13.11.1957, D. 1957, 721; Cass. 2nd Civ. 13.3.1975, D. 1975, IR, 124.
53 *Viney/Jourdain* (fn. 15) 245 f.

provide that, "Where damage is caused by an indeterminate member of a group, all the identified members of that group will be jointly liable, unless any particular member can establish that he was not the cause."[54]

3. Force majeure

In French law a "force majeure" is defined as an event which is of such importance that it renders any other event contributing to the damage insignificant. In that case, the defendant will avoid liability. Such an event must be one which is impossible to prevent and it will normally be required that the event be unforeseeable.[55] **26**

Where a natural event, such as a storm, flooding or an earthquake contributes to causing the victim's damage, but, rather than rendering any other factors insignificant, it is considered a concurrent cause along with the defendant's act or omission, then the Cour de cassation has held that the occurrence of the natural event has no effect on the liability of the defendant.[56] **27**

4. Threshold to prove causation

As has been described, the claimant is required to prove the certainty of the causal link. However, a degree of flexibility is allowed here, and in reality the courts in many cases accept probability as a sufficient level of proof.[57] **28**

5. Special rules on causation

The new French Law on GMOs, adopted in 2008, imposes a strict liability regime on GM farmers for pure economic loss caused to neighbouring farmers through GM contamination. To obtain compensation under the scheme, a non-GM farmer whose crop has been contaminated is not required to prove a causal link between the contamination of his crop **29**

54 Art. 1348 (fn. 28).

55 *Viney/Jourdain* (fn. 15) 279.

56 Cass. 2nd Civ. 11.2.1970, D. 1970, summary, 95; Cass. 2nd Civ. 30.6.1971, Bull.Civ., II, no. 240, 170. *Viney/Jourdain* (fn. 15) 297; *J. Flour/J-L. Aubert/E. Savaux*, Droit Civil. Les obligations. 2. Le fait juridique (12th ed. 2007) 169 f.

57 *Flour/Aubert/Savaux* (fn. 56) 164.

and an act or omission of the GM farmer, nor that the contamination actually came from the farmer's land or produce. All he need show is that the claimant's contaminated product comes from a parcel of land near (*à proximité*) a parcel where GM crops are cultivated, and that as a consequence of the contamination the product now has to be given a GM label.[58]

IV. Types of liability

1. Fault liability

(a) Special rules governing fault

30 A tort action may be brought on the basis of fault. However, in many cases the notion of "fault" may be interpreted by the French courts in such a way as to impose strict liability in practice.

31 Hence, French courts recognise the breach of a contractual obligation as constituting a delictual fault with respect to third parties under art. 1382 Civil Code.[59] This principle can be used to impose strict liability in tort on producers and sellers. Hence, a farmer or seed producer, by supplying a product which is defective or that does not comply with description, will be in breach of contract to the buyer, and will thereby be considered to have committed a fault and thus also be liable in tort to anyone not in the contractual chain who suffers loss as a consequence of the defect.[60] With respect to product liability claims, such an interpretation of the notion of fault will probably fall foul of the interpretation of the 1985 Directive by the European Court of Justice. We deal with this point later.[61]

32 Any violation of regulations governing GM farming will automatically be considered a fault leading to liability under art. 1382 Civil Code.[62] The victim's task of proving fault will thus be facilitated in such cases since it will be limited to demonstrating the violation of the statutory rule. Fault liability is therefore destined to play an important role here. It would appear likely that the development of the precautionary principle in civil and

---

58 Law no. 2008–595 of 25 June 2008, art. 8, now art. L 663–4 and L 663–5 Rural Code.

59 Cass. 1st Civ. 18.7.2000, Bull.Civ., I, no. 221; JCP 2000, II, 10415, report *P. Sargos*; *Viney/Jourdain* (fn. 15) 394 f.

60 Claimants in the contractual supply chain will have an action in contract.

61 Infra no. 38 ff.

62 G. *Viney*, Les principaux aspects de la responsabilité civile des entreprises pour atteinte à l'environnement en droit français, JCP 1996, I, 3900 ff.

administrative law will also have the effect of increasing situations where the defendant will be potentially liable for fault.[63]

The mere fact that the defendant has respected regulations will not necessarily prevent him from being found liable for fault.[64] 33

The importance of criminal law in the French system should also be noted here. The existence of a criminal infraction automatically constitutes a civil fault.[65] Where the defendant's fault constitutes a crime, for example where he has cultivated GM crops without seeking the relevant authorisations, then the victim will be able to bring his action before the criminal courts, or join an action already brought by the public prosecutor.[66] In such cases, he will benefit from the fact that the action will be principally financed by the state instead of from his own purse. The criminal court in such cases will apply tort law principles in establishing the defendant's liability to the claimant. Since environmental legislation creates more and more criminal infractions, this facilitates the finding of a civil fault. 34

(b) Impact of specific rules of conduct

See above. 35

2. Product liability

(a) Impact of specific rules of conduct

The development risks defence is included in the French product liability legislation transposing the 1985 Directive.[67] The French version of the defence follows the wording of the Directive, but excludes from its scope cases where the damage has been caused "by elements of the human body or products derived from such elements". Where applicable, the development risks defence will be applied in accordance with its interpre- 36

---

63 G. *Martin*, Le principe de précaution et la renaissance de la responsabilité pour faute, JCP Cahiers de droit de l'entreprise 1999, no. 1, 3 ff.; M. *Deguergue*, Les avancées du principe de précaution en droit administratif français, Revue internationale de droit comparé (RIDC) 2006, 621.

64 *Prieur* (fn. 4) 920. Cass. 2nd Civ. 14 June 1972, D. 1973, note *E. Lepointe*. Also art. L 514–19 Environment Code: "Licences are granted subject to the rights of third parties".

65 Cass. 2nd Civ. 16.11.2006, Env. 2007, note 63, observations *Boutonnet*.

66 For detailed discussion of this point, see G. *Viney*, Introduction à la responsabilité (2nd ed. 2006) 187 ff.

67 Art. 1386–11 Civil Code.

tation by the European Court of Justice in its judgment in the case *Commission v UK* on 29 May 1997.[68] A producer will therefore only be able to rely on the defence where he can show that the defect was one which could not be discovered despite the highest available level of knowledge, independent of the defendant's sector of industry, and even if the opinions of scientists were discordant.[69]

37 There was considerable debate in France as to whether to incorporate the development risks defence in the transposing legislation, and this to a large extent explains the long delay in passing the French legislation, which was only adopted in 1998 and came into force on 23 May of that year. The Law states that its provisions apply to products put into circulation after that date. As a consequence of the late transposition of the Directive, relatively few cases have been decided by the courts on this basis since many product liability claims still relate to products put into circulation before the transposing legislation came into force. In cases where the product was put into circulation between 30 July 1988 (the deadline set for Member State implementation of the Directive) and 22 May 1998, the pre-existing law (art. 1147 Civil Code for contractual liability and art. 1384-1 Civil Code for tort liability) is interpreted in the light of the provisions of the Directive. The question arises as to whether the courts, in interpreting the pre-existing law in compliance with the Directive, should interpret this law as including a development risks defence since its inclusion was optional for Member States. In a 2007 decision, the Cour de cassation held that the pre-existing rules, which imposed strict liability on the producer without permitting a development risks defence, should continue to be interpreted in this way.[70] This approach is criticised by at least one commentator.[71] If the interpretation by the Cour de cassation were to be followed by courts in the future, the defence would only be available to producers where the case relates to a product put into circulation subsequent to the introduction of the transposing legislation in May 1998.

---

68 ECJ C-300/95, *Commission v UK* [1997] European Court Reports (ECR) I-2649.

69 See the *Distiblène* case in France, where liability was based on the parallel civil liability regimes: Cass. 1st Civ., 7.3.2006, JCP G 2006, IV, 1734.

70 Cass. 1st Civ. 15.5.2007, D. 2007, 1592, note *I. Gallmeister*.

71 JCP 2006, II, 10082, note *L. Grynbaum*.

(b) Alternative routes

French law has various alternative routes for compensation of loss caused by agricultural products. However, following the ruling handed down by the European Court of Justice on 25 April 2002,[72] there is uncertainty over which of these alternatives remain available to the victim, and over the conditions of their application. **38**

The ECJ stated that the 1985 Directive put in place a liability regime based on the proof by the victim of damage, the defect in the product, and the causal link between that defect and the damage, and that the Directive precludes the application of national liability regimes based on the same grounds. However, the ECJ also confirmed that the Directive does not preclude "the application of other systems of contractual or non-contractual liability based on other grounds, such as fault or a warranty in respect of latent defects."[73] Therefore, a victim will presumably not be able to rely on actions based on the "obligation de sécurité" (safety obligation) created by the French courts on the basis of art. 1147 and 1384–1 Civil Code. These articles were adapted by the courts in order to transpose into French law the regime introduced by the Directive in anticipation of the legislative implementation through the 1998 Law. Since these actions, which imposed liability based on proof of damage, defect and a causal link, are clearly founded on the same basis as the European regime, they will logically not survive the 2002 ruling.[74] However, the decision of the ECJ would seem to permit the victim to continue to rely on art. 1641 (contractual liability for latent defects) or on art. 1382 (liability for fault in tort). **39**

In some respects, the action in contract for latent defects will be more restrictive than the action available under the Directive since, under the contractual regime, the defect must be "hidden". In addition, the action is available only to those with a contractual link. However, the contractual action for latent defects offers some significant advantages. Firstly, the action in contract would in some cases enable the victim to avoid the ten year cut-off period imposed by the Directive, the action in contract being subject simply to a limitation period of two years from the date of the discovery of the defect.[75] Secondly, a flexible approach to privity of contract **40**

---

72 ECJ C-183/00, *Gonzalez Sanchez v. Medicina Asturiana SA* [2002] ECR I-3901.

73 Ibid. par. 31.

74 D. 2003. summary 463, obs. *D. Mazeaud*; *J. Calais-Auloy*, Menace européenne sur la jurisprudence française concernant l'obligation de sécurité du vendeur professionnel, D. 2002. chron. 2458.

75 Art. 1648 Civil Code, replacing the previous "brief delay".

has been adopted by the French courts, who have allowed the buyer to sue anyone in the contractual chain. In this respect, the action in contract for latent defects is more favourable to the victim than is the Directive since the former has the effect of imposing strict liability on suppliers without entitling them to avoid liability by identifying other members of the supply chain. Given the importance placed by the ECJ on the channelling of liability to the producer, there would seem to be a distinct possibility that such an action against a supplier be considered by it as contrary to the Directive.

41 Where there is no contractual link between the producer or suppliers and the victim, the claimant will be able to rely on art. 1382 Civil Code, which imposes liability for fault. Again, using liability for fault will enable victims to avoid the ten year cut-off period imposed by the Directive. The limitation period for tort liability is also longer than that of the Directive, being ten years from the consolidation of the damage.[76] French courts in the past have employed a rather creative interpretation of "fault" to facilitate the claims of victims of defective products by stating that simply putting a defective product into circulation or supplying it constituted a fault for third party victims. Some French commentators have argued that the courts could adopt this approach once more as a way of overcoming the restrictions imposed on the victim's claim by the Directive.[77] Such an interpretation of "fault" by the courts would have the effect of imposing liability on the basis of a defect. In reality, it would therefore be an action based on the same grounds as the Directive and, in this form, will presumably therefore fall foul of the ECJ's vision of permissible parallel regimes. What is more, with respect to the liability of suppliers, there must again be doubts as to whether such an interpretation would meet with the approval of the ECJ if, as it has stated, the aim of the Directive is to channel liability to the producer in order to lighten the insurance burden of suppliers and limit recourse actions.[78]

(c) Impact of compliance with rules and regulations

42 Under the liability regime introduced by the 1985 Product Liability Directive, the fact that the defendant has obeyed all rules and regulations governing his production process will not exonerate him from liability,

---

76 Art. 22–26 Civil Code.

77 G. *Viney*, L'interprétation par la CJCE de la directive du 25 juillet 1985 sur la responsabilité du fait des produits défectueux, JCP G 2002, I, 117; *J.-S. Borghetti*, La responsabilité du fait des produits. Etude du droit comparé (2004) par. 602.

78 ECJ C-402/03, *Skov v Bilka* [2006] ECR I-199.

unless he can establish that the defect itself is due to his compliance with mandatory rules or regulations.[79] Even under traditional fault rules, the respect of rules and regulations will not necessarily provide an effective defence for the producer.

3. Environmental liability

(a) Implementation of the Environmental Liability Directive

The Environmental Liability Directive was recently implemented by the Law on Environmental Liability of 1 August 2008, incorporated in the Environmental Code.[80] The French legislator did not take up the opportunity presented by art. 16 of the Directive to enact "more stringent provisions in relation to the prevention and remedying of environmental damage". The Law adopts the option proposed by the Directive to allow the operator to avoid bearing the cost of remedial actions by demonstrating that he was not at fault or negligent and that an emission or activity was not considered likely to cause environmental damage according to the state of scientific and technical knowledge at the time when the emission was released or the activity took place.[81] The implementing legislation introduces criminal penalties for non-compliance with required remedial action.[82] 43

The legislation itself does not deal with the question of financial guarantees and no obligation is introduced to insure against environmental liability. The implementing legislation indeed does not include the provision in art. 14 of the Directive which provides for Member States to take measures to encourage the development of instruments and markets providing financial guarantees. It was considered that there was insufficient visibility and experience relating to the risks to be covered, and the public authorities took the view that at this stage it was difficult to estimate the type of guarantee which would be adapted to the diversity of possible situations.[83] A number of specific insurance policies now exist covering liability for damage caused by pollution, which can even include the cost 44

79 Art. 1386–11 Civil Code.
80 Law no. 2008-757 of 1 August 2008 *relative à la responsabilité environnementale.*
81 Art. 162-27 Law no. 2008-757 of 1 August 2008.
82 Chap. 3 of the Law.
83 Cour de cassation: responsabilité environnementale: le projet de loi relative à la responsabilité environnementale, www.courdecassation.fr/IMG/file/pdf_2007/24-05-2007/24-05-2007/-agostini.pdf.

of preventative work.[84] It is foreseen that insurers will develop policies to cover the broader liability introduced by the Directive.[85]

(b) Environmental liability regime beyond the scope of the Directive

45 There is no such liability regime.

(c) Claimants in cases of environmental harm

46 The Cour de cassation has to date shown some reluctance to award damages for harm to the environment as such,[86] and the Conseil d'Etat has rejected claims for damages for environmental damage brought before the administrative courts.[87]

47 The principal obstacle to such claims is that, under traditional liability rules, a claimant will only have locus standi where he is able to show that he has suffered personal loss as a result of the harm to the environment.

48 However, a way of overcoming this obstacle is by allowing claims for environmental harm by associations. The possibility exists in French law for associations to bring actions in defence of the interests of their members. Associations for the protection of group interests are required by French law to have a specific licence to take court action. In the environmental context, a law introduced in 1995 provides that associations for the defence of the environment have locus standi to take action to defend

---

84 See the examples provided in *M.-L. Demeester*, L'assurance et les techniques alternatives de financement des dommages environnementaux, in: B. Dubuisson/G. Viney (eds.), Les responsabilités environnementales dans l'espace européen. Point de vue franco-belge (2006) 489.

85 See the website of the French Ecology Ministry (updated only to June 2007), which gives the example of the insurance company Ace Europe which provides a policy covering a broad range of liability covered by the Directive, but is limited to pollution over a maximum ten year period and a maximum annual liability of € 20 million www.ecologie.-gouv.fr/Directive-responsabilité.html.

86 Cass. Crim. 6.2.1969, JCP 1970, II, 16528, for a fisherman complaining of the pollution to a river; Cass. 3rd Civ. 12.2.1974, Bull.Civ. III, no. 72; JCP 1975, II, 18106, note *M. Despax*, claim for moral damage by residents of a riverbank for pollution to the river rejected. *Viney/Jourdain* (fn. 15) 68 ff.; *Jourdain* (fn. 25) 161; *M.-J. Littman-Martin/C. Lambrecht*, La spécificité du dommage écologique, in: Société française pour le droit de l'environnement, Le dommage écologique en droit interne, communautaire et comparé (1992) 45 ff.

87 CE, 12.7.1969, Ville de Saint-Quentin; CE, 26.10.1984, Fédération des associations de pêche et de pisciculture de la Somme.

collective interests which fall within their stated objects provided that the action concerns facts which constitute an infraction against environmental protection legislation.[88] The 1995 Law appears to require that the act or omission constitutes an infraction of criminal law,[89] and most claims are indeed brought before the criminal courts, where, under French procedural rules, civil damages can be obtained. However, it appears that the courts may now be willing to take a more flexible approach as to the conditions imposed for claims brought by associations for the protection of the environment, notably by not insisting that the pollution constitutes a criminal offence.[90]

Likewise, the 1995 Law confers on certain public authorities such as the Environment Agency and the Organisation for the Conservation of the French Coastline the right to take action to defend the interests they have been established to defend.[91] Such a right is equally conferred on local authorities provided that they have specific responsibility for ensuring the environmental protection of an area. **49**

It should also be noted that at a more symbolic level, the Environment Charter, adopted by Parliament on 28 February 2005, and elevated to a constitutional principle,[92] declares that "everyone has the right to live in a balanced and healthy environment", and that "where a person has caused damage to the environment, he must contribute to repairing that damage in the way defined by legislation". These provisions are extremely vague and difficult to apply as such,[93] but they at least demonstrate that the public authorities have expressed a willingness to develop environmental liability. **50**

This increased consciousness of the importance of environmental protection seems to be beginning to be reflected in court decisions on environmental liability. Although decisions specifically awarding damages for environmental harm as such are still relatively rare, a recent study[94] reveals that there are an increasing number of successful claims in the **51**

88 Law no. 95-101 of 1 February 1995 *relative au renforcement de la protection de l'environnement*, now art. L 142–2 Environment Code.

89 *Viney/Jourdain* (fn. 15) 71.

90 Cass. 3rd Civ. 26..9.2007, RTDCiv 2008, chroniques, *P. Jourdain*; D. 2008, 2894, note *P. Jourdain*.

91 Art. L 132-1 Environment Code.

92 Since it is now referred to in the preamble to the French Constitution.

93 They will be difficult to apply on behalf of individuals as long as the right to a healthy environment is not recognised as a right giving rise to a legal action (*droit subjectif*).

94 *L. Neyret*, La réparation des atteintes à l'environnement par le juge judiciaire, D. 2008, 170.

lower courts, even if for the moment there are few decisions from the Cour de cassation to confirm the strength of this tendency.

52 Many of these cases award damages to associations for environmental harm indirectly by compensating moral harm. Hence, as early as 1982, a judgment of the Cour de cassation confirmed the decision of a first instance court which had awarded damages for "moral harm" to an association for the protection of birds following the shooting of a rare bird by hunters.[95] There have been an increasing number of awards of damages for moral harm in the case of harm to the environment from lower courts.[96] Courts have awarded associations damages for moral harm as a result of water pollution[97] and pollution of the sea by hydrocarbons.[98] Judges have also defined the destruction of animals belonging to protected species as moral harm.[99] Moral harm has also taken the form, for example, of damage to the image of a region,[100] and damage to the reputation of coastal tourist resorts.[101]

53 Other decisions have more directly recognised environmental harm as a specific head of damage. In 1988, a first instance court defined the harm suffered by an association for the protection of water quality in Brittany through the destruction of fish due to river pollution as "direct and certain from a biological perspective".[102] This "biological harm" was specifically listed as a head of damage separately from moral harm. More recently, an association was awarded damages for the harm inflicted on the environment by the spillage of hydrocarbons,[103] and the Bordeaux Court of Appeal awarded compensation to several associations "for the

95 Cass. 1st Civ. 16.11.1982.

96 TI Tournon, 28.4.1981, Gaz Pal 1981, 2, 560, note *E. Alauze*; RTDCiv 1981, 853, obs. *G. Durry*; Cass. 1st Civ. 16.11.1982, Bull.Civ. I, no. 331; *C. Huglo*, La réparation du dommage écologique au milieu marin: les affaires Montedison et Amoco-Cadiz, Gazette du Palais (Gaz. Pal.) 1992, 2, doc. 582 ff.; Cass. Crim. 25.10.1995, Bull.Crim., no. 322; Revue Juridique de l'environnement, 1996-2, 196: compensation awarded to an association of fishermen and a local fishing federation for damage caused to fish stocks; *Viney/Jourdain* (fn. 15) 68 ff.; also see Cass. Crim. 23.3.1999, P, no. 98-564, compensating fishing associations for the moral harm caused by pollution of a river.

97 T. corr. Rennes, 26.6.1992, no. 210992; see also T. pol. Guingamp, 5.1.2006, no. 06/00005; *Neyret*, D. 2008, 170 ff.

98 CA Rennes, 26.10.2006, 18.4.2006, 23.3.2006.

99 CA Pau, 17.3.2005, no. 00/400632; CA Aix-en-Provence, 13.3.2006; *Neyret*, D. 2008, 170 ff.

100 CA Aix-en-Provence, 25.7.2006; TGI Bastia, 4.7.1985.

101 CA Rennes, 18.4.2006.

102 T. corr. Brest, 4.11.1988.

103 T. corr. Libourne, 29.5.2001.

harm inflicted on aquatic plants and invertebrates.[104] In 2007 the first instance court in Narbonne awarded damages for the harm caused to a regional nature reserve caused by emptying chemicals into the water, again specifically listing environmental harm as a head of damage.[105]

The most notable recent example is a decision of the Paris first instance court concerning the pollution caused by the sinking of the oil tanker, the Erika, off the Brittany coast in December 1999.[106] The defendants, amongst others the ship owners and the petroleum company Total, were ordered to pay damages totalling € 192,000,000 to 70 victims. This was mostly stated to be for "moral" damage suffered by the state, regions, counties, towns, villages and associations for the defence both of the environment and of consumers. Whilst it is true that only a small proportion of this sum was allocated to the compensation of environmental damage as such, the decision again demonstrates the tendency, at least at the level of first instance courts, towards the compensation of this form of harm. The Organisation for the Protection of Birds was awarded the sum of € 300,000, to compensate for the cost of encouraging new nests and breeding birds to replace those lost, and the Brittany county Le Morbihan was awarded € 1,000,000 damages for harm inflicted on a clearly defined environmentally fragile area.[107] 54

(d) Special liability regime for losses sustained by individuals

Individuals may be able to sue for "troubles anormaux de voisinage" (nuisance), which constitutes the principal basis in French law for the recovery of individual loss from harm to the environment. In order to succeed on this basis, the claimant must establish the existence of an unreasonable level of nuisance caused by a neighbour. Courts initially based liability on art. 1382 of the Civil Code but now recognise "troubles du voisinage" as an independent legal principle. The claimant is not required to estab- 55

104 CA Bordeaux, 15.2.2005.

105 D. 2007, AJ 2731.

106 T. corr. Paris, 16.1.2008, JCP 2008, act. 88, note *K. Le Couviour*; *L. Neyret*, Naufrage de l'Erika: vers un droit commun de la réparation des atteintes à l'environnement, D. 2008, chron. 2681.

107 According to the judgment in the *Erika* case, in order to be entitled to compensation, local authorities must have "specific responsibilities with regard to environmental protection", which will be the case for county councils (*departements*), but not regional authorities. In addition, the authority must provide a detailed picture of the environmental harm caused to a specific area.

lish any fault.[108] The disturbance or nuisance must be continual or at least repetitive, and it must be considered by the court to be unreasonable or excessive. Liability can be imposed even though the defendant has obtained authorisation from the relevant administrative authorities for his activity.[109]

56 In some cases the judge assesses the unreasonable nature of the trouble by reference to the damage caused, and on other occasions by reference to the behaviour itself.[110]

57 Examples illustrate the relevance of these principles to the case in hand. Thus the Paris court of appeal found a farmer liable for "troubles du voisinage" where a treatment of crops using hormones had led to the deterioration of neighbouring lettuce crops.[111] In the same way, a cement manufacturer was found liable for the damage to neighbouring crops where the leaves of the claimant's crops were found to be covered with a fine film of grey dust, which prevented efficient photosynthesis and thus restricted growth.[112]

(e) Cartagena Protocol

58 The Cartagena Protocol was ratified by France in 2002 and came into force in September 2003. France has also ratified the International Plant Protection Convention of 1951 and the Aarhus Convention which was transposed into French law in 2004.

4. Other strict liability regimes

59 We have already mentioned liability for "troubles anormaux de voisinage" which equates to a strict liability regime. In addition, art. 1384-1 of the Civil Code, which imposes strict liability for harm caused by inanimate objects, provides an alternative action for victims of damage caused by GMOs. Under this provision, the defendant will be liable where he has

---

108 Cass. Civ. 23.3.1982, D. 1983, IR, 18, obs. *A. Robert*; Cass. 2nd Civ. 9.11.1986, Bull.Civ. II, no. 172.

109 Cass 2nd Civ. 22.10.1964, a manufacturer of castor oil was found liable despite the fact that he had obtained the relevant administrative authorisation for his activity. (*Viney/Jourdain* (fn. 15) 1217 f.).

110 *Prieur* (fn. 4) par.1159.

111 CA Paris 8th chamber, 26.6.1980, jurisdata no. 098444.

112 CA Montpellier, 11.5.1983, jurisdata no. 600730.

control ("garde") of the object, and the claimant shows that there is a causal link between that object and the damage. This principle has been used by the courts to impose liability without fault, provided that the object (in this case the genetically modified organism) has had an active role in the damage caused. A GM farmer could therefore be considered as the person in control ("le gardien") of the genetically modified organisms which have caused damage to the neighbouring crops by contaminating them and thus reducing their economic worth. The courts have in this way imposed liability on defendants for damage caused by pollution. Hence, a company producing chemicals was found liable on the basis that it had control over the gas that was emitted from its factory,[113] and another was found liable on the same basis for pollution caused by emissions of cadmium and lead particles.[114] In general however, the French courts have been reluctant to impose liability on this basis in cases of environmental pollution. This reluctance is perhaps due to the fact that the rules are too favourable to the victim since he need establish neither the fault of the defendant, nor the presence of an abnormal level of disturbance or interference.[115]

The distinction between the "gardien de la structure" (control of the structure of the thing) and the "gardien du comportement" (control of its behaviour) invented by the courts in order to impose liability on the producer of the product even though he no longer had physical control of the product itself, would no longer appear to be valid following the introduction of a specific product liability regime and the interpretation of that regime by the European Court of Justice.[116] **60**

V. Vicarious liability

1. Scope of vicarious liability

Art. 1384-5 of the Civil Code provides for vicarious liability of the "master" for the harm caused to a third party by the tortious act of his "servant" where the act is carried out in the performance of the functions for which the "servant" is employed. In order to engage the liability of the master, traditionally the victim had to show that a link of subordination existed between the master and the servant. However, today the courts take a **61**

113 Cass Civ. 17.12.1969.
114 CA Douai, 25.4.1991; *Prieur* (fn. 4) 921.
115 *Viney*, JCP 1996, 41 ff.
116 *Flour/Aubert/Savaux* (fn. 56) 289; see supra no. 38 ff.

broad view of what constitutes a link of subordination. The courts will recognise vicarious liability where the "master" retains a supervisory role: it is he who fixes the objective to be realised and the means to be employed to reach that goal.[117] In the great majority of cases, vicarious liability will concern the liability of employers for the acts of employees. In most cases with independent contractors, the terms of the contract will leave a degree of independence to the contractor which will be incompatible with vicarious liability.[118] Thus, the French courts have generally refused to consider an independent contractor as having the required "master-servant" link.[119]

2. Liability for people further up the food or feed production chain

62 Those in the distribution chain will be liable to others in that chain in the law of contract. No tort action will be possible between these parties where an action in contract is available due to the rule of "non cumul". As we have seen earlier in our discussion on the nature of fault, French tort law recognises a breach of contract as constituting a fault towards a third party victim. Therefore, the mere fact of supplying a product which does not conform to description or which is not of the quality required will mean that the supplier will be potentially liable to a third party victim.

3. Can someone further down the feed or food chain include someone further up it in a trial against him/herself?

63 A defendant may join a supplier further up the chain as a party in the action brought by the claimant where, as in the example given, the supplier (here a farmer) has a claim himself against the other supplier (here a seed producer).[120] In such cases, the farmer will remain the defendant to

117 *Flour/Aubert/Savaux* (fn. 56) 218 f.

118 *Viney/Jourdain* (fn. 15) 987 ff.

119 Cass. Crim. 17.4.1930, Gaz. Pal. 1931, 2, 37; Cass. Civ. 15.1.1954, D. 1954, 169. However, there are exceptions where the court has been willing to find a sufficient degree of supervision. Hence, in 1996, the Cour de cassation imposed vicarious liability for acts of an independent contractor: Cass. 2nd Civ. 11.12.1996., Resp. civ. et assur. 1997, com., no. 83, as it did in 1988 in the case of an independent craftsman: Cass. Crim. 22.3.1988, Bull.Crim. no. 142, RTDCiv 1988, 774, obs. *P. Jourdain*; *Viney/Jourdain* (fn. 15) 989.

120 Art. 331 new civil procedure code.

the initial action, but he will be able to shift his loss to the seed producer where appropriate.

Where the defendant is being sued under art. 1386-1 to 1386-18 Code Civil (product liability), the seller can avoid liability where he identifies his own supplier or the producer within three months of being notified of the claimant's action.[121] **64**

VI. Multiple tortfeasors

French law applies joint and several liability in the case of multiple tortfeasors. This will be the case irrespective of the basis of the liability of the different tortfeasors. The tortfeasor who is found liable may then take a recourse action against the other tortfeasors. The courts will calculate the share of liability here on the basis of the degree of fault of the various parties.[122] A defendant liable for fault will be unable to seek a contribution from another person who is liable without fault.[123] **65**

VII. Defences

1. Licence/permission to grow GM material

The existence of a licence to grow GM material (including for testing) will not in itself serve as a defence against liability triggered by its use. The liability rules under the 2008 Law on GMOs apply to licensed crops. Likewise, the obtaining of a licence will not constitute a defence for potential liability under art. 1384.1 Civil Code for damage caused by things. For liability for fault under art. 1382 Civil Code, the obtaining of a licence will not enable the defendant to avoid liability where other faults have been committed. Finally, with respect to liability for "troubles du voisinage", a long line of court decisions confirm that a licence to perform a certain activity does not constitute a defence.[124] **66**

---

121 Art. 1386-7.

122 Cass. 1st Civ. 22.10.1996, Bull.Civ. I, no. 365; Cass. 2nd Civ. 26.10.1967, Bull.Civ. II, no. 302.

123 Cass. 2nd Civ. 19.11.1970, JCP 1971, II, 16748; *Viney/Jourdain* (fn. 15) 308 ff.

124 Cass. 2nd Civ. 25.10.1978, Bull.Civ. II, no. 220; Versailles, 30.11.1989, D. 1990, IR, 18; Cass. 3rd Civ. 20.7.1994, Bull.Civ. III; RCA 1994, com. no. 375; *Viney/Jourdain* (fn. 15) 1217 f.

67 The only difference that a licence to grow crops will make is that the civil courts will, due to the principle of separation of administrative and judicial powers, be unable to order a remedy that would go directly against an administrative decision, such as ordering the cessation of the activity.[125]

2. Consent/assumption of risk

68 The consent or assumption of risk by the victim will not result in the reduction of liability of the defendant unless the victim's conduct can be considered a fault. Thus, if the fact that the victim knowingly consumed GM products can be considered manifestly imprudent behaviour, this would be sufficient to reduce the liability of the defendant. This would presumably require the victim not only to know that he was consuming GM products, but also to be aware that there was a significant risk to his health in doing so. The liability of the defendant will be reduced in proportion to the seriousness of the fault of the victim. The consent or assumption of risk by the victim will not exonerate the defendant totally from liability unless the victim's role can be considered as constituting a force majeure, which, as we have seen, requires that the victim's act be "unforeseeable and unavoidable"[126] and as having the effect of rendering it impossible for the defendant to prevent the resulting damage so that we can no longer consider the defendant as having a causal role. This could be the case where the victim ingested GM products despite clear warnings that they were unsafe for human consumption.

69 Another question to raise here is whether a GM farmer could use as a defence to a claim for nuisance the fact that an organic farmer has chosen to situate his production in fields neighbouring land which is already being used for GM crops. In French law, the fact of coming to a risk will not constitute a defence for an action for "troubles du voisinage".[127] However, if the polluting neighbour is a state-owned organisation, then administrative law principles establish that no compensation will be ordered, or at least its level will be reduced, when the nuisance, at its current level, already existed at the time that the claimant came to the land.[128]

---

125 G. *Viney* Traité de droit civil. Les effets de la responsabilité (2nd ed. 2001) par. 45.

126 Ass. Plén. 14.4.2006, Bull.Civ. AP no. 6, D. 2006, 1577, note *P. Jourdain*.

127 Cass. Civ. 4.12.1935, S. 1936, 1, 208; Cass. 1st Civ. 20.2.1968, D. 1968, 350; *Viney/Jourdain* (fn. 15) 1221.

128 CE, 19.3.1915, Rec., 84; *Prieur* (fn. 4) 935.

3. Third-party influence

Where both the act or omission of the defendant and the behaviour of a third party have had a causal role in the victim's loss, and the behaviour of the third party is entirely independent of the defendant, then both will be jointly and severally liable towards the victim. If the defendant is found liable by the court for the entire loss, he may then be able to recover a proportion of the sum paid from the third party in a recourse action if the sharing of liability has not already been decided through joining the third party to the original action. **70**

The third-party behaviour will only enable a defendant to avoid liability entirely where that behaviour constitutes a force majeure. **71**

The position is rather different where the defendant is the State, perhaps here in a case where the administrative authorities are being sued for failing to follow the correct authorisation or control procedures. Here, the Conseil d'Etat has a less firm position, and in some cases will allow the liability of the State to be reduced where its fault and the fault of a third party are in question. In such cases, the Conseil d'Etat has allowed liability to be shared ab initio.[129] **72**

4. Prescription

Recent legislation has introduced important changes to limitation periods.[130] Where the liability action has arisen due to an event which has caused personal injury, then a limitation period of ten years from the date of the consolidation of the damage will apply.[131] Where the claim relates to damage to property or financial loss without personal injury, then a civil claim will be subject to a five year limitation period, from the date that the claimant knew or ought to have known the facts giving rise to the claim.[132] For financial liability with respect to environmental damage, a longer limitation period of thirty years from the date of the act causing the damage is imposed.[133] **73**

129 CE, 15.10.1976; 5.10.1977; 28.10.1977; 14.6.1978, JCP 1980, II, 19319, note *Y. Briard; Viney/Jourdain* (fn. 15) 299.

130 Law no. 2008–561 of 17 June 2008 portant réforme de la prescription en matière civile.

131 Art. 2226 Civil Code.

132 Art. 2224 Civil Code.

133 Art. L 152–1 Environment Code.

74 Where the claim is being brought for product liability under the regime introduced by the 1985 Directive, then the three year limitation period, together with the ten year pre-emption period will apply.[134]

5. Other defences

75 None.

VIII. Remedies

1. Pecuniary compensation

(a) Bodily harm

76 There are no special rules. The regular remedies will apply.

(b) Property losses

77 Again, there are no special rules. The regular remedies will apply.

(c) Economic losses

78 Art. 8 of the Law of 25 June 2008 relating to genetically modified organisms provides special rules on the compensation of economic loss caused to organic farmers through contamination by neighbouring GM crops.[135] The economic loss will be calculated on the basis of the reduction in value of the claimant's crop resulting from the difference in sale price of a product which has to carry a GMO label under European community legislation compared to the sale price of the same product without a GM label. It is also provided under the new legislation that an alternative to financial compensation can be an exchange of products between the parties.

79 French law applies the principle that the victim is entitled to compensation for the entire loss that he has suffered. Unlike in English law, there is no principle that the victim is under an obligation to mitigate his loss,[136] and the Cour de cassation frequently rejects decisions of lower

134 Art. 1386-16 and 1386-17 Civil Code.
135 Art. L 663-4 Rural Code. See *Taylor* (fn. 2).

courts that attempt to modify the level of damages to which the victim is entitled based on factors other than a simple calculation of his loss.[137] Indeed, in a decision in 2003 the Cour de cassation specifically stated that "the victim is not required to limit his loss in the interests of the defendant".[138] Consequently, no reduction in the level of compensation payable should be made for the seasons where an organic farmer could have grown conventional crops on his land. It should however be noted that the "avant-projet" on the reform of the French law of obligations proposes to introduce a duty to mitigate. Art. 1373 of the proposals provides that "where the victim had the possibility, by reasonable means, of reducing the extent of his loss... his failure to do so will be taken into account through a reduction in his compensation". Thus, if this provision were to be enacted, it may result in a reduction in damages being imposed in the circumstances envisaged in the questionnaire.

(d) Harm to animals

The French courts will seek to compensate the victim for the entire loss **80**
suffered and to return him to the position he would have been in if the damage had not occurred. It is rather difficult to provide a precise response as to exactly how the French courts will calculate the loss since there is no obligation on the judges to explain the elements on which they base their calculation.[139] In principle, in the example provided, the farmer will be compensated for the cost of replacement of any cattle lost at the market value at the date of the damage, and the cow's potential for producing milk or meat should be taken into account in the calculation. If the cow's value is merely reduced by the harm caused by the contaminated feed, then the farmer will be entitled to compensation for the difference between the present market value and the market value if the cow had not been harmed. A reduction in the capacity to produce milk should also be compensated, as will the cost of veterinary treatment and any loss of subsidies.[140] If the mere fact that the cow has eaten contaminated feed

---

136 *Viney* (fn. 125) 121.

137 Cass. Civ. 21.10.1946, JCP 1946, II, 3348, note *P.L.P.*

138 Cass. 2nd Civ. 19.6.2003, Bull.Civ. II, no. 203, D. 2003, juris. 2326, note *Chaza*; *Flour/Aubert/Savaux* (fn. 56) 425 ff.

139 Cass. Soc. 17.3.1961, Bull.Civ. 1961, IV, 292; Cass. 2nd Civ. 15.2.1962, Bull.Civ. II, 130; *Viney* (fn. 125) 126.

140 TGI Angers, 13.1.1997, no. 9500885; Collection Lamy, Droit de la responsabilité (2007) par. 370–49.

has the effect of reducing its market value, then the owner will be entitled to compensation for this reduction in value.

(e) Are the costs of disposing of contaminated production/ animals fed with GM feed recoverable?

81 Yes, again, French law seeks to compensate for the entire loss suffered, and this will include the costs of disposal.

2. Non-compensatory damages

82 French law does not recognise punitive, exemplary or any other form of non-compensatory damages. However, the "avant-projet" for the reform of the law of obligations envisages the introduction of punitive damages in certain circumstances. Hence, art. 1371 of the proposals states that "the author of a manifest and deliberate fault, and more particularly of a fault which has had beneficial financial consequences for him, may be made liable ... for punitive damages, which the court may make payable in part to the public treasury."

3. Other remedies

83 On occasion measures are ordered requiring the return of land to the condition it was in before pollution took place. This has included, for example, an order to remove trees which had fallen on neighbouring land,[141] and to create a watertight zone preventing further pollution which also had the effect of cleaning polluted land.[142] However, in practice, the courts will rarely order the defendant to carry out work to return land to its previous condition. Most often, an order will be made for the defendant to reimburse the victim for the cost of repairs.[143] Two recent decisions by the Cour de cassation have also allowed the claimant to recover compensation for the cost of work aimed to prevent the risk of future damage.[144]

84 Certain statutes impose criminal offences which permit the court to order the polluter to take remedial or preventive measures for specific forms of

141 Cass. 2nd Civ. 5.2.2004, Bull.Civ. II, no. 50.
142 Cass. 2nd Civ. 21.2.2002, no. 98-19.338.
143 *Jourdain* (fn. 25) 171.
144 D. 2008, 2900, chron. *P. Brun*.

pollution, such as waste.[145] The courts also have the power under the Civil Procedure Code to order, in summary proceedings, any measures necessary, including an order to repair damage and to ensure the cessation of an illegal activity.[146] The newly enacted Law on Environmental Liability, implementing the 2004 Directive, introduces an obligation on polluters to remedy damage in defined circumstances.

There is, however, a major restriction to these potential remedies, since no order for remedial work or cessation of activity can be made by a court where this would be incompatible with an administrative authorisation,[147] although there is some degree of flexibility here since the civil courts have on occasion ordered the suspension (rather than the cessation) of authorised activities.[148] Thus, the civil courts could not order the destruction of a polluting GM crop where the cultivation has been licensed, although they may potentially order the suspension of that activity in order for preventative work to be done. It should, however, be noted that the French Rural Code now provides that the administrative authorities can order the total or partial destruction of GM crops where the GM farmer has failed to respect the technical conditions imposed on cultivation by the licensing authority. In this case, the costs incurred in the destruction will be borne by the offending farmer.[149] **85**

The new liability regime for economic loss under the 2008 Law on GMOs now provides that an alternative to financial compensation for the reduction in market value of contaminated crops can be an exchange of products between the parties.[150] **86**

4. Costs of pursuing a claim

(a) General cost rule

The general rule is that the loser pays his own costs and the costs of his adversary.[151] This rule relates to defined types of costs which are considered as essential to the trial and are listed in art. 695 Civil Procedure **87**

---

145 For example, Law no. 75–633 of 15 July 1975 (art. L 541-46, II, Environment Code).
146 Art. 809, 849 and 873 New Civil Procedure Code.
147 Tribunal des conflits, 23.5.1927, S. 1927, 3, 94. *Jourdain* (fn. 25)167; *Viney* (fn. 125) 93.
148 Cass. Civ. 13 July 2004, Environnement 2004, comm. 111.
149 Art. L 663-3 Rural Code, inserted by Law no. 2008–595 of 25 June 2008 relative aux organismes génétiquement modifiés.
150 Art. L 663-4 Rural Code.
151 Art. 696 New Civil Procedure Code.

Code. These "essential costs" include for example, court fees and witnesses' expenses. They do not, however, include major costs such as lawyers' fees. In order to recover these costs, the winning party may make an application to court for an order for costs against the losing party[152] under art. 700 Civil Procedure Code. Under this article, the judge has discretion as to whether to award costs and as to the level of the award. In practice, the costs ordered payable will only cover part of the winner's expenses.[153]

(b) Costs of establishing causation

88 Such costs are not included in the list of "essential costs" under art. 695 Civil Procedure Code. The winner of the trial can apply to the court for an order for these costs to be made, but it is likely that only a proportion of these costs would be recovered.

5. Advance cover

89 Art. 8 of the Law on Genetically Modified Crops introduces into the French Rural Code[154] an obligation on all GM farmers where the product is authorised to be put on the market to take out liability insurance against the potential liability imposed by the Law.

IX. Cross-border issues – Conflict of laws

1. Conflict rules applicable before (or instead of) Rome II

90 According to French conflict of law rules, the law applicable will be the law of the place where the tort has occurred (lex loci delicti).[155] However, in cases of cross-pollination, such a rule appears ambiguous since it is not clear whether it refers to the place where the tortious act ("fait générateur") took place or to the place where the damage occurred. Recent decisions of the Cour de cassation indicate that either the place of the tortious act or the place of the damage may apply, depending on the circumstances of the case. The court will choose the place which has the greatest link or

152 Or the party against whom the order for essential costs was made.
153 *L. Cadiet/E. Jeuland*, Droit judiciaire privé (5th ed. 2006) 37.
154 Art. L 663-4 Rural Code.
155 Cass. Civ. 25.5.1948, rev. cr. 1949, 89.

connection with the events. This will most likely be the place where the damage (i.e. here, the contamination) occurred,[156] although certain authors observe a tendency amongst first instance judges to prefer French law as soon as one of the elements of the tort has its origin in this country.[157]

The Rome II regulations therefore will provide clarification here, by providing that in principle the law applicable will be that of the country where the damage occurred irrespective of where the tortious act was committed.[158] **91**

2. Special regime for cross-border claims

The 1973 Hague Convention on the Law Applicable to Products Liability applies in France to cases of product liability outside the case of direct sales.[159] The law applicable will normally be that of the place where the damage occurred[160] or the place of residence of the victim.[161] Where the place of residence of the victim is also the place either where the product was acquired or where the defendant has his principal place of business, then the place of residence of the victim takes priority over the place where the damage occurred. **92**

Even after the entry into force of the Rome II Regulation, according to its art. 28 the Hague Convention will prevail and continue to apply in France. **93**

156 Cass. 1st Civ. 11.5.1999; Cass. 1st Civ. 28.10.2003. *T. Vignal*, Droit international privé (2005) 216.

157 *Y. Loussouarn/P. Bourrel/P. de Vareillers-Sommières*, Droit international privé (9th ed. 2007) 550.

158 Art. 4 general rule, art. 7 for damage to the environment.

159 Actions against the direct seller of the product by the purchaser are excluded from the scope of the Convention.

160 Provided that this is also the place of residence of the victim, the principal place of business of the defendant or the place where the product was acquired.

161 Provided that this is also either the principal place of business of the defendant or the place where the product was acquired by the victim.

X. Cases

1. Due to the adventitious presence of GMOs in a field, maize which is normally sold as conventional contains GMOs beyond the legal labelling threshold. This is not discovered before the final stage of the food production chain by the producer of taco chips. The whole production is lost since the supermarket chains refuse to accept delivery from the producer.

(a) Please solve the case by cross-referring to your general statements above. Who can sue along the chain of distribution – the taco producer, the wholesaler of the maize, etc.?

94 The taco producer would have a claim for breach of a contractual obligation of conformity with description where the product he has acquired was specifically sold as conventional.[162] There may alternatively be a claim for contractual breach of an implied condition as to quality under art. 1641 Civil Code where the defect renders the product unsuitable for its normal use. The supplier would then have a recourse action available in contract against his own supplier. Liability could therefore pass up the distribution chain.

95 In addition, the taco producer and any other intermediate supplier suffering loss as a result of the contamination will be able to sue any supplier further up the chain or the producer, since the courts consider that those in the supply chain have a direct action in contract against those higher in the chain ("action directe"). No claim in tort will be available for those in the supply chain due to rule of *non-cumul*, the principle that once the claimant has a claim in contract, he may not claim in tort.[163] This applies equally to recourse actions.

96 If the liability climbs all the way back up the distribution chain to the farmer, he may have a claim against a neighbouring GM farmer to obtain recovery of his economic loss, notwithstanding that the presence of GMOs in his field was adventitious. The new liability regime introduced by the Law on GMOs 2008 would entitle the farmer to recover compensation from neighbouring GM farmers provided that he can establish the criteria

162 Art. 1604 Civil Code.
163 *Viney/Jourdain* (fn. 15) 788 ff.

required for liability.[164] He would also have possible claims against a neighbouring farmer for nuisance ("troubles du voisinage") and under art. 1384-1 Civil Code for damage caused by things.[165] Unlike under the new statutory regime, under the two latter actions the farmer would have to establish the causal link between the neighbouring production of GM crops and his loss.

(b) Would the case be solved differently if the GMO content was below the labelling threshold?

It may be that in such cases the product would comply with the sales **97**
description, or not be considered defective, so as to not constitute a breach of contract, thus depriving the taco producer of an action against his suppliers. If the GMO content was below the labelling requirements imposed by European community rules, then no action would be available to the farmer against neighbouring GM farmers based on the liability regime under the 2008 Law on GMOs.[166]

(c) Would the case be solved differently if the admixture was not adventitious, but occurred due to the disregard of segregation rules, for example?

The answer here will depend on whether the farmer who has failed to **98**
respect segregation rules is the seller of the contaminated maize, or a neighbouring farmer whose GM crop has caused the contamination. In the former case, then the claim by those lower in the supply chain will be in contract, as described above. Since liability is strict, the fact that the farmer did not respect segregation rules will have no effect on the claim. Where the contamination is due to the act or omission of an intermediate supplier, that supplier's liability will also be based on contract. In the latter case, where the offending farmer is not part of the distribution chain, then the non-respect of segregation rules would constitute a fault under art. 1382 Civil Code,[167] and would therefore give rise to an action in tort under this article. Under French principles, there is nothing to prevent an action in tort for the compensation of pure economic loss. Any supplier

164 Supra, no. 29.
165 Supra, no. 55 ff.and no. 59 f.
166 Law no. 2008-595 of 25 June 2008, art. L 663-4 Rural Code.
167 Supra no. 32.

along the distribution chain who has suffered loss could therefore sue the farmer who had failed to respect the segregation rules in tort. The claimant would still of course face the obstacle of having to establish the causal link between the defendant's fault and his loss. The court would likely be satisfied with the proof of the probability of a causal link.[168] Where there is proof of fault, the court may be willing to employ techniques to facilitate the proof of causation, such as imposing liability for the creation of a risk. Since the non-respect of segregation rules by a GM farmer will in principle amount to a criminal offence,[169] then an action could be brought by anyone who has suffered loss in the criminal courts to recover their loss, or by joining a case brought by the public prosecutor. This would have the advantage for the claimant of enabling him to benefit from greatly reduced costs and also more generous rules on the obtaining of evidence.

99 The non-respect of segregation rules may also make it easier to establish a common law nuisance ("troubles du voisinage") in a claim for compensation by the neighbouring non-GM farmer whose crop is contaminated.

100 It may also be possible to invoke the liability of the administrative authorities where they have negligently failed to enforce segregation rules.[170]

101 Where segregation rules have not been respected, then the administrative authorities would be entitled to order the partial or total destruction of the offending crop, at the expense of the GM farmer.[171]

(d) Would the case be solved differently if the GMO found was not admitted for production in your jurisdiction?

102 The liability regime for pure economic loss introduced by the 2008 Law on GMOs relates only to GMOs which are authorised to be put into circulation.[172] Where the GM farmer was illegally producing the unauthorised crop, fault will automatically be established for art. 1382 of the Civil Code. The farmer will also incur criminal penalties, and any victim would have the possibility of initiating or joining an action brought in the criminal courts. If a foreign producer was found to have caused the contamination, then according to French conflict of law rules, the applicable law would be

168 Supra no. 18 ff.
169 Art. 671-15 Rural Code.
170 Supra no. 3.
171 Art. L 663-3 Rural Code.
172 Art. L 663-4 Rural Code.

that of the locus delicti, which in this case would most likely be interpreted as the place the damage occurred, thus enabling French law to be applied.[173]

(e) If the admixture had occurred on a non-GM field and it transpires that the GM seeds were blown from: (i) neighbouring fields; or (ii) a truck passing by, would the farmer of the affected field be liable for all or part of the loss caused further down the distribution chain?

The farmer of the affected field would be liable in contract to those further down the distribution chain, as discussed above. In such circumstances, French law applies the principle of joint and several liability.[174] The farmer of the affected field will therefore be liable for the entire loss of those further down the chain, but he will be able to recover contributions through a recourse action from the neighbouring farmers or the truck owners if he can establish their liability. If the non-GM farmer is liable without fault, he will be able to recover the entirety of the sums paid from the third party whose fault is established. He will also have the possibility of joining these parties in the action brought by distributors further down the chain.[175] **103**

2. Twenty years after the sale of GM maize used for food products, it turns out that it has certain disadvantageous health effects for humans.

(a) Can the producers be held liable at this point for risks unknown at the time of growing the maize? Who would be liable – the seed producer/farmer/food producer/distributor/ etc.?

The liability regime introduced by the Product Liability Directive has a cut-off period of ten years from the date the product was put into circulation, and will therefore not be applicable in this situation. However, the victim may be able to rely on other liability rules. The European Court of Justice has ruled against the possibility of relying on general liability rules **104**

173 Supra no. 90 f.
174 Supra no. 65.
175 Supra no. 63 f.

which have the same "grounds" as the Directive. As has been explained earlier,[176] this would seem to exclude the possibility of basing a claim on an "obligation de sécurité" under art. 1147 for contractual liability or art. 1384–1 for liability in tort.

105 The victim may therefore be able to fall back on a claim in contract for latent defects against his immediate supplier under art. 1641 Civil Code. Art. 1648 provides that the action must be brought within two years of the discovery of the defect. According to art. 1645, the victim will be able to claim damages only where the seller knew of the defect, but the French courts hold that a professional seller is considered to know of the defect. Although judges have discarded the use of art. 1641 in cases of product liability in favour of art. 1147, thereby restricting the former to its intended role of guaranteeing the purpose for which the goods were acquired,[177] they may return to using this action now that a claim under art. 1147 seems to be excluded by the ECJ. French courts allow victims to sue anyone in the supply chain using the contractual action for latent defects, and this would have the effect of imposing liability on the producer or suppliers despite the fact that the risk was unknown at the time. However, as has been stated, it may well be that a rule imposing strict liability on intermediate suppliers and the producer in this way will, in its turn, be condemned by the European Court of Justice as being actions in reality based "on the same grounds" as the Directive.

106 A victim who has no action in breach of contract against the defendant may be able to rely on a tort action under art. 1382 Civil Code. A ten year limitation period applies to such claims from the date of the consolidation of the damage.[178] The courts may be willing to consider that the distribution of a defective product equates a fault, thus enabling liability to be imposed upon any member of the chain of distribution irrespective of the knowledge of the risk.[179] Again, such an interpretation of "fault" risks being condemned by the European court as being in reality an action "on the same grounds" as the Directive.

107 In the absence of such possibilities, there remains the possibility of an action for "genuine" fault[180] which could be brought in contract (art. 1147 Civil Code) or in tort (art. 1382 Civil Code) depending on the cir-

176 Supra no. 39.
177 *Viney/Jourdain* (fn. 15) 464 ff.
178 Supra no. 73 f.
179 Supra no. 31.
180 By this we mean behaviour falling below the standards of a reasonable man.

cumstances. Indeed, a recent decision of the Cour de cassation imposes on the producer an obligation to survey the safety of his product after it has been put into circulation, and liability for fault will be imposed on him where a risk becomes apparent at a future date and he fails to take appropriate action to avoid this risk. This will be the case even if there is a diversity of opinion amongst experts on the existence of the risk.[181] The decision relates to producers of pharmaceutical products, but there seems no reason in principle why such a duty should not also be imposed on seed and food producers.

(b) Can compensation already be claimed at a point when the negative health effects have not yet materialised, but are expected to according to scientific expertise/mere rumours?

French law requires damage to be certain. In principle, therefore, no com- **108**
pensation could be claimed where the negative health effects are yet to materialise. However, in some cases the courts have accepted compensating for the fear or risk of loss.[182] These have tended to be cases where the risk of future loss is imminent and based on firm scientific evidence, although lower courts have on two occasions recognised claims based on much flimsier evidence where local residents feared the effect to their health of a mobile telephone installation.

(c) Would it make any difference if the GM maize had only been in use for feed, causing harm to the animals, which may or may not cause harm to humans consuming the meat as well?

In this case the farmer would be able to claim for the harm to his animals **109**
and for his resulting loss of profits against the feed supplier and possibly further up the chain as in the previous examples.[183] Where consumers have suffered harm but it is not established for certain that this harm is due to the GM contamination of the meat, then it will be difficult for the victim to establish the causal link, which should in principle be shown with certainty.[184] However, the courts are often willing to take a rather flexible view as to what amounts to a causal link. Proof can result from

181 Case *Distiblène*, Cass. 1st Civ. 7 March 2006, RCA 2006, commentary 164, *Ch. Radé*.
182 Supra no. 7 f.
183 Supra no. 80.
184 Supra no. 18 ff.

presumptions, provided that these are "serious, precise and concordant". In the case in hand, a mere suspicion that the GM contamination of food may cause damage to humans would not be sufficient in itself to establish liability. However, if there is some scientific evidence of a causal link between GM contamination and the human illness in question, supported by other circumstantial evidence, such as the fact that the victim fell ill shortly after consuming the contaminated food, and no other explanation can be found for the illness, then the courts may accept that a causal link has been established. Recent cases concerning the liability of clinics to victims who contracted the hepatitis C virus after having undergone a blood transfusion are examples of how courts are willing to accept proof of causation by presumptions and the absence of any other apparent causal factor. The causal link will be much easier to establish where there is clear scientific evidence of the link between the GM feed and harm to humans, although the recent cases concerning the hepatitis B vaccine and multiple sclerosis show that the lack of clear scientific proof of a link may not necessarily prevent the claimant from succeeding provided that there is circumstantial evidence on the facts.[185]

110 Even where there is a causal link between the meat consumed and the victim's illness, there will be difficulties where the supplier or producer is one of any number in the sector. In such cases, it may be very difficult to establish the liability of any particular supplier using normal principles of causation.[186] This difficulty could be bypassed if the courts are ready to accept that all the suppliers or producers are liable collectively. For this to be the case, French courts would appear to require that the suppliers acted as a concerted group (as in the hunting cases referred to earlier). An application of this technique would therefore seem unlikely unless a broader approach can be adopted to enable a presumption of liability on the producers and/or suppliers.

185 3 decisions of the Cour de cassation: Cass. 1st Civ. 22.5.2008, D. 2008, 1544, note *I. Gallmeister*.

186 Supra no. 25.

3. The driver of a food logistics company discovers that a farmer, from where he regularly picks up agricultural products fails to obey mandatory segregation rules or food or feed hygiene standards, which may lead to the admixture of GM and non-GM produce sold and packaged separately by that farmer. Does he or his employer have a duty to warn, i.e. warn the recipient of the allegedly "non-GM" produce?

The French Commercial Code[187] provides that the recipient is automatically considered a party to a delivery contract between the seller and the transporter of the goods. Any potential liability of the transporter for his own, or his employee's, failure to warn would be based on an action in contract by the recipient for breach of a possible contractual duty to provide information,[188] an action in tort being unavailable due to the rule of non-cumul.[189] 111

With respect to the potential liability of the lorry driver on his own account, an important decision of the Cour de cassation in 2000[190] established that an employee will not be liable to a third party, irrespective of whether he has committed a fault, where he has acted within the limits of the mission that he was given by his employer. The exact criteria for identifying when an act goes beyond the employee's mission are yet to be clarified by the French courts, and authors consider that these may be based on the link between the act and the employee's mission, on the severity of the fault, on the presence of a desire by the employee for personal profit or on any combination of these three.[191] It would seem in any case very difficult to argue that the employee has gone beyond his mission in the case envisaged here by omitting to warn the recipient of the produce of the failure by the farmer to respect segregation guidelines. 112

187 Art. L 132-8 Commercial Code. *P. Le Tourneau*, Droit de la responsabilité et des contrats (2008) 1020.
188 A discussion on the merits of a potential claim lies outside the scope of this report.
189 The principle of *non-cumul* provides that where the defendant is in breach of a contractual obligation, the other contracting party may not rely on an action in tort with respect to that breach.
190 Assemblée plénière de la Cour de cassation, 25.2.2000, Bull.Civ. Ass. Plén, no. 2, JCP G 2000, II, 10295, conclusions *Kessous*.
191 For an overview, see *Flour/Aubert/Savaux* (fn. 56) 238.

Damage Caused by GMOs under German Law

Jörg Fedtke

I. General overview

1. Special liability or redress scheme for GMOs

1 In 1990 Germany enacted a special regime for GMOs, the *Gentechnikgesetz* (GenTG), or Act on Genetic Engineering,[1] which provides the regulatory framework for activities in this area. The Act also addresses some of the losses that GMOs can potentially cause. Two other public law regimes, the *Umweltschadensgesetz* (USchadG), or Environmental Damage Act,[2] and the *Arzneimittelgesetz* (AMG), or Pharmaceutical Products Act,[3] may also apply. In the area of private law, liability can follow from the *Produkthaftungsgesetz* (ProdHaftG), or Product Liability Act,[4] and from the general rules found in the *Bürgerliches Gesetzbuch* (BGB), the German Civil Code.

2 Liability depends, first, on whether a particular GMO may be circulated freely[5] or not. In the latter case, losses of life, health, or property are covered by the *Gentechnikgesetz,* which establishes strict liability for harm caused by *research and development.*[6] This does not exclude claims brought

1 Gesetz zur Regelung der Gentechnik of 20 June 1990, Bundesgesetzblatt (BGBl.) 1990 I 1080 ff. Important amendments were made in December 1993 (BGBl. 1993 I 2066 ff.), December 2004 (BGBl. 2005 186 ff.), March 2006 (BGBl. 2006 534 ff.) and April 2008 (BGBl. 2008 I 499).

2 Gesetz über die Vermeidung und Sanierung von Umweltschäden of 10 May 2007 (BGBl. 2007 I 666).

3 Gesetz über den Verkehr mit Arzneimitteln of 24 August 1976 as amended in December 2005 (BGBl. 2005 I 3394).

4 Gesetz über die Haftung für fehlerhafte Produkte of 15 December 1989 (BGBl. 1989 I 2198) as subsequently amended.

5 So-called *Umgang*, see § 3 no. 6a GenTG.

6 §§ 32 ff., 3 no. 2 GenTG. See Drucksachen des Deutschen Bundestages (BT-Drs. 11/5622 at 33; *W. Wellner* in G. Schlegelmilch (ed.), Geigel, Haftpflichtprozess (25th ed. 2008), chap. 14, 363.

under the general rules of the BGB,[7] in particular nuisance[8] and the fault-based general rules of tort law.[9] Losses caused by *licensed* GMO-based products fall within the ambit of the Product Liability Act.[10] Licensed pharmaceuticals containing GMOs are subject to the Pharmaceutical Products Act.[11] Any losses caused by the licensed use of GMOs are again also covered by the Civil Code.

A second determinative factor is the type of harm caused. Environmental damage in general is covered by the *Umweltschadensgesetz* of 2007, which establishes a no-fault regime for the protection of particular species, habitats, water resources, and soil.[12] The Act applies to research and development[13] *and* licensed GMOs.[14] Limits, however, exist. Damage or imminent danger to one of the protected environmental interests will *not* attract liability under the Act if caused by armed conflict,[15] exceptional, inevitable and uncontrollable natural phenomena,[16] or if the situation falls within the ambit of a number of specified international agreements.[17] 3

It is important to note that despite the above distinction between strict liability on the basis of the GenTG and USchadG, on the one hand, and the application of the general rules of the BGB, on the other, a number of standards which influence liability under the Code are now defined by public law. This is most obvious in the case of nuisance. The provisions of the BGB which protect property interests of landowners are based on three flexible notions. There is, for one, the level of interference with land, which needs to be of *sufficient weight* if a claim for equitable compensation in money is to succeed. A neighbour will, secondly, not be able to demand that a disturbing activity be terminated or invoke general tort law to claim damages if the activity *corresponds to local custom* and, thirdly, if the interfer- 4

---

7 § 37 (3) GenTG.
8 §§ 903 ff. BGB.
9 §§ 823 ff. BGB.
10 § 37 (2) GenTG. See *Wellner* (fn. 6) 372; C. *Godt*, Rückabwicklung von Inverkehrbringensgenehmigungen und Haftung für gentechnische Produkte, Neue Juristische Wochenschrift (NJW) 2001, 1167 at 1171 f.
11 § 37 (1) GenTG.
12 See Gesetz zur Umsetzung der Richtlinie des Europäischen Parlaments und des Rates über die Umwelthaftung zur Vermeidung und Sanierung von Umweltschäden of 10 May 2007, BGBl. 2007 I 666 ff. for its application to GMOs. For a general overview of the Act see *J. Fedtke*, Germany, in H. Koziol/B. Steininger (eds.), European Tort Law 2007 (2008) 288 f.
13 § 3 and Schedule 1 no. 10 USchadG.
14 § 3 and Schedule 1 no. 11 UschadG.
15 § 3 (3) no. 1 USchadG.
16 § 3 (3) no. 2 USchadG.
17 § 3 (3) nos. 3–5 USchadG.

ence cannot be avoided by *economically reasonable* measures.[18] In practice, the Code thus requires a considerable amount of interpretation.[19]

5 Concern regarding a wider legalised use of GMOs and the protection offered by the BGB in cases of cross-fertilisation between crops, the contamination of wild honey, or the labelling of foodstuffs has prompted further legislative activity in the past few years. The *Gentechnikgesetz* was amended in 2004 to include, in particular, § 36a GenTG, a provision which provides general standards for the interpretation of § 906 BGB, the most important rule of property law in the context of GMOs. Further changes were introduced in 2008.[20] The safety standards for the handling of GMO-based products contained in § 16b GenTG are now set out in more detail in an executive ordinance (*Gentechnik-Pflanzenerzeugungsverordnung*, GenTPflEV).[21]

6 As is generally true for German strict liability statutes, the special regime for GMOs is thus merely an *addition* to the general rules of the Civil Code, and provides an advantageous alternative for the victims of particular losses. The link between the two is, however, unusually strong due to the fairly detailed (public law) regulation of duties of care regarding GMOs.

2. State liability

7 German public authorities are themselves subject to the legal framework set out above. State liability, which will require a tort committed by an official in the exercise of public duties, is given under § 839 BGB and Art. 34 of the German Constitution. § 839 BGB prevails over the applica-

18 § 906 BGB declares: (1) The owner of a piece of land is not entitled to prohibit the intrusion of gases, vapours, smells, smoke, soot, heat, noises, shocks and similar interferences emanating from another piece of land to the extent that the interference does not or only immaterially prejudices the use of his piece of land. (2) The same applies insofar as a substantial prejudice is caused by the user of another piece of land in conformity with local custom and it cannot be prevented by measures, the financing of which can reasonably be expected of users of this kind. If by virtue of this, the owner must tolerate an interference, he may demand from the user of the other piece of land an appropriate settlement in money, if by the interference in conformity with local custom the use of, or income from, his piece of land is prejudiced over and above the expected degree.

19 For details see nos. 30 ff.

20 Erstes Gesetz vom 1.4.2008 zur Änderung des EG-Gentechnik-Durchführungsgesetzes (BGBl. I S. 497) and Gesetz vom 1.4.2008 zur Änderung des Gentechnikgesetzes, zur Änderung des EG-Gentechnik-Durchführungsgesetzes und zur Änderung der Neuartige Lebensmittel- und Lebensmittelzutaten-Verordnung (BGBl. I S. 499 ff.).

21 Verordnung über die gute fachliche Praxis bei der Erzeugung gentechnisch veränderter Pflanzen (Gentechnik-Pflanzenerzeugungsverordnung, GenTPflEV) of 7 April 2008, BGBl. 2008 I 655. For details see no. 35 below.

tion of other tort provisions contained in the BGB and is more limited in scope because liability is excluded where the breach of a duty is due to negligence only *and* the victim is able to obtain redress from another source[22] or where the victim did not make use of legal remedies which could have mitigated the loss.[23] A third hurdle is the requirement that the duty in question be designed to protect the very legal interest that was infringed in the case (*drittbezogene Amtspflicht*). That said, § 839 BGB offers the advantage of imposing liability for any damage suffered through the unlawful conduct of a public official; claims for negligently inflicted pure economic loss may therefore succeed under this heading.[24]

II. Damage

1. Recoverable losses

The answer to this question depends on the applicable legal regime. Victims receive compensation for death, physical injury,[25] harm to property (both land and movable things such as animals or crops *after* they are harvested) and pain and suffering under the *Gentechnikgesetz*.[26] Particular species, habitats, water resources, and soil are covered by the *Umweltschadensgesetz*.[27] The Product Liability Act provides compensation for death, physical injury, harm to movable things, and pain and suffering.[28] The Pharmaceutical Products Act covers death, *serious* physical injury, and pain and suffering.[29] All of these losses are also recoverable under the general rules of tort law, albeit subject to proof of fault and less favourable rules concerning the burden of proof. Land, including crops *before* they are harvested, receives additional protection under nuisance. 8

Recoverable losses following death include medical costs, any disadvantages resulting from diminished earning capacity and special needs in the 9

22 § 839 (1) BGB.
23 § 839 (3) BGB.
24 For more details see *B.S. Markesinis/H. Unberath*, The German Law of Torts (4th ed. 2002) 893 ff. and *B.S. Markesinis/J. Fedtke*, Authority or Reason? The Economic Consequences of Liability for Breach of Statutory Duty in a Comparative Perspective, European Business Law Review (2007) 5 ff.
25 German law in general distinguishes between bodily harm (*Körperverletzung*) and the impairment of health (*Gesundheitsverletzung*). This text will refer to both types of loss as *physical injury*.
26 § 32 (1), (5) GenTG.
27 § 2 USchadG.
28 §§ 1 (1) and 8 ProdHaftG.
29 §§ 84 (1) and 87 AMG.

phase leading up to death, funeral costs, and loss of maintenance.[30] Physical injury will attract compensation for medical costs, disadvantages resulting from diminished earning capacity, and pain and suffering.[31]

10 Loss to property, at present the most likely kind of harm caused by GMOs, is always defined by recourse to the general rules of the Civil Code,[32] which aim at full indemnification in kind or (if restitution *in natura* is either impossible, insufficient, or possible only at unreasonable cost) in money.[33] Damages will thereby include the loss of future profits insofar as they would probably have been accrued. Losses caused by research and development will usually render contaminated crops or foodstuffs completely unmarketable. Their value will therefore have to be fully compensated. The cost of decontaminating land will also be recoverable. If crops remain marketable despite contamination, the victim will have to reduce the loss by selling them in accordance with any rules requiring specific labelling. Any depreciation following from the fact that crops cannot be marketed in the form originally envisaged will thereby be recoverable by taking into account the market price which could have been realised through private contractual agreements (e.g. with food producers). Losses caused by recalling products from the market will also have to be compensated. Finally, liability will cover indirect costs such as increased overheads to develop new outlets for goods or, more importantly, to regain a specific producer status.

2. Pure economic loss

11 German tort law protects pure economic interests only in exceptional circumstances.[34] Four avenues exist. § 823(2) BGB – which provides compensation for losses caused by the infringement of a statute designed to safeguard, inter alia, the interests of the victim (*Schutzgesetzverletzung*) – can lead to the compensation of pure economic loss if the violated rule is designed to protect such economic interest. § 826 BGB also covers pure economic loss but requires the tortfeasor to act intentionally and in a man-

30 §§ 32 (4) GenTG; 86 AMG; 7 ProdHaftG. These losses are also recoverable under general tort law.

31 §§ 32(5) GenTG; 87 AMG; 8 ProdHaftG. These losses are also recoverable under general tort law.

32 §§ 249 ff. BGB.

33 Note, however, that under § 32 (7) GenTG high costs for decontamination will not justify (mere) compensation in money if the loss caused by GMOs constitutes an impairment of the environment (*Natur oder Landschaft*).

34 *U. Magnus*, Damages under German Law, in U. Magnus (ed.), Unification of Tort Law: Damages (2001) 95.

ner *contra bonos mores*. § 824 BGB protects pure economic loss in the context of statements calculated to endanger the creditworthiness of the victim, or to affect his earnings or prospects in some other manner. Finally, pure economic loss is also recoverable under § 823(1) BGB if the right of an established and operating business is affected. The invasion must, however, be in some way directed against the business *as such* and may not simply affect rights and interests which are separable from the business as a functioning unit.[35] The effect must also be of considerable significance and duration.[36] The only way to recover pure economic loss caused by GMOs under these circumstances is arguably § 823(2) BGB in conjunction with the rules aimed at preventing cross-fertilisation between GM-crops and conventional plants or the provisions of the *Lebensmittel- und Futtermittelgesetzbuch* (LFBG), which regulates the safety of foodstuffs, feed, and household supplies[37]

3. Mere fear of a loss

Mere *fear of losses* caused by GMOs will not attract compensation under German law. With few exceptions, *measurable loss* – broadly defined as the difference between a victim's economic position after the damaging event and its estimated status without the intervention – is required.[38] Consumers who fear that they might develop an illness sometime in the future will thus not be able to claim damages. The situation can be different, however, if a victim knows that a particular loss (e.g. a serious medical condition) *will* eventually materialise and develops a recognised psychological illness as a result. **12**

4. Standard of proof

In general, courts will award damages only if no reasonable doubts about the loss remain.[39] Future losses can be taken into account if the facts allow **13**

35 For more details on this judge-made right see *Markesinis/Unberath* (fn. 24) 71 ff.

36 BGH NJW 1983, 812 at 813.

37 Lebensmittel- und Futtermittelgesetzbuch in der Fassung der Bekanntmachung vom 26. April 2006 (BGBl. 2006 I 945), zuletzt geändert durch Artikel 12 des Gesetzes vom 26. Februar 2008 (BGBl. 2008 I 215). On the application of § 823(2) BGB in this context see *S. Förster*, Internationale Haftungsregelungen für schädliche Folgewirkungen gentechnisch veränderter Organismen (2007) 60.

38 *H. Lange/G. Schiemann*, Schadensersatz (3rd ed. 2003) 26 ff.

39 § 286 Zivilprozessordnung (ZPO), or Code of Civil Procedure. See Entscheidungen des Bundesgerichtshofs in Zivilsachen (BGHZ) 53, 245; BGHZ 61, 169; BGHZ NJW 1993, 935.

the judge to make a sufficiently precise estimate on the last day of the hearing (*letzte mündliche Tatsachenverhandlung*).[40] Procedurally, victims can secure potential future losses against prescription by seeking a declaratory judgment on the factual basis of the claim only (*Festellungsklage*).[41] Otherwise, victims will have to pursue compensation once losses materialise.[42]

5. Nominal losses

14 Nominal/symbolic losses are not recognised under German law. Without measurable loss, there can be no compensation.[43]

6. Mass losses

15 German law does not provide special rules on collective or mass losses. The issue attracted some attention a decade ago[44] but specific reform proposals have thus far not been put forward.

III. Causation

1. Uncertainty of merely potential causes

16 In general, § 830 BGB provides that several persons who have jointly caused harm by an unlawful act will each be responsible for the entire loss. The same applies *if it remains uncertain which of several participants has actually caused the harm*. The Code also determines that, in the absence of a specific rule, those liable will have to come up with an equal share of the required compensation; internal recourse is open to any party which pays the full amount.[45] German law thus goes some way towards providing an answer for cases in which *ultimate* responsibility for a loss remains unclear (*Urheberzweifel*).[46] The application of this rule is, however, subject to a number of caveats. Most importantly, each participant must qualify as

40 § 287 ZPO.
41 *Magnus* (fn. 34) 90.
42 *Lange/Schiemann* (fn. 38) 45.
43 *Magnus* (fn. 34) 90.
44 See C. *von Bar*, Empfehlen sich gesetzgeberische Massnahmen zur rechtlichen Bewältigung der Haftung von Massenschäden?, 62. Deutscher Juristentag (1998).
45 §§ 840 (1), 426 BGB.
46 BGH NJW 1994, 932 at 934.

(potential) tortfeasor, which means that all other conditions of liability – except for causation – must be proven for each one.[47] Each separate action must, in particular, be capable of having caused the *entire* loss (even if each participant might actually only be responsible in part).[48] Secondly, § 830 BGB is not applicable if the loss could also have been caused by the victim. The provision is designed to ease the victim's burden of proof if several potential tortfeasors exist, but does not shift responsibility away from causes lying within the victim's own sphere.[49] Finally, recourse to § 830 BGB is excluded if one of the participants clearly caused the whole loss.[50]

The same approach is established by § 36a(4) GenTG for claims based on **17**
nuisance.[51] If it is impossible to identify as the true source one of several neighbours who could have individually caused the contamination in question due to the cultivation of the same GM crop in the area (alternative causation), each neighbour will be deemed jointly and separately responsible for the whole interference unless their respective contributions were in fact limited and particular shares can be apportioned according to § 287 ZPO. Liability is not channelled in this case; farmers will, in particular, not be able to avoid the payment of compensation despite the fact that they have adhered to the safety measures recommended by the producer of their GM seed. The same principle is applied in cases where several tortfeasors can safely be identified but the extent of each contribution remains uncertain.[52]

A special rule specifically addressing doubts about causation is also found in **18**
the Pharmaceutical Products Act, which can be relevant if GMOs are used in the production of licensed pharmaceuticals. § 84(2) AMG contains a legal presumption that a particular loss was indeed caused by the pharmaceutical in question if that product is in principle capable of causing the symptoms experienced by the patient. This presumption is rebutted if the producer can point to an alternative cause. This alternative cause may *not*, however, be another pharmaceutical product used at the same time, even it is equally capable of causing the symptoms in question, unless its use excludes liability under § 84 AMG for reasons other than the question of causation (e.g. if the combined use of these products was not *lege artis*).

---

47 BGH NJW 1989, 2944.
48 Ibid.
49 BGHZ 60, 177.
50 BGHZ 72, 355 at 361 ff.
51 § 906 (2) BGB.
52 *W. Lülling/G. Landsberg* in W. Eberbach/P. Lange/M. Ronellenfitsch, Recht der Gentechnik und Biomedizin (44. Ergänzungslieferung 06/2004), § 32 GenTG nos. 105 f.

19 The Environmental Damage Act is not applicable to situations of unclear causation (or, in the language of the Act, "a pollution not clearly defined or limited"[53]) unless a causal link between the harm and the activities of single responsible polluters can be established.[54]

20 Joint and several liability is also established, finally, by § 32(2) GenTG (which covers liability for research and development of GMOs) if the same loss is caused by more than one tortfeasor. The internal distribution of the burden depends on the respective shares of responsibility. Recourse is again possible on the basis of the general rules of the BGB. The same mechanisms are also used in the Product Liability Act,[55] the Pharmaceutical Products Act,[56] and the Environmental Damage Act.[57]

2. Complex causation scenarios

21 Complex cases such as these will have to be resolved on the basis of the rules outlined above – subject, of course, to special contractual arrangements which may exist between food producers and the farming industry.

3. Force majeure

22 The only statute in this area which excludes liability for losses caused by force majeure is the Environmental Damage Act. Losses caused by armed conflict or exceptional, inevitable and uncontrollable natural phenomena are excluded from the ambit of the USchadG.[58]

4. Threshold to prove causation

23 Within the general conditions of causation (*condition sine qua non* test) and its limitations (theory of adequate causation and considerations concerning the protective purpose of the norm), German courts will require a high degree of certainty that a particular loss was caused by the defendant. As noted above, some uncertainties concerning alternative causes are addressed in

53 "(E)ine nicht klar abgegrenzte Verschmutzung...".
54 § 3 (4) USchadG.
55 §§ 5 and 6 (2) ProdHaftG.
56 § 93 AMG.
57 § 9 (2) USchadG.
58 § 3 (3) USchadG.

the relevant strict liability statutes. These rules will, however, again require that judges are convinced that the actions of more than one defendant were in principle capable of causing the entire loss. The only statute which actually lowers the threshold to some extent is the Pharmaceutical Products Act. § 84(2) AMG identifies a number of indicators which – if given – trigger a legal assumption that the product in question did indeed cause the loss. These are the composition and individual dosage of the pharmaceutical, the way it was applied, the length of its prescribed use by the patient, the proximity between its application and the development of the harmful symptoms, the harm suffered and the general state of the patient's health at the time, as well as any other factors which might be relevant in *confirming* or *excluding* a causal link under the circumstances of the case. The sum of these considerations amounts to a "more likely than not" assessment.

5. Special rules on causation

Claimants bringing a case under the GenTG will have to prove, usually **24** with the help of an expert opinion and testing (at their own expense), both loss and causation through a GMO. It will then be presumed that the loss was the result of its modified characteristics.[59] This presumption is refutable by proof that the loss in question was in fact caused by the *un*modified genes of that GMO.[60] The GenTG thus provides only a limited degree of protection from the typical difficulties of proving causation in such cases. Some support, however, is given by discovery rights. Operators of facilities in which GMOs are developed, tested, produced, or otherwise handled must disclose details of the entire technical process so that potential victims can better ascertain whether claims based on the GenTG actually exist.[61] In the case of tests on open land, information contained in the (compulsory) public register – the specific type of crop, its modified features, and the exact location and size of the field(s) – will also be accessible.[62] Additional details will be disclosed to anyone with a legitimate interest (e.g. potential victims who can show that their property was subject to an interference by GMOs).[63] Similar disclosure rights against pharmaceutical companies and regulatory authorities can be found in § 84a AMG.

59 So-called *Ursachenvermutung*, § 34 (1) GenTG.
60 § 34 (2) GenTG.
61 § 35 GenTG.
62 § 16a (2) GenTG.
63 § 16a (5) GenTG.

IV. Types of liability

1. Fault liability

(a) Special rules governing fault

25 As explained above, cases of the kind envisaged by this study can fall within the ambit of a strict liability regime; in parallel, the general fault-based rules of German tort law will, however, always remain applicable. Thus far, it has not become apparent that courts will apply these rules in any special way when dealing with GMOs.

26 That said, the rules of nuisance – a tort in common law jurisdictions but part of property law in Germany – were modified in order to address difficulties arising from GM farming. The standards set out in the *Gentechnikgesetz* also have an indirect effect on the scope of general tort law by defining what constitutes an "unlawful" interference with neighbouring land.

27 As indicated at the outset of this report, farmers raising crops from licensed genetically modified seed will be subject to rules protecting the property of their neighbours.[64] Crops are thereby considered part of immovable property (the farmland) until the point of harvest.[65] Such farmland is, in principle, also protected by the general provisions of tort law, albeit only within the limits of any special rules which oblige the neighbour to tolerate certain levels of interference. Cross-fertilisation between modified and conventional crops was confirmed as a potential interference by introduction of § 36a GenTG. Three scenarios need to be distinguished:

28 According to § 906(1) BGB, interference which does not adversely affect a neighbouring piece of land – or which affects it only marginally – must be tolerated[66] and is thus not regarded as illegal within the scope of §§ 823 ff. BGB.[67] Neither tort nor property law (nuisance) will offer a remedy.

29 If land is used in a way *customary* in that particular region[68] and *does* impair neighbouring land *significantly*, such influence is again *not* illegal within the meaning of tort law[69] and must be tolerated if the negative effect can-

64 §§ 903 ff. BGB.

65 § 94 BGB.

66 So-called *Duldungspflicht*.

67 BGHZ 90, 255 and 92,148. *O. Jauernig* in O. Jauernig (ed.), Bürgerliches Gesetzbuch (11th ed. 2004), § 906 no. 8.

68 So-called *ortsübliche Nutzung*.

69 BGHZ 117, 110.

not be prevented by the other party through economically reasonable measures. Here § 906(2) BGB will, however, entitle the neighbour to equitable compensation.

Only if these conditions do *not* apply – i.e. if land is *not* used in a customary way or if the other party *could* prevent such significant impairment through economically reasonable measures but *fails* to do so – can neighbours demand termination of a current interference or seek injunctions if further nuisance is imminent.[70] Compensation for damage caused to crops can then be claimed on the basis of tort law or in analogous application of § 906(2) BGB.[71] **30**

These rules of property law require a considerable amount of interpretation. What level of contamination constitutes a "substantial" interference? Can the use of GM crops currently be regarded as "customary" at all? And, finally, what are the "economically reasonable" safety precautions which a farmer using GM crops must take in order to keep them from contaminating neighbouring land? § 36a GenTG now provides guidance for the application of the BGB to interference through GM crops. The rule (1) establishes a standard for "substantial" interference; (2) defines what measures can "reasonably" be expected in order to avoid disturbing others; and (3) clarifies the notion of "use according to regional custom in this context." **31**

Contamination of crops with GMOs constitutes a *"substantial" interference* within the meaning of § 906 BGB if, contrary to the intentions of the neighbour, those crops may subsequently not be sold at all,[72] may be sold but only subject to labelling ("genetically modified") as prescribed by law,[73] or may not be marketed with a certain label (e.g. "organic") as intended by the owner.[74] Any contamination with *non-approved* GMOs will thus always constitute a substantial interference since it renders the contaminated crop unmarketable. If crops are contaminated with *approved* GMOs (which leaves them affected but still potentially marketable), any thresholds contained in specific legislation will be directly applicable under § 906 BGB, in order to determine the extent of the interference. **32**

70 § 1004 BGB.

71 *Jauernig* (fn. 67) § 906 no. 9. These general rules also apply to non-licensed GMOs (in addition to the special liability regime of the *Gentechnikgesetz*), see § 37(3) GenTG.

72 § 36a (1) no. 1 GenTG.

73 § 36a (1) no. 2 GenTG.

74 § 36a (1) no. 3 GenTG (e.g. as "ecological" within the meaning of EEC Council Directive 2092/91 of 24. 6. 1991; see the explanatory memorandum of the amendment to the GenTG, BT-Drs 15/3088, 31). See also the standards required by the Verordnung zur Durchführung gemeinschaftsrechtlicher Vorschriften über neuartige Lebensmittel und Lebensmittelzutaten (NLV) of 29.2.2000.

Community law thus requires that food containing or consisting of GMOs, or produced from or containing ingredients produced from GMOs, be labelled unless it contains, consists of, or is produced from GMOs in a proportion no higher than 0.9% of the food ingredients considered individually or food consisting of a single ingredient, provided that this presence is adventitious or technically unavoidable.[75] Regulation (EC) 1829/2003 establishes the same threshold for feed.[76]

33 Interferences below these levels will not be considered "substantial" while higher percentages of contamination provide a basis for equitable compensation under § 906(2) BGB. It is, however, unclear at this point whether the 0.9% threshold could also be invoked if a GM-free farmer were to show that his contractual agreements with particular food producers are based on more severe standards, e.g., if his produce was supposed to be labelled as "free of GMOs" (see no. 36 below). It should also be pointed out that mill operators and food or feed producers who wish to market conventional products below the labelling threshold will in practice require farmers to provide material with far less than 0.9% contamination in order to ensure that their production output, based on several providers, also stays below that mark.

34 The safety measures contained in § 16b(2) and (3) GenTG (so-called *gute fachliche Praxis*) are declared *"economically reasonable"* within the meaning of § 906(2) BGB and thus provide the standard of care for GMO farming.[77] The user of genetically modified plants is obliged to avoid as far as possible (the statute uses the term *vermeiden*) cross-fertilisation (both with other crops and the environment in general) by, e.g., the maintenance of a safety corridor between his crops and surrounding land, the selection of appropriate seed, the use of techniques to counteract the intrusion of alien plants onto his land, and the use of natural barriers. Both the use of GMOs (which includes fertilizer) and safety measures must be adequately documented.[78] In a similar vein, contamination of other products by GMOs must be prevented (here the statute uses the term *verhindern*) through the use of separate storage facilities and the proper maintenance of such facilities or other equipment used in the production process.[79] Finally, the user of GMOs must also prevent the contamination of other products in transit by – again – using separate transport facilities (e.g.

75 Art. 12 (2) Reg. (EC) No. 1829/2003.
76 Art. 24 (2).
77 § 36a (2) GenTG.
78 § 16b (3) no. 1 GenTG.
79 § 16b (3) no. 3 GenTG.

trucks) and proper maintenance of any equipment.[80] The person who markets GMOs (e.g. the seed producer)[81] must provide instructions on the handling of the product which aim to meet these safety standards and give farmers guidance on how to adhere to § 16b(2) and (3) GenTG.

35 As mentioned earlier,[82] these safety standards are now set out in more detail in an executive ordinance. Duties include timely and comprehensive informing of regulatory authorities and neighbours;[83] safe storage[84] and transportation;[85] careful use of machinery;[86] the removal of alien plants from the field;[87] and comprehensive documentation.[88] More importantly, the exact size of safety corridors designed to prevent (or at least minimalise) cross-fertilisation have now been defined in an annex to the GenTPflEV. At present, only a single GM crop, corn of the variety Mon 810, is licensed for general use. Farmers wishing to grow this plant must leave a corridor of 150m to any neighbour with conventional plants and 300m to any neighbour farming organically.[89] These distances can be reduced by private agreements;[90] this will, however, require the neighbour to label his own produce as "genetically modified" even if the level of contamination (unless completely insignificant) remains below 0.9%. This strict approach results from the fact that German law – like EC rules – only accepts low contamination levels without a warning if such levels are accidental or technically unavoidable ("*zufällig oder technisch nicht zu vermeiden*").[91] The corridors defined by the ordinance now establish a legal assumption that cross-fertilisation *was* in fact avoidable if a neighbour freely accepts a smaller distance between his own crops and GM farming.

36 Finally, the rules on labelling have also been tightened. Food may not be sold as "free of GMOs" if genetically modified feed was used at some point in the production process, either by the producer of a product or at any earlier point in the production chain. This rule applies, e.g., to dairy pro-

80 § 16b (3) no. 4 GenTG.
81 § 16b (V) GenTG.
82 See above no. 4.
83 §§ 3 and 5 GenTPflEV.
84 § 6 GenTPflEV.
85 § 7 GenTPflEV.
86 § 9 GenTPflEV.
87 § 10 GenTPflEV (so-called *Durchwuchs*).
88 § 12 GenTPflEV.
89 See nos. 1 and 2 of the annex to the GenTPflEV concerning genetically modified corn. Different safety corridors may in the future be prescribed for other plants or varieties.
90 § 16b (1) GenTG.
91 § 17b (3) GenTG.

ducts if animals were fed GM corn or soya. Past use of GM feed is only allowed within certain limits. More importantly, § 3a EGGenTDurchfG prescribes that foodstuffs must be *completely free* of GMOs under this label, which is likely to translate into a threshold of less than 0.1% in practice.[92]

37 *"Customary use"* of land, finally, is not to be defined with respect to the predominant use of either GMOs or traditional production methods.[93] This provision seeks to safeguard the initial use of GMOs in a particular region.

38 §§ 36a and 16b GenTG, together with the new ordinance, thus clarify that GMOs are capable of interfering with neighbouring property rights and subject to the regime established by § 906 BGB. The criteria for the application of the BGB to GMOs, as set out by the GenTG, are thereby fairly strict and have led to much debate about the viability of GMO agriculture in Germany. The safety measures established by § 16b GenTG are costly and cannot be avoided; even if they are taken, equitable compensation will have to be paid for higher levels of contamination. §§ 36a and 16b GenTG will equally affect tort claims based on § 823(1) BGB by defining the standard of care which farmers using GMOs will have to comply with. If the threshold of a "substantial" interference with neighbouring property is crossed (constituting an infringement), farmers will thus have to show that they have met the requirements of § 16b GenTG if they wish to escape tortious liability. Evidence showing that they have followed the instructions given by their own seed supplier may not be sufficient to escape liability, but could at least provide the basis for subsequent contractual action if these instructions turn out to be insufficient.[94]

(b) Impact of specific rules of conduct

39 As explained in the previous paragraph, legislation enacted over the past five years has sought to determine standards for safe GM farming. Violations of these rules impact directly on liability under the rules of nuisance and general tort law.

---

92 §§ 3a and 3b Gesetz zur Durchführung der Verordnungen der Europäischen Gemeinschaft auf dem Gebiet der Gentechnik und über die Kennzeichnung ohne Anwendung gentechnischer Verfahren hergestellter Lebensmittel (EGGenTDurchfG) as amended by the Act of 1 April 2008 (see fn. 20). For a detailed analysis of these new rules see *G. Wagner/S. Bresser*, Natur und Recht (NuR) 2008, 695 ff.

93 § 36a (3) GenTG.

94 *T. Dolde*, Gesetz zur Neuordnung des Gentechnikrechts, Zeitschrift für Rechtspolitik (ZRP) 1/2005, 25 at 27.

2. Product liability

(a) Development risk defence

40 German legislation incorporated the development risk defence in § 1(2) no. 5 ProdHaftG by allowing producers to show that a particular defect could not be discovered on the basis of the scientific and technical knowledge available at the time a product was put into circulation. Following an amendment of the GenTG in 2000,[95] GM-based products that are made subject to a licensing requirement are currently, however, *excluded* from the ambit of this rule.[96] This is also true for agricultural products.[97] The effect is twofold. The exclusion only applies, firstly, to producers who sought and received the necessary permit *themselves*;[98] and, secondly, only covers those defects which result from genetic engineering (so-called *gentechnische Arbeiten*).[99] Farmers who are regarded as "producers" within the meaning of § 4(1) ProdHaftG will therefore be able to invoke the development risk defence[100] whereas developers or producers of GM seed will not. Courts apply an objective standard when interpreting the defence; available scientific and technical knowledge thus also includes novel research which is not yet "mainstream" accepted opinion.[101]

(b) Alternative routes

41 As explained above, GM products fall within the scope of product liability if they are subject to a licensing requirement as prescribed by §§ 37(2), 16(2), 14(2) GenTG. Such requirements are set out, e.g., in the *Pflan-*

---

95 Gesetz zur Änderung produkthaftungsrechtlicher Vorschriften of 2 November 2000, BGBl. 2000 I 1478.

96 § 37(2) GenTG. See *E. Deutsch*, Haftung und Rechtsschutz im Gentechnikrecht, Versicherungsrecht (VersR) 1990, 1041 at 1046; *Wellner* (fn. 6) 372.

97 Agricultural produce which has not yet been processed in some form was exempted from the Act until 30 November 2000 (see § 2 sent. 2 ProdHaftG before 1 December 2000) but is now subject to product liability. See *G. Wagner* in Münchner Kommentar zum Bürgerlichen Gesetzbuch V (5th ed. 2009) § 2 Produkthaftungsgesetz no. 11.

98 *Wellner* (fn. 6) 372.

99 §§ 37 (2), 3 no. 2 GenTG. See *G. Hirsch/A. Schmidt-Didczuhn*, Die Haftung für das gentechnische Restrisiko, VersR 1990, 1193 at 1204; *E. Deutsch*, Produzentenhaftung im Gentechnikrecht, Produkthaftpflicht International (PHi) 1991, 75 at 77 f.

100 *Wagner* (fn. 97) § 15 Produkthaftungsgesetz no. 14; *B. Wolfers/M. Kaufmann*, Grüne Gentechnik: Koexistenz und Haftung, Zeitschrift für Umweltrecht (ZUM) 2004, 321 f.; Drucksachen des Deutschen Bundesrates (BR-Drs.) 387/89, 36 f.

101 *H.J. Kullmann*, Produkthaftungsgesetz (4th ed. 2004), 62 ff.

zenschutzgesetz (PflSchG), or Protection of Plants Act,[102] and – under Community law – the Novel Food Regulation,[103] and will cover *most* but *not necessarily all* genetically modified (agricultural) products. Products for which no licensing requirements exist (e.g. if mere notification under the Novel Food Regulation suffices[104]) will thus continue to attract liability under the – arguably stricter – standards of the *Gentechnikgesetz*. The Act, however, came into force *after* enactment of Directive 85/374/EEC and does not fall under the privilege of its Article 13. It is therefore unlikely that the continued application of the GenTG to GM products which can be sold without a licence would survive scrutiny by the ECJ under its "exclusive standard" approach.

42 Another alternative route is general tort law. These rules are arguably not excluded by Directive 85/374/EEC,[105] though it remains to be seen whether highly distinct systems of case law regarding (fault-based) product liability will pass muster (e.g. the German jurisprudence under § 823 BGB).

(c) Impact of compliance with rules and regulations

43 Two defences of the ProdHaftG can potentially apply under these conditions. First, the defendant might be able to show that the defect in question was an immediate consequence of a *binding* rule or regulation governing this specific production process.[106] Such prescriptions are, however, rare in Germany.[107] The rules of the *Gentechnikgesetz* on GM farming do not fall into this category as they are designed to protect neighbours of GM farming *during the process* and not intended to affect the production *outcome*. The defence could, however, become relevant if regulatory authorities involved in the licensing of GM seed or foodstuffs were to prescribe specific genetic configurations of these products. The second possible defence in this scenario is the development risk argument already discussed above.[108] This defence is not available to producers who have themselves applied for the licence necessary to market a product which has a defect caused by

---

102 Pflanzenschutzgesetz in der Fassung der Bekanntmachung vom 14 Mai 1998 (BGBl. 1998 I 971, 1527, 3512 as subsequently amended.

103 Regulation (EC) 258/97 of the European Parliament and of the Council of 27 January 1997 concerning novel foods and novel food ingredients, OJ L 43, 14.2.1997, 1–6.

104 Art. 3 (IV) Novel Food Regulation.

105 ECJ C-52/00, *Commission v. France* [2002] ECR I-3827, par. 22; ECJ 154/00, *Commission v. Greece* [2002] ECR I-3879, par. 18; *Wagner* (fn. 97) § 15 Produkthaftungsgesetz no. 2.

106 § 1 (2) no. 4 ProdHaftG.

107 *Kullmann* (fn. 101) 57 ff.

108 § 1(2) no. 5 ProdHaftG.

genetic engineering; others making subsequent use of these products in their own production process may, however, be able to invoke the development risk argument.

On a more general level, note should be made of the fact that the ProdHaftG can only be invoked by victims if the product in question was designed for *private* use or consumption and was (mainly) used in that way.[109] Liability under this heading will thus not play a role in disputes between entrepreneurs and/or farmers.[110] 44

3. Environmental liability

(a) Implementation of the Environmental Liability Directive

Germany implemented Directive 2004/35/EC in 2007 with the Environmental Damage Act.[111] The Act is designed to prevent and combat existing health risks and threats to biodiversity caused by (industrial) occupational activities. It establishes liability under public law and addresses damage and threats to water resources, soil, protected species, and natural habitats.[112] In line with Community law, liability is limited to individual polluters and incidents; the Act is not instrumental to combating pollution in general.[113] Personal injury, loss to private property, and pure economic loss of third parties (even if consequential) are not covered;[114] private claims thus continue to find their most important basis in the Environmental Liability Act of 1990.[115] Note, however, that the latter regime (which attaches liability to particular industrial facilities, so-called *Anlagenhaftung*) is unlikely to cover losses caused by GMOs. 45

Polluters are required to inform the authorities about incidents that adversely affect the environment,[116] must take all necessary steps to limit or prevent damage to natural resources,[117] and restore as far as possible 46

109 § 1(1) sent. 2 ProdHaftG.
110 *L. Stökl*, Die Gentechnik und die Koexistenzfrage: Zivilrechtliche Haftungsregelungen, ZUM 2003, 274 at 277.
111 See the general remarks at no. 2 above.
112 § 2 no. 1 USchadG.
113 § 3 (4) USchadG.
114 See *D. Greinacher*, Bahnbrechend Neues oder alles wie gehabt? – Umsetzung der Umwelthaftungsrichtlinie in deutsches Recht, PHi 2007, 2 ff.
115 Umwelthaftungsgesetz (UmweltHG) of 10 December 1990, BGBl. 1990 I 2634.
116 § 4 USchadG.
117 § 5 USchadG.

the status quo ante at their own expense.[118] A schedule identifies the occupational activities which fall within the scope of the Act.[119] These include the handling, storage, and disposal of waste;[120] the legal use of water resources;[121] and the handling, storage, use, or transport of hazardous substances.[122] The statute establishes a no-fault regime.

47 The *Umweltschadensgesetz* is linked to other core environmental statutes, in particular the Environmental Protection Act,[123] the Water Resources Act,[124] the Act on Genetic Engineering, and the Soil Protection Act,[125] each of which defines specific types of environmental damage in greater detail. The more protective rules will prevail in case of an overlap. The Act does not create a completely new regime but will serve as an important default mechanism and could in the future lead to a higher level of intervention by public authorities especially in the area of long-term restorative measures.

48 The *Umweltschadensgesetz* does not prescribe financial guarantees.[126]

(b) Environmental liability regime beyond the scope of the Directive

49 The *Umwelthaftungsgesetz*,[127] mentioned in the previous section, is the private law counterpart to the *Umweltschadensgesetz* in that it regulates the claims of individuals in case of death, bodily harm, impairments of health, pain and suffering, or material loss. The Act does not establish strict liability in favour of the environment itself, but some degree of indirect protection is nevertheless achieved due to the fact that § 1 UmweltHG requires some kind of influence on soil, air or water which in turn leads to an impairment of the individual.[128] As indicated above, the Act does not, however,

118 §§ 6, 9 USchadG.
119 Anlage 1 USchadG.
120 No. 2 Anlage 1 USchadG.
121 No. 3–6 Anlage 1 USchadG.
122 No. 7 and 8 Anlage 1 USchadG.
123 Gesetz über Naturschutz und Landschaftspflege (Bundesnaturschutzgesetz, BNatSchG) of 25 March 2002, BGBl. 2002 I 1193 as subsequently amended.
124 Wasserhaushaltsgesetz (WHG) in der Fassung der Bekanntmachung vom 19.08.2002, BGBl. 2002 I 3245 as subsequently amended.
125 Gesetz zum Schutz vor schädlichen Bodenveränderungen und zur Sanierung von Altlasten (Bundes-Bodenschutzgesetz, BBodSchG) of 17 March 1998, BGBl. 1998 I 502 as subsequently amended.
126 On this see *G. Wagner*, Das neue Umweltschadensgesetz, VersR 2008, 565 at 578 ff.
127 See fn. 115.
128 *J. Fedtke/U. Magnus*, Germany, in B.A. Koch/H. Koziol (eds.), Unification of Tort Law: Strict Liability (2002), 160.

cover facilities for genetic engineering. More important in this context is the Water Resources Act,[129] which protects all natural water resources (surface water such as rivers, ponds and lakes, costal waters, and ground water) against any impairment of their water quality.[130] § 22(1) WHG establishes strict liability for harm caused by the introduction of substances or any other (legal or illegal) activity affecting the physical, chemical, or biological composition of water resources; § 22(2) WHG attaches strict liability to facilities designed for the production, storage, or transport of substances if these reach natural water resources. Both cases of § 22 WHG are not applicable if harm is caused by the use of natural water resources within the limits of a permit issued by the competent water authority;[131] here, victims can claim compensation only from the state.[132]

(c) Claimants in cases of environmental harm

Damage and threats to water resources, soil, protected species, and natural **50**
habitats are covered by the *Umweltschadensgesetz*.[133] The USchadG is purely regulatory in nature and does not provide a basis for individual claims against polluters; it is rather the competent public authority which forces polluters to take the necessary protective or restorative measures even if private property is affected.[134] That said, private individuals and environmental organisations may bring cases to the attention of the authorities and request intervention.[135]

(d) Special liability regime for losses sustained by individuals

Nuisance, part of the property regime of the BGB, is extremely relevant for **51**
the recovery of losses sustained by individuals in the course of damage to the environment as such.[136] Additional protection is offered by the Environmental Liability Act[137] and the Water Resources Act.[138]

---

129 See no. 124.
130 Now more closely defined by § 22a WHG.
131 § 11 (1) WHG.
132 §§ 22 (3); 10(2) WHG. For more details on the WHG see *J. Fedtke/U. Magnus* (fn. 128) 158 ff.
133 See no. 3, 45 ff.
134 See *Wagner* (fn. 126) 565 ff.
135 §§ 10, 11 USchadG.
136 See no. 1 ff., 16 ff. and 25 ff.
137 See no. 49.
138 Ibid.

(e) Cartagena Protocol

52 Germany is a party to the Cartagena Convention on Biological Diversity (CBD)[139] and was host to the Conference of the Parties to the Convention in May 2008. The country is also part of "The Economics of Ecosystems and Biodiversity" (TEEB) of the G8+5 Group, an initiative designed to initiate through an international study the process of analysing the global economic benefit of biological diversity, the cost of a loss of biodiversity, and the failure to take protective measures versus the costs of effective conservation. It would seem that there is currently a strong commitment to such international co-operation in Germany.[140]

4. Other strict liability regimes

53 All German strict liability regimes relevant in this context are set out above.

V. Vicarious liability

1. Scope of vicarious liability

54 In German tort law the most general rule of vicarious liability is that a master or employer[141] is liable for losses caused by servants or employees[142] within the scope of their employment *unless* he can prove that they were carefully chosen and supervised.[143] It is important to note that this defence[144] is not available under contract law.[145] These rules do not apply to persons who make use of independent contractors *who do not act under their direction*. In these cases, persons may nevertheless be liable due to a violation of their *own* duty of care regarding the choice and supervision

---

139 The Convention was ratified by Germany in November 2003; the country's membership became effective in February 2004.

140 See the Bonn Agenda on Global Biodiversity set out in the closing statement of Siegmar Gabriel, currently Germany's Environmental Minister, at the Conference of the Parties to the CBD on 30 May 2008.

141 So-called *Geschäftsherr*.

142 So-called *Verrichtungsgehilfen*.

143 § 831 BGB.

144 So-called *Exculpationsbeweis*.

145 See *J. Fedtke*/*U. Magnus*, Liability for Damage Caused by Others under German Law, in J. Spier (ed.), Unification of Tort Law: Liability for Damage Caused by Others (2003), 105. This contribution also covers other forms of vicarious liability in Germany which are not relevant in the present context.

of contractors. Contractual ties to the injured party also lead to vicarious liability under § 278 BGB.[146]

2. Liability for people further up the food or feed production chain

In purely tort law terms, each player would be liable only for his own actions. There is no ground for vicarious liability in any of the relationships envisaged here. Defective feed or food can lead to full liability on each level, though players on the lower rungs of the ladder may attempt to recover any loss from those further up. **55**

3. Can someone further down the feed or food chain include someone further up it in a trial against him/herself?

German law allows (but does not require) defendants to include third parties in a lawsuit by third party notice, or *Streitverkündung*, if they can expect to recover from that party a loss resulting from a negative outcome of the initial case against them.[147] The third party may join the proceedings on the side of the defendant but can also choose to simply await the outcome. The judgment will in both cases be binding on the third party for the purposes of any subsequent action.[148] There is no obligation of the defendant to include a third party in this way nor can he escape liability by doing so. **56**

VI. Multiple tortfeasors

Several persons who are together responsible for the loss arising from an unlawful act are liable as joint debtors.[149] This rule covers situations in which a loss is jointly caused by several persons,[150] cases of uncertain/alternative causation,[151] and vicarious liability.[152] Absent any specific rules concerning the internal distribution of responsibility, § 426(1) BGB **57**

146 For more details see *J. Fedtke/U. Magnus*, ibid. at 119 ff.
147 § 72 ZPO.
148 § 74 (3) ZPO.
149 § 840 (1) BGB; see also §§ 32 (2) GenTG, 36a (4) GenTG (for nuisance), § 5 ProdHaftG.
150 § 830 (1) sent. 1 BGB.
151 § 830 (1) sent. 2 BGB.
152 *M. Fuchs*, Deliktsrecht (6th ed. 2006), 225.

determines that those liable will have to come up with an equal share of the required compensation. Recourse is possible on the basis of § 426(2) BGB if one of the parties comes up with the full amount. Besides referring to these rules of the BGB, some strict liability regimes contain additional provisions relating to multiple tortfeasors. § 32(3) GenTG thus specifies that the operator of a facility in which GMOs are modified remains liable beside any other person who might have caused the damage.

VII. Defences

1. Licence/permission to grow GM material

58 German law requires a licence to grow GM material and/or circulate products containing GMOs. Such licence, which may be granted under the GenTG or a comparable legal regime,[153] excludes liability under §§ 32 ff. GenTG but will not in itself constitute a defence for losses caused by the use of GMOs under product liability, pharmaceutical liability, or general tort law.[154] Specific forms of *use* (feed, food, pharmaceuticals) are, moreover, subject to separate licensing requirements, which are distinct from the permission to simply *grow* GM material. GMOs which are only approved for testing fall within the ambit of the GenTG. A licence to grow such material does not cover any other use or circulation.

2. Consent/assumption of risk

59 Consent or assumption of risk can, in principle, only become relevant as a factor limiting liability in situations which lie beyond the legalised use of GM products. An example would be the use of GM-based pharmaceuticals by test persons before these are admitted to the market (so-called *klinische Prüfung*).[155] Strict requirements have to be met in these cases, which include provision of comprehensive information to the test person. A special case of consent is specifically regulated by § 16b GenTG. The provision allows farmers with conventional plants to waive their statutory rights vis-à-vis neighbours with GM-crops. The effect of such consent is threefold – liability for cross-fertilisation is excluded; the produce of the consenting farmer will have to be labelled as "genetically modified" even if contami-

---

153 §§ 14 (2), 16 (2) GenTG.
154 § 37 GenTG.
155 §§ 40 ff. AMG.

nation levels stay below 0.9%; and consenting farmers will have to take all necessary measures to protect the interests of third parties.[156]

3. Third-party influence

Third-party intervention is not a defence under the relevant German strict liability regimes[157] but will subsequently provide defendants with a cause of action. Liability for fault is excluded by third-party behaviour unless negligence by the defendant contributed to the loss.[158] **60**

4. Prescription

German strict liability statutes follow, with slight modifications, the general rule contained in § 195 BGB which bars delictual claims after three years.[159] This *regular* prescription period starts at the end of the year in which the act occurred *and* the victim has – or could reasonably have – gained knowledge of both the facts giving rise to the claim and the identity of the tortfeasor.[160] Different *absolute* time limits exist for certain types of claims. Compensation for death, physical injury, the impairment of health, or the infringement of liberty is prescribed after 30 years regardless of any knowledge.[161] Claims following from the loss of property or economic loss are prescribed after 10 years following an injury[162] or 30 years following the act which causes the loss.[163] All other claims are barred after 10 years following an injury.[164] The regular prescription period for claims brought under the Product Liability Act starts at the time the victim gains knowledge of the loss, the defect, and the identity of the responsible producer.[165] In line with Art. 11 of Directive 85/374/EEC, rights under the Product Liability Act extinguish upon the expiry of a period of 10 years from the date on which the producer put into circulation the **61**

156 § 16b (1) GenTG; see no. 33 f.
157 *Fedtke/Magnus* (fn. 128) 164.
158 *G. Knerr* in G. Schlegelmilch (fn. 6), chap. 1 at no. 29 ff.
159 §§ 32 (8) GenTG; 17 UmweltHG; 9 USchadG (for internal recourse among several responsible polluters).
160 § 199 (1) BGB.
161 § 199 (2) BGB.
162 § 199 (3) no. 1 BGB.
163 § 199 (3) no. 2 BGB.
164 § 199 (4) BGB.
165 § 12 (1) ProdHaftG.

actual product which caused the damage unless the victim has in the meantime instituted proceedings.[166]

5. Other defences

62 §§ 16a and 36a GenTG as well as the new ordinance on GM-farming establish guidelines for the interpretation of the general rules on nuisance. Adherence to these standards will serve as a defence to claims for equitable compensation based on § 906 BGB.[167]

VIII. Remedies

1. Pecuniary compensation

(a) Bodily harm

63 German strict liability statutes contain specific provisions which determine the scope of compensation for death and bodily harm. These rules do not, in substance, deviate from the regular remedies.[168] That said, strict liability is usually limited by caps. § 33 sent. 1 GenTG thus limits compensation to € 85 million for all types of loss envisaged by § 32(1) GenTG. Several victims of the same event will only receive a quota if the total loss exceeds the cap.[169] Loss of life or physical injury under the Pharmaceutical Products Act will be compensated to a total of € 600,000 or an annual pension of € 36,000 for each victim; the total amount of compensation for all losses suffered by the same product is limited to € 120 million or annual pensions of € 7.2 million, which will be distributed *pro rata* if necessary.[170] Compensation for death or physical injury under the Product Liability Act is limited to a total of € 85 million for a single defective product or several products with the same defect; victims will again receive compensation *pro rata* if the total loss exceeds this amount.[171]

166 § 13 ProdHaftG.
167 See no. 34.
168 §§ 32 (4)–(6) GenTG; 86 f. AMG; 7–9 ProdHaftG.
169 § 33 sent. 2 GenTG.
170 § 88 AMG.
171 § 10 ProdHaftG.

(b) Property losses

The regular remedies will again apply subject to specific caps. The GenTG establishes a common overall cap for all types of losses mentioned in § 32(1) GenTG, which includes loss to property.[172] Compensation for loss to property is unlimited under the Product Liability Act but victims will have to bear the first € 500 themselves.[173] This rule, a consequence of Directive 85/374/EEC, was heavily criticised in Germany. Victims will often attempt to recover losses below this threshold on the basis of general tort law.[174] Loss under the WHG is unlimited. **64**

A rule addressing the compensation of property losses caused by GMOs specifically is § 32(7) GenTG. If a loss constitutes an impairment of nature or the landscape (*Beeinträchtigung der Natur oder der Landschaft*), restoration (i.e. compensation in kind) cannot be denied simply because the cost involved would exceed the value of the property. This is a modification of the general rule established by § 251(2) BGB. Proportionality between the cost of restoration and the value of the property will nevertheless be one of the factors in determining whether compensation can be restricted to money.[175] **65**

(c) Economic losses

The regular remedies apply. Both the *Gentechnikgesetz* and the BGB aim at full compensation either in kind or money.[176] Losses include future profits as far as they would probably have been accrued in the absence of the interference, the cost of any necessary decontamination, the cost of recalling products from the market, and increased expenditure caused by the search for new outlets or measures required to regain a particular producer status. Victims are generally required, however, to mitigate a loss.[177] This will include selling contaminated but marketable crops for a period of time or growing conventional crops until contaminated land is again suitable for organic farming again. These are, however, difficult issues as restoration of GMO-contaminated land – if possible – may take more than a decade. **66**

---

172 See no. 10.
173 § 11 ProdHaftG.
174 *Kullmann* (fn. 101) 175.
175 For more details see *Förster* (fn. 37) 79.
176 See no. 10.
177 *Schadensminderungspflicht*, § 254 BGB. See *Knerr* (fn. 158) chap. 3 at no. 40 ff.

(d) Harm to animals

67 Contaminated feed can lead to property loss if the animals suffer physical harm (e.g. death, illness, infertility, or lower productivity of milk)[178] and/or if it impairs the ability of the owner to use them in a particular way.[179] The latter can be the case if the owner intended to sell meat or produce (milk) under a particular label.[180] As noted above,[181] labelling requirements have recently been tightened. Food may thus not be sold as "free of GMOs" if GM feed was used at some point in the production process. This rule is relevant for dairy products if animals were, e.g., fed contaminated corn or soya. Past use of GM feed is allowed within certain limits but the owner will suffer a recoverable loss if his current or future production is affected. Calculation of an animal's value will include all market relevant factors (e.g. breed, age, sex, reproductive capacity or productivity).[182]

(e) Costs of disposal

68 The costs of disposing in some special form of contaminated produce or animals fed with GM feed are recoverable if such disposal is required by law or administrative order. This can be the case, in particular, if the GMOs in question were not yet licensed for general use.[183] The owner will, in any case, be obliged to limit the loss by using the cheapest form of disposal (which could simply be safe storage for a particular amount of time).

2. Non-compensatory damages

69 German law does not recognise punitive, exemplary or other forms of non-compensatory damages in this context.[184]

---

178 *Substanzbeeinträchtigung.*

179 *Beeinträchtigung der bestimmungsgemässen Verwendbarkeit.*

180 *Förster* (fn. 37) 62 f.; G. *Landsberg*/W. *Lülling*, in W. Eberbach/P. Lange/M. Ronellenfitsch (eds.), Recht der Gentechnik und Biomedizin (Band 1, Stand 3/2004), § 32 GenTG no. 47.

181 See no. 35 f.

182 On the calculation of patrimonial damage in general see U. *Magnus* (fn. 34) 96.

183 See, e.g., OLG Frankfurt am Main, Case 2 U 128/07 of 6 February 2009.

184 *Magnus* (fn. 34) 92.

3. Other remedies

The primary method of compensation under German law is restitution in kind (*Naturalrestitution*). § 249 sent. 1 BGB thus requires the tortfeasor to restore the state of affairs that would have existed without the intervention that led to the loss in question. In practice, however, damages are regularly awarded in money since most victims choose to be paid the amount that is necessary for restitution in kind (§ 249 sent. 2 BGB). **70**

4. Costs of pursuing a claim

(a) General cost rule

The losing party has to bear the costs of a civil trial.[185] Costs outside the context of a trial are recoverable as far as they are reasonable to pursue justified, or to defend against unjustified, claims.[186] Costs might also be split into percentages, depending on the outcome of a case. **71**

(b) Costs of establishing causation

The reasonable costs of establishing causation, e.g., by use of sampling/testing methods, expert opinions, or even private investigators, are considered part of the recoverable damage under § 249 BGB if the claimant wins the case.[187] **72**

5. Advance cover

The GenTG places potential tortfeasors under an obligation to guarantee the payment of possible future damages (so-called *Deckungsvorsorge*) caused by particularly dangerous facilities[188] or the setting free of GMOs for testing.[189] This obligation can be met either by third party insurance[190] or an indemnification guarantee/warranty provided by the state on the federal **73**

---

185 § 91 ff. ZPO.
186 *Knerr* (fn. 158) chap. 3 at no. 115.
187 Ibid. at nos. 112, 114 ff.
188 § 7 (1) nos. 2–4 GenTG.
189 § 36 (1) GenTG.
190 § 36 (2) no. 1 GenTG.

or provincial level.[191] These provisions of § 36 GenTG are currently dormant and have to be activated by ordinance. Farmers are not subject to these obligations, which do not cover the use of GMOs licensed for general circulation. A similar mechanism was under discussion in the context of the USchadG but not included in that Act.[192] Other strict liability regimes relevant in this context do not prescribe financial guarantees.

IX. Cross-border issues – Conflict of laws

1. Conflicts rules applicable before (or instead of) Rome II

74 Regulation (EC) 864/2007 is in force in Germany.[193] The Regulation also applies to claims based on nuisance.[194]

75 Previously, courts regularly applied the law of the country in which the tortfeasor had *committed the act*, as opposed to the general rule established by Art. 4(1) of Regulation (EC) 864/2007;[195] victims could, however, also opt for the law of the country *in which the loss materialised*.[196] This would in practice happen if the latter was more favourable to the plaintiffs.[197] The choice was limited to the early phases of the first instance trial.[198] If both parties had their regular place of residence in the same country at the time the act was committed, the law of that country would be determinative.[199] This corresponds to Art. 4(2) of Regulation (EC) 864/2007. As in Art. 4(3) of Regulation (EC) 864/2007, an escape clause allowed courts to deviate from these principles if the case in question featured a particularly close connection to that law.[200] The parties were, additionally, allowed to agree on the law applicable to the case after the event.[201] This corresponds to Art. 14(1)(a) of Regulation (EC) 864/2007. A rule allowing free choice of

191 So-called *Freistellungserklärung* or *Gewährleistungsverpflichtung*; see § 36 (2) no. 2 GenTG.

192 *Wagner* (fn. 126) 578.

193 Gesetz zur Anpassung der Vorschriften des Internationalen Privatrechts an die Verordnung (EG) Nr. 864/2007 of 10 December 2008, BGBl 2008 I 2401; Art. 3 Einführungsgesetz zum Bürgerlichen Gesetzbuch (EGBGB), BGBl 1994 I 2494; 1997 I 1061 as subsequently amended. In general see *A. Junker*, Die Rom II-Verordnung: Neues Internationales Deliktsrecht auf europäischer Grundlage, NJW 2007, 3675 ff.

194 Art. 44 EGBGB.

195 Art. 40 (1) sent. 1 EGBGB.

196 Art. 40 (1) sent. 2 EGBGB.

197 *K. Haag* in G. Schlegelmilch, Geigel. Der Haftpflichtprozess (23rd ed. 2001), 1551.

198 Art. 40 (1) sent. 3 EGBGB; §§ 275 f. ZPO.

199 Art. 40 (2) EGBGB.

200 Art. 41 EGBGB.

201 Art. 42 EGBGB.

the applicable law *before* the event if all the parties are pursuing a commercial activity[202] was not available under the EGBGB.

The *amount of compensation* enforceable under foreign law in German courts was limited by the basic (*ordre public*) principles of German tort law.[203] This is not possible under Regulation (EC) 864/2007. **76**

The law of the country in which the defendant's property was located was regularly applied to claims resulting from nuisance. Victims could again opt for the law of the country in which their own property was located.[204] This was, essentially, a reversal of the regular regime and opt-out rule established by Art. 7 of Regulation (EC) 864/2007. **77**

German law did not feature a separate cross-border regime for cases arising under product liability. **78**

2. Special regime for cross-border claims

At present there is no such special regime for bilateral or multilateral cross-border claims in Germany. **79**

X. Cases

1. Due to the adventitious presence of GMOs in a field, maize which is normally sold as conventional contains GMOs beyond the legal labelling threshold. This is not discovered before the final stage of the food production chain by the producer of taco chips. The whole production is lost since the supermarket chains refuse to accept delivery from the producer.

(a) Who can sue along the chain of distribution?

It is assumed that the producer of the taco chips (TC) took all possible steps to mitigate the loss by offering to label the chips and attempting (but failing) to sell them in this form to his regular (or other) customers. TC's loss might include lost profit (if he can prove that he would have otherwise sold the chips to supermarkets at a particular price), the produc- **80**

202 Art. 14 (1) (b) Regulation (EC) 864/2007.
203 Art. 40 (3) EGBGB. For details see *K. Haag* in G. Schlegelmilch (fn. 6), chap. 43 no. 68.
204 Art. 44 EGBGB in its pre-Rome II version.

tion costs, the cost of cleaning or decontaminating his production facilities (which may involve lost production time), the cost of attempting to sell the (labelled) batch to his regular or other customers, and, finally, the cost of disposal if there is no other way of utilising the chips. Other parties in the production line include the maize wholesaler (W) and any other intermediaries (OIs) involved in refining and bringing the maize from the farmer (F) to TC (e.g. farming co-operatives, mills, or independent freight companies). W and any OIs, too, might have to decontaminate their facilities or equipment. Finally, it is assumed that the GMOs in question are *allowed* in food products (given that TC attempted to sell the contaminated batch) and that the terms of the agreement between TC and W specified the delivery of conventional maize *below* the labelling threshold.

81 Contractual claims aside, the only basis on which TC, W, or any OIs can attempt to recover their losses are the fault-based rules of general tort law. Strict liability under § 32 GenTG is not available in this case – with the exception of hypothetical 1(d) below – since these GMOs may be used in food products.[205] Strict liability on the basis of the Product Liability Act is excluded in all variations of this case. The maize as such was not intended for private consumption. The taco chips did not reach any private consumer.[206]

82 § 823(1) BGB provides a delictual claim for the loss of TC against W on the basis of (general) product liability principles. The maize delivered by W, which would in practice be a mixture of unprocessed material *from a number of sources* rather than an isolated batch coming *only from the contaminated field*, was defective because it required labelling as GM produce. As things currently stand in Germany, it is very difficult to sell products which require such labelling. This is why food producers and wholesalers take great care to avoid high GMO contamination in all phases of a production/delivery chain. W must, however, have been at fault. A duty of care which was obviously violated here was effective quality control regarding the contamination of the maize. Case law of the Federal Supreme Court (*Bundesgerichtshof*, BGH) in other areas of fault-based product liability suggests a duty of both parties to ensure effective quality control, especially if the final product is potentially dangerous for the consumer. The buyer (TC) may, however, rely on the other party to perform adequate quality control if the seller's particular knowledge and facilities indicate that such controls will have

205 See no. 2.
206 See no. 44.

regularly been performed.[207] W is a professional wholesaler operating in a very difficult market (Germany) as far as GM products are concerned. On balance, it must thus be assumed that W was at fault in delivering the maize without performing sufficient quality control. The fact that the contamination was adventitious in this first hypothetical is not relevant since the GMO level was so high as to render labelling necessary.

The above principles will also apply to other parts of the production chain. **83**
F as the ultimate producer of the maize will eventually have to bear the loss caused by his produce.

(b) Would the case be solved differently if the GMO content was below the labelling threshold?

It is unlikely that there would be a "case" at all if the level of contamina- **84**
tion had remained below the ordinary labelling threshold unless TC had intended to sell the taco chips under a "GM-free" label. TC would be able to market the taco chips in the usual way and suffer no loss. W and any OIs would still have violated their duties to monitor the level of GMO contamination vis-à-vis their respective buyers but this would all the same not translate into any loss on subsequent levels of the production chain.

(c) Would the case be solved differently if the admixture was not adventitious, but occurred due to the disregard of segregation rules, for example?

A disregard for segregation rules on any level would leave liability unaf- **85**
fected from that point on but exonerate parties which handled the maize at earlier stages of the production chain. Assuming that a neighbour of F (N) had violated segregation rules concerning GM farming (e.g. safety corridors between his crops and F's property were too narrow), F (as the very first link in the production chain) would ultimately have a claim against N based on nuisance or § 823(1) BGB.[208]

---

207 BGH VersR 1975, 922 at 923; OLG Köln NJW-RR 1990, 414.
208 See no. 24 ff.

(d) Would the case be solved differently if the GMO found was not admitted for production in your jurisdiction?

86 While the above answers would hold true for this hypothetical, an additional basis for claims would be § 32(1) GenTG. Liability is strict in this case, which would circumvent the fault requirement discussed above. The operator of the facility which caused the contamination of F's field would ultimately be liable for any losses along the production chain.

(e) If the admixture had occurred on a non-GM field and it transpires that the GM seeds were blown from:
(i) neighbouring fields; or (ii) a truck passing by, would the farmer of the affected field be liable for all or part of the loss caused further down the distribution chain?

87 In terms of segregation rules, German law does not differentiate between a contamination caused by neighbouring fields or a truck passing by, though liability is arguably slightly stricter in the latter case.[209] The legal basis of claims against the neighbour or freight company would be different (nuisance or § 823(1) BGB and § 823(1) BGB respectively) but F would in any case be able to recover his loss. Obviously, the various scenarios discussed above (contamination above or below the labelling threshold? are the GMOs admitted for production or not?) would also affect this hypothetical. In general, liability for losses caused further down the production chain would ultimately rest with the neighbour or the freight company who/which violated segregation rules.

2. Twenty years after the sale of GM maize used for food products, it turns out that it has certain disadvantageous health effects for humans.

(a) Can the producers be held liable at this point for risks unknown at the time of growing the maize? Who would be liable?

88 The focus in this hypothetical is on losses caused to consumers. A possible defence in this case is the development risk argument established under

209 Cross-fertilisation must be *avoided* whereas contamination of other products must be *prevented* (see no. 34).

(strict) product liability.[210] This defence is not available to the person who applied for the licence allowing the general use of a particular GMO, which would usually be the seed producer.[211] Given that all the other conditions of a claim under § 1 ProdHaftG are fulfilled, liability would be strict. All other links in the production chain would be able to escape (strict) product liability by invoking the development risk defence. It seems unlikely that victims would be able to prove that any of the parties involved was at fault, which excludes liability under the general rules. Strict liability under the GenTG is not engaged as the GM maize will have been brought into circulation under a licence at the time.

(b) Can compensation already be claimed at a point when the negative health effects have not yet materialised, but are to expected to according to scientific expertise/mere rumours?

Any loss must have materialised; "early" claims based on scientific prognosis or mere rumours will not succeed under German law.[212] That said, victims could seek declaratory judgment on the merits of their (future) claims.[213] The fear of future negative health effects – if objectively established and of sufficient weight – could also trigger a recognised psychological illness and give rise to a (separate) claim under (strict) product liability against the producer of the seed. **89**

(c) Would it make any difference if the GM maize had only been in use for feed, causing harm to the animals, which may or may not cause harm to humans consuming the meat as well?

The uncertainty of negative health effects on consumers eating the meat of such animals would exclude claims under any of the headings discussed above until medical evidence proving harm emerges. The harm caused to the animals themselves would not be covered by (strict) product liability unless they are intended and predominantly used for *private* consumption.[214] In this case, farmers producing such feed for others would be able to invoke the development risk defence while seed producers would **90**

210 See no. 39 ff.
211 § 37(2) GenTG.
212 See no. 12.
213 See no. 13.
214 See no. 44.

in principle be liable. It seems unlikely that the owners of affected animals would be able to prove that any of the possible defendants were at fault, which excludes (fault-based) product liability under § 823(1) BGB.

3. The driver of a food logistics company discovers that a farmer, from where he regularly picks up agricultural products fails to obey mandatory segregation rules or food or feed hygiene standards, which may lead to the admixture of GM and non-GM produce sold and packaged separately by that farmer. Does he or his employer have a duty to warn, i.e. warn the recipient of the allegedly "non-GM" produce?

91 This is a question of liability for an omission/failure to act. Omissions can trigger civil liability under German law if the person in question is under a statutory duty to protect the interests of third parties; if the person accepted a position or function which involves the duty to protect such interests; if the person is in a close factual relationship with the potential victim; or if the person creates or has control over a source of particular danger for others.[215] None of these conditions are met here. The specific duties relating to GMOs specified by the GenTG do not include the duty of third persons to report the infringement of segregation rules by those handling GMOs. Finally, there is no obligation of the driver or his employer to report the matter under Art. 19 of Regulation (EC) 178/2002[216] or German laws regulating the safety of foodstuffs, feed, and household supplies.[217]

---

215 *Knerr* (fn. 158), chap. 1 no. 4.

216 Regulation (EC) No. 178/2002 of the European Parliament and of the Council of 28 January 2002 laying down the general principles and requirements of food law, establishing the European Food Safety Authority and laying down procedures in matters of food safety, OJ L 31, 01.02.2002, 1.

217 See no. 10.

Damage Caused by GMOs under Greek Law

Eugenia Dacoronia

I. General overview

1. Special liability or redress scheme for GMOs

In Greece there is no special liability regime which exclusively or specifically addresses the liability of GMOs. The relevant matters are dealt with under Law 1650/1986 on the protection of the environment, given that both art. 17 of the Joint Ministerial Decision H.Π.: 11642/1943/2002[1] (issued in implementation of Council Directive 98/81/EC, which modified Council Directive 90/219/EC on terms and conditions for the contained use of genetically modified micro-organisms, and of Council Directive 2001/204/EC) as well as art. 33 of the Joint Ministerial Decision 38639/2017/2005[2] (which implemented Directive 2001/18/EC on the deliberate release into the environment of genetically modified micro-organisms)[3] include a provision stating that the civil sanctions provided by art. 29 of l. 1650/1986 are imposed on any person who, by acting or omitting to act, violates the provisions of the said Ministerial Decisions.[4] 1

Art. 29 of l. 1650/1986 provides that: "Whoever, physical person or legal entity, provokes pollution or other degradation of the environment is 2

---

1 Government Gazette (FEK) Issue B? 831/2002.

2 FEK Issue B'1334/2005.

3 For the implementation and application of the Community rules on GMOs in the Greek legal order see *Ath. Takis*, The legal status of GMOs in the European Union and elements of the adaptation of the Greek law, Armenopoulos (Arm) 60 (2006) 1552–1556.

4 Also art. 12 of the Joint Ministerial Decision 95267/1893/1995, which implemented Council Directive 90/219/EC as amended by Council Directive 94/51/EC (FEK Issue B'1030/1995), subsequently replaced and abolished by art. 21 of the Joint Ministerial Decision H.Π.: 11642/1943/2002, contained a similar provision. The Joint Ministerial Decision 278787/2005 on Necessary Complementary Measures for the Implementation of Regulations (EC) No. 1829/2003 and (EC) No. 1830/2003 of the European Parliament and the European Council (FEK Issue B? 998/2005), however, includes only penal and administrative sanctions.

liable for damages, unless he proves that the damage is due to an act of God or it was the result of a third party's wilful act." Pursuant to said article, in order to establish liability, it suffices that there is an unlawful act or omission causing pollution or environmental degradation, damage and causation between the said act or omission and the damage. The defendant may assert the defences of act of God or the malicious act of a stranger, in order to be discharged of liability.

3 For more details see *E. Dacoronia*, Economic Loss Caused by GMOs in Greece, in: B.A. Koch (ed.), Economic Loss Caused by Genetically Modified Organisms (2008) 233–241, no. 1–24.

2. State liability

4 The Introductory Law of the Greek Civil Code (IntrLGCC) contains provisions (art. 104–106) concerning State liability and liability of legal persons of public law in general. Art. 104 IntrLGCC stipulates that the State is liable, according to the provisions of the Greek Civil Code (GCC) on legal persons, for acts and omissions of its organs regarding private law legal relations or related to the State's private patrimony. Art. 105 and 106 IntrLGCC provide that the State is liable for the damage that the organs of the State cause during the execution of their duties.

5 More particularly in art. 105 IntrLGCC, it is stipulated that the State is liable to pay damages for illegal actions or omissions of its organs during the exercise of the public authority that has been entrusted to them, unless the action or omission was made in breach of a provision existing in favour of the general interest. The culpable person is also jointly liable in parallel to the State, with the reservation of special provisions for the liability of ministers. According to art. 106 IntrLGCC, art. 104 and 105 IntrLGCC also apply for the liability of municipalities, communities or other legal persons of public law for actions or omissions of organs in their services.

II. Damage

1. Recoverable losses

6 Personal injury, damage to property, and moral damage can be recovered according to Greek liability rules. As to the extent of the recoverable damage, art. 297 ff. GCC states that damage includes the reduction of the value of the existing estate of the affected party (positive damage, damnum emergens), as well as the loss of profit (lucrum cessans). That which can be expected as probable profit in the usual course of events or by reference to the special circumstances and particularly to the preparatory measures taken, shall be reckoned as loss of profit (art. 298 GCC).[5] Given that it is highly likely that the GMO admixture may initially remain undetected and the consequences of the use of GMOs may come about in the future, it is accepted that future and indirect damage is also compensated for according to art. 29 of l. 1650/1986.[6]

Regardless of the compensation for damage to property, the court may award reasonable pecuniary compensation for emotional stress and strain, for damages to goods such as life, health, physical integrity, freedom, honour, etc. 7

Compensation, in principle, is paid in money (art. 297, subpar. 1 GCC). 8 Provision, however, is made, by way of exception, for the possibility of its payment in natura. Thus, sub-par. 2 of art. 297 GCC lays down that the court may, taking into consideration any special circumstances, order, in lieu of compensation in money, the restoration of the former state of affairs (status quo ante) if this is not contrary to the interests of the creditor. In the case of ecological damage, the provision of art. 297 GCC provides the legal basis so that the restitution in natura of the impaired element of the environment, to the extent that is possible, is achieved.

Economic damage also includes money expended in diminishing damage 9 by the person who sustained it.[7]

2. Pure economic loss

Pure economic loss is recoverable in Greek tort law, provided that the gen- 10 eral requirements for liability are fulfilled, namely that a causal link between the loss and the act or omission generating liability is established.[8]

5 For the notion of positive damage and loss of profit in Greek law (in English) see *M. Stathopoulos*, Contract Law in Hellas (1995) no. 305.

6 See *I. Karakostas*, Environmental Law (in Greek) (2nd ed. 2006) 522.

7 *Karakostas* (fn. 6) 521.

8 About civil liability for pure economic loss in Greece see (in English) *Th. Liakopoulos/G. Mentis*, Civil liability for pure economic loss in Greece, Revue Hellénique de Droit International (RHDI) 1/1998, 61 ff.; *K. Christodoulou*, Pure Economic Loss: Aspects of an Anglo-American Legal Issue under Greek Law, RHDI 2/1998, 599 ff.; *id.*, in: M. Bussani/ V. Palmer (eds.), Pure Economic Loss in Europe (2003) and (in Greek) *E. Zervogianni*, Pure economic loss of third persons, Kritiki Epitheorissi (KritE) 1/2003, 246 ff.

3. Mere fear of a loss

11 The Greek jurisprudence has not dealt yet with issues relating to GMOs, but we are of the opinion that if a case concerning GMOs were to be brought before the Greek courts, the latter would decide after taking into consideration the precautionary principle, as a large majority of them have done in several cases dealing with the risk of exposure to electromagnetic radiation emitted by mobile telephone base stations.[9] For example, in the injunction order 4531/2004 of the First Instance Court of Athens,[10] the plaintiffs living in the area of "Stathmos Larisis", which is one of the most densely populated areas of Athens, in their petition asked for an injunction order for an immediate removal of the mobile telephone base stations. The injunction order was granted on the basis of the precautionary principle. The plaintiffs had stated that they were suffering feelings of fear, worry and mental distress for the consequences the daily exposure to electromagnetic radiation, emitted by the mobile telephone base stations in question would have on their mental health and their environment, since, apart from their homes, antennas were also located in the vicinity of the base stations, at schools and colleges of their children.

12 Apart from the application of the precautionary principle in cases of daily exposure to electromagnetic radiation, it has recently been held[11] that the said principle, combined, however, with the principle of proportionality, should also apply to cases of non-checked toxic plant drugs, where a possibility of harm to inhabitants' health and in particular to the health of pupils at the adjacent vineyard, when sprayed with the toxic drugs, might exist.

---

9 Single Member Court of First Instance of Thessaloniki 13776/2002 Perivallon kai Dikaio (PerDik) 2002, 360 (followed by a note of M. *Kotzaivazoglou*); 16242/2003 Arm 2005, 1202; 9069/2005 published at the database NOMOS; 10165/2005 (not published); 10252/2005 (not published); 17599/2005 (not published); Multimember Court of First Instance of Thessaloniki 26223/2005 PerDik 4/2005, 614 (followed by a note of *Ap. Sinis*); Single Member Court of First Instance of Larissa 3867/2005, Dikografia (Brief) 2005, 557. The decisions of the One Member Court of First Instance of Patras 1558/1998 PerDik 2/2001, 247; 3421/2000 PerDik 1/2001, 88, of the One Member Court of First Instance of Herakleion 802/2003 Nomiko Vima (NoV) 2003, 1458 and the decision of the Court of Appeal of Patras 182/2001 PerDik 2/2001, 249 (followed by a note of *T. Nikolopoulos*) were the first decisions, not explicitly mentioning the precautionary principle, but actually implying it, as they founded their judgment on the probability of risks to human health from the electromagnetic radiation.

10 Arm 2005, 467.

11 Single Member Court of First Instance of Korinthos 2449/2008 Chronika Idiotikou Dikaiou (ChrID) Θ/2009, 124 followed by a note of *Z. Tsolakidis*).

Though fear of a loss triggered by GMOs (e.g. fear of developing an illness in the future) could easily, according to the above, lead to the issuing of an injunction prohibiting the use of GMOs, an action for damages for the mere fear of a loss triggered by GMOs without proof of actual damage would not be easily admitted according to Greek tort law, which requires an actual loss with respect to person or property in order for a claim for damages to be accepted. Therefore, mere fear of a loss triggered by GMOs will not easily constitute compensable damage.[12] This traditional approach, however, has begun to be criticised in Greece, as various diseases related to environmental degradation appear after decades, there is scientific uncertainty and it is extremely difficult if not impossible for the plaintiffs to establish the causal relationship between the disease and the exposure to dangerous substances; it is suggested thus that the only way out is to consider the risk per se as damage and to compensate it.[13] 13

4. Standard of proof

According to the general rules of the Greek Code of Civil Procedure (GCCP), the burden of proof lies with the plaintiff: the plaintiff is burdened with proving the elements of the rule of law he invokes. Causation is one of the preconditions required for the application of art. 29 of l. 1650/1986, which normally should be proved by the plaintiff. Therefore, if the GCCP is to be applied, the plaintiff should have to prove that the damage he sustained is the consequence of the presence of GMOs. As such a proof is difficult in cases of environmental damage, a reversal of the burden of proof is possible by adopting the position of the doctrine 14

12 Mere fear, however, has been taken into consideration for the adjudication of pecuniary satisfaction in a case of product liability. More particularly, the Athens Court of Appeal, in a decision confirmed by the Supreme Court (Areios Pagos [AP] 1051/2004 Arm 2006, 596), held that the pins found by the plaintiff in a standardized frozen cheese pie had intruded into the product during its production process, that there was no evidence that this was due to sabotage by any of the producer's employees, and that the only damage the plaintiff sustained causally connected to the product defect was the reasonable fear of eventual damage to her health. The Court consequently held that this fear caused the plaintiff mental pain and discomfort, due to which the producer, being strictly liable according to the provisions on product liability, had to pay pecuniary satisfaction amounting to approximately € 3,000 (GDR 1,000,000) after having taken into consideration the conditions of the tort and the general social and economic condition of the litigants.

13 *G. Balias*, The Risk per se: towards a widening of the civil environmental liability?, Lecture delivered at the Congress organized by the Piraeus Bar and the Piraeus Prefecture and held in Piraeus on 26 and 27 June 2009 under the general title "Environmental Liability: Doctrinal Development and Issues of Application".

according to which cases of ecological harm must be treated in the same way as cases of product liability as to the burden of proof.

15 For more details about how the burden of proof is distributed see *E. Dacoronia* (supra no. 3) no. 52–55.

16 As the presence of actual damage needs to be established by the plaintiff under Greek law, the plaintiff risks having his claim rejected by the Court as vague and unfounded should he not be able to prove the exact quantum of damage or if its proof is unreasonably difficult, as in the case of future losses.

5. Nominal losses

17 Nominal losses are not recognized.

6. Mass losses

18 There are no special provisions governing mass losses.

III. Causation

1. Uncertainty of merely potential causes

19 L. 1650/1986 has no special provision regulating the above issues. With regard to the matter of multiple causes, the general rules of the GCC apply and in particular art. 926 and 927 thereof, for which see in detail *E. Dacoronia* (supra no. 3) no. 56–61.

20 The GCC provides in its art. 300 that if the injured party has contributed by his own conduct to the causing or the extent of the damage he has sustained, the court may, at its free discretion, either not award compensation or reduce its sum. Art. 300 GCC applies in any case of damage caused either because of non-performance of a contract or as a result of a delictual act. Art. 29 of l. 1650/1986 does not expressly include the contributory negligence of the victim in the reasons for exculpation of liability, but it is accepted in doctrine[14] that art. 300 GCC also applies to risk liability cases.

---

14 *Karakostas* (fn. 6) 524 f.

2. Complex causation scenarios

Here also the above-mentioned art. 926 and 927 of the GCC would apply. There are no other special rules. 21

3. Force majeure

Art. 29 of l. 1650/1986 provides that: "Whoever, physical person or legal entity, provokes pollution or other degradation of the environment, is liable for damages, unless he proves that the damage is due to an act of God (force majeure) or it was the result of a third party's wilful act." Thus, the presence of force majeure excludes the liability of the tortfeasor. The definition of force majeure, however, is left to legal theory. 22

Two German theories regarding the range of force majeure have also been introduced in Greece: the objective (or absolute) theory on the one hand, which recognises a narrower range of force majeure by regarding only those events as force majeure that are by their very nature impossible to avert such as war, earthquake, revolution and other similar occurrences, i.e. events which are "external" to the defendant and the subjective (or relative) theory on the other hand, which widens the range of events of force majeure by including "internal" occurrences as well, i.e. also events that could not be foreseen and averted even by measures of extreme care and prudence on the part of the perpetrator.[15] Though the subjective theory seems to be prevailing in Greek jurisprudence, there are tendencies, particularly among scholars, to favour the objective theory. This applies in particular in cases of environmental liability, where it is suggested[16] that it is fairer that the keeper of a source of risk is exculpated only for events that lie outside of his sphere of influence. 23

4. Threshold to prove causation

Under Greek law, as a rule, the proof provided by the plaintiff must fully convince the court that the defendant caused the injury (full proof), i.e. so that no logical and learned man can seriously doubt it.[17] Probability is a 24

15 See in details M. *Stathopoulos*, Law of Obligations, General Part (in Greek) (2004) § 6 IV 2 no. 101 ff.; *id.* (fn. 5) no. 259.

16 *Karakostas* (fn. 6) 524.

17 *I. Tentes* in: K. Kerameus/D. Kondylis/N. Nikas (eds.), Code of Civil Procedure. Article-by-Article Commentary, vol. I (in Greek) (2000) art. 347 no. 1.

sufficient level of proof only as an exception, when e.g. provisionary measures are sought.[18]

5. Special rules on causation

25 See answer to Question 4 under II (no. 14–15).

IV. Types of liability

1. Fault liability

(a) Special rules governing fault

26 For an answer to this question see E. *Dacoronia* (supra no. 3) no. 62 f.

(b) Impact of specific rules of conduct

27 For an answer to this question see E. *Dacoronia* (supra no. 3) no. 64 f.

2. Product liability

(a) Development risk defence

28 The development risk defence was implemented into art. 6 § 8 e of l. 2251/1994 on the protection of the consumer. Thus, liability of the producer is excluded if he/she is able to show that the properties of the product could not be identified as defective according to the state of science and technology. There is no relevant Greek jurisprudence to our knowledge.

29 The EC Product Liability Directive (85/374/EC) was initially implemented by l.1961/1991. Instead of the development risks instituted by the Directive as an exculpation reason for the producer, the Greek legislator introduced, in l.1961/1991, a provision which contained an important deviation from the Community law. Despite the fact that the letter of the Community provision favours the application of objective criteria only, the letter

18 See, among others, G. *Rammos*/N. *Klamaris*, Synopsis of Procedural Law, vol. 1 (in Greek) (1998) 103; D. *Kranis* in: K. Kerameus/D. Kondylis/N. Nikas (eds.), Code of Civil Procedure. Article-by-Article Commentary, vol. II (in Greek) (2000) art. 690 no. 6.

of the relevant provision of l. 1961/1991 regarding the exculpation ("he excusably did not know neither could he have known"), referred to fault-based liability. Thus, the producer was not held liable if, without fault, he was not aware of the defect, which meant that culpability constituted a negative prerequisite for his liability.[19] The above provision of l. 1961/1991 breached the primary and fundamental principle of the liability of the manufacturer of defective products without culpability and was thus amended by l. 2251/1994, which introduced the development risk defence.

In order for the exculpation grounds in art. 6 § 8 e of l. 2251/1994 to be 30
applied, the producer must have been objectively unable – given the scientific and technological knowledge of the time – to discover the defect of the product at the time of its circulation. The allegation on the part of the producer that he could not have discovered the defect, because of his own diminished scientific and technological knowledge, is not excusable and does not constitute a ground for his exculpation.[20] If the defect could have been discovered given the international level of scientific and technological knowledge of the particular production branch, the producer is not excused for not having discovered the defect, provided, of course, that the international level of scientific and technological knowledge was accessible to the representatives of that particular branch.[21]

(b) Alternative routes

Whenever products which are already in circulation cause a loss, the gen- 31
eral tort law is applicable in addition to l. 2251/1994, as well as the provisions on sale in the cases where a contractual relationship between the producer and the victim exists. Every legal basis presents its own advantages and disadvantages. The general tort law regime on the one hand covers a broader field of application (it applies not only to consumers but to everybody, it compensates for damage to the defective product as well, etc.). The regime implementing Directive 85/374/EC on the other hand is more lenient to the consumer as it provides for strict liability, whilst the general tort law regime, even with the reversal of the burden of proof regarding causality and liability, is based on fault and the general burden of proof for all requirements of the claim lies with the plaintiff.[22]

19 *I. Karakostas*, Product liability (in Greek) (2008) 270.
20 Ibid.
21 Ibid., 271.
22 For the advantages of each regime (tort law regime and the one implementing Directive 85/374/EC) as well as for the concurrence of claims deriving from the producer's liability

(c) Impact of compliance with rules and regulations

32 The fact that the defendant has obeyed all rules and regulations governing his production process will not exonerate him from liability. Even under traditional fault rules, the respecting of rules and regulations will not necessarily provide an effective defence for the producer.

33 For more on the subject-matter see *E. Dacoronia* (supra no. 3) no. 64 f.

3. Environmental liability

(a) Implementation of the Environmental Liability Directive

34 The Environmental Liability Directive (2004/35/CE) has been implemented in Greece with the Presidental Decree (P.D.) 148/2009 (FEK Issue A'190/ 29.9.2009), long after prescribed in the Directive term for its implementation. Because of this late implementation the European Court of Justice, in its decision dated 19.05.2009, condemned the Greek State for not having implemented the said Directive within the prescribed term.[23]

35 The P.D. 148/2009 includes an article on financial security, in conformity with art. 14 § 1 of the Directive which provides for Member States to take measures to encourage the development of instruments and markets providing financial guarantees. More particularly, it provides in its art. 14 § 1 that the operators of the professional activities under the decree can use financial guarantees (private insurance as well as other forms of financial guarantees) in order to cover their liability under this decree, via appropriate economic and financial operators, including financial mechanisms in case of insolvency. In such a case the operator of the professional activity should take care that the terms of the insurance and of the other forms of financial guarantees safeguard the best possible coverage of the restitution of the environmental damage presented as probable at a reasonable cost and under reasonable preconditions.

36 According to art. 14 § 2 of the P.D. as from 1 May 2010 all professional activities covered by its Annex III shall be compulsorily submitted to a financial security system (private insurance and other forms of financial guarantees), via appropriate economic and financial operators, including

---

and from the provisions on sale see *I. Karakostas*, Law of Consumer's Protection (in Greek) (2004) 242 ff.

23 ECJ C-368/08, *Commission v. Hellenic Republic*, [2009] ECR I-00089.

financial mechanisms in case of insolvency. For each one of the above activities, the timetable of the above-mentioned compulsory submission, taking into consideration the European Commission's report, as provided by art. 14 § 2 of Directive 2004/35/EC as well as the ability of the economic market to provide for an integrated and economically possible disposition of the relevant financial security will be precisely specified by means of ministerial decisions.

According to art. 14 § 4 of the P.D., in the cases mentioned in art. 14, the **37**
Ministry of Economy and Economics, being the competent authority according to the law, specifies the amount of the financial security. This specification does not, in any case, determine the liability of the operator under this decree, and is made according to the extent, the kind and the amplitude of the damage that each activity may cause. The method for the specification of the above amount will be set out by a common decision of the Minister of Economy and Economics and the Minister for the Environment, Physical Planning and Public Works, and should rely on technical criteria which will guarantee an homogeneous estimation of the risk scenarios as well as of the relevant restitution expenses and should also assure a uniform delimitation of the essential insurance cover for each activity.

About the regime in force governing the insurance cover for environmen- **38**
tal damage in Greece see *E. Dacoronia* (supra no. 3) no. 77 f.

(b) Environmental liability regime beyond the scope of the Directive

There is no liability regime which exceeds the scope of the Directive and **39**
which covers environmental harm as such without the need to prove the infringement of a private interest as well.

Greek private law has not recognized the environment, at least explicitly, **40**
as a distinct legally protected individual right. The environment is not included among the "social goods", such as health, physical integrity or property, explicitly recognised and protected by the provisions of the Greek Civil Code. Thus, legal aid for an offence to the environment as such without an infringement at the same time of one of the above-mentioned legally protected private goods cannot be sought.[24] The environment may benefit only indirectly from civil law provisions, that is only

24 *I. Karakostas/E. Dacoronia*, Environment and Private Law in Greece, Annuaire International des Droits de l' Homme, vol. I/2006, 479.

within the scope of an infringed individual right whose remedy coincides with the objectives of environmental protection. In other words, it is through the infringement of specific individual rights that civil liability arises and, once these individual rights refer to environmental goods, the environment is also protected.[25]

(c) Claimants in cases of environmental harm

41 Environmental harm is only recoverable if it can be qualified as damage to the plaintiff's property pursuant to general tort law.

(d) Special liability regime for losses sustained by individuals

42 Reflecting a period when pollution of the environment was not a vital problem, the GCC did not include provisions especially devoted to the protection of the environment. Nevertheless, its provisions regarding:

43 a) the law of neighbours (art. 1003 ff. GCC),

44 b) the protection of common things, such as the air and the sea, and of things of common use, such as big lakes, rivers, etc., and

45 c) the protection of the personality (art. 57–59 GCC),

46 all of them read in the light of the Greek Constitution of 1975, which in its art. 24 introduces an express right respecting the environment, prove to be an adequate ground for the solution of legal problems arising from the pollution of the environment. Art. 1003 GCC states that: "The owner of an immovable shall be bound to tolerate the emission of smoke soot exhalations, heat vibrations or other similar side effects originating from another immovable to the extent that such effects do not prejudice substantially the use of the owner's immovable or result from a use which is common to the immovables of the region in which the offending immovable is situated." This article has to be interpreted in the light of the Constitution (art. 24) so as to be read in the following way: "Emissions even usual for the area are only then considered legal when they do not injure elements of the vital area".[26] On the grounds of protection of ownership as a right of enjoyment of a land in a healthy vital area,[27] the owner of such

---

25 *I. Karakostas/I. Vassilopoulos*, Environmental Law in Greece (1999) § 207.
26 See *Karakostas* (fn. 6) 416.
27 See *Karakostas* (fn. 6) 412.

land may be protected by means of art. 1003 GCC in combination with art. 1108 and 989 thereof. So the owner of land can sue his neighbour for emissions bringing against him the negatory action of art. 1108 GCC, on the basis of which he will ask the latter to cease the impairment and not to repeat the offence in the future, provided that the emissions severely impair the use of the land and are not usual for the area according to its economic characterization (for example, the area is not an industrial one). A similar action is given to the owner in his capacity as the possessor of the land according to art. 989 of the GCC. A parallel action for damages and compensation for moral harm may also be available under the tort provisions of the GCC.[28]

(e) Cartagena Protocol

The Cartagena Protocol of 2000 on Biosafety to the Convention on Biological Diversity was ratified by l. 3233/18.2.2004. Greece is generally eager to ratify international agreements regarding the protection of the environment and of biodiversity. **47**

4. Other strict liability regimes

There are no other strict liability regimes. **48**

V. Vicarious liability

1. Scope of vicarious liability

Art. 922 GCC introduces an exception to the rule that delictual liability is based on one's own fault, by holding the master or a person who has assigned to another the task of performing a service strictly liable for the prejudice caused illegally to a third party by the servant or the person assigned while performing his service.[29] It is long established by jurispru- **49**

---

28 See ibid., 441.

29 See relatively *Ap. Georgiadis*, Law of Obligations – General Part (in Greek) (1999) § 62 III no. 26–41; *P. Kornilakis*, Law of Obligations – Special Part I (in Greek) (2002) § 91 2, 533 ff.; *M. Stathopoulos*, Law of Obligations – General Part (in Greek) (3rd ed. 1998) § 7 II, 136 and the recent monography of *Z. Tsolakidis*, Contractual and delictual liability for assistants (in Greek) (2008). See also (in English) *E. Dacoronia*, Mass Torts: a Greek approach, RHDI 47 (1994) 88.

dence[30] that a contract between the master and the servant or the person assigned while performing his service is not necessary in order to establish liability according to art. 922; vicarious liability can be based on a purely factual or friendly or parental relationship or it can take place occasionally for only one particular act. Thus, it is of no importance how the auxiliary person has been employed or if his employment has been decided by another person and not by the principal. It suffices that the latter acted under the guidance and orders of the principal on how he must fulfil his duties.[31]

50 According to the Greek Supreme Court,[32] illegal and faulty behaviour on the part of the auxiliary person, which is causally connected with the resulting damage and an interior causal relation between the above-mentioned behaviour and the execution of the service assigned to the auxiliary person must exist for the establishment of vicarious liability according to art. 922 GCC. A relation between the principal and the auxiliary person exists when in the frame of a juridical or another, as mentioned above, relation between these two persons, the principal appoints the auxiliary person to perform a function and the latter is subject to the control or at least to the general instructions of the principal. Thus, this special relation may also exist in case of a contract for independent services.

51 Greek scholars, however, point out[33] that the application of the element of dependence may lead to unacceptable results, because either it narrows the field of application of art. 922 GCC or it unreasonably broadens it. As a result it might be better to use the criterion of the integration of the auxiliary performer in the field of the professional or social activity of the principal.

---

30 For which see *Kornilakis* (fn. 29) § 91 2 II 1, 537, fn. 13.

31 AP 121/2002 ChrID B/2002, 322, 323. For a brief summary of the facts and the judgment of the Court (in English) see *E. Dacoronia* in: H. Koziol/B.C. Steininger (eds.), European Tort Law 2002 (2003) 237 f., no. 21–23.

32 AP 1226/2007 Theory and Practice of Civil Law (Efarmoges Astikou Dikaiou, EfAD) 2008, 60 f.; AP 1362/2007 EfAD 2008, 62 f. For a brief summary (in English) of the facts and the judgment of the said decisions which referred to vicarious liability of medical centres for the liability of doctors see *E. Dacoronia* in: H. Koziol/B.C Steininger (eds.), European Tort Law 2007 (2008) 317 f., no. 26–29.

33 *M. Stathopoulos*, Law of Obligations. General Part I (in Greek) (1978) § 7 III, 220; *Georgiadis* (fn. 28) § 62 no. 34; *Kornilakis* (fn. 29) § 91 2 II 1,540, 541; *E. Poulou*, note under AP 1226/2007 EfAD 2008, 61 f.

2. Liability for people further up the food or feed production chain

Vicarious liability cannot be easily established within the feed and food production chain, as the latter consists of independent contractors not acting under the guidance of each other. However, if the respective independent contractor actually was called in by the principal in the performance of his contractual duties, vicarious liability may arise. 52

3. Can someone further down the feed or food chain include someone further up it in a trial against him/herself?

In principle, someone further down the feed or food chain is not able to bring someone further up into a lawsuit filed against himself, neither is he obliged to include himself, nor can he escape liability by shifting the loss. Provided that vicarious liability is applicable, the principal who has compensated the plaintiff is able to take recourse against the agent according to art. 927 GCC regulating the recourse among multiple tortfeasors.[34] It is accepted in Greek theory and jurisprudence[35] that this recourse can be exercised by one of the tortfeasors even before the payment of the compensation, through the mechanism of art. 69 par. 1 GCCP. By this mechanism, the said tortfeasor can, while the trial for damages against him is still pending, introduce his co-tortfeasor as an additional litigant to the opened trial. This is achieved by the filing of a relevant petition parallel to the filing of the recourse collateral action. The reason for this procedural possibility is that, with the exercise of the recourse right, the co-tortfeasor is considered to be a procedural guarantee of the tortfeasor who was the initial defendant in the trial for damages. 53

34 *Kornilakis* (fn. 29) § 91 2 III 1, 546.

35 See *Ap. Georgiadis* in: Ap. Georgiadis/M. Stathopoulos, GCC, vol. IV (in Greek) (1982) art. 927 no. 27; *Ath. Kritikos*, Damages from Car Accidents (in Greek) (4th ed. 2008) 181, no. 45; *N. Nikas* in: K. Kerameus/D. Kondylis/N. Nikas (eds.), Code of Civil Procedure. Article-by-Article Commentary vol. I (in Greek) (2000) art. 88 § 1, 201; AP 1138/2007 NoV 55, 2444; Patras Court of Appeal 68/2006 Achaiki Nomologia (AchNom) 23 (2007) 618; 259/2004 AchNom 21 (2005) 245; Athens Court of Appeal 1969/2003 Arm 2005, 359; Athens Court of Appeal 7038/2002 Elliniki Dikaiosini (EllDni) 2007, 516.

VI. Multiple tortfeasors

54 According to art. 926 GCC, if several persons unlawfully and culpably caused harm, or if several persons are responsible for certain harm, they are liable jointly and severally. The same applies if more have acted simultaneously or successively and it cannot be ascertained whose action caused damage.

55 According to art. 927 GCC, the tortfeasor who has paid compensation to the injured person according to the previous article has a right of recourse against the other tortfeasors. The court allocates the damage among the tortfeasors according to the degree of the fault of each single tortfeasor. If this degree cannot be ascertained, the damage is divided between all equally.

VII. Defences

1. Licence/permission to grow GM material

56 As there is no specific rule regarding GMOs, the general rules apply. It is clear in the Greek legal order that abiding by administrative provisions does not suffice for the exclusion of fault and, therefore, for the exoneration from liability.[36] The bearer of a possible source of risk for the environment must take all measures of precaution and safety required and not only the ones that are specifically prescribed by administrative provisions. The latter merely define the minimum standards to which the said bearer must comply and, therefore, compliance with them does not result in exoneration from liability.[37]

57 If the polluter has acted in conformity with the law and has also taken all measures of providence and care, then he is not liable for any damage which may occur.[38] In cases of release of GMOs, however, if the releaser has acted in conformity with the Ministerial Decisions on GMOs, he

36 See relatively AP (in full bench) 146/1940 Themis NA', 417; AP 343/1968 NoV 16, 943.

37 *Karakostas* (fn. 6) 475 f., 524.

38 It must be noted here, however, that it has been suggested (*E. Dacoronia*, Emissions and damage to the environment from the operation of an enterprise under licence from the competent authority (relation of art. 1003, 914 of the GCC), PerDik 1997/1, 22 f.) that in such a case, the damage must be covered for reasons of equity, by analogy with other provisions of the GCC (e.g. art. 387, 675 § 2, 918 etc.) which provide for such a possibility (i.e. they recognise a claim for reasonable damages at the discretion of the Court if this is dictated by good faith and equity, even if the activities that caused the damage are legal).

must be held liable for any damage which may occur even if he has taken all measures of providence and care due to the uncertainty of the risk caused by the GMOs.

2. Consent/assumption of risk

If the victim was aware that there was a significant risk to his health in consuming GM products and, nevertheless, he knowingly consumed them, then art. 300 GCC on contributory negligence might apply. For the time being, however, GMOs and GM products are considered to have no adverse effects on human health, thus the knowing consumption of these products by the victim will have no impact on the tortfeasor's liability. **58**

3. Third-party influence

Art. 29 of l. 1650/1986 which, as stated above in the general overview, also applies to GMOs cases, provides that: "Whoever, physical person or legal entity, provokes pollution or other degradation of the environment, is liable for damages, unless he proves that the damage is due to an act of God (force majeure) or it was the result of a third party's wilful act." Thus, if the operator invokes and proves that the damage was caused by a wilful act of a third party, the causal relation between the dangerous operation and the damage is interrupted and the operator is discharged of liability.[39] **59**

4. Prescription

Where the claim is being brought for product liability under the regime introduced by the 1985 Directive, then the three year limitation period, together with the ten year pre-emption period, will apply (art. 6 § 13 of l. 2251/1954). The same limitation period will also apply if the claim is being brought under tort law, as art. 14 § 5 of l. 2251/1994 provides that the provisions of ordinary law regarding prescription and pre-emption which are more favourable to the consumer do not apply. However, in case of claims accumulation, which are not part of the harmonised law, such as the consumer's claims for the satisfaction of his moral harm or **60**

39 *Karakostas* (fn. 6) 524.

pain and suffering, or the claim for restitution of the damage which was caused to the defective product itself, the prescription is to be calculated exclusively according to the relevant tort law provision of the GCC.[40]

61 The harmonisation sought with the Directive and the law that implemented it does not allow the shortening of the above-mentioned cut-off period by choosing as applicable law the one that provides a shorter prescription period. Thus, the same deadlines of art. 6 § 13 of l. 2251/1954 apply for the compensation claim of art. 6 even in case of concurrence with the contractual liability of the producer-salesman for which a two-year prescription period from the delivery of the product to the purchaser is provided (art. 554, 555 GCC as replaced by art. 1 § 1 of l. 3043/2002).

5. Other defences

62 According to art. 29 of l. 1650/1986, if the operator invokes and proves that the damage was due to an act of God (force majeure), the causal relation between the dangerous operation and the damage is interrupted; thus, the operator is discharged of liability.[41]

63 Regarding how force majeure is understood under Greek law, see answer to Question 3 under III (no. 22 f.).

VIII. Remedies

1. Pecuniary compensation

(a) Bodily harm

64 There are no special rules. The regular remedies apply.

(b) Property losses

65 Same answer as above.

---

40 *I. Rokas*, Product Liability (in Greek) (1997) 304 f.; *Karakostas* (fn. 19) 282 f.; Thessaloniki Multi Member Court of First Instance 23048/2004 Arm 2005, 1035.

41 *Karakostas* (fn. 6) 524.

(c) Economic losses

There are no special rules on compensating economic losses caused by GMOs. 66

(d) Harm to animals

According to Greek law (art. 948, 950 GCC), animals are non-fungible movable things[42] in the sense of property law. Consequently, if an animal is hurt or killed by a third party, the owner of the animal is entitled to claim compensation according to the general rules governing property damage. The Greek courts seek to compensate the victim for the entire loss suffered. The animal's (e.g. the cow's) potential to produce milk, meat and offspring is included in the calculation of the commercial value of the animal. The estimated cost, however, of feeding the animal during its usual life expectancy, that the farmer no longer has to bear because of the damaging event, is also calculated and reduces accordingly the damages awarded.[43] In case of death of animals (e.g. sheep) for which State subsidies are given, the relevant amounts lost for each animal are also recoverable.[44] For the loss of the animals, a reasonable amount for compensation of the moral harm of the farmer is also adjudicated.[45] 67

If the animal's value is reduced by the harm caused by the contaminated feed, then the farmer will be entitled to compensation for the difference between the present market value and the market value if the animal had not been harmed. 68

The mere fact that an animal eats contaminated feed does not constitute harm to the animal or, therefore, recoverable damage for its owner. However, if it can be proven that its market value is reduced for this reason, then the owner will be entitled to compensation for this reduction in value. 69

42 See F. *Dimakou* in: Ap. Georgiadis/M. Stathopoulos (eds.), GCC, vol. V (in Greek) (1985) art. 950 no. 4.
43 Patras Court of Appeal 852/2005 Achaiki Nomologia (Ach Nom) 22 (2006) 741.
44 Patras Court of Appeal 919/2004 Ach Nom 21(2005) 641; 852/2005 (fn. 43) 741.
45 AP 901/2008 NoV 56, 2466; Patras Court of Appeal 852/2005 (fn. 43).

(e) Costs of disposal

70 The Greek courts seek to compensate the victim for the entire loss suffered. Consequently, the costs sustained by disposing of the contaminated production/animals fed with GM feed would be recoverable, as a causal link between the damage (disposal costs) and the wrongful behaviour of the tortfeasor can certainly be established.

2. Non-compensatory damages

71 As it is a general principle of Greek law that damages have an exclusively compensatory character, punitive damages are not recognised. However, the Supreme Court held,[46] though not unanimously but with two dissenting views, that a judgment of a US court by which punitive damages were adjudicated on grounds of contractual liability, due to the fact that the debtor fraudulently breached the contract, did not violate the national public order and could be declared enforceable in Greece.

3. Other remedies

72 Compensation, in principle, is paid in money (art. 297, subpar. 1 GCC). Provision, however, is made, by way of exception, for the possibility of its payment in natura. Thus, subpar. 2 of art. 297 GCC lays down that the court may, taking into consideration any special circumstances, order, in lieu of compensation in money, the restoration of the former state of affairs (status quo ante), if this is not contrary to the interests of the creditor. In the case of ecological damage, the provision of art. 297 GCC provides the legal basis for the restitution in natura of the impaired element of the environment, to the extent that is possible, to be achieved.

46 AP 17/1999 (in full bench) EllDni 40, 1288 = DEE (Dikaio Etaireion kai Epicheiriseon, Law of Companies and Enterprises) 2000, 181.

4. Costs of pursuing a claim

(a) General cost rule

Art. 173–193 GCCP regulate how the expenses related to the litigation are to be borne by the litigants. As a general rule, costs are recoverable on the basis of the "loser pays" principle (art. 176 GCCP).[47] 73

Litigation expenses are characterised as judicial, encompassing the Court expenses (payable to the State), and extra-judicial, encompassing the amounts payable to lawyers, bailiffs etc. as fees and expenses.[48] The claim for the payment of the expenses is a claim of substantive law, which depends, however, on the fulfilment of the conditions set out by the above-mentioned relevant provisions of GCCP.[49] Greek jurisprudence consistently holds[50] that said expenses cannot be recuperated on the basis of art. 914 GCC but only on the basis of the above-mentioned provisions of the GCCP. 74

The Greek Supreme Court has held[51] that it derives from art. 914 GCC that in case of a tort, the tortfeasor is obliged to restitute the damage which is causally connected to the event that caused it. The expenses the party that sustained the damage voluntarily incurs in order to support his civil claims against the tortfeasor, such as the judicial and extra-judicial expenses before and during the litigation, are included in this damage, as long as they fall within the protective ambit of art. 914 GCC; the judicial expenses that can be claimed on the basis of art. 176, 178, 181, 184 and 189 GCCP cannot be recuperated on the basis of art. 914 GCC. Otherwise, according to the Court, the relevant provisions of the GCCP which regulate the matter of judicial expenses, to the exclusion of other provisions which could ground a relevant claim, would be set aside. 75

---

47 For an analysis of art. 176 GCCP see, among others, G. *Orfanidis* in: K. Kerameus/D. Kondylis/N. Nikas (eds.), Code of Civil Procedure. Article-by-Article Commentary, vol. I (in Greek) (2000) 412 ff.

48 See, among others, *Orfanidis* (fn. 47) Introductory Remarks to art. 173–193 GCCP no. 1.

49 *Orfanidis* (fn. 47) Introductory Remarks to art. 173–193 GCCP no. 3.

50 Athens Court of Appeal 6590/2003 Episkopissi Emporikou Dikaiou (EpiskEmpD) 2004,162; 4027/1978 Arm 1979, 21; Piraeus Court of Appeal 1290/1996 Epitheorissi Dikaiou Polykatoikias (EDPol) 1998, 176; Creta Court of Appeal 133/1971 Arm 1971, 1001.

51 AP 1345/2008 and AP 1609/2007 published at database NOMOS.

(b) Costs of establishing causation

76 As mentioned hereinabove, the extra-judicial expenses before and during the litigation are included in the damage, as long as they fall within the protective ambit of art. 914 GCC. Thus, the costs of establishing causation (e.g. the costs of detecting GMOs at farming level) could be recovered under Greek law, provided they were necessary (art. 189 GCCP).

5. Advance cover

77 There is no provision which obliges the operator who uses or releases GMOs to obtain advance cover, be it by contracting third party liability insurance or by other appropriate means. Neither the provisions of the Joint Ministerial Decision Η.Π.:11642/1943/2002 nor those of the Joint Ministerial Decision 38639/2017/2005, the only provisions regulating GMOs in Greece, provide for an obligation of the operator who uses or releases GMOs to obtain such advance cover.

IX. Cross-border issues – Conflict of laws

1. Conflicts rules applicable before (or instead of) Rome II

78 According to art. 4 § 1 of the Rome II Regulation which entered into force on 11 January 2009 (art. 32 thereof), an international claim for damages based on general tort law is to be treated pursuant to the law of the state where the damage occurred (lex loci damni). Under the Greek conflict of laws rules (art. 26 GCC), the law of the state where the tort was committed (lex loci delicti commissi) had to be applied.[52]

2. Special regime for cross-border claims

79 There is no GMO fund regime foreseeing provisions for cross-border claims.

---

52 For art. 26 GCC and the Rome II Regulation see *Sp. Vrellis*, Private International Law (3rd ed. 2008) 249 ff.

X. Cases

1. Due to the adventitious presence of GMOs in a field, maize which is normally sold as conventional contains GMOs beyond the legal labelling threshold. This is not discovered before the final stage of the food production chain by the producer of taco chips. The whole production is lost since the supermarket chains refuse to accept delivery from the producer.

(a) Who can sue along the chain of distribution?

According to Greek law, without the proof of actual damage there is no compensation claim. In the case above the only party who actually would suffer a loss along the chain of distribution is the taco producer, because he would not receive any money for his production from the supermarket chains which justifiably refuse to accept delivery of the contaminated taco chips. Delivery of the contaminated taco chips would constitute a defective or wrong delivery according to the contract between the supermarket chains and the producer which provided for delivery of conventional maize. As the burden of proof lies in principle with the plaintiff, the taco producer would have to find out who originally caused the contamination of the maize with GMOs and to prove his fault if damages are sought according to tort law rules. This might not be easy or even possible to achieve in the case of adventitious presence of GMOs in a field. As a result, the taco producer might not receive any compensation for his loss at all. **80**

(b) Would the case be solved differently if the GMO content was below the labelling threshold?

In such a case the product would not be considered defective, the maize would be treated as conventional and the supermarket chains would violate their contractual obligations if they refused delivery of the taco chips. **81**

If the affected farmer would face difficulties in selling the maize or would have to sell it at a lower prize because of the existence of GMOs even below the labelling threshold, then he could seek compensation provided, as above, that the person who originally caused the contamination of the maize with GMOs could be identified and his fault proven, which is difficult or even impossible to achieve in the case of adventitious presence of GMOs in a field. **82**

(c) Would the case be solved differently if the admixture was not adventitious, but occurred due to the disregard of segregation rules, for example?

83 Yes, the case would be solved differently if the admixture was not adventitious but occurred due to the disregard of segregation rules. Fault-based liability would be applicable and the taco producer would be able to sue the person who was liable for the disregard of segregation rules under general tort law, provided that he succeeded in establishing the tortfeasor's fault.

(d) Would the case be solved differently if the GMO found was not admitted for production in your jurisdiction?

84 No, as explained under (a), the supermarket chains would again refuse delivery of the taco production, the taco producer would not receive any money for his production and compensation under the fault-based tort law regime would depend on the taco producer's ability to establish the fault of the person who caused the contamination of the maize with the GMO.

(e) If the admixture had occurred on a non-GM field and it transpires that the GM seeds were blown from: (i) neighbouring fields; or (ii) a truck passing by, would the farmer of the affected field be liable for all or part of the loss caused further down the distribution chain?

85 The farmer would not be liable for all or part of the loss caused further down the distribution chain under tort law because of lack of fault on his part. Under the strict liability regime of art. 29 of l. 1650/1986 he would not be held liable either if the above incidents were characterised as force majeure events.

2. Twenty years after the sale of GM maize used for food products, it turns out that it has certain disadvantageous health effects for humans.

(a) Can the producers be held liable at this point for risks unknown at the time of growing the maize? Who would be liable – the seed producer/farmer/food producer/distributor/ etc.?

There would be no liability of the producers under l. 2251/1994 regulating product liability for two reasons: a) the development risk defence (art. 6 §8 e of l. 2251/1994) and, in any case, b) the period of limitation (ten years after the product has been put into circulation according to art. 6 § 13 of l. 2251/1994). There would be no liability under the general tort law provisions either, as explained (see hereinabove under VII answer to Question 4 no. 60), the ten year limitation of l. 2251/1994 applies here also; anyway, the answer would have been the same even if art. 14 § 5 of l. 2251/1994 did not exist and the limitation period of art. 937 GCC had applied, as the said article provides for a 20-year limitation period starting from the commission of the tort. **86**

(b) Can compensation already be claimed at a point when the negative health effects have not yet materialised, but are expected according to scientific expertise/mere rumours?

As Greek law requires actual damage in order for damages to be awarded, in principle no compensation could be claimed as long as the negative health effects have not yet materialised. **87**

For more on the issue of whether mere fear can constitute compensable damage see answer to Question 3 under II (no. 13). **88**

(c) Would it make any difference if the GM maize had only been in use for feed, causing harm to the animals, which may or may not cause harm to humans consuming the meat as well?

In this case the farmer would be able to claim only for the harm to his animals and for his resulting loss of profits. A claim for compensation for damage to health is grounded only if there is proof of actual damage to health. **89**

3. The driver of a food logistics company discovers that a farmer, from where he regularly picks up agricultural products fails to obey mandatory segregation rules or food or feed hygiene standards, which may lead to the admixture of GM and non-GM produce sold and packaged separately by that farmer. Does he or his employer have a duty to warn, i.e. warn the recipient of the allegedly "non-GM" produce?

90 The behaviour of a person, being a requisite of tortious liability, can be either a positive act or an omission. An omission gives rise to an obligation to pay damages when it is illegal,[53] i.e. when the tortfeasor omitted to act even though he was obliged to by law, by contract, by good faith according to the prevailing social understanding, because of his previous behaviour or by the general spirit of law. The principle of good faith – encountered in art. 200, 281, 288 GCC – entails the straightforwardness in transactions displayed by an upright and prudent person.[54] In the case presented here, however, it would seem very difficult to argue that the principle of good faith imposes on the employee or his employer any duty to warn the recipient of the produce of the failure by the farmer to respect segregation rules, because, in principle, only the person who by *his/her own* behaviour causes a dangerous situation is obliged, according to the principle of good faith, to take all the necessary protective measures, according to the circumstances, in order to avoid the occurrence of damage.[55]

53 For the illegal omission see, among others, *Georgiadis* (fn. 29) § 60 III no. 28–33; *Stathopoulos* (fn. 15) § 15 IV 3 no. 44.

54 See indicatively AP 821/2004 ChrID D/2004, 985.

55 See *Stathopoulos* (fn. 15) § 15 IV 3 no. 44.

Damage Caused by GMOs under Hungarian Law

Attila Menyhárd

I. General overview

1. Special liability or redress scheme for GMOs

The most comprehensive genetic technology related legislation in Hungary is the Act No. XXVII of 1998 on genetic technology activity as it is amended[1] by the Act no. CVII (further referred to as the Genetic Technology Act). This Act provides for a special liability regime for genetic technology activity in general[2] as well as for liability for damage caused as a result of incomplete segregation of GM and traditional crop production. As a general rule, § 27 of the Genetic Technology Act provides that as genetic technology activity may imply considerable hazards, the strict liability regime for abnormally dangerous activities (§ 345 ff. of the Hungarian Civil Code) shall be applied to liability for damage caused by genetic technology activity. A similar regime is established for liability for incomplete segregation. § 21/D subparagraph 5 and 6 of the Genetic Technology Act provide that for liability for damage caused as a result of incomplete segregation of GM and traditional crop production, § 345 and § 346 of the Hungarian Civil Code (the strict liability regime for dangerous activities) are to applied. Under this strict liability regime that is to be applied for liability for GMOs in the course of genetic technology activity in general, the tortfeasor (the person considered operator of the genetic technol- 1

1 The amendment, which came into effect on 22 December 2006, establishes the special liability regime for incomplete segregation of GM crop production from traditional crops. The main aim of this amendment was the implementation of the European Union Recommendation no. 2003/556/EEC of 23 July 2003 and provision of proper regulation for coexistence of GM and traditional crops.

2 I.e. damage caused by genetic technology activities such as establishing an institution (e.g. a laboratory) that performs genetic technology activity, modification of genes, utilisation of gene-manipulated micro-organisms in closed systems, emissions, export, import, putting the output of genetic technology activity into circulation and elimination.

ogy activity) has to prove that the cause of the damage was an unavoidable one (i.e. force majeure) that fell outside the scope of the genetic technology activity in order to exonerate himself from liability.

2 If, however, the victim as the owner or user of the neighbouring land has consented in written form to the growing of genetic plants according to § 21/C of the Genetic Technology Act, the general fault-based liability regime provided in §§ 339–342 and § 344 of Hungarian Civil Code is to be applied.

3 Thus, it is not the losses caused by GMOs as such that are covered by specific liability regimes but rather the losses caused by genetic technology activity. Losses caused by genetic technology activity are covered by the special strict liability regime of the Hungarian Civil Code for liability for dangerous activities, except liability for incomplete segregation if the victim as the owner or user of the neighbouring land has given his written consent to the growing of genetic plants according to § 21/C of Genetic Technology Act, as for such cases the general rules of fault-based liability are to be applied. The specific regime is provided for genetic technology activities but not for damage caused by GMOs in general. Where damage has been caused by GMOs, but not through activity defined as genetic technology activity according to the provisions of the genetic technology regulation, the "normal" fault-based liability is to be applied except for the product liability regime.

4 Under the regime of strict liability for abnormally dangerous activity covering liability for losses caused by GMOs in the course of genetic technology activity too, the operator of the abnormally dangerous activity shall be liable for damage caused by such an activity, and the operator may exonerate himself only by proving that the cause of the damage fell outside the scope of the dangerous activity and was unavoidable or that the victim was the one who caused the damage wrongfully. § 346 of the Hungarian Civil Code (subpars. 1–4) provides that if damage is caused by two or more persons through activity that involves considerable hazard, the general rules and regulations governing liability shall apply to their relationship with one another. If the cause of damage is not attributable to either of the parties, but derives from a malfunction that occurred within the realm of activity involving considerable hazard performed by one of the parties, that party shall be liable for paying damages. If the cause of damage is a malfunction that occurred within the sphere of both parties' activity involving considerable danger and, furthermore, if such malfunction cannot be attributed to one of the parties, each party shall, since individual responsibility cannot be established, bear liability for his own loss.

According to the basic rule of liability establishing the fault-based liability regime, (§ 339 of the Hungarian Civil Code), the person causing damage unlawfully to another person shall be liable for damages and he can exonerate himself from liability by proving that he acted as was generally expected under the given circumstances. The "under the given circumstances generally expected behaviour" (i.e. the required standard of conduct) is a flexible standard allowing great leeway for the courts in risk allocation. This regime – except in cases of vicarious liability – makes the person causing the damage liable. The strict liability regime for abnormally dangerous activities provided in § 345 of the Hungarian Civil Code and applicable to liability for genetic technology activity too, implies six basic differences as compared to the fault-based liability regime. Beyond these differences, the general rules and principles of – fault-based – liability are to be applied to liability for abnormally dangerous activities too and they may also be applied together in the same case.[3] **5**

The most striking difference appears in the prerequisites of exoneration. While under the fault-based liability regime the tortfeasor may exonerate himself by proving that he acted in compliance with the required standard of conduct, under the strict liability regime for abnormally dangerous activities the operator running the abnormally dangerous activity may exonerate himself from liability only by proving that the cause of the damage fell outside the scope of the dangerous activity *and* was unavoidable or that the victim was the one who caused the damage wrongfully. **6**

The second difference is that the person liable under the strict liability regime for abnormally dangerous activities is the operator of the activity, who is not necessarily the direct tortfeasor. The concept of the operator is a flexible one; there is no general definition of operator, instead the courts use a system of criteria for assessing and establishing, case by case, who the operator is. Normally, but not necessarily, the owner is held to be the operator, as the owner is assumed to exercise the control over a thing or an activity, but if the permanent control[4] of the activity was exercised by another person, this person may be deemed to be the operator and in certain cases the question of in which person's interest (including the power of making decisions concerning it)[5] a certain activity has been performed is decisive. These criteria, i.e. ownership, control and interests are applied in court practice in order to allocate and shift the risks of the given activity **7**

3 Statement no. PK. 38. of College of Civil Law of the Supreme Court.
4 Statement no. PK. 40. of College of Civil Law of the Supreme Court.
5 Supreme Court, P. törv. III. 20 970/1986. sz. – BH 1987., 437. sz.

and in each case they may be accorded different weight. The decisive factor in court practice, however, seems to be the power of control; liability for abnormally dangerous activity is generally shifted to the person who had the opportunity of preventing harm by making decisions concerning the activity or intervening in it generally.[6]

8 The third difference is that in order to establish the liability of the operator under the strict liability regime for abnormally dangerous activities, the link of causation must be established between the harm that occurred and the dangerous activity itself. Thus, the operator shall be liable if the harm was the consequence of the abnormally dangerous nature of the activity (i.e. resulted from it).[7] In special forms of liability, such as the liability for abnormally dangerous activities, the *conditio sine qua non* is not an explicit precondition of liability, insofar as the liability is not established by natural causation but is shifted according to an implied obligation of certain persons, for example, to keep the dangerous activity safe for others. The liable person is defined through the regulation – e.g. the person who shall be treated as running the dangerous activity – without the general test of causation.[8] The attempt of the theory to provide a general and unified explanation of liability – reflected also in the Civil Code – lays stress upon legal causation instead of natural causation because, in these special forms of liability, natural causation in a strict sense (*conditio sine qua non* or but-for test) is not a precondition of liability. Causation shall be established between the dangerous character of the activity and the damage.

9 The fourth difference is that while normally the Civil Code allows the contractual limitation of liability and also exclusion of liability in tort,[9] under the strict liability regime for abnormally dangerous activities, § 345 subpar. (3) of the Hungarian Civil Code makes any exclusion or limitation of liability null and void with the exception that this prohibition shall not apply to damage caused to a thing.

10 The fifth difference is the specific regulation provided in § 346 of the Hungarian Civil Code and specified above in no. 2 for allocating damage if two

6 *K. Benedek*, in: Gy. Gellért (ed.), Commentary to the Civil Code (Complex, 2008) § 345 no. 6.

7 *K. Benedek* (fn. 6) § 345 no. 7.

8 Causation is established in these special forms of liability only on the level of theoretical explanations [e.g. the caretaker is a cause of the harm caused by the child insofar as she did not act in caretaking as was generally expected. *F. Petrik*, A kártérítési jog (1991) 30].

9 With the limits that any contractual clause shall be null and void if it limits or excludes in advance liability for damage proceeding from wilful or gross negligence; injury to life, physical integrity, or health; or the consequences of a crime (§ 342 subpar (1) of the Hungarian Civil Code).

or more abnormally dangerous activities coincided and the damage shall be apportioned among them in their relationship.

11 The sixth difference is the special limitation period of three years provided in § 345 subpar (4) of the Hungarian Civil Code for claims established on the basis of strict liability for abnormally dangerous activities. If the passing of this limitation period of three years prevents the victim from claiming damages under the strict liability regime for abnormally dangerous activities, he may still have a claim on the basis of fault-based liability on the ground of the same damaging event provided the general limitation period of five years also applicable for claims of damages under the fault-based liability regime have not passed yet.[10]

12 The same rules and principles are suggested in the New Hungarian Civil Code concerning strict liability for especially dangerous activities except that the new text would specify the criteria to be applied in the course of establishing who the operator (i.e. the person liable under the strict liability regime for especially dangerous activities) of the dangerous activity shall be. According to the New Hungarian Civil Code,[11] the person (i) maintaining the dangerous activity, (ii) having disposal over the dangerous activity, (iii) in the interests of whom the dangerous activity was performed, (iv) the person obliged to implement specific preventive measures concerning the dangerous activity, (v) the person qualified by the law as operator of the dangerous activity, shall be deemed the operator of the dangerous activity.[12] This suggested provision does not attempt to change the existing law but only to implement the criteria developed in court practice for determining the operator also on the level of the Civil Code.[13] It also has been suggested that if there are more than one operators of the dangerous activity, they shall be jointly and severally liable *vis-à-vis* the victim.[14]

13 The scope of regulation provided by the Genetic Technology Act including the provision concerning liability covers the production and distribution of GM-products as genetic technology activity in general as well as damage caused in the course of growing genetically modified crops beside tradi-

10 *K. Benedek* (fn. 6) § 345 no. 13.

11 As a result of overall revision of the Civil Code of 1959 (the basis of existing private law today) the New Hungarian Civil Code has been passed as the Act no. CXX of 2009 on the Civil Code. It is still not in force and the date of its coming into force is uncertain as the separate act which is to put the New Hungarian Civil Code into force is under a revision before the Constitutional Court of the Hungarian Republic.

12 § 5: 490 subpar (1) of the New Hungarian Civil Code.

13 Motivation to § 5: 490 of the New Hungarian Civil Code.

14 § 5: 490 subpar (2) of the the New Hungarian Civil Code.

tional crop production. Risks resulting from actual or feared GMO presence in non-GM crops are covered by this liability rule only insofar as they are the result of incomplete segregation of GMO crops from neighbouring traditional plants cultivated. There is no other compensation regime that covers these kinds of risks.

14 As the application of the strict liability regime for abnormally dangerous activities is imposed on those who are engaged in genetic technology activity as defined by the regulation, in cases where damage was caused by GMOs but the tortfeasor was not engaged in genetic technology activity (e.g. producer of traditional crops selling products containing GMOs a result of adventitious mixture of GM and traditional seeds) the liability for GMOs is to be covered by the general fault-based liability regime. Under this regime, in order to be exonerated from liability, the tortfeasor shall prove that he acted as was generally expected under the given circumstances (i.e. complied with the required standard of conduct).

2. State liability

15 In Hungary there is no specific state liability regime to be applied for such losses; the normal rules and principles concerning state liability established in court practice are to be applied for losses covered by the liability regime of genetic technology regulation too, this practically leads to sovereign immunity. The Hungarian Civil Code provides regulation for public liability as a specific form of vicarious liability. Liability for damage caused within the jurisdiction of government administration shall be established only if common legal remedies could not abate the damage or if the aggrieved person resorted to the ordinary legal remedies in order to avoid damage without success. Unless otherwise provided by legal regulation, these provisions shall also be applied to liability for damage caused by judges and public prosecutors within their official activity (§ 349 of Hungarian Civil Code). Neither this approach nor the system seems to be changed in the New Hungarian Civil Code. The New Hungarian Civil Code would preserve this system. §§ 5:524–5:526 of the New Hungarian Civil Code specify three cases of liability for exercising public power including (i) liability for improper decision-making in the course of exercising executive power, (ii) liability for failure of judges and public prosecutors and (iii) liability for unfair or unreasonably slow court procedure. These forms of liabilities are limited to exercising executive power and are basically fault-based. Hungarian court practice rejects the claims for damages against legislative bodies and refuses to establish liability for improper

legislation.[15] From this it follows that the state as such shall not be the subject of strict liability for abnormally dangerous activities and the state cannot be liable for compensating damage caused by market players through the application of genetic technology.

II. Damage

1. Recoverable losses

As there is no specific regulation covering recoverable losses under the strict liability regime for abnormally dangerous activities to be applied to liability for genetic technology activities or in the context of liability for losses caused by GMOs in general, the general rules covering compensable losses under liability in tort are to apply. In the Hungarian system of tort law, the concept of damage is tightly linked with the concept of damages. According to § 355[16] of the Hungarian Civil Code, the tortfeasor who is responsible for the damage shall be liable for restoring the original state, or, if this is not possible or if the aggrieved party refuses restoration on a reasonable ground, he shall compensate the aggrieved party for pecuniary and non-pecuniary damage. Compensation must be provided for any depreciation in value of the property belonging to the aggrieved person (*damnum emergens*) and any pecuniary advantage lost due to the tortfeasor's conduct (*lucrum cessans*) as well as compensation of the costs necessary for the attenuation or elimination of the pecuniary and non-pecuniary loss suffered by the victim. From the principle of full compensation generally accepted in Hungarian tort law, it follows that all damage must be compensated regardless of the nature of damage (i.e. whether it was *damnum emergens* or *lucrum cessans,* or whether the harm was caused in property, in person or it was an economic loss) or the degree of fault (provided it was attributable to the tortfeasor). This corresponds to the traditional view according to which[17] damage is the depreciation in the wealth of the aggrieved person, including actual damage (depreciation of value in property), lost profit, the costs of attenuation or elimination of damage and non-pecuniary damage. **16**

As far as compensation of non-pecuniary loss is concerned, there are strong arguments in Hungarian legal theory that, in the context of non-pecuniary damages, one cannot speak of damage in the traditional sense **17**

15 Supreme Court, Legf. Bír. Pf. V. 24.375/1998. sz. – BH 2002., 264. sz.
16 Subpar (1) and subpar (4).
17 G. *Eörsi,* A polgári jogi kártérítési felelősség kézikönyve. (1966) no. 205.

of this word. The theoretical starting point of this idea – followed by the courts as well[18] – is that awarding non-pecuniary damages is a special sanction for wrongful interference with others' personality rights. That is why in the context of non-pecuniary damages, the damage as a precondition of liability is replaced with the fact of wrongful interference with personality rights of the victim and the aggrieved person shall be entitled to non-pecuniary damages without proving any actual harm, costs or loss in the traditional sense of the word.[19]

18 The New Hungarian Civil Code[20] does not bring any changes or amendments to the previous provisions in either in its structure of regulation or in its regulatory method (i.e. in its definition of damage and its definition of damages as compensation for damage) and – with the exception of small clarifications and the abandonment of the concept of non-pecuniary damages replacing it with a special form of indemnity for unlawful interference with personality rights – keeps the same regulatory system.[21] One of the most significant amendments that has been suggested in the draft of the New Civil Code as regards the existing law is the above-mentioned abolition of non-pecuniary damages and the replacement of this special kind of damages with a special indemnity as a direct monetary sanction of wrongful interference with personality rights (similar to the German *Schmerzensgeld*).[22]

2. Pure economic loss

19 Pure economic loss as an independent category of damage is not known in the Hungarian legal system, even though it is a very important type of loss in tort law. The main conceptual feature of pure economic loss is that it is a loss without antecedent harm to the plaintiff's person or property, which is not a consequential loss in the same patrimony in which property has been damaged and which is not the loss of the plaintiff, who as person has been injured.[23] Pure economic loss is a "harm not causally consequent

---

18 Supreme Court, Legf. Bír. Pf. III. 26.339/2001. sz. – EBH 2003. 941. sz. [EBH = Official Collection of the Decisions of the Supreme Court of the Republic of Hungary. Decisions published in this collection are selected and highlighted decisions of the Hungarian Supreme Court.].

19 *T. Lábady*, A nem vagyoni kártérítés újabb bírói gyakorlata (1992) 31 and *F. Petrik*, Kártérítési jog (2002) 74.

20 Fn. 11.

21 § 5:478 and § 5:480 of the New Hungarian Civil Code.

22 § 2:90 of the New Hungarian Civil Code.

23 *M. Bussani/V. Palmer*, The notion of pure economic loss and its setting in: M. Bussani/ V. Palmer (eds.), Pure Economic Loss in Europe (Cambridge 2003) 5.

upon an injury to the person (life, body, health, freedom or other rights to personality) or to property (tangible or intangible assets)."[24] The problem of pure economic loss is the problem of limitation of liability and mostly it is a problem of legal causation in case law and in legal theory. In Hungarian legal theory – summarizing the impact of limitation doctrines in the court practice – *Eörsi* exposed that the court may dismiss a claim for damages or refuse full compensation if the harm was unforeseeable to the tortfeasor (foreseeability doctrine); it was beyond rational probability, was atypical or unique (adequate causality); was beyond the normal consequences and was too unexpected; if the harm as the consequence of the tortfeasor's conduct was too remote; if in the causal link the interference of an unexpected cause altered the normal foreseeable consequences and contributed to the causing of the harm; if the damage is within the normal risk imputed to the aggrieved party; or if it would be disproportionate considering the amount of damages and the degree of fault.[25] This theoretical starting point of *Eörsi,* the main authority in tort law in Hungary until today, suggests that pure economic loss shall not be recoverable and this seems to be the mainstream position in Hungarian tort law even though in certain cases the Supreme Court seems to tend to allow compensation for certain types of pure economic loss.

"Ricochet loss"[26] was, according to the illustration provided by *Eörsi,*[27] a **20**
loss not to be compensated on the ground of its being too remote but nonetheless a relatively recent judgment of the Hungarian Supreme Court accepted this type of claim. A sales representative suffered a car accident which was caused negligently by another car driver. The sales representative was on his way to conclude a prepared contract in the name of his employer (the plaintiff) with a business partner of theirs. The concluding

24 *H. Koziol,* Compensation for Pure Economic Loss from a Continental Lawyer's Perspective, in: W. H. van Boom/H. Koziol/C. A. Witting (eds.), Pure Economic Loss (2004) 141 ff.

25 *Gy. Eörsi,* A közvetett károk határai, in: Emlékkönyv Beck Salamon születésének 100. évfordulójára (Budapest, 1985) 62 ff.

26 Cases where a "physical damage is done to the property or person of one party and that loss in turn causes the impairment of the plaintiff's right." This is a three-players' scene where a "direct victim sustains physical damage of some kind, while the plaintiff is a secondary victim who incurs only economic harm."

27 *Eörsi* (fn. 25) 62. In this the employer sends his employee (a mechanic who has special skills in repairing certain machines) to a factory located in another part of the country. On his way to the railway station, a car runs down the mechanic. According to Eörsi, the driver of the car shall be liable to the employee to compensate his lost earnings (salary, etc. for the period he is unable to work) but not to the employer for the loss resulting from the stoppage of the factory because of the failure or further delay of repair. The reason for the limitation here is that the economic loss suffered by the employer is beyond the normal consequences and too unexpected for the tortfeasor.

of the contract failed because the accident prevented the sales representative from getting to the place of contracting. The Supreme Court ascertained that if the sales representative had concluded the contract in the name of his employer, his employer would have had a certain income. The Court held that the unrealised net income, which the employer would have had from the performance of the contract if the contracting had not been frustrated through the accident, is an economic loss of the employer. The driver who caused the accident of the sales representative has caused this economic loss. On this ground, the Court held the driver liable for the economic loss of the employer and ordered the defendant (the insurer of the driver who caused the accident) to pay the lost net income as compensation to the plaintiff.[28]

21 To another category of pure economic loss called "transferred loss" belong those cases where the harm caused to the primary victim is shifted to another person (the secondary victim). In these cases, it is the contractual or statutory obligation which results in the secondary victim taking the loss of the primary victim upon himself.[29] In Hungarian court practice and literature, it is not a special tort situation. If the party is obliged to bear the loss (e.g. on the grounds of insurance) of another, the right of recourse is usually statutorily (if the obligation is imposed by statute) or contractually available to him.

22 The speciality of a further type of pure economic loss, "closures of public markets, transportation corridors and public infrastructures", is that here the loss "arises without a previous injury to anyone's property or person" and usually public restraints are involved in these cases.[30] There are not too many precedents for these types of cases, but both the theory and the practice seem to be ready for the limitation of liability. In a case from 1964, a Hungarian court dismissed the claim of a plaintiff who claimed compensation for his additional costs from the use of a longer route when a road had been closed because of a car accident that had been caused by the defendant. The court pointed out that the defendant could not

28 Supreme Court, Legf. Bír. Pfv. VIII 20.295/1999. sz. – BH 2001. no. 273. It is notable that the defendant was the liability insurer of the tortfeasor and there is a tendency in court practice that the courts are more willing to order compensation if the risk is shifted to an insurance company.

29 *M. Bussani/V. Palmer* (fn. 23.), 12.

30 *M. Bussani/V. Palmer* (fn. 23.) 12. Van Boom categorises these cases as "interference with resources" and attaches here also the cable cases. See *W. H. van Boom*, Pure Economic Loss: A Comparative Perspective, in: W. H. van Boom/H. Koziol/C. A. Witting (eds.), Pure Economic Loss (2004) 26.

reckon with the possibilities of this harm.[31] In a decision from 1989, the Supreme Court[32] established that the owner of a roadside restaurant who suffered loss as a result of a decrease in the number of guests because of the closing of a route shall be entitled to claim damages to compensate the loss from the operator of the car which caused the accident making the closure of the route necessary. In another decision,[33] however, the Supreme Court rejected the claim of a sport club for compensation of the costs of replacing the sportsman they lost when he died in a car accident caused by the defendant's vehicle.[34] According to an illustration of *Eörsi,* if a bridge is wrecked because of the conduct of the defendant, the plaintiff shall not be compensated for the loss he suffers because of the traffic detour. The reason for the limitation in these cases is that traffic may be detoured for very different reasons and its occurrence is a normal risk which everyone has to bear on his own. *Eörsi* seems to share the view that in these cases the defendant shall be liable for causing the risk itself but not for the realisation of it.[35]

There are cases where the aggrieved party suffers harm as a result of reli- **23**
ance on data, information or professional services given by the tortfeasor in a situation where the tortfeasor provides such on the basis of a contract with another party but not with the plaintiff.[36] According to Hungarian tort law, these cases seem to fall under the normal liability test without special limitations, at least as far as only a limited number of possible plaintiffs are involved. If a lawyer causes harm, e.g. by composing an invalid contract, he shall be liable towards his clients for breach of contract; towards other parties the lawyer shall be liable on the ground of torts.[37] On the basis of the proportionality doctrine, the compensation would be limited if the person who provided the false information or otherwise

31 The decision was not a Supreme Court decision but a first instance decision, which has not been appealed by the plaintiff and as such may be taken into account only with reservation as reference. *F. Petrik* (fn. 8) 31.

32 Not reported.

33 Not reported.

34 *K. Benedek* (fn. 6) § 345 no. 7.

35 *Eörsi* (fn. 25) 65. If the person who has to use the diversion suffers damage in an accident he cannot be compensated on the basis that he would not have been involved in an accident if he had not been forced to use the alternative route, because the link of causation is abnormal and is outside of the ordinary probability.

36 *M. Bussani/V. Palmer* (fn. 23) 13; see also: *van Boom* (fn. 30) 19. A typical illustration here may be the facts of the English cases *Hedley Byrne & Co. v. Heller & Partners* [1964] AC 465 on the liability for negligent misstatements (See for detailed analysis: B.S. Markesinis/S.F. Deakin (eds.), Tort Law (3rd ed. 1993) 86 ff.) and on the liability for professional services toward third parties *White and another v. Jones and others* [1993] All ER (CA) 481.

37 *Eörsi* (fn. 17) no. 254.

caused harm to third parties outside the contractual relationship acted with only a low degree of negligence. Presumably this would be the case if investors and market operators sued the accountant who provided falsely calculated and published balance sheets which the buyers relied on before they decided to buy shares.

24 To sum up, pure economic loss is not a special type of damage (or loss) in Hungarian tort law and cases of pure economic loss are not addressed under a common heading in Hungarian tort law theory and practice. The cases known as "pure economic loss" are addressed as causation problems and one cannot say that there is a general policy for restricting or rejecting claims in pure economic loss cases, albeit in theory and literature a strong limitation is suggested and such approach seems to be followed – but not in every respect – by the courts in general too. The court practice is, however, neither clear nor consequent in this respect as there are some decisions where the Supreme Court has awarded compensation for pure economic loss. It is not clear to what extent the examples where the Supreme Court awarded compensation for pure economic loss result in changes of the original position of theory and practice of rejecting such claims. Court practice is, on the other hand, consistent in that the amount of damages must be proven by the plaintiff,[38] and there is only the possibility to award so-called general damages if it is *per se* impossible to prove the amount of damages. If the plaintiff fails to prove the amount of damages (the exact loss) despite it being objectively possible, the claim will be rejected.

3. Mere fear of a loss

25 Hungarian court practice does not seem to accept that fear itself may be a compensable harm. Mere fear may not be considered a material damage and the generally accepted starting point is that wrongful interference with personality rights may be a ground for non-pecuniary damages with the only exception being compensation for the lost chances of recreation and the lost holiday experience as well as the missed entertainment and relaxation of a holiday abroad as a result of breach of a travel contract.[39] From this it follows that in order to award non-pecuniary damages as

38 Supreme Court, Legf. Bír. Gf. VI. 30.036/2002. sz. – BH2003. 249. sz. Supreme Court, Legf. Bír. Gfv. II. 30. 016/2001. sz. – EBH2001. 544. sz., Supreme Court, Legf. Bír. Pfv. III. 23.402/1998. sz.- BH2000. 541. sz.

39 Supreme Court, Legf. Bír. Pfv. VIII. 23.243/1996. sz. – BH 1998. 278. sz.

compensation for mere fear, fear would have to be pre-qualified as wrongful interference with personality rights of the victim. Hungarian court practice – even though there are decisions of lower courts showing some inclination to move to this direction[40] – does not seem to accept such extension of the concept of personality rights or the concept of non-pecuniary loss. If the fear results in a normally compensable loss, e.g. depreciation of value of land (e.g. because of assumed radiation from a working mobile phone transmission tower built on the neighbouring land)[41] or in a psychiatric illness (as non-pecuniary loss),[42] this loss shall be compensated but mere fear itself has not been accepted as a compensable loss in the practice of the Hungarian Supreme Court so far.

Thus, according to the principles which seem to be outlined in the court practice, mere fear of a loss triggered by GMOs is not a recoverable loss. If this fear, however, results in an otherwise compensable material or non-pecuniary loss, this otherwise recoverable loss shall be recovered. **26**

4. Standard of proof

There is no specific standard of proof with respect to establishing losses envisaged by this report. The main principle is that the plaintiff has to prove the fact of the loss he suffered as well as the extent and amount of it according to § 164 subpar (1) of the Hungarian Civil Procedure Act. Normally, this means that – apart from awarding general damages as addressed below – the court will award damages only if the loss has been proven completely by the plaintiff. As, however, under the system of evidence in Hungarian civil procedure law – with the main exception of certain qualified documents – evidence presented to the court by the parties shall be evaluated on the basis of free judicial discretion, the courts may establish the result of evidence on a relatively flexible basis. This is supported by an explicit rule concerning claims for damages provided in § 206 subpar (3) of the Hungarian Civil Procedure Act, which gives the courts a special discretionary power in establishing the sum to be paid as damages if the loss suffered by the victim cannot be established according to the proof (expert opinions, etc.) presented to the court. In such cases, the judges may establish the sum to be paid as compensation as they think **27**

40 As the decision of the Supreme Court refers to the decision of the court of 2nd instance in Legf. Bír. Pfv. III. 20.911/2007. sz. – EBH 2007. 1691. sz.

41 Supreme Court, Legf. Bír. Pfv. III. 21.543/2007. sz. – BH 2008. 211. sz.

42 Supreme Court, Legf. Bír. Pfv. III. 21.334/2007. sz. – EBH 2007. 1694. sz.; Supreme Court, Legf. Bír. Pfv. III. 21.147/2005. sz. – BH 2007. 6. sz.

fit. Under this regime of free weighing up and considering of evidence provided in § 206 subpar (1) and (3) of the Hungarian Civil Procedure Act, the courts may establish the result of evidence on probabilities[43] in the past as well as in the future. The authorisation provided to the courts in § 206 subpar (1) and (3) of the Hungarian Civil Procedure Act for a very wide discretionary power in weighing up, considering and establishing the results of evidence as well as in establishing the sum to be paid as compensation if (and only if) the results of evidence do not result in certainty, is very clear. Even if there are some decisions suggesting that certain types of loss or damage – like an allowance as compensation of salary to be lost in the future[44] – may be compensated only if the future loss had been established without any doubts, there are not enough reported decisions to allow us establish the limits of the discretionary power provided in § 206 subpar (1) and (3) of the Hungarian Civil Procedure Act.

28 According to § 359 of the Hungarian Civil Code, if the extent of the loss – at least partly – cannot be precisely calculated, the person responsible for causing the damage can be compelled by court to pay a general indemnification that would be sufficient to provide the aggrieved person with full compensation. General damages shall be awarded only if – and insofar as – it is impossible *per se* to prove the loss suffered by the victim. The fact that under the given circumstances it is complicated, difficult or expensive to prove the loss suffered shall not be enough ground for awarding general damages. Awarding general damages shall not be used as a tool for relieving the victim of the burden of proof of the loss he claims.[45] The rationale behind awarding general damages is to prevent the tortfeasor from being relieved of liability simply by the fact that he caused a loss that cannot be proven in its amount.

5. Nominal losses

29 Nominal or symbolic losses are not recognised so far in Hungarian tort law. There are no provisions of the Hungarian Civil Code or the Civil Procedure Act addressing such losses (either as such or as damages) but it has been established in court practice that symbolic damages shall not be awarded under Hungarian law. The plaintiff is entitled to claim less than full compensation but symbolic or nominal damages shall not be awarded

---

43 *Eörsi* (fn. 17) no. 301.

44 Supreme Court, Legf. Bír. Pfv. VI. 20. 545/1994. sz. – BH 1995. 639. sz.

45 Statement no. PK. 49. of College of Civil Law of the Supreme Court.

to relieve the victim of the burden of proof of the loss he suffered. If the victim cannot prove the loss he suffered, he shall not be entitled to claim symbolic or nominal damages either.[46]

6. Mass losses

There are no special rules in Hungarian tort law for mass losses and the court practice has not developed such guidelines or principles either. 30

III. Causation

1. Uncertainty of merely potential causes

According to the general liability rule of the Hungarian Civil Code (§ 339), the wrongdoer shall be liable if he causes the harm. Thus, according to the general rule of liability, the but-for test or *conditio sine qua non* is accepted as the first necessary step for establishing liability.[47] If a link of natural causation cannot be established, a precondition of liability is missing. One cannot be liable for damage which one did not cause. Theory and practice use the but-for test[48] which requires that, in order to establish liability, the harm could not have occurred without the tortfeasor's conduct. Establishing the causal link is a question of facts of the case which have to be proven by the plaintiff seeking compensation for the loss he suffered, even though legal theory and practice focus on problems of determining the relevant cause and risk allocation instead of natural causation. Within the general form of liability, if the aggrieved person (the claimant) cannot prove[49] the causal link between the defendant's conduct and the harm, the court will dismiss the claim. According to the general (or basic) rules of liability, one cannot be liable if one's activity or omission was not a necessary cause (a *conditio sine qua non*) of the harm suffered by the claimant. 31

Thus, the starting point is that the burden of proof of the causal link between the defendant's conduct and the harm suffered rests on the claim- 32

46 Supreme Court, Legf. Bír. Pf. V. 24.697/2002. sz. – BH 2004. 464. sz.

47 *G. Eörsi*, Kötelmi jog általános rész (8th ed. 1988) 269.

48 *F. Petrik*, A kártérítési jog (fn. 8.) 27.

49 As far as the preconditions of liability are concerned, Hungarian court practice takes a very strict line on the burden of proof. It must be taken into account that the civil procedure rules are based on an unbound system of evidence where the court is not bound in the evaluation of the proofs. The conviction of the court is decisive.

ant. If the victim could have suffered the damage alternatively by virtue of an actor or by a hazard and the claimant cannot prove that the actor caused the victim's loss, his claim shall be rejected. The burden of proof is regarded very strictly by Hungarian courts. If the cause of the harm is unknown, the claimant cannot establish the liability of the defendant on the ground of general rules of liability. That may be why this problem is neither treated in the literature nor arises in the court practice.[50]

33 The courts in certain cases may solve the problem by establishing multiplicity of tortfeasors and holding possible multiple tortfeasors as jointly and severally liable *vis-à-vis* the victim but only where the danger was an indirect cause of the damage, i.e. the but-for test can be satisfied relating to the danger. The starting point of the regulation and practice is that the claimant has to prove that the defendant caused the harm. In cases of possible alternative tortfeasors', however, it would be impossible to prove who caused the harm and the damage would result in no compensation.[51] This need for proof or *Beweisnotstand* would leave the damage uncompensated in cases where the damage and fault are obvious and the narrow circle of possible tortfeasors is known. The principle of prevention could also be undermined through rejecting the claim in such cases. The court may solve this problem by declaring the defendant one of the multiple tortfeasors who are jointly and severally liable.[52] Under the strict liability regime for abnormally dangerous activities – to be applied as a general rule for liability for genetic technology activity – the causal link shall be established between the activity "operated" by the defendant and the damage suffered by the victim.

2. Complex causation scenarios

34 There are no special rules to be applied for such scenarios. The court may shift the risk of unknown producer to the party transferring the product to the market – e.g. to the seller or other person utilising the product – and may establish joint and several liability of those who created the risk, but cannot shift the ultimate risk to a person whose liability has not been

50 Supreme Court, Legf. Bír. Pfv. IV. 21. 910/2001. sz. – EBH 2002. 626. sz.

51 The courts at the first and second instances in this case held that the claimant could not be relieved of the burden of proof since he could not prove who the tortfeasors were and rejected the claim. According to these decisions, the claimant should have proven the causal link between the harm and the defendant's conduct, which he had not been able to do.

52 Supreme Court, Legf. Bír. Pf. VI. 21. 525/1993. sz. – BH 1995. 214. sz.

established by a proven causal link. Market share liability is not applied in Hungarian court practice.

One decision of the Supreme Court from 1993 may be worthy of mention here. A child wanted to buy hydrochloric acid in a shop at the request of her mother. Because of a defect in the bottle, the acid splashed onto her face and blinded her. The claimant sued the shop and the manufacturer. The court rejected the claim against the manufacturer because it turned out that the shop had bought hydrochloric acid from two manufacturers and it was impossible to decide which one of the two manufacturers had produced the defective product. The court found only the shop liable on the grounds of liability for abnormally dangerous activities.[53] **35**

3. Force majeure

There are no specific doctrines concerning force majeure in the context of causation in Hungarian tort law. In the context of this report, however, force majeure may be relevant as a ground for exoneration under the fault-based liability regime, as one – but not the only – component of exoneration[54] from liability under the strict liability regime for abnormally dangerous activities covering the cases of liability for genetic technology activity in general or as an event cutting the causation link under the general liability regime to be applied for damage caused by GM crops to the neighbour who consented in written form to the growing of GM crops. **36**

4. Threshold to prove causation

The starting point is that a causal link shall be established between the damage and the tortfeasor's conduct and this shall mean certainty but the courts have a great discretionary power in assessing and evaluating evidence according to § 206 of the Civil Procedure Act and deciding what *they hold* to be certainty. Although there are no explicit policies, guidelines, fixed principles or established doctrines concerning the threshold of proof for the judges, I would say that it must be at least very close to certainty. **37**

53 Supreme Court, Legf. Bír. Pf. III. 21 046/1992. sz. – BH 1993. 678. sz.

54 The other is that the cause fell outside the scope of the abnormally dangerous activity. If the cause of the damage did not fall into the scope of the abnormally dangerous activity or it was not unavoidable, the operator shall be relieved of liability only insofar as the damage has been caused wrongfully by the victim himself.

5. Special rules on causation

38 There are no specific rules or doctrines on causation in Hungarian tort law that would be specifically relevant or apply otherwise especially to GMO cases.

IV. Types of liability

1. Fault liability

(a) Special rules governing fault

39 The fault-based liability regime shall be applied *vis-à-vis* the owner or user of the land neighbouring GM cultivation if he has consented in written form to the growing of genetically modified plants according to § 21/C of the Genetic Technology Act and in cases where the damage has been caused by GMOs outside of genetic technology activity. This system is a system of reversed burden of proof. Thus, the tortfeasor has to prove that he acted as was generally expected under the given circumstances in order to exonerate himself from liability. This is a general rule to be applied to liability for GMOs under the fault-based liability regime too, as well as a general structural principle to be applied for all the special forms of liability in Hungarian tort law. Under the strict liability regime to be applied for liability for GMOs in general (that is cases where the damage has been caused by genetic technology activity) the tortfeasor (the person held to be the operator of the genetic technology activity) has to prove that the cause of the damage was an unavoidable one (i.e. force majeure) that fell outside the scope of the genetic technology activity in order to exonerate himself from liability.

40 The system of strict liability for abnormally dangerous activities covering liability for genetic technology activity too does not make it necessary to apply a specific standard of care under the fault-based liability regime to be applied in the clearly distinguishable cases of written consent of the neighbour.

(b) Impact of specific rules of conduct

41 Such rules may establish the required standard of care very clearly but I do not think that it can otherwise make any difference if such rules have been violated.

2. Product liability

(a) Development risk defence

The Hungarian Product Liability Act incorporates the development risk defence. In Hungary, the Product Liability Directive has been implemented by the Act X of 1993 on Product Liability. According to § 7 subpar (1) d, of the Hungarian Product Liability Act, the manufacturer shall be relieved from liability defined in the Product Liability Act upon proving that at the time the product was put into circulation the defect could not have been discovered according to the current state of scientific and technological developments. As there are only two reported decisions of the Hungarian Supreme Court[55] regarding the application of the provisions of the Product Liability Act and they also deal only with marginal problems, it is impossible to give an analysis or even shallow overview on how this aspect is or might be handled in practice. 42

(b) Alternative routes

There is no special regime for compensating losses caused by products. The alternatives to the product liability regime are the general rules of liability as well as special forms of liability which covered product liability cases before the specific product liability regulation came into force in Hungarian law. § 12 of the Product Liability Act does not preclude such alternatives as it provides that the scope of the Product Liability Act shall not extend to the claim enforcement options of the injured party, based on the rules regarding liability for damages due to any breach of contract, or as caused outside the scope of a contract, or those defined in a separate legal regulation. In the light of the interpretation given by the European Court of Justice and holding the Directive to provide maximum protection,[56] providing such alternatives as § 12 of the Hungarian Product Liability Act does is hardly compatible with European law. 43

55 Supreme Court, Legf. Bír. Pfv. VI. 23.641/1998. sz. – BH 2000. 350. sz. and Supreme Court, Legf. Bír. V. Pfv. VII. 20.620/2004. sz. – BH 2005 354. sz.

56 ECJ C-183/00 *Maria Victoria González Sánchez v. Medicina Asturiana SA* [2002] ECR I-3901; ECJ C-52/00 *Commission v. French Republic* [2002] ECR I-3827; and ECJ C-154/00 *Commission v. Hellenic Republic* [2002] ECR I-3879. See for more detailed comments *B. A. Koch*, Report on European Union, in: H. Koziol/B. C. Steininger (eds), *European Tort Law 2002* (2003) 448 ff.

(c) Impact of compliance with rules and regulations

44 Obeying all rules and regulations governing the production process would not affect liability. In Hungarian legal theory, practice and tort law regulation there seems to be a Chinese wall between public law regulation and civil law. This wall is built of the autonomous concept of unlawfulness and fault in tort law which are independent from administrative regulation. From the autonomous concept of unlawfulness generally accepted in Hungarian law, follows that even in the absence of infringement of a statutory provision the tortfeasor shall be held liable and – on the other hand – the compliance of the tortfeasor's conduct with statutory provisions or administrative permission in itself does not prevent the tortfeasor from being held liable.[57] The Chinese wall between administrative law and private law concerning liability in tort means that the lawfulness of the tortfeasor's conduct in public law does not mean that it shall be deemed lawful in tort law as well. The lawfulness of the tortfeasor's conduct in public law in itself does not amount to a permission to cause damage to others.[58] The Hungarian legal system does not allow a regulatory permit defence. Compliance with statutory or individual permission makes the tortfeasor's conduct lawful in public law but does not make it lawful in tort law. The permission itself does not constitute an exemption for the tortfeasor from civil law liability. In a very recent case, the Supreme Court decided – confirming well established practice – that the neighbour of a cell phone transmission tower shall be entitled to claim damages according to the general rule of liability in tort for the depreciation in value of the land and the mobile phone company shall be liable even if the necessary permissions were given to build the tower in that certain place.[59] From the inner logic of the prevailing theoretical "Chinese Wall" approach, it follows that if the regulation itself makes causing harm lawful and exempts – explicitly or implicitly – the tortfeasor from the obligation to provide compensation, causing damage shall be deemed lawful and the tortfeasor does not have to pay compensation. If the regulation allows the tortfeasor to cause the damage but provides the obligation to pay compensation, causing damage shall be deemed lawful under the obligation to pay compensation.

---

57 *B. Lenkovics*, A környezetszennyezés polgári jogi szankciói (The civil law sanctions of environmental damage) in: L. Asztalos/K. Gönczöl (eds.), Felelősség és szankció a jogban (Liability and sanction in the law) (1980) 317 ff., 324.

58 Supreme Court, Legf. Bír. Pfv. I. 23.084/1998. sz. – BH 1999. 449. sz.; Supreme Court, Legf. Bír. Pfv. X. 21.156/1999. sz. – BH 2000. 244. sz.

59 Supreme Court, Legf. Bír. Pfv. III. 20.852/2005. sz. – BH 2006. 184. sz.

3. Environmental liability

(a) Implementation of the Environmental Liability Directive

Directive 2004/35/CE of the European Parliament and of the Council of 21 April 2004 on environmental liability with regard to the prevention and remedying of environmental damage (OJ L 143, 30.4.2004, 56–75) has been implemented in Hungarian law by Act no. XXIX. of 2007 on Amending Environmental Liability Legislation which came into force on 30 April 2007.[60] One of the most significant features of this legislation may be the implementation of the concept of environmental damage *per se* in private law liability as well, which is provided in Art. 2.2. of the Directive. Art. 2.2. of the Directive provides that under "damage" shall be understood a measurable adverse change in a natural resource or measurable impairment of a natural resource service which may occur directly or indirectly. Even if Act no. LIII. of 1996 on the Protection of Nature defines damage not only as the actual damage (*damnum emergens*), the lost profit (*lucrum cessans*) and the costs of the remedial measures, but also the non-pecuniary damage resulting from the reduction of quality of the natural environment or worsening of the living conditions of individuals or group of individuals in society, this definition of damage rests on the traditional concept of material and immaterial damage to persons (§ 81 subpar (1)). As a result of the amendment, § 4 point 13. of Act no. LIII. of 1995 on General Rules of Environmental Protection (Environmental Protection Act) defines environmental damage as a measurable, significantly adverse change, occurring directly or indirectly in the environment or in natural resources and the significant, measurable impairment of a natural resource service occurring directly or indirectly, respectively. This new provision of the Environmental Protection Act extends the criminal, civil law and public law liability provided already in the Act for such environmental damage as well. This amended legislation implementing the Directive brings a new concept of damage into Hungarian tort law reaching beyond the traditional definition of damage covering actual damage, lost profit and costs of loss avoidance or restoring the original state. The impacts of implementing the concept of environmental damage *per se* are not yet clear in Hungarian law. 45

This new legislation has reformulated § 102 of the Environmental Protection Act as well. This section of the Act, in its amended form, now estab- 46

60 *B.A. Koch*, European Union, in: H. Koziol/B. C. Steininger (eds.), European Tort Law 2004 (2005) 595 f.

lishes – until the contrary is proved – the joint and several liability of persons who, following the occurrence of environmental damage, owned, possessed or used the immovable where the activity damaging or endangering the environment had been pursued. The owner may exempt himself from liability by specifying the actual user of the immovable and by proving without a doubt that he himself is not liable. If users of the environment establish an economic organisation in order to merge their former similar or complementing activities, this new economic organisation shall – within the scope of duties and obligations regarding environmental protection – be deemed legal successor of these users and this economic organisation shall be jointly and severally liable with the users establishing it under the Environmental Protection Act. As a case for "lifting the corporate veil," the liability of members (shareholders) and managers of a company, that passes a resolution or takes measures although they knew or they should have known that executing or implementing such would cause environmental damage, shall be joint and several and unlimited in case of liquidation of the company insofar as the company did not perform its obligations to provide compensation and to restore the environment to its original state. Members (shareholders) and managers of the company who refrain from passing the resolution or taking the measures, who voted against the resolution, or who protested against taking certain measures shall be exempted from liability.

47 As a result of the amendment, the new § 102/B of the Environmental Protection Act – implementing the requirements of Art. 8 of the Directive – provides that in order to get financial guarantees for the costs of restoration, the Hungarian State shall be entitled to establish restraint on alienation and encumbrance on the lands of the user of the environment if the environmental harm has already been declared. On movables, collateral may be established under the same conditions. This restraint on alienation and encumbrance established on lands and collateral established on movables are to provide the financial guarantees required by Art. 8 of the Directive.

(b) Environmental liability regime beyond the scope of the Directive

48 Liability for environmental damage is also covered by liability for abnormally dangerous activities according to § 345 of the Hungarian Civil Code (see above). This is explicitly provided in the second sentence of § 345 subpar (1) of the Hungarian Civil Code and also referred to in § 103 of the Environmental Protection Act (subpar (1)) which also provides in

§ 103 subpar (2) that if the direct victim does not claim for damages *vis-à-vis* the person liable for the environmental damage, the minister shall be entitled to do this for the benefit of the Environmental Protection Fund of the State.

(c) Claimants in cases of environmental harm

The principle of full compensation shall be applied here too; above the normal losses – i.e. pecuniary and non-pecuniary damage in the traditional sense – environmental damage per se also shall be compensated. The environmental protection regulation presupposes that there is a victim who suffers environmental damage in the traditional sense and – except for the case mentioned above where the victim does not submit a claim and the minister does it for the benefit of the Environmental Protection Fund – does not provide specific rights to other persons to claim compensation for environmental damage. **49**

The Environmental Protection Act in its § 109 subpar (2) entitles the public attorney to seek an order prohibiting the tortfeasor from carrying on the activity threatening environmental damage. The entitlements of the public attorney do not cover claims to damages; they only give the public attorney the right to seek a prohibitory judgment.[61] **50**

(d) Special liability regime for losses sustained by individuals

There is no special liability regime covering losses sustained by individuals in the course of damage to the environment as such. Such cases fall under the system of strict liability for abnormally dangerous activities to be applicable for environmental liability too. Neighbourhood or nuisance regimes provide claims for *actio negatoria* under property law regulation in this context but not a special liability regime. **51**

(e) Cartagena Protocol

Hungary is a party to most of the important international conventions on environmental protection and protection of biology and biodiversity, **52**

61 Supreme Court, Legf. Bír. Gfv. IV. 32.530/1999. sz. – EBH 2000. 321. sz.; Supreme Court, Legf. Bír. Pfv. III. 22.798/1999. sz. – BH 2003. 419. sz.

among them the Cartagena Protocol[62] and the Rio de Janeiro Convention on Biodiversity.[63]

4. Other strict liability regimes

53 N/A.

V. Vicarious liability

1. Scope of vicarious liability

54 Under the title "Damage Caused by Employees, Members of Cooperatives, Representatives, and Agents", the Hungarian Civil Code addresses three special forms of vicarious liability: the liability of employers for employees, the liability of principals for agents, and the liability of the state for public officers. The relevant forms of vicarious liability under the regime of the Hungarian Civil Code may be the following.

55 (i) If an employee causes damage to a third person in connection with his employment, unless otherwise provided by law, the employer shall bear liability towards the injured person. This provision shall also be applied for liability for members of co-operatives causing damage to a third person in connection with such membership (§ 348 of the Hungarian Civil Code). The provisions of the Labour Code and the provisions of separate legal regulations shall be applied to such claims between employees and employers and between cooperatives and their members (§ 350 subpar (5) of the Hungarian Civil Code). (ii) Liability for damage caused by exercising governmental administrative power including judiciary power and the power of public prosecutors shall be established only if common legal remedies cannot abate the damage or if the aggrieved person resorted to the ordinary legal remedies for the abatement of damage without success (§ 349 of the Hungarian Civil Code). (iii) A principal shall be subject to joint and several liability with his agent for any damage caused to a third person by the agent in this capacity. The principal shall be relieved of liability if he proved that he was not negligent in choosing, instructing, and supervising his agent. This provision shall not apply to the liability of per-

---

62 Act no. CIX. of 2004 on Proclamation of the Cartagena Protocol on Biosafety to the Convention on Biological Diversity.

63 Act no LXXXI. of 1995 on Proclamation of the Convention on Biological Diversity of Rio de Janeiro.

sons who perform activities involving abnormally dangerous activities. In respect of permanent agency or if the principal and the agent are economic organisations, the court may apply the regulation governing employer's vicarious liability to the relationship between the victim and the principal. A client shall not be liable for damage caused by his legal representative. An agent shall be liable to the principal for any damage he causes, as shall a representative not employed by a client to that client. The principal shall be entitled to demand reimbursement from the agent for compensation paid by the principal to a third person (§ 350 of the Hungarian Civil Code).

The policy underlying this solution in the Hungarian Civil Code is the 56 idea that the agent and the employee are acting primarily in the interests of their employer and not in their own interests. The liability of the employer and the principal rests on this consideration. The main difference between the employer/employee and the principal/agent relationship (which is a question of legal qualification of the legal relationship according to the content of the rights and obligations of the parties) is the permanent character of the employment relationship and the incidental one of the agency. That is why the Code provides the possibility of applying the rules of employers' liability to cases where agency has a permanent character. The damage caused by the employee, as emphasised in the motives to the Draft of the Civil Code of 1959, may be seen as if it had been caused by the enterprise itself. Thus, the activity of the employee shall be attributed to the corporate entity.

In Hungarian court practice the decisive factor seems to be the qualifica- 57 tion of the contractual relationship between the employer and the employee. The qualification is based on the normal tests drawing the line between contracts for work on goods, relationships of agency and employment (labour) contracts. The main distinguishing factors are the right to and degree of control and instructions, the ownership of the means that are necessary to pursue the given activity, the management of work and the way pricing or remuneration functions. The employee is only the party to an employment contract; for an agent § 350 of the Hungarian Civil code is applicable (joint and several liability or vicarious liability in case of permanent agency or if the principal and the agent are economic organisations). An independent contractor shall not be deemed an employee. Establishing vicarious liability according to § 348 of the Hungarian Civil code requires that the employee acted within the scope of his employment relationship. This formulation is relatively wide as it extends employer's vicarious liability also to cases where the employee acted outside of his strictly interpreted sphere of employment activity or worked outside of the control of his employer.[64] If, for instance, the employee exceeds his authorisation and as false procurator without the right to act in the name of his employer and with false documents, also committing a crime, orders goods, the employer shall be liable vicariously towards the other party with whom the contract could not be concluded but who entered into it knowing nothing about the false representation.[65] If a doctor, on the other hand, was acting during his holiday outside the territory of his practice without remuneration, his employer shall not be vicariously liable for him because his activity did not fall within the scope of his employment.[66]

58 Thus, according to the Hungarian Civil Code, someone cannot be held liable for an independent contractor hired by him/herself. The New Hungarian Civil Code suggests that vicarious liability for independent contractors shall be extended only insofar as the party employing someone as an independent contractor under a private law relationship shall be liable for this independent contractor if this person is not known to the victim and the party employing the independent contractor shall be relieved from liability if he identifies the person (i.e. the independent contractor) to the victim.[67]

2. Liability for people further up the food or feed production chain

59 Under the tort (non-contractual) liability regime, none of these persons shall be held vicariously liable for the others as they are presumably deemed independent contractors. The concept of the operator in the context of strict liability for abnormally dangerous activities, to be applicable to liability for damage caused in the course of genetic technology activity with GMOs, may be wide enough to cover those persons who did contribute directly to the damage. The primary interest in pursuing the activity as being the source of the damage may be a ground for establishing the position of operator and the liability for the abnormally dangerous activity. On this ground, e.g., the customer or buyer of a service may also be held to be an operator of the relevant activity, as in case of spreading chemicals to the land from an airplane, the person who ordered the chemical

64 G. *Eörsi*, Kötelmi jog, általános rész (1981) 298.

65 Supreme Court, Legf Bír. Gf. I. 31 500/1993. sz. – BH 1994. 96. sz.

66 Supreme Court, Legf. Bír. Pf. V. 20.063/1995. sz. – BH 1996. 89. sz.

67 § 5:524 of the New Hungarian Civil Code.

spraying shall be deemed operator if the chemical spread causes damage to the crops or animals of other persons.[68]

3. Can someone further down the feed or food chain include someone further up it in a trial against him/herself?

If someone has a legitimate interest in participating in a civil lawsuit between other parties (e.g. on the ground that one of the parties might have a claim against him depending on the result of the civil lawsuit), he can intervene in the lawsuit in order to support one of the parties but he cannot be forced to do so if he does not want to intervene. The farmer cannot bring his seed producer into a lawsuit filed against him by a customer if the producer does not want to intervene. If the farmer has been found liable, he may have a regress claim *vis-à-vis* the producer but he cannot shift the liability to the producer if he was the one being held to be the operator of a genetic technology activity. 60

VI. Multiple tortfeasors

The Hungarian Civil Code provides special regulation for damage caused by multiple tortfeasors. According to § 344 of the Civil Code, if the damage is caused jointly by two or more persons, their liability shall be joint and several towards the aggrieved person, while their liability towards one another shall be divided in proportion to their respective degree of responsibility. Liability for damages shall be divided in equal proportions among the responsible persons if the degree of their responsibility cannot be established. The court shall be entitled to declare joint and several liability and oblige the persons having caused the damage to bear the compensation in proportion to their respective contributions provided doing so would not jeopardise or considerably delay the compensation for damage or if the aggrieved person has himself contributed to the occurrence of the damage or has procrastinated in enforcing his claim without any excusable reason. In the literature and in practice there is a controversy about whether the two or more persons should have acted with a certain degree of common intention or whether they can act independently in order to be held jointly and severally liable for the damage. In academic literature there are opinions according to which the common intent is a 61

68 Supreme Court, Legf. Bír. Gfv. XI. 30.293/2006. sz. – BH 2007. 301. sz.

necessary requirement for establishing joint and several liability.[69] This view is not in accordance with the motives behind the draft of the Civil Code which explicitly states that common intention of more than one tortfeasors is not a precondition for treating them as joint or multiple tortfeasors within the meaning of § 344 of the Civil Code. More authentic interpretations also stress the objective character of the assessment and that the common intent is not a precondition of common liability; the object of the tortfeasors' conduct is irrelevant. If, for instance, two cars collide and as a result of the accident someone who is travelling in one of the cars is injured, the two car drivers shall be treated as multiple tortfeasors and are jointly and severally liable[70] and this example may be generalised: if two abnormally dangerous activities contributed to the damage, they may be held to be multiple tortfeasors jointly and severally liable. If only one of the contributing parties shall be liable under the strict liability regime, this does not preclude establishing – as multiple tortfeasors – joint and several liability but each of the tortfeasors shall be liable according to the system governing his liability.

62 Mere interdependence in causation is, however, not always enough for establishing common liability. If someone negligently fails to fulfil his obligation and this makes it possible for someone else to cause harm, he also shall be jointly and severally liable with the tortfeasor who caused the harm directly. The two main principles for rendering joint and several liability are the prevention and the provision of a better chance of compensation for the plaintiff. The distinction between – jointly and severally liable – multiple tortfeasors and several independently liable tortfeasors can be found in terms of causation: the tortfeasors are jointly and severally liable multiple tortfeasors if the behaviour of each is a *condition sine qua non* but the tortfeasors shall not be jointly and severally liable if there is not any causal interdependence between their harmful conduct or if the interdependency is too remote. If, for instance, someone causes a car accident and the injured person suffers an injury which is not fatal but dies because the surgeon is negligent, the two tortfeasors are not jointly and severally liable[71] and this approach also can be generalised.

---

69 *B. Kemenes/L. Besenyei,* A kártérítés általános szabályai, in: G Gellért (ed.), A Polgári Törvénykönyv Magyarázata (2002) 110 ff., 1120.

70 *K. Benedek/M. Világhy,* A Polgári Törvénykönyv a gyakorlatban (1965) 349.

71 A Magyar Népköztársaság Polgári Törvénykönyve – az 1959. évi IV. törvény és a törvény javaslatának miniszteri indokolása [Motivation for the Hungarian Civil Code] (1963) The motivation for § 344.

VII. Defences

1. Licence/permission to grow GM material

From the autonomous concept of unlawfulness and fault – explained supra no 43 in more detail – it follows that the regulatory permit defence is not accepted in Hungarian tort law and the conduct held to be lawful in context in public law may still be unlawful in private law. Thus, the licence/permission to grow GM material does not serve as a defence against liability triggered by its use. I do not think that it would make any difference if the GMOs are approved for testing only. **63**

2. Consent/assumption of risk

According to § 342 subpar (2) of the Hungarian Civil Code, as a general rule, the victim shall not be entitled to get compensation if the damage was caused with his consent and this does not violate or endanger any social interest. According to § 345 subpar (2), as a specific rule to be applied to liability for abnormally dangerous activities, damage shall not be compensated to the extent that it originates from a conduct attributable to the victim. From this follows that the victim shall not be compensated for a risk he undertook knowingly, provided that it was not contrary to social interest and I think that normally this consent in the cases covered with this questionnaire would not involve unacceptable interference with social interests. **64**

Thus, the impact of consent/assumption of risks on liability is that – provided extreme circumstances do not make it contrary to public interests – harm, in so far as it was covered by the consent, is not to be held unlawful and the defendant shall not be liable for such. The consent shall, however, be narrowly construed as it results in the victim losing his right to claim damages and according to § 207 subpar (4) of the Hungarian Civil Code, judicial acts resulting in a waiver of rights shall always be construed narrowly. **65**

3. Third-party influence

Behaviour of third parties would influence liability only in so far as it may result in exoneration of the operator from liability, e.g. if the contribution of the third party to the occurrence of the damage was an unavoidable **66** cause falling outside the scope of the abnormally dangerous activity (under application of the strict liability regime for abnormally dangerous activities) or results in establishing that the tortfeasor had acted as was generally expected under the given circumstances (the prerequisite for exoneration under the fault-based liability regime). I do not think that behaviour of third parties in itself would be relevant in the course of establishing liability but in the context of considering whether the tortfeasor (operator) could be exonerated from liability, such behaviour shall be considered and taken into account just as any other relevant circumstance.

4. Prescription

67 The general limitation period is five years and this shall be applicable in tort law in general. There is, however, a special limitation period of three years for claims under the strict liability regime for abnormally dangerous activities (to apply to genetic technology activity too) and a three year limitation period is also provided for claims under the product liability regime. As far as strict liability for abnormally dangerous activities (which is the most relevant aspect in the cases covered with this study) is concerned, the expiration of the three years period does not bar the victim from claiming damages under the fault-based liability regime (within the standard limitation period of five years) from the operator of the abnormally dangerous activity.

68 As provided in § 326 of the Hungarian Civil Code, the period of limitation commences upon the due date of the claim. If, however, the obligee is unable to enforce a claim for an excusable reason, the claim shall remain enforceable for one year from the time when said reason is eliminated or, in respect of a period of limitation of one year or less, for three months, even if the period of limitation would otherwise already have lapsed. This provision shall also apply if the obligee has granted a respite for performance after expiration. § 327 of the Hungarian Civil, however, provides that a period of limitation shall be suspended by a written notice for performance of a claim, the judicial enforcement of a claim, the amendment of a claim by agreement (inclusive of composition), and the acknowledgment of the debt by the obligor. The period of limitation shall recommence after suspension or following the non-appealable outcome of a suspension proceeding. If a writ of execution is issued in the course of a suspension proceeding, the period of limitation shall be suspended only by the acts of enforcement.

5. Other defences

I do not think that there are any other relevant defences not yet mentioned in this report. 69

VIII. Remedies

1. Pecuniary compensation

(a) Bodily harm

There are no special rules on compensating personal injury caused by GMOs; the general rules are to be applied also for bodily harm, but as awarding non-pecuniary damages is a special sanction of wrongful interference with personality rights, in this context there is a difference between damage to "persons" and damage to property. Non-pecuniary damages shall not be awarded as a compensation of damage to property while it is a *per se* consequence of wrongful interference with personality rights including bodily harm. There are no liability caps for such damages. 70

(b) Property losses

There are no caps or other special rules on compensating property losses caused by GMOs; the regular remedies are to apply. 71

(c) Economic losses

There are no special rules or principles on compensating economic losses caused by GMOs; the normal remedies apply also for economic losses. The application of these general rules, however, may mean that the organic farmer cannot recover the full costs of restoring a field that was contaminated with GMOs to meet organic standards because deductions would apply for the time he could have grown conventional crops there. According to § 340 subpar (1) of the Hungarian Civil Code, in order to prevent or mitigate damage, the victim shall act in a manner that can generally be expected in the given situation and no liability shall apply to the portion of the damage incurred due to the failure of the aggrieved person to perform this obligation. If it was established that the farmer growing organic crops might have mitigated the loss by shifting to growing con- 72 ventional crops, this may result in such deductions. This, however, shall be considered according to the required standard of conduct to be established in the context of the given circumstances of the case and the answer for this may not be generalised.

(d) Harm to animals

73 There are no specific rules or principles that apply for this case. The damage would be calculated according to the depreciation in value of the animal; calculating this, the cow's potential for producing milk or meat should also be considered; at least this should be the case based on the principle of full compensation. I do not think that the mere fact that an animal eats GM contaminated feed would constitute recoverable harm in Hungarian tort law. As animals do not possess personality rights, damages shall be awarded in such a case only if the fact that the animal eats GM contaminated feed results in damage to property or economic loss.

(e) Costs of disposal

74 Such costs are recoverable insofar as they were manifested as damage in the owner's property, the economic loss to him or insofar as they are to be deemed costs for mitigation or attenuation of the damage/loss suffered by the owner.

2. Non-compensatory damages

75 Hungarian tort law does not recognise punitive, exemplary or any other forms of non-compensatory damages, either on a regulatory level or in court practice. In Hungarian tort law, the concept of damages is tightly linked with the concept of damage. According to § 355[72] of the Hungarian Civil Code, the tortfeasor responsible for the damage shall be liable for restoring the victim to her original state, or, if this is not possible or if the aggrieved party refuses restoration on a reasonable ground, the tortfeasor shall compensate the aggrieved party for any pecuniary and non-pecuniary damage suffered by the latter. Compensation must be provided for any depreciation in value of the property belonging to the aggrieved

72 Subpar (1) and subpar (4).

person (*damnum emergens*) and any pecuniary advantage lost due to the tortfeasor's conduct (*lucrum cessans*) as well as compensation of the costs required for the attenuation or elimination of the pecuniary and non-pecuniary loss suffered by the victim. These provisions not only dictate the principle of full compensation, they also draw the limits of liability as they define damages as compensation for compensable loss. A general principle of Hungarian tort law – which is similar to other continental tort law systems – is that no one shall be enriched by her own damage. It seems that restitutionary damages or claims for the benefit gained to be shifted from the tortfeasor to the victim thus depriving the tortfeasor of the profit she gained from the wrongful conduct at issue (*Gewinnabwehr*) should be held to be compatible with the principles and policies underlying tort law.[73] It is, however, a generally accepted principle that the victim should be prevented from making a profit on her own loss. This principle is generally accepted but not expressly declared in the current Civil Code, although it clearly follows from the concept of damage (i.e. that damage includes the actual loss, lost profits and the costs of prevention and avoidance of the loss) and restitution of unjust enrichment. According to this principle, in the course of calculating the sum of damages to be awarded, the amount of damages shall be reduced by the sum the victim earned or saved as a result of the damage (e.g. payments under a national health care system[74] or an increase of value in the property of the victim as a result of the event which caused damage). In line with the principle of full compensation, the plaintiff shall be compensated for all the losses she suffered but cannot be paid more.[75] Neither in its structure of regulation nor in its regulatory method (i.e. in its definition of damage and its definition of damages as compensation for damage) does the draft of the new Hungarian Civil Code suggest changes or amendments to the previous provisions and – with the exception of small clarifications and the abandonment of the concept of non-pecuniary damages replacing it with a special form of indemnity for unlawful interference with personality rights – it keeps the same regulatory system.

73 *Marton* (fn. 42) nr. 117. § 2:88 of the Draft of the New Hungarian Civil Code (deprivation of the profit gained through wrongful interference with personality rights).

74 Supreme Court Legf. Bír. Mfv. I. 10.244/2002/3. sz. – EBH 2002. 695. sz.; Supreme Court Legf. Bír. Mfv. I. 10.744/2006. sz. – BH 2007. 354. sz.; Supreme Court Legf. Bír. Mfv. I. 10.697/2006. sz. – BH. 2007. 274. sz.

75 *A. Harmathy* in G. Gellért (ed.), A Polgári Törvénykönyv Magyarázata (7th ed., CompLex, 2007) § 355 no. 4.

76 Non-pecuniary damages in Hungarian tort law are primarily of a compensatory character[76] even if they also imply a redress which helps to prevent, in general, unlawful behaviour in the future and helps to avoid interferences with the human dignity of others in society.[77] This function, of providing redress beyond compensation, has been stressed in a decision of the Hungarian Supreme Court, establishing that the victim shall be entitled to non-pecuniary damages even though she was in coma which prevented her from enjoying any kind of reparation that non-pecuniary damages could have brought to her.[78] The new form of indemnity as a special sanction of wrongful interference with personality rights, which is proposed in the draft for the new Hungarian Civil Code to replace non-pecuniary damages, would not change the functions of non-pecuniary damages. The idea behind this amendment of the Civil Code and the introduction of an indemnity instead of non-pecuniary damages was to relieve the plaintiff of the burden of proving some form of detriment suffered as a result of the defendant's wrongful conduct and in this way, it makes access to indemnification easier for the plaintiff. This, however – according to the underlying policy of the suggested amendment – would not deprive the sanction of its compensatory function, although it may inevitably put an emphasis on its repressive character.[79]

3. Other remedies

77 As has been mentioned although not emphasised above, as a primary remedy the tortfeasor shall be obliged to restore the original state and only as a secondary remedy shall he be obliged to pay damages. As provided in § 355 of the Hungarian Civil Code, the tortfeasor shall primarily be liable for restoring the original state, and if this is not possible or if the aggrieved party refuses restoration on a reasonable ground, he shall provide compensation as damages for the material and immaterial harms suffered by the victim.

76 Constitutional Court of the Hungarian Republic, decision of 34/1992. (VI. 1.) AB hat. On the revision of § 354 of the Hungarian Civil Code from the point of view of its compatibility with the Constitution.

77 *T. Lábady*, A nem vagyoni kártérítés újabb bírói gyakorlata (1992) 40.

78 Supreme Court P. törv. III. 20 703/1989. sz. – BH 1990. 15. sz. The compensatory function has been stressed, however, in another decision, the Supreme Court subordinated the repressive and preventive function of non-pecuniary damages to the compensatory function of civil law liability. Supreme Court Legf. Bír. Pfv. IV. 20.419/2006. sz. – BH 2006. 318. sz. (EBH 2006. 1398. sz.).

79 Motivation to the Draft of the New Hungarian Civil Code. Motivation to § 2:90.

As a main rule, damage shall be indemnified in cash, unless compensation in kind (as a special form of damages here) is justified under the given circumstances. Compensation in kind may be justified particularly if the object of indemnification is produced by the person responsible for the damage or is otherwise at his disposal. 78

4. Costs of pursuing a claim

(a) General cost rule

Regulation for recoverability of costs of a civil trial is provided in the Civil Procedure Act (Act no. III of 1952 on Civil Procedure). § 78 subpar (1) of the Civil Procedure Act provides that – as a main rule – the costs of the procedure are to be borne by the losing party. There are numerous exceptions to this rule but all of them are attached to specific situations like when the defendant did not give cause for the procedure and admitted the plaintiff's claim at the first hearing, then the costs are to be borne by the plaintiff; costs created by delay or which were induced superfluously are to be borne by the party creating them independently of the result of the procedure; if the party submits a claim in spite of the result of agreement made under a mediation procedure, the costs are to be borne by him; in case of partial losing, the losing party shall bear the costs only to the extent which they lost.[80] There is an important exception concerning claims for damages: if the claim is of a nature pending in its amount – even in case of success – on judicial discretion (as typically claims for damages are), the court may oblige the party who lost the procedure only partially to bear the whole costs if the claim was not obviously exaggerated.[81] 79

(b) Costs of establishing causation

I think that such costs are recoverable either according to § 75 subpar (1) of the Civil Procedure Act defining recoverable costs as all the costs that occurred during or before the procedure in connection with their necessary procedural acts in good faith or as a part of their claim for damages. 80

80 §§ 80–81 of the Hungarian Civil Procedure Act.
81 § 81 subpar (2) of the Hungarian Civil Procedure Act.

5. Advance cover

81 In Hungary, no such redress schemes have been introduced and the introduction of such a system planned (as far as I know).

IX. Cross-border issues – Conflict of laws

1. Conflict rules applicable before (or instead of) Rome II

82 § 32 of the Law-Decree no. 13 of 1979 on the Private International Law of Hungary provides for a complex regulation of applicable law to claims for damages under the non-contractual liability regime. As a main rule, for liability in tort, the law applicable at the place and time of the activity or omission resulting in damage (*lex loci delicti commissi*) shall be applied (first alternative as the main rule). If, however, it is more favourable to the victim, the law of that state in the territory of which the damage occurred shall apply (*lex loci damni*, second alternative). A third alternative is that if the places of residence of the tortfeasor and the victim are in the same state, the law of that state shall be applied. § 32 (4) of the Private International Law Law-Decree also provides that if, according to the law of the place of the activity or omission causing the damage, culpability is a condition of liability, the capacity of culpability may be established either according to the personal law of the party causing the damage, or according to the law of the place of the violation of law. As far as multi-state torts are concerned, Hungarian court practice and legal literature are uncertain how they should be handled under the application of the Hungarian Private International Law Law-Decree. It is suggested that in such cases the second alternative (*lex loci damni*, if it is more favourable to the victim) shall be applied in a creative and flexible way having regard to all circumstances of the case.[82]

83 The Rome II Regulation – as far as its general provisions are concerned – seems to be based only partly on the same foundations as the Hungarian private international law and also according to other preferences. While the primary connecting principle in Hungarian private international law is the *lex loci delicti commissi*, the Rome II Regulation has the *lex loci damni* as a primary rule (which is only a secondary option according to the Hungarian Private International Law Code) and the connecting principle provided in Art 4. subpar (3) of the Rome II Regulation – the principle of

---

82 *F. Mádl/L. Vékás*, Nemzetközi magánjog és nemzetközi gazdasági kapcsolatok joga. (2004) 388.

strongest connection – would be completely new in the context of torts in Hungarian Private International Law.

2. Special regime for cross-border claims

I do not know any of such relevant regimes to apply to resolve bilateral or multilateral cross-border claims for compensation. 84

X. Cases

1. Due to the adventitious presence of GMOs in a field, maize which is normally sold as conventional contains GMOs beyond the legal labelling threshold. This is not discovered before the final stage of the food production chain by the producer of taco chips. The whole production is lost since the supermarket chains refuse to accept delivery from the producer.

(a) Who can sue along the chain of distribution?

The taco producer has a choice either to sue his wholesaler on a contractual 85
basis or to sue the farmer or the wholesaler on a non-contractual basis.[83] If the taco producer chooses to sue the wholesaler on a contractual basis claiming damages as a remedy for breach of contract, the basic rule of fault-based liability shall be applied, whereby the wholesaler could be exonerated from liability by proving that the damage had been unavoidable – according to court practice the standard of conduct required is stricter (up to unavoidable causes) than in cases of tort.[84] As the damage was caused outside of the genetic technology activity, the liability for it shall be assessed under the fault-based liability regime. If the taco producer sues the wholesaler of the maize or the farmer producing it, the wholesaler or the farmer is to be liable under the general rules of fault-based liability and they can be exonerated by proving that they complied with the generally required standard of conduct i.e. they acted as was generally expected under the given circumstances. The producer of the original (traditional) crops is liable under the fault-based liability regime.

---

83 G. *Eörsi*, Elhatárolási problémák az anyagi felelősség körében (1962) 124 ff.

84 I. *Kemenes*, A gazdasági szerződések követelményei és az új Ptk. Polgári Jogi Kodifikáció (2001) 9 ff.

(b) Would the case be solved differently if the GMO content was below the labelling threshold?

86 If this would mean that the supermarket chain could not refuse delivery – in other words they did so unlawfully – this would result in no liability for the wholesaler etc., but compliance with statutory regulation could not be enough defence in itself (see answer to Q.IV. 2. c.).

(c) Would the case be solved differently if the admixture was not adventitious, but occurred due to the disregard of segregation rules, for example?

87 If the producer growing GM crops disregards the segregation rules, he is liable under the strict liability regime for abnormally dangerous activities and he would hardly be able to prove that the cause of the damage had fallen outside the scope of his activity and that the cause of the damage was unavoidable (in fact he was at fault), so he should be liable. If the producer of the traditional crops disregards the segregation rules, he is liable under the fault-based liability regime (as he did not pursue genetic technology activity) and he would hardly be able to prove that he acted according to the generally required standard of conduct, so he should be liable. If the defendant is not the producer of the crops, the result would not be different and the case would not be solved differently.

(d) Would the case be solved differently if the GMO found was not admitted for production in your jurisdiction?

88 This may be relevant under the fault-based liability regime in establishing incompliance with the required standard of conduct (= fault) of the producer (which would presumably be established in such a case) but not under the strict liability for abnormally dangerous activities covering genetic technology activities or fault-based liability of others in the distribution chain who could not have known the origin of GMO. So otherwise, I do not think that the case would be solved differently.

(e) If the admixture had occurred on a non-GM field and it transpires that the GM seeds were blown from: (i) neighbouring fields; or (ii) a truck passing by, would the farmer of the affected field be liable for all or part of the loss caused further down the distribution chain?

I think so, as there is no rule or doctrine of limiting liability for such cases. **89** The farmer, however, may have a regress claim *vis-à-vis* the producer or carrier of the GM seeds.

2. Twenty years after the sale of GM maize used for food products, it turns out that it has certain disadvantageous health effects for humans.

(a) Can the producers be held liable at this point for risks unknown at the time of growing the maize? Who would be liable?

According to the strict liability regime for especially dangerous activities, **90** the producers shall be liable as they would hardly be able to prove that the cause of the damage had fallen outside the scope of their activity which is one of the prerequisites for exoneration from liability under the strict liability regime for abnormally dangerous activities that is to apply to GM activities too (see Q.I.1.). As far as, however, such a case falls under the product liability regime, the scientific development defence shall be applied because cases falling under the product liability regime shall not be brought under other regimes more favourable to the victim according to the ECJ case law (see Q.IV.2.b.).

I do not think that under Hungarian tort law, the person to be liable could **91** be described in such an abstract way. None of the persons in the production and distribution chain may be excluded from the circle of potentially liable persons.

(b) Can compensation already be claimed at a point when the negative health effects have not yet materialised, but are to expected to according to scientific expertise/mere rumours?

According to the concept of damage (see Q.II.1.), an injury that did not **92** occur yet in itself may not be a compensable material damage. Damage which may occur in the future probably would result in a health condition of being exposed to risk which in itself which may – depending on the proximity of risk – be a wrongful interference with the personality rights of the person exposed to such a risk and thus provide a basis for claiming non-pecuniary damages.

(c) Would it make any difference if the GM maize had only been in use for feed, causing harm to the animals, which may or may not cause harm to humans consuming the meat as well?

93 No, I do not think that this would make any difference either concerning *(a)* or *(b)*.

3. The driver of a food logistics company discovers that a farmer, from where he regularly picks up agricultural products fails to obey mandatory segregation rules or food or feed hygiene standards, which may lead to the admixture of GM and non-GM produce sold and packaged separately by that farmer. Does he or his employer have a duty to warn, i.e. warn the recipient of the allegedly "non-GM" produce?

94 Not only the requirement of good faith and fair dealing but also a general duty to cooperate is established in § 4 subpar (1) of the Hungarian Civil Code. There is not, however, a settled court practice on how this duty shall be construed outside of the context of contractual relationships. In the course of preparatory works for the New Hungarian Civil Code, it was established that such a duty to cooperate shall be provided for contractual relationships only (including the pre-contractual phase) but not in general.[85] Thus, I would conclude here that – except when specific circumstances would establish a proximate relationship between the driver or the employer of the driver and the recipient – although there is a general duty to cooperate provided in the Civil Code which might be a source of such a duty of disclosure, this duty would not be established in court practice.

---

85 *L. Vékás* in: L. Vékás (ed.), Szakértői javaslat az új Polgári Törvénykönyv tervezetéhez (2008) 69.

Damage Caused by GMOs under Italian Law

Alberto Monti and Federico Fusco

I. General overview

1. Special liability or redress scheme for GMOs

A special liability and compensation regime addressing liability for GMOs and covering the economic damage resulting from GMO admixture in non-GM products has been enacted in Italy in the form of general principles that still require detailed implementation and specification at regional and local level. Pending implementation, the cultivation of GM crops is prohibited in Italy, subject to criminal sanctions. Given this situation, no case law has been developed on liability for GMOs. Therefore, any issue relating to this type of liability must be addressed according to the few and general provisions enacted so far, which have to be interpreted in light of general rules of tortious liability. 1

Decree-law no. 279 of 22 November 2004 (hereafter: Dl 279/04), as amended by Law no. 5 of 28 January 2005, defines the minimal normative frame of reference for coexistence. In particular, art. 5 of Dl 279/04 provides for a special fault-based liability regime for damage resulting from GMO admixture in non-GM products as a consequence of the violation of coexistence measures, with a reversal of the burden of proof. Pursuant to art. 5 par. 1-*bis*, of Dl 279/04, a farmer who suffers damage resulting from other farmers' inobservance of the measures stipulated in the local coexistence plan or in the mandatory business management plan is entitled to compensation. 2

At present, it is not clear whether this liability regime should be deemed an exclusive one or whether it could overlap with the general tortious liability regime laid down in art. 2043 ff. of the Italian Civil Code (hereafter: CC). The main difference between this special liability regime and the general tort law regime consists in the reversal of the burden of proof. Indeed, according to art. 5 of Dl 279/04, the burden of proving 3 full compliance with all the applicable coexistence measures lies with the defendant.[1]

4 In any event, no special provision has been enacted in Italy with regard to losses caused by GMOs to consumers or to the environment as such.

2. State liability

5 No specific rules on state liability are contemplated by the liability regime mentioned under Question 1. However, pursuant to art. 5, par. 1-*ter* of Dl 279/04, the implementing decree of the Minister of Agriculture and Forestry, which has never been promulgated, should have regulated the access of injured farmers to the existing National Solidarity Fund, set up by Legislative Decree no. 102 of 29 March 2004 and aimed at preventing and restoring losses suffered by agriculture as a result of natural catastrophes and calamities.

6 As far as losses to third parties (other than farmers cultivating non-GM crops) are concerned, some authors have suggested that the public authorities which grant the relevant licences to cultivate GMOs should be held jointly liable for damage resulting from the commercialisation of GM food and feed.[2]

II. Damage

1. Recoverable losses

7 According to the Guidelines for the enactment of regional provisions on coexistence, every Region ought to set up a fund to cover any damage that cannot be compensated under the rules of tortious liability.[3] According to the Guidelines, the losses that can be compensated by the regional funds are: 1) the depreciation of contaminated products, 2) costs associated with sampling and testing of GMO presence in other products, 3) lost chances, such as the chance to enter a given sales channel.[4]

1 For a more detailed description of this special liability regime see *A. Monti/F. Fusco*, Economic Loss Caused by GMOs in Italy, in: B. Koch (ed.), Economic Loss Caused by Genetically Modified Organisms (2008) 299 ff.

2 *L. Prati/F.Massimino*, Organismi geneticamente modificati, danno alla salute e danno ambientale, Danno e responsabilità 2001, 337 ff.

3 "Linee Guida per le normative regionali di coesistenza tra colture convenzionali, biologiche e geneticamente modificate", approved on October 18, 2007.

4 See chap. 12.3 of the Guidelines.

Aside from this, however, the general rules of tort law should be applicable to losses caused by GMOs to third parties. The Italian Civil Code does not provide for a general definition of "damage", but the term is generally understood as designating something injurious to an interest or, more narrowly, something detrimental (to property or person) resulting from "injury to an interest". Art. 2056 CC, which sets out the tests to assess the magnitude of the damage inflicted, refers to art. 1223 ff. CC, pursuant to which the assessment of recoverable damage shall include both the loss sustained and the lost profits (economic detriment), insofar as they are the direct and immediate consequence of the wrongful act or omission. **8**

Italian law distinguishes between pecuniary losses, which are recoverable under the general rule of art. 2043 CC and include all losses that can be evaluated economically, and non-pecuniary losses, which are recoverable under art. 2059 CC only in cases provided for by the law, for example in cases of damage caused by criminal activities (art. 185 Criminal Code) or damage to interests protected by the Constitution (the most important example being the so-called "damage to health").[5] Both types of losses may be recoverable in cases of damage caused by GMOs. **9**

2. Pure economic loss

The category of pure economic loss is unknown under Italian tort law, as no distinction is made between physical harm and economic losses. Of course, the latter may be recoverable if they can be qualified as wrongful under the general rule of liability provided by art. 2043 CC,[6] and if they satisfy the criteria set forth in art. 1223 CC. **10**

3. Mere fear of a loss

Dl 279/04 does not provide specifically for feared admixture; however, it seems reasonable to conclude that only losses deriving from actual admixture would be recognized as compensable, according to the general rules of tort law. **11**

---

5 For a comment on the most recent developments in the case law with regard to non-pecuniary losses see *C. Castronovo*, Danno esistenziale: il lungo addio, Danno e responsabilità 1/2009, 5 ff.

6 See *M. Bussani/V.V. Palmer*, The liability regimes of Europe – their façades and interiors, in: M. Bussani/V.V. Palmer (eds.), Pure Economic Loss in Europe (2003) 133 ff.

12 In general, feared losses are not recoverable under Italian tort law. Recently, the Supreme Court has confirmed that "neither serenity nor safety represent, in themselves, fundamental rights of the person protected at a constitutional level, whose violation allows for compensation of non-pecuniary losses".[7]

13 However, in the famous "Seveso" case, the Court of Appeal of Milan stated that "the environmental pollution deriving from a criminal act, which causes citizens to be exposed to undefined quantities of toxic substances, resulting in prescriptions and limitations to their freedom of action and life due to the changed relationship with a contaminated environment, gives them a right to be awarded moral damages, regardless of the existence of a proven health impairment that constitutes biological damage, as it gives rise to a fear syndrome which humiliates and conditions the inhabitants of the area".[8]

4. Standard of proof

14 General rules of tort law are applicable, according to which losses must be actual in order to be recoverable. Of course, lost profits include the future effects of a damaging event. Nonetheless, they are recoverable only if they are linked to an actual infringement of an interest of the injured party, who needs to prove the existence of a certain damage.

5. Nominal losses

15 Nominal losses are not recognized.

6. Mass losses

16 There are no special rules dealing with mass losses.

III. Causation

1. Uncertainty of merely potential causes

17 Causation at law is a complex subject and cannot be dealt with satisfactorily in a short answer. For the purposes of this Questionnaire, however, it

7 Cass. civ., 12.02.2008, no. 3284.

8 App. Milano, 15.04.1994, in Giurisprudenza Italiana (Giur. It.) 1994, I, 2, 961.

is important to note that Italian courts and scholars have acknowledged that even if it cannot be shown that the defendant's conduct was with certainty a cause of the victim's injury, civil liability can be imposed based on a probabilistic assessment of the causal link validated by scientific evidence.[9] If causation is established, the victim is entitled to recover the loss sustained.

2. Complex causation scenarios

If the probabilistic test of causation mentioned supra, in no. 17, is not satisfied, liability cannot be imposed. **18**

3. Force majeure

Under Italian law, *force majeure* can interrupt the causal link between the conduct of the wrongdoer and the damage. **19**

4. Threshold to prove causation

In civil matters, the threshold in court practice is coming closer and closer to "more likely than not", especially in the field of medical malpractice. In criminal matters, however, the standard is much more stringent. **20**

5. Special rules on causation

There are no specific rules on causation that apply to GMO cases. **21**

IV. Types of liability

1. Fault liability

(a) Special rules governing fault

Art. 5 of Dl 279/04 provides for a fault-based liability regime with a reversal of the burden of proof towards the defendant. **22**

9 See Court of Cassation, no. 21619/2007.

23 Fault arises as a consequence of the mere breach of the provisions contained in the regional coexistence plans and in the business management plans, which means that liability for GMOs will arise upon breach of the measures on coexistence, provided that all the other requirements of tort liability are met (existence of damage, causal link between the conduct of the agent and the damage, and the capacity of the tortfeasor).

24 As the burden of proof is reversed by operation of law, the defendant must provide evidence that he/she has acted in full compliance with all the applicable measures on coexistence, otherwise he/she will automatically be considered at fault.

25 Pursuant to art. 5 par. 2, of Dl 279/04, the farmer who proves that he/she used only GMO-free seeds – certified by the public authority and by the producer – is always exempted from liability.

26 Aside from the reversal of the burden of proving fault contemplated by the special liability regime provided for by Dl 279/04, we are not aware of any other deviation from general fault rules in connection with losses caused by GMOs.

(b) Impact of specific rules of conduct

27 See answer to Question (a) above. In general, the breach of statutory or regulatory provisions is relevant to the assessment of fault. Fault is defined, indeed, by art. 40 of the Italian Penal Code (hereafter: PC), according to which fault may derive from negligence, imprudence or lack of skill on the part of the agent, or from the agent's breach of statutes, regulations, orders or guidelines.

2. Product liability

(a) Development risk defence

28 The EC Product Liability Directive was implemented in Italy by Presidential Decree no. 224 of 24 May 1988, which was subsequently repealed by art. 146 of Legislative Decree no. 206 of 6 September 2005. The provisions on product liability are now contained in the Italian Consumer Code[10] (art. 114 ff.).

10 Legislative Decree 06.09.2005 no. 206.

The development risk defence is contained in art. 118(e) of the Consumer Code, according to which liability is excluded "if the state of scientific and technical knowledge, at the time when the producer put the product into circulation, did not yet enable assessment of the product as defective". **29**

In order to be exempted from liability according to said provision, the producer must prove that, at the time when he put the product into circulation, he adopted both the control measures (to ascertain the existence of the defect) required by laws and regulations in force, and the measures that could be taken according to the scientific and technical knowledge available at that time, even if the latter were not mandatory and regardless of their cost and complexity. In other words, as the ECJ has clarified, this defence "does not refer to the practice and safety standards in use in the industrial sector in which the producer is active but, with no restrictions, to the state of the scientific and technical knowledge, including its most advanced level, existing at the time when the product in question was marketed. Moreover, this controversial exemption does not refer to the state of knowledge of which the producer in question was aware or could be actually or subjectively aware, but to the objective state of the scientific and technical knowledge of which the producer is presumed to be aware".[11] **30**

As far as losses caused by GMOs are concerned, the development risk defence might not be easily applicable, given the great deal of opinions circulating in the scientific community as to the potential danger of genetically modified organisms. In particular, the development risk defence would conflict with the so-called precautionary principle,[12] which implies a state of uncertainty as to the safety of GM products and requires continuing controls on the side of the producer, even after the product has been put into circulation.[13] In our modest view, the development risk defence should not be applicable in case of losses caused by GM products. **31**

---

11 ECJ C-300/95, *Commission v. United Kingdom* [1997] ECR I-2649.

12 With specific regard to GMOs, the precautionary principle was defined by the EC Commission in its Communication of 02.02.2000 COM (2000)1. A general precautionary principle as to the safety of products is also contained in art. 107 of the Italian Consumer Code.

13 Some French authors have called this continuing obligation to monitor the products "obligation de suivi". See P. *Kourilsky*/G. *Viney*, Le principe de précaution. Rapport au Premier ministre (2000).

(b) Alternative routes

32 Italian Courts may consider the cultivation of GM crops a "dangerous activity" pursuant to art. 2050 CC, which would entail the application of a quasi-strict liability regime. Indeed, pursuant to art. 2050 CC, whoever causes damage in the performance of a dangerous activity is liable to pay compensation if he does not prove that all adequate measures aimed at preventing the damage have duly been taken.

33 Case law has interpreted the notion of dangerous activity in broad terms, extending the notion of "danger" beyond the definitions provided by special laws with regard to specific activities. Even though it may be observed that the cultivation of GM crops in accordance with applicable laws and coexistence plans may not be deemed "dangerous" as such, art. 2050 CC may nonetheless be applicable on account of the precautionary principle, which raises the threshold of the duty of care and may justify the reversal of the burden of proof towards the defendant as provided for by art. 2050 CC.

34 The application of art. 2050 CC to losses caused by agricultural products is, to some extent, more favourable for the victim, as long as the development risk defence is not applicable to liability for dangerous activities. From this point of view, liability for dangerous activities under art. 2050 CC may be attacked in light of the principle affirmed by the ECJ in C-183/00, according to which national liability regimes cannot be more favourable than the liability regime set forth in Directive 85/374/EC.[14] In light of the ECJ's ruling, it might be argued that art. 2050 CC would be legitimately applicable in case of losses caused by the cultivation of GMOs, where liability for defective products would not be applicable. On the other hand, it would not be applicable in case of damage resulting from the commercialisation of the final GM products, where product liability should be exclusively applicable. However, we believe it is unlikely that a liability regime more favourable for the victim will be considered illegitimate by the ECJ, especially if it is examined from the perspective of the highly debated field of GM products, where the precautionary principle plays a dominant role.

---

14 ECJ C-183/00, *María Victoria González Sánchez v. Medicina Asturiana SA* [2002] ECR I-03901.

(c) Impact of compliance with rules and regulations

Under general Italian tort law, a tortfeasor can be held liable even if he acted in compliance with all applicable rules and regulations governing his activity if breach of the general standards of diligence and care can be demonstrated by the injured party. The same holds true also for product liability and liability for dangerous activities.[15] 35

3. Environmental liability

(a) Implementation of the Environmental Liability Directive

The Environmental Liability Directive (2004/35/EC) was implemented in Italy with the enactment of Legislative Decree no.152 of 3 April 2006, also known as the Single Act on the Environment (*Testo Unico Ambientale* or *Codice dell'Ambiente*, hereafter: Single Act). No financial guarantees are currently required pursuant to the relevant provisions contained in Part VI of the Single Act. 36

(b) Environmental liability regime beyond the scope of the Directive

In addition to the rules implementing the Directive, the Single Act contains other provisions establishing a broader liability regime. In particular, art. 311 of the Single Act states that, in case of environmental harm, the Italian Minister of the Environment can alternatively: (a) file a civil liability lawsuit against the liable party to obtain either *restitution in integrum* (specific performance) or monetary compensation (damages); (b) make use of the provisions contained in Part VI of the Single Act (i.e. the rules implementing the Directive). In the event that the Minister elects to file the civil liability claim, the liable party would be whoever committed a wrongful act, or omitted to act when legally required, in breach of any law, regulation, or administrative provision, and through negligence caused harm to the environment, destroying, deteriorating or altering it in whole or in part. Pursuant to art. 313, moreover, the Minister may also issue an administrative order compelling the liable party – in the event that he/she did not voluntarily undertake the reparation measures set forth in 37

15 See *A. Monti/F.A. Chiaves*, Tort and Regulatory Law in Italy, in: W.H. van Boom/M. Lukas/C. Kissling (eds.), Tort and Regulatory Law (2007) 181 ff.

the Directive – to proceed with the *restitutio in integrum* within a predetermined timeframe. If it is impossible or too costly to proceed in whole or in part with the *restitutio in integrum*, then the Minister enjoys the power to issue an injunction to obtain from the liable party the payment, within sixty days, of a sum of money equal to the economic extent of the damage.

(c) Claimants in cases of environmental harm

38 See answer to previous question.

(d) Special liability regime for losses sustained by individuals

39 Loss sustained by individuals in the course of damage to the environment can be recovered on the basis of the general rule in art. 2043 CC. Art. 2050 CC, envisaging a quasi-strict liability regime, would apply in case of loss caused by dangerous activities.

(e) Cartagena Protocol

40 Italy ratified the Convention on Biological Diversity through Law no. 124 of 14 February 1994. It then ratified the Cartagena Protocol on 22 June 2004. Italy, as an EU Member State, is contributing to implementing at regional and national level the "EU Action Plan to 2010 and beyond" annex I to EU Council conclusions "Halting the loss of Biodiversity by 2010", adopted in December 2006. Furthermore, Italy was the first Party to commit as a Government to the IUCN/CE initiative "COUNTDOWN 2010" (www.countdown2010.net). The initiative was launched during the Malahide conference on biodiversity in Europe, under the Irish EU presidency in April 2004. The official Italian engagement was signed on the occasion of the first meeting of the AHOEWG-PA1 in Montecatini. In the last year, Italy has especially been focusing on the elaboration, by 2010, of a national strategy for biodiversity through a participative process with national and local institutions, as well as other stakeholders coming from the private sector and civil society.[16]

---

16 Information taken from the official website of the Convention on Biological Diversity (http://www.cbd.int/).

V. Vicarious liability

1. Scope of vicarious liability

41 Pursuant to art. 2049 CC, "masters and employers are liable for the damage arising from the torts of their servants or employees in the functions for which they have been employed". In other words, vicarious liability under art. 2049 CC arises whenever a servant/employee commits a tort (under the general rule of fault liability of art. 2043 CC) in the performance of the business activity of a third person (the master/employer) to whom the tortfeasor is bound by an employer-employee relationship.

42 Art. 2049 applies regardless of whether the relationship between master and servant is a formal employment relationship, whether it is permanent or temporary, and whether it is remunerated, as long as the master is in a position to direct and control the servant. However, agents and independent contractors are not covered by this rule of liability, as they are considered to act in full autonomy. Principals may be held liable for the acts of their agents and independent contractors under the rule of art. 2043 CC when it can be demonstrated that they breached the general standard of diligence in selecting such agents (*culpa in eligendo*) or they were directly involved in the performance of the agents' activity, so as to deprive the agents of their autonomy and independence.

43 In the area of product liability, it must be noted that art. 3 of the Consumer Code now includes in the definition of "producer" also the intermediaries. The notion of "intermediary" should be equivalent to that of "producer of any raw material or the manufacturer of a component part" contained in art. 3 of Directive 85/374/EC, which establishes the liability of all the operators in the production chain. According to some authors, "intermediaries of the producer" are all the operators involved in the distribution chain, whose activity is apt to affect the safety features of the products.[17] From this perspective, the operators involved in transportation and stocking activities of raw materials may also be liable under the rules of product liability. However, the final producer is always jointly liable with its intermediaries, save for the cases where the product has become defective after it was put into circulation by the producer (art. 119 Consumer Code).

17 *E. Bellisario*, in: G. Alpa/L. Rossi Carleo (eds.), Codice del Consumo. Commentario (2005) 749.

44 In the field of contractual liability, art. 1228 CC provides that the debtor is liable for the acts of the third parties (the "auxiliaries") he uses to perform his obligation, unless it is otherwise agreed between the parties.

2. Liability for people further up the food or feed production chain

45 Under the general rules of vicarious liability, the different operators in the production chain would hardly ever be liable for damage caused by other independent operators in the same production chain. However, contractual liability aside, the rules of product liability may apply. As long as crops containing GMOs are considered to be "defective products", people in the production chain (e.g. the crop retailer) may be found to be jointly liable with people further up in the same production chain (e.g. the farmer).

46 Please note that the special liability regime provided for by art. 5 of Dl 279/04 also applies to the suppliers of technical means of production and to the other operators in the primary production chain. These operators, however, should be independently liable for their own conduct and, in turn, the farmer who proves that he/she used only GMO-free seeds – certified by the relevant public authority and by the producer – would always be exempted from liability.

3. Can someone further down the feed or food chain include someone further up it in a trial against him/herself?

47 See answer to Question 2 above.

VI. Multiple tortfeasors

48 Pursuant to art. 2055 CC, "if the harmful event can be ascribed to multiple tortfeasors, they are all jointly and severally liable" but the tortfeasor who has compensated for the whole amount of damage may recover from the other tortfeasors, in proportion to their respective degree of fault and to the consequences resulting therefrom. However, if the percentage of each tortfeasor's fault is not clear, it is presumed to be equal.[18] The same rule applies to product liability, pursuant to art. 121 Consumer Code.

---

18 For a more detailed analysis of the issues raised under Italian law by cases of multiple tortfeasors see *F.D. Busnelli/G. Comandé*, Multiple Tortfeasors under Italian Law, in: J. Spier (ed.), Unification of Tort Law: Multiple Tortfeasors (2004) 117 ff.

VII. Defences

1. Licence/permission to grow GM material

As already mentioned, under Italian law compliance with all the applicable rules and regulations covering a given sector may not suffice to exclude liability if breach of the general standard of diligence can be demonstrated by the injured party. This principle should be all the more valid if applied to GM products, given the uncertainty as to their safety. Accordingly, the licence to grow GM material should not be, in our opinion, a defence against liability triggered by its use, especially where such material is approved for testing only. 49

2. Consent/assumption of risk

Pursuant to art. 122 par. 2, Consumer Code, "compensation is not due when the victim was aware of the defect of the product and of the danger resulting therefrom but he nonetheless exposed himself to it". In theory, this provision would apply in case of losses caused by GM products. However, it must be taken into account that the commercialisation of GM products implies an assumption of risk also on the side of the producer, who knowingly put into circulation products whose safety is not at all certain. In this regard, it may be argued that the interest of the victim should prevail. 50

3. Third-party influence

Third-party behaviour, like fortuitous events, may break the chain of causation between the conduct of the tortfeasor and the damage, so that liability has to be excluded. However, there might be cases in which third-party behaviour is coupled with negligence on the side of the tortfeasor. For instance, a farmer cultivating GM crops who fails to take adequate segregation measures, thus facilitating sabotage or improper conduct by others, may be held liable notwithstanding the intervention of a third party who, in turn, may be held jointly liable as concurring in the causation of the damage. 51

4. Prescription

52 In cases where art. 2043 or art. 2050 CC are applicable, the general rule laid down in art. 2947 CC would apply, according to which the right to compensation for damage arising from an unlawful fact is prescribed five years from the date on which the fact occurred. Since the late 1960s, courts and scholars have developed the principle of *conoscibilità del danno* (actual or constructive knowledge of the damage), according to which the limitation period begins to run only when the damage has manifested itself, which may be some time after the date of the unlawful act or omission. The Italian Court of Cassation has recently confirmed this principle, clarifying that prescription starts to run only from the moment at which the damage can be perceived by the victim as the consequence of someone's conduct, taking into account ordinary diligence and available scientific knowledge.[19]

53 For cases to which product liability applies, the limitation period is three years from the day on which the victim became aware, or should reasonably have become aware, of the damage, the defect, and the identity of the producer. In case of aggravation of the damage, the limitation period does not start to run before the day on which the victim became aware, or should reasonably have become aware, of damage serious enough to justify legal proceedings (art. 125 Consumer Code). In any event, the right to compensation is forfeited ten years after the product causing the damage was put into circulation (art. 126 Consumer Code).

54 For cases of environmental liability, where art. 304 or art. 305 of the Single Act on the Environment are applicable, if the operator does not comply with the duty to adopt the required preventive or remedial actions, or if it cannot be identified, the Minister of the Environment may take such actions directly, bearing the necessary expenses. The reimbursement of such expenses can be claimed from the liable parties if they are identified within five years of the date of the payment.

5. Other defences

55 N/A.

19 Cass. S.U. 11.01.2008, nos. 581 and 583. See F. *Fusco*, Commencement of the Prescription Period in Case of Damage Caused due to Omissions, in: H. Koziol/B.C. Steininger (eds.), European Tort Law 2007 (2008) 79 ff.

VIII. Remedies

1. Pecuniary compensation

(a) Bodily harm

As of now, regular remedies would apply. 56

(b) Property losses

As of now, regular remedies would apply. 57

(c) Economic losses

As of now, regular remedies would apply. Restoration costs would presumably be included in the loss deemed to be sustained by the organic farmer, who would nonetheless be under an obligation to mitigate damages pursuant to art. 1227 par. 2 CC. 58

(d) Harm to animals

In the absence of any specific provision, the owner of the contaminated animals could recover their commercial value or the diminution thereof, as long as he demonstrated that the animals have actually lost all or some of their full commercial value or that they can only be destined to less remunerative uses. 59

(e) Costs of disposal

Once again, in the absence of any specific provision, disposing costs would only be recoverable if the owner could demonstrate that the animals had lost their full commercial value and must be put down, if still living. 60

2. Non-compensatory damages

Punitive damages are not contemplated in the Italian legal system, which is governed by the general principle that a victim may only be redressed for the damage that he actually suffered and not a cent more. 61

3. Other remedies

62 The general rule under Italian tort law provides for monetary compensation, equivalent to the loss sustained. According to art. 2058 CC, however, the victim may claim "specific restitution" when such is possible in whole or in part. But the judge can exclude specific redress when it is too costly for the tortfeasor.

4. Costs of pursuing a claim

(a) General cost rule

63 Pursuant to art. 91 of the Italian Code of Civil Procedure, the costs of the proceedings must be paid by the losing party. In practice, judges often use their discretionary power to offset the costs sustained by the parties, who may, therefore, end up paying their own procedural costs.

(b) Costs of establishing causation

64 In general, the costs of establishing causation are included in the losses sustained by the victim, and are recoverable as long as they are a direct and immediate consequence of the wrongful conduct of the tortfeasor (in line with the provision contained in art. 1223 CC). Accordingly, a food producer should be able to recover the costs of proving that his products were contaminated by the presence of GMOs at farming level if this investigation was actually needed in order to establish causation.

5. Advance cover

65 N/A.

IX. Cross-border issues – Conflict of laws

1. Conflicts rules applicable before (or instead of) Rome II

66 The Rome II Regulation came into force in Italy in January 2009. It replaces art. 62 and 63 of Law no. 218 of 31 May 1995, dealing with the

law applicable respectively to tortious liability in general and to product liability. In a nutshell, the main differences are as follows.

Art. 62, subs. 1, of Law no.218/1995 provides that the law applicable to tortious liability is the law of the country where the damage occurred. However the victim can claim application of the law of the country where the fact causing the damage took place; art. 4, subs. 1, of the Rome II Regulation states, instead, that "the law applicable to a non-contractual obligation arising out of a tort/delict shall be the law of the country in which the damage occurs irrespective of the country in which the event giving rise to the damage occurred and irrespective of the country or countries in which the indirect consequences of that event occur". The wording of the Regulation aims at eliminating any misinterpretation of the distinction between "damaging event" and "damage", stating very clearly that the relevant criterion, determining the applicable law, is where the damage (i.e. the direct damaging consequences of the tort) occurred. **67**

Art. 62, subs. 2, of Law no. 218/1995 states that wrongful facts involving only citizens of the same country, having their domicile there, are governed by the law of that country. Art. 4, subs. 2, of the Rome II Regulation contains the same rule, the only difference being the specification that the persons involved must have their domicile (they need not be citizens) in the same country at the moment when the damage occurred. Art. 4, subs. 3, of the Regulation also adds that "where it is clear from all the circumstances of the case that the tort/delict is manifestly more closely connected with a country other than that indicated in paragraphs 1 or 2, the law of that other country shall apply. A manifestly closer connection with another country might be based in particular on a pre-existing relationship between the parties, such as a contract, that is closely connected with the tort/delict in question". **68**

Art. 63 of Law no. 218/1995 states that the law applicable in cases of product liability can be chosen by the victim from between (i) the law of the country where the producer is domiciled and (ii) the law of the country where the product was purchased. Instead, art. 5 of the Rome II Regulation provides that the applicable law is that of the country in which the victim had his habitual residence when the damage occurred if the product was marketed in that country; or, failing that, that of the country in which the product was acquired if the product was marketed in that country; or, failing that, that of the country in which the damage occurred if the product was marketed in that country. However, the law applicable is that of the country in which the producer is habitually resident if he could not reasonably foresee the marketing of the product, or a product of the **69** same type, in the country the law of which is applicable under the three previous rules. In any event, where it is clear from all the circumstances of the case that the tort/delict is manifestly more closely connected (e.g. pre-existing contract between the parties) with a country other than that indicated so far, the law of that other country applies.

2. Special regime for cross-border claims

70 N/A.

X. Cases

1. Due to the adventitious presence of GMOs in a field, maize which is normally sold as conventional contains GMOs beyond the legal labelling threshold. This is not discovered before the final stage of the food production chain by the producer of taco chips. The whole production is lost since the supermarket chains refuse to accept delivery from the producer.

(a) Who can sue along the chain of distribution?

71 Neither product liability nor environmental liability seem to be applicable to this case, which would be solved primarily according to the rules of contractual liability. The producer of taco chips would presumably sue the wholesaler of the maize who, in turn, would sue the farmer cultivating the maize for breach of contract. The producer of taco chips would allege that it has lost profits due to the impossibility of selling the taco chips to supermarket chains. Eventually, either the wholesaler or the farmer would be held liable for the presence of GM maize in products which were expected to be GMO-free. According to general rules of contract law, indeed, their fault would be presumed and, in order to escape liability, they would have to prove that the delivery of contaminated maize was due to fortuitous events or, in any event, it was not imputable to them. If the farmer cultivating the maize was held liable, he would have an interest in tracing the cause of the admixture. If he believed that any neighbouring farmer or operators in the primary production chain were responsible for such contamination, he would likely bring an action against them under the special liability regime set forth by Dl 79/04.

(b) Would the case be solved differently if the GMO content was below the labelling threshold?

If the GMO content was below the labelling threshold, the supermarket chains would have no legitimate reason to refuse delivery from the taco chips producer. 72

(c) Would the case be solved differently if the admixture was not adventitious, but occurred due to the disregard of segregation rules, for example?

As mentioned above, if the farmer cultivating the maize was held liable, he would have an interest in tracing the cause of the admixture. If he could detect the source of the contamination and trace it back to a neighbouring farmer or to another operator in the primary production chain, the latter could be held liable for their breach of segregation rules, if any, under Dl 279/04. 73

(d) Would the case be solved differently if the GMO found was not admitted for production in your jurisdiction?

In such a case, the farmer or operator who was found to be liable for admixture would also be subject to criminal sanctions if it was ascertained that he cultivated or dealt with GMOs prohibited in Italy. If the source of the GMO was traced to an operator in a neighbouring country, the Italian farmer who sold the contaminated maize – and who was found to be liable for the loss sustained by the wholesaler and/or by the taco producer – could sue the foreign operator under Italian law, according to the Rome II Regulation. 74

(e) If the admixture had occurred on a non-GM field and it transpires that the GM seeds were blown from: (i) neighbouring fields; or (ii) a truck passing by, would the farmer of the affected field be liable for all or part of the loss caused further down the distribution chain?

The farmer of the affected field would be held liable for breach of contract unless he can prove fortuitous event (unprecedentedly strong winds blowing GM seeds from remote sites) or third parties' conduct, not detectable 75 using ordinary diligence. In the latter case, the third parties to whom admixture is attributable could ultimately be held liable under Dl 279/04 or under art. 2050 CC.

2. Twenty years after the sale of GM maize used for food products, it turns out that it has certain disadvantageous health effects for humans.

(a) Can the producers be held liable at this point for risks unknown at the time of growing the maize? Who would be liable?

76 Even if any actual damage occurs, an action for product liability would be time barred pursuant to art. 126 of the Consumer Code. The victims could nonetheless bring an action under art. 2043 and/or art. 2050 CC within five years from the moment at which they perceived the damage as the consequence of someone's conduct, taking into account ordinary diligence and available scientific knowledge. Hence, the alleged tortfeasors (most likely the seed producer and the food producer) could be held jointly liable for the damage caused to the victims' health, in proportion to their respective contribution.

(b) Can compensation already be claimed at a point when the negative health effects have not yet materialised, but are to expected to according to scientific expertise/mere rumours?

77 Under Italian law, only actual damage can be recoverable. See, however, footnote 8.

(c) Would it make any difference if the GM maize had only been in use for feed, causing harm to the animals, which may or may not cause harm to humans consuming the meat as well?

78 Once again, only actual damage can be recoverable.

3. The driver of a food logistics company discovers that a farmer, from where he regularly picks up agricultural products fails to obey mandatory segregation rules or food or feed hygiene standards, which may lead to the admixture of GM and non-GM produce sold and packaged separately by that farmer. Does he or his employer have a duty to warn, i.e. warn the recipient of the allegedly "non-GM" produce?

Food logistics companies are not under a specific obligation to warn, unless the conduct of the farmer constitutes a crime, in which case they must inform the competent public authority. As a general rule, mere carriers do not have a duty to police the quality of the products they transport. However, pursuant to art. 104 Consumer Code, producers and distributors have a duty to market only safe products and to properly inform consumers about their features. In particular, distributors must act diligently to ensure that only safe products are put into circulation, and are under a duty not to distribute products which they know to be dangerous, or should have known to be dangerous according to information available to them in their capacity as professional operators. Moreover, distributors must take part in the safety controls, providing the competent authorities with any relevant information as to the risks of the products. **79**

Damage Caused by GMOs under Liechtenstein Law

Märten Geiger

I. General overview

1. Special liability or redress scheme for GMOs

1 The current Liechtenstein legal order contains only one specific provision as regards the liability regime for losses caused by GMOs. Art. 24 of the Act of 17 December 1998 on handling genetically modified or pathogenic organisms (GPOG)[1] provides for a strict liability regime. Exculpation is possible only in case of force majeure or gross negligence of the injured or another third party. In addition, the provisions of general tort law in the Civil Code[2] are applicable, especially its §§ 1293 ff.

2 This provision is strongly inspired by the former art. 59a of the Swiss Federal Act on the protection of the environment,[3] as in force between 1 July 1997 and 1 January 2004. Consequently, Swiss materials concerning this provision are of interest as well when reflecting on the Liechtenstein provision.

3 In March 2009, the Liechtenstein Government presented a draft bill on a new Act on handling organisms to the Liechtenstein legal order.[4] This draft bill was submitted to public consultation until the beginning of July 2009. Based on the comments received, it is expected that the Govern-

---

1 Law of 17 December 1998 über den Umgang mit gentechnisch veränderten oder pathogenen Organismen (GPOG), LR 816.1 (LR is an abbreviation for "Systematische Sammlung der liechtensteinischen Rechtsvorschriften" which means the systematic compendium of the Liechtenstein law, http://www.gesetze.li).

2 Allgemeines Bürgerliches Gesetzbuch (ABGB) of 1 July 1811, LR 210.0.

3 Federal Law of 7 October 1983 über den Umweltschutz (Umweltschutzgesetz, USG), SR 814.01.

4 Vernehmlassungsbericht der Regierung betreffend die Schaffung eines Gesetzes über den Umgang mit Organismen: http://www.llv.li/amtsstellen/llv-rk-amtsgeschaefte-vernehmlassungen/llv-rk-amtsgeschaefte-vernehmlassungen-abgelaufene_2009.htm.

ment will present its final legislative proposal in late spring 2010. If legislative procedure is not delayed, the new Act could be dealt with by Parliament in summer 2010 and enter into force at the end of the same year.

The draft bill provides a number of new provisions on liability issues, retracing the respective developments of the last years in Swiss law. Concretely, the proposed Act reflects almost word by word the provisions in the Swiss Federal Act relating to non-human gene technology,[5] in force since 1 January 2004 and replacing the former art. 59a of the Swiss Federal Act on the protection of the environment, mentioned above (no. 2). Therefore, once the new Act on handling organisms is in force in Liechtenstein, the respective comments in this report relating to the Swiss provisions will be of relevance for Liechtenstein as well. 4

However, as already mentioned in no. 1, art. 24(4) GPOG as well as art. 57(8) of the proposed new Act on handling organisms contain a *general* reference to the general tort law of the Liechtenstein Civil Code, which is very strongly inspired by the Austrian Civil Code. In the Swiss archetypes of the Federal Act on the protection of the environment as well as of the Federal Act relating to non-human gene technology, there is only a reference to some selected provisions of the Swiss Civil Code. Consequently, this might possibly lead to the situation that there be different interpretations of the more or less identical provisions in Liechtenstein and in Switzerland as soon as the general tort law is in play. As the Liechtenstein courts never had to deal with cases based on the GPOG – at least no such case has ever been published – this is a purely academic statement until now. 5

2. State liability

The state acting as a private person, for example as a farmer growing GMO crops, is submitted to the same rules outlined above as any other private actor. Accordingly, it can be held liable for losses caused by GMOs under the same conditions as any other private actor. 6

Besides this, in execution of its sovereign tasks, the state is submitted to the special regime of the Act on government liability.[6] In addition, due to its membership of the European Economic Area (EEA), extending the 7

5 Federal Law of 21 March 2003 über die Gentechnik im Ausserhumanbereich (Gentechnikgesetz, GTG), SR 814.91 (SR is an abbreviation for "Systematische Rechtssammlung" which means the systematic compendium of the Swiss federal law, http://www.admin.ch/ch/d/sr/sr.html).

6 Law of 22 September 1966 über die Amtshaftung, LR 170.32.

internal market of the European Union (EU) to the three members of the European Free Trade Association (EFTA) Iceland, Norway and Liechtenstein, the community rules on state liability are applicable as well.[7]

II. Damage

1. Recoverable losses

8 As there is no special definition of damage in the context of the GPOG, the general regime in §§ 1293 ff. of the Civil Code applies. Accordingly, damage to person and property is covered, including lost profit. In addition, contrary to the Swiss archetype in the former art. 59a of the Swiss Federal Act on the protection of the environment,[8] liability under the GPOG also includes environmental damage, as explicitly stated in the Government's explanation to its bill.[9]

2. Pure economic loss

9 Under the liability regime of the Civil Code, pure economic loss is only recoverable in case of liability *ex contractu*. The GPOG does not contain any deviating special provision, extending the protection of legal interests to pure economic loss.[10]

3. Mere fear of a loss

10 The mere fear of a loss triggered by GMOs is not recoverable, as long as there is no materialised damage. This might be the case if the price of land decreases after use of GMOs in the area or in case of farmers whose customers fear – even mistakenly – that the products are no longer GMO-free.

---

7 In parallel to the judicial development in the EU, the EFTA Court introduced the principle of state liability into the EEA in case E-9/97, *Erla María Sveinbjörnsdóttir* [1998] EFTA Court Report 97.

8 See as well *H.R. Trüeb* in: Vereinigung für Umweltrecht and *H. Keller* (eds.), Kommentar zum Umweltschutzgesetz (2nd ed. 2004) no. 68 ff.

9 Bericht und Antrag (BuA) der Regierung an den Landtag des Fürstentums Liechtenstein zur Schaffung eines Gesetzes über den Umgang mit gentechnisch veränderten oder pathogenen Organismen, No. 74/1998, 43.

10 For the same result regarding the former art. 59a of the Swiss Federal Act on the protection of the environment, see *M. Brülhart*, Gentechnik und Haftpflicht (2003) 145.

4. Standard of proof

Due to the fact that the GPOG does not contain any special provision, the general tort law of the Civil Code applies. Accordingly, it is the claimant's task to prove his losses, including future losses. In order to convince a court, a very high level of probability, close to certainty, is required. See also infra no. 18. **11**

5. Nominal losses

Nominal/symbolic losses are not recognised in the Liechtenstein jurisdiction. The existence of a real loss is one of the preconditions of liability. **12**

6. Mass losses

There are no explicit special rules for mass losses in Liechtenstein law. **13**

III. Causation

1. Uncertainty of merely potential causes

Due to the fact that the GPOG does not provide for a specific rule as regards potential causes, one has to refer to general tort law. Consequently, all actors are jointly and severally liable, unless one actor proves that he has not caused the damage.[11] The tortfeasor who compensated the loss has a right of recourse against the other tortfeasor. **14**

The same applies as well in cases where causes possibly lie within the victim's own sphere. Based on § 1304 Civil Code, the injured party will see the amount of recoverable damage being reduced according to the seriousness of its own contributory misconduct. However, predominant fault on one side can justify full liability by one of the involved parties. **15**

11 *R. Reischauer* in: P. Rummel (ed.), Kommentar zum Allgemeinen bürgerlichen Gesetzbuch (3rd ed. 2007) § 1302 no. 12.

2. Complex causation scenarios

16 In absence of special rules, the general principles outlined above are applicable.

3. Force majeure

17 Force majeure interrupts the causal link. As art. 24(2) GPOG states explicitly, a defendant is exempted from liability if he or she can prove that the damage was caused by an Act of God or through gross misconduct of the injured or a third party.

4. Threshold to prove causation

18 In general, a very high level of probability, close to certainty, is required in order to convince judges that the defendant caused the injury. This is the general rule of tort law, to which the GPOG does not contain any specific amendment.

19 However, in the draft bill on a new Act on handling organisms proposed by the Liechtenstein Government in March 2009, art. 61 provides for the possibility to accept lower degrees of likelihood in certain situations.

5. Special rules on causation

20 Under the current regime of the GPOG, there are no such special rules. As just mentioned above, this might change with the new legislation as recently proposed by the Liechtenstein Government.

IV. Types of liability

1. Fault liability

(a) Special rules governing fault

21 In principle, the burden of proof for all elements of liability is with the plaintiff. In certain situations, the plaintiff only has to establish *prima facie* evidence, especially if the defendant has violated a protective legal provision that was designed to protect against the kind of damage that has

occurred (§ 1311 Civil Code). As far as the GPOG – and the future act on handling organisms – can be qualified as protective law, the burden of proof will be eased for the plaintiff.

The standard of care changes for experts in the sense of § 1299 Civil Code. **22** As GMO-farmers or other actors dealing with GMOs will most probably be considered such experts, an objective standard of fault is applied.

(b) Impact of specific rules of conduct

In case of violation of such specific rules, it will be more difficult for a **23** defendant to exculpate himself, especially if he is qualified as an expert in the sense of § 1299 Civil Code.

2. Product liability

(a) Development risk defence

Liechtenstein is not member of the European Union but of the European **24** Economic Area (EEA), extending the Internal Market to the three EFTA-States Iceland, Norway and Liechtenstein. Directive 85/374/EC has been taken over into the EEA Agreement and has been implemented in Liechtenstein law by the Act dated 12 November 1992 on product liability.[12]

The development risk defence in the sense of art. 7 lit. e of the Directive **25** mentioned above has been incorporated into Liechtenstein Law in art. 7(1)(e) of the Act on product liability. As there is no published case law as regards this provision, its handling in practice in Liechtenstein is still undefined. However, due to the fact that the Liechtenstein legislator was strongly inspired by the Austrian act on product liability, future Liechtenstein case law will take into account the Austrian practice.

(b) Alternative routes

The extension of the scope of Directive 85/374/EC to agricultural products **26** through Directive 1999/34/EC has been incorporated into the EEA Agreement as well. Accordingly, the respective exemption in art. 5(2) of the Liech-

12 Law of 12 November 1992 über die Produktehaftpflicht, LR 215.112.2.

tenstein Act on product liability has been deleted. Besides this, there are no further routes for compensating losses caused by (agricultural) products.

(c) Impact of compliance with rules and regulations

27 As long as the product as such is defective, the fact that the defendant has obeyed all rules and regulations governing his production process does not affect his product liability which is not based on fault.

3. Environmental liability

(a) Implementation of the Environmental Liability Directive

28 The Directive has been implemented by the Act on the protection of the environment,[13] especially its art. 11–13. There is no obligation as regards financial guarantees and/or assurance.

(b) Environmental liability regime beyond the scope of the Directive

29 There is no such liability regime in Liechtenstein.

(c) Claimants in cases of environmental harm

30 As already mentioned supra no. 1, liability under the GPOG also includes environmental damage, as explicitly stated in the Government's explanation to its bill.[14] This departs from the Swiss archetype in the former art. 59a of the Swiss Federal Act on the protection of the environment. However, in the absence of relevant case law, the limits of the range of losses which can be recovered remain unclear.

31 All those who are affected in their individual pecuniary interests can recover losses. This excludes all environmental non-governmental organi-

13 Umweltschutzgesetz (USG) of 29 May 2008, LR 814.01.

14 Bericht und Antrag (BuA) der Regierung an den Landtag des Fürstentums Liechtenstein zur Schaffung eines Gesetzes über den Umgang mit gentechnisch veränderten oder pathogenen Organismen, No. 74/1998, 43.

sations from the group of actively legitimated actors. They only have the possibility to request the competent state authority to take action, as foreseen in art. 12 of the Environmental Liability Directive 2004/35/EC.

(d) Special liability regime for losses sustained by individuals

There is no such liability regime in Liechtenstein law. **32**

(e) Cartagena Protocol

International agreements in the field of environmental protection and **33** shipment of dangerous goods – such as the Cartagena Protocol – are in most cases signed by Liechtenstein. However, due to the small size of the Liechtenstein administration, ratification sometimes only follows years later.

As regards the Cartagena Protocol on Biosafety, it has not been signed by **34** Liechtenstein. Nevertheless, the Protocol is applied in Liechtenstein indirectly, based on the Customs Union Treaty with Switzerland.[15] The implementing Swiss Ordinance on cross-border transfer of GMOs[16] has been included in the list of Swiss provisions which are applicable in Liechtenstein due to the Customs Union Treaty.[17] Hence, companies must obtain authorisation from the importing country prior to shipment of GMOs and are obliged to provide detailed information on the product.

4. Other strict liability regimes

There are no further strict liability regimes applicable in cases of harm **35** caused by GMOs.

15 Vertrag vom 29. März 1923 zwischen der Schweiz und Liechtenstein über den Anschluss des Fürstentums Liechtenstein an das schweizerische Zollgebiet, LR 0.631.112.

16 Verordnung vom 3. November 2004 über den grenzüberschreitenden Verkehr mit gentechnisch veränderten Organismen (Cartagena-Verordnung), SR 814.912.21.

17 Kundmachung vom 23. Juni 2009 der aufgrund des Zollvertrages im Fürstentum Liechtenstein anwendbaren schweizerischen Rechtsvorschriften, LR 170.551.631, 51.

V. Vicarious liability

1. Scope of vicarious liability

36 Art. 24 GPOG attaches liability to the tenant of a business or an installation dealing with GMOs. This liability also encompasses liability for auxiliary persons, including independent contractors.[18]

2. Liability for people further up the food or feed production chain

37 In principle, every actor is liable for those incidents occurring in his or her sphere of influence only. However, art. 24(3) GPOG explicitly states that the person responsible for the first placing on the market of a GMO is also liable for those losses which are caused to third parties even if all rules of conduct have been respected. According to the Government's explanatory note to the GPOG,[19] this provision aims at protecting all those actors producing GMO-free products which suffer from a contamination further up the production chain.

3. Can someone further down the feed or food chain include someone further up it in a trial against him/herself?

38 Referring to art. 24(3) GPOG mentioned above, a farmer is explicitly enabled to bring someone further up the production chain into a lawsuit in order to escape liability.

VI. Multiple tortfeasors

39 Cases of multiple tortfeasors have to be dealt with according to general tort law. The GPOG does not contain specific provisions. Hence, according to § 1302 Civil Code, multiple tortfeasors who have caused damage concertedly and intentionally are jointly and severally liable. Among themselves, they have a right of recourse based on § 896 Civil Code.

---

18 *Trüeb* (fn. 8) 27, no. 53.

19 BuA No. 74/1998 (fn. 9) 44.

If the tortfeasors have not acted in a concerted manner or acted by negligence only, the injured party first has to try to apportion the damage among the tortfeasors according to their share in the damage. If this is not possible, he is entitled to claim full damages from each defendant. Again, they are jointly and severally liable with a right of recourse among themselves. 40

VII. Defences

1. Licence/permission to grow GM material

Art. 24 GPOG provides for a strict liability regime. Exculpation is possible only in case of force majeure or gross negligence of the injured or another third party. Consequently, a licence to grow or to test genetically modified material does not serve as a defence against liability triggered by its use, as the damage is the realisation of the risk the GPOG is focusing on. 41

2. Consent/assumption of risk

As mentioned supra no. 1, there are only three possibilities for exculpation: force majeure and gross negligence of the injured or of another third party. Consequently, consent/assumption of risk does not affect liability deriving from the strict liability regime of the GPOG. 42

3. Third-party influence

Third-party behaviour is the basis of an exculpation of the defendant as far as it can be qualified as grossly negligent (art. 24(2) GPOG). Lower degrees of fault by third parties lead to situations of joint and several liability.[20] 43

4. Prescription

In absence of special provisions in the GPOG, the general statutes of limitation of § 1489 Civil Code apply, as for most other tort cases. Concretely, this provision provides for a relative prescription period of three years and an absolute one of thirty years. 44

20 *Trüeb* (fn. 8) 39, no. 90.

5. Other defences

45 In certain cases, the defence might be based on the above-mentioned art. 24(3) GPOG. It states explicitly that the person responsible for the first placing on the market of a GMO also remains liable for those losses which are caused to third parties further down the production chain even if all rules of conduct have been respected. This might be helpful for an actor producing GMO-free products suffering from a contamination further up the production chain.

VIII. Remedies

1. Pecuniary compensation

(a) Bodily harm

46 There are no special rules, the regular remedies apply.

(b) Property losses

47 There are no special rules, the regular remedies apply.

(c) Economic losses

48 In Liechtenstein law, there are no special rules on compensating economic losses caused by GMOs. The regular system applies and the loss which has to be recovered is the concrete loss the injured party is faced with.[21] The situation existing before the damaging incident shall be re-established. Consequently, an organic farmer can recover the full costs of restoring a contaminated field.

(d) Harm to animals

49 Harm to animals is qualified as harm to the pecuniary interests of their owner.[22] As mentioned supra no. (c), the loss to recover is the concrete loss the injured party is faced with. If the animal was used for the production of milk or meat, the value of the animal includes this potential as well.

---

21 Rummel/*Reischauer* (fn. 11) § 1293 no. 2b.
22 Ibid., § 1332a no. 2.

It is difficult to imagine that a court would qualify the mere fact that an animal eats GM contaminated feed as harm to the animal. However, if the owner of the animal is an organic farmer producing meat, this would constitute damage to his or her pecuniary interests, which can be recovered because the animal can no longer be used as foreseen. **50**

(e) Costs of disposal

As the costs of disposing of contaminated production/animals fed with GM feed affect the pecuniary interests of the injured party, these costs fall under the damage which can be recovered. **51**

2. Non-compensatory damages

The Liechtenstein jurisdiction does not recognise punitive, exemplary or any other form of non-compensatory damages. **52**

3. Other remedies

Beside pecuniary compensation, the Liechtenstein tort law foresees restoration in kind (§ 1323 Civil Code). **53**

4. Costs of pursuing a claim

(a) General cost rule

According to § 41 of the Code of Civil Procedure,[23] the general rule of the "loser pays" principle applies and the costs of a civil trial have to be borne by the losing party. **54**

(b) Costs of establishing causation

In general, all those costs which have been necessary in order to prove a party's position in court are recoverable. Their necessity has to be assessed by the competent judge on a case-by-case basis. **55**

23 Law of 10 December 1912 über das gerichtliche Verfahren in bürgerlichen Rechtsstreitigkeiten (Zivilprozessordnung), LR 271.0.

5. Advance cover

56 Art. 25(1) GPOG obliges all tenants of enterprises or installations dealing with GMOs to contract insurance covering their liability. Par. 2 of the same provision enables the Government to define, if necessary,[24] the exact scope of this insurance coverage. However, when deciding the Ordinance on the GPOG,[25] no such specification has been included.

IX. Cross-border issues – Conflict of laws

1. Conflicts rules applicable before (or instead of) Rome II

57 In the absence of special rules in the GPOG, situations of conflict of laws concerning Liechtenstein have to be resolved on the basis of the statute on international private law.[26] According to its art. 52, in case of claims for damages based on general tort law, the right of the state where the damaging conduct occurred has to be applied.

2. Special regime for cross-border claims

58 There is no such special regime in Liechtenstein law.

X. Cases

1. Due to the adventitious presence of GMOs in a field, maize which is normally sold as conventional contains GMOs beyond the legal labelling threshold. This is not discovered before the final stage of the food production chain by the producer of taco chips. The whole production is lost since the supermarket chains refuse to accept delivery from the producer.

(a) Who can sue along the chain of distribution?

59 All actors along the chain of production having suffered damage can sue the tenant of the enterprise or installation dealing with GM-material responsible for its presence in the normally GMO-free field.

---

24 Explicitly stated in the Government's explanatory memorandum to the GPOG, BuA No. 74/1998 (fn. 9) 44.

25 Verordnung vom 20. April 1999 zum Gesetz über den Umgang mit gentechnisch veränderten oder pathogenen Organismen, LR 816.11.

26 Law of 19 September 1996 über das internationale Privatrecht (IPRG), LR 290.

(b) Would the case be solved differently if the GMO content was below the labelling threshold?

If the GMO content was below the labelling threshold, there would be no damage to allow the taco producer or others to claim for damages. 60

(c) Would the case be solved differently if the admixture was not adventitious, but occurred due to the disregard of segregation rules, for example?

If the disregard of segregation rules is qualified by the competent court as grossly negligent behaviour, the tenant of the enterprise or installation dealing with the genetically modified material can be exculpated on the basis of art. 24(2) GPOG. Accordingly, the person who acted wrongfully becomes liable. 61

(d) Would the case be solved differently if the GMO found was not admitted for production in your jurisdiction?

No, the liability aspect would not change. However, the actor having introduced the GMO which was not admitted for production would have to face additional criminal responsibility pursuant to art. 29 GPOG. 62

(e) If the admixture had occurred on a non-GM field and it transpires that the GM seeds were blown from: (i) neighbouring fields; or (ii) a truck passing by, would the farmer of the affected field be liable for all or part of the loss caused further down the distribution chain?

The farmer of the affected non-GM field does not fall under the scope of the GPOG, thus is not submitted to its liability provisions. In order to protect him from liability in cases as described above, art. 24(3) GPOG explicitly states that the person responsible for first putting the GMO on the market remains liable in such situations. 63

2. Twenty years after the sale of GM maize used for food products, it turns out that it has certain disadvantageous health effects for humans.

(a) Can the producers be held liable at this point for risks unknown at the time of growing the maize? Who would be liable?

64 Under the regime of the GPOG, the producer can – most probably – be held liable for development risks. Due to the fact that this was not clearly stated in the Swiss archetype for the GPOG, there was an intensive discussion in Switzerland about this question.[27] In the meantime, the legal situation in Switzerland has been amended in order to clarify that development risks are covered as well. With the new Liechtenstein Act on handling organisms, currently with the legislator, this development on the Swiss side will be reproduced.

(b) Can compensation already be claimed at a point when the negative health effects have not yet materialised, but are to expected to according to scientific expertise/mere rumours?

65 Compensation can only be claimed once damage has materialised.

(c) Would it make any difference if the GM maize had only been in use for feed, causing harm to the animals, which may or may not cause harm to humans consuming the meat as well?

66 No, this would not make any difference as long as concrete damage materialises, e.g. negative health effects to humans after consumption of the meat.

27 See *Trüeb* (fn. 8) 45, no. 104 ff.

3. The driver of a food logistics company discovers that a farmer, from where he regularly picks up agricultural products fails to obey mandatory segregation rules or food or feed hygiene standards, which may lead to the admixture of GM and non-GM produce sold and packaged separately by that farmer. Does he or his employer have a duty to warn, i.e. warn the recipient of the allegedly "non-GM" produce?

Beside eventual duties to warn *ex contractu*, there is no general provision **67** foreseeing such an obligation in the context of the GPOG. However, if liability cases arise at a later stage in action by the food logistics company against the farmer and it appears that the driver knew about the disregard of the mandatory segregation rules, this might be qualified as contributory misconduct in the sense of § 1304 Civil Code. Consequently, the food logistics company will see the amount of recoverable damage reduced according to the seriousness of its own contributory misconduct. However, depending on the concrete circumstances, the faulty behaviour of the farmer might still be qualified as predominant fault, justifying full liability of him alone.

Damage Caused by GMOs under Luxembourg Law

Patrick Goergen

I. General overview

1. Special liability or redress scheme for GMOs

1 In the Grand-Duchy of Luxembourg, the legal framework for the use and marketing of GMOs is regulated by the Law dated 13 January 1997[1] as amended by the Law dated 13 January 2004[2] (hereafter named "the GMO Law"). This Law contains provisions of general importance (subject matter, definitions, technical means of genetic modification, worthiness and professional qualification requirements), provisions regarding the use of GMOs (exclusions, classification of GMOs and their use, principles, risk evaluation, authorisation request, public consultation, authorisation modalities, principles of good micro-biological practice, registration), provisions regarding intentional dissemination and placing on the market of GMOs (risk evaluation, principles, authorisation request, public consultation, administrative decision, intra EU information exchange) as well as miscellaneous provisions (ministry committee, confidentiality, preventive measures, measures in the case of accidents, cooperation with the EU Commission and other Member States, liability, inspection, withdrawal of authorisations, court actions, control powers, criminal sanctions). Sev-

---

1 Loi du 13 janvier 1997 relative au contrôle de l'utilisation et de la dissémination des organismes génétiquement modifiés, Mémorial A 1997 of 24 January 1997, 10 and 13 February 1997, 584 (rectificatif).

2 Loi du 13 janvier 2004 modifiant la loi du 13 janvier 1997 relative au contrôle de l'utilisation et de la dissémination des organismes génétiquement modifiés, Mémorial A-5 of 23 January 2004, 22 (Parliament document no 4913). This Law implements Dir. 98/91/EC and 01/18/EC.

eral executive regulations have been issued for the purpose of implementing the GMO Law.[3]

This legal framework is completed by a Law dated 18 March 2008[4] (hereafter named "the Coexistence Law") which replaces the Law of 9 November 1971 regulating trade in seeds and plants[5] and introduces, on the basis of Art. 26 para. 1 of Directive 2001/18,[6] measures to manage the coexistence of genetically modified crops and conventional and organic crops. 2

3 See for example, Règlement grand-ducal du 4 avril 2005 portant sanction et exécution des dispositions – du règlement (CE) N° 1829/2003 du Parlement Européen et du Conseil du 22 septembre 2003 concernant les denrées alimentaires et les aliments pour animaux génétiquement modifiés et – du règlement (CE) N° 1830/2003 du Parlement Européen et du Conseil du 22 septembre 2003 concernant la traçabilité et l'étiquetage des organismes génétiquement modifiés et la traçabilité des produits destinés à l'alimentation humaine ou animale produits à partir d'organismes génétiquement modifiés, et modifiant la Dir. 2001/18/CE. Mémorial A-52 of 20 April 2005, 828 – Règlement grand-ducal du 18 avril 2004 déterminant les exigences en matière de surveillance des organismes génétiquement modifiés ayant fait l'objet d'une autorisation, Mémorial A-56 of 27 April 2004, 870 – Règlement grand-ducal du 18 avril 2004 déterminant les lignes directrices pour l'établissement des rapports d'évaluation en matière de demandes de mise sur le marché d'organismes génétiquement modifiés, Mémorial A-56 of 27 April 2004, 871 – Règlement grand-ducal du 18 avril 2004 déterminant les principes applicables à l'évaluation des effets néfastes potentiels sur la santé humaine et l'environnement lors des demandes d'autorisation de projets de dissémination volontaire et celles de projets de mise sur le marché d'OGM, Mémorial A-56 of 27 April 2004, 872 – Règlement grand-ducal du 17 octobre 2002 modifiant le règlement grand-ducal du 6 décembre 1999 fixant les critères de classement des organismes génétiquement modifiés et de leurs utilisations et définissant les mesures de sécurité et les modalités de confinement relatives à ces utilisations, Mémorial A-120 of 31 October 2002, 2892 – Règlement grand-ducal du 5 octobre 2001 déterminant les informations que doivent contenir les demandes d'autorisation de projets d'utilisation confinée d'organismes génétiquement modifiés. Mémorial A-128 of 18 October 2001, 2591 – Règlement grand-ducal du 10 mars 2000 déterminant les informations que doivent contenir les demandes d'autorisation de projets d'utilisation confinée d'organismes génétiquement modifiés, Mémorial A-24 of 17 March 2000, 638 – Règlement grand-ducal du 12 juin 1998 concernant l'étiquetage et l'emballage des produits consistant en organismes génétiquement modifiés ou en contenant, Mémorial A-46 of 22 June 1998, 698- Règlement grand-ducal du 17 avril 1998 déterminant les informations que doivent contenir les demandes d'autorisation de projets de dissémination volontaire d'OGM et de projets de mise sur le marché d'OGM, Mémorial A-32 of 28 April 1998, 458 (amended on 18 April 2004, Mémorial A-56 of 27 April 2004, 876) – Règlement grand-ducal du 6 décembre 1999 fixant les critères de classement des organismes génétiquement modifiés et de leurs utilisations et définissant les mesures de sécurité et les modalités de confinement relatives à ces utilisations, Mémorial A-143 of 14 December 1999, 2590.

4 Loi du 18 mars 2008 sur la commercialisation des semences et plants ainsi que sur la coexistence des cultures génétiquement modifiées, conventionnelles et biologiques, Mémorial A-32of 27 March 2008, 446.

5 Loi du 9 novembre 1971 portant réglementation du commerce des semences et plants, Mémorial A-77 of 17 November 1971, 2030 (Parliament document 1492).

6 Directive 2001/18/EC of the European Parliament and of the Council of 12 March 2001 on the deliberate release into the environment of genetically modified organisms and repealing Council Dir. 90/220/EEC, OJ L 106, 17.04.2001, 1.

3 Compulsory segregation distances between GMO and conventional cultures are, by implementation of the Coexistence Law, set at 600 metres for maize, 100 metres for beet and 50 metres for potatoes.[7]

4 Since the entry into force of the Coexistence Law, the Grand-Duchy of Luxembourg has applied a strict liability regime for everyone growing genetically modified seed and plants, imposing liability for economic damage caused by the adventitious presence of the genetically modified organism of this variety in the produce of neighbouring plots and in the yields of honey or pollen proceeding from neighbouring apiaries.[8]

5 The latest piece of specific legislation is the Law of 20 April 2009 on the prevention and remedying of environmental damage[9] (hereafter named "the Environmental Liability Law"). The Environmental Liability Law implements EU Directive 2004/35/EC of 21 April 2004,[10] as amended by Directive 2006/21/EC,[11] and aims to establish a framework of environmental liability based on the "polluter pays principle", to prevent and remedy environmental damage.[12]

6 Given the absence of any special regime before the adoption of the Environment Liability Law, standard tort law rules applied to the compensation of environmental damage. As the civil liability rules of the Luxembourg Civil Code only protect people and private property, environmental damage might only have been compensated if it also constituted personal injury or damage to private property. As the environment is part of *res nullius*, environmental damage outside of private property rights was not likely to be compensated under traditional tort law mechanisms. The

---

7 Règlement grand-ducal du 22 juillet 2008 fixant les distances d'isolement et les conditions techniques à respecter lors de la mise en culture de semences et plants génétiquement modifiés, Mémorial A-121 of 20 August 2008, 1824.

8 Art. 15 of the Coexistence Law, which, in its French version, reads as follows: "Quiconque met en culture des semences et des plants génétiquement modifiées est responsable de plein droit du préjudice économique résultant de la présence fortuite de l'organisme génétiquement modifié de cette variété dans la production de parcelles avoisinantes portant des cultures non génétiquement modifiées et dans la production de miel ou de pollen provenant de ruchers avoisinants".

9 Loi du 20 avril 2009 relative à la responsabilité environnementale en ce qui concerne la prévention et la réparation des dommages environnementaux, Mémorial A-82 of 27 April 2009, 968 (Parliament document no 5877).

10 Directive 2004/35/EC of the European Parliament and of the Council of 21 April 2004 on environmental liability with regard to the prevention and remedying of environmental damage, OJ L 143, 30.04.2004, 56.

11 Directive 2006/21/EC of the European Parliament and of the Council of 15 March 2006 on the management of waste from extractive industries and amending Dir. 04/35/EC, OJ L 102, 11.04.2006, 15.

12 Art. 1 of the Environmental Liability Law.

Environmental Liability Law, which faithfully implements the European Directive, has now introduced either strict liability or fault-based liability, depending on the activity that caused the environmental damage or the threat of such damage.

A strict liability regime would apply to dangerous or potentially danger- **7**
ous occupational activities as listed in Annex III of Directive 2004/35/EC. This list mainly comprises installations subject to permit pursuant to Council Directive 96/61/EC,[13] discharges into inland surface water and all discharges of substances into groundwater, waste management operations, installations releasing polluting substances into the air as well as any contained use, including transport, involving genetically modified micro-organisms. Such a regime allows for the imposition of liability on the operator even in the absence of any fault.

Fault-based liability would, on the other hand, apply to any occupational **8**
activities other than those listed in Annex III, but only in the case of damage caused to protected species and natural habitats.

The Environmental Liability Law only applies to activities carried out in **9**
the course of an economic activity, a business or an undertaking. For any activities which are not of an occupational character, standard tort law rules shall apply.[14]

The Law does not cover environmental damage or any imminent threat of **10**
such damage arising from an incident in respect of which liability or compensation falls within the scope of the international conventions ratified by the Grand-Duchy of Luxembourg.[15] Damage caused by pollution of diffuse character requires the establishment of a causal link between the damage and the activities of individual operators.[16]

Luxembourg lawmakers, using the option foreseen by Art. 8 para. 4. of **11**
Directive 2004/35, have exempted the operator from bearing the cost of remedial actions where he demonstrates that he was neither at fault nor negligent and that the environmental damage was caused by an emission or activity or any manner of using a product in the course of an activity which the operator demonstrates was not considered likely to cause envi-

---

13 Council Directive 96/61/EC of 24 September 1996 concerning integrated pollution prevention and control, OJ L 257, 10.10.1996, 26, last amended by Reg. (EC) no. 1882/2003.

14 Opinion of the Council of State of 23 September 2008, see Parliament document 5877-5 of 2 October 2008, 3.

15 Art. 5.2. of the Environmental Liability Law.

16 Art. 5.5. of the Environmental Liability Law.

ronmental damage according to the state of scientific and technical knowledge at the time when the damage occurred.[17] The burden of proof (which is a double proof, as the operator has to demonstrate that he was not at fault or negligent on the one hand, and that the environmental damage was not likely to be anticipated, on the other hand) lies on the operator.

12 The Luxembourg Environmental Liability Law will only apply to environmental damage caused after its entry into force,[18] that is after 1 May 2009.[19] The prescription period is set at thirty years.[20]

13 The Law of 20 April 2009 does not affect legal and regulatory provisions which may confer a right of compensation for environmental damage or an imminent threat of such damage.[21]

14 Legal actions initiated by victims in GMO matters will be based on the Coexistence Law or on standard tort law rules based on Art. 544, 1382, 1383 and 1384 of the Luxembourg Civil Code. State liability can only be pursued under the State Liability Law.[22] Environmental damage would be governed by the Environmental Liability Law.

2. State liability

15 In the Grand-Duchy of Luxembourg, the liability of the State and other public authorities is regulated by the law dated 1 September 1988[23] (hereafter named "the State Liability Law"). According to case-law, however,[24] Art. 1 of the State Liability Law[25] does not establish a specific liability sys-

---

17 Art. 9.4. of the Environmental Liability Law.

18 Art. 15.2. of the Environmental Liability Law.

19 Art. 15.1. of the Environmental Liability Law.

20 Art. 4, par. 2, of the Environmental Liability Law.

21 Art. 4, par. 4, of the Environmental Liability Law, which, in its French version, reads as follows:"La présente loi n'affecte pas les dispositions légales ou réglementaires susceptibles de fonder une indemnisation à la suite d'un dommage environnemental ou d'une menace imminente d'un tel dommage".

22 See infra no. 15 ff.

23 Loi du 1ier septembre 1988 relative à la responsabilité civile de l'Etat et des collectivités publiques, Mémorial A-51 of 26 September 1988, 1000. This law entered into force on 30 September 1988.

24 Cour de Cassation, 24 April 2003, 32, 368 ("L'article 1ier, alinéa 1 de la loi du 1er septembre 1988, sans instaurer un régime spécifique, ne fait qu'appliquer aux personnes morales de droit public dans une terminologie adaptée à celles-ci, le principe de la responsabilité civile délictuelle de droit commun qui se fonde sur le concept de la faute").

25 The original French version of this provision reads as follows: "L'Etat et les autres personnes morales de droit public répondent, chacun dans le cadre de ses missions de ser-

tem, but rather applies the general rules of standard tort law which are grounded on the concept of fault.

In order to hold the State, or other public authorities, liable for any losses caused by GMOs, the victim has to prove both fault and a causal link between the fault and the damage. 16

The State may, within this framework, be seen in the context of its duties as State-administrator, State-judge or State-legislator. In the third case, a fault could consist in the fact that the State has omitted to exercise a normally diligent law-making power.[26] It will be up to a court to decide, on a case by case basis, if such omission has been proved. 17

The State is not involved in the special regimes set up by the Coexistence Law and the Environmental Liability Law. 18

II. Damage

1. Recoverable losses

Under standard tort law rules, any material and non-material damage is recoverable.[27] 19

The Coexistence Law of 18 March 2008 imposes liability for economic damage caused by the adventitious presence of a genetically modified organism of this variety in the produce of neighbouring plots and in the yields of honey or pollen proceeding from neighbouring apiaries. 20

Under the Product Liability Law,[28] recoverable damage includes any damage other than (i) nuclear injury or damage covered by international treaties in force in Luxembourg; (ii) damage to the defective property itself; (iii) damage to, or destruction of, any item of property provided that the item of property is of a type ordinarily intended for private use or consumption and was used by the injured person mainly for his own private use or consumption.[29] Luxembourg applies a € 500 deduction for damage caused to items of property.[30] 21

---

vice public, de tout dommage causé par le fonctionnement défectueux de leurs services, tant administratifs que judiciaires, sous réserve de l'autorité de la chose jugée".

26 See Justice of the Peace Luxembourg, 29 April 1996, no. 1937/96, declaring the principle of State liability in an environment case.

27 See our report "Economic loss caused by GMOs in Luxembourg", no. 36–38, 332.

28 See question IV.2. below.

29 Art. 2 (4) of the Product Liability Law.

30 Art. 2, final sentence, of the Product Liability Law.

22 The Environmental Liability Law applies to environmental damage and imminent threat of such damage. Damage is defined as a measurable adverse change in a natural resource or measurable impairment of a natural resource service which may occur directly or indirectly.[31] This definition does not include damage to protected species and natural habitats which result from an act by an operator which was expressly authorised by the relevant authorities on the basis of the Law of 19 January 2004.[32] The Law of 20 April 2009 does not provide for the compensation of material damage, economic loss or non-material damage. Therefore, it does not provide a valid ground for legal action to be brought by victims.

2. Pure economic loss

23 Pure economic loss is recoverable, on the one hand under standard tort law rules,[33] on the other hand under the Coexistence Law and the Product Liability Law.

24 Under the Coexistence Law, which foresees a strict liability regime for everyone growing genetically modified seed and plants, the victim has to prove the presence of the GMO seeds or plants in his field, as well as the reality of the economic loss.

25 Under the Product Liability Law[34] as well as under standard tort law rules, the injured person shall be required to prove the damage (no special provisions apply to economic losses), the defect and the causal relationship between the defect and damage.

3. Mere fear of a loss

26 In general, potential damage is not compensated under Luxembourg law. Judges have to consider only future elements which present a sufficient degree of certainty and are likely to be assessed. In no case will judges consider a possible future change of situation which only constitutes a hypothetical event, as such a case may not be compensated.[35] A mere fear of loss triggered by GMOs is therefore difficult to recover.

---

31 Art. 2 of the Environmental Liability Law.

32 Loi du 19 janvier 2004 concernant la protection de la nature et des ressources naturelles, Mémorial A-10 of 29 January 2004, 147, as amended. In fact, this includes projects, plans and constructions implemented for imperative reasons of major public interest.

33 See our report "Economic loss caused by GMOs in Luxembourg", no. 43, 333.

34 Art. 3 of the Product Liability Law.

35 Luxembourg Appeal Court, 26 February 1997, no. 19083.

The precautionary principle is slowly appearing in Luxembourg law, at least in legal writing. According to a famous author,[36] the principle would apply to any person who has the power to start or to stop an activity likely to present a risk for others. It must, therefore, be applicable to any decision-makers and tends to be applicable to any professionals and also private persons. Such principle would be appealed to in the case of human activities presenting, currently or in the future, adverse effects which science is currently not able to determine with certainty. Mobile phone radiation is one example of the scope of this principle. The exposure to such risk has been qualified as neighbourhood nuisance under Luxembourg law, and would allow those affected to seek compensation under Art. 544 of the Luxembourg Civil Code.[37] 27

4. Standard of proof

Pursuant to Art. 58 of the Luxembourg New Civil Code on Procedure,[38] the burden of proof rests with the party who invokes a legal or factual point to validate his claim or defence. Evidence is produced to explain, support and confirm the party's claim or defence.[39] 28

The burden of proving losses is not reversed in GMO matters. Future losses may be compensated, provided they are certain and may be assessed.[40] 29

5. Nominal losses

Luxembourg courts do not recognize nominal or symbolic losses as capable of being compensated. 30

6. Mass losses

Luxembourg law does not foresee any special rules for mass losses. 31

36 G. *Ravarani*, La responsabilité civile des personnes privées et publiques, Pasicrisie luxembourgeoise (2nd ed. 2006) 65.
37 See question IV.3. (d) below.
38 In the French version, this provision reads as follows: "Il incombe à chaque partie de prouver conformément à la loi les faits nécessaires au succès de sa prétention".
39 See our report "Economic loss caused by GMOs in Luxembourg", no 15–21.
40 See also question II.3. here above.

III. Causation

1. Uncertainty of merely potential causes

32 Luxembourg law requires the fulfillment of three conditions to establish civil liability (fault, damage and causation), whether liability is contractual or tortious. Every plaintiff who can prove a fault/act[41]/negligent conduct/imprudence,[42] damage and a direct link between this fault/act/negligent conduct/imprudence and damage, can claim compensation.

33 In GMO matters, a plaintiff would therefore have to prove the causal link between the damage and the presence of the GM crop concerned.[43]

34 With regard to the general rules of the Luxembourg Civil Code, the damage suffered by the plaintiff must be the direct and immediate consequence of unlawful conduct, i.e. the violation of a contractual or legal provision or a tort (fault/act/negligent conduct/imprudence) committed by the defendant.

35 In assessing the direct link between the damage and the unlawful conduct, Luxembourg courts apply the theory of adequate causality (*causalité adéquate*). According to this theory, the court will assess whether the fault, act or imprudence in question could be considered a cause which would, normally and according to life experience, have led to the alleged damage. Any potential causes which might have contributed to the damage cited by the plaintiff to the court are analysed by the court pursuant to such principle.

36 In looking for the adequate cause of the damage, Luxembourg courts operate a selection procedure within all the factors which have contributed to the damage, and eliminate those appearing exceptional.

37 Uncertain causes will likely not pass the examination by the judge and be eliminated, as they may not be considered as having contributed to the damage. Where it is unclear, for example, who violated segregation rules or good farming practice, such causes will not be taken into consideration.

38 The causal link, according to case-law,[44] is broken at the moment when an initiative taken by a victim or a third party interrupts the individual links

41 Art. 1382 of the Luxembourg Civil Code: "Tout fait quelconque de l'homme, qui cause à autrui un dommage, oblige celui par la faute duquel il est arrivé, à le réparer".

42 Art. 1383 of the Luxembourg Civil Code:"Chacun est responsable du dommage qu'il a causé non seulement par son fait, mais encore par sa négligence ou par son imprudence".

43 See our report "Economic loss caused by GMOs in Luxembourg", no. 11–14.

44 Luxembourg Appeal Court, 20 February 2002, no. 24911, B.I.J. 2003, 16.

that constitute the chain of events from the initial fact until the damage. There is a general principle, well anchored in Luxembourg case-law, that each victim is obliged to moderate its damage.

2. Complex causation scenarios

Where the damage has been caused by an undetermined member of a group of specified persons whose identity, however, is known (as in the case where several producers have delivered maize to a food producer), the victim will not be able to prove the exact cause of the damage and will not be compensated. Luxembourg law does not know a principle of collective liability of group members. **39**

French case-law, which might be referred to by Luxembourg judges in the absence of specific Luxembourg case-law,[45] foresees in such cases, in order to help victims, a collective fault of the group members, which results in liability *in solidum* if all group members have taken part in the collective fault. Such a collective fault can, however, not be established if no fault may be charged to some members or if their participation in the fault cannot be proven. **40**

Several producers held liable under the Product Liability Law for a single loss shall be solidarily liable.[46] **41**

In the case given in the example above, where it remains unclear which batch was contaminated, a collective liability of all producers presupposes that it can be proven that all producers participated in a collective fault. Under the Product Liability Law, it must be proven that the batches of all the producers were defective. If several batches have materially intervened in the production of a single loss, case-law provides that each of the guardians of the different batches is deemed to have caused the damage in its entirety; the victim is however obliged to prove the material intervention of different inanimate things in the production of the damage. This seems difficult to handle before court. **42**

45 *Ravarani* (fn. 36) no. 931.
46 Art. 6 of the Product Liability Law.

3. Force majeure

43 Force majeure has a role to play in the case of presumption of liability, where liability is presumed on the grounds of the intervention of inanimate things or where a contract foresees a duty of result.

44 Someone presumed liable cannot exonerate himself by proving the absence of fault or by proving that the cause of the damage remains unknown. He has to prove that the damage has an extraneous cause.

45 An extraneous cause premises the fulfilment of certain criteria. These criteria are the criteria of force majeure and have to be considered *in abstracto*[47] i.e. the judge will assess the fault by referring to the concept of a normally diligent, prudent and wise person (*homme normalement diligent, prudent et avisé, le bon père de famille*).

46 Firstly, if it is a question of liability for items under custody, the cause has to be exterior to the person of the custodian and to the item in custody.

47 The other two criteria required for force majeure as cause are that it is irresistible and unpredictable. It is sufficient that the fact which has given rise to the damage could not reasonably be foreseen and humanly avoided. There was no reason to think that such event would happen.[48] All three conditions have to be concurrently fulfilled.

48 Case-law considers extraneous causes for natural events, for acts of the victim and for acts of a third person.[49]

4. Threshold to prove causation

49 Causation is established by presumptions which are subject to the judges' understandings and caution. Luxembourg Civil Code requires that such presumptions shall be serious, precise and corroborating.[50] For one famous author, it will be sufficient if such presumptions are very strong.[51] In tort

---

47 Luxembourg District Court, 21 May 1980, *CNAMO v. Rech*.

48 Luxembourg District Court, 6 January 2004, no. 2/2004; 9 March 2004, no. 70/2004; 22 February 2005, no. 38/2005.

49 See our report "Economic loss caused by GMOs in Luxembourg", no. 32.

50 Art. 1353 of the Luxembourg Civil Code, stating in its French version: "Les présomptions qui ne sont point établies par la loi, sont abandonnées aux lumières et à la prudence du magistrat, qui ne doit admettre que des présomptions graves, précises et concordantes, et dans les cas seulement ou la loi admet les preuves testimoniales, à moins que l'acte ne soit attaqué pour cause de fraude ou de dol".

51 *Ravarani* (fn. 36) 694.

law, proof of a fault or negligent conduct or imprudence may be brought by any legal means and in particular by simple presumptions. Judges have a discretionary power to assess these presumptions. Even if such presumptions have no absolute certainty, they realize a complete proof without any limit or conditions, provided they persuade the judges.[52]

Luxembourg law would therefore tend towards an "almost certainty" standard. 50

5. Special rules on causation

The Coexistence Law establishes a strict liability regime for any GMO user where GMO crops are present in neighbouring fields. This does not mean that the behaviour of the GMO user is deemed to constitute a fault, but that he is presumed to be at the origin of the damage. It is the causal link between such behaviour and the damage which is presumed. 51

IV. Types of liability

1. Fault liability

(a) Special rules governing fault

Art. 1382 and 1383 of the Luxembourg Civil Code provide that compensation is due for any fault, act, negligent conduct or imprudence committed by the author. Bad faith is not required. Any fault/act/negligent conduct/imprudence will be analysed *in abstracto*, i.e. the judge will assess the fault by referring to the concept of a normally diligent, prudent and wise person (*homme normalement diligent, prudent et avisé, le bon père de famille*). Notwithstanding the objective analysis, the judge has to take into consideration the external circumstances, i.e. the judge compares the behaviour of the author of the act with that of any wise individual if confronted with a similar situation.[53] These rules also will apply to GMO cases handled under Luxembourg standard tort law. 52

Under the Coexistence Law, which establishes a strict liability regime for any GMO user where GMO crops are present in neighbouring fields, the producer of GMO seed and plants is presumed to be at the origin of the 53

52 Luxembourg Appeal Court, 20 October 1959, 18, 11.
53 See our report "Economic loss caused by GMOs in Luxembourg" no. 27.

damage. He is obliged to prove an extraneous cause in order to exonerate himself from liability.

(b) Impact of specific rules of conduct

54 Any fault/act/negligent conduct/imprudence will be analysed *in abstracto*, i.e. the judge will assess the fault by referring to the concept of a normally diligent, prudent and wise person (*homme normalement diligent, prudent et avisé, le bon père de famille*). Notwithstanding the objective analysis, the judge has to take into consideration the external circumstances, i.e. the judge compares the behaviour of the author of the act with that of any wise individual if confronted with a similar situation.

55 The fact that specific rules governing GM or non-GM farming have been violated would be considered by the judge to constitute behaviour which is not that of a normally diligent, prudent and wise person.

2. Product liability

(a) Development risk defence

56 The Grand-Duchy of Luxembourg has transposed the EC Product Liability Directive (85/374/EC)[54] into Luxembourg law by the legislative act of 21 April 1989[55] (hereafter named "the Product Liability Law").

57 The development risk defence has not been incorporated by the Product Liability Law, despite the option offered by the Directive. The Government, followed by the legislator, argued in favour of this decision by referring to the fact that the development risk defence was unknown in Luxembourg law including its case-law in the field of tort law and contractual liability.[56]

---

54 Council Directive 85/374/EEC of 25 July 1985 on the approximation of the laws, regulations and administrative provisions of the Member States concerning liability for defective products, OJ L 210, 07.08.1985, 29, as amended by Directive 1999/34/EC of the European Parliament and of the Council of 10 May 1999, OJ L 141, 04.06.1999, 20.

55 Loi du 29 avril 1989 relative à la responsabilité civile du fait des produits défectueux, Mémorial A-25 of 28 April 1989, 522 (Parliament document 3287), as amended by the law of 6 December 1989 (Loi du 6 décembre 1989 concernant la juridiction du travail, Mémorial A-83 of 27 December 1989, 1632).

56 Parliament document 3287, 5.

It seems that, due to the restrictive provisions of the Product Liability Law, the financial threshold and the competing standard tort law rules,[57] which are more favourable to the victim of damage caused by a defective product, the Product Liability Law has given rise to almost no litigation in Luxembourg courts.[58] **58**

(b) Alternative routes

Agricultural products are included in the scope of the Product Liability Law[59] and damage caused by such products shall be compensated according to the regime of that Law. **59**

An alternative route to obtain compensation for losses caused by agricultural products is recourse to the standard tort law regime, explained in other sections of this contribution and – in our view – not subject to invalidation by ECJ case-law. **60**

(c) Impact of compliance with rules and regulations

By virtue of Art. 1 of the Product Liability Law, the producer shall be liable for damage caused by a defect in his product. **61**

This constitutes liability without fault on the part of the producer. The characteristic element of the producer's liability is the defect of the product, which is the case when the product does not provide the safety which a person is entitled to expect, taking all circumstances into account, including the presentation of the product, the use to which it could reasonably be expected to be put, and the time when the product was put into circulation.[60] **62**

The lack of compliance does not matter, and the observance of rules governing the production process does not matter either. **63**

---

57 Art. 8 of the Product Liability Law, stating in its French version: "Les dispositions de la présente loi ne portent pas atteint aux droits dont la victime d'un dommage peut se prévaloir au titre du droit commun de la responsabilité contractuelle ou extracontractuelle ou au titre d'un autre régime spécial de responsabilité".

58 *Ravarani* (fn. 36) no. 512.

59 Art. 2 of the Product Liability Law, stating in its French version: "Pour l'application de la présente loi, on entend par: 1° "produit": tout bien mobilier, même s'il est incorporé dans un autre meuble ou dans un immeuble; le terme "produit" désigne également l'électricité; . . .".

60 Art. 2, 3 par. 1, of the Product Liability Law.

64 The defect of the product is evaluated *in abstracto*, that is, compared to the common expectation of an average user.

65 The Product Liability Law offers the producer several means to escape liability, which are additional to those under standard tort law rules. The producer shall not be liable if he proves that he did not put the product into circulation,[61] or that the product was neither manufactured by him for sale or any form of distribution for economic purpose nor manufactured or distributed by him in the course of his business.[62] No liability shall be incurred either if, having regard to the circumstances, it is probable that the defect which caused the damage did not exist at the time when the product was put into circulation by him or that this defect came into being afterwards.[63] The producer shall also escape liability if the defect is due to compliance of the product with mandatory regulations issued by the public authorities.[64]

3. Environmental liability

(a) Implementation of the Environmental Liability Directive

66 The Environmental Liability Directive 2004/35/EC was implemented in the Grand-Duchy of Luxembourg by the Environmental Liability Law of 20 April 2009,[65] adopted by the Luxembourg Parliament on 19 March 2009.[66]

67 The Environmental Liability Law faithfully implements Directive 2004/35/EC. It names as competent authority the governmental Administrations for the Environment (*Administration de l'Environnement*), Water and Forests (*Administration des Eaux et Forêts*) and Water Management (*Administration de la Gestion de l'Eau*), each acting within the frame of their legal mandate, as well as the members of Government having these three administrations in their responsibility.[67]

---

61 Art. 4 a) of the Product Liability Law.
62 Art. 4 c) of the Product Liability Law.
63 Art. 4 b) of the Product Liability Law.
64 Art. 4 d) of the Product Liability Law.
65 See supra fn. 9.
66 The European Court of Justice found on 24 March 2009 that the Grand-Duchy of Luxembourg had failed to fulfil its obligations under Dir. 04/35/EC introduced by the European Commission on 18 July 2008. See judgment of 24 March 2009, *Commission of the European Communities v. Grand-Duchy of Luxembourg*, C-331/08.
67 Art. 2 (17) and (18) of the Environmental Liability Law.

The Law of 20 April 2009 does not consider the first case of financial exoneration as foreseen by Art. 8.4 (a) of Directive 2004/35/EC. Transposing Art. 8.4 (b), it therefore only rules that the operator does not have to bear the cost of remedial actions where he demonstrates that he was not at fault or negligent and that the environmental damage was caused by an emission or activity or any manner of using a product in the course of an activity which the operator demonstrates was not considered likely to cause environmental damage according to the state of scientific and technical knowledge at the time the damage occurred.[68] **68**

Contrary to the provisions of the Directive, the Luxembourg Environmental Liability Law does not include future amendments of pertinent international conventions and instruments as applicable within the framework of the Environmental Liability Law.[69] **69**

Regarding preventive action, the Environmental Liability Law foresees that, in case of an imminent threat of environmental damage, the operator shall take the necessary preventive measures within seven days after the threat arose.[70] **70**

Luxembourg did make use of the option allowed by Art. 12.5 of the Directive to decide not to apply modalities of the request for action to cases of imminent threat of damage. **71**

The Environmental Liability Law, at any time when it deals with significant risk of adverse effects on human health, assimilates such risk into the risk of negative impact on the environment in protected habitats of Community interest, in those of national and local interest within the meaning of the amended law of 19 January 2004.[71] **72**

Luxembourg did not take measures to encourage the development of financial security instruments and markets by the appropriate economic and financial operators, including financial mechanisms in case of insolvency, with the aim of enabling operators to use financial guarantees to cover their responsi- **73**

68 Art. 9.4 of the Environmental Liability Law, which reads as follows: "Le coût des mesures de réparation ne peut être mis à charge de l'exploitant s'il apporte la preuve qu'il n'a pas commis de faute ou de négligence et que le dommage à l'environnement résulte d'une émission, d'une activité ou de tout mode d'utilisation d'un produit dans le cadre d'une activité dont l'exploitant prouve qu'elle n'était pas considérée comme susceptible de causer des dommages à l'environnement au regard de l'état des connaissance scientifiques et techniques au moment du fait générateur du dommage.".

69 See Art. 5.2., 5.3. and 5.4. of the Environmental Liability Law (corresponding to Art. 4 subpar. 2, 3 and 4 of Dir. 2004/35/EC).

70 Art. 6.1. of the Environmental Liability Law.

71 See Art. 2.15. and Annexe II (point 2).

bilities, as required by Art. 14.1 of Directive 2004/35/EC. The lawmakers explained this omission by pointing to the fact that it would not be appropriate, for evident reasons, to establish such a regime on the level of solely national law, and opted to wait for Community initiative on this matter.[72]

(b) Environmental liability regime beyond the scope of the Directive

74 Luxembourg law does not know any liability regime covering environmental harm which exceeds the scope of Directive 2004/35/EC.

(c) Claimants in cases of environmental harm

75 In case of harm to biodiversity or other harm to the environment as such, as foreseen under the Environmental Liability Law, it is up to the competent Government Member to recover, inter alia, via security over property or other appropriate guarantees, from the operator who has caused the damage or the imminent threat of damage, the costs it has incurred in relation to the preventive or remedial actions taken under the Environment Liability Law. Competent authorities may decide not to recover the full costs where the expenditure required to do so would be greater than the recoverable sum or where the operator cannot be identified.[73]

76 Cost recovery proceedings against the operator, or if appropriate a third party who has caused the damage or the imminent threat of damage in relation to any measures taken, have to be initiated within five years from the date on which such measures were completed or on which the liable operator, or third party, was identified, whichever is later.[74]

77 Natural or legal persons affected or likely to be affected by environmental damage or with a sufficient interest in environmental decision-making relating to the damage or, alternatively, alleging the impairment of a right, are entitled to request action by the Government.[75]

78 Recoverable losses are the costs for the preventive and remedial actions taken pursuant to the Environmental Liability Law.[76]

---

72 Parliament document 5877 of 8 May 2008, Exposé des motifs, 16.
73 Art. 9.2. of the Environmental Liability Law.
74 Art. 11 of the Environmental Liability Law.
75 Art. 12 of the Environmental Liability Law.
76 Art. 9.1. of the Environmental Liability Law.

(d) Special liability regime for losses sustained by individuals

We refer to our report "Economic losses causes by GMOs in Luxembourg".[77] 79

(e) Cartagena Protocol

The Cartagena Protocol, signed by Luxembourg on 11 July 2000, was approved and transposed into Luxembourg legislation by the law of 29 May 2002.[78] 80

The Luxembourg Parliament, at the time of discussing the ratification of the Cartagena Protocol, recognised that the Protocol creates an important legal framework for new technology which has great potential, but also presents some risks. The Protocol was recognised as an important step towards assuring biological security, by establishing a balance between commercial and environmental preoccupations and by assuring foreseeability for exporters and importers. It constitutes the most recent effort made by the international community to establish rules in the interest of the protection of human health and environment without impeding the international rules of free and fair trade.[79] 81

4. Other strict liability regimes

There are no (other) strict liability regimes applicable in cases of harm caused by GMOs. 82

V. Vicarious liability

1. Scope of vicarious liability

According to Art. 1384 para. 3, of the Luxembourg Civil Code, masters (*maîtres*) and employers (*commettants*) shall be liable for the damage caused by their servants (*domestiques*) and employees (*préposés*) in the func- 83

77 No. 34–35.

78 Loi du 29 mai 2002 portant approbation du Protocole de Cartagena sur la prévention des risques biotechnologiques relatif à la Convention sur la diversité biologique, fait à Montréal, le 29 janvier 2000, Mémorial A-59 of 13 June 2002, 1496.

79 Report of the parliamentary committee for the Environment of 11 April 2002, Parliament document 4842–3 of 23 April 2002, 7.

tions for which they have been employed.[80] This constitutes non-contractual liability.

84 Two conditions are required in order to engage the liability of the employer for his employee's act: the hierarchical superiority of the employer over the employee, and a harmful act committed by the employee, which is in relation with his functions.

85 The hierarchical superiority is normally defined as a link of subordination, being a result of the employer's authority over the employee, authority which comprises the right to give orders and instruct a person on the manner of how to fulfil her or his duties.[81] The hierarchical relation is characterised by the fact that the employee participates in the employer's activity under the employer's command.

86 Situations which are characterised by the independence of the person working on another person's account do not, however, give rise to such a hierarchical link. This is the case with respect to contractor agreements, except if the other contracting party has reserved for himself the full and absolute direction of the work to be done.[82]

87 There is no link of subordination constituting a hierarchical link between the work-master and the contractor,[83] nor between the main contractor and a sub-contractor.[84]

88 The vicarious liability based on Art. 1384 para. 3, of the Luxembourg Civil Code is a strict liability. Once the harmful act by the employee and the link with said employee's functions are established, the employer may not escape liability by proving that he was not at fault[85] or by demonstrating that the employee's act was an extraneous cause as far as the employer is concerned.

89 The employer can, however, escape liability by proving that the harmful act was an extraneous cause presenting the character of force majeure as far as the employee is concerned.

80 In its French version, this provision reads as follows: "On est responsable non seulement du dommage que l'on cause par son propre fait, mais encore de celui qui est cause par le fait des personnes dont on doit répondre ... Les maîtres et les commettants, du dommage cause par leurs domestiques et préposés dans les fonctions auxquelles ils les ont employés".

81 Diekirch District Court, 15 February 1900, 7, 172; Luxembourg District Court 20 January 1954, 16, 86; Luxembourg Appeal Court 12 November 2003, no. 26868.

82 Luxembourg Appeal Court, 27 June 1893, 3, 216.

83 Luxembourg Appeal Court, 14 January 2004, no. 26 658.

84 Luxembourg Appeal Court, 14 January 2004, no. 26 658.

85 Luxembourg Appeal Court, 26 May 1916, 9, 559.

2. Liability for people further up the food or feed production chain

Considering that there is no link of subordination or hierarchy between the crop retailer and the crop wholesaler, or between the crop wholesaler and the farmer, or between the farmer and the seed producer, people "further up" in the feed and food production chain cannot be held vicariously liable. **90**

3. Can someone further down the feed or food chain include someone further up it in a trial against him/herself?

Under vicarious liability rules, one cannot bring someone further up the hierarchical chain into a lawsuit. **91**

This is, however, possible under fault-based liability rules. Thereunder a farmer could (but is not obliged to) bring his seed producer into a lawsuit filed against him by a customer. If the farmer can prove the producer's fault, and his own fault is proved by the customer, the farmer and the producer will be liable *in solidum*. Each of the liable persons may take recourse against the jointly liable person in order to have the question of their respective shares of liability settled by the court. **92**

Under the Product Liability Law, the product supplier may inform the injured person of the identity of the producer or of the person who supplied him with the product.[86] The supplier may thus escape liability in cases where the producer was not identified by the victim. **93**

VI. Multiple tortfeasors

If damage has been caused by several faults, each of these faults shall be considered to have caused the entire damage.[87] The solution will be the same in the event that several things have intervened in the production of a sole and identical loss. **94**

The different authors shall be liable *in solidum* towards the victim. Each of them shall be obliged to repair the entire damage, even if an appeal between the jointly liable persons is already impossible for a factual or legal reason. **95**

---

86 Art. 2 (2) of the Product Liability Law.

87 Luxembourg Appeal Court, 25 October 1961, 18, 387; Luxembourg District Court, 18 October 1974, no. 17625; Cour de cassation 26 June 1975, 25, 116; Luxembourg District Court 9 November 1982, no. 763/82 I, 26 January 2004, no. 443/04 XV, B.I.J. 2005, 25.

96 The respective seriousness of the faults does not matter much. To escape liability *in solidum*, being able to measure the respective importance of the committed faults will not be sufficient. Apportionment of liability shall only affect reciprocal relations between the co-authors as well as any recourse actions between themselves. Luxembourg law does not know partial causality which would authorise the court to mitigate the liability according to the causal role played by the different factors in the production of a single and identical loss, even if none of them, on its own, would have caused such loss or at least not to the same extent.

97 This principle is applicable only in the case of a single loss. If it is possible to divide the damage and if the damage caused by each author is known, the authors shall not be jointly liable. If distinct facts cause divisible damage, where it is possible to attribute each part to its author, each of the liable persons shall compensate only the damage which is attributable or attachable to it.[88]

VII. Defences

1. Licence/permission to grow GM material

98 Considering that the Coexistence Law has established strict liability for the user of authorised crops, the permission to grow GM material cannot serve as a defence against liability triggered by its use. It does not make any difference whether the GMOs are approved for testing only.

99 The authors of the Environmental Liability Law have ruled out an exemption based on permission, as such exemption by the ordinary proof that the conditions of such permission were fulfilled is not likely to be justified considering the "principle of polluter-pays".

2. Consent/assumption of risk

100 According to the traditional position of Luxembourg case-law, someone who has taken part in an activity while knowing or having to know that this activity presented an abnormal risk, i.e. unforeseeable consequences, has to be considered as having accepted such risk.[89] By taking a risk exceeding normal standards, he is thus at fault or unwise, which is

88 Luxembourg Appeal Court, 3 October 2002, no. 25643.
89 Luxembourg Appeal Court, 23 March 2005, no. 28774.

deemed to be a contribution to any damage occurring and which must, consequently, partially exempt the author of such damage from liability.[90]

To constitute an abnormal risk, the danger to which the potential victim is exposed must be sufficiently characterised in such a way that the realisation of the harmful event appears, in the eyes of all, if not certain then at least reasonably foreseeable, not to say likely, it being understood that a simple possibility of damage will not be sufficient.[91] **101**

The concept of abnormal risk certainly does not mean rare or exceptional risk, but risk which is not specific or outside the activity in question, and which a priori is not likely to happen. **102**

The extent of the acceptance of such risk is, as a matter of course, a function of the specific nature of the activity. If in such case the damage was foreseeable, the presumption of liability for things in the victim's custody (Art. 1384 para. 1 and Art. 1385) is set aside. If the harmful act was beyond reasonable prediction, the presumption applies. **103**

As regards fault-based liability, the author may partially escape liability if he proves that the victim contributed to the damage by accepting the risk of damage in full awareness of such. **104**

If the victim knowingly consumed GM products, it will be up to the judge to assess whether the harmful act was within the bounds of reasonable prediction and only in such case will the judge decide that the victim has contributed to the damage. **105**

3. Third-party influence

As fault-based liability is concerned, the author of damage may not exempt himself partially from liability by proving a third person's fault which has contributed to the damage. The third party's fault cannot be invoked as far as liability to the victim is concerned. If the third party has contributed to the damage by a personal fault, he shall be liable *in solidum* with the defendant. In order to be unburdened, even partially, from liability towards the victim, the defendant must take recourse against this third party. **106**

---

90 Luxembourg Appeal Court, 20 December 1967, 20, 426; Luxembourg District Court, 15 May 1991, no. 359/91, 10 December 1997, no. 18190; 29 October 1998, no. 658/98 VI.

91 Luxembourg District Court, 7 March 1983, no. 170/83 I, confirmed by the Luxembourg Appeal Court on 26 November 1984; Luxembourg Appeal Court, 20 January 2000, no. 22151.

107 If the third party's act does not constitute a fault, recourse will not be possible, as fault is a condition of such recourse.

4. Prescription

108 There are no special statutes of limitation, deviating from standard tort cases, which would apply to the cases envisaged by this study.

5. Other defences

109 There are no other defences that might be relevant in a GMO liability case.

VIII. Remedies

1. Pecuniary compensation

(a) Bodily harm

110 Luxembourg GMO legislation does not foresee special rules on compensating bodily harm caused by GMOs. The aim of the Environment Liability Law is not to compensate victims. The Coexistence Law only foresees compensation of material damage.

111 The regular remedies will therefore apply.

(b) Property losses

112 Luxembourg GMO legislation does not foresee special rules on compensating property losses caused by GMOs. The aim of the Environment Liability Law is not to compensate victims. The Coexistence Law only foresees compensation of material damage.

113 The regular remedies will therefore apply.

(c) Economic losses

114 Luxembourg GMO legislation does not foresee special rules on compensating economic losses caused by GMOs. The Environment Liability Law

does not aim to compensate victims. The Coexistence Law only foresees compensation of material damage.

The regular remedies will therefore apply. **115**

An organic farmer may recover the full costs of restoring a field that was **116** contaminated with GMOs to meet organic standards. The question whether deductions would apply for the seasons that he could have grown conventional crops there is left up to interpretation by the court.

(d) Harm to animals

Under Luxembourg law, in order to be reparable, damage has to be legiti- **117** mate, certain, direct and individual. Luxembourg law knows a very broad conception of reparable damage, and no damage, even the most unseen, is generally excluded from assessment and compensation. Full compensation is concretely assessed and the courts tend towards compensation point by point.

Under such conception, the value of an animal would include, for exam- **118** ple, a cow's potential for producing milk or meat. In assessing possible damage, the judge takes into consideration any factor which, even in the future, presents a sufficient degree of certainty and can be assessed. However, he does not take into consideration any possible future change of the situation which only constitutes a hypothetical event.[92]

Case-law actually recognises the recoverability of both non-material and **119** material damage suffered by indirect victims. Luxembourg's highest court has already decided that the loss of a thing may cause to its owner subjective and emotional damage for which he may claim compensation.[93] Referring to French case-law and finding the same solution for loss of animals or harm caused to them, it is likely that Luxembourg courts would not generally exclude harm suffered by animals as a cause of damage to the animal's owner.

---

92 Luxembourg Appeal Court, 22 November 1994, no. 419/94 V; Luxembourg District Court, 3 May 1995, no. 5/95 I.C. 51; Luxembourg Appeal Court, 7 April 1998, no. 140/98 V; Luxembourg District Court 15 November 2005, no. 203/2005 VIII.

93 Cour de cassation, 10 May 1990, 28, 37.

(e) Costs of disposal

120 According to the broad conception of reparable damage in Luxembourg, basically no damage is excluded from assessment and compensation. The costs of disposing of contaminated production/animals fed with GM feed may therefore be recoverable.

2. Non-compensatory damages

121 Luxembourg does not recognize non-compensatory damages such as punitive or exemplary damages.[94] Luxembourg law does not allow the adjustment of damage compensation according to the seriousness or non-seriousness of the fault that caused the damage.[95]

3. Other remedies

122 Luxembourg case-law operates the principle that any damage is to be repaired in kind. If this is impossible or not appropriate for some reason, the compensation will be by the allocation of damages.

123 Case-law considers that compensation in kind is the only way to eliminate the damage suffered by the victim and that reparation in kind must be ordered whenever victims request such compensation.[96]

124 Reparation in kind may be impossible under certain circumstances. It may be materially impossible, legally impossible or impossible due to the debtor's refusal. It may also be too expensive or not appropriate. In these cases, damages will be allocated.

4. Costs of pursuing a claim

(a) General cost rule

125 Luxemburg law states that any losing party will be ordered to bear the judicial costs, but reserves for the court the option to take a special and

94 *Ravarani* (fn. 36) no. 1095.

95 See our report "Economic losses caused by GMOs in Luxembourg" no 39.

96 Luxembourg Appeal Court, 24 March 1994, no. 14195; 26 October 1994, no. 15242; Luxembourg District Court 19 January 2005, no. 11/2005 XVII.

reasoned decision to order the other party to bear the costs, in full or partially.[97]

It also allows the court to order a party to bear the costs of the proceedings.[98] This generally comprises the costs not included in the judicial costs, and in particular lawyer fees as well as transport and office costs. **126**

(b) Costs of establishing causation

According to the broad conception of reparable damage under Luxembourg law, and the principle that compensation must include everything which is necessary to restore the victim to the state before the damage, the costs of establishing causation may be recoverable. **127**

Only useless and needless costs will not be compensated. The victim is also obliged to take any reasonable measures to that effect, and the burden of proof lies on the author of the damage. **128**

5. Advance cover

In Luxembourg, a mandatory insurance coverage or contribution to an officially recognised fund covering liability for GMO production is required by the Coexistence Law.[99] **129**

IX. Cross-border issues – Conflict of laws

1. Conflicts rules applicable before (or instead of) Rome II

General rules of jurisdiction and choice of law have been discussed in our report "Economic losses caused by GMOs in Luxembourg".[100] The general rule stated in Art. 4.1 of the Rome II Regulation[101] is in line with the prevailing Luxembourg case-law. **130**

---

97 Art. 238 of the New Code of Civil Procedure.
98 Art. 240 of the New Code of Civil Procedure.
99 Art. 15 (2) of the Coexistence Law.
100 No. 67–68.
101 Regulation (EC) No. 864/2007 of the European Parliament and of the Council of 11 July 2007 on the law applicable to non-contractual obligations (Rome II), OJ L 199, 31.07.2007, 40.

131 Apart from product liability which continues to be governed by the Hague Convention on the Law Applicable to Products Liability, Rome II special rules will apply in fields where Luxembourg law[102] currently does not foresee conflict of law clauses.

2. Special regime for cross-border claims

132 Under Luxembourg law, bilateral or multilateral cross-border claims for compensation are not dealt with by any special regime.

133 The Environmental Liability Law only foresees cooperation procedures between Member States pursuant to Art. 15 of Directive 2004/35/EC.

X. Cases

1. Due to the adventitious presence of GMOs in a field, maize which is normally sold as conventional contains GMOs beyond the legal labelling threshold. This is not discovered before the final stage of the food production chain by the producer of taco chips. The whole production is lost since the supermarket chains refuse to accept delivery from the producer.

(a) Who can sue along the chain of distribution?

134 Anyone suffering economic losses due to the adventitious presence of GMO crops in the production of neighbouring plots may sue the producer of the GMOs. The victim has to prove the presence of the GMO in the neighbouring plot and, in order to benefit from the strict liability regime introduced by the Coexistence Law, must establish the economic damage. The action may be introduced by anyone along the chain of distribution, i.e. the farmer of the neighbouring field, the maize wholesaler, the maize detailer or the taco producer.

102 For unfair competition and acts restricting free competition: the Law of 30 July 2002; for environmental damage: the Environment Liability Law of 20 April 2009; for infringement of intellectual property rights: the Law of 18 April 2001; for industrial action: general tort law rules, for unjust enrichment, negotiorum gestio and culpa in contrahendo: Luxembourg Civil Code.

The GMO farmer, in order to exonerate himself from liability, has to prove that the damage has an extraneous cause, fulfilling the criteria of force majeure, possibly natural events, an act by the victim or a third party. **135**

The maize wholesaler, the maize retailer and the taco producer could also bring a legal action against the farmer of the affected field who provided GMO affected maize. The basis of the wholesaler's action would be contractual liability (due to the contract concluded with the farmer concerned), the basis of the detailer's and taco producer's actions would be unintentional tort liability. Fault, negligent conduct or imprudence has to be proved. **136**

In such a case, the farmer of the affected field could sue the GMO producer in order for the latter to be held jointly liable by the court (if the farmer of the affected field has himself committed a tort), i.e. liable *in solidum* with the farmer of the affected field for compensation of the damage incurred, or solely liable (if the farmer of the affected field has not committed a wrong). If there is liability *in solidum*, the farmer of the affected field could initiate a recourse action with the purpose of apportioning liability between the two jointly liable parties. In order to succeed in the action against the GMO producer, the farmer of the affected field would have to prove the GMO producer's fault, negligent conduct or imprudence. If the action is grounded on the Product Liability Law, the defect of the product would have to be established as well. **137**

(b) Would the case be solved differently if the GMO content was below the labelling threshold?

The GMO content threshold has no impact on the resolution of the case if it is based on the Coexistence Law. **138**

Under standard tort law, the fact that the GMO content was below the labelling threshold would be considered by the judge in assessing the fault, negligent conduct or imprudence of the GMO producer. **139**

(c) Would the case be solved differently if the admixture was not adventitious, but occurred due to the disregard of segregation rules, for example?

The Luxembourg Coexistence Law expressly deals with adventitious admixture of GMOs in a field. General tort law rules will, however, inter- **140** vene in the event that the admixture was not adventitious, but for example occurred due to the disregard of segregation rules.

141 The disregard of segregation rules would be considered to constitute fault, negligent conduct or imprudence by the GMO producer and would be a ground for the victims' right to compensation of their damage.

(d) Would the case be solved differently if the GMO found was not admitted for production in your jurisdiction?

142 The solution would be the same as under (c) above, as the culture of GMO crops not admitted for production in Luxembourg would be considered to constitute fault, negligent conduct or imprudence by the GMO producer.

(e) If the admixture had occurred on a non-GM field and it transpires that the GM seeds were blown from: (i) neighbouring fields; or (ii) a truck passing by, would the farmer of the affected field be liable for all or part of the loss caused further down the distribution chain?

143 If it transpires that the GM seeds were blown from neighbouring fields, the farmer of the affected field cannot be held liable if no fault, negligent conduct or imprudence is proven against him. If such fault, negligent conduct or imprudence is proven (unlikely in this case as the GM seeds were blown from the neighbouring field), he would in any case have a possible recourse action against the farmer of the GM seeds with whom he is declared jointly liable. The farmer of the affected field could take action against the farmer of the GM seeds on the basis of the strict liability regime established under the Coexistence Law.

144 If the GM seeds were blown from a truck passing by, such a recourse action would not succeed as the cause is uncertain and as the owner of the GM seeds transported by the truck is unknown.

2. Twenty years after the sale of GM maize used for food products, it turns out that it has certain disadvantageous health effects for humans.

(a) Can the producers be held liable at this point for risks unknown at the time of growing the maize? Who would be liable?

While, generally speaking under standard tort law rules, producers and/or distributors may not be held liable for risks unknown at the time of growing the maize, the Product Liability Law has not implemented the development risk defence foreseen by Directive 85/374/EEC. Under Luxembourg law, proof that the state of scientific and technical knowledge at the time when the product was put into circulation was not such as to enable the existence of the defect to be discovered is not open to the producer as a defence. **145**

The Product Liability Law requires, however, that proceedings for the recovery of damages be initiated within three years, which begin to run from the day on which the plaintiff became aware, or should have become aware, of the damage, the defect and the identity of the producer. **146**

In the first line, the producer of the defective product would be held liable. If the producer cannot be identified, each supplier would be held liable, unless he informed the injured person of the identity of the producer. **147**

(b) Can compensation already be claimed at a point when the negative health effects have not yet materialised, but are to expected to according to scientific expertise/mere rumours?

Potential damage is not subject to compensation. Only damage which presents a sufficient degree of certainty and can be feasibly assessed will be considered by Luxembourg courts. **148**

(c) **Would it make any difference if the GM maize had only been in use for feed, causing harm to the animals, which may or may not cause harm to humans consuming the meat as well?**

149 Causing harm to an animal may, under certain circumstances, ground a right for compensation for the animal's owner, on the basis of subjective and emotional damage, not to speak of the material damage.

150 In assessing the meat consumers' right to take action, the courts will apply the adequate causality theory, and only consider a cause which would, normally and according to life experience, have led to the alleged damage, and eliminate exceptional causes. Due to the lack of any published case-law in this context, it is rather difficult to foresee the magistrates' position.

3. **The driver of a food logistics company discovers that a farmer, from where he regularly picks up agricultural products fails to obey mandatory segregation rules or food or feed hygiene standards, which may lead to the admixture of GM and non-GM produce sold and packaged separately by that farmer. Does he or his employer have a duty to warn, i.e. warn the recipient of the allegedly "non-GM" produce?**

151 Under Luxembourg law, denunciation is not mandatory.

152 Nonetheless, the driver and/or his employer could be sued by the victim asserting that the logistics company was aware of the admixture of GM and non-GM products. If the victim establishes the causal link of this fault/negligent conduct/imprudence to the damage sustained, the driver and/or his employer could incur liability. Due to the lack of any published case-law in this context, it is rather difficult to foresee the magistrates' position.

Damage Caused by GMOs under Maltese Law

Eugène Buttigieg

I. General overview

1. Special liability or redress scheme for GMOs

One must distinguish between economic losses caused by GMOs to private individuals such as consumers, farmers and other third parties and harm caused to the environment itself. 1

As far as the former is concerned, the position has remained as described in the first report, namely that there is no special liability or other compensation regimes in Malta for GMOs but the general tort regime found in the Civil Code[1] is applicable. These provisions and related case law were discussed in depth in the first report. 2

On the other hand, where the GMOs cause harm to the environment, the Prevention and Remedying of Environmental Damage Regulations, 2008[2] which transpose Directive 2004/35/EC[3] into Maltese law establish a framework of environmental liability, based on the polluter-pays principle, to prevent and remedy environmental damage caused *inter alia* by GMOs. 3

These Regulations apply to environmental damage (or any imminent threat of such damage) caused by any of the occupational activities listed in Schedule III to the Regulations through the fault or negligence of the operator. The activities listed in this Schedule include the contained use, including transport, of genetically modified micro-organisms and the deliberate release into the environment, transport and placing on the mar- 4

1 Chap. 16 of the Laws of Malta, Art. 1029-1051A.

2 LN 126 of 2008.

3 Directive 2004/35/EC of the European Parliament and of the Council of 21 April 2004 on environmental liability with regard to the prevention and remedying of environmental damage, OJ L143, 30.04.2004, 56–75.

ket of genetically modified organisms. The Regulations, of course, have a wider scope of application because these activities are just two out of a long list of activities that are all covered by the Regulations.

5 However, the Regulations make it clear that they do not give private parties a right of compensation for such environmental damage or the threat of such damage. They merely provide for preventive and remedial action in the interest of the community at large, not for individual redress.[4]

6 The Regulations include under the definition of 'environmental damage' 'land damage, which is any land contamination that creates a significant risk of human health being adversely affected as a result of the direct or indirect introduction, in, on or under land, of substances, preparations, organisms or micro-organisms'.[5]

7 These Regulations do not apply where the damage is caused by a natural phenomenon of exceptional, inevitable and irresistible character and where the damage is caused by pollution of a diffuse character; the Regulations are applicable only where it is possible to establish a causal link between the damage and the activities of individual operators (i.e. the person operating or controlling the activity concerned).

8 Apart from providing for preventive action where there is an imminent threat of environmental damage in order to avert such danger, the Regulations also provide for remedial action where such damage has occurred. In the latter case, the operator must, without delay, inform the Malta Environment and Planning Authority of all relevant aspects of the situation and he must take all the practicable steps necessary in order to immediately control, contain, remove or otherwise manage the relevant contaminants and any other damage factors in order to limit or to prevent further environmental damage and adverse effects on human health or further impairment of services. Moreover, he has to take all the necessary remedial measures prescribed by the Regulations. If he does not act accordingly, the Authority is empowered to require him to comply with these obligations and if he fails to do so or if the operator responsible cannot be identified, the Authority may take these measures itself, as a means of last resort.

9 As stated above, the operator must identify the potential remedial measures in accordance with the common framework set in the Schedule to the Regulations and submit them to the Authority for its approval, unless the Authority has taken action itself as aforesaid. The Authority decides

4 Reg. 4 (3).
5 Reg. 3.

which remedial measures shall be implemented in accordance with the framework set in this Schedule and with the cooperation of the relevant operator. Natural or legal persons affected or likely to be affected by environmental damage, or having a sufficient interest in environmental decision-making relating to the damage, and in particular the persons on whose land the remedial measures would be carried out, would be invited to submit observations and the Authority is obliged to take these into account.

The operator would bear the costs for these remedial actions. If the **10** Authority has taken any actions itself, it would recover, *inter alia* via security over property or other appropriate guarantees from the operator who has caused the damage, the costs that it has incurred in relation to the remedial actions taken under these Regulations, unless the operator cannot be identified or the expense involved in recovering the costs would be greater than the recoverable sum. However, the operator would not be required to bear the cost of the remedial actions taken if he can prove that the environmental damage was caused by a third party and occurred despite the fact that appropriate safety measures were in place; or resulted from compliance with a compulsory order or instruction emanating from a public authority other than an order or instruction consequent upon an incident caused by the operator's own activities. Likewise, the operator would not bear the cost of the remedial actions taken where he demonstrates that he was not at fault or negligent and that the environmental damage was caused either by an event expressly authorised by, and fully in accordance with the conditions of, an authorisation conferred by or given under the laws or regulations specified in the Regulations as applied on the date of the event, or by an activity or any manner of using a product in the course of an activity which the operator demonstrates was not considered likely to cause environmental damage according to the state of scientific and technical knowledge at the time when the activity took place.

The Authority is entitled to initiate cost recovery proceedings against the **11** operator, or if appropriate a third party who has caused the damage, in relation to any preventive or remedial measures taken in pursuance of these regulations within five years from the date on which those measures have been completed or the liable operator, or third party, has been identified, whichever is the later date.

12 These Regulations should be read in conjunction with the Contained Use of Genetically Modified Micro-Organisms Regulations 2008,[6] which in Regulation 17(4) provide that in the event of an incident involving a significant and unintended release of genetically modified micro-organisms in the course of their contained use, which may present an immediate or delayed hazard to human health or the environment, the Malta Environment and Planning Authority, in collaboration with the Occupational Health and Safety Authority, may require the user to defray or contribute towards any or all of the costs incurred by the Authority arising from such an incident.

13 Moreover, as pointed out in the first report, concurrently, Article 24 of the Environment Protection Act[7] provides that, without prejudice to the civil law provisions on damages, any person who causes damage to the environment is liable to pay to the Environment Protection Fund, set up by the same Act, such sum as may, in the absence of agreement, be fixed by the court *arbitrio boni viri* to make good the damage caused by the non-observance of any law or regulation by such person or by his negligence or wilful act or lack of skill in his art or profession to the environment and suffered by the community in general. This action would be instituted by the Chairman of the Fund on behalf of the Government.

2. State liability

14 The same tort provisions of the Civil Code referred to above are applicable to state liability. Likewise, the Prevention and Remedying of Environmental Damage Regulations, 2008 described above are applicable irrespective of whether the operator is a private person or a public person/entity.

II. Damage

1. Recoverable losses

15 The general tort regime does not specifically envisage losses caused by GMOs.

6 LN 127 of 2008.
7 Chap. 435 of the Laws of Malta.

2. Pure economic loss

As explained in the first report, the loss that is recoverable under the general tort regime is the actual loss directly caused by the tortious act, the expenses that the injured party may have incurred in consequence of the damage and the loss of actual or future earnings suffered as a result of the tortious act.[8] Details concerning the applicable rules and principles may be found in the first report. **16**

3. Mere fear of a loss

As shown in the first report, in order to be recoverable under the Civil Code provisions the economic loss must be certain in the sense that it is inevitable either because it has already been suffered or because the cause exists that will inevitably produce such damage. So future damage is covered provided it is inevitable. **17**

Non-material damage or moral damage (i.e. non-pecuniary loss) is not recoverable. Under Maltese law, as a general rule non-material damage may not be recovered except in those instances where the law expressly prescribes otherwise; in the case of tort actions under the Civil Code the law does not provide for such damages. **18**

4. Standard of proof

The standard of proof in civil proceedings, including tort actions under the Civil Code like the ones under discussion, is that the plaintiff's claim must be proved on a balance of probabilities. On the other hand, as shown in the first report, the tort provisions of the Civil Code are silent on the question of who bears this onus of proof. However, the provisions on court procedure in the Code of Organisation and Civil Procedure[9] prescribe that as a general rule, unless provided otherwise in any law, the burden of proving any fact rests in all cases on the party alleging it.[10] Moreover, it is an established principle in case law that the onus of proving the causal link between the damage and the unlawful act rests solely with the party claiming damages, i.e. the plaintiff.[11] The law does not envisage any cir- **19**

8 Art. 1045 of the Civil Code.
9 Chap. 12 of the Laws of Malta.
10 Art. 562.
11 *David Gatt* v. *Peter Calleja*, Civil Court, First Hall, judgment of 30 May 2002.

cumstances where there might be a reversal of the burden of proof. Even where the tortious act or omission constitutes a breach of a duty imposed by law, though the plaintiff might not need to prove negligence (there were cases such as *S. Caruana* v. *Kaptan E. Skapinakis noe*[12] where the court held that the non-observance of regulations is *prima facie* proof of negligence), he would still be required to prove the causal relationship between the act or omission constituting the breach of duty and the alleged damage.[13]

5. Nominal losses

20 N/A.

6. Mass losses

21 N/A.

III. Causation

1. Uncertainty of merely potential causes

22 As far as causation is concerned, as shown in the first report, case law relating to the general tort provisions of the Civil Code has established that as a general rule it must be shown that the tortious act was the immediate and direct cause of the damage.[14] However, occasionally, Maltese courts have also held that, provided the nexus is not too remote, it may suffice if the tortious act was only the indirect cause of the damage; provided that is, that the act led to a state of affairs which would not have existed were it not for the act concerned.[15]

2. Complex causation scenarios

23 N/A.

---

12 Vol. XXXV, Pt II, 548.

13 Art. 1033.

14 *G. Cefai* v. *G. Attard*, Vol. XXV, Pt. 1, 811.

15 *Brookes* v. *Saré*, Vol. XL, Pt. II, 815 and *A. Mallia* v. *C. Moore* Vol. XXXVIII, Pt. I, 249.

3. Force majeure

Force majeure is a defence under the Civil Code as damage caused by *force majeure* is borne by the victim; Article 1029 provides that: 'Any damage which is produced by a fortuitous event, or in consequence of an irresistible force, shall, in the absence of an express provision of the law to the contrary, be borne by the party on whose person or property such damage occurs'. **24**

4. Threshold to prove causation

The threshold is weighed more towards an 'almost certainty' test as the law prescribes, as shown above, that only loss that is 'directly caused' to the injured party by the tortious act is recoverable.[16] **25**

5. Special rules on causation

N/A. **26**

IV. Types of liability

1. Fault liability

(a) Special rules governing fault

As explained above, there are no specific rules dealing with damages actions in the case of GMOs. Liability under the general tort provisions is based on the concept of fault and the onus of proving this fault, the damage and the causal link between them is on the injured party and, as stated above, the law does not envisage any circumstances where there might be a reversal of this burden of proof. **27**

(b) Impact of specific rules of conduct

Since there is as yet no law or case law dealing specifically with such issues, one may only answer this question hypothetically. As stated in the first report, under the general tort provisions there is liability for the ensuing **28**

16 Art. 1045 of the Civil Code.

damage when a person, even without the intent to injure, voluntarily or through negligence, imprudence or lack of attention acts or fails to act, in breach of a duty imposed by law.[17] This implies that a person who acts voluntarily or negligently in breach of specific statutory rules governing GM or non-GM farming would automatically be liable for the damage caused but the injured party would still have to prove the damage and the causal relationship.

29 The Civil Code also provides that liability for damages would subsist if a person, lacking the necessary skill, undertakes any work or service and causes damage to others through his lack of skill.[18] So again here this implies that if GMO-related legislation were to require a person to have a certain specialised skill, damage caused as a result of lack of such skill would give rise to liability. Yet again, however, the damage and the causal relationship would have to be proved.

2. Product liability

(a) Development risk defence

30 The Consumer Affairs Act that implements the EC Product Liability Directive incorporates the development risk defence by reproducing *verbatim* the wording of the relative provision of the directive.[19] There is no case law yet to show how this defence would be interpreted by the court if invoked in court.

(b) Alternative routes

31 The general tort provisions of the Civil Code discussed above would of course provide an alternative route, but since they are based on fault liability not strict liability this alternative route would have to survive the scrutiny of the ECJ in terms of the relevant case law to date.

---

17 Art. 1033 of the Civil Code.

18 Art. 1038 of the Civil Code.

19 Chap. 378 of the Laws of Malta, Art. 56–71A. The defence is incorporated in Art. 62 (e).

(c) Impact of compliance with rules and regulations

The Consumer Affairs Act incorporates the defence provided by the EC Product Liability Directive that a producer would not be liable if he proves that 'the defect in question is due to compliance with a mandatory requirement imposed by law or by a public authority'.[20] This means, however, that for the defence to succeed it does not suffice that he shows he followed the rules and regulations governing the production process but he must also prove specifically that the defect leading to the damage is directly due to his observance of these rules and regulations with which he was obliged to comply. **32**

3. Environmental liability

(a) Implementation of the Environmental Liability Directive

As stated above, the Environmental Liability Directive was implemented via the Prevention and Remedying of Environmental Damage Regulations, 2008.[21] Regulation 9(2) of these Regulations obliges the Malta Environment and Planning Authority to recover the costs it incurred in relation to the preventive and remedial actions it had to take in respect of the environmental damage caused by the operator's activities or the imminent threat of such danger, *inter alia*, by obtaining appropriate financial guarantees from the operator who caused the damage or the imminent threat of damage including by means of security over his property. **33**

(b) Environmental liability regime beyond the scope of the Directive

As mentioned above, the Environment Protection Act[22] and the Contained Use of Genetically Modified Micro-Organisms Regulations, 2008[23] also provide for liability for environmental harm. **34**

---

20 Art. 62 (d) of the Consumer Affairs Act.
21 LN 126 of 2008.
22 Chap. 435 of the Laws of Malta.
23 LN 127 of 2008.

(c) Claimants in cases of environmental harm

35 Under the Prevention and Remedying of Environmental Damage Regulations, 2008[24] it is the Malta Environment and Planning Authority that can order remedial measures in case of environmental damage and recover the cost of such measures if they are undertaken by the Authority itself, as explained above. Likewise, under the Contained Use of Genetically Modified Micro-Organisms Regulations 2008,[25] in the event of an incident involving a significant and unintended release of genetically modified micro-organisms in the course of their contained use which may present an immediate or delayed hazard to human health or the environment, it is again the Malta Environment and Planning Authority, in collaboration with the Occupational Health and Safety Authority, which may require the user to defray or contribute towards any or all of the costs incurred by the Authority arising from such an incident. Under the Environment Protection Act,[26] it is the Chairman of the Environment Protection Fund on behalf of the Government who, in the absence of agreement, may institute an action against any person who causes damage to the environment for a court ruling obliging such person to pay to the Environment Protection Fund a sum of money to make good the damage caused to the environment, as detailed above.

(d) Special liability regime for losses sustained by individuals

36 N/A.

(e) Cartagena Protocol

37 Malta ratified the Cartagena Protocol on 5 January 2007 and it entered into force in Malta on 5 April 2007. The Malta Environment and Planning Authority carries out all the functions pursuant to this Protocol.

4. Other strict liability regimes

38 N/A.

---

24 LN 126 of 2008.
25 LN 127 of 2008.
26 Chap. 435 of the Laws of Malta.

V. Vicarious liability

1. Scope of vicarious liability

The general tort provisions of the Civil Code provide that where a person for any work or service employs another person who is incompetent, or in respect of whom he has no reasonable grounds to expect competence, he would be liable for the damage which this person, through incompetence in the performance of such work or service, causes to others.[27] 39

2. Liability for people further up the food or feed production chain

Such a situation would not arise as under Maltese law these persons are each responsible only for the damage that may result from their own culpable action once they are all operating independently of each other. It is only under the product liability provisions of the Consumer Affairs Act[28] that someone in the supply chain may be held liable in lieu of the producer for damages resulting from a defective product – and this only in the event that the supplier fails to disclose the identity of the producer.[29] 40

3. Can someone further down the feed or food chain include someone further up it in a trial against him/herself?

Only if he can show that the other person was also partly (or even solely) to blame for the damage caused. Articles 1049–1050 of the Civil Code provide for joint and several liability where two or more persons have maliciously caused any damage. On the other hand, where some of them have acted with malice and others without malice, the former would be jointly and severally liable, while each of the latter would only be liable for the part of the damage that he/she may have caused. However, where the part of the damage which each has caused cannot be ascertained, the injured party may file for the whole damage to be made good by any one of the persons concerned, even though all or some of them have acted without malice, but the defendant would have the right to seek relief from the other or the others. In such a case, the defendant may demand 41

27 Art. 1037.
28 Chap. 378 of the Laws of Malta.
29 Art. 59.

that all the persons who caused the damage be joined in the proceedings and the court apportion among them the sum fixed by way of damages, in equal or unequal shares, according to the circumstances, saving always the right of the injured party to claim the whole sum from any one of the persons concerned who would be held jointly and severally liable.

VI. Multiple tortfeasors

42 See paragraph 40.

VII. Defences

1. Licence/permission to grow GM material

43 Neither the product liability provisions nor the general tort provisions seem to accommodate (by analogy) such a defence.

2. Consent/assumption of risk

44 Both the product liability provisions and the general tort provisions provide for contributory negligence. Article 1051 of the Civil Code prescribes that if the injured party through his 'imprudence, negligence or want of attention' contributed to or occasioned the damage, the court, in assessing the amount of damages payable to him, would determine, in its discretion, the proportion of damage to which he has so contributed or given occasion to and the amount of damages payable to him by the defendant will be reduced accordingly.[30]

45 Moreover, the defectiveness test under the product liability regime is based on consumer expectations – 'A product is defective if it fails to provide the safety which a person is entitled to expect'.[31] So if the product clearly warns the purchaser that it is a genetically modified product and the average consumer considers a GM product to be generally safe notwithstanding the risks involved, the product would not be deemed 'defective' under the product liability regime.

30 Art. 65 of the Consumer Affairs Act makes Art. 1051 of the Civil Code applicable also to product liability cases.

31 Art. 60 of the Consumer Affairs Act.

3. Third-party influence

Since the product liability provisions of the Consumer Affairs Act are based on strict liability, Article 66 thereof provides that 'the liability of the producer shall not be reduced when the damage is caused both by a defect in the product and by the act or omission of a third party'. On the other hand, under the general tort provisions where liability is fault-based, the court in order to establish the degree of fault of the defendant in respect of the damage and the relative causal link would take into account third-party behaviour. **46**

4. Prescription

Since there is no specific legislation dealing with claims as envisaged by the study, the applicable limitation periods are those applicable in standard damages actions. **47**

The limitation period for an action for damages in tort is two years[32] but if the tortious act or omission constitutes a criminal offence (e.g. under the Food Safety Act),[33] the limitation period applicable even for the civil action would be the one prescribed for the criminal action.[34] **48**

On the other hand, the limitation period under the product liability provisions of the Consumer Affairs Act is three years from the day when the injured party became aware, or should reasonably have become aware, of the damage, the defect and the identity of the producer; however, in the latter case, the action for damages would be extinguished upon the lapse of ten years from the date on which the producer put into circulation the actual product which caused the damage.[35] **49**

In the case of proceedings under the Prevention and Remedying of Environmental Damage Regulations, 2008,[36] the Malta Environment and Planning Authority is entitled to initiate cost recovery proceedings within five years from the date on which the preventive or remedial measures in respect of the environmental damage have been completed or the responsible operator, or third party, has been identified, whichever is the later date. **50**

32 Art. 2153 of the Civil Code.
33 Chap. 449 of the Laws of Malta.
34 Art. 2154 (1) of the Civil Code.
35 Art. 67.
36 LN 126 of 2008.

51 As for proceedings under Article 24 of the Environment Protection Act,[37] which empowers the Chairman of the Environment Protection Fund to institute an action in court to recover money from the person responsible in order to make good the damage caused to the environment, these are prescribed by the lapse of eight years.

5. Other defences

52 N/A.

VIII. Remedies

1. Pecuniary compensation

(a) Bodily harm

53 There are no special rules, so the regular remedies apply.

(b) Property losses

54 There are no special rules, so the regular remedies apply.

(c) Economic losses

55 There are no special rules, so the regular remedies apply.

(d) Harm to animals

56 As stated above, under the general tort regime, the damage that is recoverable is the actual loss that the tortious act has directly caused to the injured party together with the expenses that the latter may have incurred in consequence of the damage and the loss of actual earnings as a result of the tortious act (collectively referred to as *damnum emergens*) as well as the loss of future earnings as a result (*lucrum cessans*). The underlying principle is the *restitutio in integrum* principle, namely that the claimant should be put

37 Chap. 435 of the Laws of Malta.

in the position he would have been in, had the tortious act or omission not occurred.

The courts have developed rules on the quantification of damages only in the sphere of personal injury and traffic accidents, as these are the most common compensation-related cases to come before them. So one may only speculate what the court might decide in the scenario hypothesised here, on the basis of the above-mentioned principles. I would assume that the plaintiff would be able to recover the market value of the animal harmed by the contaminated feed including the loss of future earnings from the produce of the harmed animal if as a result of the feed the animal is no longer of any commercial use to its owner. In the absence of specific case law, one cannot say whether the court will consider the mere fact that an animal eats GM contaminated feed as constituting harm to the animal and thereby as giving rise to recoverable damage to its owner. 57

(e) Costs of disposal

On the basis of the above-mentioned principles governing recoverable damages and in particular the *restitutio in integrum* principle, I would assume that the costs of disposing of contaminated production/animals feed with GM feed would be recoverable. 58

2. Non-compensatory damages

As noted above, only compensatory damages are recoverable under the general tort regime as such is based on the principle that the claimant should be put in the position he would have been in, had the tortious act or omission not occurred. So exemplary damages are not contemplated by this law. 59

On the other hand, Article 24 of the Environment Protection Act[38] is wider in scope and does allow for non-compensatory damages, providing that, *without prejudice to the civil law provisions on damages*, any person who causes damage to the environment is liable to pay to the Environment Protection Fund such sum as may, in the absence of agreement, be fixed by the court *arbitrio boni viri* to make good the damage caused to the environment and suffered by the community in general by the non-observance of any law 60

38 Chap. 435 of the Laws of Malta.

or regulation by such person or by his negligence or wilful act or lack of skill in his art or profession. Hence the law here contemplates the payment of a contribution to the Fund by the person responsible for the damage over and above the compensatory damages that he would have paid to the injured party.

61 Moreover, Article 26 of the same Act provides that where the person's action or omission causing the environmental damage amounts to a criminal offence, the person may pay to the Fund a contribution (compromise penalty) as may be agreed with the Malta Environment and Planning Authority *in lieu* of undergoing criminal proceedings and his criminal liability with respect to that offence would be extinguished. This compromise penalty would not however extinguish his civil liability to make good the damage caused to any injured party; so again this compromise penalty is a contribution that is over and above any compensatory damages that are due.

3. Other remedies

62 As noted in the first report, there is a general provision in the Civil Code that might be applicable to a situation where the GMO cultivation in one field is causing damage to non-GM crops in neighbouring fields. Article 539 of the Civil Code provides that 'where any person has reasonable cause to apprehend any serious and impending damage to a tenement or other thing possessed by him, from any building, tree or other thing, he may bring an action demanding, according to the circumstances, either that the necessary steps be taken to obviate the danger, or that the neighbour be ordered to give security for any damage the plaintiff may suffer therefrom'.

4. Costs of pursuing a claim

(a) General cost rule

63 The general rule followed by Maltese courts is to allocate all costs borne by the party winning the case to the party losing the case. However, at law the courts enjoy wide discretion in apportioning these costs. In fact, Article 223 of the Code of Organisation and Civil Procedure,[39] while providing that 'every definitive judgment shall award costs against the party cast', states that 'in all cases, it shall be lawful for the court to order that the

39 Chap. 12 of the Laws of Malta.

costs shall not be taxed as between party and party, when either party has been cast in some of the points at issue, or when the matter at issue involves difficult points of law, or when there is any other good cause'.

Moreover, where an *ex parte* expert witness is produced by any of the parties, Article 223 leaves it up to the court to determine how the costs of this expert witness are to be apportioned between the parties. Where two or more persons are ordered to bear costs, in terms of Article 224, each person would be deemed liable *in solidum* or in proportion to his interest in the cause according to the decision on the merits. **64**

(b) Costs of establishing causation

In view of the wide discretion enjoyed by the courts in apportioning costs, as shown above, and given the absence of any case law concerning allocation of the costs of testing or of other means to establish causation, it is not possible to say how a Maltese court might allocate the costs incurred in proving causation in a case involving GMOs. **65**

5. Advance cover

There is no general or specific duty at law to obtain liability insurance or to provide for other advance cover for potential liability nor are there any general compensation schemes available under Maltese law. **66**

IX. Cross-border issues – Conflict of laws

1. Conflicts rules applicable before (or instead of) Rome II

Before the entry into force of the Rome II Regulation in Malta on 11 January 2009, the choice of law principle followed by Maltese private international law in relation to actions in tort was that the applicable law was the *lex loci delicti commissi* so that in relation to GMO related actions, the applicable law would have been the law of the country where GMO contamination had occurred. **67**

Article 4 of the Regulation now lays down the new general conflicts rule for tort actions as being that the applicable law is the law of the country in which the damage occurs irrespective of the country in which the event giving rise to the damage occurred and irrespective of the country or coun- **68** tries in which the indirect consequences of that event occur. There are, however, two exceptions to this rule: (i) where the person allegedly responsible for the tortious act and the person sustaining the damage both have their habitual residence in the same country at the time when the damage occurs, the applicable law is then the law of that country; and (ii) where it is clear from all the circumstances of the case that the tortious act is manifestly more closely connected with a different country (e.g. deriving from a pre-existing relationship between the parties such as a contractual relationship), then the law of that other country would apply.

69 Article 4 of the Regulation has therefore radically changed Malta's choice of law rule. However, for our purposes this is not entirely so because Article 7 of the Regulation lays down that where the tortious act gives rise to environmental damage or damage sustained by persons or property as a result of such damage, the person seeking compensation for damage may choose to depart from this general rule and base his or her claim on the law of the country in which the tortious event giving rise to the damage occurred; in other words, in the case of environmental damage the *lex loci delicti commissi* principle that governed Malta's conflicts rule before the coming into force of the Regulation subsists.

70 Moreover, also relevant for our purposes is Article 5 of the Regulation that departs from the Regulation's general conflicts rule as described in paragraph 68 above in the case of product liability cases by laying down that the applicable law in product liability cases would be (i) the law of the country in which the person sustaining the damage had his or her habitual residence when the damage occurred, if the product was marketed in that country; or, failing that, (ii) the law of the country in which the product was acquired, if the product was marketed in that country; or, failing that, (iii) the law of the country in which the damage occurred, if the product was marketed in that country. However, the applicable law would be the law of the country in which the person allegedly responsible for the product is habitually resident if he or she could not reasonably foresee the marketing of the product, or a product of the same type, in the country the law of which is applicable under (i), (ii) or (iii). Nevertheless, as in the case of the general conflicts rule, the said rules concerning product liability cases do not apply (a) where the person allegedly responsible for the defective product and the person sustaining the damage both have their habitual residence in the same country at the time when the damage occurs, as in this case the applicable law is the law of that country; or (b) where it is clear from all the circumstances of the case that the tort is manifestly more closely connected with a different country, as in this case the law of that other country would apply.

In addition, there is the overarching rule in the Regulation that respects the parties' freedom of choice. Article 14 provides that, subject to certain limitations, the parties may agree to submit non-contractual obligations to the law of their choice by entering into an agreement before (where all the parties are pursuing a commercial activity) or after the event giving rise to the damage occurred. 71

2. Special regime for cross-border claims

N/A. 72

X. Cases

1. Due to the adventitious presence of GMOs in a field, maize which is normally sold as conventional contains GMOs beyond the legal labelling threshold. This is not discovered before the final stage of the food production chain by the producer of taco chips. The whole production is lost since the supermarket chains refuse to accept delivery from the producer.

(a) Who can sue along the chain of distribution?

Since Malta's tort regime is fault-based, no action for damages would be possible unless it can be shown that although the admixture is adventitious, someone along this chain of events was negligent (using the reasonable person standard) or lacked the necessary skill to undertake the work concerned and his negligence or lack of skill led to the contamination.[40] However, there is an important factor here, namely that in selling the maize the farmer had breached a statutory duty – that of ensuring that the conventional maize being sold does not contain GMOs beyond the threshold set by law. Under the Civil Code, a person is *automatically* liable for damages when, even if without the intent to harm, he voluntarily, negligently or through imprudence or lack of attention acts in breach of a 73

40 Art. 1029 of the Civil Code states that 'any damage which is produced by a fortuitous event ... shall ... be borne by the party on whose person or property such damage occurs'.

duty imposed by law.[41] So the taco producer (only he can sue as only he suffered damage) might be able to sue the farmer successfully by showing that the farmer negligently breached a statutory rule by selling maize that contained GMOs beyond the legal labelling threshold as conventional and that there is a direct causal link between this tort and the damage he suffered. However, if the farmer had taken all necessary measures and care (according to a reasonable person standard) to ensure that the GMO content did not exceed the limits but his product nonetheless breached these limits due to fortuitous events beyond his control, no liability can be imputed to the farmer.

(b) Would the case be solved differently if the GMO content was below the labelling threshold?

74 If the GMO content was below the labelling threshold, it would be impossible for the taco producer to prove fault on the part of the farmer as the GMO content was within the limit.

(c) Would the case be solved differently if the admixture was not adventitious, but occurred due to the disregard of segregation rules, for example?

75 Yes, because clearly in this case there would be a clear intentional breach of a statutory duty and so there would be automatic liability. Of course, the plaintiff would still have to prove the damage and the causal link between said breach and the damage.

(d) Would the case be solved differently if the GMO found was not admitted for production in your jurisdiction?

76 Not necessarily, because although this is also a breach of a statutory duty, it does not necessarily follow that this breach directly led to the damage sustained by the taco producer; there would be no direct causal link as there is in the case of a breach of the segregation rules.

---

41 Art. 1033.

(e) If the admixture had occurred on a non-GM field and it transpires that the GM seeds were blown from: (i) neighbouring fields; or (ii) a truck passing by, would the farmer of the affected field be liable for all or part of the loss caused further down the distribution chain?

In such circumstances no fault and hence no liability can be imputed to the farmer. **77**

2. Twenty years after the sale of GM maize used for food products, it turns out that it has certain disadvantageous health effects for humans.

(a) Can the producers be held liable at this point for risks unknown at the time of growing the maize? Who would be liable?

Under the tort regime, there can be no liability as no fault can be established where the risks were unknown at the time. Under the product liability regime (applicable here because it is a question of personal injury), a damages action would not be possible because of the ten-year extinction period – the GM maize would have been out on the market for more than ten years. Moreover, if scientific knowledge at the time of production was not advanced enough to make such risks detectable, the development risks defence would also be a defence against liability. **78**

(b) Can compensation already be claimed at a point when the negative health effects have not yet materialised, but are to expected to according to scientific expertise/mere rumours?

Under the tort regime, future damage can be compensated provided it is inevitable because the cause exists that will inevitably produce such damage. So it all depends on how conclusive the scientific expertise is; certainly mere rumours do not suffice. So far, court cases involving future damage have related solely to the calculation of the loss of future earnings resulting from an incapacity induced by a tortious event and never to the calculation of damage resulting from yet-to-materialise negative health effects; thus, one can only speculate about the court's approach in such a scenario. **79**

(c) Would it make any difference if the GM maize had only been in use for feed, causing harm to the animals, which may or may not cause harm to humans consuming the meat as well?

80 Yes, because then the nexus between the tortious act and the harm to humans might be too remote to establish liability.

3. The driver of a food logistics company discovers that a farmer, from where he regularly picks up agricultural products fails to obey mandatory segregation rules or food or feed hygiene standards, which may lead to the admixture of GM and non-GM produce sold and packaged separately by that farmer. Does he or his employer have a duty to warn, i.e. warn the recipient of the allegedly "non-GM" produce?

81 Under the Product Safety Act,[42] the distributor (a term which would include a 'food logistics company' which I assume is a company that transports agricultural products but whose activity does not affect the safety properties of the product) is required to act with due care to ensure compliance with the general safety requirements established by law, in particular by not supplying products which he knows or should reasonably know, on the basis of the information in his possession as a practitioner in the relevant business, do not comply with such requirements. It would be a criminal offence for any distributor to supply products which he knows to be unsafe. This Act does not establish vicarious liability, so the obligation to warn would rest on the employer of the driver if the latter informs the former about the farmer's non-compliance with the segregation rules.

42 Chap. 427 of the Laws of Malta, Art. 7.

Damage Caused by GMOs under Dutch Law

Ingrid Greveling and Willem H. van Boom[1]

I. General overview

1. Special liability or redress scheme for GMOs

There are no specific rules on liability or compensation of damage relating 1
to GMO crops.

2. State liability

No specific rules apply to state liability for GMO crops. Therefore, the normal 2
rules apply. Under Dutch law, public authority liability is a complex field of the law in which tortious liability under civil law (basically a negligence-based liability for wrongful state intervention) interacts and sometimes clashes with public authority "liability" under administrative law. Moreover, the concept of *égalité devant les charges publiques* may compel the state to alleviate the financial burden resting upon specific well-defined groups in society.[2]

II. Damage

1. Recoverable losses

According to art. 6:95 *Burgerlijk wetboek* (Civil Code, BW), damage consists of 3
patrimonial damage and non-patrimonial damage. Patrimonial damage

1 Ingrid Greveling is Assistant and Willem van Boom is Professor at the Rotterdam Institute of Private Law, Erasmus School of Law, The Netherlands. For further information, please visit www.ripl.eu.

2 Further on this topic, e.g., *W.H. van Boom/A. Pinna*, Liability for Failure to Regulate Health and Safety Risks; Second-Guessing Policy Choice or Showing Judicial Restraint? in: H. Koziol/B.C. Steininger (eds.), European Tort Law 2005 (2006) 1 ff.

includes costs incurred, loss suffered and loss of profit (art. 6:96 Civil Code). Death, personal injury, property damage and pure economic loss are on an equal footing in this regard.

4 With regard to non-pecuniary loss, the following is relevant. The injured party may only claim non-patrimonial damage in one of the situations mentioned in art. 6:106 Civil Code: firstly, if the liable party had the intention to cause immaterial damage; secondly, if the injured party incurred a physical injury, if his reputation or his honour is damaged, or if his person is damaged in any other way; thirdly, if the reputation of a person who has passed away is damaged (only if that person, were he alive, would also have had the right to compensation for damage to his reputation).

2. Pure economic loss

5 As such, pure economic loss is not special under Dutch law. If the conduct of the respondent is held to be wrongful and all the requirements laid down in art. 6:162 BW have been met, then there is liability. Liability may include pure economic loss. No specific thresholds apply with regard to pure economic loss. Having said that, it may well be possible that the court may consider the respondent not to have acted tortiously vis-à-vis the claimant on the basis that the claimant's interest was of a purely economic nature. This depends on the case at hand.

6 Take for instance a case where consumers lose trust in certain agricultural produce for fear of GMO contamination. Although there are no court decisions on this matter, we feel that the loss sustained by farmers who suffer from a diminishment of consumer trust in their produce will not be compensated easily under tort law. A court would certainly require proof of a wrongful act or omission leading to admixture or contamination. A GMO-farmer may possibly be held liable, for instance, for not informing neighbouring farmers of his GMO-activities – thus depriving them of the chance to take precautionary measures. In that case, liability can also cover pure economic losses such as sudden drop in turnover. Dutch law does not lay down actual admixture or interference as a formal prerequisite for liability, so in effect the adjudication of compensation for pure economic loss is feasible. Whether compensation is granted may depend on the specific facts of the case.

3. Mere fear of a loss

Non-pecuniary loss is only compensated under Dutch law under very specific conditions. The injured party can only claim non-pecuniary loss in one of the situations mentioned in art. 6:106 Civil Code, i.e., (1) if the liable party had the intention to cause non-pecuniary loss, (2) if the injured party suffered personal injury, damage to his reputation or honour (either alive or deceased!), or if his person is "personally damaged in any other way". Mere fright over future events causing pure economic loss as such is unlikely to generate claims for non-pecuniary loss unless filed under "personal injury" or "personal damage in any other way". **7**

4. Standard of proof

Concerning evidence of damage, we need to distinguish three phases: evidence that the claimant has suffered loss (1), evidence that the loss was caused by the defendant (2), and evidence of the extent of the loss (3). **8**

Regarding proof that the claimant suffered loss, the normal rules of burden of proof apply. As a starting point, the burden of proof lies on the claimant. This rule is laid down in art. 150 *Wetboek van Burgerlijke Rechtsvordering* (Code of Civil Procedure). The claimant has to prove the facts underpinning his claim regarding the obligation to compensate. To some extent this includes the fact of loss, but there is no need for precise evidence of the extent of the damage. Courts are allowed to make a rough estimate of the damage. **9**

Evidence of *conditio sine qua non* (2) will be dealt with infra, no. III.4. Calculating the exact extent of damage (3) is not subject to strict rules of evidence. Courts have considerable room for calculating or even roughly estimating the pertinent amount. **10**

As far as future damages are concerned, the courts are allowed to award damages either as a lump sum or as a periodic allowance (art 6:105 BW). In personal injury legal practice, both injurer and injured party generally prefer the payment of a lump sum (partly for the purpose of avoiding income tax). The payment for future damages by means of a lump sum is calculated on the basis of reasonable projections on how the future would have evolved if the injury had not occurred. **11**

5. Nominal losses

12 Nominal/symbolic losses are not acknowledged as such.

6. Mass losses

13 The *Wet Collectieve Afwikkeling Massaschade* (WCAM), the 2005 Collective Settlement Mass Damage Act, supports efforts to settle claims for mass losses with a minimum of judicial intervention. The main focus of the legislator was on designing an efficient mechanism for the settlement of events causing mass personal injury but it seems that the Act is of more practical relevance for securities litigation.

14 The WCAM 2005 allows parties to a voluntary collective settlement contract to request the Amsterdam Appeals Court to declare this settlement binding on all victims. Art. 7:907 Civil Code provides that an amiable settlement "concerning the payment of compensation for damage caused by an event or similar events concluded between a foundation or association with full legal competence and one or more other parties which have committed themselves by this agreement to pay compensation for this damage may, at the joint request of the parties that concluded the agreement, be declared binding by the court on persons to whom the damage was caused, provided the foundation or association represents the interests of these persons pursuant to its articles of association".

15 The Amsterdam Court will consider the settlement according to specific standards of fairness. Moreover, if the Court declares the settlement binding on all victims, individual victims may opt out during a certain period (art. 7:908 Civil Code).

III. Causation

1. Uncertainty of merely potential causes

16 In the case of multiple uncertain causes, art. 6:99 Civil Code (alternative causation) provides as follows. When the damage may have resulted from two or more events for each of which a different person is liable and it has been determined that the damage may have been caused by at least one of these events, each one of these persons is jointly and severally liable and therefore liable to repair the damage, unless he can prove that the damage is not a result of the event for which he is liable.[3] Note that if it is unclear

who *violated* certain rules, art. 6:99 will not come into play: for joint and several liability it must be ascertained that the defendant did in fact violate the relevant rules – as others did – but remain uncertain which of these violations *caused* the damage.

As far as uncertain causation is concerned in cases where the defendant is responsible for potential cause A and the claimant is responsible for potential cause B, there may be room for awarding damages proportionate to the likelihood that cause A was in fact the sole cause. In a recent decision (*Nefalit v Karamus*), the Dutch Supreme Court ruled that where an employee who had been exposed to asbestos dust by his negligent employer and had also smoked for years could not prove which of these potential carcinogenics caused his lung cancer, the employer was liable in proportion to the (statistical) likelihood that the asbestos exposure had in fact caused the cancer.[4] The basis for this proportionate liability was found in the special legal relationship between employer and employee, which in the eyes of the Court justified the extensive application by analogy of art. 6:101 (1) Civil Code (contributory negligence).[5] 17

2. Complex causation scenarios

As mentioned supra, in scenarios comparable to DES, the Dutch Supreme Court applies joint and several liability. The example given in the question, however, is not necessarily such a case. Joint and several liability can only be applied if there is evidence that the defendant acted wrongfully and his wrongful act was adequate to cause the damage at hand. If it remains unclear which batch was contaminated, the first question to answer is whether the defendant did in fact act tortiously. If it is clear 18

3 Hoge Raad (HR) 9 October 1992, Nederlandse Jurisprudentie (NJ) 1994, 535 (DES). Cf. *W.H. van Boom*, Multiple Tortfeasors under Dutch Law, in: W.V.H. Rogers (ed.), Unification of Tort Law: Multiple Tortfeasors (2004) 135–150.

4 HR 31 March 2006, Rechtspraak van de Week (RvdW) 2006, 328.

5 Art. 6:101, par. 1, BW reads: "When the damage is partly caused by an occurrence that can be imputed to the injured party, the obligation to pay compensation is reduced by apportioning the damage between the injured party and the liable party in proportion to the degree in which the occurrences that can be imputed to the parties have contributed to the damage, provided that account is taken of the disparity of the seriousness of the respective faults, or other circumstances of the case, to decide whether equity demands that an alternative apportionment or full recovery takes place or that the obligation to pay lapses." On art. 6:101 BW, see, e.g., *W.H. van Boom*, Contributory Negligence under Dutch Law, in: U. Magnus/M. Martín-Casals (eds.), Unification of Tort Law: Contributory Negligence (2004) 129–148.

that he in fact acted wrongfully, then imputation to him of the uncertain causation is possible within the framework of art. 6:99 Civil Code.

3. Force majeure

19 A prima facie wrongful act is considered not to be wrongful whenever force majeure justified it (art. 6:162 BW). As far as causation is concerned, force majeure in the sense of external causes other than the wrongful act may mitigate liability altogether (for lack of *conditio sine qua non* link between wrongful act and damage) or reduce the obligation to compensate damage (art. 6:101 BW).

4. Threshold to prove causation

20 According to Dutch law, a two-stage test must be applied. First, the well-known *conditio sine qua non* ("but for") test is applied. According to this requirement, there is a causal link between the damage and the wrongful act if the act was a necessary condition for the existence of the damage. In other words: without the act there would not be any damage. Unsurprisingly, Dutch courts do not apply a "scientifically approved" threshold as far as the degree of likelihood is concerned but rather an open-textured standard of the court's conviction based on all the evidence put forward and ultimately "on the balance of all probabilities".

21 With regard to the burden of proof concerning causation, the Dutch Supreme Court (Hoge Raad) has in recent years developed the so-called *omkeringsregel*, the "reversal rule". In a number of decisions the Hoge Raad has stated that if an act which constitutes a wrongful act is known to create the risk that a specific damage will occur, and if this risk subsequently materialises (so the damage occurs), the causal link between the damage and the act is presumed unless the respondent proves otherwise. In recent cases, the scope of this rule has been limited to cases in which the risk that materialised was of a certain specific nature that could be associated easily with the wrongful act. Hence, the rule is easily applied to contamination of a neighbouring crop if the contaminating substance is easily associated with a specific GMO crop in the area. It is unlikely, however, that it can be applied in a case where a GMO-farmer has acted wrongfully by not taking precautionary measures against migrating pollens and a drop in profits is experienced by all corn-producing farmers after negative publicity. Although there may be evidence of the intermediate cause of the negative publicity reflecting on corn as such, the market price mechanisms ruling corn trade are far too complicated to say that a drop in profits in corn farming is typically associated with negligent GMO-farming.

Obviously, the *conditio* requirement is too extensive; without any further delimitation too many causal links between act or omission and damage would be seen as cause of the damage. Therefore, if the first stage of the test has been satisfied, a second is applied: the imputation test. The test is laid down in art. 6: 98 BW, which reads: **22**

> "Compensation can only be claimed insofar as the damage is related to the event giving rise to liability in such a fashion that the damage, also taking into account its nature and that of the liability, can be imputed to the debtor as a result of this event."

This test was developed in case law. For instance, the Dutch Supreme Court decided (HR 20 March 1970, NJ 1970, 251, Waterwingebied) that for the establishment of the causal link it was also necessary that the damage was reasonably imputable to the act (or omission as the case may be). This requirement was thus called the requirement of "reasonable imputability". For a specific damage caused (in the sense of *conditio sine qua non*) by an unlawful action to be imputable, there are a number of relevant factors that have to be weighed up. In general, the damage should not be too exceptional a result of the unlawful action, nor in such a distant relation to it that it cannot reasonably be imputed to the liable person. **23**

The aforementioned case law has been codified in art. 6:98 BW. However, art. 6: 98 BW identifies only two of many factors that decide imputation: the nature of the damage and the nature of the liability. Although foreseeability of the damage is not mentioned in art. 6:98 BW, it is surely an important factor as well. As far as the nature of the damage suffered is concerned, both case law and doctrinal writings are inclined to stretch the limits of causal connection very far whenever bodily harm is involved, somewhat less far when damage to property is involved, and the least far in the case of loss related to neither of the two former categories (i.e. pure economic loss). **24**

It must be stressed that before the "reasonable imputability" test can be invoked, in principle the *conditio sine qua non* test should be met first. There are, however, specific conditions under which the requirement of *conditio sine qua non* does not apply, for instance in case of alternative causation or in case of two independent concurring causes that each would constitute an adequate condition to bring about the entire damage. These were dealt with supra III. **25**

5. Special rules on causation

26 There are no specific rules relating to GMO cases, but it seems that the rules of art. 6:98 BW are most relevant in such cases. We refer to supra, no. 22.

IV. Types of liability

1. Fault liability

(a) Special rules governing fault

27 No specific statutory rules or case law are applicable. Therefore, the general principles apply. As a starting point the burden of proof lies on the claimant.

28 In general, the injured party has to prove a) the facts that give rise to liability,[6] and b) the causal connection between these facts and the damage incurred (*conditio sine qua non*).[7] Thus, the claimant has to prove the facts underpinning his claim regarding the wrongful act committed. According to art. 150 of the Dutch Code of Civil Procedure, the burden of proof may be reversed by the court if a special (statutory) rule so requires or if reasonableness and fairness demand such reversal.[8]

(b) Impact of specific rules of conduct

29 Violation of rules – either statutory or generally accepted in the line of GMO business – helps the case of the claimant. Fault-based liability for unlawful acts is based on art. 6:162 BW. Fault-based liability consists of four elements: there must be an unlawful act, the act must be imputable to the actor, there must be damage and there must be a causal link between the damage and the unlawful act.

30 Firstly, there must be a wrongful act. Art. 6:162 Civil Code defines three types of wrongful act: the infringement of a subjective right, an act or omission violating a statutory duty (e.g., importing a banned GMO-product), or "conduct contrary to the unwritten standard of conduct seemly in society",

---

6 Effectively, the legal qualifications of "wrongful", "imputable", etc. do not need proof in the strict sense. See *I. Giesen*, Bewijs en aansprakelijkheid (2001) 14 f.

7 Cf. *Giesen* (fn. 6) 112 f.

8 *Giesen* (fn. 6) 98 f. Note that exact proof of the *calculation* of the amount in damages is not required.

the so-called "maatschappelijke betamelijkheid". This latter category is the most important one. It can be considered a residual category: whenever the injured party cannot base his claim on either of the first two categories, the third provides a comfortable fall-back option. Because of its open texture, many claims are based on this category. Violation of a statutory rule that purports to protect the interest that was in fact damaged by the violation constitutes a wrongful act vis-à-vis the damaged person. Violation of a customary rule may constitute violation of the standard of conduct seemly in society and may thus constitute a wrongful act in its own right.

2. Product liability

(a) Development risk defence

With regard to product liability, the Dutch legislator fully implemented the European Directive on Product Liability. The Dutch legislature has chosen to allow the "state of the art" defence (see art. 6:185 (1) (e) Civil Code), and not to exclude non-pecuniary loss. **31**

(b) Alternative routes

Apart from the strict liability on the basis of the European Directive, manufacturers' liability for defective or unreasonably unsafe products can in most cases also be based on the general fault liability for wrongful acts. The Dutch Supreme Court has stipulated some "hard and fast rules" in this respect. First, there is the very general rule that a manufacturer acts wrongfully if he markets a product that causes damage when it is used in a normal fashion and in accordance with its purpose. Second, there is the rule that a manufacturer is at fault if he does not assure himself of the absence of possible unsafe characteristics and flaws in his product.[9] These two rules in combination provide a strong basis for fault-based liability in most of the actual product liability cases, where lack of inspection or lack of utmost care in production methods provide the main sources of danger. **32**

As a side note, we do not believe that the strict and formal application by the ECJ of the full harmonisation effect of Directive 85/374/EEC really stands in the way of continuation of the Dutch domestic tortious liability regime. Firstly, it should be noted that fault-based liability was already a basis for **33**

9 HR 6 December 1996, NJ 1997, no. 219, and HR 22 October 1999, NJ 2000, no. 159.

product liability under Dutch law before promulgation of the Directive and therefore it seems to constitute an exception as referred to in art. 13 of the Directive. Moreover, although the Court has ruled that adding compliance with the recall duty to the conditions under which a manufacturer can exempt himself from liability was contrary to art. 7 and 15 of the Product Liability Directive, it did not rule that construing a recall duty under national law as a local remedy for tortious breach of the material duties under the GPSD was contrary to art. 13 of the Directive. Moreover, it could be argued that "after sales duties' under general tort law principles (for instance, recall duty) are altogether outside the scope of the Product Liability Directive. If, for instance, a public authority decides that a manufacturer should recall the defective product and the manufacturer refuses to do so, this may constitute both a criminal or administrative offence under the GPSD (depending on the domestic implementation of the GPSD) and a tortious breach of statutory duty vis-à-vis any consumers who sustain injuries after such a breach. We would argue that since breach of statutory duties as such surely constitutes a source of tortious liability under domestic legal systems in the sense of art. 13 of Directive 85/374/EEC (i.e. art. 13 allows continuation of tortious liability for breach of statutory duties if such liability predates the Directive), the GPSD can be enforced through common tort law rules. Moreover, breach of the recall duty does not necessarily injure the same consumers as the defective products as such do, hence enforcing the GPSD through tort law does not constitute a competing system of liability for defective products.[10]

(c) Impact of compliance with rules and regulations

34 As a rule, compliance with regulatory standards is relevant but does not preempt liability. It delivers evidence that there was compliance with regulatory law and that the defendant did not act wrongfully in that respect. However, civil law standards may go beyond the level of precaution demanded by regulatory standards. Therefore, there may be cause for a court to rule that a duty of care under the general standard of "conduct seemly in society" was breached even if all statutory rules were complied with.[11] This depends, inter alia, on the nature of the details and the drafter's intent with the regulation.

10 On the interpretation of art. 13 in light of the ECJ rulings, see, e.g., *S. Whittaker*, Form and Substance in the Harmonisation of Product Liability in Europe, Zeitschrift für Europäisches Privatrecht (ZEuP) 2007, 865 ff.; *J.-S. Borghetti*, La Responsabilité du Fait des Produits (2004) 563 ff.

3. Environmental liability

(a) Implementation of the Environmental Liability Directive

The Environmental Liability Directive (2004/35/EG) has been implemented in the *Wet Milieubeheer* (Environmental Management Act), art. 17.6–17.18, 18.2g.[12] 35

(b) Environmental liability regime beyond the scope of the Directive

There is no specific liability regime that specifically covers environmental harm and that exceeds the scope of the Directive. However, the strict liability for hazardous and noxious substances (art. 6:175 Civil Code) may be relevant with regard to environmental damage. The concept of damage is nevertheless restricted to the traditional heads of damage (property damage, reasonable clean-up costs, etc). On art. 6:175 Civil Code, see, infra no. 41. 36

(c) Claimants in cases of environmental harm

The Dutch legislator has fully implemented Directive 2004/35/EC in the public law Environmental Management Act. To the extent that harm to biodiversity and harm to the environment are recoverable under the regime of the Directive, a designated competent authority is authorised to claim expenses under public law. Private parties are not competent to claim under the Directive, but neither are their rights under private (tort) law affected. Consequently, proprietors of contaminated soils can still claim compensation from tortfeasors under common tort law principles. 37

11 See, e.g., HR 10 March 1972, NJ 1972, 278. Cf. *R.J.P. Kottenhagen/P.A. Kottenhagen-Edzes*, in: W.H. van Boom/M. Lukas/C. Kissling (eds.), Tort and Regulatory Law (2007) 197, no. 59 ff. See also *E. Bauw*, Privaatrechtelijke betekenis van zorgplichten in de milieuwetgeving (1996) 96–102.

12 Parliamentary Proceedings 2007–2008, 30 920. Staatsblad 2008, 166, 178.

(d) Special liability regime for losses sustained by individuals

38 No such special regime exists in The Netherlands. The common rules on wrongful acts and omissions (art. 6:162 BW) apply.

(e) Cartagena Protocol

39 The Cartagena Protocol has been ratified. The Dutch government considers agreements such as the Cartagena Protocol as important instruments that deserve ratification and implementation, provided that they lead to a workable balance for both importing and exporting nations.[13]

4. Other strict liability regimes

40 Here, there may be two relevant sources of liability. Vicarious liability (art. 6:170 Civil Code) and strict liability for hazardous substances (art. 6:175 Civil Code).[14] For vicarious liability see infra Question V.

41 Art. 6:175 Civil Code defines the liability for hazardous and noxious substances. Liability rests on anyone who uses or keeps the dangerous substance in his profession or business. As follows from the criteria of art. 6:175, non-professional possessors cannot be held strictly liable. Art. 6:175 Civil Code may be relevant if it is generally acknowledged that the GMO crop poses a specific, inherent and serious threat to life and limb and this risk materialises. Hence, this strict liability can only be applied to inherent dangers of substances which are scientifically proven at the time of the damaging event or exposure. This is not (yet) the case.

42 Art. 6:175 creates strict liability for dangerous substances used or kept in the course of business or trade. The article defines a dangerous substance as a substance of which it is known that it has such properties as to pose a special danger of a serious nature to persons or things. Such a "special danger" is posed in any case (according to the article) by substances which are explosive, oxidative, flammable, or poisonous as defined in specific public law legislation. We do not think that, according to the current state of science, GMOs as such can be considered dangerous substances. This may depend, however, on the specific case and the specific dangers the GMO may pose to persons or things. The Ministry of Justice has taken

---

13 See Parliamentary Proceedings 2005–2006, 26 407, 25.

14 See generally *W.H. van Boom/C.E. du Perron*, The Netherlands, in: B.A. Koch/H. Koziol (eds.), Unification of Tort Law: Strict Liability (2002) 227–255.

the position that GMO crops are unlikely to fall under "dangerous substances" in the sense of art. 6:175 Civil Code.[15] Whether this will also be the courts' position, remains to be seen.

Liability arises if the "special danger" materialises. Since the danger is defined as being "to persons or things", compensation of *pure* economic loss cannot be based on this article. Hence, we believe that even if a GMO was to be considered a dangerous substance under art. 6:175 Civil Code, a mere drop in turnover as a result of the absence of consumer confidence in crops neighbouring a GMO crop would not be deemed compensable damage. **43**

According to art. 6:178, liability on the basis of art. 6:175-177 is excluded, inter alia, in the following situations: **44**

- the damage is the result of armed conflict, civil war, revolt, riots, insurgence or mutiny;
- the damage is the result of a natural event of an exceptional, unavoidable and irresistible nature;
- the damage is solely caused by following an order or regulation of the government;
- the damage is intentionally caused by a third party;
- the damage is (the result of) a nuisance, pollution or any other consequence for which no liability would have existed on the basis of the general principles of tort law if the defendant had caused it intentionally (so the damage is considered an ordinary burden that one has to bear).

V. Vicarious liability

1. Scope of vicarious liability

We understand the concept of vicarious liability to reflect a liability without wrongfulness or fault of one person for the tortious acts of another person. Hence, we would consider this to be a strict liability for others' **45**

15 See *Notitie Ministerie van Justitie* – Aansprakelijkheid voor schade in het kader van coëxistentie van gg-gewassen en conventionele en biologische gewassen, in: Coëxistentie Primaire Sector – Rapportage van de tijdelijke commissie onder voorzitterschap J. van Dijk; commissiepartijen: Biologica, LTO Nederland, Plantum NL en Platform Aarde Boer Consument (2004) B57.

torts. There are typically two sources of such liability: art. 6:170 and art. 6:171 Civil Code.[16]

46 Art. 6:170 Civil Code defines the strict liability of the employer for tortious acts committed by employees. According to this article, strict liability for tortious acts of employees lies on the person in whose service the employee fulfils his duties if the risk of committing a mistake was increased by the assignment to fulfil the duty and the employer had control over the conduct of the employee. The scope of art. 6:170 Civil Code is wide in the sense that not only labour contracts are covered but also more flexible contract forms which constitute some form of hierarchy between the provider of a service and the client.

47 In addition to art. 6:170 there is art. 6:171 Civil Code. This article deals specifically with independent contractors. If such an independent service provider is hired by a client and performs activities in the exercise of the client's business, the client is liable vis-à-vis third parties for torts committed in the course of these activities. Note that art. 6:171 imposes vicarious liability only if the independent contractor was actually or seemingly a part of the business process of the client. If, however, an outsider – such as the potential victim – can easily see that the independent contractor is not part of the business process of the client, then there is no liability under art. 6:171 Civil Code.[17] Obviously, this substantially restricts the ambit of art. 6:171.

2. Liability for people further up the food or feed production chain

48 If we disregard contractual liability, the question becomes one of imputation of tortious conduct of others. Liability for upchain torts is rather limited under Dutch law, although there are cases in which downward chains were recognised under a duty of care vis-à-vis end-users to inspect the quality of the semi-products that were delivered before processing these into end-products.[18] Moreover, the Product Liability Directive seems to impose joint and several liability to a large extent on the entire chain for upward negligence. See art. 6:185 (1) (b), art. 6:187 (2) and (4), art. 6:189

16 Note that there is also art. 6:172 Civil Code, imposing vicarious liability on the principal for torts committed by the agent in the execution of his duties as an agent.

17 See HR 10 January 2001, NJ 2002, no. 75.

18 See, e.g., HR 6 December 1996, NJ 1997, no. 219, and HR 22 October 1999, NJ 2000, no. 159.

Civil Code, the implementation of art. 1, 3, 5, 7a–7f, 8 lid 1 of the EC Product Liability Directive (85/374/EEC).

3. Can someone further down the feed or food chain include someone further up it in a trial against him/herself?

No specific rules apply and therefore the defendant can summon any third party in an ancillary proceeding (*vrijwaring*) aimed at shifting the loss onto this third party in the event that the defendant fails in his defence in the main proceeding. Neither the court nor the defendant can compel third parties to join the main proceedings. 49

VI. Multiple tortfeasors

Joint and several liability of multiple tortfeasors is dealt with in art. 6:102 (1) Civil Code. This article states: 50

> "If two or more persons are each obliged to compensate the same damage, they are jointly and severally liable. In order to determine their contribution as amongst themselves on the basis of article 10, the damage is apportioned amongst them by applying the standard set in article 101, unless another division is demanded by statute or juristic act."

The subject of *hoofdelijke aansprakelijkheid* covers a wide range of concurrent liabilities. According to the present law of obligations, there is no fundamental distinction – that is, from a dogmatic point of view – between joint tortfeasorship (*Mittäter*, *Gehilfe*, etc.) and concurrent tortfeasorship (*Nebentäter*). Both types of tortfeasor are covered by the same flexible system of joint and several obligations. 51

Where the acts of A and B combine to cause harm to V, each is liable to V for the whole loss. So, in principle both A and B are liable in full. According to the general rules of tort law, tortfeasor A is liable in full for the damage that is caused by his act in a *conditio sine qua non*-sense, unless it is unreasonable to impute the resulting harm to the act that he committed (the legal causation as laid down in art. 6:98 BW).[19] Possibly, either A or B is not to be held liable at all if the resulting harm cannot be imputed to 52

19 On the topic of causation according to Dutch law, see *W.H. van Boom*, The Netherlands, in: B.A. Koch/H. Koziol (eds.), Compensation for Personal Injury in a Comparative Perspective (2002) 213, no. 9 f.

the act that was committed. This may be the case, e.g., when the damage caused was unforeseeable to either A or B. However, the mere fact that another cause was a necessary condition for the harm to materialise is not a sufficient reason for not applying the rule of liability in full.

VII. Defences

1. Licence/permission to grow GM material

53 As explained supra no. 34, the mere fact of having a licence or permission from a public authority does not as such render the operator of any risky activity immune from liability. It may depend on the circumstances whether the operator has breached other rules or acted in a negligent manner.

2. Consent/assumption of risk

54 Again, this depends on the specific case. Consent and assumption may be valid defences, either in the sense of justification for wrongful behaviour or in the sense of contributory negligence mitigating the obligation to compensate damage. Whether such a defence holds will depend on the specific case.

3. Third-party influence

55 As a rule, third-party behaviour will be considered to cause joint and several liability rather than to be a full or partial defence against liability. See art. 6:102 Civil Code, which we dealt with supra, no. VI. In some cases, however, third-party behaviour leads to a full release from liability. For instance, sabotage may be a full defence against strict liability for hazardous and noxious substances under art. 6:175 Civil Code. See art. 6:178 (a) and (e), as mentioned supra no. IV.4.44.

4. Prescription

56 Limitation periods (prescription) in the cases envisaged by this study do not deviate from the normal periods that apply in tort cases. Briefly explained, these periods are:

- In case of death and personal injury, a period of five years starting from the day that the claimant had cognizance of both the damage and the liable person.[20]
- In case of soil or water contamination and air pollution, a combined period of (1) five years starting from the day that the claimant had cognizance of both the damage and the liable person and (2) thirty years running from the day of the most recent or final occurrence.
- In other cases, a combined period of (1) five years starting from the day that the claimant had cognizance of both the damage and the liable person and (2) twenty years running from the day of occurrence.

The product liability prescription period consists of a combined three and ten years period (art. 6:191 Civil Code = art. 10, 11 Directive). **57**

5. Other defences

There are no specific defences that we are aware of. **58**

VIII. Remedies

1. Pecuniary compensation

(a) Bodily harm

There are no specific rules, so the regular remedies apply. **59**

(b) Property losses

There are no specific rules on liability or compensation of damage relating to GMO crops. **60**

(c) Economic losses

There are no specific rules on liability or compensation of damage relating to GMO crops. Therefore, the common rules of private tort law apply. Whether **61**

20 This new prescription period for death and personal injury was inserted in the Civil Code in 2004.

an organic farmer can claim the full costs of restoring a field rather than the full costs minus the return from conventional crops is a matter of the court's discretion. The court is allowed to calculate the damage in accordance with the nature of the damage (art. 6:97 Civil Code). As agriculture is mostly a commercial activity, it seems logical to require the farmer to mitigate his damage (art. 6:101 Civil Code) by farming conventional crops in the meantime.

(d) Harm to animals

62 If a cow is harmed by feed to the extent that it needs replacement, the cost of such replacement is awarded. The cost of replacement will usually be the market value of the animal in question. The market value will usually reflect its potential for producing milk or meat. The mere fact that an animal eats contaminated feed will – as far as we can foresee – not constitute harm to the animal. It could constitute pure economic loss in the sense that the produce will have a lower market value. If there is liability, such pure economic loss may be compensated within the framework of art. 6:162 and 6:98 Civil Code.

(e) Costs of disposal

63 These costs are recoverable within the framework of art. 6:162 and 6:98 Civil Code.

2. Non-compensatory damages

64 The Dutch legal system does not avail itself of punitive or exemplary damages.

3. Other remedies

65 Art. 6:103 Civil Code gives the court a discretionary power, at the request of the injured party, to award damages in any other form. This enables the imposition of a duty of reparation in kind, provided that is the appropriate remedy in the specific circumstances.

4. Costs of pursuing a claim

(a) General cost rule

The general rule is a diluted "loser pays" rule: the loser has to pay the costs of the opposing party according to a standardised valuation method. This method does not reflect the true cost of litigation and thus merely compensates to some extent. This standardisation does not apply, however, to pre-trial costs (i.e. all relevant legal costs, expert costs, etc. accrued prior to serving and related to pre-trial legal activities). Therefore, if the court sustains the claimant's assertion that the defendant is liable vis-à-vis the claimant, then the claimant can also claim reimbursement of these pre-trial costs (art. 6:96 Civil Code). **66**

(b) Costs of establishing causation

There are no specific rules concerning the covering of sampling and testing costs. Costs associated with sampling and testing of GMO presence in other products may be considered to be pecuniary loss (art. 6:96 (2) (b) Civil Code) if these costs are accrued to assess damage and liability. Hence, sampling and testing of GMO presence in products constitute part of the loss suffered by the injured party. Such costs may even be recoverable under tort law – even if the test does not prove actual GMO presence – provided that the liability of the GMO farmer is established. For example: a farmer has used some GMO in his crops in breach of a statutory ban, and consequently the GMO crop is suspected of having contaminated other crops of an adjacent farmer. The farmer pays for testing his crop and he claims the cost of these tests from the GMO-farmer. The test reveals that no admixture has occurred and customers have continued purchasing the products of the claimant. Hence, the farmer does not suffer any damage, but the GMO farmer is still liable for breach of a statutory provision. If the test proves GMO presence but no admixture, the respondent GMO farmer can be held liable for the expenses incurred in connection with the test. The basis for this claim is art. 6:96 Civil Code: the claimant is to be reimbursed for the reasonable cost of assessing liability and possible damage even if the wrongful act turns out not to have caused damage.[21] **67**

21 See HR 11 July 2003, NJ 2005, no. 50.

5. Advance cover

68 There is no general or specific statutory duty on "operators" to take out liability insurance, although specific public law legislation does enable local authorities to oblige some operators to take out some form of insurance or a bank guarantee for clean-up costs related to ultra-hazardous activities.[22] In practice, this does not seem to apply to GMO-farmers.

IX. Cross-border issues – Conflict of laws

1. Conflicts rules applicable before (or instead of) Rome II

69 To a large extent, the Dutch conflict law regime concerning cross-border GMO torts is identical to the Rome II Regulation. From a formal point of view, there seems to be one big difference between these two regimes. The Dutch 2001 Conflict of Laws (Tort Cases) Act (*Wet conflictenrecht onrechtmatige daad*, WCOD, Stb. 2001, 190) starts from the *lex loci delicti* rule (with a number of exceptions) whereas the Rome II Regulation seems to start from the *lex loci damni* rule (again with a number of exceptions). In practice, the differences between the two regimes will not be very substantial, with the exception of environmental damage. Art. 7 Rome II Regulation gives the victim of environmental damage the option to choose from (at least) two legal systems, whereas art. 4 of the WCOD exclusively points to the *lex loci damni*.[23]

2. Special regime for cross-border claims

70 We are not aware of any such special regime, apart from the private international law regime under the 1973 Hague Convention on the Law Applicable to Product Liability.

---

22 Besluit financiële zekerheid milieubeheer, in: Staatsblad 2003, no. 71, based on art. 8.15 Wet milieubeheer.

23 On the conflict between the WCOD and the Rome II Regulation, see, e.g., *L. Strikwerda*, Van 'lex loci delicti' naar 'lex loci damni', Weekblad voor Privaatrecht, Notariaat en Registratie (WPNR) 2008, no. 6780, 993 ff.

X. Cases

1. Due to the adventitious presence of GMOs in a field, maize which is normally sold as conventional contains GMOs beyond the legal labelling threshold. This is not discovered before the final stage of the food production chain by the producer of taco chips. The whole production is lost since the supermarket chains refuse to accept delivery from the producer.

(a) Who can sue along the chain of distribution?

Assuming that the supermarket chain is allowed to refuse the delivery, the producer of taco chips may try to claim compensation from his counterpart (in contract) or any of the previous links in the chain that can be held accountable (in tort).[24] Although the damage sustained by the taco producer is in part purely economic, this as such does not preclude liability in tort. If any of the businesses in the production chain was aware of the problem or should have had procedures in place that could have prevented the distribution and processing of the maize (e.g., arrival and/or exit inspections), then the taco producer may state a claim on the basis of negligence ("conduct contrary to the unwritten standard of conduct seemly in society"; art. 6:162 Civil Code; see supra no. 30). Note that, as a rule, the links in the trading or production chain are not responsible in tort for acts and omissions elsewhere in the chain. 71

The mere fact that the legal labelling threshold was surpassed may in itself constitute a breach of a statutory duty owed by the farmer to the taco producer. Whether the duty was indeed owed to the producer depends on statutory interpretation and the analysis of the protective scope of this specific piece of regulation. In practice, however, this issue seems less decisive as Dutch law avails itself of the broad concept of negligence ("conduct contrary to the unwritten standard of conduct seemly in society"; art. 6:162 Civil Code; see supra no. 30). As a result, even if the scope of the regulation was not intended to protect the producer, a court may still find that the farmer owed a duty of care under this broad concept of negligence. 72

24 Note that the case is not covered by the strict liability of the Product Liability Directive.

(b) Would the case be solved differently if the GMO content was below the labelling threshold?

73 This depends on the specific circumstances: if the taco producer was allowed to expect the raw materials to be fully GMO-free, then obviously something has gone wrong in the production chain and those businesses negligently omitting to implement precautionary measures can be held liable in tort.

(c) Would the case be solved differently if the admixture was not adventitious, but occurred due to the disregard of segregation rules, for example?

74 This may shift the focus from one tortfeasor to another. Again, all depends on the exact facts of the case. These facts will point towards the exact link that failed within the chain. Disregard of segregation rules may constitute negligent breach of the duty of care under the rule of "conduct contrary to the unwritten standard of conduct seemly in society" (art. 6:162 Civil Code; see supra no. 30).

(d) Would the case be solved differently if the GMO found was not admitted for production in your jurisdiction?

75 Depending on the exact cause of the admixture, this may shift the focus from one tortfeasor to another. Again, all depends on the exact facts of the case. These facts will point towards the exact link that failed within the chain.

(e) If the admixture had occurred on a non-GM field and it transpires that the GM seeds were blown from: (i) neighbouring fields; or (ii) a truck passing by, would the farmer of the affected field be liable for all or part of the loss caused further down the distribution chain?

76 If we concentrate on tortious liability and exclude possible claims of contractual counterparts, then the question is whether the farmer committed a wrongful act. What did he do wrong? Possibly, he could be held liable if there is a custom among farmers of his kind to perform "exit inspections".

If not, it seems unlikely that the farmer would be liable for this external cause.

2. Twenty years after the sale of GM maize used for food products, it turns out that it has certain disadvantageous health effects for humans.

(a) Can the producers be held liable at this point for risks unknown at the time of growing the maize? Who would be liable – the seed producer/farmer/food producer/distributor/ etc?

This is ultimately a matter of interpretation of the Product Liability Directive, specifically the "state of the art" or "development risk" defence. At face value, it seems unlikely that risks unknown at the time of distribution of the produce into the trade chain will be imputed to any of the links in the chain. **77**

(b) Can compensation already be claimed at a point when the negative health effects have not yet materialised, but are expected to according to scientific expertise/mere rumours?

This is ultimately a matter of interpretation of the Product Liability Directive, specifically the "state of the art" or "development risk" defence. At face value, it seems that scientific expertise may be decisive in shifting the development risk from the consumer to the chain. **78**

(c) Would it make any difference if the GM maize had only been in use for feed, causing harm to the animals, which may or may not cause harm to humans consuming the meat as well?

If we interpret the question to be one of scientific uncertainty concerning causation of health impairment, this would be an obstacle for the consumer's claim to compensation. In principle, the consumer has to prove causation (art. 6:188 Civil Code = art. 4 Product Liability Directive). **79**

3. The driver of a food logistics company discovers that a farmer, from where he regularly picks up agricultural products fails to obey mandatory segregation rules or food or feed hygiene standards, which may lead to the admixture of GM and non-GM produce sold and packaged separately by that farmer. Does he or his employer have a duty to warn, i.e. warn the recipient of the allegedly "non-GM" produce?

80 As a rule, the mere fact that someone has knowledge of failure to comply with specific regulation does not put him under a duty to inform other interested parties. Assessing the existence of a duty to warn under the rule of "conduct contrary to the unwritten standard of conduct seemly in society" (art. 6:162 Civil Code) involves a number of factors. According to case law, a great many factors determine impropriety in any concrete case, e.g., foreseeability of the loss (also described as the chance of a loss occurring as a result of the act), the degree of blameworthiness, the costs of avoiding the loss, the nature of the damage, and the relationship between the injured party and the injurer.[25]

81 It seems obvious that if concrete lives are at stake, the duty to warn or even intervene is more likely to arise than if there is a mere commercial interest at stake.

---

25 Most of these criteria originate from the landmark decision HR 5 November 1965, NJ 1966, no. 136. See further on the subject: *J. Spier*, The Netherlands – Wrongfulness in the Dutch Context, in: H. Koziol (ed.), Unification of Tort Law: Wrongfulness (1998) 94 f.

Damage Caused by GMOs under Norwegian Law

Bjarte Askeland

I. General overview

1. Special liability or redress scheme for GMOs

In 1993 Norway enacted a special act concerning genetic technology, the Norwegian Act on Genetic Technology (*Lov om framstilling og bruk av genmodifiserte organismer* [genteknol.] 2 April 1993 no. 38). This Act comprises a number of regulatory provisions designed to ensure that the production and use of genetically modified organisms is conducted in an ethical manner that is moreover prudent in the light of societal interests. 1

In addition, the Act features a general liability clause in § 23. A translation of the clause reads as follows: 2

> "One who is responsible for activity under the scope of this statute is liable without fault when the activity by placing or emitting genetically modified organisms in the environment causes damage, inconvenience or loss. In addition, the rules enacted in the Pollution Act (Forurl.) chapter 8 on liability for pollution apply as far as they are appropriate."

This provision will cover the situation where non-GM crops are contaminated by GM crops. The provision establishes strict liability for damage caused by GMOs. Without this provision, one would have had to resort to the general uncodified rule on liability for negligent behaviour or the strict liability regime of the Pollution Act (*Lov om vern mot forurensning og avfall* 13 March 1981 no. 6, Forurl. § 55). Taking into consideration the fact that the GMO damage will be covered by the strict liability regime of pollution damage in the Pollution Act, it is fair to say that the two Acts overlap when it comes to the legal basis of strict liability for GMO damage. According to information gathered from the Norwegian Ministry of Agriculture, there has (over the past few years) been some work towards designing rules on the co-existence of GMO and non-GMO crops. No such rules have, however, 3 been produced so far and there are no official documents available on the subject. For the time being, one must therefore resort to the rules described above.

2. State liability

4 The state may be liable for any such losses provided that the state is involved in the enterprise in the same way as other liable persons. No kind of sovereign immunity clause applies. State liability only deviates from the general tort law regime as far as the special doctrine of a lenient culpa standard applies. This special doctrine is explained in the Norwegian report on tort law and regulatory law.[1]

II. Damage

1. Recoverable losses

5 The general regime, whereunder pure economic loss, personal injuries and damage to things are compensated, will apply.

2. Pure economic loss

6 Pure economic loss is recoverable provided that there is adequate causation between the wrongful GMO-activity and the loss.

3. Mere fear of a loss

7 The mere fear of a loss is not recoverable under Norwegian law.

4. Standard of proof

8 To deem someone liable requires in principle that it is proved beyond 50% certainty that the alleged tortfeasor has caused the harm.[2] This standard is applied also to future losses.[3]

---

1 B. *Askeland*, Tort and regulatory law in Norway, in: W. van Boom/C. Kissling/M. Lukas (eds.), Tort and Regulatory Law, 205–221 no. 16 f.

2 See for example N. *Nygaard*, Skade og ansvar (6th ed. 2007) 339 ff.

5. Nominal losses

Nominal/symbolic losses are not recognised. 9

6. Mass losses

There are no special rules for mass losses. The general reduction clause in *skadeserstatningslov* § 5–2 may, however, apply. The main criterion for any reduction is that paying in full would be "unreasonable" for the tortfeasor. The fact that there are many claimants with claims that taken together represent a large sum is relevant to the question of whether paying in full is "unreasonable".[4] 10

III. Causation

1. Uncertainty of merely potential causes

The general principle (mentioned supra no. 8) that the decisive facts of the case must be proved more likely to have occurred than not to have occurred, applies. This test must also be applied where good practice has been violated. The judges have, however, according to Norwegian Supreme Court practice, the competence within certain limits to shift the burden of proof on a discretionary basis.[5] The fact that 1) the neighbour uses GM crops and 2) that the claimant has got GM-contamination in his crop *may* be sufficient for the judge to reverse the burden of proof so that the neighbouring GM crop holder must prove that the contamination was *not* caused by him. But whether or not judges will use their competence in this manner is not easy to tell in advance. 11

As for GMO cases, the Pollution Act and the special rule on possible polluters may apply, see Askeland (fn. 7) no. 9–13. 12

3 See for example *M. Strandberg*, Skadelidtes hypotetiske inntekt (2005) 90.

4 This interpretation follows from the wording "urimelig tyngende" ("unreasonably burdening"). It is also consistent with the preparatory works for the reduction clause; see Innstilling fra Erstatningslovkomiteen, 1 May 1971, 46–48.

5 Cf. *Nygaard* (fn. 2) 341 f.

2. Complex causation scenarios

13 There are no special rules for this situation apart from the special rule on possible polluters, see supra no. 13 and Askeland (fn. 7) no. 9–13.

3. Force majeure

14 In the general doctrine of causation within tort law, the notion of force majeure is seldom addressed. The phenomenon of force majeure is brought more into the discussion as a limitation of the general uncodified rule of strict liability. The prevailing opinion is that one can not claim damages where the damage is caused by force majeure such as war or extreme weather.[6] This exception is codified in some of the Norwegian acts that have provisions on strict liability, see for example the Act on Atomic Energy (*Atomenergiloven* 12 May 1972 no. 28) § 24. In some acts the occurrence of force majeure does not eliminate liability totally, but rather paves the way for a reduction of the award, see The Act on Liability for Railways (*Jernbaneansvarsloven* 10 June 1973 no. 73.) § 12, second section. The fact that damage is caused by force majeure may in some circumstances render the alleged tortfeasor's contribution to the damage unimportant compared to the forces of nature or other types of force majeure. When a *condition sine qua non* is considered relatively unimportant, it may not constitute liability.[7] One should mention that traditionally the question of force majeure under Norwegian law is first and foremost considered to be part of the law of obligations.

4. Threshold to prove causation

15 Under Norwegian law the general degree of likelihood is "more likely than not". This question follows the general regime on standards of proof, see supra no. 8.

---

6 See P. *Lødrup*, Lærebok i erstatningsrett (4th ed. 1999) 275–278.

7 See B. *Askeland*, Norwegian report, in: B.A. Koch (ed.), Economic Loss Caused by Genetically Modified Organisms, 361–372, no. 24 with reference to Retstidende (Rt.) 1992, 64.

5. Special rules on causation

There are no special rules on causation. The special rule on liability due to lack of evidence in forurl. § 59 should be mentioned. This provision is presented in Askeland (fn. 7) no. 9–13. **16**

IV. Types of liability

1. Fault liability

(a) Special rules governing fault

Because of the fact that there is an established strict liability regime for GMO damage (see supra no. 3–4), fault rules will not apply. If general fault rules were to be applied, it would be hard to find a legal basis for reversing the burden of proof or to establish special standards of care. **17**

(b) Impact of specific rules of conduct

The fact that a regulatory provision has been violated will in general point towards liability, Askeland (fn. 1) no. 2. The same goes for the fact that good farming practice has been violated.[8] **18**

2. Product liability

(a) Development risk defence

The Norwegian jurisdiction did not incorporate the development risk defence. A reason for this choice was that liability for development risks was consistent with Norwegian tort law in general. Moreover, the Norwegian lawmakers were of the opinion that the risk should be borne by the producer rather than the victim.[9] There have not been any cases dealing with this specific problem. **19**

8 *Nygaard* (fn. 2) 206.

9 See the preparatory works for the Norwegian Act on Product liability, Odelstingsproposisjon (Ot. prp., “Proposition to Parliament”) no. 48 (1987–88) 40–42.

(b) Alternative routes

20 The special rules on gene technology described in the Askeland (fn. 7) report will of course apply, see supra no. 1–4. There is, however, a slight possibility that the liability regime in the genteknol. Act will not survive the scrutiny of the ECJ. The liability rule in the genteknol. Act § 23 prescribes an *unconditional strict liability*, whereas the Product Liability Directive apparently does not go that far: The criterion in art. 6, "does not provide the safety which a person is entitled to expect" at least theoretically provides a possibility for the producer/farmer to escape liability unlike under the regime of the genteknol.[10] It may, however, be debatable whether the Product Liability Directive prescribes something less than unconditional strict liability.[11] In all circumstances, it is probably possible to interpret genteknol. in accordance with the demands of the Product Liability Directive. This would probably be the most sensible alternative given the fact that genteknol. so far has never been applied due to the lack of GMO production in Norway. Hence the Norwegian court practice has the opportunity to choose a path within the boundaries of the Product Liability Directive.

(c) Impact of compliance with rules and regulations

21 In general we have no doctrine that excludes liability due to the fact that the defendant has obeyed all rules and regulations. He may still be liable if the requisite of culpa is fulfilled. The fact that all rules are obeyed will, however, narrow the possibility of proving culpability on the part of the defendant. For a further elaboration of "the regulatory permit defence" under Norwegian law, see Askeland (fn. 1) no. 37–38.

10 The Norwegian theorist N. Nygaard discusses the specialities of product liability in a way that implies that it is something other than unconditional strict liability, see *Nygaard* (fn. 2) 461–467. A debate on consequences in the aftermath of the ECJ cases on the exclusiveness of the Product Liability Directive in 2000 also presupposes that the product liability rules are not as strict as traditional uncodified strict liability under Norwegian law; see on this debate *A. Stenvik*, Erstatningsrettens internasjonalisering, Tidskrift for Erstatningsrett (TfE) 2005, 33–61, 44 f.

11 Comparative legal theory perceives the Directive to describe a strict liability regime; see e.g. *C. van Dam*, European Tort Law (2006) 343.

3. Environmental liability

(a) Implementation of the Environmental Liability Directive

The EEA Committee has decided that the Directive (2004/35/EC) shall be included in the EEA Agreement, Appendix XX (Environment).[12] One has found that the Norwegian Act on Pollution (forurl.) and other Acts satisfied the demands of the parts of the Directive that are comprised by the EEA agreement.[13] The rules on limitation reach, however somewhat further than the Norwegian rules. The Directive in art. 10 prescribes a five year period of limitation whereas the Norwegian general rule in *Foreldelseslov* (Norwegian Limitation Act, fel.) 18 May 1979 no. 18 § 2 prescribes a limitation period of only three years. Norwegian statutory law will have to be changed on this point, preferably with a special provision on limitation in forurl. (the Norwegian Pollution Act) and genteknol. (the Norwegian Act on Gene Technology).[14] As one can observe, Norway has on this occasion not met the deadline of implementation, 30 April 2007.[15] The fact that there already exists a proposition to the Parliament makes it, however, probable that the mentioned changes soon will be enacted. There are no indications of any planned new financial guarantees. 22

(b) Environmental liability regime beyond the scope of the Directive

The Pollution Act (forurl.) § 55 cfr. 57 provides strict liability for environmental harm, but also for harm to things (property) and persons and the causing of pure economic loss by pollution. The Norwegian parliament has consented to this by decision of 5. June 2009. This way, the Norwegian protection reaches further than the Directive, see preamble no. 14 which excludes personal injury and damage to property and economic loss. The fact that the government has already recognised that the act will have to be changed, makes it probable that changes soon will be enacted. On other points also, the protection rules probably reach further than the Directive.[16] 23

---

12 See EEA-Committee decision 17/2009, 5 February 2009.

13 See Stortingsproposisjon ("Proposition to Parliament", St. prp.) no. 62 (2008–2009) 5.

14 See St.prp. no. 62 (2008–2009) 5.

15 See Directive 2004/35/EC, art. 19.

16 Hence, Norwegian officials have stressed that the Directive is a minimum Directive, For the purpose of the Questionnaire, I assume that there is no need for a detailed investigation on this point.

(c) Claimants in cases of environmental harm

24 The local municipality may claim damages where the public enjoyment of the land has deteriorated, see forurl. § 58. According to this provision, one may be liable for damage affecting the common usage of nature. This may be the case where a subject "hinders, makes difficult or puts limitations on" the common enjoyment of nature. The value of common usage must, however, be converted into an economic loss, such as expenses incurred to re-establish the level of enjoyment of the environment. An example of such expenses would be the costs of deployment of fish in lakes. The assessment of damages will in this way be based on a reflection of the expenses incurred. The authority responsible for supervising pollution within the municipality has the competence to sue the defendant on behalf of the municipality. If the pollution affects different counties, the special pollution supervising entity of the state (*Statens forurensningstilsyn*) will sue on behalf of the state of Norway.[17]

(d) Special liability regime for losses sustained by individuals

25 The above-mentioned provision in forurl. § 58 is probably the best legal basis for claiming such damages, see supra no. 24. The Act on Neighbour Relationships (*Grannelova* 16 June 1961 no. 15, gl.) also has special rules on liability. In § 2 of this Act it is stated that no one may unreasonably or unnecessarily take measures that damage the neighbouring land or the people living there. In § 9 there is a legal basis for claiming economic compensation for such damage. The criterion of economic loss may, however, make it difficult to get compensation for damage to the environment as such. One strategy may be to incur expenses in order to restore the environment, but such a strategy has in principle been turned down by the Supreme Court, see Rt. 1980, 309. In this case, the Court did not allow the claimant to convert non-economic loss (noise) into a compensable expense (by building a wall to eliminate the noise). The court held that the claimant could not better his own legal position by incurring expenses. This decision has, however, been criticised by legal commentators.[18] It is uncertain whether the solution would be the same were the case decided today.

---

17 The competence to claim on behalf of the state or municipalities is given in the provision in forurl. § 58 subpar. 2.

18 *H.Chr. Bugge*, Forurensningsansvaret (1999) 386; *E. Stavang*, Erstatningsrettslig analyse (2007) 155.

(e) Cartagena Protocol

Norway is a party to the Cartagena Protocol.[19] Many official documents produced by the government refer to the obligations thereunder as an element in our environmental policy. Norway is also a party to many other multi- and bilateral agreements concerning the environment.[20] 26

4. Other strict liability regimes

N/A. 27

V. Vicarious liability

1. Scope of vicarious liability

The scope of vicarious liability is in general limited to the employees of an employer, see the Norwegian Act on Compensation (*Skadeserstatningsloven* 13 June 1969 no. 26, skl.) § 2–1. As for damaging acts committed by an independent contractor, the main principle is that an employer is *not vicariously* liable.[21] There are very few exceptions to this principle and the principle is also applied in the Pollution Act (forurl. § 55). 28

2. Liability for people further up the food or feed production chain

Under the Norwegian private law regime, there is no identification between lower levels and higher levels. The crop retailer may be liable for exposing a third party to a defective crop, but there is no vicarious liability with regard to the crop wholesaler. 29

---

19 The Cartagena Protocol was signed by Norwegian representatives on 24 May 2000 and came into force on 11 September 2003.

20 A search for international agreements concerning the environment produced 24 hits in a relevant database.

21 See B. *Askeland*, Erstatningsrettslig identifikasjon (2002) 202 ff. and 144 ff.

3. Can someone further down the feed or food chain include someone further up it in a trial against him/herself?

30 Within the Norwegian system there is no way to include other possibly liable subjects in the litigation process by force. The Norwegian approach is normally that the farmer – if he is found liable – will have a right of recourse against the producer provided that the requisites of liability are met.

VI. Multiple tortfeasors

31 Under Norwegian law there is a general rule on solidary liability for multiple tortfeasors, see skl. § 5-3. This rule resembles very much the content of the rule on solidary liability in the Principles of European Tort Law art. 9:101. There is no tradition of proportional liability under Norwegian tort law. Historically, one has perceived solidary liability as a natural consequence of the doctrine of causality based on *condition sine qua non*.[22]

VII. Defences

1. Licence/permission to grow GM material

32 The permission does not in itself serve as a defence against liability. The fact that the defendant has permission will, however, make it harder to prove that he is at fault. For a general elaboration of the regulatory permit defence, see Askeland (fn. 1) note 37–38.

2. Consent/assumption of risk

33 Under Norwegian law, a plaintiff may lose his claim because he consented to a risk. This was a major point in a case where a smoker sued a tobacco company, claiming damages after having suffered from cancer generated by the smoking.[23] As for consent to a GM product that it is perfectly legitimate to sell, one would probably not find the risk profound enough to

---

22 See B. *Askeland*, Plurality of Liable Persons and Prescription of Recourse Claims, in: H. Koziol/B.C. Steininger (eds.), European Tort Law 2007 (2008) 94 ff., no. 54 f. with further references.

23 See Rt. 2003, 546, reported in H. Koziol/B.C. Steininger (eds.), European Tort Law 2004 (2005) 453 f.

deprive the plaintiff of his claim. If the risk of consuming the GM product were comparable to continuous smoking, there might be at least a reduction of the claim according to the Norwegian rules on contributory negligence, see skl. § 5-1.

3. Third-party influence

If the acts of the third party amounted to being the dominating cause of the damage, the GMO farmer might be acquitted of liability. This would, however, only be the case if the farmer's contribution to the damage could be considered "unsubstantial" or "unimportant" in the total chain of events that lead to the damage.[24] **34**

4. Prescription

The general Norwegian Act on Limitation (*Foreldelsesloven* 18 May 1979 no. 1, fel.) § 9 applies. This provision states that a claim must be filed within three years after the day when the claimant knew or ought to have known of the tortfeasor and the act or activity that led to liability on the part of the tortfeasor. **35**

The general rules on limitation for non-contractual obligations also stipulate a long stop period of 20 years, see fel. § 9 no. 2. Hence, the claimant will have no claim 20 years after the harmful act regardless of whether he has gained knowledge of the tortfeasor. **36**

5. Other defences

No defences apart from the general question of assumption of risk. **37**

VIII. Remedies

1. Pecuniary compensation

(a) Bodily harm

The regular remedies apply. **38**

24 Rt. 1992, 64, 70: The crucial requisite is that the party's contribution is so unsubstantial that "it is not natural" to make the party liable for the damage.

(b) Property losses

39 The regular remedies apply. No caps.

(c) Economic losses

40 There are no special rules on compensation.

(d) Harm to animals

41 The value of an animal is assessed in the same way as other things.[25] The market value is decisive. Hence, a cow's potential for producing milk or meat is integrated in the market value of the cow. The mere fact that the animal has eaten GM contaminated feed does not constitute damage unless the market value of the animal has dropped.

(e) Costs of disposal

42 As far as the act of contaminating the production/feeding the animals with the GM feed is considered a wrongful act, all economic loss that stems from the act is to be recovered. The extent of expenses compensated is governed by the rules of adequate causation. As for the above-mentioned acts, the incurring of costs in getting rid of production or animals must be considered to be a foreseeable and adequate consequence. Hence, such costs are recoverable.

2. Non-compensatory damages

43 Norwegian law does not recognise punitive, exemplary or any other form of non-compensatory damage. Compensation for non-pecuniary damage in personal injury cases will sometimes comprise "an element of punitive damages", namely that the gravity of the harmful act may be reflected in the compensation.[26] This is the closest to non-compensatory damages under Norwegian tort law.

---

25 *Nygaard* (fn. 2) 81 ff.

26 See *B. Askeland*, Scandinavian report, in: H. Koziol/V. Wilcox (eds.), Punitive Damages in Europe (2009) 115–122, no. 4.

3. Other remedies

In the Norwegian tort law system there is no tradition of compulsory restitution in kind or reparation. Ordinary compensation in money is the dominant remedy. Community contributions have so far not been part of the possible remedies. There is, however, a legal basis for fines or penalties in the Pollution Act, see forurl. § 78. A person who causes pollution may be fined provided that the pollution is committed with negligence or intent and deemed unlawful according to the Pollution Act. **44**

4. Costs of pursuing a claim

(a) General cost rule

The general rule on recoverability of costs is that the winning party has a right to full compensation for expenses incurred because of the case, see *Tvisteloven* (the Norwegian Act on Litigation) 17 June 2005, no. 90 § 20-2 (1). **45**

(b) Costs of establishing causation

The general rule on recoverability mentioned in no. 44 applies. A special provision on assessment of recoverable costs is found in tvistel. § 20-5 (1). The main requirement is that the expenses incurred have been necessary for taking care of the winning party's interests connected to the case. Whether or not the above-mentioned costs are recoverable will be a concrete decision left to the judge's discretion. **46**

5. Advance cover

There are so far no official documents indicating that there will be initiatives taken in the direction of establishing advance cover of any kind. **47**

IX. Cross-border issues – Conflict of laws

1. Conflicts rules applicable before (or instead of) Rome II

The Rome II Regulation will not come into force within the Norwegian jurisdiction. Except for the general principle of *lex loci delicti*, there are no other specific provisions designed for cross-border cases. *Lex loci delicti*, however, obviously does not provide any good answers for cross-border cases because it is doubtful whether *lex loci delicti* is the place where the **48** damaging activity took place (*lex loci actus*) or where the damage occurred (*lex loci injuriae*). One should note that Norwegian writers within the field of private international law hold that Norwegian solutions should copy the solutions within Rome II even though the Regulation does not directly apply. They point to the value of harmonised rules throughout Europe.[27] Hence, it is held that *lex loci injuriae* should apply to cross-border cases.[28]

49 There are no special jurisdictional or conflict of law rules in force or planned, except for a general rule concerning damage covered by forurl., namely forurl. § 54. This provision states that forurl. applies to pollution damage that occurs on Norwegian territory, forurl. § 54, par. 1, litra a. Moreover, it may apply to pollution damage outside Norway provided that the activity that causes the damage took place in Norway. The claimant is entitled to require that the question of compensation shall be decided in the country where the polluting activity took place, cf. forurl. § 54, para. 4. Because of the fact that genteknol. § 23 prescribes that the forurl. should apply to gene technology cases as far as appropriate, the said provisions will apply to GMO-cases.

2. Special regime for cross-border claims

50 There are no special regimes. However, there is relevant regulation within the Nordic Convention on the Protection of the Environment.[29] This Convention was incorporated into Norwegian law by *Lov om gjennomføring i norsk rett av miljøvernkonvensjon mellom Norge, Danmark og Sverige, undertegnet 19. februar 1974* (the Act on Implementation of the Nordic Convention between Norway, Denmark, Sweden and Finland, signed 19 February 1974), an Act of 9 April 1976 no. 21. Art. 3 of the said Convention reads that a claimant that has been exposed to pollution damage in his own country has the right to sue or apply administrative remedies against the tortfeasor in the country where the polluting activity took place. The claimant is entitled to a compensation regime that is as much to his advantage as the compensation regime in the country where the polluting activity took place.

27 L.A. *Heimdal*, Erstatning ved grenseoverskridende personlighetskrenkelser (2008) 52, in the same direction also G. *Cordero Moss*, Nye EF-Forordninger om lovvalg, Nytt i Privatretten 2008, no. 10, 10–11.

28 *Heimdal* (fn. 27) 52.

29 Convention 19 February 1974, *Miljøvernkonvensjon mellom Danmark, Finland, Norge og Sverige av 19 februar 1974*.

X. Cases

1. Due to the adventitious presence of GMOs in a field, maize which is normally sold as conventional contains GMOs beyond the legal labelling threshold. This is not discovered before the final stage of the food production chain by the producer of taco chips. The whole production is lost since the supermarket chains refuse to accept delivery from the producer.

(a) Who can sue along the chain of distribution?

Both the maize producer, the wholesaler of maize and retailers can sue the farmer in contract. There are rules providing for a direct action in contractual chains, see Kjøpsloven 13 May 1988 no. 27 (the Norwegian Act on Sale of Goods, kjl.) § 84.[30] **51**

(b) Would the case be solved differently if the GMO content was below the labelling threshold?

The case would probably not be solved any differently because GM food is considered controversial in Norway. The customers would want to know whether the food is GMO-contaminated. As a consequence, the presence of GMO-contamination would probably in itself constitute a breach of contract within the law of sales. **52**

(c) Would the case be solved differently if the admixture was not adventitious, but occurred due to the disregard of segregation rules, for example?

For the same reason mentioned above in no. 52, the case would probably not be solved any differently. **53**

(d) Would the case be solved differently if the GMO found was not admitted for production in your jurisdiction?

As a matter of fact, no GMOs have so far been admitted for production in Norway; this fact will make the case even more serious and hence the case **54**

30 For an elaboration of such direct actions under Norwegian law, see *A. Bjøranger Tørum*, Direktekrav (2007).

of the claimant would be strengthened. In principle, nonetheless, the case would be solved in the same way.

(e) If the admixture had occurred on a non-GM field and it transpires that the GM seeds were blown from: (i) neighbouring fields; or (ii) a truck passing by, would the farmer of the affected field be liable for all or part of the loss caused further down the distribution chain?

55 This question challenges the scope of the codified strict liability rule in the Act on Genetically Modified Organisms (Genteknologiloven 2 April 1993 no 38, genteknol.) § 23. In this case, the farmer would probably not be liable according to this provision because the requirement that the damage must be caused *by deployment* of GMO (*ved utsettelse eller utslipp*) would hardly be fulfilled. It is more doubtful whether the farmer escapes the strict liability clause in forurl. § 55. One line of defence may, however, be that it is not the farmer who *caused* (*volder*) the pollution. The link between the genteknol. and forurl. (see supra no. 48) provides for an argument against liability. Forurl. is to be applied as far as the act fits the cases on GMOs. It would therefore be a bit peculiar if forurl. had liability rules reaching further than the liability clause in gentekonol.

56 In the case presented above we may assume that there is no negligence on the part of the farmer. Accordingly, he would not be subject to non-contractual liability of any kind, He would, however, probably be liable in contract simply because he delivered a good of a different quality than promised, see supra no. 52. Because of the special rule on direct claims in kjøpsl. 84, as mentioned above in no. 50, the farmer may also be liable further down the distribution chain.

2. Twenty years after the sale of GM maize used for food products, it turns out that it has certain disadvantageous health effects for humans.

(a) Can the producers be held liable at this point for risks unknown at the time of growing the maize? Who would be liable?

57 The producer may in principle be held liable for risks unknown at the time of growing the maize. As explained above in no. 19, Norway did not incorporate the development risk defence when implementing the Product Liability Directive by enactment of the Norwegian Product Liability Act (pal.) in 1988. Hence, the liability clause in pal § 2–1 will apply.

The general limitation rule is that a claim of compensation is precluded three years after the time when the claimant discovers or ought to have discovered the damage and the tortfeasor.[31] This limit may, however, be prolonged as a result of the fact that the claimant did not discover the damage. The claim has, however, to be filed within a period of 20 years, see the "long stop"- provision in fel. § 9 no. 2. **58**

According to the Norwegian Product Liability Act (pal.) § 2–1 cfr. § 1–3 no. 1, the producer of the maize (i.e. the farmer) is liable. The food distributor may only be liable if the producer is hard to identify, see pal. § 1–3 no. 1 c). If the maize is imported from a foreign country, the importer is liable, see pal. § 1–3 no. 1 e). **59**

(b) Can compensation already be claimed at a point when the negative health effects have not yet materialised, but are to expected to according to scientific expertise/mere rumours?

No, compensation may not be claimed at such an early point. One has to prove actual economic loss. The mere fear of suffering non-pecuniary loss does not qualify for compensation under Norwegian tort law. **60**

(c) Would it make any difference if the GM maize had only been in use for feed, causing harm to the animals, which may or may not cause harm to humans consuming the meat as well?

As a consequence of the fact that the situation described under (b) does not qualify for compensation, the situation sketched out under (c) would not make any difference. The situation under (c) is even further from qualifying than the situation under (b). **61**

31 See fel. (The Norwegian Limitation Act) § 9 and 3.

3. The driver of a food logistics company discovers that a farmer, from where he regularly picks up agricultural products fails to obey mandatory segregation rules or food or feed hygiene standards, which may lead to the admixture of GM and non-GM produce sold and packaged separately by that farmer. Does he or his employer have a duty to warn, i.e. warn the recipient of the allegedly "non-GM" produce?

62 This question is governed by the general standard of culpability. The point of departure is that a person who has no connection to the victim or the tortfeasor has no duty to interfere.[32] As for the driver, he is part of the chain of activities that places the potentially damaging agricultural products on the market. Moreover, he is employed in a company that has a contractual relationship to the potential victims. Presumably, these two elements of connection to the damaging activity is sufficient to generate a duty to warn the victim. Duty in this sense implies that the driver may be found to have acted culpably if the duty to warn is not heeded.

32 See e.g. *Nygaard* (fn. 2) 185 f.

Damage Caused by GMOs under Polish Law

Ewa Bagińska

I. General overview

1. Special liability or redress scheme for GMOs

In the Law of 22 June 2001 on Genetically Modified Organisms (consolidated text published in the Journal of Laws 2007, no. 36, item 233), hereafter Law on GMOs, there is a special provision in art. 57 (chap. 7 of the Law) addressing the civil liability of GMO users. 1

According to art. 57 sec. 1 Law on GMOs, a GMO "user" is liable under civil law for damage to persons, damage to property and damage to the environment caused by the carrying out of a contained use of GMOs or of a deliberate release of GMOs into the environment, including the placing of GMOs on the market. 2

The liability is strict, i.e. it is based on the principle of risk. There are three exculpatory grounds with respect to this liability: 1) *force majeure*, 2) the exclusive fault of the injured party, or 3) the exclusive fault of a third person for whom the GMO user is not liable. The fact that the activity that caused damage was carried out on the basis of and within the scope of an administrative decision does not exempt the user from the liability envisaged in art. 57 sec. 1 Law on GMOs (art. 57 sec. 3 of the Law). 3

The same rules of liability apply to the transit of GMO products through the territory of Poland. The liability is channelled to the person obliged to obtain a licence according to art. 51 Law on GMOs. 4

The reference to the civil law in art. 57 Law on GMOs corresponds with the traditional approach of Polish law, which has always used the Civil Code (*kodeks cywilny*, KC) liability rules in the field of environmental protection, regardless of any special legislation. 5

6 As pointed out in art. 57 Law on GMOs, the losses that are compensable under this special regime include damage to the environment. In this case, however, the claim for compensation may be brought by the State Treasury, a local authority or an ecological organisation (art. 57 sec. 2 Law on GMOs).

7 This type of damage is specifically addressed by the Environmental Protection Law (EPL) of 27 April 2001[1] (art. 322–328). The EPL provides that unless the statute stipulates otherwise, the Civil Code provisions apply to damage incurred through influence on the environment. The relations between the Law on GMOs and the EPL are not determined by the respective regulations. The application of the classical rules of statutory interpretation and the fact that both regulations refer to the application of the Civil Code lead to the conclusion that the regime of civil liability for GMO activities constitutes *lex specialis* with regard to the rules of civil liability for environmental damage envisaged in EPL.

8 Finally, release of GMOs is also covered by the Act on Prevention and Remedying of Environmental Protection, adopted on 13 April 2007, implementing Environmental Liability Directive 2004/35. The Act applies only to environmental damage that occurred after 30 April 2007. The said Act complements both the Law on GMOs[2] and the EPL (see art. 7a EPL) with regard to the liability for damage to the environment.

9 There is yet another bill (of August 2008) aiming to replace the Act of 22 June 2001 with a new Law on GMOs, however, it has not met with political approval to this date. It slightly modifies the provisions on liability for damage, while reiterating the principle of risk as the basis of the claim. Pursuant to art. 218 of the draft, a "user" (any person who undertakes any GMO activity regulated by the law, also a farmer) is liable under civil law for damage to persons, damage to property and damage to the environment caused by a contained use, a deliberate release or placing on the market of a GMO as a product or in a product and by a cultivation of GMO crops, unless the damage is attributed to *force majeure*, the exclusive fault of the injured party or the exclusive fault of a third person for whom the user is not liable. In addition, art. 218 sec. 3 of the draft Act provides for joint and several liability of users whose conduct (activity) caused the damage. In the case of damage to the environment, the claim for compensation may be brought by the State Treasury or a local authority. However,

1 Dziennik Ustaw (Journal of Laws, Dz.U.) 2001, no. 62, at 627 as amended.

2 Although the first draft of the Polish law of 13 April 2007 provided for an explicit reference in the Law on GMOs to the new Act with respect to the liability for damage stemming from GMO-linked activities, the final version of the law abandoned that reference (since it was considered by the parliamentary commission to be redundant).

in this case any award is to be transferred to the benefit of the National Fund of Environmental Protection and Water Management.

2. State liability

The regulation of public liability in Poland is very recent and modern. It has always been situated in the civil law and regarded as a special tort regime. Nowadays, almost all cases of liability of public authorities are embraced by the rules contained in the Civil Code. Every statutory regulation relating to state liability must be construed in compatibility with art. 77 sec. 1 of the Polish Constitution (1997) which states: "Everyone shall have the right to compensation for any harm done to him by any action of an organ of public authority contrary to law". Correspondingly, the Civil Code provides for the liability of the State and other public bodies for the illegal exercise of public authority (art. 417 KC). This is a case of strict liability where wrongfulness of the *acte de pouvoir* suffices to establish liability, provided that the damage and causation are proved. The liability rests on the institution (structure), not on its agents (functionaries), and is connected with the exercise of public authority, i.e. of the prerogatives and official powers. All legal persons (corporations) exercising such prerogatives are subject to the same rules, irrespective of public/private ownership.[3] If the exercise of a task falling within the scope of public authority was delegated to a local authority or other legal person by way of an administrative arrangement, both the performer of the task and the State Treasury, or the local authority who delegated the task, are liable and their liability is joint and several. **10**

The State (local community and any other corporation) is liable for non-governmental conduct as would be any private person, i.e. on the grounds of fault or of risk (art. 427, 430, 435, 436 ff. KC). Consequently, two sets of rules exist: one for acts of a governmental nature and the other for proprietary acts. The distinction between what is and is not "exercise of public liability" is not entirely clear. Historically, Polish courts have avoided the troubling governmental/proprietary distinction, gradually widening the scope of the corporate (economic) activity of public authorities for which the liability in tort could arise. Thus the State was successfully sued on a basis that was more favourable to the victim than the ordinary vicarious liability rules. **11**

3 See *E. Bagińska*, Poland, in: H. Koziol/B.C. Steininger (eds.), European Tort Law 2004 (2005) no. 1–10.

12 From the above presentation, it follows that the first and most probable case for holding the State liable for damage linked to GMO activities is where a public administration organ issues a wrongful decision allowing for GMO activity and damage thus ensues to the party who undertook the GMO activity or third parties (provided that the "normal causation" requirements are met). In this scenario, compensation may be demanded only after having declared the illegality of the final decision in proper proceedings.[4] Similary if the damage resulted from omission to render a decision, when a provision of law provides for such a duty, compensation may be demanded only after having the illegality of the omission declared in proper proceedings, unless special rules provide otherwise (art. 417^1 § 3 KC).

13 The State may be held liable pursuant to art. 417 KC for failure to regulate or for failure to supervise, albeit in that case only direct victims are indemnified. Lack of proper monitoring seems to create a possible ground for public authority liability in the GMO context. It may reasonably be doubted that the State should be liable for failure to legislate or for unconstitutional regulation in this field.[5]

14 Finally, it should be added that the claimant will still have a general cause of action, provided that he proves all conditions of liability. In the ruling of 5 March 1999 (I ACa 1386/98),[6] the Court of Appeals in Warsaw held that even in the absence of any special provisions in environmental law, the State Treasury might still be liable according to the general rules, either in tort or in contract, for damage done to the plaintiff's fishing farm by otters. This judgment represents the practice of the lower courts and the traditional approach of applying tort law rules to redress damage caused by negative impacts on the environment.

4 The same applies to a case where a final decision has been rendered on the basis of a legislative act not conforming with the Constitution or an international agreement or a statute (art. 417^1 § 2 KC).

5 If the damage has been caused through enactment of a legal act, compensation may be demanded only when in proper proceedings the legal act has been declared void on the ground of its unconstitutionality or non-conformity with an international agreement or a statute (art. 417^1 § 1 KC). If the damage resulted from an omission to issue a legal act, where there is a legal duty to issue such act, the court that hears the claims for compensation determines the illegality of the omission (art. 417^1 § 4 KC).

6 OSA 2001/3, item 17. See E. *Bagińska*, Poland, in: H. Koziol/B.C. Steininger (eds.), European Tort Law 2001 (2002) no. 27–29.

II. Damage

1. Recoverable losses

Art. 57 of the Law on GMOs refers to liability "as provided by civil law" and points to the three heads of damage: damage to persons, damage to property and damage to the environment. Accordingly, the scope of liability is determined by the following rules: art. 361–363 KC, and art. 444–449 KC (with respect to personal injury). **15**

In Polish law, damage is considered to be every wrong inflicted upon an interest protected by law, be it property or personality interests,[7] suffered by a person against her will. All kinds of damage, pecuniary and non-pecuniary, must be compensated. Material damage is to be compensated in every case, while non-pecuniary loss is compensable when it is permitted by the law. Damages comprise both (real) losses (*damnum emergens*) and lost profits (*lucrum cessans*). They must be established in accordance with the theory of differentiation. Moreover, the adequate causation test (art. 361 § 1 KC) relates to both types of damage. **16**

As regards damage to the environment see IV.3(b)61. Polish academic writers do not classify "damage to the environment" as a new type of damage under civil law. Based on the interpretation of the international conventions they see this damage as an aggregate notion for all losses incurred as a result of negative impacts on the environment.[8] This type of damage is characterised by its source and not by its effects (similarly to nuclear damage). **17**

2. Pure economic loss

Pure economic loss is not distinguished as a separate head of damage in Polish law. It is commonly understood by doctrine as exclusively financial damage arising extra-contractually without any previous harm to the person or property of the claimant.[9] In general, only persons who are directly **18**

7 See A. *Szpunar*, Ustalenie odszkodowania w prawie cywilnym (1975) 36, A. *Szpunar*, Odszkodowanie za szkode majatkowa (1998) 22–24.

8 See M. *Pyziak-Szafnicka*, Odpowiedzialność cywilna prowadzacego przedsiebiorstwo za szkode ekologiczna, in: Studia z prawa prywatnego. Ksiega pamiatkowa ku czci Profesor Biruty Lewaszkiewicz-Petrykowskiej (1997) 247 f., compare W. *Katner*, Prawo czowieka do korzystania z wartości środowiska to prawo obywatelskie czy takze podmiotowe prawo cywilne, Studia Prawno-Ekonomiczne 1990, 35.

9 See T. *Pajor*, Poland, in: V.V. Palmer/M. Bussani (eds.), Pure economic loss. New Horizons in Comparative Law (2008) 258.

injured (i.e. by the wrongful action directed against them) may claim compensation. The loss suffered by a ricochet victim is distinguished from the harm of the directly injured. The exclusion of claims of persons injured only indirectly, even where their damage has a normal causal connection with the tort, is derived by doctrine from art. 446 § 1 KC.[10] Pursuant to this provision, if as a result of a bodily injury the injured person dies, the person obliged to redress the damage shall refund the costs of treatment and the funeral to the person who has incurred them.

19 The Polish courts award damages for pure economic loss when they find that in the particular case such loss falls within the category of *lucrum cessans*. The proof of high probability of lost profits is required, as well as a normal causal link between the tort and the scope of the economic loss claimed by the plaintiff. The courts emphasise that damage to property makes up the difference between the present property standing of the injured and the standing which would have existed if the event causing the damage had not occurred".[11] Thus, property damage is seen as one loss which may comprise different elements, for example the loss of value of property, loss of income that the property was to produce, etc. The principle of full compensation, subject to statutory or contractual limitations, entails that the amount of damages may not exceed the scope of damage or enrich the injured party.

3. Mere fear of a loss

20 Unquestionably, mere fear of economic loss will not be compensated.

21 Fear of developing a disease is compensated under the heading of non-pecuniary loss. By way of example, one could quote the judgment of the Regional Court in Wroclaw of 20 September 1999 (IC 708/96).[12] In this case, the plaintiff claimed compensation for non-pecuniary loss for the stress she had been suffering from and the constant threat to her life posed by the needles in her heart that had been left there during surgery about 15 years earlier. There is no actual need to remove the needles since they remain rigid in the plaintiff's body. Awarding PLN 80,000 as damages, the Court emphasised that the plaintiff was reasonably anxious concerning the possibility of negative consequences appearing in the future and this had led to her permanent mental trauma.

10 Ibid., 262.

11 Sad Najwyzszy (Polish Supreme Court, SN) 22 November 1963, III PO 31/63, Orzecznictwo Sadów Polskich I Komisji Arbitrazowych (OSPiKA) 1964, at 147.

12 Affd. Court of Appeals 1 June 2000, I A Ca 323/00, See *Bagińska* (fn. 6) no. 46–48.

There are no reported cases concerning claims arising from the fear of mobile phone radiation or other consequences of modern technology. I doubt that the courts would recognise liability in the absence of personal injury or of a violation of a personal interest.[13] 22

4. Standard of proof

As regards pecuniary loss, the proof of actual damage or of a high probability of loss of profits is required. 23

The loss of expected profits is usually more difficult to evaluate because a hypothetical situation must be taken into consideration. The plaintiff must show that the loss of profits has actually been brought about. The subjective expectations and hopes of the plaintiff will not meet with this requirement. Gaining *lucrum cessans* must be objectively feasible and real (e.g. products have been contracted for). The claimant has to establish that they have lost a profit which was to be obtained with certainty or at least with a high degree of probability. 24

In personal injury cases, compensation should also cover future harm which is certain to occur as well as harm which can be foreseen at the moment of adjudication. If at the time the damage cannot be assessed precisely, the court may issue a declaratory judgment that would proclaim the defendant's liability for future damage.[14] This is particularly justified in the case of minors. 25

The rule provided in art. 322 Code of Civil Procedure (*kodeks postepowania cywilnego*, KPC) may be viewed as the relaxation of the burden of proof concerning consequential losses. Pursuant to this rule, if in the case of a claim for damages, a court deems that it is impossible or extremely difficult to prove accurately the amount of the claim, it may adjudicate a relevant amount of money in accordance with its evaluation based on the consideration of all circumstances of the case. Therefore, it is sufficient to prove that the plaintiff had an actual possibility to gain certain profits and that he did not gain these due to the consequences of the damaging event. 26

13 The Polish courts cannot award non-pecuniary damages unless it is permitted by a statutory provision. Apart from personal injury cases, non-financial wrongs are also recognised by special legal provisions (such as violation of patients' rights or copyright infringement).

14 SN 7 August 1969, Orzecznictwo Sadu Najwyzszego (OSN) 1970, at 217.

5. Nominal losses

27 Nominal losses are not recognised in the Polish jurisdiction. The symbolic PLN 1 as compensation for non-pecuniary damage is disputable. The established case law emphasises that damages for non-pecuniary loss serve the aim of compensation, thus they may not be symbolic or excessive in relation to the inflicted damage. Conversely, if the adjudicated award is symbolic (e.g. PLN 1), whether in civil or criminal proceedings, the injured party still has a valid claim for non-pecuniary loss, unless the judgment states that the symbolic award makes up fully for the damage sustained.[15]

28 As regards material damage, nominal losses are non-existent in Polish law unless we include the problem of liquidated contractual damages in the discussion. In the latter case, the obligee does not need to prove the damage in order to claim liquidated damages.

6. Mass losses

29 In the general tort law there are no special rules concerning recovery for mass losses but as of 19 July 2010 they may be a subject of class actions.

30 In Polish atomic law and in maritime law, there are financial limits to liability for mass injuries. If the claims exceed the limit, a special fund must be created. The establishment and distribution of the fund is carried out pursuant to the rules of the Maritime Code relating to the limitation of liability for maritime claims.[16]

III. Causation

1. Uncertainty of merely potential causes

31 In Polish law there are no special rules concerning alternative, potential or uncertain causation. The problem is solved through the application of the general principle of adequate causation (art. 361 § 2 KC). The causal connection may be direct or indirect, as long as all the consequences of the causes meet the test of being adequate (normal) consequences of an event.[17]

---

15 SN 23 November 1966, OSN 1967, at 94; SN 24 November 1966, OSN 1967, at 47.

16 See *E. Bagińska*, Economic Loss Caused by GMOs in Poland, in: B.A. Koch (ed.), Economic Loss Caused by Genetically Modified Crops. Liability and Redress for the Adventitious Presence of GMOs in Non-GM Crops (2008) no. 62.

17 See *E. Bagińska* (fn. 16) 11–16.

It is accepted by doctrine that if a "personally attached group of persons" as a whole created a dangerous situation leading to damage and where the actual actor (perpetrator) cannot be established, all the members of the group may be considered joint tortfeasors. This is explained by the adequate causal link between the creation of the dangerous condition and the harm inflicted (art. 361 § 1 KC).[18] Conversely, if *any* of the possible wrongdoers could have inflicted the harm, but only one actually did, the answer could be different. Polish law does not recognise so-called alternate concurrent causation. It remains an open question in Polish legal academic writing whether in such a situation the liability of concurrent actors should be joint and several.[19] **32**

Polish doctrine strongly supports the concept of *prima facie* evidence used by the courts in the cases of uncertain causation involving contracting contagious diseases (HBV virus, HIV or staphylococcus).[20] By way of example, the Court of Appeals in Cracow[21] ruled: "A health care institution is obliged to exercise due diligence aimed at the protection of patients from the danger of becoming infected with a contagious disease. If an infringement of this duty causes increased risk of infection, and the patient's contagious disease is related thereto, the health care institution bears liability for the damage incurred by the patient unless it presents evidence that the damage is a consequence of other circumstances. Requiring from the injured person that he/she proves direct causality between the conduct which increased the risk of infection and the infection with the contagious disease, would bring insurmountable evidential difficulties for the injured person". The Court of Appeals in Wroclaw[22] held that a plaintiff had proven causation by showing that when entering the hospital he was not infected with any virus and that he could not have contracted it anywhere else because of the poor quality of his personal life. Where a patient was infected with a contagious disease, the defendant hospital may exclude its liability for damages only by proving a high probability of a different cause of that damage. **33**

18 See B. *Lewaszkiewicz-Petrykowska*, Wyrzadzenie szkody przez kilka osób (1978) 77 ff.

19 See W. *Czachórski*, in: Z. Radwański (ed.), System prawa cywilnego, vol. III, part 1 (1981) 264, for joint and several liability *Lewaszkiewicz-Petrykowska* (fn. 18) 78.

20 See M. *Nesterowicz*, Medical Law (8th ed. 2007) 62.

21 Judgment of 14 October 1992, I ACr 374/92, OSA Kr r II, item 44.

22 Judgment of 28 April 1998, I Aca 308/98, Prawo i Medycyna 2002/12, 147.

2. Complex causation scenarios

34 Lacking a Polish example of a DES-type case, it is worth mentioning the Supreme Court judgment of 23 October 2002,[23] dealing with complex causation in a product liability case. In this case the plaintiff, who owned a chicken farm, bought defective feed from producer D1. Because a veterinarian advised changing the feed, the claimant bought another feed from producer D2 one week later, but this feed was also defective. The flawed product of both manufacturers caused an excessively high death rate of chickens, a weakening of the young chicks and underweight chickens despite their prolonged feeding. The claimant sued both producers for full compensation. The lower courts awarded damages, as it was proven that the feed was defective. The Court of Appeals emphasised that the damage inflicted by both manufacturers of flawed feed is a single loss whose elements cannot be separated and so both defendants are liable jointly and severally. Since the claimant did not obtain full recovery, he appealed to the Supreme Court. The Court held that this is a case of a cumulative concurrence of causes which excludes neither a possibility of the valuation of the causes from the viewpoint of adequate causation nor a possibility of estimation of the scope of the damage imputable to just one wrongdoer, subject to the rule of art. 322 KPC. Each producer should bear responsibility just for the part of the damage that he had inflicted. In the case of several causes of harm, their precise separation is usually not possible, therefore it is sufficient to establish the influence of each cause on the infliction of the damage in the light of an adequate causal relation and with a great degree of probability. Therefore, the Supreme Court reversed and remanded the case. It held that the determination of the damage caused by each of the producers would require a more thorough analysis of the expert opinion as to the possibilities of separation of the consequences of the serving of the defective feed of one producer from the results of the subsequent serving of the defective feed of another producer.

35 On the facts of this case one might argue that the damage has been inflicted by several successive events. In fact, the problem here concerns the damage inflicted from the moment of the "intervention" of the second feed manufacturer. It is certain that both producers caused the damage, but it is impossible to establish which part of it was caused by them individually. Therefore, the Court of Appeals ruled on their joint and several liability (art. 441 KC), but the Supreme Court tried to divide the damage, based on the degree of probability of each of the causing factors' contribution to it. The latter view would appear to stem from the fact that Polish law grants no basis for the

23 II CKN 1185/00, not published.

apportionment of damages *pro rata parte* in the case of multiple tortfeasors. Thus, in fact, the Supreme Court wants to break down the damage in order to establish which part of it was inflicted by each producer. This would also eliminate their joint and several liability because each would compensate for the individual damage they caused. It would seem that the Court of Appeal's decision was correct. The expert had already stated in his opinion that a separation of the damage caused by one feed and then by the second one was not possible. Therefore, the Court correctly held that the claimant's loss was one and the same, and its parts were not separable. After all, the same chickens were fed with the feed coming from both defendants.

3. Force majeure

Force majeure is defined objectively and refers to an extraneous event, the effects of which could not be foreseen or prevented. The fact that it was foreseeable does not matter for the qualification of the phenomenon as *force majeure* (e.g. the arrival of a tornado, flood or avalanche). **36**

In tort law it plays the role of a fact exonerating from liability. For the contractual liability regime, the Civil Code does not use the term "force majeure", but speaks of "circumstances for which the obligor bears no responsibility" (art. 471 KC). **37**

Legal authorities vary as to who carries the burden of proof of *force majeure* and causation, in particular in cases based on art. 435 KC. According to some authors, it is sufficient that the plaintiff proves the cause in fact (*conditio sine qua non*) between the operation of the enterprise and the damage.[24] Another distinguished professor of law is of the opinion[25] that there is a presumption of a causal connection between the activity of an enterprise and the damage, which can be rebutted by the proof of an exonerating fact. The dominant view is that an event fulfilling the conditions of *force majeure* exempts the obligee from strict liability and at the same time it defies causation between his activity (the operation of the enterprise) and the damage sustained.[26] The courts agree with this approach, but they emphasise that any one of the three possible exonerating facts must be the exclusive cause of the damage in order to deny the liability.[27] Hence, no **38**

---

24 See *M. Safjan* in: K. Pietrzykowski (ed.), Kodeks cywilny. Komentarz (4th ed. 2005) 1233; *T. Dybowski* in: Z. Radwański (ed.), System prawa cywilnego, vol. III, part I (1981) 270.

25 See *B. Lewaszkiewicz-Petrykowska*, Odpowiedzialność cywilna prowadzacego przedsiebiorstwo wprawiane w ruch przy pomocy si przyrody (art. 435 KC) (1967) 130.

26 See *M. Nesterowicz* in: J. Winiarz (ed.), Kodeks cywilny z komentarzem, vol. I (2nd ed. 1988) 428. Cf. *W. Dubis* in: E. Gniewek (ed.), Kodeks cywilny. Komentarz (3rd ed. 2008) 770.

other causes, whether attributable or not to any person, that would pass the test of adequate causation may exist in a given situation. But the question still remains: who has the burden of proof? Typically, as soon as it is established that the alleged damage is linked to the operation of the defendant's enterprise and he submits evidence of the exonerating fact, the courts will also require from him evidence that no other causes within the defendant's sphere exist (or if any come into play – that they are not in the normal causal connection with the damage).[28] See infra no. 44.

4. Threshold to prove causation

39 The rule in art. 6 KC imposes the burden of proving the causation on the injured person ("The burden of proof relating to a fact rests on the person who attributes legal consequences to that fact"). Generally, courts require "a probability bordering on certainty", i.e. the judge must be convinced beyond reasonable doubt. In some cases an evidence of high probability of the causal connection between the fact and the inflicted loss will suffice. One can also find cases where the court states that the plaintiff has proved causation with "sufficient degree of probability".

40 In general, in order to establish causation, a court may avail itself of any evidence, both direct and absolutely certain and indirect with a significant degree of probability. In accordance with art. 231 KPC, the court may recognise as established the facts which have crucial significance for the adjudication of the case if such a conclusion may be drawn from other established facts. This competence has particular significance in medical malpractice cases.[29]

5. Special rules on causation

41 No such rules are to be found in the positive law.

42 According to the Supreme Court,[30] in a case where two persons may be held liable for damage on the basis of strict liability (for the creation of risk – art. 435 KC) and there is no way to determine which person's activity was the actual cause of the damage, the two persons are jointly and severally liable

27 SN 27 July 1973, II CR 233/73, OSPiKA 1974/9, at 190, cmt. *A. Rembieliński*.

28 Ibid. and SN 28 November 2007, V CSK 282/07, OSNC-ZD 2008/2, at 54.

29 See *M. Nesterowicz/E. Bagińska/A. den Exter*, Poland, Medical law, in: International Encyclopaedia of Laws (2nd ed. 2007) no. 91.

30 SN 4 July 1985, IV CR 202/85, not published.

(art. 441 KC) because neither of them is able to prove any defence to strict liability. The evolution of Polish case law in this respect is quite interesting.

Earlier case law took the view that the duty to redress damage arose only in the case of such negative emissions by an industrial plant as exceeded the commonly suffered effects of the deterioration of the environment in a given area (see art. 144 KC and infra no. 69).[31] That interpretation, limiting the notion of risk (hazard) imposed on an enterprise set in operation by natural forces, has no ground in the current law. The position was changed in the Supreme Court judgment of 6 October 1976.[32] In this case, the claimants demanded recovery for personal injuries and damage to their agricultural and farming businesses that stemmed from the emission of harmful lime-stone dust by the adjacent defendant enterprise. In the area where the claimants lived and ran their businesses, various industrial facilities which discharged damaging smoke were operating. The claimants sued only one enterprise, the one they deemed to be the greatest polluter and cause of the damage. The defendant contested his liability claiming that the discharge from his chimneys and from the chimneys of the other facilities was never measured. The Supreme Court held that the causal link between a given disease of the injured party and the operation of an industrial enterprise which emits harmful substances should be considered to be established as soon as it is proven that the victim was exposed to the damaging pollution released by such enterprise if his disease may be a normal consequence thereof. In addition, the Supreme Court indicated that the fact that the damage could arise only in the case of aggravation of harmful substances emitted by different industrial units does not exclude the strict liability of an enterprise (art. 435 KC). Each industrial enterprise can and should be aware of the fact that any additional source of harmful air pollution worsens living conditions in a given area and, as a result of the accumulation of different effluences, it may cause certain harms, even though the emissions originating in that particular enterprise did not themselves exceed the environmental standards prescribed in the relevant legal provisions. Because the defendant asserted that the density of the lime-stone dust emitted by him was not measured, the Supreme Court held that the law imposes the obligation to measure the density of polluting gases on the enterprises which emit such gases. It is the defendant enterprise, and not the claimant, who should bear the 43

31 SN 7 April 1970, III CZP 17/70, OSPiKA 1971/9, at 169; SN 3 July 1969, II CR 208/69, OSPiKA 1971/5, at 87.

32 IV CR 380/76, OSN 1977/5-6, at 93.

burden of proving that the harmful dust discharged by the former had no impact on the damage suffered by the latter.

44 The position taken by the Court was approved by doctrine.[33] Since then the Supreme Court has typically ruled that a defendant enterprise is responsible for the entire damage unless it supplies counter evidence supporting the conclusion that its activities did not cause the claimant's harm or caused only a small part of it.[34]

45 The case above indicates that strict liability is crucial in relation to any environmental damage caused by any activity. This is also true for damage caused by GMO-linked activities. The complexity of establishing the identity of the party who actually causes environmental damage through a GMO activity validates the shift in the burden of proof of another source of the damage.

IV. Types of liability

1. Fault liability

(a) Special rules governing fault

46 The cases envisaged here will not typically be based on fault, although fault is not entirely excluded as a ground for liability due to the possibility of the concurrence of grounds of liability in Polish law.[35]

47 The rule in art. 6 KC imposes on the plaintiff the burden of proving all facts on which his right to recover may be based. The burden of proving fault is sometimes shifted, thus a presumption of fault is established. This is true for cases of vicarious liability (e.g. art. 429 KC).

48 There are no impediments, other than practical ones, to using the fault regime rather than the strict liability one. Polish civil law permits concurrence of causes of actions (art. 443 KC). This notwithstanding, one should recall that in order to prove fault, the plaintiff has to first establish unlawfulness of the defendant's conduct. This means that a violation of legal rules, or of administrative limits on pollution (contained in licences or other legal acts) or soft-law standards must be found by a court.

---

33 See *J. Panowicz-Lipska/Z. Radwański*, Przeglad orzecznictwa Sadu Najwyzszego, Nowe Prawo 1979/6, 87.

34 See in particular SN 24 February 1981, IV CR 17/81, OSPiKA 1982/5-6, at 64, cmt. *J. Skoczylas*.

35 See *Bagińska* (fn. 16) 68–70.

A court's approach similar to the doctrine of *res ipsa loquitur* may be helpful in establishing certain elements of liability, and in particular causation and negligence. Understandably, a court may apply presumptions of facts (art. 231 KPC) or rely on *prima facie* evidence. 49

Also, one should not overlook the fact that pursuant to art. 57 sec. 5 Law on GMOs, the plaintiff has a right to ask the court to oblige the defendant to supply all the information necessary to establish the scope of his liability. 50

(b) Impact of specific rules of conduct

It may make a difference in fault-based liability. The violation of customary or statutory standards is wrongful conduct, and in the case of a professional it also means negligence.[36] 51

2. Product liability

(a) Development risk defence

Under the product liability regime (art. 449^1–449^{11} KC), the producer and other persons liable like a producer can make use of the development risk defence. No reported court cases are available in order to show how this defence operates in practice. 52

(b) Alternative routes

There is no alternative system of compensation for losses caused by (agricultural) products. The general fault basis can always be applicable. 53

(c) Impact of compliance with rules and regulations

Under the product liability regime, if the product conformed to the compulsory technical standards set by an appropriate state agency as mandatory law, the manufacturer may use this as a defence against strict liability (art. 449^3 § 2 *in fine* KC). 54

It follows that the plaintiff could only question the level of the legal standards, by trying to prove them to be too old or not to meet the ordinary 55

36 See ibid., 82–84.

consumer expectation test, and, consequently, base his cause of action on fault (art. 415 in conjunction with art. 449^{10} KC).[37]

3. Environmental liability

(a) Implementation of the Environmental Liability Directive

56 The Environmental Liability Directive was implemented by the Act on Prevention and Remedying of Environmental Protection (PREP), adopted on 13 April 2007. The Act uses in part the terms from the Directive as well as its own terms.

57 The Polish Act covers all natural habitats and protected species protected under national law,[38] provided that fault of the operator has been established. The state of the art defence and the permit defence are not allowed. PREP introduces one exception in the application of the Act which is contrary to the Directive. It relates to the "forest management carried out in compliance with the rules of sustainable forest management as referred to in the Forest Act of 1991".

58 As regards financial guarantees, they are not compulsory. The competent authority is entitled (but not obliged) to require that the operator must arrange for some kind of financial security when granting an emission permit (art. 187 EPL). This security may be in the form of a bank guarantee, insurance guarantee or policy, or a money bank deposit. The possibility for introducing the financial security is, therefore, limited only to the activities causing emission. The Minister of the Environment is entitled to issue an executive regulation listing categories of installations (not activities) requiring compulsory security. No such regulation has been issued so far.

59 However, financial security may be required by a licensing organ in a decision allowing for the release of GMOs (art. 25–30 and 36 Law on GMOs). The security that is meant to cover damage to the environment should be provided in the form of a deposit to a special bank account set up by the licensing organ, a bank guarantee or insurance policy.[39]

---

37 See *E. Letowska*, Prawo umów konsumenckich (2nd ed. 2002) 123.

38 The Act defines the damage to protected species and habitats as damage to all species and habitats protected under the Polish Nature Conservation Act of 2004 and not only to those protected under the Habitat and Birds Directives (the Act uses the option foreseen by art. 2.3 (c) of Directive 2004/35/EC).

39 See *Bagińska* (fn. 16) 52, 56–60.

(b) Environmental liability regime beyond the scope of the Directive

Art. 12 sec. 2 and 3 PREP establishes joint and several liability of a holder of land and the actual polluter in the case where the environmental damage or threat of this damage occurred with the consent and to the knowledge of the land holder. **60**

With respect to the damage to the environment, account should be taken of the provisions of the EPL. Although the latter does not define "damage to the environment", the general doctrinal view has been rather consistent with what is now considered environmental damage under Directive 2004/35. The reparation of damage to the environment is regulated in the EPL. The compensation will in general comprise the costs of prevention and restitution.[40] **61**

Pursuant to art. 324 EPL, if the damage was caused by an enterprise operating at an aggravated or major risk (which is defined by art. 248 EPL and the relevant ministerial regulations), art. 435 KC (i.e. strict liability) applies even if the enterprise is operated "without the use of natural forces". It is important to note that under Polish law, strict liability can arise in relation to environmental damage caused by any activity of an enterprise in the meaning of art. 435 KC and art. 324 EPL.[41] However, farming has not been considered by courts to fall within the scope of art. 435 KC unless the activities conducted on a farm and their effects directly depended on the use of machines set in operation by natural forces (ergo creating a risk).[42] Hence, the harmful effects on the environment were not relevant for the application of art. 435 KC. Therefore, the combined legal basis of the claim (art. 324 EPL and art. 435 KC) seems to fill this gap. **62**

Under the Law on GMOs, strict liability is linked with all activities involving the production or release of GMOs and covers any risk posed by the GMO products that have been released into the environment. This will include farming business that uses GMO products. For these reasons, the Law on GMOs may be seen as a regime exceeding the scope of the EPL, but not the scope of Directive 2004/35/EC (which covers GMO activities). **63**

40 See ibid. 63 f.
41 See *Lewaszkiewicz-Petrykowska* (fn.25), *Pyziak-Szafnicka* (fn. 8) 251–262.
42 See SN 15 February 2008, I CSK 376/07, OSNC – ZD 2008/4, at 117.

(c) Claimants in cases of environmental harm

64 Traditionally, in Polish tort law the actor's conduct must violate someone's subjective right for the liability to arise. However, in the case of damage to the environment, neither the Polish doctrine nor the courts have accepted the concept of the subjective right to the environment.[43] Consequently, the environment is protected as a common good (value).

65 Both the Law on GMOs and the EPL restrict legal standing with regard to claims based on the harm to the environment. Hence in the case of environmental damage, the claim for compensation may be brought by the State Treasury, a local authority or an ecological organisation (art. 57 sec. 2 Law on GMOs, art. 323 sec. 2 EPL).[44]

66 Pursuant to the EPL, the person who has repaired the damage to the environment has a right of indemnity towards the actor who caused it; however, the indemnity is limited to the reasonable expenses incurred in the restoration of the previous state (art. 326 EPL).

(d) Special liability regime for losses sustained by individuals

67 Art. 18 PREP allows the land holder to demand compensation if the preventive or remedial measures that were carried out on his land caused him (individual) damage.

68 In accordance with art. 323 EPL, *anyone* who is exposed to the risk of damage, or who incurred damage through another's illegal influence on the environment, may demand from the person liable that he restores the lawful state and takes preventive measures, such as installing safety appliances or machines. Where the preventive action is impossible or unreasonably difficult, the claimant may demand that the actor abstains from the infringement.

69 The relations between art. 323 EPL and the "law of neighbours" in the Civil Code are debatable. The opinion that the EPL broadens the remedies available in the Civil Code, quite rightly, prevails.[45] Conversely, the nature of liability in the law of neighbours is not clear either. There are many

43 See *Pyziak-Szafnicka* (fn. 8) 251–248, see broadly *W. Radecki*, Odpowiedzialność cwyilna w ochronie środowiska (1987).

44 Both provisions refer to the damage to the environment "as a common value".

45 See *J.J. Skoczylas*, Odpowiedzialnosc cywilna na podstawie ustawy – prawo ochrony srodowiska, Przeglad Sadowy (PS) 4/2003, 68.

plausible arguments for qualifying the liability as delictual. Indeed, this is the dominant view in the legal writings.[46] *Actio negatoria* (art. 222 § 2 KC) and a claim to abstain from infringement on the basis of art. 144 KC may be brought against any person who infringes upon the environment. According to art. 144 KC, the owner of real estate should, in exercising his rights, refrain from activities which would disturb the utilisation of neighbouring real property above an average degree, stemming from the socio-economic purpose of the real property and local conditions. This claim belongs to owners and persons enjoying other proprietary rights. An activity that would disturb the neighbouring land beyond the acceptable local standards is illegal, thus it creates a claim for damages based on the general rules. On the other hand, the illegality of the activity carried out on the neighbouring land would not be relevant for the action for compensation based on art. 435 KC (strict liability of an enterprise), even if the claim based on art. 144 KC could not be raised.

A slightly different aspect of the law of neighbours applies in the case of **70**
mining and geological damage (the Act of 4 February 1994 – Geological and Mining Law).[47] The land owner (or other parties endangered by mining plant operations) may not object to the threats caused by mining plant operations if such operations take place in accordance with the rules specified in the Act. They may demand reparation of the damage caused by these operations in accordance with provisions of the Act (art. 91). These damages should be repaired by way of restoration of the former state or by the provision of land, constructions, devices, premises, water or other goods of the same kind. Damaged afforested and agricultural lands should be subject to cultivation. If the restoration of the former state is not feasible or the cost thereof would vastly exceed the loss incurred, the compensation may be paid in money. If the mining operations violated the respective legal rules, the entrepreneur shall be responsible for damage according to the rules specified in the Civil Code (ergo, the claimant has a choice, although not unlimited,[48] between restitution in kind and pecuniary damages).

46 See W. *Katner*, Ochrona wasności nieruchomości przed naruszeniami pośrednimi (1982) 130; D. *Maśniak*, Ubezpieczenia ekologiczne (2003) 99 ff.

47 Dz.U. no. 27, at 96 with later amendment.

48 According to art. 363 § 1 KC, redress of damage must take place, either through restoration of the former state or through payment of an appropriate sum of money, at the option of the injured party. If, however, restoration of the former state would be impossible, or would entail excessive difficulty or cost to the person liable, the injured person's claim is limited to pecuniary payment.

(e) Cartagena Protocol

71 Poland has ratified multiple international treaties, including the Biodiversity Convention (1996), as well as the Cartagena Protocol in 2003. It may be said that the minimal international standards are met (at least imposed, if not actually enforced) by Polish environmental legislation. The governmental approach seems to be a minimalist one, since it takes action only with respect to treaties that have to be ratified because of the obligations arising from EU membership.

V. Vicarious liability

1. Scope of vicarious liability

72 To begin with, one should say that there is a substantial difference between vicarious liability in tort and vicarious liability in contract. The latter is stricter. An obligor is liable *ex contractu* for the action and omission of persons with whose help he performs the obligation (subcontractors and assistants) as if for his own action or omission (art. 474 KC). Hence, the assessment of their conduct (breach of duty) is carried out according to the standard of due care required from the debtor as if he had performed the obligation personally. The liability is strict and may not be excluded by furnishing the proof of lack of negligence in choosing the subcontractor (*culpa in eligendo*) or in supervising (*culpa in custodiendo*). The subcontractors and assistants need not be subordinate to the main contractor, and in their relationship with him there is no requirement of dependency.

73 In the tort regime, the possibility of raising an independent contractor immunity as a defence is dependent on the determination of whether we have an independent contractor's or rather a servant's (employer's[49]) liability case. In order to answer this question one has to look at the scope of power of management and control over a person's activities. Art. 429 KC provides that the person who entrusts an independent contractor with the performance of a task is liable for the damage he may cause while doing so. In contrast to employer's liability (art. 430 KC), however, this lia-

49 Assuming we determine the direct actor is a independent contractor and not an employee. See M. *Nesterowicz*/E. *Bagińska*, Liability for damage caused by others under Polish Law, in: J. Spier (ed.), Unification of Tort law: Liability for damage caused by others (2003) 190–194.

bility is based on fault. The defendant may escape liability if he refutes the presumption of *culpa in eligendo* in either of these two ways:

- he was not negligent in choosing the contractor; or
- he entrusted the performance of the transaction to a person, enterprise or institution which performs such transactions within the scope of their professional activity.

The master is still personally liable for any other negligence.[50] In that case both the master and the independent contractor are liable jointly and severally for damage according to art. 441 § 1 KC. 74

2. Liability for people further up the food or feed production chain

Since both the regime of liability for GMOs and the product liability regime are of strict liability, it could reasonably be argued that the former will prevail with respect to every person who puts a product containing a GMO on the market and may be considered a GMO user conducting activities as defined by the Law on GMOs (in art. 3). Therefore, I would incline to the view that the question of the liability for people further up the feed and food production chain should be answered in the light of art. 57 Law on GMOs. Each person in the production and distribution chain that can be considered "a GMO user" will be strictly liable pursuant to art. 57 Law on GMOs. They may be liable jointly and severally liable for damage according to art. 441 § 1 KC. In order to determine all the persons who should be allocated joint and several liability, it is critical that the law imposes tort liability on them, whether as a direct tortfeasor, or through vicarious liability, or for aiding, abetting or consciously taking benefit from damage caused to other persons. Hence, all of the persons involved in the feed and food production chain may be responsible for the whole damage. See also no. 80, 81. This interpretation is in favour of the injured consumer, due also to the fact that the GMO law allows the defendant only three exculpatory grounds. One of them is the fact that the damage was caused exclusively by the fault of a third person for whom the GMO user 75

50 Example: A employs B, a building contractor, to build a house. During the work, B causes damage to X, a passer-by. A will not be liable for damage caused to X by building contractor B if he proves the defence ad 1) or ad 2). However, A is still liable for any other negligence (e.g. defects in construction design or erroneous description of the estate's boundaries), which led to X's damage (art. 415 KC).

is not liable. In practice, it is hard to meet the requirements for this exculpatory ground. See the remarks and court cases at supra no. 43.

76 As already mentioned, the draft Law on GMOs introduces the express basis of the joint and several liability of users whose conduct/activity caused the damage.

3. Can someone further down the feed or food chain include someone further up it in a trial against him/herself?

77 On the assumption that both the farmer and the seed producer are jointly and severally liable the following answer can be offered:

78 In principle, an obligee may sue co-obligors simultaneously or successively. A joint and several debtor (e.g. the farmer) may not raise the defence that not all liable parties have been sued. The customer may decide to bring successive proceedings against the seed producer. However, if in the first law-suit against the farmer, the judgment is entered in favour of the defendant because the court has considered a defence common to all the debtors (for example, *force majeure*, contributory negligence) and partially or entirely dismissed the action, then in a separate trial against the other co-debtor (the seed producer) the action must be dismissed accordingly.[51] In the successive trial, the court is bound by the previous evaluation of the defences raised. Even if it deems the first ruling incorrect, it still cannot enter a judgment against the other co-obligor.[52]

79 If the customer sues the farmer alone, the latter may notify the seed producer of the proceedings and call on him to join the case as an intervenor (art. 84 KPC). The producer, however, is not obliged to join the defendant farmer. If he decides not to join him in the trial, this decision will not militate against him. Thus, this is not the way for a co-obligor to escape liability. In the following contribution lawsuit between the co-obligors, the producer will be allowed to contest the facts established at the first trial, "whether they concerned fault or the scope of damage inflicted in (the accident").[53] This is because the findings of the court in the original proceedings are not binding on the court in the action over contribution brought by the obligor(s) who performed.[54]

---

51 SN 23 July 1961, 4CR 909/60, Państwo i Prawo 1962/5-6, 948.
52 SN (7 judges panel) 26 August 1965, III Co 9/65, OSPIKA 1968/6, 114.
53 SN 15 January 1963, 3CR 1008/61, Nowe Prawo 1964/5, 548.
54 SN 30 April 1957, 3 CR 340/56, OSPIKA 1958/4, 98.

VI. Multiple tortfeasors

Art. 441 § 1 KC lays down a general rule of joint and several liability in all cases where more than one person is liable for the damage arising from a tort.[55] The basic requirement of art. 441 § 1 KC is the delictual infliction of damage. Secondly, the damage for which all the persons are liable must be one (indivisible) loss. The third requirement is the causal relation between the damage and the event (tort), for which each of the joint tortfeasors bears liability. The causal relation should be viewed in the light of art. 361 § 1 KC.[56] **80**

Art. 441 § 2 KC controls the determination of contribution claims between the tortfeasors who are liable on the same legal basis (here: strict liability). The scope of indemnity depends on the circumstances of the case, in particular on the degree to which each tortfeasor contributed to the injury. **81**

VII. Defences

1. Licence/permission to grow GM material

The fact that the activity that caused damage was carried out on the basis and within the scope of an administrative decision does not exempt the user from the strict liability for GMOs (art. 57 sec. 3 Law on GMOs). There is no distinction between types of possible decisions, hence it does not make a difference if the GMOs are approved for testing only. The exclusion of the "permit defence" corresponds with the traditional approach of the jurisprudence towards liability for environmental damage in Poland. **82**

2. Consent/assumption of risk

Polish law contains no specific regulation on the assumption of risk. **83**

As a rule, consent of the injured person will not always exclude strict liability.[57] According to art. 437 KC, the liability defined in art. 435 KC (enter- **84**

55 See *M. Nesterowicz/E. Bagińska*, Multiple Tortfeasors under Polish Law, in: W.V.H. Rogers (ed.), Unification of Tort Law: Multiple Tortfeasors (2004) 151–155, 162.

56 See no. III.1.32. and *Bagińska* (fn. 16) 74–76, 79 f.

57 The judicature indicates that an injured person assumes the risk of loss when, contrary to the prohibition sign, he bathes in a dangerous section of a lake or when he enters a military field during exercises. However, in such situations Polish courts would rather find an exclusive fault of the victim or no fault of the defendant (if his liability is based on fault).

prise liability) and art. 436 (liability of possessors of vehicles) cannot be excluded or limited in advance. In other cases, a person may not consent to bodily harm or infliction of a health disorder if the wrongdoer's conduct is unlawful. Such consent would be held contrary to public policy (art. 58 § 2 KC).

85 Art. 362 KC stipulates that "if the injured person contributed to the occurrence or increase in the damage, the obligation to redress it shall be correspondingly reduced according to the circumstances, and in particular to the degree of fault of both parties." As follows from art. 362 KC, a contribution to the increase of damage is subject to the same principles as a contribution to causing damage in the first place.[58] The former occurs if the dimensions of damage would have been less significant but for the injured person's conduct. A distinct duty of the injured person to act so as to reduce damage under pain of denial of compensation is provided for exclusively in art. 826 § 1 KC which refers to property insurance.

3. Third-party influence

86 Only the exclusive fault of a third person for whom the GMO user is not liable will be a complete defence (art. 57 sec. 3 Law on GMOs). The judicature emphasises that the fault of a third person, which exonerates from the strict liability, is to be understood as applying when the fault is the exclusive cause of the damage. See the remarks and court cases referred to supra no. 43.

4. Prescription

87 Since art. 57 sec. 1 Law on GMOs refers to liability according to civil law and no statutes of limitation are to be found in the Law on GMOs, the standard rules of limitation of tort claims will apply.[59] They do not change the interpretation of art. 449^8 KC which is a special rule on prescription in the product liability regime and must be construed in conformity with Directive 85/374/EEC.

---

58 See: M. *Nesterowicz*/E. *Bagińska*, Contributory Negligence under Polish Law, in: U. Magnus/M. Martín-Casals (eds.), Unification of Tort Law: Contributory Negligence (2004) 151–155.

59 The revised rules on prescription entered into force on 24 May 2007.

Pursuant to the Civil Code, a claim for redress of the damage due to a tort expires after the lapse of three years from the day on which the injured person learned about the damage and about the person obliged to redress it. However, that time may not exceed ten years from the day on which the event that caused the damage occurred (art. 442^1 § 1 KC). If the damage was due to a crime or misdemeanour, the claim for its redress expires twenty years after the day on which that crime or misdemeanour was committed, regardless of when the person who suffered the damage learned about it and about the person obliged to redress it (art. 442^1 § 2 KC). It should be remembered that a civil court decides on the criminal nature of the actor's conduct. No criminal court verdict is required. **88**

In a personal injury case, a claim may not expire before the lapse of three years from the day on which the injured person learned about the injury and about the person obliged to redress it (art. 442^1 § 3 KC). Hence, the absolute periods of ten or twenty years are "suspended" in personal injury cases. Nevertheless, the periods running *a tempore facti* may prove more convenient for the injured parties as regards evidence. The claims of a minor for redress of the damage stemming from personal injury may not expire earlier than after the lapse of two years from the day of their maturity. **89**

5. Other defences

There are no specific defences available. **90**

One should not overlook the principle of *compensatio lucri cum damno*. The reduction of damages may occur only with respect to the gains flowing from the same event. Moreover, any gains obtained from the tortfeasor are to be taken into account when determining the amount of damages. In several decisions, the Supreme Court has related to collateral sources such as the relation between damages sought from the tortfeasor and insurance benefits. **91**

VIII. Remedies

1. Pecuniary compensation

(a) Bodily harm

See supra no. 17.[60] **92**

60 See also *Bagińska* (fn. 16) no. 34 and 42.

(b) Property losses

93 The regular remedies apply to the compensation of property losses.[61]

(c) Economic losses[62]

94 An organic farmer can recover the full costs of restoring a field that was contaminated with GMOs to meet organic standards. Deductions for the seasons that he could have grown conventional crops there depend on whether the court decides that growing the conventional crops is conduct that should have been undertaken in order to reduce the scope of the loss (see supra no.86).

(d) Harm to animals

95 The animal's value will probably be calculated according to the market value. The farmer's damage comprises: 1) the difference between the full market value of an animal fed with healthy food and the value of the animal fed with GM contaminated feed, and 2) the consequential economic damages (provided that he meets the high standard of proof), i.e. the cow's potential for producing non-contaminated milk or meat will determine the loss of profits by the farmer (the depreciation of the products actually sold or rejected by his contractors due to the contamination).

(e) Costs of disposal

96 The cost of disposing of contaminated production/animals should be seen as the farmer's real loss that is in an adequate causal relation with the contamination (pursuant to the principle of full compensation).

2. Non-compensatory damages

97 No punitive, exemplary or any other forms of non-compensatory damages are recognised.

---

61 See ibid., 47–50.
62 See ibid., 46–50.

3. Other remedies

N/A. 98

4. Costs of pursuing a claim

(a) General cost rule

The Polish civil procedure follows the "loser pays" principle (art. 98 KPC). 99
It is mitigated by art. 100 KPC that permits a proportionate distribution of the costs if the claimant won only partially. Moreover, in situations justified on the ground of justice (e.g. in a serious personal injury case), the court may award only parts or none of the costs from the losing party (art. 102 KPC).

(b) Costs of establishing causation

Such costs may be recoverable under the theory of difference.[63] 100

5. Advance cover

It is possible to ask for advance cover in personal injury cases. Pursuant to 101
art. 444 § 1 KC, upon demand by the injured person, the person obliged to redress the damage must deposit in advance a sum necessary for medical costs, and if the claimant becomes an invalid, also the sum necessary to meet the costs of acquiring another professional skill.

IX. Cross-border issues – Conflict of laws

1. Conflicts rules applicable before (or instead of) Rome II

Polish law declares the law of the place of the harmful event as governing 102
in the situation of conflict of laws (pursuant to art. 31 § 1 of the Act on International Private Law). This rule applies, *inter alia*, to compensation of the damage arising out of a tort. Thus, in comparison to the Rome II

63 See *Bagińska* (fn. 16) 99 f.

Regulation, it is the place of the tort, not the place of damage, which determines the applicable law.

103 However, in line with the Regulation, if a tort was committed in one country, the law of another country could be applied pursuant to art. 31 § 2 of the Act, when both parties are citizens of that other country and reside there (*lex patriae et domicilii*). The rule is also applicable to relations between legal persons (the place of incorporation and seat of the legal person replaces the requirement of citizenship and residency) as well as between legal and physical persons.

104 According to doctrine, the relationship between § 1 and § 2 of art. 31 of the Act on International Private Law is vague. However, it appears plausible that the two paragraphs form two conflicts of law rules which are not in a *legis generalis–legis specialis* relation. Thus, in order to establish the applicable law, one should first consider the combined personal link (*lex personalis communis–lex patriae et domicilii*), i.e. the nationality and residence of the parties, and only in the second phase the place of a delict. One of the links should exist at the moment in time of the event causing damage, so any later changes to the residency or citizenship or to the person of the defendant are irrelevant.

105 The place of a tort may be either the place of the actor's conduct or the place of the result. No case law is available with regard to this matter. There are certain grounds for supposing that all connecting factors may be used, depending on the facts and type of the tort committed. According to the dominant doctrinal view, if there are several losses and the conduct took place in several countries, a court has freedom to decide which law to apply. The suggested criteria to be used by the court are the following: (i) the most favourable law to the plaintiff, (ii) the closest link between the tort and a given jurisdiction.

2. Special regime for cross-border claims

106 N/A.

X. Cases

1. Due to the adventitious presence of GMOs in a field, maize which is normally sold as conventional contains GMOs beyond the legal labelling threshold. This is not discovered before the final stage of the food production chain by the producer of taco chips. The whole production is lost since the supermarket chains refuse to accept delivery from the producer.

(a) Who can sue along the chain of distribution?

The persons in the production and distribution chain can be considered GMO users in the meaning of the Law on GMOs if they fulfil the statutory definitions which require a conscious and deliberate GMO activity. The accidental presence of GMOs in a maize field does not automatically make any of these persons a GMO user. Conversely, if we found the GMO user from which the contamination of the field originated (e.g. a laboratory), we could apply the regime of art. 57 Law on GMOs to his liability. See also (e) below. **107**

The chips producer who only sells produce made of conventional maize may sue the maize wholesaler on a fault basis, since the contamination is beyond the legal labelling threshold. He must prove that the defendant had a duty to test the maize for the adventitious presence of GMOs and the causal link between the damage and the negligent breach of duty.[64] **108**

(b) Would the case be solved differently if the GMO content was below the labelling threshold?

The fact that the GMO content is below the labelling threshold means only that a conventional product is not officially regarded as a GMO product. Thus there is no wrongfulness of the party responsible for the labelling. **109**

Pursant to Law on GMOs the answer is the same as supra no. 109. **110**

Under Polish law there is no labelling threshold and all GMO products put on the Polish market must be labelled as GMO products. The answer could vary if we discussed an imported GMO product that is circulating freely in the EU market. **111**

64 With regard to the duty of testing and sampling, see *Bagińska* (fn. 16) 96 f.

(c) **Would the case be solved differently if the admixture was not adventitious, but occurred due to the disregard of segregation rules, for example?**

112 If the wholesaler distributes both non-GMO and GMO products and is in breach of segregation rules, he is strictly liable for the damage as a GMO user. He must be considered as a person who deliberately releases GMOs into the environment (which includes the placing of GMOs on the market).

113 Everyone who is bound to apply the segregation rules can be held liable for negligence. See no. 50, 51.

(d) **Would the case be solved differently if the GMO found was not admitted for production in your jurisdiction?**

114 The same answer as supra no. 108.

(e) **If the admixture had occurred on a non-GM field and it transpires that the GM seeds were blown from: (i) neighbouring fields; or (ii) a truck passing by, would the farmer of the affected field be liable for all or part of the loss caused further down the distribution chain?**

115 Under Polish law the domestic case described in (e) is highly hypothetical. As of 2 July 2006, the Act on Seeds of 23 June 2003[65] banned any genetically modified seed from the Polish market (art. 57 sec. 3 of the Act).

116 The farmer would not be liable on general grounds as it cannot be said that he was at fault in either of the two situations. Nor may he be considered a GMO user, thus allowing the strict liability regime provided for in art. 57 Law on GMOs to apply.

117 The possibility of applying the product liability regime to the farmer is also excluded because of the kind of losses sought.

118 If we assumed that the ban on GM seed was annulled (because of its incompatibility with the EU law), then we might argue that the cause of action could be based on the farmer's (hypothetical) failure to inform (or misinformation to) his contractors about the nature of crops grown on neighbouring fields.

65 Consolidated text Dz.U. 2007, no. 41, item 271 with later amendment.

It is valid to speculate that any action for damages will fail on the grounds of lack of adequate causation between the losses and the conduct of the farmer (one cannot require someone to take measures against the wind blowing in the direction of his field). 119

As to the truck passing by – see liability for the transportation of GMO products art. 57 sec. 4 Law on GMOs. See no. 4. 120

2. Twenty years after the sale of GM maize used for food products, it turns out that it has certain disadvantageous health effects for humans.

(a) Can the producers be held liable at this point for risks unknown at the time of growing the maize? Who would be liable?

As to the liability of the persons in the feed and food chain see V.2. 121

As to the prescription of claims – see no. 88, 89. 122

All persons in the distribution chain that are not GMO users, if sued in the regime of product liability, could take advantage of the risk development defence, were the claim not prescribed (extinguished) under the special prescription regime (art. 449^8 KC). 123

As to the liability of the state for permitting the sale – see no. 12. 124

As to the exonerating facts – see no. 3. 125

It should be noted that the licence for distributing GMO products on the Polish market is given for a limited time, not longer than ten years. Thus, the potential effects on human health are subject to research and tests after ten years at the latest. 126

(b) Can compensation already be claimed at a point when the negative health effects have not yet materialised, but are to expected to according to scientific expertise/mere rumours?

For the admissibility of the claim for redress of personal injury, some negative effects have to materialise. As to the fear of negative health effects – see no. 21. 127

128 With respect to damages for non-pecuniary loss, compensation also covers future harm which is certain to occur or which can be foreseen at the moment of adjudication. See no. 25.

(c) Would it make any difference if the GM maize had only been in use for feed, causing harm to the animals, which may or may not cause harm to humans consuming the meat as well?

129 It makes a difference in establishing causation when the cause is uncertain. See III.1 above.

3. The driver of a food logistics company discovers that a farmer, from where he regularly picks up agricultural products fails to obey mandatory segregation rules or food or feed hygiene standards, which may lead to the admixture of GM and non-GM produce sold and packaged separately by that farmer. Does he or his employer have a duty to warn, i.e. warn the recipient of the allegedly "non-GM" produce?

130 It may reasonably be doubted whether the Polish courts would recognise such a duty to warn. However, the broad definition of wrongfulness has a long tradition in Polish civil law. According to the courts,[66] some duties are justified by life experience and the professional nature of the defendant's activity. The general duty not to cause damage to anyone (*neminem laedere* rule) requires that all necessary action be taken in order to avoid damage to persons or things. Failure to do so can be wrongful, depending on the circumstances.

66 SN 2 December 2003, III CK 430/03, Orzecznictwo Sadów Polskich (OSP) 2005/2, at 21, reported by *E. Bagińska*, Poland, in: H. Koziol/B.C. Steininger (eds.), European Tort Law 2005 (2006) no. 6–15.

Damage Caused by GMOs under Slovenian Law

Rok Lampe

I. General overview

1. Special liability or redress scheme for GMOs

There is no specific liability or other compensation/redress regime in Slove- 1
nian jurisdiction. The existing liability and compensation/redress regime is regulated through the Code of Obligations, which sets out the general compensation/redress regime. However, there have been attempts on the part of the legislator also to cover a special liability regime in the new Act on Genetically Modified Organisms. The draft actually intends to regulate the special liability regime but on the basis of the existing general regime provided by the Code of Obligations. So the legislator is trying to do something in this respect, but rather unsuccessfully because the special regime which would have to be created is based on the existing general one.

2. State liability

It can be argued that, according to Art. 13 of the Act on Treatment with 2
Genetically Modified Organisms 2005, the State takes the position of subsidiary liability. This article provides that, in cases where the State has responsibility to reduce the harmful impact which results from working with a GMO in a closed system, the intentional spreading of GMOs into the environment or placing a product on the market, the Ministry (for the Environment) must ensure preparation and executing measures.

II. Damage

1. Recoverable losses

3 It could be argued that the general compensation scheme set out in the Code of Obligations would cover pure economic loss (*damnum emergens*) and loss of profit (*lucrum cessans*).

2. Pure economic loss

4 Pure economic loss is recoverable according to the general provisions of tort law. The Code of Obligations sets out in Art. 132 that damage is any diminishment of property. It is recoverable if the damage is proven and if there is a causal link between the damage and the tortious act. The tortfeasor must prove that he did not act with fault (Art. 135). On the other hand, liability can also be strict. In this case, according to Art. 149 and 150 of the Code of Obligations, the causal link is presumed and the defendant can only exculpate himself by proving one of the grounds for exculpation set out in Art. 153.

3. Mere fear of a loss

5 I would argue that mere fear cannot be the basis for compensation in such cases.

4. Standard of proof

6 The general standard of proof is applicable in this respect. But of course the main question is whether a loss should be proven with certainty, or whether the standard likelihood is applicable. Unfortunately there is no case law or scholarly materials that would defend either position.

5. Nominal losses

7 It could be argued that there is legal ground also for nominal and symbolic losses – the general provision on damage. However, there is no support in case law or in legal literature.

6. Mass losses

There is a compensation scheme with established funds for victims of asbestos, for crime victims and for victims of violations of Art. 6 of the European Convention on Human Rights. 8

III. Causation

1. Uncertainty of merely potential causes

To answer this question I found a case dealing with alternative causation in Slovenian law. In this case,[1] the plaintiff demanded cessation of the operation of a bakery that presumably produced illegal emissions. She argued that her health conditions had been aggravated – she suffers from asthma – since the defendant started his bakery production. The defendant's business presumably emitted irritants (hydrogen sulfide, nitric oxides, etc.) above the legally permissible level and above what is customary in the community. 9

The district court rejected the plaintiff's demand for injunctive relief for nuisance. It also rejected the plaintiff's claim for compensation for non-pecuniary loss. The appellate court confirmed the district court's decision, highlighting the fact that the plaintiff failed to prove the existence of a causal link between the bakery's emissions and her health problems. It also rejected the appeal for injunctive relief and the decision became final under the Act on Civil Procedure. 10

The plaintiff challenged the appellate decision at the Supreme Court. She argued that the deterioration of her health was caused by emissions which were presumably above the legally permissible level. She presumably had not suffered any asthmatic crisis before the bakery started its production. She also stressed that the lower courts had reached their decisions contrary to the testimonies of expert witnesses. One of them presumably testified that flour contains one of the irritants that caused the plaintiff's health deterioration. If she was allergic to other irritants, then she would have suffered the deterioration earlier. The Supreme Court did not accept this argument and rejected the appeal. It stressed that the crucial question would be whether emissions produced by the defendant were above the legal limits. Expert witnesses testified negatively. They also could not 11

1 Judgment of the Supreme Court, II Ips 76/2003.

establish with certainty that irritants contained in the bakery's emissions caused the plaintiff's health conditions. Therefore, the plaintiff failed to prove the existence of a causal link between said emissions and her health deterioration.

12 It is impossible to conclude that an irritant in the emissions was the cause, even if the plaintiff's state of health worsened at the time when the production started. The Court stressed the following formula for discovering causal links in toxic tort cases – if it is not possible to single out at least one of the factors among internal and external causes (emissions) which would cause the existing health conditions, then the causal link is not proven. In this particular case, the expert witness left open a possibility that one of the irritants contained in the defendant's flour could, among other internal and external factors, cause the plaintiff's state of health. He did not give any assurance that this irritant was the actual cause. I decided to include the case in this category because it is a question of alternative causation. The victim did indeed suffer from asthma from the time the defendant started his bakery production. However, it is impossible to prove that this illness was caused by the emissions from the bakery.

2. Complex causation scenarios

13 There are no special rules.

3. Force majeure

14 In Slovenian tort law, *force majeure* is traditionally a question of fault and one of the grounds for exculpation in strict liability regimes. Thus, it is not directly connected with causation.

4. Threshold to prove causation

15 A "certainty" rule is generally applicable in causation issues.

5. Special rules on causation

16 There is a two-step regime which is applicable to general rules on causation in toxic tort cases. The first one is the general step – where it must be scientifically shown that a link is established between cause A and result B and the second step is a factual one, where it must be shown with certainty that this precise cause gave rise to the concrete result. So, for example in asbestos cases, first the general scientific causal link between asbestos and asbestosis must exist, and then it is a question of whether the particular asbestosis patient contracted the disease because of the concrete asbestos plant.

IV. Types of liability

1. Fault liability

(a) Special rules governing fault

If the fault principle applies, then the burden of proof is reversed according to the general rule set in Art. 131 of the Code of Obligations – he who causes damage to another must compensate it, except if he proves that he is not at fault. This requires that the defendant shows that he acted with due care either as a prudent person or as a prudent professional. 17

(b) Impact of specific rules of conduct

The rules on the burden of proof are the same; in such case only the standard of care due is set out in the statute or by the customary rule. According to the standard set out in the statute, the standard makes it easier to define the care due, which is now a technical standard. Art. 16 of the Act on GMOs requires that the risk assessment must be made before using the GMO in a closed system. Approval by the responsible Ministry should reduce the risk of an accident. However, when it comes to the question of liability for damage, it is the standard of care due for a prudent professional which applies to the defendant. 18

2. Product liability

(a) Development risk defence

There are no data available to answer how this aspect would be handled in practice.

(b) Alternative routes

19 N/A.

(c) Impact of compliance with rules and regulations

20 In such case the defendant cannot be held liable because there is no illegality in his action.

3. Environmental liability

(a) Implementation of the Environmental Liability Directive

21 Slovenia was actually held liable by the European Court of Justice because it failed to implement the Environmental Liability Directive.

(b) Environmental liability regime beyond the scope of the Directive

22 There is only the general one.

(c) Claimants in cases of environmental harm

23 In cases where there is an environmental impact, the anthropocentrical approach is still the only one. According to the Slovenian Constitution, everyone has a right to a healthy environment. Being another personality right, violations of this right to a healthy environment are remedied. According to this approach, the victim whose personality right is infringed is entitled to compensation.

(d) Special liability regime for losses sustained by individuals

24 There is a nuisance regime under the Property Act (Art. 5). According to this rule, neighbours must refrain from causing nuisance to one another. The standard of nuisance is measured according to the local custom. According to case law, there is still some doubt whether governmental approval overrides the local custom regime.

(e) Cartagena Protocol

Slovenia ratified the Cartagena Protocol on Biosafety to the Convention on Biodiversity in 2002. As a ratified multilateral treaty it is directly applicable in Slovenian law, according to Art. 8 of the Constitution. **25**

4. Other strict liability regimes

No.

V. Vicarious liability

1. Scope of vicarious liability

The general provision on vicarious liability is regulated in Art. 147 of the Code of Obligations. This article is entitled "Liability of employers". However, this article is not only applicable to employers but also to other vicarious relations. The type of liability in these cases is strict. The employer is strictly liable for the damage caused by his employee unless the employer proves that the employee acted with due care. The victim can also demand compensation from the employee directly if the employee acted intentionally. **26**

2. Liability for people further up the food or feed production chain

According to the general provision of Art. 186 par. 3 of the Code of Obligations, all of them are liable jointly and severally if it is not proven that one of them contributed more to the final damage than people further up. **27**

3. Can someone further down the feed or food chain include someone further up it in a trial against him/herself?

I can only speculate on the answer to this question. It could be argued that the farmer can involve the seed producer; he could exculpate himself if he proves that he acted according to the standard of care, not knowing about the "bad seed" produced by the seed producer. **28**

VI. Multiple tortfeasors

29 The Slovenian Code of Obligations defines multiple tortfeasor liability in Art. 186.[2] It rules that all participants are to be held jointly liable for the damage if this damage was caused by multiple tortfeasors who acted together. *Ex lege* accomplices, inciters (instigators, agitators) or any person who assisted the tortfeasors to avoid discovery are jointly liable for the damage. Art. 186 – joint liability is also applicable in cases when:

- tortfeasors acted independently of each other and their contributions to the damage can not be assessed exactly;
- there is no reasonable doubt that the damage was caused by at least one tortfeasor within a connected group of persons, although it can not be determined exactly which of them it was.

VII. Defences

1. Licence/permission to grow GM material

30 There is no support in practice to give a positive answer.

2. Consent/assumption of risk

31 According to the general rule on consent (Art. 140, par. 1, of the Code of Obligations), one who gives approval for his own damage to an act of the tortfeasor cannot then claim compensation for the damage incurred. However, par. 2 of Art. 140 provides that such consent is not valid if the damaging act is prohibited by law.

2 Art. 186 (Joint liability):
"(1) Participants are to be held jointly liable for the damage if this damage was caused by multiple tortfeasors who acted together.
(2) Accomplice, inciter, or any person who assisted the tortfeasors not to be caught are jointly liable for the damage.
(3) Joint liability is imposed on those who caused the damage and acted independently from each other if it is not possible to determine their share on the caused damage.
(4) When there is no doubt that the damage was caused by one of the two or more persons who acted in any connected way, and it is not possible to determine who caused the damage, then joint liability is imposed."

3. Third-party influence

The act of a third party is one of the exculpation reasons according to Art. 153. Therefore, the defendant cannot be held liable if he proves that the damage occurred exclusively because of an act by a third party. The condition is that the defendant could not have expected the consequences, nor have excluded or removed them. **32**

4. Prescription

There are no specific statutes that would set limitation periods. The absolute time limit is five years from the time the damage was produced, and the subjective time limit is three years. **33**

5. Other defences

There are no other relevant defences. **34**

VIII. Remedies

1. Pecuniary compensation

(a) Bodily harm

There are no special rules. **35**

(b) Property losses

There are no special rules. **36**

(c) Economic losses

There are no special rules. **37**

(d) Harm to animals

38 There are no special provisions on that, so general tort rules would apply. It could be argued that the loss of a cow, e.g., would result in a loss of a profit issue. However, there is no support to argue that contaminated feed constitutes recoverable damage.

(e) Costs of disposal

39 There are no special rules and no support in general tort law practice.

2. Non-compensatory damages

40 No specific non-compensatory damages are recognised in tort law practice.

3. Other remedies

41 N/A.

4. Costs of pursuing a claim

(a) General cost rule

42 The loser pays principle applies, resulting also in shared costs (for example, 80:20) especially in tort law cases, for example due to contributory negligence.

(b) Costs of establishing causation

43 There is no relevant case law that would support the answer.

5. Advance cover

44 N/A.

IX. Cross-border issues – Conflict of laws

1. Conflicts rules applicable before (or instead of) Rome II

The conflicts of law regime remained the same as before – as described in the next answer. 45

2. Special regime for cross-border claims

There is no special bilateral or multilateral cross-border compensation scheme in Slovenian jurisdiction. Slovenia being a party to the Hague Convention on the Law Applicable to Products Liability will continue to apply that regime (cf. Art. 28 Rome II). 46

In conflicts of laws, the general rule of tort law set out in the Act on International Private Law previously applied. Art. 30 of that Act set out that in respect of tortious liability, the law of the country where the damage was caused is applicable. If such works in favour of the defendant, the applicable law could also be the law where the tortious consequence resulted. However, this did not apply if the tortfeasor could not foresee the place of the damage. 47

According to Art. 55 of the Act on International Private Law, Slovenian courts have jurisdiction in tort cases if the tortious act at issue was committed in Slovenia or if (as mentioned above) the tortious consequence resulted in Slovenia. 48

X. Cases

1. Due to the adventitious presence of GMOs in a field, maize which is normally sold as conventional contains GMOs beyond the legal labelling threshold. This is not discovered before the final stage of the food production chain by the producer of taco chips. The whole production is lost since the supermarket chains refuse to accept delivery from the producer.

(a) Who can sue along the chain of distribution?

I can not imagine that this relationship would not be contractually regulated. Therefore, I see this case as a potentially contractual case and not a tort law case. If we concentrate only on tort law, both the final producer 49 and the wholesaler are potential applicants. Both could claim loss of profit.

(b) Would the case be solved differently if the GMO content was below the labelling threshold?

50 No.

(c) Would the case be solved differently if the admixture was not adventitious, but occurred due to the disregard of segregation rules, for example?

51 Yes, in this case the liable person would be the party who violated the segregation rules.

(d) Would the case be solved differently if the GMO found was not admitted for production in your jurisdiction?

52 No.

(e) If the admixture had occurred on a non-GM field and it transpires that the GM seeds were blown from: (i) neighbouring fields; or (ii) a truck passing by, would the farmer of the affected field be liable for all or part of the loss caused further down the distribution chain?

53 In this case Art. 153 of the Code of Obligations would apply: specifically par. 2, according to which the maize grower (if such is the defendant) can exculpate himself if he proves that the organism was genetically modified due to an act of a third party.

2. Twenty years after the sale of GM maize used for food products, it turns out that it has certain disadvantageous health effects for humans.

(a) Can the producers be held liable at this point for risks unknown at the time of growing the maize? Who would be liable?

There is no support in general practice to answer this question. There have never even been tobacco claims brought to Slovenian Courts. The only pertinent experience is with asbestos. In this context, the State passed a law with a special compensation regime which allows the applicants to recover through an out-of-court mechanism. **54**

(b) Can compensation already be claimed at a point when the negative health effects have not yet materialised, but are to expected to according to scientific expertise/mere rumours?

As I argued the two-step approach must be taken into consideration. The first step is scientific. Going back to the tobacco cases, although there is scientific proof of the damaging effect of smoking, no claims have yet been brought to the court. Maybe the negative health effect has not materialised yet, or the patients are not aware of the possibilities. Therefore, it can be argued that merely scientific proof is not enough to claim compensation. Moreover, the second step must also be fulfilled, namely that the particular reason caused the concrete consequence. **55**

(c) Would it make any difference if the GM maize had only been in use for feed, causing harm to the animals, which may or may not cause harm to humans consuming the meat as well?

No. **56**

3. The driver of a food logistics company discovers that a farmer, from where he regularly picks up agricultural products fails to obey mandatory segregation rules or food or feed hygiene standards, which may lead to the admixture of GM and non-GM produce sold and packaged separately by that farmer. Does he or his employer have a duty to warn, i.e. warn the recipient of the allegedly "non-GM" produce?

57 In Slovenian law, this question would more likely be solved within the framework of criminal law. According to Art. 184 of the Penal Code, the production and marketing of "food and other products which could be damaging to health" is a criminal offence. Hence, anyone who produces, sells or in some other way places on the market food which could be dangerous and thus causes danger to human lives or health can be sentenced to up to three years imprisonment. If such an action is taken negligently, then the penalty is up to one year imprisonment.

58 Since such an action is a criminal offence, it could be argued that the duty to warn is based on a general duty to prevent eventually punishable actions. Everyone should report a criminal offence they have witnessed. On the other hand, the same rule would also apply in civil law, that everyone who is aware of a risk of future harm should warn and try to prevent it. But in such a case, I would argue that there cannot be any civil law consequence for the driver if he did not warn the farmer.

Damage Caused by GMOs under Spanish Law

Miquel Martín-Casals and Albert Ruda*

I. General overview

1. Special liability or redress scheme for GMOs

As is well known, the cases of pollution of conventional crops with GM material are creating problems for the farmers in those countries where the use of such material is allowed. Spain is no exception. It was the first European country to authorise commercial farming of GM crops by allowing several varieties of GM maize to be planted since 1998. It is also the only European country where Bt maize is farmed on a significant scale with commercial aims, occupying approximately 11% of the cultivated area.[1] Although it seems that during the years there have been no remarkable problems of coexistence, several cases of genetic pollution have been detected and the affected farmers have allegedly suffered substantial losses. To date there is no specific liability regime concerning this type of pollution and the affected farmers seem to have the feeling that they are being neglected and lack the protection of the authorities.[2] However, the litigation derived from these problems, if any, has still not reached the Supreme Court and there does not seem to be claims in tort for damage caused by GMOs on a relevant scale. In spite of this, it seems that there have been some agreements out of court whereby seed producers have agreed to pay compensation for damage caused to farmers by defective 1

* This paper was drafted within the framework of the Spanish Ministry of Science and Technology grant for the project 'The Principles of European Tort Law, beyond the so-called "Common Frame of Reference" (CFR)' (FFI2008-00647 R & D), which is directed by Prof Dr Miquel Martín-Casals (Girona).

1 According to *J. Messeguer/E. Melé*, Coexistencia de cultivos tradicionales y genéticamente modificados, in: E. Muñoz (ed.), Organismos modificados genéticamente (2006) 267.

2 As *F. Casero Rodríguez*, En defensa de la agricultura ecológica, in: R. Herrera Campos/M.J. Cazorla González (eds.), Sociedad de consumo y agricultura biotecnológica (2006) 165 explains.

seed. The Public Administration and farmers unions would have acted as mediators in the conflict between producers and farmers.[3]

2 As explained in our previous report,[4] under Spanish law, the legal framework on GMOs is provided by the Act 9/2003, of 25 April, on the legal regime of the confined utilisation, voluntary release and commercialisation of genetically modified organisms.[5] This Act – also known as the Biosecurity Act, hereafter the GMO Act – was developed by the Government through a so-called General Regulation which was passed on 30 January 2004.[6] The Act and the Regulation transposed the Directives 1998/81/EC and 2001/18/EC into Spanish law.

3 The GMO Act establishes the distribution of statutory competences between the State and the Autonomous Regions or Communities; the requirements and procedures to carry out activities of confined utilisation, voluntary release and commercialisation of GMOs; the rules on information, surveillance and control of these activities; the violations and penalties, as well as the composition and competence of the Inter-Ministerial Council on GMOs and the National Commission of Biosecurity, among other things. As regards liability, the Act lays down a duty to restore damage in the case of infringement of its provisions. If the liable person does not restore damage on their own, the Public Administration is entitled to carry out the restoration at the offender's expense (Art. 38). The Act adopts a public law approach, since it does not give standing to private parties to enforce its provisions. Also, there is no definition of damage, but a rather imprecise reference to restoration of the previous state of affairs.[7] Therefore, in theory it is possible that the Public Administration force the offender to restitute damage to private interests. The administrative decision may afterwards be challenged by affected private

3 According to the information provided by *J.D. Téllez de Peralta,* La responsabilidad por semillas, semilleros, y transgénesis, Revista de Derecho Agrario y Alimentario 2000, 20. However, the author does not make clear whether such agreements refer to damage caused by GM seed or not.

4 *M. Martin-Casals/A. Ruda,* Spain, in: B.A. Koch (ed.), Economic Loss Caused by Genetically Modified Crops (2008) 443 ff.

5 *Ley* 9/2003, de 25 abril, *de régimen jurídico de la utilización confinada, liberación voluntaria y comercialización de organismos modificados genéticamente* (BOE, Boletín Oficial del Estado, Official Gazette no. 100, 26 April 2003, 16214).

6 *Real Decreto* 178/2004, de 30 de enero, *por el que se aprueba el Reglamento general para el desarrollo y ejecución de la Ley 9/2003, de 25 abril, de régimen jurídico de la utilización confinada, liberación voluntaria y comercialización de organismos modificados genéticamente* (BOE no. 27, 31 January 2004, 4171).

7 Such lack of precision also affects other aspects of the legal regime on GMOs. See *M. Tarrés Vives,* Bioseguretat i normes tècniques, Revista Catalana de Dret Públic 2008, 139.

parties before the courts of justice according to the general rules. The broad concept of damage could also include damage to the environment as such, also called pure ecological damage. For instance, the use of genetic engineering to make a plant resistant to a specific kind of insect (so-called "target insect") could cause damage to other live organisms that have contact with the plant. As is known, the case usually discussed in this regard is the use of Bt maize, which is poisonous for the European corn borer but could also be detrimental to the Monarch butterfly.[8]

Therefore, the GMO Act is very broad and it is not focussed specifically on any particular kind of damage caused by GMOs. Although the Spanish Government has been working on a draft regulation on the so-called coexistence of GM and non-GM crops,[9] the liability issues are usually left aside, so there is still no specific liability regime on the pollution of crops with GM material from neighbouring crops.[10] Moreover, a similar act or omission may amount to an infraction of the GMO Act and other statutory rules, such as for instance the Act on the protection of plant varieties.[11] The latter lays down a liability rule referring only to the infringement of the rights of the holder of the variety (Art. 22). Penalties are established for specific administrative violations, such as in the case of the transfer of plant material protected by a title of plant variety which does not match the features which appear in its official description (Art. 29.2.a). However, this Act does not confer any specific powers on the Public Administration to enforce the liability rule for damage to the title holder. Therefore, the liability rules of the GMO Act may be applied to damage not covered by this Act. Also mention can be made of the Act on seeds,[12] which lays down another specific regime on administrative violations "without prejudice to the civil or criminal actions" (Art. 57). One of the violations may be the lack of indication or the lack of exactitude of the labels of the seeds containers (Art. 61.c). However, the Act does not specifically address the problems related to GMO pol- 4

8 On these risks see *F. García Olmedo*, La biotecnologia davant dels reptes de la nova agricultura, in: L. Argemí d'Abadal/M. Rodríguez Rodríguez-Zúñiga (eds.), L'agricultura moderna: de l'alimentació al medi ambient (2004) 168.

9 *Coexistence* is understood as the possibility that farmers choose between conventional production, ecological production or production of GM crops, complying with the statutory obligations on labeling and pureness.

10 See the criticism for this reason by *M.J. Cazorla González*, Comentario al proyecto de Real Decreto sobre coexistencia de cultivos modificados genéticamente con los convencionales y ecológicos, in: R. Herrera Campos/M.J. Cazorla González (eds.), Sociedad de consumo y agricultura biotecnológica (2006) 152.

11 *Ley* 3/2000, de 7 de enero, *de régimen jurídico de la protección de las obtenciones vegetales* (BOE no. 8, 10 January 2000).

12 *Ley* 30/2006, de 26 de julio, *de semillas y plantas de vivero y de recursos fitogenéticos* (BOE no. 178, 27 July 2006).

lution. Moreover, it has been suggested by some scholarly opinions that the liability rule of the Seeds Act does not meet the needs of the farmer, since restoration will arrive too late due to the usual length of the court trials.[13]

5 An approach similar to that of the GMO Act has been adopted by the Act on Environmental Liability 2006 (LRM),[14] which transposes Directive 2004/35/CE of the European Parliament and of the Council of 21 April 2004 on environmental liability with regard to the prevention and remedying of environmental damage[15] into Spanish law.[16] Private parties are explicitly denied standing to bring a claim on the basis of this Act (Art. 5). Following the Directive, the Act provides for a public law mechanism of restoration and prevention of damage to the environment. The key element is the definition of damage, which also follows the provisions of the Directive in general terms. However, the Spanish legislature has gone beyond the Directive in certain aspects. Specifically, damage to the soil is covered by the liability provisions of the Act not only in cases where pollution of the soil creates a significant risk of adverse effects on human health, (Art. 2.1.c) Directive) but also in case of significant risk of adverse effects on the environment (Art. 2.1.d) LRM). Moreover, damage to wild species and habitats falls within the new liability regime, without the requirement that they are already protected by European law (Art. 2.1.a)). In addition to damage to water, damage to the seashore and those long, narrow, tidal inlets known as "rías" also triggers liability according to the new regime (Art. 2.1.c)). In general terms, the statutory definition of damage is rather complex and unpractical. Moreover, it has to be taken into account that some subspecies of damage to the environment are defined by way of reference to a different Act or even to a European Directive, which contributes to make the new statutory regime overly intricate. This is in striking contrast with the general rule of liability for fault laid down by the Spanish Civil Code (Art. 1902), which does not actually define damage. In addition to this, for liability to be established according to the Environmental Liability Act, it is required that the damage is serious (Art. 2). This is not a condition for liability according to the Civil Code.[17]

13 In this vein *Cazorla* (fn. 10) 153.

14 *Ley* 26/2007, de 23 de octubre, *de Responsabilidad Medioambiental* (BOE no. 255, 24 October 2007).

15 Official Journal (OJ) L 143, 30.4. 2004, 56–75.

16 Summarily see *J. Ribot*/*A. Ruda*, Spain, in: H. Koziol/B.C. Steininger (eds.), European Tort Law 2007 (2008) 541 ff.

17 See *A. Ruda*, Las relaciones entre la responsabilidad ambiental conforme a la Directiva 2004/35/CE y la responsabilidad civil, in: A. Pigrau Solé (ed.), Nuevas perspectivas de la responsabilidad por daños al medio ambiente (2006) 97.

The Spanish legislature has also enlarged the scope of the new regime in a different way. The operator is obliged on a strict basis to adopt preventative measures in the presence of a threat of damage, as well as to avoid creating further damage. It is not required that the economic or professional activity which caused such threat fits within the list established by the Annex III of the Act (which is similar to the same Annex of the Directive; Art. 17.1). However, the operator will have the duty to restore damage only if his activity is included in the list (Art. 19.1). This liability rule is also strict. If the activity is not in the list, the duty to adopt restoration measures will exist only if the operator behaved in a negligent fashion (Art. 19.2). This latter rule applies in cases of environmental damage as defined by the statute, and – in contrast with the criterion of the Directive – not only in case of damage or threat of damage to wild species and habitats. It has to be kept in mind that the Spanish Civil Code lays down a liability rule for damage deriving from toxic emissions and fumes (Art. 1908). Such a provision, which has not been affected by the LRM, has been construed by the courts as laying down a general rule of strict liability for pollution damage[18] and is considered by some scholars to provide for an appropriate solution in case of traditional damage caused by environmental interferences or disturbance.[19] From this point of view, the new regime may offer less protection than general tort liability rules. Apart from this, damage caused by GMOs falls within the scope of the new statutory regime (Annex III, no. 12), as the Directive had already established.[20] However, the Act provides that damage to crops caused as a result of the release of GMOs will be compensated according to the rules of private tort law (Additional Disposition no. 4). As a matter of fact, such a provision was not necessary, since damage to crops does not fall within the definition of environmental damage laid down by Art. 2.1 LRM and – as has been said – claims by third parties cannot be founded upon LRM (Art. 5). The separate liability regime laid down by the GMO Act has therefore been left unaffected. 6

Finally, the Acts mentioned do not affect the criminal liability of the offender, which is governed by the Criminal Code (CP).[21] According to this, the manipulation, transport or possession (*tenencia*) of organisms which violate 7

18 For further details, see *A. Ruda*, El daño ecológico puro (2008) 409.

19 See for instance *M.J. Santos Morón*, Acerca de la tutela civil del medio ambiente: algunas reflexiones críticas, in: A. Cabanillas Sánchez et al. (eds.), Estudios jurídicos en homenaje al profesor Luis Díez-Picazo II (2003) 3019.

20 On this issue see now *R. Herrera*, Responsabilidad derivada de los daños producidos por la biotecnología (2007).

21 *Ley Orgánica* 10/1995, de 23 de noviembre, *del Código Penal* (BOE no. 281, 24 November 1995).

the established norms or measures of safety, putting at concrete risk the life, physical integrity or the health of persons, or the environment, amounts to a crime (Art. 349). The penalty established is imprisonment of 6 to 24 months, a financial punishment and special disqualification from a public occupation or post. Legal scholarship makes clear that although the word "organisms" has a very broad meaning, the legislature wanted to refer to behaviours related to genetic modification of organisms and the risks derived from it.[22] Otherwise the possibility would exist that the offender's behaviour was punishable because of an infringement of the Criminal Code, without at the same time having infringed the GMO Act.[23] However, the truth is that the literal wording of the provision is so broad in meaning and the reference to the "established norms or measures" so imprecise that it has even been suggested that it infringes upon the Constitution.[24] Oddly enough, the provision does not mention production and commercialisation of GMOs, which, as has been seen, are activities mentioned by the GMO Act.[25] Also, the Criminal Code does not require that the GMOs are for human consumption and neither does the GMO Act. In any case, such a provision would be a compromise from a criminal law point of view of the GMO Act. Therefore, the definition of organism in the sense of Art. 349 CP has to be arrived at by reference to this Act, which means that the "organism" which is manipulated, transported, or possessed, has already been modified genetically; this is the prevailing opinion. Therefore, the first genetic manipulation of the organism at stake falls outside of the scope of this provision. Both the national rules and the European Directives may be taken into account to apply the criminal law provision.[26] Moreover, although not explicitly required by the Criminal Code, it is understood that the risk created by the offender must affect a large and undetermined group of people.[27] If the conditions of application of the Criminal Code provision are met, not only will the fine established be imposed on the offender, but he will also have the duty to compensate for damage caused by

22 See *L. Escajedo San Epifanio*, Derecho penal y bioseguridad: los riesgos derivados de organismos modificados genéticamente, in: C.M. Romeo Casabona (ed.), Genética y Derecho penal. Previsiones en el Código Penal Español de 1995 (2001) 267 ff.

23 As *C. Villacampa Estiarte*, Comentario, in: G. Quintero Olivares (ed.), Comentarios al nuevo Código Penal (4th ed. 2005) 1831 observes.

24 See *M. Corcoy Bidasolo*, Les conductes il·lícites relatives a la genètica, in: Consejo General del Poder Judicial (ed.), Bioètica i Dret (2004) 131.

25 See the criticism by *M. Cugat Martí*, Comentario, in: J. Córdoba Roda/M. García Arán (eds.), Comentarios al Código Penal. Parte especial II (2004) 1503.

26 See *S. Romeo Malanda*, Bioseguretat: l'avaluació del Dret comunitari i la seva recepció a l'ordenament jurídic espanyol, in: Consejo General del Poder Judicial (ed.), Bioètica i Dret (2004) 336.

27 See *M. Serrano Pascual*, Comentario, in: E. Bacigalupo et al., Código Penal II (1997) 3342.

him (pursuant to Art. 109.1 and 116.1 CP). The affected party may prefer to bring his claim in tort before the private law courts instead (Art. 109.2 CP). Such a duty to compensate for damage caused includes restitution (Art. 110.1), restoration (Art. 110.2) and compensation for material and non-pecuniary losses (Art. 110.3) both to the aggrieved party and third persons (Art. 113).

2. State liability

Under Spanish law, liability of the state is governed by a special statutory regime, namely the Act on the legal regime of public administration and general administrative procedure (LRJPAC).[28] According to this, the state is liable for damage caused by either the normal or abnormal functioning of the public services, except in the case of force majeure (Art. 139). The burden of proof of force majeure lies on the Public Administration. Such liability is strict, in contrast with the general rule of tort liability which, as above-mentioned, requires fault (Art. 1902 CC).[29] Therefore, no sovereign immunity applies according to these legal regimes. This may be important since, as has also been said, it is the Public Administration that may oblige the liable party to restore damage caused by GMOs according to the GMO Act. **8**

Apart from this, it is required for liability to be established according to the LRJPAC that the affected individual brings evidence as to how damage affects him or her on an individual basis, i.e. damage has to be individualised. Therefore, collective damage does not trigger liability under this special regime. The act clearly requires that the victim proves that damage affects "his or her goods or interests" (Art. 139), which is construed as private or traditional damage.[30] In many situations it may be impossible to determine to what extent specific individuals have been harmed, unless for instance it is possible to state that the residents of a building are affected by noise produced by neighbouring industry. Damage can be individualised if it can be proved that it affects a person or a group of individuals only, but not if it affects the public as a whole.[31] Therefore, it may **9**

28 *Ley* 30/1992, de 26 de noviembre, *de Régimen Jurídico de las Administraciones Públicas y del Procedimiento Administrativo Común* (BOE no. 285, 27 November 1992).

29 Instead of many see *J. González Pérez*, Responsabilidad patrimonial de las Administraciones Públicas (3rd ed. 2004) 186.

30 See N. *Álvarez Lata*, Daños al medio ambiente, in: F. Reglero Campos (ed.), Tratado de responsabilidad civil III (4th ed. 2008) 113 and *Santos* (fn. 19) 3028.

31 See *González Pérez* (fn. 29) 211 ff. and 328.

be rather difficult, if not impossible, to file a claim for environmental damage as such on the mere basis of this provision.

10 As for the statutory regime on environmental liability, it has already been mentioned that it lays down a public law mechanism of liability where the role of pursuing the liable party and forcing him or her to restore damage is assigned to a public authority only (Art. 2.22 LRM). Liability of the state itself is not expressly excluded, so in theory it is possible that the public authority in charge start a liability procedure against a public administration on the basis of the Environmental Liability Act. This possibility is reinforced by the fact that some Public Administrations have filed claims against other Administrations for damage to the natural resources belonging to the first. For instance, this is the case with municipalities who claimed against the Ministry of Defence for damage caused by military training with fire arms.[32] Moreover, it is provided that liability according to the LRM will be excluded whenever damage derives from an activity aimed at national defence or international security, as well as activities exclusively aimed at protecting against natural disasters (Art. 3.4.c). Such an exclusion clause has been criticised on the basis that it equates a sovereign immunity which is foreign to our legal system and could even infringe the Spanish Constitution (Art. 106).[33]

II. Damage

1. Recoverable losses

11 As has been said, the GMO Act does not define damage but provides that the offender will have the duty to restore the state of affairs previous to its causation (Art. 38). Since the public administration is entitled to start proceedings against the offender arising from the violation of the statutory provisions, the probable outcome is that the administration would assess all the damaging consequences arising from such a violation. The Act lays down some criteria for the assessment of damage which will be

32 See *G. Valencia Martín*, El impacto (favorable) de la Directiva 2004/35/CE en el "sistema" español actual de responsabilidad por daños ambientales, in: J. Jordano Fraga/J.F. Alenza García/E. Orteu Berrocal/G. Valencia Martín/R. Márquez Molero/Á. Ruiz de Apodaca Espinosa, Estudios sobre la Directiva 2004/35/CE de responsabilidad por daños ambientales y su incidencia en el ordenamiento español (2005) 136.

33 In connection to the Environmental Liability Directive see *J. Jordano Fraga*, La responsabilidad por daños ambientales en el Derecho de la Unión Europea, in: Jordano/Alenza/Orteu et al., Estudios sobre la Directiva 2004/35/CE de responsabilidad por daños ambientales y su incidencia en el ordenamiento español (2005) 27.

applicable whenever it is difficult to assess, namely, the theoretical cost of restitution and restoration, value of the damaged goods, cost of the project or activity which caused damage, and benefit derived from the infringing activity. These criteria may be applied as a whole or separately (Art. 38.1 par 2). Such criteria are in accordance with other statutory provisions which lay down similar liability rules, pursuant to which the person who infringes a statutory provision may be obliged by the Public Administration to restore damage.[34] In these cases, the Public Administration usually links the administrative violation with the liability of the offender and issues a decision which includes both aspects.[35] This may cause some concern since both liability to the Public Administration itself and damage to private parties may be assessed by the Administration according to such criteria. However, claims by private parties should be the object of a judicial procedure.[36] Nevertheless, the damage assessment carried out by the Public Administration can be reviewed by the courts of justice, since the person who disagrees may challenge by filing a judicial claim according to the general rules. Actually, the GMO Act follows the trend already set by previous law, in the sense that the Public Authority has an expanding power in connection with the protection of the environment and the possibility to request the restoration of natural resources regardless of whether they are privately owned or not.

The same can be said in connection with the provisions of the Environ- **12**
mental Liability Act. Assessment of damage is again carried out by the public authority (Art. 45.2.b). However, such an assessment does not aim at establishing a sum to be paid by the liable party in compensation of damage caused. As a matter of fact, the primary objective of the Act is restoration in kind (so-called primary restoration, pursuant to Annex II LRM). If restoration in kind is incomplete, it will be complemented by other measures in connection with natural resources or natural services (so-called complementary restoration). No financial compensation will be paid to the public for interim losses, but other measures will be adopted in the same natural resources affected or in other areas.

As for general liability rules as established by the Spanish Civil Code, any **13**
kind of damage is recoverable provided that all the conditions of liability – and fault in particular where needed (Art. 1902) – are met. Personal injury, damage to things, economic and non-pecuniary losses are recoverable,

34 See *J. Conde Antequera*, El deber jurídico de restauración ambiental (2004) 85 ff.
35 See *S. González-Varas Ibáñez*, La reparación de daños causados a la Administración (1998) 228.
36 See *Martín-Casals/Ruda* (fn. 4) 446 no. 8 with further references.

regardless of whether they are directly or indirectly caused by the liable party. If he did not have the intention to create damage, liability will be limited and he will only be liable for those consequences of the activity which were foreseeable (Art. 1107 CC).

2. Pure economic loss

14 Spanish law does not contemplate a separate category of pure economic loss. As already explained, the concept of damage used by the general provisions on tort liability (Art. 1902 ff. CC) is not limited on the basis of the nature or scope of the protected rights or interests. In principle, any kind of rights or interests are protected by tort law, provided that they are legitimate. Only very recently has Spanish legal writing started to refer to the concept of pure economic loss, as a concept used in other legal systems and foreign to Spanish law, but with reference only to some very specific issues.[37] Therefore, pure economic loss can be recovered under the general rules of tort law and under the same conditions. However, the courts seem to be very demanding as to the proof of the causal link between alleged pure economic loss and the conduct of the tortfeasor.[38]

3. Mere fear of a loss

15 With regard to non-pecuniary loss, Spanish law has adopted a quite broad approach, probably due to the influence of the French Civil Code on the Spanish one. In contrast to other systems which establish restrictions concerning compensation of such kind of loss, the courts consider mental trauma a recoverable damage, regardless of whether it amounts to a psychiatric illness or not. In connection with nuisance consisting in noise affecting neighbouring land, anguish, the feeling of impotence, anxiety or mere uneasiness have amounted to compensable harm.[39] Compensation for non-pecuniary loss does not require a bodily or mental injury. On the contrary, it covers damage that does not medically qualify as an illness. In nuisance cases, Spanish courts have even admitted compensation for harm consisting only in the environmental detriment which impairs

37 See M. *Martín-Casals*/J. *Ribot*, Compensation for Pure Economic Loss under Spanish Law, in: W.H. van Boom/H. Koziol/Ch.A. Witting (eds.), Pure Economic Loss (2004) 62.

38 See A. *Ruda*, Spain, in: M. Hinteregger (ed.), Environmental Liability and Ecological Damage in European Law (2008) 475.

39 See *Ruda* (fn. 38) 575 with further references.

the quality of life of the claimants.[40] With regard to oil pollution, it has even been suggested that suffering caused by the viewing of the consequences of an oil spill may amount to non-pecuniary loss, although obviously in these cases difficulties of damage assessment may arise.[41]

Courts have not yet dealt on a relevant scale with tort claims for the mere fear of suffering harm before physical damage has become manifest. Certainly there have been some decisions in cases of pleural plaques caused by asbestos fibres, but non-pecuniary loss for the anguish or fear of falling ill has not yet been discussed by the courts. Legal scholarship has recently started to pay attention to these cases in the light of developments in other countries.[42] An opinion has suggested that the claim for non-pecuniary loss would have been dismissed anyway on the basis that such damage is not certain.[43] However, some authors consider that compensation would be awarded in cases of mere fear of suffering future damage as well.[44] This view is more consistent with the inclusive approach of Spanish courts, since even the unrest, sorrow, pain, anxiety or mental suffering, stress, difficulties to sleep or memorise, loss of mental concentration or difficulties to communicate orally, feelings of impotence, uneasiness or the lack of quality of life are enough for compensation to be awarded on the basis of non-pecuniary loss.[45] Remarkably enough, the Supreme Court has stated that the mere situation of latent risk for human health caused by pollution amounts to non-pecuniary loss (STS 16 January 1989 [RJ 1989, 101]).[46] In any case, non-pecuniary loss has to be proved. **16**

4. Standard of proof

In order to obtain compensation, damage has to be certain, i.e. the victim is required to bring forward evidence as to its existence. In the case of future damage, compensation is only awarded provided that it is certain **17**

40 See *J. Ribot*, Spain, in: H. Koziol/B.C. Steininger (eds.), European Tort Law 2004 (2005) 545.

41 As *M.P. García Rubio*, A indemnización dos danos causados polo *Prestige*, in: E.J. Torres Feijó (ed.), Prestige: a catástrofe que despertou a Galiza? (2003) 118, points out.

42 See *A. Azagra Malo*, La tragedia del amianto y el derecho español (2007) 92 ff.

43 *A. Azagra Malo*, Placas pleurales, angustia e incremento de riesgo, InDret 2008, 15 ff.

44 *M. Martín-Casals/J. Ribot/J. Solé*, Spain, in: W.V.H. Rogers (ed.), Damages for Non-Pecuniary Loss in a Comparative Perspective (2001) 202.

45 See, in connection with nuisance, *Ruda* (fn. 18) 216 with references to case law.

46 However, the court finally dismissed the claim on purely procedural grounds, because this issue had not been raised before the lower courts. See *A. Ruda*, La tipología de los daños causados por el *Prestige*, in: M.P. García Rubio/S. Álvarez González (eds.), La responsabilidad por los daños causados por el hundimiento del Prestige (2007) 63.

that damage will occur in the future. Therefore, the judge will take into account the loss of earnings that the victim will experience in the future and will award compensation for such losses. To this end, it is required that there is a high probability that such losses will take place, so the certainty requirement can be met.[47]

5. Nominal losses

18 Courts have the duty to explain the criteria according to which assessment of damage is carried out and how the compensation award is established. Sometimes the courts merely refer to prudent and reasonable criteria without fully explaining the reasoning which leads to the amount of damages awarded. However, the Constitutional Court has reminded the courts of justice on several occasions that their decisions have to be sufficiently reasoned (Art. 120.3 Spanish Constitution), so they should specify in an explicit manner the criteria which led to an amount being set as compensation. The lack of a sufficient grounding of damage assessment by the court has recently been discussed in connection with symbolic damages. In particular, the Constitutional Court blamed the Supreme Court for having granted damages which could well be qualified as symbolic or merely nominal.[48] Such damages are not allowed under Spanish law, not even in the case of compensation for non-pecuniary loss, since they are insufficient to restore the loss suffered by the victim (STC no. 300/2006, 23.10.2006).[49]

6. Mass losses

19 There are some special rules in connection with a sort of class action (*acciones de clase*) provided by Spanish law, although they are actually not entirely coincident with the class actions of US federal law.[50] Several conditions have to be met to file such an action. First, damage has to have been caused to consumers and users (Art. 15.1 Civil Procedure Act [LEC]),[51] i.e. persons

47 See *E. Vicente Domingo*, El daño, in: F. Reglero Campos (ed.), Tratado de responsabilidad civil I (4th ed. 2008) 318.

48 See *J. Ribot*/*A. Ruda*, Spain, in: H. Koziol/B.C. Steininger (eds.), European Tort Law 2003 (2004) 389.

49 See *A. Ruda*, Spain, in: H. Koziol/B.C. Steininger (eds.), European Tort Law 2006 (2008) 438 ff.

50 See the paper by *A. Ruda*/*J. Solé Feliu*, Spain, in: K. Oliphant (ed.), Aggregation and Divisibility of Damage (2009).

51 *Ley* 1/2000, de 7 de enero, *de Enjuiciamiento Civil* (BOE no. 7, 8 January 2000).

who act in a sphere other than an entrepreneurial or professional activity, pursuant to Art. 3 of the General Act for the Protection of Consumer and Users (LGDCU).[52] This requirement prevents class actions from being used to claim compensation for collective damage other than damage to consumers or users, such as collective environmental damage or the like.[53] The second condition for a class action to be filed is that a damaging event (*hecho dañoso*) has occurred (Art. 15.2 LEC). This includes damage in tort as well as contractual damage. The mere fact that the circumstances of damage suffered by every single victim may vary does not prevent them from filing a class action, since damage derives from their having consumed a good or service and is therefore considered to be common to all of them.[54] Also any kind of damage may be the basis of a class action, be it economic or personal damage, including non-pecuniary loss. There are also no thresholds or liability caps.[55]

The decision must have a separate and explicit pronouncement on the actions filed within the collective process by consumers or users (Art. 221.3 LEC). As regards the class action, whenever the judge finds for the claimants, the decision has to define the consumers or users who will be beneficiaries of the decision on an individual basis (Art. 221.1 LEC). If it is not possible to determine the affected persons on an individual basis, the decision will have to establish the data, features and requirements needed to be entitled to the payment of the compensation award at the procedural stage of the execution of judgment (Art. 221.2 LEC). This is a key aspect of the class action, since it allows the judge a certain degree of indeterminacy as to the subjects entitled to compensation as well as to what they will be entitled to receive.[56] The judge will verify at the execution stage whether the individual consumers meet the criteria already established in the decision. This is also an important difference with individual procedures, where the judge is required to establish at least the basis which will allow the exact compensation award to be determined through a mere arithmetical operation in the execution stage (Art. 219 LEC). Therefore, a collective claim filed for an apparently scant amount for many may eventually turn into an **20**

52 *Real Decreto Legislativo* 1/2007, de 16 de noviembre, *por el que se aprueba el texto refundido de la Ley General para la Defensa de los Consumidores y Usuarios y otras leyes complementarias* (BOE no. 287, 30 November 2007).

53 See *Ruda* (fn. 18) 518.

54 See *M.I. González Cano*, La tutela colectiva de consumidores y usuarios en el proceso civil (2002) 41 ff.

55 See *J.J. Marín López*, Las acciones de clase en el derecho español, InDret 2001, 4.

56 See *C. Senés Motilla*, Las partes del proceso civil, in: I. Díez-Picazo Giménez (ed.), Disposiciones generales relativas a los juicios civiles en la nueva Ley de enjuiciamiento civil (2000) 37 ff.

execution for millions. Probably, there will be a fluid recovery of damages, which may infringe the principle of full compensation.[57]

III. Causation

1. Uncertainty of merely potential causes

21 Spanish tort law lacks a general rule for the situation where it is uncertain which of several defendants of a group caused damage to the victim but it is sure that one of them actually caused it. Certainly some statutory rules refer to such scenario but it is in the field of hunting accidents or development of buildings, so there is no equivalent to, for instance, the German § 830 I(2) Bürgerliches Gesetzbuch (BGB), in the Spanish Civil Code. Therefore, case law has played a major role in this area by developing a rule according to which all the potential tortfeasors are held liable on a joint and several basis unless they prove that they did not cause damage. Such doctrine, which is coincident with the criterion established by the aforementioned statutory rules, is usually known as *improper joint and several liability*.[58] This is also the prevailing view in legal scholarship.[59] However, case law has never formulated a general definition of the concept of group in this context.[60]

22 As can be observed, the Spanish courts adopt an "all-or-nothing" approach with regard to alternative causation. The same approach is adopted with regard to proof of the causal link in general. As a rule, it is the claimant who bears the burden of proof of such a link (Art. 217 LEC). Therefore, the victim will have to bring evidence that, for instance, a GM product caused damage. In the case of product liability, proof of the defect is also required, as well as proof that the defect caused damage. For instance, it should be proved that the lack of labelling of a product as containing GMOs was the cause of damage suffered by the victim, such as an allergic reaction. However, it may happen that the exact cause of the allergic reaction cannot be proved or it cannot be attributed to the GMO.[61] In the case

57 For this reason the rule has been criticised by *Á. Carrasco Perera/M.C. González Carrasco*, ¿Acciones de clase en el proceso civil? Aranzadi Civil 2001, 1908.

58 See *Ruda* (fn. 38) 428.

59 See *J. Solé Feliu*, Pluralidad de causantes del daño y solidaridad, Revista de Derecho Privado (RDP) 2008, 13 ff.

60 See *J. Ribot/A. Ruda*, Spain, in: B. Winiger/H. Koziol/B.A. Koch/R. Zimmermann (eds.), Essential Cases on Natural Causation (2007) 374.

61 See *J. Valls Prieto*, La utilización comercial de organismos modificados genéticamente (OMG) y el Derecho penal, in: R. Herrera Campos/M.J. Cazorla González (eds.), Aspectos legales de la agricultura transgénica (2004) 118.

of damage allegedly suffered by a farmer as a result of defective seed, it should be established that the seed did not offer the safety which could be expected from it on a legitimate basis, taking into account all the circumstances, in particular the presentation, the foreseeable reasonable use and the moment of putting the product into circulation (Art. 137.1 LGDCU). The claimant will thus bear the burden of proving that such a defect caused damage (Art. 139 LGDCU). This may be extremely difficult in the case of products which are perishable or are destroyed because of being consumed.[62] Moreover, proof may be complicated due to the incomplete scientific understanding or knowledge available on GMOs and related issues.[63] In general, Spanish courts are very reluctant to shift the burden of proof to the defendant in connection with the causal link.[64] However, it seems that in practice proof of the defect is eased in this context by not requiring a direct proof. Instead, an indirect proof or proof by presumptions is considered to be sufficient in many cases.[65]

The courts have adopted a very strict stance as to the standard of proof and require that proof of the causal link is categorical (*terminante*), so the threshold that the claimant has to reach here is an indisputable certainty. Mere conjectures, speculations or probabilistic estimations are insufficient to establish causation.[66] However, some scholars have advocated a less demanding criterion, such as proof of a "sufficient probability" that the causal link exists.[67] The use of mere statistics and probabilities is also allowed in other contexts, such as with regard to the application of the precautionary principle in order to shift the burden of proof.[68] **23**

---

62 As pointed out by *J. González García/D. Jiménez Liébana*, Aspectos preventivos y resarcitorios relacionados con los productos alimenticios transgénicos, in: R. Herrera Campos (ed.), Homenaje al profesor Bernardo Moreno Quesada II (2000) 711.

63 Among others see *A. Kemelmajer de Carlucci*, Responsabilidad civil, principio de precaución y transgénicos, in: C.M. Romeo Casabona (ed.), Principio de precaución, Biotecnología y Derecho (2004) 349.

64 See *Ruda* (fn. 49) 449 and *P. del Olmo*, Tort and Regulatory Law in Spain, in: W.H. van Boom/M. Lukas/Ch. Kissling (eds.), Tort and Regulatory Law (2007) 283.

65 According to case law analysis by *J.J. Marín López*, Daños por productos: estado de la cuestión (2001) 82.

66 See *A. Ruda*, Fairchild v. Glenhaven Funeral Services Ltd and others. Spanish case note, European Review of Private Law (ERPL) 2004, 251 ff.

67 In this vein for instance *J.F. Alenza García*, El régimen público de responsabilidad por daños ambientales en la legislación española y en la Directiva de responsabilidad ambiental, in: Jordano/Alenza/Orteu et al., Estudios sobre la Directiva 2004/35/CE de responsabilidad por daños ambientales y su incidencia en el ordenamiento español (2005) 98.

68 See *Advisory Committee on Ethics of Scientific and Technical Research*, Report/Genetically Modified Organisms in Agriculture and Food (2005) 61 and *F. Garrido Peña*, La teoría de la elección racional y biotecnología de transgénicos, in: I.F. Benítez Ortúzar (ed.), Genética humana en el tercer milenio. Aspectos éticos y jurídicos (2002) 266 ff.

24 Certainly, proof by presumption is admitted under procedural law (Art. 386 LEC) but its availability is made dependant by case law upon strict requirements.[69] Apart from this general rule, the Environmental Liability Act also lays down a rule according to which the causal link may be presumed in connection with environmental damage (Art. 3.1 par. 2). According to such rule, it is presumed that an economic or professional activity included in an Annex to the Act has caused damage or the imminent threat of damage, provided that such an activity is appropriate to cause it, with regard to the intrinsic nature of the activity or the way it has been carried out.

25 This presumption can be rebutted by the offender if he or she brings so-called "contrary evidence" (*prueba en contrario*). It should be open to discussion whether it is enough that the offender points out another potential offender, or that he or she proves that he or she did not carry out the activity at stake, or that he or she proves he or she did not cause such damage or threat of damage.[70] It also remains unclear whether the presumption rule is applicable to all kind of activities or only to those included in the Annex to the Act. Actually, a different rule provides that the Act will also be applicable to environmental damage or threat of damage caused by economic or professional activities different than those included in the Annex (Art. 3.2). However, as has been said, the presumption rule makes reference to activities included in the Annex only. It has been suggested that the rule should not be extended to other activities, since the liability rule established by the LRM is not, properly speaking, a tort liability rule. Moreover, it is thought that activities not included in the Annex create a lesser risk for the environment.[71] However, the difficulty of proving causation may exist in both situations, regardless of whether the activity at stake is included in an Annex or not. Therefore, the possibility cannot be completely excluded that the presumption rule will also be applied to activities not included in the Annex, by analogy or on the basis of the general procedural rules. In any case, it seems doubtful whether the presumption rule introduced by the Environmental Liability Rule was really necessary. As a matter of fact, causation has not been an insurmountable hurdle in environmental trials and proof by presumptions was already provided for by general procedural rules as already referred to above.[72]

---

69 See *Ribot/Ruda* (fn. 60) 413.

70 See *A. Ruda*, Comentario a la Sentencia de 2 de noviembre de 2007, Cuadernos Civitas de Jurisprudencia Civil (CCJC) 2008, 1135.

71 In this vein *J. Esteve Pardo*, Ley de responsabilidad medioambiental. Comentario sistemático (2008) 62.

72 Similarly, see *Santos* (fn. 19) 3020 and *E. Cordero Lobato*, Derecho de daños y medio ambiente, in: L. Ortega Álvarez (ed.), Lecciones de Derecho del medio ambiente (2nd ed. 2000) 447.

2. Complex causation scenarios

26 Some legal scholars have suggested that alternative causation scenarios such as those created by DES cases should be solved by means of so-called market share liability.[73] However, as has been said it would seem very difficult to apply such a rule in Spanish law because of the all-or-nothing approach already referred to above.[74] Moreover, solidary liability seems to provide more protection for the victim so it is unlikely that courts will replace it with liability in proportion to the defendants' market share.

27 As regards alternative causation in general, joint and several liability is only established provided that the claimant shows that all the members of the group have probably caused damage to him or her, i.e., that they belong to the group of potential tortfeasors. Solidary liability will probably be established if it is not possible to determine to what extent each of the tortfeasors contributed to damage, but it is clear that all of them did. With regard to multiple tortfeasors, the legal regime on product liability provides that they will be held liable on a joint and several basis (Art. 132 LGDCU). However, the rule requires that all the tortfeasors are liable and does not particularly address the issue of alternative causation. Moreover, it requires that the liable tortfeasors have caused the same damage. Also, the product liability regime is applicable to damage to things aimed at private consumption or use. For this reason, the farmer will not be able to found his claim in tort for damage to his crop on the product liability provisions.[75] Therefore, it is probable that the victim of contamination in the hypothetical would need to resort to the case law doctrine already mentioned in connection with alternative causation and found his claim on the general tort law provisions. As a result, the victim would have to prove that it is probable that the contaminated batch was produced by one of the defendants. However, an isolated opinion has suggested that such a doctrine would not be applicable in this context. It is argued that the product liability regime does not contain a specific rule on alternative causation and, therefore, the victim could not file a claim against the potential tortfeasors.[76]

73 See for instance *Marín* (fn. 65) 100 ff.

74 See *A. Ruda*, La responsabilidad por cuota de mercado a juicio, InDret 2003, 1 ff.

75 See *D. Jiménez Liébana*, Riesgos y responsabilidades civiles en los cultivos y productos alimentarios, in: R. Herrera Campos/M.J. Cazorla González (eds.), Sociedad de consumo y agricultura biotecnológica (2006) 280.

76 See *González/Jiménez* (fn. 62) 712.

3. Force majeure

28 According to the Spanish Civil Code, force majeure excludes liability (Art. 1105). Force majeure is not defined by the Spanish Civil Code but it is considered a cause of extinction of the obligation to give a specific and determined thing (Art. 1182 CC) as well as a cause of extinction of liability for breach of an obligation (Art. 1105 CC). Such a rule is included in a general set of rules concerning all kinds of obligations so it is also applicable to obligations based on tort. Force majeure also excludes liability of the possessor of an animal which causes damage (Art. 1905 CC). If harm is partly caused by the liable party and partly due to force majeure, the former will be held liable for the part corresponding to his contribution to damage.[77] Force majeure is also a defence in most strict liability rules outside the Civil Code.[78]

29 With regard to liability of the Public Administration, force majeure is also established as a defence (Art. 106 Spanish Constitution and Art. 139 LRJPAC). Courts usually define force majeure in this context as an event which is generally unforeseeable and is always unavoidable. It is also required that such an event is foreign to the functioning of the public service.[79] Usually, force majeure in this context refers to meteorological phenomena which could not be foreseen by the Public Administration, such as river flooding caused by very heavy rain. In some cases, a strike by public officials has been considered as force majeure, provided that the minimum services established by the law are fulfilled.[80]

30 As regards the Environmental Liability Act, liability is excluded if damage is caused by a natural phenomenon of an exceptional, irresistible and unavoidable character (Art. 3.4.b) or an armed conflict, hostilities, civil war or an insurrection (Art. 3.4.a). Such clauses are interpreted as excluding liability for force majeure.[81] Usually case law refers to force majeure whenever damage is caused by war, naval blockade, or intervention of the public authority.[82] Finally, it is usually said that force majeure breaks or interrupts the causal link between damage and the liable party.[83]

---

77 Instead of many, *E. Roca,* Derecho de daños (4th ed. 2003) 75.

78 See *M. Martín-Casals/J. Ribot/J. Solé*, Spain, in: B.A. Koch/H. Koziol (eds.), Unification of Tort Law: Strict Liability (2002) 306 ff.

79 See *B. Soro Mateo*, La responsabilidad ambiental de las Administraciones Públicas (2005) 93.

80 See *González Pérez* (fn. 29) 417 ff.

81 See *Jordano* (fn. 33) 27.

82 See *L. Reglero Campos/L. Medina Alcoz*, El nexo causal, in: F. Reglero Campos (ed.), Tratado de responsabilidad civil I (4th ed. 2008) 857.

83 Among many others, see *Cordero* (fn. 72) 448.

4. Threshold to prove causation

Some legal scholars have emphasised that establishing the causal link between damage and GMOs may be extremely difficult, due to the innovative nature of GMOs and the incomplete knowledge available regarding their consequences.[84] It has to be taken into account that the degree of certainty required by Spanish courts in order to consider that the causal link has been established is very high. Some court decisions even require absolute certainty, so the causal link can be established surely (e.g. STS 28 June 1979 [RJ 1979/2553] and 2 March 2000 [RJ 2000/1304]). Therefore, a mere probabilistic approach is not sufficient. However, other decisions have stated that it is not required that the claimant attains mathematical exactitude (STS 25 September 1999 [RJ 1999/7275]). Instead, it is sufficient that the causal link is proved with a qualified probability (STS 26 July 2001 [RJ 2001/84236]). However, mere speculations, conjectures or pure probabilistic estimations are deemed insufficient (STS 20 February 2003 [RJ 2003/1174]). As a result, the criterion adopted by Spanish courts implies that certainty or almost certainty has to be achieved in connection with proof of the causal link. 31

Having said that, it must be pointed out that in certain very exceptional cases the Spanish Supreme Court has adopted a more flexible approach. As is well known, in the famous colza oil case it was not proved before the court which specific molecule of the toxic oil had actually caused damage. However it seemed clear to the Supreme Court that the oil had indeed caused damage, so following the German doctrine in the *Lederspray* case, it considered that the causal link had been sufficiently established (STS 2nd Chamber 23 April 1992 [RJ 1992/6783]). It has sometimes been suggested that following this doctrine the court could adopt a subjective standard, whereby it could establish the causal link on the basis of a subjective belief different from a mere suspicion.[85] However such an exceptional decision has to be seen in the light of the extreme difficulty of the case.[86] 32

5. Special rules on causation

As has been said, the specific statutory rules on liability for damage caused by GMOs are very simple. The GMO Act does not include any special rule 33

84 See *Ruda* (fn. 18) 301.

85 See F. *Pérez Álvarez*, Alimentos transgénicos y Derecho penal. Apuntes para una reflexión, in: C.M. Romeo Casabona (ed.), Genética y Derecho penal. Previsiones en el Código Penal Español de 1995 (2001) 330.

86 See on this decision *M. Martín-Casals/J. Solé Feliu*, Defectos que dañan, InDret 2000, 2.

on causation, so in principle general rules on evidence apply. However, the Environmental Liability Act is also applicable to environmental damage caused by GMOs. In particular, the confined utilisation, including transport, of genetically modified microorganisms – in the sense of the GMO Act – is included as an activity to which the Environmental Liability Act will be applicable (Annex III no. 11 in connection with Art. 3.1). Moreover, the Act has also included in the Annex any intentional release into the environment, transport or commercialisation of GMOs according to the GMO Act (Annex III no. 12). The activities included in this list are governed by a more stringent set of rules in comparison to other activities which may cause environmental damage. Regarding these rules, there is the causal presumption referred to above. However, such a presumption is only applicable to environmental damage as defined by the Environmental Liability Act. According to its provisions, non environmental damage caused to crops as a result of the release of GMOs will be restored by means of compensation of damage according to private law rules (Additional Disposition no. 4 LRM). Therefore, proof by presumptions admitted by private procedural law (Art. 386 LEC) will be applicable to such damage in theory, as explained above.

IV. Types of liability

1. Fault liability

(a) Special rules governing fault

34 Environmental damage caused by GMOs is governed by a strict liability rule according to the Environmental Liability Act (Art. 3.1). Liability under such a rule does not require that an administrative infraction has been committed. Therefore, the obligation to restore damage could be imposed on the liable party in an administrative procedure where a penalty or punishment will also be established, but must not be.[87] This is in contrast with the liability rule previously established by the GMO Act, which apparently made liability dependant on the fact that such an infraction had taken place (Art. 38). This is the criterion generally adopted by the regimes of liability such as this one, which confer on the public authority the power to oblige the liable party to restore damage. In such kinds of legal regimes, only exceptionally is the administrative infraction not a requirement for liability

87 See *Alenza* (fn. 67) 71 and 91.

to be established.[88] Therefore, liability under the latter provision requires at least fault on the part of the defendant in order to establish liability, since an administrative violation requires fault of the offender (e.g. an explicit provision in Art. 58.1 Seeds Act).[89] This means that the liability rule laid down by the GMO Act is less stringent than the private law regime in connection with nuisance (Art. 1908 CC), as explained above.[90] Also it has to be taken into account that over the last decades the Spanish courts have reversed the burden of proof of fault and increased the required standard of care in many instances of damage caused by tort. Therefore, even if the aforementioned rules were not applicable, the proof of fault would be eased by such a reversal and it would be more difficult for the defendant to escape liability.[91] However, the recent decisions of the Spanish Supreme Court seem to adopt a more restrictive stance as to the reversal of the burden of proof of fault, although not specifically in connection with GMOs.[92]

(b) Impact of specific rules of conduct

As already explained in the reply to the previous question, violation of **35** statutory rules or regulations is required by the GMO Act in order to establish liability for damage caused by GMOs, inasmuch as liability is derived from an administrative infraction. As regards the Environmental Liability Act, this is no longer a requirement, since only damage is required and liability for damage caused by GMOs' confined utilisation – including transport – is strict (Art. 3.1 in connection with Annex III no. 11).[93] However, the violation of other rules such as those referred to in the question would not be entirely irrelevant. As has been said, the Environmental Liability Act lays down a presumption that damage has been caused by an activity included in the statutory list, provided that such an activity is adequate (*apropiada*) to cause damage (Art. 3.1 2nd part). To establish such adequacy, the public authority has to take into account the intrinsic nature of the activity as well as the way it has been carried out (pursuant to the same provision). Therefore, it may be suggested that the violation of specific rules on GMOs may be taken into account by the public authority as an

88 See *A. Ruda*, En tierra de nadie. Problemas de delimitación del nuevo daño medioambiental, RDP 2009, 21 ff.
89 Among others, see *Valencia* (fn. 32) 127 ff.
90 See on this difference of legal treatment *Martín-Casals/Ruda* (fn. 4) 452 ff.
91 See *Martín-Casals/Ribot/Solé* (fn. 78) 291 ff.
92 See *Ribot/Ruda* (fn. 16) 555 ff.
93 Similarly see *Álvarez* (fn. 30) 45.

element indicating that the activity at stake is adequate to cause damage, according to the circumstances of the case.

36 Apart from the above, the infringement of statutory or administrative standards could be taken into account as an element which indicates that the defendant behaved in a negligent way. This would ease the proof of fault pursuant to the general rule of liability in tort (Art. 1902 CC). However, such a violation is not required in order to establish liability according to this regime. For instance, to date there is no regulation on the coexistence of GM and conventional crops. Should this exist, a violation of the rules on distances between crops could be used by the claimant as evidence that fault existed. Nevertheless, the mere fact that the correct distances were observed does not necessarily exclude fault.[94] Moreover, it has to be borne in mind that liability for damage caused by nuisance is strict (pursuant to Art. 1908.2 CC).[95]

2. Product liability

(a) Development risk defence

37 The development risk defence was introduced in the Spanish Product Liability Act 1994 (Art. 6.1.e) and applied to all products except medicines, food and food products for human consumption (Art. 6.3). Such rules have been incorporated into the new Consumer Protection Act 2007 (Art. 140),[96] which repeals the previous Act. Legal scholarship has suggested that such a defence is not applicable in the case of production defects. Instead, the defence should be construed in a restrictive way. Moreover, the defence excludes liability for risks which could not be known when the product was put into circulation. However, liability is not excluded if harm is the result of risks which were known or potentially existing but impossible to discover or eliminate.[97] As regards food and food products, as has been said, the development risk defence is not applicable. Therefore, the legal situation of the farmer who produces food is worse in comparison to the one who produces other products such as fodder, live animals (unless they are prepared for human con-

---

94 See *Jiménez* (fn. 75) 290. See also (c) below (no. 46 ff.).

95 See no. 6 above.

96 *Real Decreto Legislativo* 1/2007, de 16 de noviembre, *por el que se aprueba el texto refundido de la Ley General para la Defensa de los Consumidores y Usuarios y otras leyes complementarias* (BOE no. 287, 30 November 2007).

97 See *J. Solé Feliu*, El concepto de defecto del producto en la responsabilidad civil del fabricante (1997) 532 ff.

sumption), or plants before the crop is done.[98] Legal scholarship has pointed out that such a rule is sufficient and appropriate from the point of view of the protection of the consumer or final user of the product.[99]

The defence has also been introduced in the legislation on the liability of the Public Administration (Art. 141.1.2 LRJPAC as amended in 1999).[100] The amendment incorporated a second defence to liability of the Public Administration in an already existing provision which was devoted to force majeure. Nevertheless, most legal scholars have criticised the poor legal drafting of the reform, and have accordingly distinguished force majeure from development risks. The issue arose in connection with damage caused by transfusion of blood infected with the HIV virus. Even if the point of view of force majeure was adopted, it is considered that such harm is not due to a circumstance external to the functioning of the public service. Therefore, this is not properly speaking an instance of force majeure, so liability of the Public Administration cannot be excluded on the basis of such defence. Instead, the development risk defence is applicable.[101] **38**

This defence has also been introduced in the Environmental Liability Act. **39** According to it, the operator will not be obliged to bear the cost of the measures of restoration whenever two conditions are met. First, he must bring evidence that he did not behave with fault or intent. Second, he also has to prove that environmental damage was caused by an activity, an emission, or the utilisation of a product which, at the moment of being carried out or utilised, were not considered potentially harmful to the environment according to the state of scientific and technical knowledge existing at that time (Art. 14.2.b). If both conditions are fulfilled, the operator will indeed be obliged to adopt the measures of prevention, avoidance and restoration of environmental damage. However, he will be able to recover the costs of the measures of restoration by way of recoup against the State Fund for Restoration of Environmental Damage (Art. 14.3 in connection with Art. 15.2). Such Fund is managed by the Ministry of the Environment and receives its funding from the general budget of the state (Art. 34.1). Therefore, the operator will have to bear the cost of the measures of prevention and avoidance of damage, which are not refundable from this Fund.

---

98 See *Jiménez* (fn. 75) 280.

99 Thus for instance *A. Aguado de Maetzu*, Las responsabilidades en Derecho alimentario, in: V. Rodríguez Fuentes (ed.), El Derecho agroalimentario (2003) 231.

100 See *Martín-Casals/Ribot/Solé* (fn. 78) 309.

101 See *J.M. Busto Lago*, La responsabilidad civil de las Administraciones Públicas, in: F. Reglero Campos (ed.), Tratado de responsabilidad civil III (4th ed. 2008) 961 ff.

40 As for the practical application of this defence, it has to be noted that no cases are known where it has been applied in connection with GMOs. The Environmental Liability Act establishes that the defence requires that the activity, emission or utilisation of the products "were not considered" to be potentially harmful to the environment according to the state of the art (Art. 14.2.b). Therefore, the criterion is objective, since it does not depend on the actual knowledge of any specific operator. The statutory rule refers to the fact that the activity, emission or product was not considered harmful. However, legal scholarship states that liability is excluded by the impossibility of discovering the defect. Therefore, it is not enough that knowledge about the potentially harmful character of the activity, emission or product exists. It is also required that such knowledge is accessible.[102]

41 The development risk defence has been criticised by some scholars in general on the basis that it weakens the precautionary principle and the strict liability rule.[103] It has also been specifically criticised in connection with damage caused by GMOs on the basis that it is hard to imagine harm caused by GMOs to which such a defence is not applicable.[104] Moreover, it may be very difficult for the operator to bring evidence that the harmfulness of the activity, emission or product was completely unknown, since this implies that a negative fact has to be proved.[105] Apart from that, it may seem surprising that the rule requires for the defence to be applicable that the operator did not behave with "fault, intent or negligence" (Art. 14.2). If he has behaved in a negligent manner, this means that damage was foreseeable so it is logically not possible to apply the development risk defence.[106] Finally, some legal scholars have pointed out that it is quite probable that manufacturers of GMO products will rely on the risk development defence to escape liability. Therefore, it may be rather difficult for the victim to obtain compensation in these cases.[107]

(b) Alternative routes

42 In theory, agricultural activities could cause environmental damage as governed by the Environmental Liability Act, since they require an admin-

102 See *P. Salvador Coderch/J. Solé Feliu*, Brujos y aprendices. Los riesgos de desarrollo en la responsabilidad de producto (1999) 45.
103 See *Alenza* (fn. 67) 95. With regard to product liability, see *Marín* (fn. 65) 115 ff.
104 Thus with regard to the Environmental Liability Directive, *Jordano* (fn. 33) 32 ff.
105 Similarly *Esteve* (fn. 71) 79 ff.
106 See *Ruda* (fn. 18) 427.
107 See *González/Jiménez* (fn. 62) 709.

istrative authorisation to carry out the activity pursuant to the Act on Integrated Pollution Prevention and Control[108] (Annex III no. 1 LRM). Industries of so-called "food and agriculture" (*agroalimentación*) or animal production are included in an Annex to the latter Act (Annex I no. 9), and therefore governed by the Environmental Liability Act under a strict liability rule (Art. 3.1 LRM). According to the latter Act, if damage or threat of damage is caused by the use of a product, the operator may claim compensation from the producer, the importer or the provider of the costs that the operator incurred as a result of the application of the Environmental Liability Act. For this claim to be possible, it is required that the operator has carried out his or her activity whilst observing in an strict manner the conditions established for the use of the product as well as the legal rules existing at the moment when the emission or the fact which caused environmental damage took place (Art. 16.2 LRM). In theory, there is a risk of a lack of internal coherence, since the development risk defence is provided for by the Environmental Liability Act as described above, but the product liability regime excludes such a defence in connection with food and food products.

As a matter of fact, the Consumer Protection Act – which includes the legal **43**
regime on product liability, as has been said – does not lay down any special rules for agricultural products. Such products fall within the scope of application of the product liability regime, provided that they are defective and meet the other conditions of liability pursuant to such regime (Art. 128 ff. LGDCU). Those products resulting from manipulation through genetic engineering can be considered transformed products, regardless of whether they are produced for human consumption or not.[109] However, nowadays both transformed and non-transformed products are governed by the product liability regime, as a result of the statutory reform motivated by the "mad cow" crisis. Moreover, the concept of product is interpreted in a broad way. It could be a seed or an embryo, but also the plant obtained from that seed or the animal resulting from such an embryo, as well as the fruits obtained from them.[110] Therefore, there is no specific statutory regime on liability for harm caused by agricultural products apart from that. The possibility of applying the general rules of the Spanish Civil Code on tort liabi-

108 *Ley* 16/2002, de 1 de julio, *de prevención y control integrados de la contaminación* (BOE no. 157, 2 July 2002).

109 See *C. Vattier Fuenzalida*, Responsabilidad por alimentos defectuosos y seguridad alimentaria, Cuadernos de Derecho Agrario 2004, 51. Similarly *Téllez*, Revista de Derecho Agrario y Alimentario 2000, 12.

110 See *González/Jiménez* (fn. 62) 710.

lity remains open.[111] As already stated, the development risk defence is not applicable to damage caused by food or food products for human consumption (Art. 140.3 LGDCU).

44 Apart from that, there is some controversy as to what knowledge has to be taken into account to decide whether the risk was known or not. In general terms, the prevailing opinion has suggested that the appropriate standard is a global one.[112] Therefore, the potentially liable parties should take into account not only the information available in their country but also that available abroad. However, it has also been suggested that such a standard may be inappropriate for some food producers. Probably, the Spanish courts will take into account the economic dimension and the capacity to carry out research by the specific activity, since the R&D which can be carried out by a multinational corporation is not the same as that of a small business.[113] So, for instance, a court decision on a DES case considered that the defendant should not be held liable because the state of the art *in Spain* did not facilitate knowledge of the negative effects of the drug.[114] Similarly, a non-global, geographically reduced standard has been suggested as the more appropriate with regard to farmers of small dimensions, in particular those located in the Spanish mountains.[115] In a similar vein, it has sometimes been pointed out that holding a small farmer liable for damage caused by development risks may be excessive, whereas this is more justified in the case of big companies.[116]

45 As regards administrative violations (Art. 49 ff. LGDCU), there is a specific regulation.[117] The Consumer Protection Act expressly provides that this regulation will be taken into account when dealing with administrative sanctions for infractions of consumer protection rules (Final Disposition

---

111 See *José A. Navarro Fernández*, Los agricultores y las formas de organización de la empresa agraria, in: Navarro (ed.), Introducción al Derecho agrario (2005) 366 ff.

112 See *Salvador/Solé* (fn. 102) 55 and *M.P. García Rubio*, Los riesgos del desarrollo en la responsabilidad por daños causados por los productos defectuosos. Su impacto en el Derecho español, Actualidad Civil 1998, 859.

113 In this regard see *F. Pantaleón Prieto*, Cómo repensar la responsabilidad civil extracontractual (también la de las Administraciones Públicas), in: J.A. Moreno Martínez (ed.), Perfiles de la responsabilidad civil en el nuevo milenio (2000) 452.

114 On this decision see *Ruda*, InDret 2003, 4 ff.

115 Thus *Vattier* (fn. 109) 58.

116 See *G. de Castro Vítores*, Tendencias actuales en materia de seguridad alimentaria, y su repercusión en obligaciones y responsabilidades, in: P. de Pablo Contreras/Á. Sánchez Hernández (eds.), Régimen jurídico de la seguridad y calidad de la producción agraria (2002) 181.

117 *Real Decreto* 1945/1983, de 22 de junio, *por el que se regulan las infracciones y sanciones en materia de defensa del consumidor y de la producción agroalimentaria* (BOE no. 168, 15 July 1983).

no. 3 LGDCU). However, this aspect only affects liability before the Public Administration, so it does not affect tort liability rules with regard to claims filed by private parties. As a rule, administrative penalties are compatible with tort liability.[118]

(c) Impact of compliance with rules and regulations

If the defendant has obeyed every single rule and regulation governing his production process, liability will probably be excluded under the general rule of tort liability based on fault (Art. 1902 CC). The reason is that in such a case it will be very difficult to prove that the defendant did something wrong and behaved in a negligent way. However, the mere fact that existing regulations or statutory rules are observed while carrying out the damaging activity does not free the defendant from liability.[119] The applicable rules may be incomplete or out of date. Therefore, the standard of care may be more stringent than provided by such regulation or statutory rules. Furthermore, for fault to be established it is not required under Spanish law that a violation of a regulation is proved. Liability based on fault may exist even though all the administrative requirements have been carefully observed by the defendant. Moreover, as has been said, Spanish courts have increased the required level of care. If the defendant is a professional, he or she will be subject to an increased level of care according to the so-called *lex artis ad hoc* or standard of care applicable to that profession.[120] However, the courts may take existing regulations into account as one of the criteria – among many others – to decide whether the behaviour of the defendant was lawful or not.[121] For instance, if the defendant has exceeded the administrative limits established by a regulation – in connection with noise – for instance, the disturbance will be more readily considered excessive, risky or detrimental to the victim than if no administrative limit was exceeded.[122] **46**

As for the Environmental Liability Act, it differs in part from the situation which has been described above. According to Art. 14.1.b), the fulfillment **47**

118 See *Del Olmo* (fn. 64) 257.
119 Among others see *González/Jiménez* (fn. 62) 706. Also see no. 36 above.
120 See *Roca* (fn. 77) 68.
121 As pointed out by *J. Egea Fernández*, Relaciones de vecindad, desarrollo industrial y medio ambiente, in: J. Esteve Pardo (ed.), Derecho del medio ambiente y administración local (2006) 398.
122 See *J. Almagro Nosete*, Tutela procesal frente al ruido, in: I. Sierra Gil de la Cuesta (ed.), Responsabilidad civil medioambiental (2005) 34.

of an order or a compulsory instruction given by the competent public authority is a liability defence available to the operator. Should this be the case, the operator would not be obliged to bear the cost of the measures of prevention, avoidance and restoration of environmental damage or threat of damage, provided that such damage or threat was caused exclusively by such fulfillment. This means that the operator will be obliged to adopt measures of restoration, prevention and avoidance of environmental damage, but in the end he will be able to shift the costs incurred thereby to the Public Administration which issued the order or instruction. To this end, the operator will have to file a claim against the Public Administration (Art. 15.1). An exception is provided where the order or instruction is issued to deal with an emission or an incident previously generated by the operator's own activity (Art. 14.1.b) par. 2). The projects approved by the Public Administration will not be considered orders or instructions in the sense of the previous rule (Art. 14.1.b) par. 3). If environmental damage is caused as a result of a vitiating factor in the project prepared by the Public Administration in connection with a works contract or a production supply contract, the operator will not be obliged to bear the cost of the measures adopted (Art. 14.1.b) par. 4).

48 As regards the obligation to bear the costs of restoration of environmental damage, the Act also establishes that the operator will not have to bear them if he or she proves that he or she did not behave in a negligent way or with intent to cause the damage. However, it is additionally required that the emission or the fact which is the direct cause of environmental damage is the explicit and specific object of an administrative authorisation issued in conformity with the rules applicable to the activities included in Annex III to the Act (Art. 14.2.a). This implies that the Environmental Liability Act allows the so-called "licence or permit defence" provided that some very specific requirements are met.[123] As has already been seen, not every licence allows the operator to escape liability since it is required that he or she behaved in a careful way. Moreover, it is required that the operator has strictly observed the conditions and content established in the authorisation as well as the normative provisions applicable thereto at the moment when the emission or the fact which caused environmental damage took place (Art. 14.2 par. 2). It has to be taken into

123 This is a limited admission of such a defence. See *E. Orteu Berrocal/R. Márquez Molero*, La transposición de la Directiva 2004/35/CE, del Parlamento Europeo y del Consejo, de 21 de abril, sobre responsabilidad medioambiental en relación con la prevención y reparación de daños ambientales, in: Jordano/Alenza/Orteu et al., Estudios sobre la Directiva 2004/35/CE de responsabilidad por daños ambientales y su incidencia en el ordenamiento español (2005) 59.

account that if the authorisation has expressly taken into consideration the hypothesis that environmental damage will result from the authorised activity, the authorisation itself will stipulate the appropriate measures to restore such damage. In such cases, the authorisation is issued assuming that the activity may quite probably cause environmental damage. Therefore, the operator will not have to bear the costs of restoration on the basis of the Environmental Liability Act but he may be forced to comply with the specific terms of the authorisation according to public law rules. This is not an innovation of the Environmental Liability Act but actually follows the criterion already existing in practice before it entered into force. As a rule, the mere fact of having a licence is not a defence and the private law courts cannot decide on whether the licence has been issued correctly or not, as this is, properly speaking, an administrative issue.[124] However, private law courts usually oblige the liable party to adopt measures to prevent damage from occurring in the future.[125] It has also been suggested that if the contrary solution was followed and the operator was forced to bear the restoration costs, this would harm Spanish industry.[126]

Apart from that, the Environmental Liability Act also lays down a liability rule for damage based on fault, applicable to environmental damage caused by economic or professional activities not included in Annex III to the Act (Art. 3.2). This rule will probably be construed in a way similar to the interpretation of Art. 1902 CC as described above. **49**

3. Environmental liability

(a) Implementation of the Environmental Liability Directive

The Environmental Liability Directive was transposed by an Act of the Spanish Parliament, the Environmental Liability Act referred to above. This is a *de minimis* statute, in the sense that other statutory rules may establish more stringent rules of liability in connection with environmental damage (Additional Disposition no. 2 LRM). It is expressly provided in this Disposition that the Autonomous Communities may pass more rigorous provisions on environmental liability or in connection with specific activities. This is a consequence of the constitutional distribution of com- **50**

124 See *Cordero* (fn. 72) 438.
125 This possibility is endorsed by legal scholarship. Among others see *Almagro* (fn. 122) 19.
126 So M. *Zubiri de Salinas,* El seguro de responsabilidad civil por daños al medio ambiente (2005) 228.

petences among the Spanish state and the Autonomous Communities. The Spanish Constitution lays down that the Communities may assume competences in the field of management of the protection of the environment (Art. 148.1.9). The Spanish state has an exclusive competence to pass so-called basic legislation on environmental protection though (Art. 149.1.23). However, the latter provision expressly allows the Autonomous Communities to establish "additional rules of protection". In principle, this does not necessarily mean that the Autonomous Communities have the possibility to develop the rules adopted by the state, but they may introduce statutory rules which are more protective than those passed by the state.[127] In spite of this, the Environmental Liability Act has gone a step further by providing that both the statutory development and the execution of the Act fall within the competence of the Autonomous Communities in whose territory the damage or threat of damage is located (Art. 7.1).

51 The Environmental Liability Act takes into account the competences of the Autonomous Communities within their territory in other senses as well. As has been said, the standing to file a claim is given by the Act to the public authority. However, taking into account the distribution of competence described above, it is laid down that the competent authority will be the one to which such competence is conferred in the corresponding sphere (i.e. within the General Administration of the Spanish state, the Autonomous Communities or the autonomous cities of Ceuta and Melilla, Art. 2.22). Therefore, the Autonomous Communities will, as a rule, start proceedings on the basis of the Environmental Liability Act for damage to natural resources located in their territory. With regard to damage caused to natural resources belonging to the Spanish state – such as waters and coasts according to their specific statutory rules – it is provided that the General Administration of the State will apply the Environmental Liability Act (Art. 7.3 LRM). This means that if waters belonging to the Spanish state but located in an Autonomous Community are damaged in the sense of the Environmental Liability Act, the General Administration of the State has the standing to claim against the operator. This may entail that the provisions of the Act are construed in different ways depending on which Autonomous Community starts proceedings against the operator and on the practical application of the Act by the Autonomous Communities and the state. If damage affects several Autonomous Communities or both an Autonomous Community and the State are obliged to

127 See *L. Ortega Álvarez*, La organización administrativa del medio ambiente, in: Ortega (ed.), Lecciones de Derecho del medio ambiente (2nd ed. 2000) 103.

act to protect the environment, the affected Public Administrations will establish mechanisms of cooperation (Art. 7.4).

Apart from that, in general terms, the Environmental Liability Act closely follows the provisions of the Environmental Liability Directive. However, the Spanish Ministry of the Environment very soon criticised that the Directive fell short in certain aspects and that the Ministry would prepare a more rigorous Draft regime. Among the Directive's shortcomings, the most evident for the Ministry was the lack of financial guarantees. Such guarantees are provided for by Chapter IV of the Act. Basically, the operators of the activities included in Annex III to the Act have the obligation to provide a financial guarantee which allows them to face environmental liability derived from the activities they want to carry out (Art. 24.1). Such a guarantee will only cover environmental liabilities derived from the economic or professional activity of the operator (Art. 25.1) and will be independent from other guarantees that the operator may have (Art. 25.2). **52**

The sum to be covered by the guarantee will be established by the competent authority. Such a sum is not a liability cap but refers only to the cover by the financial guarantee. Therefore, the liability of the operator may be beyond the sum covered by the guarantee. To establish the sum covered, attention will be paid to the intensity and extent of damage which the activity may cause, according to the criteria established by a regulation (Art. 24.2). Such regulation is the Royal Decree 2090/2008 for the partial development of the Environmental Liability Act (RLRM).[128] The Regulation established that an assessment of risks will be carried out (Art. 33.2). Such an assessment will identify the scenarios of an accident and establish the probability that each scenario occurs (Art. 33.2.a). Also the assessment will establish the extent of environmental damage associated with each scenario (Art. 33.2.b). **53**

According to the Environmental Liability Act, the competent authority will justify why a particular amount has been established (Art. 24.3). The method for establishing such an amount will be based on technical criteria which guarantee a homogeneous assessment of the risk scenarios and the restoration costs associated with each of them. Also a uniform definition of the cover for each activity or facility has to be ensured (Art. 24.3 2nd part). **54**

There are three different possible modalities of financial guarantee (Art. 26 LRM). The operator may choose between them and he or she may **55**

128 *Real Decreto* 2090/2008, de 22 de diciembre, *por el que se aprueba el Reglamento de desarrollo parcial de la Ley 26/2007, de 23 de octubre, de Responsabilidad Medioambiental* (BOE no. 308, 23 December 2008).

also combine them. First, the operator may choose an insurance contract – within the terms provided by the Insurance Contract Act[129] – with an insurance company authorised to operate in Spain (Art. 26.a). As a matter of fact, it was often pointed out before the Environmental Liability Act was passed that environmental insurance has scarcely taken roots in Spanish practice due to the difficulties of assessing the risks in this area.[130] Spanish insurers have always been very sceptical as to insurability of pure ecological damage, due allegedly to problems of definition and lack of experience in this field – although this line of reasoning is in fact a *circulus vitiosus*. Moreover, this insurance is different from the compulsory insurance which the operator is obliged to provide according to other statutory rules such as the Nuclear Energy Act (Art. 55 ff. LEN) or the Waste Act (Art. 21.2 [LR]),[131] among others. Remarkably, it seems that previous insurance policies covering harm for environmental disturbance excluded damage caused by genetic mutations or GMOs.[132]

56 The second possibility is that the operator obtains a guarantee (*aval*) from a financial entity authorised to operate in Spain (Art. 26.b) LRM). Third, the operator may create a so-called "technical provision" through an *ad hoc* fund. Such a fund has to materialise in financial investments supported by the public sector (Art. 26.c). In principle, this possibility is addressed to big companies with important financial resources, a part of which can be used to guarantee environmental liability.[133]

57 As for the persons whose liability will be covered by the financial guarantee, the Act establishes that the primary person is the operator. However, it is also possible to include the subcontractor and the professionals who collaborate with the operator to carry out the authorised activity in the cover (Art. 27 LRM).

58 Not every operator in the sense of the Environmental Liability Act is obliged to provide a financial guarantee, indeed some of them are expressly excluded from such an obligation. In particular if it is assessed that an activity carried out by an operator may cause damage which may be restored for less than € 300,000, such an operator will be freed from the obligation to provide a guarantee (Art. 28.a) LRM). This is also the case of the operators of activities which may cause damage, the restoration

129 *Ley* 50/1980, de 8 de octubre, *de Contrato de Seguro* (BOE no. 250, 17 October 1980).
130 Among others, see *González-Varas* (fn. 35) 248 and *Del Olmo* (fn. 64) 277.
131 *Ley* 10/1998, de 21 de abril, *de residuos* (BOE no. 96, 22 April 1998).
132 According to the information provided by *Zubiri* (fn. 126) 233.
133 As *Orteu/Márquez* (fn. 123) 61 observe, this possibility would be out of the reach of small or medium companies.

of which is assessed at costing between € 300,000 and € 2,000,000. For such an exemption to apply, these operators must prove that they have adhered to the European Eco-Management and Audit System (EMAS) or the environmental management system UNE-EN ISO 14001:1996 on a permanent and continued basis. Finally, operators who carry out the utilisation of phytosanitary products and biocides with agropecuary and forest aims are also excluded from the obligation to provide a financial guarantee (Art. 28.c) LRM in connection with Annex III no. 8.c) and d)). Commercialisation and utilisation of phytosanitary products and the registering, authorisation and commercialisation of biocides, are governed by two separate regulations.[134,135]

Moreover, the obligation to provide a financial guarantee is not applicable to legal persons governed by public law. In particular, the Act establishes that such an obligation cannot be required of the General Administration of the State or the public entities linked to or dependent on it. It will also not be applicable to the local entities, autonomous organs or other entities governed by public law and dependent on them (Additional Disposition no. 7.1). However, the Autonomous Communities may provide that this obligation will be applicable to their public administration and their own public organs (Additional Disposition no. 7.2). **59**

As regards the cover provided by the financial guarantee according to the Environmental Liability Act, there are some kinds of costs which have to be covered (Art. 29), namely: a) those derived from the obligations of the operator with regard to prevention and avoidance of new damage (Art. 17 LRM), provided that damage the prevention or limitation of which it is aimed at has been caused by pollution (Art. 29.a); b) costs derived from the obligations of the operator in connection with restoration of damage (Art. 19 and 20), provided that damage the prevention or limitation of which it is aimed at has been caused by pollution (Art. 29.b). If such damage affects water, wild species of habitat and coastal inlets (*riberas del mar*) and *rías,* the costs covered are limited to those which fit into the concept of primary restoration pursuant to Annex II no. 1.a) of the Act. **60**

The financial guarantee to be provided by the operator has a maximum amount, not to be exceeded, established by the Environmental Liability **61**

134 *Real Decreto* 2163/1994, de 4 de noviembre, *por el que se implanta el sistema armonizado comunitario de autorización para comercializar y utilizar productos fitosanitarios* (BOE no. 276, 18 November 1994).

135 *Real Decreto* 1054/2002, de 11 de octubre, *por el que se regula el proceso de evaluación para el registro, autorización y comercialización de biocidas* (BOE no. 247, 15 October 2002).

Act, which is € 20,000,000 (Art. 30.1). This amount will be applied as a limit per event and annuity (Art. 30.2). A deductible can be established to a maximum of the 0.5% of the amount to be guaranteed. In that case, the operator will have to bear such amount. As regards the definition of event, all the claims for environmental damage which derive from the same emission, happening or incident will be considered as amounting to one and the same event, regardless of the number of affected people and even though the claims may take place at different moments. The amount per event and annuity of the insurance established in the guarantee will be applicable to such unity of event or event in series (Art. 30.2 final part).

62 In any case, the guarantee must be provided from the date when the authorisation required to carry out the activity at stake has effect (Art. 31.1). The operator has the obligation to maintain his or her guarantee in force during all the time the activity is being carried out. The public authority may request the information needed from the financial entities and the operators in order to supervise such obligation.

63 The Environmental Liability Act allows that temporal limits are set to the financial guarantee (Art. 32). In particular, it is allowed to set a temporal limit so that the guarantee cover includes liabilities provided that the following circumstances are met: first, that the beginning of the emission which causes pollution or the beginning of the situation of imminent risk of pollution is identified and it is proved that it occurred within the period of the guarantee. Second, that the first verifiable manifestation of pollution has taken place within the period of the guarantee or within the three months following the termination of the guarantee. The first manifestation is defined as the moment when it is first discovered that there has been pollution, regardless of whether it is considered dangerous or not at that moment. Third, that the claim against the operator for the pollution has been filed within the period of the guarantee or within three years from its termination.

64 In connection with the temporal limitation which has been described, the generating fact (*hecho generador*) is considered to be the pollution which is caused in an accidental and fortuitous (*aleatorio*) way, i.e., that has an extraordinary character and has not been generated by any of the following causes: a) intent;[136] b) a normal and foreseen consequence of the possession of buildings, facilities or equipment for the service of the authorised activity; c) a consequence of a fact which has been foreseen and consented to by the operator, which has occurred within the premises

136 Liability for intentional acts of the insured parties is not insurable under Spanish law according to the Insurance Contract Act (Art. 19).

where such an activity is carried out or in the geographical sphere for which the activity has been authorised; d) an infringement of the rules applicable to the authorised activity – be they concerning environmental aspects or not – provided that such an infringement is known by the insured party or he or she could not have ignored it; e) the conscious bad utilisation of the facilities or mechanisms and their components, or the lack or inadequacy of their maintenance; f) the abandonment or prolonged lack of use of the facilities, without having adopted the appropriate measures to prevent the protection or safety conditions from deteriorating; g) the result of popular disturbances, riots, strikes, internal disturbances, sabotage, terrorist acts or acts by armed groups. With regard to terrorist attacks, it has to be borne in mind that there is insurance coverage provided by a public entity, the *Consorcio de Compensación de Seguros*.[137] Moreover, insurance policies usually exclude the liability of the insurer for damage caused by terrorist attacks, war, sabotage and similar circumstances amounting to force majeure.[138]

(b) Environmental liability regime beyond the scope of the Directive

As is well known, the Environmental Liability Directive is often criticised **65** because of the use of vague concepts, the fact that it does not alleviate the burden of proof of causation, its restricted scope and many other shortcomings.[139] As a result, the Directive may be less protective in comparison to the liability provisions previously existing in several European countries. As regards Spain, the Environmental Liability Act was officially presented as the mechanism which would bring the "polluter pays" principle into practice. However, the truth is that there were – and still are – many statutory provisions which provide a level of protection comparable to that provided by the Environmental Liability Act. The environment was not completely unprotected before the entrance into force of this Act, quite the contrary. There were many statutory provisions expressly establishing the duty to restore damage to natural resources. Such restoration should be in kind and, where this is not possible, the liable party has the obligation to pay compensation in money. Usually, such statutory provisions even establish

137 See *Ribot* (fn. 40) 540 with further references.

138 See *Zubiri* (fn. 126) 227.

139 See among others *B.A. Koch*, European Union, in: H. Koziol/B.C. Steininger (eds.), European Tort Law 2004 (2005) 595 ff.

several criteria to determine the amount of the compensation sum, such as restoration costs or the benefit obtained by the wrongdoer.

66 The natural resources protected by these provisions are, for instance, water,[140] woodlands,[141] soil in case of damage caused by waste (Art. 36 LR),[142] coasts,[143] and flora and fauna.[144] Moreover, general public law rules establish that the person who commits an administrative violation has the duty to restore damage deriving from it (Art. 130.2 LRJPAC). The liable party can be forced by the Public Administration to restore the situation he or she altered and restore it back to the situation which existed before the alteration took place (*status quo ante*). Moreover, he or she may be forced to pay compensation, to be determined by the administrative authority (Art. 130.2 LRJPAC). The liable party may challenge the administrative decisions before the courts of justice. If there is a plurality of liable parties, they will be held liable on a solidary basis (Art. 130.3 LRJPAC). These liability rules have a very broad scope of application. Whenever someone commits an administrative infraction violating a rule on environmental protection, it may be considered that the Public Administration may enforce the duty to restore damage according to these provisions. Such liability according to public law does not prevent the affected private parties from filing a claim in tort if they have suffered damage as a result of the offender's behaviour.[145]

67 As a result of what has been explained, the level of protection provided by Spanish law before the Environmental Liability Act entered into force was quite high. This is not to say that it was entirely satisfactory. It could be argued that the liability rules were scattered among many different statutes and that some natural resources – such as air or the atmosphere – lacked sufficient protection. However, at least from the point of view of the natural resources protected, the Environmental Liability Act does not seem to have improved the legal protection previously existing.[146] Even with regard to air and the atmosphere, the Spanish legislature passed a specific statutory

140 Art. 118.1 of the *Real Decreto Legislativo* 1/2001, de 20 de julio, *por el que se aprueba el texto refundido de la Ley de Aguas* (BOE no. 176, 24 July 2001).

141 Art. 77 of the *Ley* 43/2003, de 21 de noviembre, *de montes* (BOE no. 280, 22 November 2003).

142 See fn. 131 above.

143 Art. 100 of the *Ley* 22/1988, de 28 de julio, *de costas* (BOE no. 181, 29 July 1988).

144 Art. 37 of the *Ley* 4/1989, *de conservación de espacios naturales y de la flora y la fauna silvestres* (BOE no. 74, 28 March 1989). This Act has been repealed by the *Ley* 42/2007, de 13 de diciembre, *del Patrimonio Natural y de la Biodiversidad* (BOE no. 299, 14 December 2007). The duty to restore damage to biodiversity is laid down in Art. 75 which is commented on under the reply to question no. IV.3.(c).

145 See *J. Conde Antequera*, El deber jurídico de restauración ambiental (2004) 181 ff.

regime which provides a liability mechanism similar to the one laid down by the Environmental Liability Act.[147] The drafters of the Environmental Liability Act thought that pollution of the air has a diffuse character and that the establishment of the causal link is far too complex to include the air as a protected natural resource in that Act.[148] In any case, as has been said, liability according to public law provisions usually assumes that the liable party has committed an administrative offence. From this point of view, the scope of the Environmental Liability Act may be broader than the other liability regimes which confer similar powers on the public authority to claim in case of damage to a natural resource.

(c) Claimants in cases of environmental harm

As is well known, the Environmental Liability Directive has not conferred standing to private parties to file a claim against the person who caused environmental damage, so harm to individuals remained outside its scope (Art. 3.3). Following this path, the Spanish Environmental Liability Act establishes in a provision on "harm to private parties" (Art. 5) that this Act will not protect the filing of claims for personal injury, damage caused to private property, or any kind of economic loss, and will not affect any right related to such kind of harm or any other economic damage which does not have the condition of environmental damage, even where damage results from the same facts which gave origin to environmental liability (Art. 5.1). Such claims will be governed by other legal provisions such as, it may be added, the Civil Code (e.g. Art. 1902 and 1908). Therefore, it may happen that the same act by an operator gives rise to two different claims. First, the competent public authority may start an administrative procedure for environmental damage, which may give rise to liability governed by the Environmental Liability Act. Second, a private party may have suffered damage to an interest of his or hers which also may trigger liability. This may give rise to a claim before the courts of justice, which will be governed not by the Environmental Liability Act but by other rules. **68**

146 See similarly *Grupo de Economía Ambiental*, Análisis económico del anteproyecto de Ley de Responsabilidad Ambiental (2006) 12 and *Ruda*, RDP 2009, note 43 and accompanying text.

147 *Ley* 34/2007, de 15 de noviembre, *de calidad del aire y protección de la atmósfera* (BOE no. 275, 16 November 2007). A synopsis of this regime can be found in *V. Manteca Valdelande*, Regulación jurídica de la calidad del aire y la protección de la atmósfera, Actualidad Administrativa 2008, 476 ff.

148 See *Orteu/Márquez* (fn. 123) 54.

69 However, so-called double recovery of costs is forbidden by the Environmental Liability Act. Private parties are not allowed to claim restoration of or compensation for environmental damage inasmuch as it has been restored by virtue of the application of said Act (Art. 5.2). If the liable party has actually been obliged to pay twice, he will be entitled to claim the corresponding sum back from the respective party.

70 The mere fact that a claim grounded in private law is filed against the operator does not prevent the public authority from starting an administrative procedure against him or her. It also does not free the operator from his or her liability according to the Environmental Liability Act (Art. 5.3). It may then happen that the affected private party files a claim against the operator before a court of justice, but the claim has to be dismissed because damage has already been restored due to the enforcement of the Environmental Liability Act by the public authority.

71 It seems open to debate whether the Environmental Liability Act excludes the possibility that private parties file a claim before the courts of justice to obtain redress for damage to the environment as such (so-called pure ecological damage). Art. 5.1 LRM could be interpreted in the sense that private parties completely lack standing to file such claims.[149] However, the provision does not refer to damage to the environment but to instances of damage to private parties. Certainly, the Environmental Liability Act does not confer private parties standing to sue for damage to the environment as such, but it also does not exclude this. In theory, the Act has left such a possibility – if it existed previously – unaffected. Indeed, the interested parties may contact the public authority if environmental damage has been caused in order to inform it that such damage has occurred (Art. 41.1). However, only the public authority has the statutory power to start an administrative procedure on environmental liability according to the Act aforementioned. Also in other ways, it seems that private claims for damage to the environment may face formidable hurdles. To start with, the claimant will have to prove that damage affects him or her in a personal way and that damage is certain. Certainly, tort liability does not require that a subjective right of the victim has been harmed. It is sufficient that damage exists and this would undoubtedly be the case if any interest of the victim was affected. However, legal scholarship has repeatedly underlined that the environmental interest is of a collective nature. Therefore, no one has legal standing to file a claim for damage to the environment as such.[150] As a

149 Apparently this is the interpretation by *Álvarez* (fn. 30) 37.
150 Among others see *Santos* (fn. 19) 3023 ff.

result, it would seem very hard to prove before a private law court that pure ecological damage affects an individual victim in a personal way.

However, on occasion some NGOs have succeeded in claiming compensation for damage to the environment as such before a criminal court.[151] It has to be taken into account that criminal procedural law does confer standing to sue on everyone as an *actio popularis* (*acción popular*). Therefore, such ecological organisations could request the court to adopt measures conducing to the restoration of the perturbed ecological equilibrium (pursuant to Art. 339 Criminal Code).[152] Apparently, other NGOs have not followed suit and this has not become a general resort for obtaining compensation for pure ecological damage under Spanish law. Moreover, *actio popularis* is also possible with regard to administrative procedures, so any person may file a request to the Public Administration to ensure that the legal system is observed. However, it is usually said that such a claim aims at protecting the observance of the legal system as such and does not allow the claimant to request compensation for damage suffered by the violation of the legal rules.[153] **72**

As regards damage to so-called "natural patrimony" and biodiversity, it has already been said that the Public Administration has the authority to oblige the liable party to restore damage according to the Act on Natural Patrimony and Biodiversity (Art. 75 LPNB).[154] The liability rule is formulated in a similar way to the aforementioned provisions of liability towards the Public Administration for administrative violations. For liability to be triggered, it is required that the liable party has committed an act or made an omission which infringes the provisions of the Act (Art. 75.1). Such liability is explicitly qualified as "administrative" (Art. 75.1 1st proposition), in the sense that the responsibility to enforce it belongs to the Public Administration. Moreover, such liability leaves any other liabilities of a different kind unaffected. Therefore, it is possible to file a claim in tort or a criminal claim for the same fact which gives rise to such an administrative liability.[155] **73**

As for the details of the duty to restore damage, the Act on Natural Patrimony and Biodiversity refers to the provisions of the Environmental Liability Act (Art. 75.2). Rather surprisingly, the Act on Natural Patrimony and Biodiversity establishes that the offender will be obliged to compensate damage which cannot be restored, according to the terms indicated **74**

151 See *Del Olmo* (fn. 64) 271 and *Ruda* (fn. 18) 517 ff.
152 As *García Rubio* (fn. 41) 138 points out.
153 See *González Pérez* (fn. 29) 211.
154 See full reference in fn. 144 above.
155 Obviously the offender cannot be punished twice for the same behavior (Art. 75.5 LPNB), because otherwise the principle of *non bis in idem* would be infringed.

in the administrative decision (Art. 75.2 final part).[156] This seems to suggest that liability according to the Act on Natural Patrimony and Biodiversity could go further than the provisions of the Environmental Liability Act. Such a possibility would be of great interest in those cases where the means of damage restoration laid down by the latter Act could be considered less satisfactory or insufficient. In theory, the administrative decision could force the liable party to pay compensation per equivalent, i.e. compensation in money. However, the Act (Art. 75.3) also establishes that damage assessment will be carried out pursuant to the provisions of the Environmental Liability Act and the provisions which develop it. A similar doubt may arise with regard to the other statutory provisions which have already been mentioned and still allow the possibility that the Public Administration sets a compensation sum to be paid by the liable party instead of forcing him to restore damage in kind. Since the Environmental Liability Act has not repealed previous legislation, it seems unclear to a certain extent whether such a possibility still remains.[157]

(d) Special liability regime for losses sustained by individuals

75 In contrast to the legal framework existing in other European countries, there is no specific statute governing nuisance at a national level in Spain. Claims under nuisance rely primarily on general doctrines of tort law and abuse of rights (Art. 7.2 CC).[158] As has been explained, an isolated rule on the liability of the person who creates toxic fumes (Art. 1908.2 CC) has been widely construed so as to forbid nuisance of any kind and lay down a strict liability rule for such damage.[159] In general, not every disturbance may amount to nuisance; it is required that its author abuses his rights (Art. 7.2 CC), be it because the activity does not create any profit or utility for him, or he produces the disturbance intentionally, or he is exceeding the normal limits of the exercise of the right of property. The conditions and the customary use of the place may be taken into consideration to decide whether the disturbance is tolerable or not.[160] Moreover, tolerability may also be defined by reference to the administrative standards.[161]

156 If there is a plurality of liable parties, and it is not possible to determine the degree each of them contributed to the administrative infraction, all of them will be liable on a solidary basis (Art. 75.4 LPNB), which actually is the general rule followed elsewhere. See *Solé*, RDP 2008, 18 ff.

157 See *Ruda*, RDP 2009, 34 ff.

158 See *Ruda* (fn. 38) 151 ff.

159 See no. 3 above.

160 Instead of many see *Cordero* (fn. 72) 441.

Such rules have been helpful in practice with regard to nuisance but it has been sometimes pointed out that they are inadequate to cope with environmental damage derived from existing technological risks. In particular, these provisions aim at the protection of individual interests and only in an indirect way may they have an effect on the protection of collective interests.[162]

In some regions – Navarra and Catalonia – specific statutory regimes exist which make the creator of the nuisance liable for damage resulting from it. However, none of these regimes establishes a liability rule for damage to the environment as such. Also in Spanish law there is no provision equivalent to par. 16 of the German *Umwelthaftungsgesetz*, which establishes that if damage to a thing implies also damage to the environment or the landscape, the costs of repairing the damaged thing will not be considered disproportionate because of the mere fact that they exceed the value of such a thing. Finally, nuisance has not been used by private parties in Spanish practice to claim compensation for pure ecological harm before the courts of justice. **76**

(e) Cartagena Protocol

Spain ratified the Cartagena Protocol in 2001.[163] The Protocol has received little attention from Spanish legal scholarship. It is usually considered that the Protocol provisions will not have a major impact on Spanish law, due to the fact that existing EU legislation already deals with the same issues.[164] At the same time, notice is usually had to the hard negotiations to provide a liability regime for damage caused by GMOs.[165] It has also been suggested that the Protocol contains certain expressions or concepts **77**

161 As *Del Olmo* (fn. 64) 285 points out.

162 See *Egea* (fn. 121) 397 and 404.

163 *Instrumento de ratificación de 10 de diciembre de 2001* (BOE no. 181, 30 June 2003, Correction of errors in BOE no. 284, 27 November 2003).

164 See *A. Fresno*, La regulación de los Organismos Modificados Genéticamente, in: E. Muñoz (ed.), Organismos modificados genéticamente (2006) 67 and *E. Barahona Nieto*, El Protocolo de Cartagena sobre bioseguridad, in: E. Iáñez Pareja (ed.), Plantas transgénicas: de la Ciencia al Derecho (2002) 265 ff. (Ms. Barahona participated in the negotiation of the Protocol as a representative of the Spanish Government).

165 See *S. Álvarez*, El protocolo de Cartagena sobre bioseguridad y la regulación internacional de los movimientos de los organismos modificados genéticamente: de las palabras a la acción, Ambienta. Revista Ministerio de Medio Ambiente 2006, 47 ff. For an overview of the Protocol see also *E. Barahona*, El protocolo de Cartagena, in: J. Gafo (ed.), Aspectos científicos, jurídicos y éticos de los transgénicos (2001) 85 ff. and *E. Barahona Nieto/M. Corrales Rodrigáñez*, El Protocolo de Cartagena sobre Bioseguridad, Gestión Ambiental 16 (2000) 1 ff.

which are rather imprecise if such liability regime is to be considered seriously. In particular, the definition of compensable damage remains unclear to a certain extent.[166] Finally, a scholarly opinion has expressed regret regarding the fact that the Protocol did not take the opportunity to determine the liable party or whether liability should be based on fault or not, among other aspects of liability.[167]

4. Other strict liability regimes

78 All the liability regimes which may be relevant in case of harm caused by GMOs have already been referred to. Specially, the GMO Act should be taken into account.[168]

V. Vicarious liability

1. Scope of vicarious liability

79 According to the Spanish Civil Code, a person is liable for the acts of others only if such person was negligent himself (Art. 1903). The wording of this provision does not recognise liability for the acts or omissions of others as vicarious liability, strictly speaking.[169] The person who is liable for another is liable due to his or her own fault (*culpa in eligendo* or *in vigilando*) and this fault is rebuttably presumed (Art. 1903.6 CC). Rebuttal requires that persons mentioned by Art. 1903 CC prove that they acted in a diligent way but such proof may be very difficult in practice. Liability is imposed upon certain persons (such as parents, guardians, owners of establishments or companies and owners of schools) who are liable for damage caused by other persons with whom they are tied by family, professional or educational relations of dependency. The main feature of these relations is that there is a hierarchical or subordinate relationship between the liable person and the other who caused harm.[170]

---

166 See *Ruda* (fn. 18) 122 ff.

167 See *G. Doménech Martínez*, La agricultura transgénica: algunas precisiones en torno a dicho concepto y su tratamiento en textos legales, in: P. de Pablo Contreras/Á. Sánchez Hernández, Régimen jurídico de la seguridad y calidad de la producción agraria (2002) 201.

168 See above no. 1 ff.

169 See M. *Martín-Casals/J. Solé Feliu*, Liability for Damage Caused by Others under Spanish Law, in: J. Spier (ed.), Unification of Tort Law: Liability for Damage Caused by Others (2003) 231 ff.

170 Ibid., 232.

As for the independent contractor, the Civil Code provides that employers are liable for the damage caused by their employees in the services of the branches in which they are employed or on account of their duties (Art. 1903.4). Liability of the employer under this provision requires a relationship of dependence between the employer and the employee, which for instance exists where both are linked by a labour contract or when the employer has retained the power to supervise, control or direct the performance of the contractor. Therefore, the employer will not be liable for damage caused by an independent contractor under Art. 1903 CC as a rule, unless such power of direction, supervision or control has been retained by the employer.[171] This leaves the possibility of filing a claim under the general rule of liability for fault (Art. 1902 CC) against the employer – provided that his fault contributed to damage caused by the independent contractor – unaffected. For this to be possible, the victim would have to prove that the employer had a duty to prevent damage – even that caused by a third party – from occurring and infringed such duty in a negligent way.[172] **80**

The solution is similar to the one adopted in case of liability of the Public Administration for environmental disturbance caused by a third party. Although the damage may not have been caused by an action of the Public Administration itself, the latter may have contributed to its causation by way of omission. This would happen for instance if the Public Administration authorised an activity which caused interference to third parties and did not carry out proper control or supervision of the activity as regards noise or vibrations.[173] The same can be said with regard to liability for damage caused by GMOs, e.g. when the Public Administration fails to supervise their commercialisation.[174] Therefore, liability may be apportioned between the Public Administration and the third party according to their respective contribution to damage.[175] **81**

171 See *Martín-Casals/Solé* (fn. 169) 249 ff. with further references.

172 See *E. Gómez Calle*, Los supuestos de la responsabilidad civil. La responsabilidad por hecho ajeno, in: F. Reglero Campos (ed.), Tratado de responsabilidad civil I (4th ed. 2008) 1033.

173 See *González-Varas* (fn. 35) 242 and *Cordero* (fn. 72) 459.

174 See *González/Jiménez* (fn. 62) 705.

175 See *González Pérez* (fn. 29) 448.

2. Liability for people further up the food or feed production chain

82 As explained above, Spanish law lays down a rule of liability for damage caused by others who are in a relationship of subordination or dependency with the liable person (Art. 1903 CC). Spanish law does not specifically provide for the possibility that a person lower on the feed and food production chain is liable for the people further up. However, the Civil Procedure Act provides, as a general rule, that a statute may allow the defendant to call a third party to court (so-called provoked intervention, *intervención provocada,* Art. 14 LEC). In this case, the third party will be given the opportunity to come to court but not in the condition of defendant. Nonetheless, he will be given the same possibilities of actuation as the defendant himself (Art. 14.1 LEC). It has to be made clear that the third party has no duty to come to court from the mere fact that he is called to it. Any intervention by a third party before the court is done on a voluntary basis, since there is no obligation to reply to the call. Technically speaking, legal scholars describe this situation not as an obligation, but as a burden (*carga*) which affects the third party as he will have to bear the negative consequences of his not coming to court after being called to do so.[176]

83 After the third party has attended the procedure, the defendant may consider that his place in the procedure should be occupied by the third party instead (Art. 14.4 LEC). Then, a so-called procedural succession may take place, i.e., a change of procedural parties, by means of which the court considers that the third party will replace the original defendant and be the defendant from then on (Art. 18 LEC). The crop retailer may accordingly argue before the court that he is not liable for the damage alleged by the victim and that the claim should be brought against someone further up the production chain instead. Nevertheless, the court may consider that no statutory rule allows for succession in such an specific situation, since neither the GMO Act or the Environmental Liability Act provide for the possibility that the defendant calls a third party to the procedure to intervene in it.

84 Apart from this, the Civil Procedure Act provides that a plurality of persons may act before the court as a single procedural party (so-called joinder of parties or *litisconsorcio,* Art. 12.1 LEC). According to the Act, if it is required for justice to be done that several people are brought to court

176 See F. *Cordón Moreno*, Comentario, in: F. Cordón Moreno (ed.), Comentarios a la Ley de Enjuiciamiento Civil (2001) 197.

together, all of them will have to be addressed in the claim as a joinder of parties, except in cases where an Act provides otherwise (Art. 12.2 LEC). This means that the defendant may reply to the claim by arguing that not only he, but also other persons, should be brought to court (so-called *litisconsorcio necesario*). The court may then consider that the claim has to be dismissed because, properly speaking, some people who should be defendants are not included, unless the claimant brings those other persons into it as defendants.

An alternative possibility for the victim is to file a claim against the manufacturer of the defective product. The Spanish statutory regime on product liability lays down a liability rule respecting the producer (Art. 138 LGDCU). Such is defined as the manufacturer of the good or the person who delivers a service or his intermediary, as well as the importer of the good or service in the territory of the EU. Moreover, any other person who appears as a producer because the product indicates his name, trademark or any other distinctive sign, be it in the container or wrapper or any other element of protection or presentation, will be considered a producer (Art. 5 LGDCU). The producer will have this condition regardless of whether he has manufactured a finished product, an element which has been integrated into a finished product, or raw material (Art. 138 LGDCU). Since liability according to this Act is strict (Art. 135), the producer may be held liable for having manufactured a product which contained a defective element manufactured by another. Therefore, the manufacturer of a product containing GM material may be held liable for damage which is due to the defective condition of the GM material. Following this line of reasoning, it cannot be excluded that the producer of the final product is held liable for the acts of the producer of the GM product or the producer of the GM seed.[177] However, it has sometimes been suggested that the traceability requirements applicable to GMOs may allow the defendants to escape liability, or at least reduce their share in the compensation amount by pointing to the contribution to damage by others in the production chain.[178] Moreover, the producer of a component of the finished product will not be held liable if he proves that the defect is attributable to the conception of the product to which such 85

177 See *G. Doménech Martínez*, La agricultura transgénica: algunas precisiones en torno a dicho concepto y su tratamiento en textos legales, in: P. de Pablo Contreras/Á. Sánchez Hernández, Régimen jurídico de la seguridad y calidad de la producción agraria (2002) 202; in the same volume, see also *L. Amat Escandell/D. Llombart Bosch*, La defensa de los consumidores y la responsabilidad civil por productos agroalimentarios defectuosos, 130 ff.

178 See *De Castro Vítores* (fn. 116) 187.

component has been incorporated or to the instructions given by the manufacturer of such product (Art. 140.2 LGDCU).

3. Can someone further down the feed or food chain include someone further up it in a trial against him/herself?

86 Please see the reply to the previous question.

VI. Multiple tortfeasors

87 As has been already pointed out, Spanish courts usually hold a plurality of tortfeasors liable on a solidary basis. This is clearly the case where a statutory rule expressly so provides, as e.g. in the case of a plurality of producers who are liable for the same damage (pursuant to Art. 132 LGDCU) or in cases where several people have an obligation concerning seeds pursuant to the Seeds Act (Art. 58.1). As a rule, the liable party will then have a recourse claim against the other producers to recover from them according to their contribution to damage (Art. 132 2nd part LGDCU and Art. 1145 par. 2 CC). Moreover, in case of damage caused by defects in the construction of polluting facilities, the author of the disturbance will be held liable on a solidary basis together with the work contractors and the technicians who plan and direct the works (Art. 1909 CC).[179] In these cases, solidarity makes it unnecessary to file a claim against all the liable parties. Therefore, the victim may choose against whom to file the claim and bring to court only one, some or all the liable parties.

88 However, the courts have extended the scope of solidary liability not only to those specific situations where a statutory rule so provides, but also to almost every scenario where there is a plurality of liable parties, and especially in those situations where it is not possible to individualise the contribution to damage of each of them but it is certain that all of them contributed to it.[180] As a matter of fact, very few court decisions rule for separate liability even when it has been possible to elucidate the share corresponding to each tortfeasor.[181] Solidary liability is also the rule in case of alternative causation, provided that it is certain that one of the defendants actually caused damage, although there is no equivalent to the provisions which so

179 See *Cordero* (fn. 72) 45.

180 See *Solé*, RDP 2008, 19 ff. and no. 27 above and *Gómez Calle* (fn. 172) 973 ff.

181 See M. *Martin-Casals*/J. *Solé*, Multiple Tortfeasors under Spanish Law, in: W.V.H. Rogers (ed.), Unification of Tort Law: Multiple Tortfeasors (2004) 190 with further references.

provide in other European Civil Codes (e.g. § 830 BGB in Germany), as has been said.[182] Such a solution is controversial due to the lack of a specific statutory basis in the Code and the alleged impossibility of applying certain special rules (such as the one in the Hunting Act, Art. 33.5) by way of analogy.[183] But this has not been an insurmountable hurdle for case law, since most decisions have chosen solidary liability in spite of all. However, there may be cases in which there is more room for such an analogical application, as in the case of damage caused by people in charge of several animals.[184] Therefore, if the claim is brought against several defendants but the cause of damage remains uncertain so it has not been established that one of such defendants actually caused damage, they cannot be held liable.[185]

Rather strangely, an exception to the rule of liability on a solidary basis **89** can be found in the Environmental Liability Act. According to it, if there is a plurality of liable parties for the same damage, liability will be separate, unless a specific statute provides otherwise (Art. 11). This entails a lack of systematic coherence within the Spanish law of torts. However, some legal scholars have pointed out the negative effects of solidary liability in connection with environmental liability.[186]

VII. Defences

1. Licence/permission to grow GM material

The GMO Act does not specifically provide for any liability defences. How- **90** ever, as has been said, liability according to its provisions depends on the fact that the liable party has committed an administrative violation (Art. 38). Since administrative liability is based on fault, as described above, any defence which excludes fault of the defendant would free him from liability according to the general rules of administrative liability.

As regards the effect of the licence or permit on liability, as a rule it is no **91** defence under Spanish tort law.[187] Licences or permits are issued as a rule even though the authorised activity may cause damage to third par-

182 See *Ruda*, ERPL 2004, 251 ff. and III above (no. 21 ff.).
183 It is usually understood that this is a strict liability rule and its analogical application is therefore refused. See *Gómez Calle* (fn. 172) 1004 with further references.
184 As *Ribot* (fn. 40) 559 suggests.
185 See *M. Martín-Casals*/*A. Ruda*, Comentario a la Sentencia de 26 de noviembre de 2003, CCJC 2004, 843 ff. and *Ribot*/*Ruda* (fn. 48) 401 ff.
186 See *Ruda* (fn. 18) 351 with further references.
187 On the special rules laid down by the Environmental Liability Act see no. 48 above.

ties. In fact, authorisations are usually issued including a clause which leaves the rights of third parties in respect of damage caused to them unaffected (*sin perjuicio de terceros*) on an explicit basis.[188] Therefore, courts usually presume that the licensee has not caused damage unless the victim proves otherwise.[189]

92 The fact that the licence has been issued for testing only does not seem to affect the rule described. However, it may be taken into account in connection with administrative liability. For instance, an infringement of the conditions established in a licence concerning the confined utilisation of GMOs amounts to a severe infraction (Art. 34.3.d) GMO Act), whereas such infringement in the case of licences for free release and commercialisation amounts to a very severe infraction (Art. 34.4.b) GMO Act).

2. Consent/assumption of risk

93 The consent of the victim excludes liability of the tortfeasor, provided that damage affects rights or interests which are at the free disposal of the victim.[190] Assumption of risk also excludes liability. This circumstance is usually taken into account to exclude liability of the organisers of risky activities such as rafting, fireworks, kart races, bullfighting and the like. Nonetheless, assumption of risk may not be taken into account with regard to risks created by the negligent behaviour of the defendant, e.g. in the case of the negligent organisation of a sport competition.[191] In some cases, assumption of risk may reduce compensation due to this negligent behaviour of the organiser.[192] However, assumption of risk is still a blurry concept under Spanish law but it is sometimes referred to, although properly speaking liability is excluded for a different reason, such as lack of fault of the defendant.[193] Therefore, if the victim was aware of the risk created by GM products, liability would be excluded. However, it is required that such a knowledge existed, which would obviously be a problem in connection with development risks.

---

188 See *Santos* (fn. 19) 3032 and *Ruda* (fn. 18) 403 with further references.

189 See *Cordero* (fn. 72) 443.

190 See *M. Martín-Casals/J. Solé Feliu*, Contributory Negligence under Spanish Law, in: U. Magnus/M. Martín-Casals (eds.), Unification of Tort Law: Contributory Negligence (2004) 180.

191 See the case commented on by *Ruda* (fn. 49) 444 ff.

192 See *Reglero/Medina* (fn. 82) 906 ff.

193 See *J. Solé Feliu*, Los perfiles borrosos de la asunción del riesgo en el Derecho comparado, in: A. Cabanillas Sánchez et al. (eds.), Estudios jurídicos en homenaje al profesor Luis Díez-Picazo II (2003) 3097 ff.

3. Third-party influence

The act of a third party usually excludes or reduces liability, depending on the contribution of the third party and the defendant to damage.[194] If the intervention of the third party is so important that the behaviour of the other party becomes irrelevant, such intervention of the third party is considered to amount to force majeure with regard to liability of the other.[195] This rule has an exception in the case of environmental liability. The operator will not be obliged to bear the costs of restoration of environmental damage provided that it is proved that the damage or threat of damage has been caused by the behaviour of a third party. However, for such an exception to be applied, it is also required that the third party does not belong to the sphere of organisation of the operator's activity and that he is independent from the same. Moreover, it is required that damage or threat of damage is caused by the third party in spite of the appropriate safety measures (Art. 14.1.a) LRM). **94**

If these requirements are met, the liable party will be obliged to adopt the measures of prevention, avoidance and restoration of environmental damage pursuant to the Act (Art. 14.3). However, inasmuch as he is not obliged to bear the costs referred to, he will have the possibility to file a claim against the third party who caused damage or threat of damage to recover the costs he incurred (Art. 15.1 and 16.1). Moreover, the public authority may oblige the third party to pay the costs incurred (Art. 15.1 par. 2). **95**

Also the statutory regime on liability for damage caused by defective products or other goods or services refers to the intervention of a third party as a circumstance which will not reduce liability of the liable party according to the Act. Therefore, the liable party – e.g. the producer (Art. 135 LGDCU) – will be liable before the victim, but will have the possibility to file a claim against the third party to recover compensation (Art. 133 LGDCU). **96**

4. Prescription

The general rule established by the Spanish Civil Code provides that limitation of actions in tort is one year after the moment when the victim knew about liability (Art. 1968.2). The same period is established with **97**

194 Instead of many see *Roca* (fn. 77) 162.
195 See *Reglero/Medina* (fn. 82) 825 and 874.

regard to liability of the Public Administration (Art. 145.2 LRJPAC). In this case, the limitation period starts to run from the moment when the damaging event took place.[196] Very often, environmental disturbance is caused by activities which go on for some time. In case of permanent or continuing damage, the limitation period starts running when the activity ceases causing damage or damage is finally caused.[197]

98 In contrast to this, the Environmental Liability Act provides that its provisions will not be applicable to environmental damage if 30 years have passed after the emission, event or incident which caused it (Art. 4 par. 1). Such period will start from the day when such emission, event or incident has completely finished or has taken place for the last time (Art. 4 par. 2). This seems to be an absolute or security period, similar to the statutes of repose existing in other countries such as the United States of America. Such kinds of periods are considered safety or "caducity" periods (*caducidad*) – i.e. periods which cannot be interrupted – and not limitation periods. Therefore, it is not possible for the victim to file a claim after such period has expired, even though the damage may still not have manifested itself.[198] The rule has been criticised on the basis that some environmental disturbances may produce effects many years later and actually, these effects may not even be known or properly understood according to present scientific knowledge.[199] The absolute period could also be seen as an additional way to exclude liability for damage caused by development risks.[200]

99 Something similar could be said in connection with the product liability regime. Following the Directive on Product Liability, a limitation period of three years from the moment when the victim suffered harm is established (Art. 143.1 LGDCU). For the period to start running it is required that the victim knows the identity of who caused him damage. The liable party has the possibility to claim against the other tortfeasors in an internal manner. Such a claim has a limitation period of one year from the day when the liable party paid compensation to the victim (Art. 143.1 2nd part LGDCU). The limitation period can in any case be interrupted according to the rules of the Civil Code (Art. 143.2 LGDCU). However, the Act also pro-

196 See *González Pérez* (fn. 29) 508 ff.

197 Among others see *Cordero* (fn. 72) 451.

198 See *L.F. Reglero Campos*, La prescripción de la acción de reclamación de daños, in: L.F. Reglero Campos (ed.), Tratado de responsabilidad civil I (4th ed. 2008) 1225 and *Ruda* (fn. 18) 430 ff.

199 See *M. Medina de Lemus*, Medio ambiente. Protección y responsabilidad (2007) 74. Similarly, see *J.M. Mora Sánchez*, Biotecnología vegetal: un enfoque legal, in: E. Iáñez Pareja (ed.), Plantas transgénicas: de la Ciencia al Derecho (2002) 235.

200 See *Jordano* (fn. 33) 36.

vides that the rights of the victim to obtain compensation according to the product liability regime are extinguished 10 years after the date when the product was put into circulation, unless the claim is filed before the court during such a period (Art. 144 LGDCU). Given the scientific uncertainty and the fact that damage caused by GMOs may arise many years after utilisation, such a caducity period seems an important hurdle for the victim to get compensation.[201]

Apart from this, the Environmental Liability Act lays down a limitation period applicable to the claim by the public authority to recover costs incurred against the liable operator, in the cases where the public authority had to intervene in a direct way to protect the natural resources (Art. 23) or where it had to intervene on a subsidiary basis due to a breach of the operator's obligations (Art. 47). Such limitation period is five years (Art. 48.2). The period will start to run from the later of the following dates: a) the moment when the public authority ended the adoption of the measures; b) the moment when it identified the liable party. This rule, which follows the Directive (Art. 10), departs from the general regime, which does not explicitly require that the victim identifies the liable party to start the limitation period running. However, legal scholarship requires such identification, taking into account the need to protect the victims of environmental harm.[202] Apart from this, the rule laid down by the Environmental Liability Act also departs from the rule in connection with goods belonging to a Public Administration (so-called *demanio* or *bienes demaniales*). Such goods cannot be acquired on the basis of adverse possession under Spanish law (Art. 132.1 CE). Therefore, case law has sometimes considered that the claims to protect such goods are not affected by limitation of actions.[203] Some statutes protecting specific kinds of natural resources even establish on an explicit basis that the liability claim can be filed regardless of the period of time that has passed since the damage occurred.[204] 100

The Environmental Liability Act also establishes the causes of interruption of the limitation period (Art. 48.3), in contrast with the Civil Code, which has no rules on this subject. Such causes are the following: a) any claim by the public authority in charge, filed with the formal knowledge of the liable party, with the aim of attributing to such party any kind of liability derived from the Environmental Liability Act or any other Act; b) the 101

201 As *Kemelmajer* (fn. 63) 351 and *Jiménez* (fn. 75) 292 point out.
202 See *Ruda* (fn. 18) 431.
203 See *Jordano* (fn. 33) 35.
204 Such as the Woodlands Act (Art. 77) and the Coasts Act (Art. 92). See also *Valencia* (fn. 32) 129.

beginning of a criminal procedure for the same facts which give rise to liability governed by the Environmental Liability Act; c) the petition of the interested parties, with the formal knowledge of the liable party (pursuant to Art. 44 LRM), and finally d) any other act of acknowledgement of liability by the liable party.

5. Other defences

102 Most relevant defences have probably already been dealt with in the replies to the previous questions. Apart from that, the Environmental Liability Act provides that in case of environmental damage or threat of damage caused by diffuse pollution, liability will only be established provided that it is possible to prove that a causal link exists between damage and the activities of specific operators (Art. 3.3). As for the statutory regime on product liability, it is provided that contributory negligence may exclude or reduce liability of the producer (Art. 145 LGDCU). This is in conformity with the general rules on tort liability. Finally, coming to the nuisance is no defence under Spanish law.[205]

VIII. Remedies

1. Pecuniary compensation

(a) Bodily harm

103 As has been explained, bodily harm lies outside of the scope of the Environmental Liability Act.[206] Liability according to this Act is unlimited, although the financial guarantees have a limited cover as has also been explained.[207] As regards the GMO Act, it merely refers to damage without distinguishing whether it is personal injury or a different kind of damage. Liability pursuant to this Act is also unlimited. Therefore, the regular remedies apply to bodily harm caused by GMOs.

---

205 See *Cordero* (fn. 72) 441 and *Egea* (fn. 121) 421.
206 See no. (c) above (c).
207 See no. 52 ff. above.

(b) Property losses

With regard to private property, the regular remedies apply to claims filed by private parties, e.g. the owner or the holder of a different interest in the thing. However, the Environmental Liability Act lays down special rules concerning restoration of damage to natural resources, which may belong to private parties or not. These remedies follow in general the provisions of the Environmental Liability Directive, in the sense that no financial compensation to the public is provided for and preference is given to restoration in kind.[208] **104**

(c) Economic losses

In the case of economic losses caused by GMOs, the regular remedies would be applicable. To the reporters' knowledge, no claims have been filed before the Spanish courts in connection with the economic loss described in the question. In that case, the Public Administration would be entitled to force the polluter to restore the damage pursuant to the GMO Act (Art. 38 ff.). This is also a kind of pollution taken into account by the Environmental Liability Act, as has been explained.[209] In theory, the affected farmer could file a claim in tort to oblige the polluter to restore the field to its previous condition or subsidiarily pay financial compensation for damage suffered, plus any other kind of damage derived from such pollution. Such a claim should be grounded upon general provisions of tort liability. However, as a rule the victim of damage has the duty to mitigate damage, since otherwise it would contribute to damage and compensation could be reduced accordingly or even be excluded.[210] Nevertheless, for such a reduction to take place it should be proved that the farmer in the hypothetical would actually have reduced damage by having grown conventional crops, which will probably depend on the circumstances of the case. **105**

(d) Harm to animals

In general, restoration of damage has to be in kind, i.e. the cause of damage has to be eliminated and so as to restore the situation which previously existed. However, in the case of an animal harmed because of hav- **106**

208 See no. 12 above.
209 See no. 6 above.
210 See *Martín-Casals/Solé* (fn. 190) 179.

ing eaten contaminated feed, restoration in kind may not be possible. In this case, the liable party has to restore damage per equivalent, by paying the amount of money at which the harmed object is assessed, or the cost of its restoration and of the loss of value.[211] To do such an assessment, the characteristics of that kind of animal would be taken into account. Therefore, it would be assumed that a cow is normally used to produce milk or meat. In the case described by the example, it is assumed that the animal will not lose all its value due to the mere fact of having eaten contaminated feed. Probably it may keep the potential to produce milk or meat, although they would not qualify as GMO free. Therefore, compensation should be paid for the loss of value, should this loss occur, and other losses arising from it.

(e) Costs of disposal

107 Such costs would be recoverable if all the conditions of liability according to the general rules on tort are met. In particular, the victim should prove that damage is certain and has been caused by the defendant. The defendant may argue that disposing of the production or animals is an excessive response to pollution with GMOs and therefore the duty to mitigate damage has been infringed by the victim.[212]

2. Non-compensatory damages

108 Some legal scholars have suggested that punitive damages should be awarded to the victim in cases of nuisance, e.g. for damage caused by noise.[213] However, punitive damages are foreign to the Spanish legal tradition and rejected by the prevailing opinion.[214]

3. Other remedies

109 The GMO Act lays down several criteria that the Public Administration may use to assess damage resulting from an administrative violation on

211 See *Roca* (fn. 77) 203.

212 See no. (c) above (c).

213 Among others see *F. J. Fernández Urzainqui*, La tutela civil frente al ruido (2003) 101. Similarly but in connection with the Environmental Liability Directive, see *Jordano* (fn. 33) 46.

214 See *García Rubio* (fn. 41) 116 and *Ruda* (fn. 18) 225 with further references.

an economic basis. In particular, such criteria are provided for situations where damage is otherwise difficult to assess and constitute the following: the theoretical cost of restitution and restoration, value of the damaged goods, cost of the project or activity which caused damage, and benefit derived from the infringing activity. These criteria may be applied as a whole or separately (Art. 38.1 par. 2).[215]

With regard to the Environmental Liability Act, no financial compensation to the public is provided for. Instead, the objective pursued by the statutory regime is that restoration be made in kind. If this is not sufficient, other measures will be adopted to compensate interim losses as well as remaining damage.[216] **110**

4. Costs of pursuing a claim

(a) General cost rule

According to the Civil Procedure Act, each party in a private law procedure has to pay his own legal expenses and costs as soon as they are incurred (Art. 241.1).[217] However, the Act makes a distinction between expenses (*gastos*) and costs (*costas*). The first are those payments which are due in a direct and immediate way to the existence of the procedure. The second category comprises several kinds of costs, among which there are the lawyer's fees (Art. 241.1.1) and the fees of other experts who may have intervened in the procedure (Art. 241.1.4). Such experts cannot request the payment of their fees before the case has been ended by a definitive judicial decision. These fees are usually included, together with the other procedural expenses, in an assessment of costs done by the Judicial Secretary once the decision has been issued (Art. 242.3). Therefore, the person who incurs expenses as a result of the evidence brought to court is not obliged to pay at that moment to satisfy the experts' fees but will have to wait for the outcome of the trial. **111**

Usually, the court decisions include a pronouncement on the costs of the trial. If such a pronouncement is lacking, the distinction between **112**

215 See *Martín-Casals/Ruda* (fn. 4) 444.

216 See no. 12 above.

217 Claimants with fewer resources can apply for so-called free legal assistance (*asistencia jurídica gratuita*). Such assistance includes, inter alia, the intervention of experts in the procedure with no cost for the beneficiary.

expenses and costs mentioned above will not be applied.[218] Such a pronouncement will depend on the outcome of the case but it generally follows the "loser pays" principle (*principio del vencimiento*). If the claimant has completely failed and none of his requests are accepted by the court, then he will have to bear the full costs of the trial (*condena en costas*). However the court may consider that the case was doubtful for factual or legal reasons. If such an exception is made, the court will not impose the costs on the claimant only. Doubtfulness of the case will be determined taking into account case law in similar cases (Art. 394.1 LEC). If the court finds for the claimant but only partially, i.e. some of his requests are accepted and others dismissed, each party in the case will have to bear the costs that it caused. Common costs will be divided between the two parties in equal shares, unless the court considers that one of them has filed the claim or opposed it in a reckless way (*litigado con temeridad*, Art. 394.2 LEC). In case of admission or acceptance of the claim by the defendant (*allanamiento*) before replying to it, the court will not impose the costs on the defendant, unless the court considers that he behaved in bad faith (*mala fe*, Art. 395.1 LEC). If the procedure ends because the claimant desists from his claim and the defendant does not accept it, the claimant will have to bear the costs (Art. 396.1 LEC). The present legal regime does not make any exception in cases of claims for GM pollution or for damage caused by a defective product. Such a framework can be very harsh for the claimant, since in case of losing the action he will have to face potentially very high expert fees. Therefore, it is suggested that several claimants file an action together or that they create an association of affected persons to share costs.[219] The associations of consumers of a broad geographical range – i.e. more than one Autonomous Region – may also apply for free legal assistance under certain conditions (Art. 37 LGDCU).

(b) Costs of establishing causation

113 Such costs may be recovered under the conditions described in the reply to the previous question. Therefore, as a rule the party who incurs these costs may have to bear them unless the court shifts them to the other party on the basis of the circumstances already mentioned.

---

218 See *J.F. Herrero Perezagua*, Comentario, in: F. Cordón Moreno (ed.), Comentarios a la Ley de Enjuiciamiento Civil (2001) 879.

219 See *Marín* (fn. 65) 96 f.

5. Advance cover

The financial guarantees have been described in the reply to a previous question.[220] 114

IX. Cross-border issues – Conflict of laws

1. Conflicts rules applicable before (or instead of) Rome II

Given its universal character (Art. 3), the Rome II Regulation will replace the applicability of the regime established by the Spanish Civil Code. According to the latter, so-called extra-contractual obligations are governed by the law of the place where the fact from which they derive has occurred (Art. 10.9 par. 1). Therefore, the criterion of *lex loci delicti commissi* is followed by the Code, as in the Rome II Regulation (Art. 4.1). This is the general criterion under Spanish law, applicable to all instances of liability in tort and functions as a subsidiary rule, applicable to those instances not falling under the scope of application of a different Spanish statute or an international convention ratified by Spain.[221] There is no specific statutory rule concerning environmental damage, in contrast to the Rome II Regulation (Art. 7). 115

With regard to liability for damage caused by defective products, this will be governed by the Hague Convention on the Law Applicable to Products Liability (1973), which has been in force in Spain since February 1989.[222] Since the application of the Convention is also universal or *erga omnes* (Art. 11), it displaces the general rule laid down by the Civil Code within the material scope of application of the Convention. As is well known, this Convention makes use of the technique of multiple points of connection, and the criterion *lex loci delicti commissi* is not considered the main connexion point. With regard to damage which may be localised in several different jurisdictions, legal scholarship has pointed out that the solution provided by the Convention is imprecise to some extent and that the criterion of *lex loci delicti commissi* should be construed as exclusively referring to the place where damage manifests itself.[223] In any case, it is also known 116

---

220 See no. 52 ff. above.

221 See *M.A. Amores Conradí*, in: J.D. González Campos/J.C. Fernández Rozas/A.L. Calvo Caravaca/M. Virgós Soriano/M.A. Amores Conradí/P. Domínguez Lozano, Derecho internacional privado. Parte especial (6th ed. 1995) 210 and *G. Palao Moreno*, Aspectos internacionales de la responsabilidad civil por servicios (1995) 58, 60.

222 *Instrumento de ratificación de 7 de noviembre de 1988* (BOE no. 21, 25 January 1989).

223 See *J.C. Fernández Rozas/S. Sánchez Lorenzo*, Derecho internacional privado (2nd ed. 2001) 613 and *M.E. Zabalo Escudero*, La ley aplicable a la responsabilidad por daños derivados

that the Rome II Regulation also provides for a rule concerning product liability (Art. 5). Nonetheless, it will not displace the application of the rules of the Convention referred to, since as is known Rome II does not prejudice the application of international conventions to which Member States are parties at the time when this Regulation was adopted (Art. 28.1).

117 If we leave this aside, it may be underlined that the new regime introduced by the Rome II Regulation is more flexible than the previously existing Spanish regime. For instance, the Civil Code lays down no exception similar to Art. 4.2 of the Regulation, according to which if the person allegedly liable and the person sustaining damage both have their habitual residence in the same country at the time when the damage occurs, the law of that country shall apply. In contrast to this, Art. 10.9 CC is rather inflexible, providing a single criterion which completely ignores the possibility taken into account by the Regulation. Moreover, it is usually considered that Art. 10.9 CC lays down a mandatory rule, so the parties are not allowed to agree on a different criterion – quite the opposite is provided by Art. 14 of the Rome II Regulation, as is also known. Therefore, most legal scholars would probably agree that the flexibility of the Regulation is preferable to the CC rule.[224] Apart from this, the Art. 10.9 CC rule may be inappropriate whenever the results of the damaging events manifest themselves in several different countries, for instance in case of infringement of personality rights through mass media, or damage arising from cross-border pollution.[225] In connection with the latter, some scholars suggest that the law of the place where the result manifests itself should be applied.[226]

118 In fact it may be very difficult to establish which is the "place where the fact occurred" as referred to by the rule laid down by Art. 10.9 CC. It may remain unclear whether the relevant place is the one where the action which caused the damage took place or the place where the result of such an action was produced.[227] This has given rise to a lively scholarly debate. According to one opinion, the solution should be determined by the courts in every single individual case, taking into account that the judge

de los productos en el Derecho internacional privado español, Revista Española de Derecho Internacional (R.E.D.I.) 1991, 90.

224 See the criticism against the latter by *Fernández/Sánchez* (fn. 223) 607 f.; *Zabalo*, R.E.D.I. 1991, 106; *P. Abarca Junco*, in: E. Pérez Vera (ed.), Derecho internacional privado II, Derecho civil internacional (1998) 339 and *A.-L. Calvo Caravaca/J. Carrascosa González*, Derecho internacional privado II (2003) 601.

225 See *A. Miaja de la Muela*, Derecho internacional privado II, Parte especial (10th ed. 1987) 398 and *Calvo/Carrascosa* (fn. 224) 584.

226 See *G. Palao Moreno*, La responsabilidad civil por daños al medio ambiente (1998) 118, 128.

227 See *Zabalo*, R.E.D.I. 1991, 77.

may apply the law which is most favourable to the victim.[228] However, it seems that the principle of protection of the victim (*favour laesi*) cannot be used in such a way as a basis for a solution since it is difficult to reconcile with both Spanish tort law and private international law.[229] On the other hand, leaving it up to the judge to decide whether to apply the law of the place of the damaging activity or the place of the damaging result seems an insecure solution which does not meet the demands of legal certainty.[230] Another opinion has held that a distributive approach should be adopted, so establishment of liability is decided according to the law of the place where the damaging behaviour – e.g. pollution – originates, whereas the obligations derived from liability, proof of the causal link and the nature and extent of compensable damage be decided according to the law of the place where the victim has suffered the result of such behaviour.[231] However, this has been criticised on the basis that applying two different laws to the same case would be against the Spanish legal rule, according to which the law (in singular) of the place where the act has occurred has to be applied (Art. 10.9 CC).[232] Finally, a different author has argued that it is better to interpret the rule laid down by Art. 10.9 CC depending on the function of the particular area of tort law and depending on policy at stake and the interests that are pursued in that area of tort law.[233]

As regards the cases where the causal sequence occurs in more than one **119**
place, i.e., where the behaviour of the tortfeasor could lead to liability taking into account his conduct in each of the places at stake, or the damaging result manifests itself in more than one place, there are no reasons to prefer one causal place in relation to others, nor for preferring one place where the results manifest themselves in relation to others. For instance, it has been pointed out by legal scholarship that if the behaviour of the tortfeasor amounts to unfair competition and the result of his conduct

228 According to *Abarca* (fn. 224) 341.
229 See *Amores* (fn. 221) 227.
230 See *J.M. Espinar Vicente*, Notas para un estudio de las nuevas perspectivas de la responsabilidad por daños en el Derecho internacional privado, Anuario de estudios sociales y jurídicos 1977, 85, and *Palao* (fn. 226) 117.
231 This opinion is held by *N. Bouza Vidal*, Aspectos de Derecho internacional público en la protección del medio ambiente, in: Asociación Española de Profesores de Derecho Internacional y Relaciones Internacionales (ed.), Problemas internacionales del medio ambiente (1985) 90, in connection with damage derived from pollution.
232 See *K. Fach Gómez*, La contaminación transfronteriza en Derecho internacional privado (2002) 182.
233 This is the opinion of *M. Virgós Soriano*, El comercio internacional en el nuevo Derecho español de la competencia desleal: un análisis del art. 4 de la Ley española de competencia desleal de 1991 (1993) 99.

produces effects in more than one jurisdiction, there will be different liabilities, one in each of the markets affected.[234]

2. Special regime for cross-border claims

120 In contrast with the GMO Act, which contains no rules in this regard, the Environmental Liability Act lays down a specific rule concerning cross-border damage (*daños transfronterizos*, Art. 8). According to the same, in the presence of environmental damage or an imminent threat that such damage occur, which may affect another Member State of the EU, the authority in charge will have to inform the Ministry of the Environment (Art. 8.1). The Ministry has the duty to provide information to the authorities of the affected Member States and cooperate with them to prevent damage from occurring, as well as to restore it. Moreover, it will have to adopt all measures needed so that the liable operators assume the costs which the authorities of the affected Member States have incurred, according to reciprocity criteria (Art. 8.2). It is also provided that the Spanish authorities may adopt some measures in case of damage or threat of damage caused by an activity carried out in the territory of a different Member State, namely, making recommendations and initiating the adoption of preventive or restoration measures (Art. 8.3). However, no specific rules concerning liability in cross-border scenarios are provided by these Acts.

X. Cases

1. Due to the adventitious presence of GMOs in a field, maize which is normally sold as conventional contains GMOs beyond the legal labelling threshold. This is not discovered before the final stage of the food production chain by the producer of taco chips. The whole production is lost since the supermarket chains refuse to accept delivery from the producer.

(a) Who can sue along the chain of distribution?

121 As a matter of fact, the GMO Act does not refer to the relationship between GMOs and agricultural food but is applicable to any GMO in general. Some

234 See *Amores* (fn. 221) 228.

legal scholars have pointed out that a specific legal regime is needed to take problems related to the release and commercialisation of plants derived from GMOs into account.[235] In this hypothetical, the maize producer might not only suffer economic loss, but even lose his certification of ecological agriculture.[236] This hypothesis is reminiscent of the case decided upon by the Spanish Supreme Court on 27 January 2007 (RJ 2007/296).[237] Due to the construction of a motorway, one of the plots of land on which the claimant developed an organic farm had to be expropriated. His other plots had to be removed from the register of agricultural farms certified by the board regulating organic farming, since it considered that the proximity of the motorway gave rise to a change in environmental conditions that prevented that sort of production from continuing. The Supreme Court stressed that organic and ordinary farming fit into a single activity and that the difference between them refers to the farming techniques used and to the profitability of the farm, but not to the nature of the activity. The Court admitted that the construction of the motorway by the Public Administration had an impact on the development of organic farming, but stated that it did not give rise to any impediment to the further continuation of ordinary farming activities by the claimant. The termination of all agricultural activity was his free decision and the economic losses resulting from it were not attributable to the Administration. On the other hand, deprivation of the opportunity to continue with organic farming does not give rise, by itself, to a real and actual harm, since it is necessary to prove that the harm is directly linked to this fact.

Apart from this case, the present statutory regime is very broad in scope **122**
and not very detailed, as has been explained. The Public Administration is entitled to force the liable party to restore damage in kind and pay monetary compensation (Art. 38).[238] However, for liability to arise under this regime, it is required that the liable party has committed an administrative violation. Therefore, it should be established in an administrative procedure that someone infringed the existing regulations–for instance, because of a lack of adoption of measures on the confined utilisation of GMOs (Art. 34.3.f), the release of GMOs into the environment without a licence (Art. 34.4.a) or the breach of the licence conditions (Art. 34.4.b).

---

235 See D. *Llombart Bosch*, La utilización confinada de microorganismos modificados genéticamente, y la liberación intencionada de OMG en el ambiente, Cuadernos de Derecho agrario 2004, 105.

236 As *Cazorla* (fn. 10) 154 observes.

237 The summary by M. *Martín-Casals*/J. *Ribot Igualada* in the Spanish report for the Digest of European Tort Law (forthcoming) is followed here.

238 See no. 3 ff. above.

Only then could liability pursuant to the GMO Act be determined. Should this requirement be met, any of the affected parties could request the Public Administration to proceed against the person or persons who caused pollution of the field by GMOs.

123 The victims would also have standing to sue according to the general rules of liability based on fault (Art. 1902 CC) and liability for nuisance (Art. 1908.2 and 1908.4 CC) as construed by the courts.[239] Such provisions provide for liability rules without reference to whether the liable party has also violated administrative standards or not. However, it could be argued that the victims should have detected GMO pollution sooner. Nevertheless, it is well known that sometimes feed is sold mixed with conventional seed, since in some countries of origin separation of seeds is not compulsory. In these cases, the victims could argue that it was not possible for them to know that the product was GM. The same happens in cases of adventitious pollution of crop fields which are close to a test field where GMOs are released. In this situation, it is very difficult for the conventional farmer to detect the presence of GMOs in his or her crop.[240]

124 As for the Environmental Liability statutory regime, private parties cannot ground their claims in tort on the provisions of the Act (Art. 5).[241] Moreover, this is not strictly speaking "environmental damage" as defined by the Act (Art. 2),[242] since presumably the victims may wish to obtain compensation for economic loss and not restore damage to a natural resource protected by that Act.

125 Therefore, the affected parties should resort to the general tort law rules already mentioned. In any case, damage to the crop itself because of GMO pollution may not be recoverable under the product liability regime if it is considered to be damage to the product itself (pursuant to Art. 142 LGDCU).[243] However, legal scholarship refers to the compensability of damage consisting in the costs of buying the defective seed, the phytosanitary treatments, the cost of the irrigation water, the journals of the employees, the interest rates of the loans contracted by the farmer to finance the production of his crop, and even the loss of market share.[244]

---

239 See no. 6 above.
240 As *González/Jiménez* (fn. 62) 713 note 131 point out.
241 See no. (c) above (c).
242 See no. 5 above.
243 See *González/Jiménez* (fn. 62) 712.
244 See *Téllez*, Revista de Derecho Agrario y Alimentario 2000, 17 ff.

If damage was caused by pollen coming from other crops or the admixture of crops, the producer of the seed would probably not be liable, or at least not to the same extent. The claim should be filed against the farmers of the crops from which the GM material comes instead. If damage is caused by an indeterminate member of a group, the solution explained in connection with alternative causation would be applicable.[245] Compensation could also be sought on the basis of other liability provisions, including the ones provided for by the law of contracts. **126**

(b) Would the case be solved differently if the GMO content was below the labelling threshold?

As has been observed, our statutory rules do not provide any specific solution for the special case of GM pollution of a conventional crop. As a rule, any damage has to be repaired (Art. 38 GMO Act) and no difference is made depending on whether it is serious or not (Art. 1902 CC). Only the Environmental Liability Act requires that a seriousness threshold is exceeded (Art. 2). However, the affected parties who bring a claim on the basis of other statutory provisions may argue that they suffered damage even though pollution does not reach a specific threshold. An apparently insignificant level of pollution may increase after it is added to the pollution coming from other sources in later stages of production. For this reason, it has been suggested that liability should arise even if the threshold of 0.9 % – established by EU legislation – was not exceeded.[246] **127**

(c) Would the case be solved differently if the admixture was not adventitious, but occurred due to the disregard of segregation rules, for example?

The disregard of segregation rules could be a signal that the defendant behaved in a negligent way. This would be relevant to establish liability based on fault under the general liability regime (Art. 1902 CC).[247] The violation of segregation rules, should they exist, could also be taken into account in connection with the presumption of the causal link pursuant to the Environmental Liability Act.[248] **128**

---

245 Thus, on the basis of an analogical interpretation, *Jiménez* (fn. 75) 290. See no. III ff. above (III).
246 See *Casero* (fn. 2) 165.
247 See no. 5 ff. above.
248 See no. 24 ff. above.

(d) Would the case be solved differently if the GMO found was not admitted for production in your jurisdiction?

129 The infringement of the rules on confined utilisation, voluntary release and commercialisation of GMOs may amount to a violation of public law and give rise to the application of the penalty or punishment established by the GMO Act.[249] This may even mean that a crime or misdemeanour has been committed by the defendant. It could be argued that the behaviour of the offender violated the rules on safety and therefore created a situation of risk pursuant to the Criminal Code.[250] Therefore, criminal liability could be established. Apart from this, the circumstance that the GMO found was not admitted for production could be taken into account to establish that the behaviour of the defendant was against the law and therefore wrongful or illegal (*antijurídico*). Although it has been questioned by some authors, it is usually understood that for liability to be established it is required that the defendant behaved in a wrongful way. Wrongfulness may be derived from the mere fact that the defendant caused damage to another and therefore infringed the duty not to harm others (*neminem laedere*).[251] Moreover, the infringement of the legal rules could be taken into account as an element which indicates fault on the side of the defendant.[252]

(e) If the admixture had occurred on a non-GM field and it transpires that the GM seeds were blown from: (i) neighbouring fields; or (ii) a truck passing by, would the farmer of the affected field be liable for all or part of the loss caused further down the distribution chain?

130 The defendant will be held liable in proportion to his contribution to damage. If it can be established that there were several people who contributed to it, the rule is that they will be held liable on a solidary basis.[253] It may happen that it cannot be established whether damage was caused by the farmer or the truck, but it is clear that it was one of them who caused damage. In this case, liability would probably be established on

249 See no. 122 ff. above.
250 See no. 7 ff.
251 See *Roca* (fn. 77) 72.
252 As explained in the reply to the previous question.
253 See no. VI ff. above.

the basis of alternative causation.[254] If GM pollen was blown by extremely strong winds, it could be argued that the damage was caused by force majeure.[255] Then, liability of the defendant could be reduced accordingly or even completely excluded, as already explained.

2. Twenty years after the sale of GM maize used for food products, it turns out that it has certain disadvantageous health effects for humans.

(a) Can the producers be held liable at this point for risks unknown at the time of growing the maize? Who would be liable?

If the damaging effect of the GM maize was unknown when it was used, **131** fault on the part of the defendant would be excluded under the general tort liability regime (Art. 1902 CC).[256] Liability pursuant to the GMO Act would also be excluded, since it requires fault on the side of the offender (Art. 38).[257] As for the Environmental Liability Act, liability of the operator is excluded if the development risk defence applies.[258] However, the defendant may be held liable under the product liability regime, since as has been said, the development risk defence is excluded in connection with damage caused by food and food products.[259] In that case, limitation periods may bar the claim, since the rights of the victim to obtain compensation according to the product liability regime are extinguished ten years after the date when the product was put into circulation.[260]

(b) Can compensation already be claimed at a point when the negative health effects have not yet materialised, but are to expected to according to scientific expertise/mere rumours?

As has been explained, for liability to be established, it is usually required **132** that damage is certain.[261] In case of mere speculations as to whether

254 See no. III ff. above.
255 See no. 2 and 3 ff. above.
256 See no. 5 and 2 above.
257 See no. IV above.
258 See no. 39 above.
259 See no. 2 above.
260 See no. 99 above.
261 See no. 16, 4, 71 ff. and (e) above.

damage exists or not, the court may consider that damage has not been sufficiently established. If damage will arise in the future with certainty, then it is recoverable.[262] Compensation for the mere fear of suffering damage in the future could probably also be recovered.[263]

(c) Would it make any difference if the GM maize had only been in use for feed, causing harm to the animals, which may or may not cause harm to humans consuming the meat as well?

133 As has been seen, the development risk defence is only excluded in case of damage caused by defective food or food products for human consumption, according to the product liability regime.[264] Therefore, if GM maize had been used only to feed animals, liability would probably be excluded on the basis of this defence, should all the conditions for its applicability be met. Apart from this, it could be considered that such harm amounts to damage to be repaired under the GMO Act provisions. Harm to farm animals would nonetheless fall outside the scope of the Environmental Liability Act, since it provides no standing for individuals for private or so-called traditional damage.[265] The claim should therefore be based on other provisions, such as the general regime of liability on fault.[266]

3. The driver of a food logistics company discovers that a farmer, from where he regularly picks up agricultural products fails to obey mandatory segregation rules or food or feed hygiene standards, which may lead to the admixture of GM and non-GM produce sold and packaged separately by that farmer. Does he or his employer have a duty to warn, i.e. warn the recipient of the allegedly "non-GM" produce?

134 According to the GMO Act, any activity by means of which the genetic material of an organism is modified, as well as the farming, storage, use, transport, destruction or elimination of a GMO, is considered to be a confined utilisation governed by this Act, provided that the activity is carried out adopting confinement measures to restrict the contact of the GMO

262 See no. 4 above.
263 See no. 16 above.
264 See no. 2 above.
265 See no. (c) above.
266 See no. 2 above.

with the public and the environment (Art. 5.1). The Act lays down a duty to inform the Public Administration in advance that a confined utilisation is going to take place (Art. 8). This duty affects any physical or legal person who wants in the first place to use specific facilities for the confined utilisation of GMOs (Art. 8.1 par. 1). However, this obligation is also applicable to any other person who wants to carry out any activity of confined utilisation of GMOs, except if the activity creates an insignificant risk or no risk at all (Art. 8.1 par. 2). Therefore, it could be argued that the driver in the hypothetical would be obliged in the first place to inform the Public Administration.

The duty to label the GMO product is referred to by the GMO Act on the **135** commercialisation of GM products. Before commercialising GM material, it is required that the interested person files a request with the Public Administration (Art. 14). Such a request must include, inter alia, a proposal for labelling and packaging (Art. 14.1.e). The Public Administration will decide on whether to authorise the commercialisation or not (Art. 16.1). The authorisation will have to specify the requirements of labelling and packaging (Art. 16.3.e). However, transport is not considered to be a commercialisation act (Art. 13.3). Therefore, it seems that these obligations are not applicable to the driver in the hypothetical. Nevertheless, the Act establishes some duties concerning traceability of the GMOs. According to it, the persons who commercialise GMOs or products which contain GMOs have the obligation to preserve and transfer the data and information established by the regulations to make control easier and to allow for the withdrawal of the GMOs or product from the market at all the stages of commercialisation (Art. 18).[267]

The Regulation which develops the GMO Act imposes a duty to warn on **136** the operators[268] who commercialise products containing or consisting of GMOs at any stage of their production or distribution (Art. 45). They must preserve in writing and transfer to the operator who receives the product the following pieces of information: a) an indication that the product contains or is made up of GMOs, and b) the unique identifier or identifiers assigned to those GMOs according to the EC procedure established

267 There are very detailed special rules concerning labelling and traceability of specific kinds of products such as meat, fish and eggs. For further details see *F. Ruiz Luque*, Etiquetado y trazabilidad en los alimentos, in: R. Herrera Campos/M.J. Cazorla González (eds.), Sociedad de consumo y agricultura biotecnológica (2006) 213 ff.

268 According to the Regulation, an operator is any person who commercialises a product or receives a product commercialised in the EU, be it from a Member State or from a third country, at any stage of its production or distribution. The final consumer is not to be considered an operator in this sense (Art. 4.e).

to this effect (Art. 45.1). In the case that the products made up of GMOs or containing GMOs are intended to be used as either food or fodder or to be processed, the information to be provided by the operator can be replaced by a so-called declaration of use (*declaración de uso*) by the operator. Such a declaration has to be joined to a list of the unique identifiers of all the GMOs which have been used to create the mixture (Art. 45.2). In any case, regardless of whether the GMO product is intended to be used as food or fodder or not, the operators have the duty to guarantee that the label will contain certain specific pieces of information (Art. 45.3). In case of pre-packaged products, the label must indicate that the product contains GMOs or the names of the GMOs it contains. And in case of non pre-packaged products offered to the final consumer, the indication that the product contains GMOs or the names of the GMOs it contains will have to appear in the presentation of the product or in the documentation accompanying it. It has sometimes been suggested that such obligations may be difficult to fulfill since in many cases determining whether food derives from a GMO or not will be very complicated from a technical point of view.[269] Moreover, it has been suggested that a clearer solution would have to expressly provide for the possibility that the producer indicates that the product is "GMO free". Otherwise it is argued that the consumer may still have some doubts as to whether a specific product contains GMOs or not.[270] In any case, these rules (pursuant to Art. 45 of the GMO Regulation) are not applicable to the traces of GMOs which are present in a proportion not exceeding the thresholds established by EU law, provided that such presence is accidental or technically unavoidable (Art. 45.5). The operators will have to prove that they have adopted the required measures so as to avoid the presence of GM material.[271] In conclusion, the driver in the hypothetical may be affected by the duty to warn as established by the present rules on traceability, but may escape such an obligation if the latter exception is applicable, having regard to the proportion of GM material and the other circumstances referred to above.

269 See G. *Morcillo/E. Cortés/J.L. Garcia*, Biotecnología y alimentación (2005) 341.

270 See C.L. *De Cuenca*, Normativa de etiquetado en alimentos y piensos modificados genéticamente, in: E. Muñoz (ed.), Organismos modificados genéticamente (2006) 73.

271 As *id.*, 76 suggests.

Damage Caused by GMOs under Swedish Law

Philip Mielnicki and David Langlet

I. General overview

1. Special liability or redress scheme for GMOs

In Sweden no special liability or compensation regime is in place, nor has any such regime been proposed. That has thus far been deemed unwarranted, as concluded by an expert commission which was instructed to analyse these issues.[1] Hence, existing liability rules shall be applied. 1

As already outlined in the first study on GMO liability,[2] there are two viable liability regimes which may apply: the *special regime of strict liability of the Environmental Code*[3] (chap. 32), or the *Tort Liability Act*[4] together with general principles of tort law. The liability under the Environmental Code particularly differs from general tort liability in that the former explicitly provides for strict liability, it offers a wider liability for pure economic loss and has a relaxed burden of proving causation (on the balance). However, it is uncertain whether that special liability regime may be applied to GMO admixture. At an official level, there seem to be two conflicting views as to the applicability of that regime: the Government presupposed the regime to be non-applicable[5] whereas the expert commission held the regime likely applicable. Notwithstanding, these views do not have any binding effect on what liability regime is applicable. The issue must therefore be decided by the courts. 2

1 Government Official Report (SOU 2007:46, Ansvarsfrågan vid odling av genmodifierade grödor), presented to the Government in June 2007.

2 *D. Langlet/M. Schultz*, Economic Loss Caused by GMOs in Sweden, in: B.A. Koch (ed.), Economic Loss Caused by Genetically Modified Organisms (2008) 475 ff.

3 Miljöbalk (1988:808).

4 Skadeståndslag (1972:207).

5 As outlined in the Government Instructions (dir. 2006:38, Ansvarsfrågan vid odling av genmodifierade grödor) requesting the aforementioned Government Official Report.

3 We shall apply the same method as used in the previous study and thus primarily base our analysis on the Tort Liability Act and general principles of tort law, but make reference to the special liability regime of the Environmental Code where it may be of interest.

2. State liability

4 State liability is particularly provided for under the Tort Liability Act for personal injury, property damage and pure economic loss caused by wrongful or negligent conduct in the course of the exercise of public authority (chap. 3 sec. 2).[6] The liability is wider for the State (including municipal authorities) than under the general rules of damages liability, particularly having regard to the fact that liability for pure economic loss is not coupled to any special restriction. Furthermore, the assessment of (wrongful conduct and) negligence is made objectively, disregarding any individual characteristics of the authority or civil servants involved. Negligence may be cumulated, in the sense that several instances of improprieties may constitute attributable negligent conduct, and negligence may also be anonymous whereby it is not necessary to attribute negligence to any particular civil servant.

5 Liability is thus triggered for decisions but also other conduct involved in exercising public authority in relation to a person. With regard to GMOs, such authority is exercised by supervision and the granting of concessions by the relevant authority. However, there is a rather limited scope for liability to arise in relation to a person claiming that damage has been caused as a result of poor supervision or control on part of the public authority. Firstly, the rules governing the public authority's activity and responsibilities must entail a protective purpose aiming at the protection of the person claiming damages.[7] This may indeed be the case with regard to the rules on precautionary measures which should be supervised by the relevant public authority. Secondly, it must be determined that the public authority was negligent in its granting of a concession or in its supervision, whereby it must be held that the duty to take precautionary measures did not rest solely upon the GMO-farmer or another person. The primary duty rests upon the GMO-farmer, and the precautionary rules for GMOs are also primarily addressed to those persons carrying out activities involving GMOs. However, if the public authority has failed to monitor a

6 See *B. Bengtsson/E. Strömbäck*, Skadeståndslagen – En kommentar (2nd ed. 2006) 88 ff.
7 *J. Hellner/M. Radetzki*, Skadeståndsrätt (7th ed. 2006) 450.

person who has applied for and been granted a GMO concession, and omitted its obligation to check if the conditions for granting such a concession were at hand, liability could arise for the State. A few cases indicate that liability may arise under such circumstances. The cases concern building concessions granted for structures which later turned out to be faulty and caused property damage and even damage suffered by a third person who had not applied for the concession but bought the property subsequently from another.[8]

State liability may also arise under special circumstances for pure economic loss caused by information or advice given by a public authority and attributable to wrongful or negligent conduct (Tort Liability Act chap. 3 sec. 3). Some information or advice given may even qualify under the general head of state liability (chap. 3 sec. 2), where such communications may have been interpreted as a directive from the public authority and are thus comparable to conduct in the course of the exercise of public authority. That may be the case where an authority conveys to a person certain information of a nature that it can be interpreted as an obligation for that person to invest in certain measures (e.g. safety equipment) which de facto and de lege are unnecessary.[9] However, other information or advice upon which a person has placed reasonable reliance may also trigger state liability under the second head (chap. 3 sec. 3). 6

Lastly, it may be mentioned that the idea of establishing a public compensation fund has been rejected thus far.[10] 7

II. Damage

1. Recoverable losses

Under the general liability rules of the Tort Liability Act, losses caused as a result of a personal injury or property damage are recoverable based on negligence, whereas pure economic loss is only recoverable under certain conditions (see no. 13 below). 8

Generally with regard to *personal injury*, compensable losses include (Tort Liability Act chap. 5 sec. 1): a) costs for medical care and other costs; 9

8 Nytt Juridiskt Arkiv (NJA; selected case-law from the Supreme Court) 1984, 340 I-III; cases concerning liability for public authorities which had granted building concessions for structures which later on turned out to be faulty causing property damage.

9 NJA 1990, 705.

10 Government Official Report SOU 2007:46 (fn. 1) 246 ff.

b) loss of income; c) and pain and suffering. However, it is unclear what, if any, personal injuries may arise from GMO admixture.

10 *Property damage* is the primary type of damage which may arise from GMO presence in a non-GM crop or other admixture of GMO. The notion of property damage not only encompasses physical damage to property as such, but also a change of property whereby consideration is taken of whether its function has been lost or deteriorated.[11] Generally, the same notion of property damage is adhered to when it comes to liability for damage, insurance- and product liability. Hence, GMO admixture to another's property – movable or immovable – does harmonise with the notion of property damage.

11 Compensable losses resulting from property damage, as far as relevant here, include (Tort Liability Act chap. 5 sec. 7): a) value compensation involving the replacement cost, e.g. if an ecological farmer or producer has no use for feed which contains GMOs; b) costs consequential to the property damage, such as for re-labelling, extra transports, cleaning of machinery and other costs; c) income loss, e.g. if a product containing GMOs can be sold but at a lower price, the difference can be recovered, and infringements on professional activities associated with losses and costs when the damaged property cannot be used in a professional activity, e.g. losses associated with the loss of a certification.[12]

12 Pure economic loss (see below) must be distinguished from recoverable consequential loss to property damage on the one hand, and third party loss consequential to another's loss on the other hand.[13] The latter is not recoverable under the general principles of tort law. However, if a person has suffered property damage, as earlier described, that person may incur liability towards another person. The cost of that liability may be recoverable for the person who suffered the property damage directly, as a consequential loss of that damage.

2. Pure economic loss

13 Pure economic loss, i.e. an economic loss arising without a connection with personal injury or property damage, is recoverable according to the

11 *J. Kleineman*, Ren förmögenhetsskada (1987) 153 ff.; NJA 2004, 566; NJA 1996, 68; NJA 1990, 80.

12 *Hellner/Radetzki* (fn. 7) 412 ff.

13 *Kleineman* (fn. 11) 180 ff.; *M. Radetzki*, Skadeståndsberäkning vid sakskada (2004) 57 ff.

Tort Liability Act chap. 2 sec. 2 if such a loss is caused as a result of criminal conduct. While this does not constitute a definite exclusionary rule on liability for pure economic loss in other instances, pure economic loss has only been acknowledged in a few types of cases without any involvement of criminal conduct, for example in the case of negligent misrepresentation and inducement to breach of contract.[14]

Nevertheless, there are other special liability regimes which recognise pure economic loss in a wider scope. The strict liability regime under the Environmental Code which may perhaps be applicable to the spreading of GMOs to another's crop,[15] in particular, recognises pure economic losses which are not of an insignificant nature (chap. 32 sec. 1 [2]). **14**

A pure economic loss which could manifest itself by diminished property value or precautionary investments in a non-GM crop, triggered by the presence of a nearby GM-crop would thus not engender a recoverable loss under the Tort Liability Act. It is also unlikely that such losses could be recoverable under the special liability regime of the Environmental Code (chap. 32), which in some instances provides liability for value diminishments and costs resulting from, e.g. noise and negative aesthetic alteration due to activities on another's property (see also no. 17).[16] **15**

3. Mere fear of a loss

With regard to pure economic loss which farmers or producers not using GMOs may suffer due to the fear felt by consumers or others if one or several non-GMO farmers or producers suffers damage from GMO admixture,[17] such a loss reflects a change of the market value which technically could be described as a third party loss consequential to another's loss. Such losses are generally barred from damages liability. **16**

A non-pecuniary loss which is the result of fear, but not consequential to an established personal injury, is not recognised as resulting in liability, although this has been discussed in connection with environmental damage and emissions (such as temporary noise disturbances).[18] **17**

14 NJA 1987, 692 and NJA 2005, 608.
15 See no. 2.
16 *Hellner/Radetzki* (fn. 7) 335 f.
17 See *Langlet/Schultz* (fn. 2) par. 37.
18 *Hellner/Radetzki* (fn. 7) 336 with references.

18 Fear which manifests itself in psychological suffering may only ground non-pecuniary damages if it is the result of criminal conduct.[19]

19 Fear or risks of certain environmental disturbances from a nearby activity may result in liability if pure economic loss can be established, although GMO presence is unlikely to qualify as such a disturbance.

4. Standard of proof

20 The burden for proving that a loss has arisen rests upon the claimant, who must fully substantiate the loss. However, under the Code of Judicial Procedure[20] chap. 35 sec. 5, the loss may be estimated to a reasonable quantum by the court if no evidence to this end can be provided or only with considerable difficulties, although this provision is rather restrictively applied.

5. Nominal losses

21 Symbolic damages are not recognised as such, but at the same time there is no general exclusionary rule barring trivial damages.[21]

6. Mass losses

22 There are no such rules.

III. Causation

1. Uncertainty of merely potential causes

23 For a general overview of causation in Swedish tort law, see the previous study.[22]

---

19 Tort Liability Act, chap. 2 sec. 3; NJA 1990, 186.

20 Rättegångsbalken (1942:740).

21 *B. Bengtsson*, Om grannelagsansvaret idag, in: B. Bengtsson (ed.), 21 uppsatser (2003) 65; Government Bill Proposition 1985/86:83, Ersättning för miljöskador, 19.

22 *Langlet/Schultz* (fn. 2) no. 7–19. See further *J. Hellner*, Causality and Causation in Law, Scandinavian Studies in Law 40 (2000) 119 ff., and *M. Schultz*, Further Ruminations on Cause-In-Fact, Scandinavian Studies in Law 41 (2001) 467 ff.

In cases where causation is difficult to establish, the plaintiff's burden of proof is sometimes alleviated. If there are two or more possible causes, the burden of proof may be fulfilled if the plaintiff can make his or her causal explanation "clearly more probable" than any other explanation, but only if the plaintiff's explanation is probable in itself. This standard has been recognised by the Supreme Court for cases where the plaintiff is confronted with considerable difficulties when it comes to presenting full evidence that other possible causes can be excluded.[23] It is unclear when this standard may be applied.[24] But having regard to the fact that several of the instances where it has been deemed applicable have involved environmental or similar damage, this standard may indeed be deemed applicable in cases involving GMO admixture. **24**

Thus, if the defendant argues that the source of the damage originates from the sphere of the victim, the victim must produce evidence showing the defendant to be the probable source of damage, which in comparison to the victim as a potential cause is deemed clearly more probable. The precise nature of this assessment cannot be defined more exactly, but must be decided on a case-by-case basis. **25**

However, if one agent has violated the precautionary GMO farming rules or otherwise acted negligently, it may indeed be contended that this constitutes a probable cause which also is clearly more probable than a cause for which no liability can be attributed (such as a cause lying within the victim's sphere or a cause attributable to casus where liability therefor is not possible). Under such circumstances, it can be established that the condition originating from the aforementioned agent was *a* cause, while it is not necessary to establish that it was *the* (only) cause.[25] If an additional cause originating from another agent, who cannot be identified also contributed to the damage, this does not change the conclusion: joint and several liability is at hand unless the damage caused can be divided.[26] **26**

Nevertheless, under the aforementioned standard of proof, a difficult situation may arise for the victim, involving instances of several uncertain but potential causes. The case NJA 1982 p. 421 revealed this predicament for the plaintiff. It concerned personal injuries sustained during X-ray examinations, where the plaintiffs claimed that the injuries had been sus- **27**

23 NJA 1981, 622; NJA 1982, 412; NJA 1993, 764; NJA 2001, 657.

24 *Bengtsson/Strömbäck* (fn. 6) 37.

25 Cf. NJA 1982, 421, (referred to no. 27 below). See further *M. Schultz*, Kausalitet (2007) 475; *Hellner/Radetzki* (fn. 7) 200 f., 213.

26 *H. Karlgren*, Skadeståndsrätt (5th ed. 1972) 211.

tained from the defendant pharmaceutical company's drug used during the examination. The defendant claimed that the injuries were sustained by other causes, inter alia, another company's drug which was also used during the examination. Each of the drugs was considered a probable cause, but as the defendant's drug was not deemed the "clearly more probable" one, damages were denied. This outcome has been criticised, since the plaintiff would likely have been left without damages even if the other drug company had also been involved in the damages actions, as no clear probability on the balance could be attributed to their drug either.[27] Thus, there is no recognised rule under general tort law which takes problems of alternative causation into consideration by granting joint and several, or proportional, liability where no particular cause can be established with a clear preponderant probability.[28]

28 The special strict liability regime under the Environmental Code (chap. 32), which may perhaps be applicable to instances of GMOs spreading from a GM crop to a non-GM crop,[29] provides for an even more relaxed burden of proof than under general tort liability. The standard thereby provided only requires that the alleged cause be probable on the balance, i.e. with somewhat greater than 50% probability (chap. 32 sec. 3 [3]).[30]

2. Complex causation scenarios

29 If the author of a causal condition cannot be identified, even under the alleviated standard of proof as presented above, a causal link to any of the potential agents cannot be established.

30 Furthermore, a similar predicament can arise if negligent conduct can be attributed to several agents but it is disputed which agent caused the damage and it cannot be established that even one of them contributed to the causation of the damage. The textbook example entails two persons who independently throw stones in the direction of a car, but it cannot be established which one actually hit the car. Strictly speaking, causation can

27 *Hellner/Radetzki* (fn. 7) 200 f., 213; B. *Dufwa*, Flera skadeståndsskydliga (1993) par. 2659 ff., 3007 f.; *Schultz* (fn. 25) 494 ff.

28 *Dufwa* (fn. 27) par. 3008.

29 See no. 2 above, and *Langlet/Schultz* (fn. 2) par. 3.

30 The provision states: "Damage shall be deemed to have been caused by a disturbance referred to in the first paragraph where, in view of the nature of the disturbance and its adverse effects, other possible causes and any other circumstances, the balance of probability indicates that the disturbance was the cause.".

not be established thereby for either one unless some collaboration between the two causes can be established.[31]

The notions of proportional liability or rebuttably presumed joint (and several) liability are foreign in that connection to Swedish tort law. **31**

However, in many instances involving GMO admixture, it should be possible to establish that at least one agent contributed to the damage, taking the relaxed burden of proof into account, and thus was one cause of the damage, giving rise to joint and several liability as described in the answer to the previous question. But difficulties of establishing causation may indeed potentially arise for the claimant, as also elaborated in the answer to the previous question, where not even one cause can be attributed to an agent. **32**

3. Force majeure

Force majeure is normally not considered in connection with the issue of causation. Force majeure does not constitute a general exemption from non-contractual liability in Sweden.[32] However, with regard to strict liability, certain causes may indeed be disregarded, i.e. not give rise to liability. Under general principles of tort law, this includes causes which cannot be subsumed under the particular risks which are affiliated with the strict liability activity. Damage or loss caused in a way which is not affiliated with the typical risk of that activity or the underlying reason behind a strict liability regime, therefore, does not give rise to liability, e.g., the strict liability regime applicable to the railway does not attract strict liability with regard to a passenger assaulted by another passenger.[33] **33**

Furthermore, the theory of adequacy may imply limitations to liability concerning causes which are not foreseeable or otherwise not deemed adequate causes.[34] **34**

With regard to the possibly applicable strict liability regime concerning the spreading of GMOs between crops – the Environmental Code (chap. 32) – it has been contended that the aforementioned delimitations of liability may not be applicable under the Environmental Code since it delimits liability with particular prerequisites.[35] **35**

---

31 *Hellner/Radetzki* (fn. 7) 201 f.; *Dufwa* (fn. 27) par. 3000 ff.
32 *J. Hellner*, Strict Liability in Sweden, in: In memoriam Jean Limpens (1987) 59.
33 *Hellner/Radetzki* (fn. 7) 176 f.
34 *H. Andersson*, Skyddsändamål och adekvans (1993) 105, 396 ff.
35 *Hellner/Radetzki* (fn. 7) 339. Cf. NJA 1997, 684 and NJA 1983, 209.

4. Threshold to prove causation

36 If the alleviated burden of proof is applied (as outlined no. 24 above), the degree of likelihood for a cause must be "clearly more probable" than another, and deemed probable in itself. This definitely involves a probability which exceeds a probability on the balance.[36]

37 If the alleviated burden of proof under the special strict liability regime of the Environmental Code is deemed applicable,[37] chap. 32 sec. 3 (3) provides for a burden of proof of probability on the balance, whereby the probability of an alleged cause must be more probable than another, having regard to the circumstances of the case and other potential causes.

5. Special rules on causation

38 No special methods or rules of causation are applicable, such as rebuttably presumed joint (and several) liability or proportional liability in cases of alternative causation, although an alleviated burden of proof may be applicable.[38]

IV. Types of liability

1. Fault liability

(a) Special rules governing fault

39 As already mentioned, we believe that general tort liability under the Tort Liability Act is primarily applicable to cases involving liability for GMOs. This regime is based on fault, or culpa as it is usually denoted in Swedish tort law, with regard to property damage and personal injury.

40 The notion of negligence under Swedish tort law has already been presented in the earlier study.[39] Negligence with regard to GMOs will be handled according to general principles of tort, which entails various considerations. However, no special delimitations, such as reversing the burden of proof or altering the standard of care, will be applicable thereby.

---

36 *Dufwa* (fn. 27) par. 2658 ff.
37 See no. 2 above, and *Langlet/Schultz* (fn. 2) par. 3.
38 See previous paragraphs.
39 See *Langlet/Schultz* (fn. 2) par. 20–25.

(b) Impact of specific rules of conduct

As already described in the earlier study,[40] the presence of statutory or customary rules may indeed influence and guide the assessment of negligence.[41] Additionally, regulations on safe GMO farming are also now in place. However, the negligence assessment can encompass several other considerations, such as general risk assessments. Nevertheless, the existing farming regulations on GMOs will certainly have a strong impact on the negligence assessment, albeit not every single violation of those rather precise regulations will conclude such assessment. **41**

2. Product liability

(a) Development risk defence

Sweden did incorporate a defence for development risks under the Product Liability Act[42] sec. 8 (4). It provides that liability is excluded if the defendant proves that it was not possible to discover the safety defect based on the scientific and technical knowledge available at the time of putting the product into circulation. The defence was incorporated at a late stage of the legislative process, based on competitive concerns for Swedish industry. The nature of the defence has thus far not been elaborated in case-law. **42**

However, if a victim's claim fails due to the defence, negligence may still be attributable to the defendant based on the duty of care to monitor defects and injuries or losses that have arisen subsequent to the release of the product on the market, or based on the defendant's negligence with respect to the observation of the development of scientific or other knowledge of risks associated with the product and thus failure to issue warnings or take other precautions.[43] **43**

Strict product liability under the Product Liability Act only covers personal injury and property damage to property intended for personal use (consumers' property), and is based on the prerequisite that damage has been caused due to a safety defect of a product. Hence, the scope of appli- **44**

40 See *Langlet/Schultz* (fn. 2) par. 22.
41 *Hellner/Radetzki* (fn. 7) 129 ff.
42 Produktansvarslag (1992:18).
43 *B. Bengtsson/H. Ullman*, Det nya produktansvaret (2nd ed. 1993) 30; *Hellner/Radetzki* (fn. 7) 312.

cation with regard to GMOs must be considered very limited, as far as present-day knowledge on the effects of GMOs is concerned.

45 Product liability outside the scope of that statute may be considered, as it has been developed in case-law, but it does not extend the scope of liability in comparison to the general tort liability referred to earlier (see further under the next question).

(b) Alternative routes

46 Consumers have an extended protection under product liability, pursuant to the Consumer Sales of Goods Act sec. 31, in the sense that liability for lack of conformity extends to damage to other property as a result of such lack of conformity.

47 Leaving aside the aforementioned Product Liability Act as well as particular consumer protection, issues of product liability in a wider sense, i.e. damage to property other than a product which has been bought, have been dealt with in case-law. This has in a few instances resulted in *strict liability* for such damage, something which must be seen as a very limited possibility for the purposes of this study, as presented below.

48 Otherwise, liability may also be based on *negligence* under the Tort Liability Act.

49 However, with regard to both those grounds of liability (case-law based strict liability and fault liability), another delimitation arises concerning *claims directed against a third party*, i.e. a party who contracted with the contractual partner of the claimant. The general rule is that such claims are barred, unless a safety defect of a product has caused damage to other property, as further explained below.

50 To begin with, damage to property other than the object of the sales contract does not result in damages liability under the Sales of Goods Act (according to sec. 67 (2) 2nd sentence). Such damage is thus subject to the law on extra-contractual liability. Hence, so-called ingredient-damage does not attract the application of contract law, i.e. where a delivered product which lacks conformity with the contract is subsequently incorporated with some other property by the buyer. However, if the buyer's additive is an insignificant component of the end-product, the rules on sales of goods may still be applicable.[44]

---

44 *Hellner/Radetzki* (fn. 7) 318.

Turning firstly to *strict liability* which has been considered in the case-law, it is unlikely that such liability could be triggered with regard to GMO admixture in a dispute between two non-consumer parties concerning property damage.[45] That is especially so where a safety defect is not at hand (see further below). 51

Liability may only be strict if the seller of the product which has caused property damage made a guarantee to the buyer with regard to the quality of the product. However, an explicit guarantee does not necessarily need to be at hand. Some cases have given rise to strict liability where the court has deemed an implied guarantee to be at hand. Such may be derived from an explicit labelling which turns out to be misguiding, or otherwise where the buyer had reason to assume a certain quality of the product.[46] However, cases based on implied guarantees involve property damage where the property has been destroyed. It is unclear if such a guarantee, de facto construed by the court, would be acknowledged with regard to all property damage, including all alterations of property which is intended to be GMO-free but still may have a market value. 52

Instead, liability for damage to property other than the product subject of the contract may be based on *negligence*.[47] Negligence can be at hand due to mistakes during the production of the product, insufficient quality monitoring, lack of information to the buyer, omission to follow safety standards or otherwise. 53

However, it should be noted that Swedish law takes a cautionary position on *direct claims against a third party* to a contract, i.e. claims directly against a party further down the contractual chain and thus other than the immediate contracting party, unless such are explicitly provided for in a statute or a contract.[48] It is considered a general rule that such direct claims, which sidestep a contractual party in a contractual chain, are barred.[49] 54

A direct claim may only be granted if the product which damaged other property was afflicted with a safety deficiency.[50] Such a deficiency implies 55

45 NJA 1982, 380 (strict liability denied concerning chicken feed which negatively impacted the taste of the eggs); NJA 1983, 118 (strict liability denied concerning horse feed which poisoned and damaged a horse used for professional harness racing).

46 NJA 1968, 285; NJA 1985, 614; NJA 2001, 309.

47 *Hellner/Radetzki* (fn. 7) 319.

48 See generally *H. Hellner*, Linked Contracts, in: R. Cranston (ed.), Making Commercial Law – Essays in Honour of Roy Goode (1997) 167 ff.

49 NJA 2007, 758 and NJA 2001, 309.

50 NJA 2007, 758; NJA 2001, 309; NJA 1983, 118; NJA 1945, 189. Cf. NJA 1986, 712.

that a significant safety feature of the product is deficient and that the deficiency involves a considerable risk of personal injury or extensive property damage.[51] It is thus questionable whether such claims could be made on the basis of GMO admixture, other than against the immediate contracting party who has delivered a product wrongfully containing GMOs.

56 It is also uncertain whether a guarantee by a third party to the injured person may suffice to grant a direct claim, at least if that guarantee is not explicitly addressed to the injured party.[52]

57 Finally, the aforementioned remedies for product liability, within its wider meaning, are rather limited and do not seem to be in conflict with the case-law of the ECJ concerning extended product liability, especially taking art. 13 of the EC Directive into account.[53]

(c) Impact of compliance with rules and regulations

58 With regard to the protection afforded to consumers under the Product Liability Act, such a defence only exists where a safety defect is the result of mandatory provisions issued by a public authority with which the product had to comply (sec. 8 [3]). This defense is considered to be of minor practical importance for delimiting liability.[54]

59 With regard to product liability falling outside the scope of the Product Liability Act, where a claim has to be based on negligence,[55] the fact that rules and regulations have been obeyed does not exculpate the defendant. The negligence assessment may involve several considerations and not be limited merely to negligence in connection with the production of the product or with regard to the material used, but also extend to omissions to inspect the product prior to delivery or to provide the buyer with adequate information about the product.[56]

---

51 NJA 1986, 712 and NJA 2007, 758.
52 NJA 1983, 118. Cf. *Hellner/Radetzki* (fn. 7) 323.
53 *Hellner/Radetzki* (fn. 7) 317; NJA 2001, 309.
54 *Bengtsson/Ullman* (fn. 43) 30.
55 See above no. 46 ff.
56 *Hellner/Radetzki* (fn. 7) 319.

3. Environmental liability

(a) Implementation of the Environmental Liability Directive

The Environmental Liability Directive has been implemented through a revision of chapter 10 of the Environmental Code on activities that cause environmental harm (before the revision it was on "polluted areas"). Primarily, the revision introduced the notion of "serious environmental harm". That comprises environmental damage which: 1. through pollution of land, constitutes a significant risk to human health; 2. by affecting a water area or groundwater has a significant negative effect on the quality of the water environment, or; 3. on a significant scale harms or impedes the preservation of an animal or plant species or the habitat of such a species, if the harm concerns i) a protected area which is listed as part of the Natura 2000 network; ii) an animal's breeding sites or resting places which have been protected by provisions based on chap. 8 sec. 1 of the Environmental Code; or iii) a species protected by provisions based on chap. 8 sec. 2 of the Environmental Code. The sections referred to implement, inter alia, the habitat directive.[57] **60**

With respect to financial guarantees, the Environmental Code enables relevant authorities to require the furnishing of a financial security for the costs of after-treatment and other restoration measures that may be necessary as a result of an activity which requires a permit, an approval or an exemption in order to be carried out.[58] **61**

(b) Environmental liability regime beyond the scope of the Directive

The Swedish Supreme Court found in 1995 that the Government was entitled to compensation for poached wolverines despite the fact that the species is protected and has no market value. The amount was set according to an estimation of the conservation measures paid for by the Swedish State which, as a result of the poaching, had come to naught.[59] A seemingly important factor was the obligation incumbent on the Swedish Government under EC and international law to protect the wolverine. Lia- **62**

57 Council Directive 92/43/EEC of 21 May 1992 on the conservation of natural habitats and of wild fauna and flora, Official Journal (OJ) L 206, 22.7.1992, 7–50.

58 Environmental Code, chap. 16 sec. 3.

59 NJA 1995, 249.

bility could arise on this basis outside the ambit of the Environmental Liability Directive, although species affected by protective measures taken by the Government are likely to also be protected under the Directive.

63 The provisions on polluted areas in chap. 10 of the Environmental Code establish an obligation to remedy polluted areas or to pay for remedial action taken by public authorities also in cases when that would not follow from the Directive. This applies to situations where land or water areas have been so polluted that they *may cause damage or detriment* to human health or the environment.[60] This should be compared to the Directive's definition of environmental damage which identifies "land damage" as land contamination that creates *a significant risk* of human health being adversely affected, whereas "water damage" means damage that *significantly adversely affects* ecological, chemical and/or quantitative status and/or ecological potential.[61] However, the liability for remedial action with respect to polluted areas that would not trigger liability under the Directive only applies to the extent that such liability is deemed reasonable considering, e.g., the length of the time that has elapsed since the pollution occurred and whether the person liable was obliged to prevent future damage.[62]

(c) Claimants in cases of environmental harm

64 Only a supervisory authority (normally a County Administrative Board or a municipal committee) may demand that remedial action be taken or paid for when the damage concerned is purely environmental, i.e. does not entail an economic loss.[63] Liability entails an obligation to carry out or pay for the remedial action that, due to the pollution damage, is needed to prevent or counteract the occurrence of damage or detriment to human health or the environment. As indicated under (b) above, the liability for remedial action which goes beyond that of the Directive only applies to the extent that it is deemed reasonable considering, inter alia, the length of the time that has elapsed since the pollution occurred and whether the person liable was obliged to prevent future damage.[64] If the damage qualifies as "serious environmental harm" (see (a) above), the responsible person(s) shall carry

60 Environmental Code, chap. 10 sec. 1.

61 Directive 2004/35/CE of the European Parliament and of the Council of 21 April 2004 on environmental liability with regard to the prevention and remedying of environmental damage, OJ L 143, 30.4.2004, 56–75.

62 Environmental Code, chap. 10 sec. 4.

63 Environmental Code, chap. 10 sec.14.

64 Environmental Code, chap. 10 sec. 4.

out or pay for the remedial action necessary to immediately prevent further damage to the environment and risk to human health. If the damage consists of polluted land which poses a considerable risk to human health, that land should no longer constitute such a risk. If it is a case of damage to a protected area or to a protected species or its habitat (see further (a) above), the environment shall be restored to the state in which it would have been had the damage not occurred. In such a case lost environmental values shall also be compensated for pending restoration. If restoration is not possible, lost environmental values shall be compensated for in some other way.[65]

(d) Special liability regime for losses sustained by individuals

To our knowledge there is no such regime relevant for those purposes. **65**

(e) Cartagena Protocol

Sweden is party to the Cartagena Protocol.[66] It is implemented primarily through EC-law in the form of Regulation 1946/2003.[67] The Swedish Environment Protection Agency is responsible for coordinating Sweden's participation in the information exchange procedure established by the Protocol.[68] **66**

V. Vicarious liability

1. Scope of vicarious liability

The Tort Liability Act (chap. 3 sec. 1) provides vicarious liability for persons engaging employees in their undertaking, with regard to damage caused by employees in the course of their engagement for that employer. The employer is accountable for personal injury and property damage, where the employee was negligent, and pure economic loss, where the employee acted criminally. **67**

65 Environmental Code, chap. 10 sec. 5.

66 Cartagena Protocol on Biosafety to the Convention on Biological Diversity, Montreal, 29 January 2000, (2000) 39 International Legal Materials (ILM) 1027.

67 Regulation (EC) No 1946/2003 of the European Parliament and of the Council of 15 July 2003 on transboundary movements of genetically modified organisms, OJ L 287, 5.11.2003, 1–10.

68 Förordning (2002:1086) om utsättning av genetiskt modifierade organismer i miljön, sec. 4a.

68 In contractual relations, a person who engages an independent contractor is generally also liable for that contractor and his employees. However, it is necessary thereby that the conduct causing damage lay within the purview of that contractual relationship, or alternatively that the independent contractor is engaged in order to fulfil a particular contractual obligation which the immediate contractor is fulfilling towards the injured person.[69] This liability is more of a contractual nature.

69 Extra-contractual vicarious liability has been acknowledged in certain instances with regard to independent contractors. Firstly, certain non-delegable duties will give rise to liability regardless of whether a person has engaged an independent contractor to fulfil those duties. Such duties primarily follow from statutes or public regulations where the duty is clearly expressed in relation to the relevant person.[70] That may indeed be the case with regard to the farming rules issued on GMO farming.

70 Otherwise, a person is, as a general rule, not liable in an extra-contractual situation for damage caused by an independent contractor.[71]

2. Liability for people further up the food or feed production chain

71 In a production chain each person is liable towards those with whom a contractual relationship exists. Hence, each party is contractually liable for his product towards the party to whom he has sold the product. Liability thus normally lies at the top of the production chain. (On third party liability, see no. 54–56 above).

3. Can someone further down the feed or food chain include someone further up it in a trial against him/herself?

72 No (see previous paragraph).

---

69 *Hellner/Radetzki* (fn. 7) 165.

70 *Hellner/Radetzki* (fn. 7) 166 f.

71 *Bengtsson/Strömbäck* (fn. 6) 79; NJA 1959, 695 and NJA 1979, 773 (liability for a subcontractor denied, having regard to the fact that the subcontractor executed the work in an independent manner, using his own machinery and without the supervision of the contracting party).

VI. Multiple tortfeasors

73 Joint liability is recognised under the Tort Liability Act (chap. 6 sec. 4), where two or more tortfeasors are liable for the same damage. Liability is actually joint and several as a general rule, according to the general principles of tort law.[72] The problems of multiple causation have already been outlined above. Joint and several liability is at hand if at least *a* cause of the damage can be attributed to one tortfeasor, as established under the relevant burden of proof, even if other tortfeasors also possibly contributed to the causation of the damage.[73]

VII. Defences

1. Licence/permission to grow GM material

74 The fact that permission has been granted to grow GM material by the competent authority does not serve as a defence against liability. As mentioned earlier, we would say that the general negligence rule will be applicable to cases involving GMOs. The characteristics of the negligence assessment have been outlined in the previous study on GMOs.[74] That assessment involves several considerations, where the rules on precautionary measures on farming with GM material is one aspect and in this context an important one. Those measures are conditions for the granting of permission, but that does not exculpate the farmer who violates them or otherwise acts negligently.

75 The fact that permission has been granted for testing and not for releasing any produce on the market, does not alleviate the farmer from observing his duty of care.

2. Consent/assumption of risk

76 Consent constitutes a defence under general principles of tort law. However, consent to a type of damage cannot serve as a defence if the harm that materialised is atypically more severe than could be expected.[75]

72 *Hellner/Radetzki* (fn. 7) 242; *Dufwa* (fn. 27) par. 4328 ff.
73 *Karlgren* (fn. 26) 211; *Hellner/Radetzki* (fn. 7) 238.
74 See *Langlet/Schultz* (fn. 2) par. 20–25.
75 *Hellner/Radetzki* (fn. 7) 123 f.

77 The defence of assumption of risk is separately discussed in Swedish literature with respect to instances where a legitimate consent to damage or risk of damage cannot be established but instead merely a conscious risk-taking, or, as it is sometimes denoted, acceptance of risk. It is generally acknowledged that instances of such risk-taking may not necessarily provide a defence which eliminates liability, but instead attract the alternative defence of contributory negligence whereby the damages are reduced, or, it may also instead be considered within the scope of assessing the tortfeasor's negligence by way of increasing the scope of acceptable risk-taking for the tortfeasor in relation to the risk taken by the plaintiff.[76] The former alternative, the one of applying contributory negligence, is generally accepted and often times preferred, as it does not necessarily give rise to a complete denial of damages for the plaintiff having regard to the ambiguity of the distinction between the defence and contributory negligence, especially concerning personal injuries which trigger particularly good protection.

78 The Damages Liability Act (chap. 6 sec. 1) stipulates stringent limits for reduction of damages for personal injuries, and even gross contributory negligence seldom renders reductions. Hence, it is fundamentally questionable whether a general defence for acting at one's own risk would be accepted, except in certain special instances where the risk is particularly clear and even being pursued, e.g. in sports.[77]

79 If a supplier delivers a product containing GM material to a person who is made aware of the presence of such material, the former may in certain instances be under the obligation (duty of care) to inform the latter of possible damage which may follow from the use of the product, as far as he is aware of or should be aware of such consequences.[78]

80 Under product liability, the producer has a duty of care to monitor defects and injuries or losses arising subsequent to the release of the product on the market, and to observe scientific developments or other knowledge regarding the risks associated with the product and thereby to issue warnings or take other precautions.[79]

---

76 *B. Bengtsson*, Skadestånd vid sport, lek och sällskapsliv (1962) 62 ff.; *A. Agell*, Samtycke och risktagande (1962) 12 ff.,176 ff.

77 *Bengtsson/Strömbäck* (fn. 6) 50.

78 *Hellner/Radetzki* (fn. 7) 319.

79 *Bengtsson/Ullman* (fn. 43) 30; *Hellner/Radetzki* (fn. 7) 312.

3. Third-party influence

81 The issue of a third party's influence on the causation of damage is usually considered within the theory of adequacy, whereas adequate causation is required in order to give rise to liability.

82 The problem dealt with usually concerns the interference of a third person in the causation of the damage. This implies a keeper of the source of risk for harm (the defendant), i.e. someone who handles a source of risk which is realised through a third person or a third person worsening an occurring damage. Adequate causation is usually considered to be at hand thereby.[80]

83 Turning more closely to GMO farming, a person carrying out such an activity is under a duty of care to take precautions. That may involve testing of the crop. Hence, third-party conduct which in some way results in damage being caused by the above-mentioned GMO farmer may still be considered to have been caused adequately by that farmer.[81]

84 Hence, even rather unforeseeable acts by a third person which influence the causation of damage do not necessarily mean that the causation of the first party's conduct is no longer adequate.

85 This is also the case under the strict liability regime of the Environmental Code (chap. 32), where third-party behaviour does not serve as a defence.[82] However, in this connection it has been argued that if a person for instance unlawfully disposes of toxic substances on another's property, which then spread to yet another's property, this can not render the owner of the former property liable if the owner does not use the property for any activity at all.[83] Such limited exclusion of liability may perhaps be at hand where a person grows GMO plants on an abandoned meadow without the permission of the owner who does not carry out any activity on that property. However, if duty of care exists with regard to the property, adequate causation may nevertheless be at hand despite third-party misconduct.[84]

80 NJA 1931, 7 is usually cited in that connection, involving a person A, who negligently allowed gasoline to leak into the waters of a harbour while emptying a container of gasoline. Another person B who observed this, wilfully threw a burning match into the water causing a fire that damaged a nearby ship. A (jointly with B) was considered liable for that damage.

81 *Andersson* (fn. 34) 421 f.

82 *Hellner/Radetzki* (fn. 7) 339.

83 Preparatory works NJA II 1986, 141.

84 Cf. NJA 1981, 622 liability recognised for the municipal keeper of a water pipe line to which a company had connected a sewer without permission. The resulting damage was considered adequately attributable to the municipality.

4. Prescription

86 The period of prescription for damages claims is ten years according to the Act on Prescriptions[85] (sec. 2 [1]), unless it is suspended whereby it starts to run anew. The general contention is that the period of prescription starts when the harmful conduct is committed, even if that conduct only results in damage much later.[86] If the harmful conduct is ongoing for a longer period of time, or, if the conduct constitutes an omission, the period of prescription starts at the point of time when damage-preventive measures could have been taken. The special strict liability regime under the Environmental Code (chap. 32) also follows this general period of prescription. A liability claim based on the Product Liability Act provides for a three year period starting when the claimant became aware of, or should have become aware of, the possibility of a claim (sec. 12). However, if based on general tort liability the general ten year period of prescription applies.[87]

5. Other defences

87 To our knowledge there are no other particular defences.

VIII. Remedies

1. Pecuniary compensation

(a) Bodily harm

88 There are no caps or other special rules in that regard.[88]

89 Damages for personal injuries come under strong protection in Swedish tort law. This may, inter alia, be identified by looking at the very restrictive approach to reducing damages on the basis of contributory fault. Damages for personal injuries are only reduced by way of exception; according to the Damages Liability Act chap. 6 § 1, they may be reduced where the injured person acted with intent or gross negligence, the threshold of gross negligence being a rather high one, based on social and humanitarian grounds.

---

85 Preskriptionslag (1981:130).
86 *Hellner/Radetzki* (fn. 7) 433 ff.
87 *Hellner/Radetzki* (fn. 7) 436.
88 With regard to general remedies, see no. 9 above.

(b) Property losses

There are no caps or other special rules in that regard.[89] **90**

(c) Economic losses

As mentioned above,[90] economic loss will primarily be compensable when resulting from property damage (Tort Liability Act chap. 5 sec. 7). The income loss suffered because of a decreased vending value of the produce caused by GMO admixture is therefore recoverable. Other consequential losses to the direct victim of property damage are, as earlier outlined, also compensable. That may include costs incurred from: necessary re-labelling, the need to keep contaminated produce separated from other produce, extra transport, the changing of the cultivation cycle. **91**

The question as to whether an organic farmer could recover damages for costs resulting from the restoration of a field to its previous state may be somewhat ambiguous. Swedish tort law does rest upon the general principle of *restitutio in integrum*. **92**

However, it should be noted, that on the one hand, the farmer would have the right to recover damages for income loss and other losses related to the disruption of his professional activity, including losses resulting from the loss of a certification for organic farming. On the other hand, the *duty to limit one's loss* must be taken into account. That duty arises especially with regard to consequential economic losses,[91] which are relevant here. Losses incurred by the injured person are generally recoverable if they would not have been incurred without the damage inflicted, or if the costs incurred have been rendered useless thereby. However, a rather strict assessment is made as to whether the losses (or costs) incurred could have been avoided or made useful by the injured person.[92] Hence, the injured person has a duty to limit his losses. **93**

In certain instances the duty to limit losses cannot be set too high as it may involve significant and unexpected sacrifices[93] and the injured person should be afforded some time to assess the necessary costs involved in lim- **94**

89 With regard to general remedies, see no. 11 f. above.
90 See no. 10.
91 *Andersson* (fn. 34) 479 ff.
92 *Hellner/Radetzki* (fn. 7) 417.
93 NJA 1911, 472 (separate opinion of Justitierådet Sjögren).

iting losses.[94] If production is halted, the duty may however be rather extensive as unnecessary production costs ought to be limited.[95]

95 Consequently, a farmer or a producer suffering from GMO contamination has the duty to limit costs arising from the damage as well as a duty to make his enterprise useful as far as possible in the state that it is in after the damage has been caused.

96 Turning back to the question of recovering costs for restoring a field to its former state in order to meet previous organic farming standards, the possibilities of making the interim produce useful and the income losses incurred by the fact that organic produce can no longer be delivered will have to be weighed against the costs of restoring the field. That is certainly the case for a period of time after the damage was caused, meaning the duty to limit loss will be rather extensive in that regard.

97 Hence, a deduction for seasons during which the farmer could have grown conventional produce will apply. If an economic loss has nevertheless been suffered, costs for the restoration may in principle be recovered, presuming that the requirements of adequate causation and accountability have been met.

(d) Harm to animals

98 In this case, property other than the product bought from the feed-supplier is damaged, and thus product liability in its wider sense is at hand because tort law can be applied with regard to the damage to that other property, as described above.[96]

99 If an animal is harmed, one may firstly consider whether an (implied) guarantee concerning the feed is at hand. As presented earlier, that conclusion may indeed be reached in the case of significant damage (or harm) to the animal, as indicated by some case-law.[97] Damages liability is then strict. Otherwise, negligence has to be attributed to the seller or the producer of the feed. The damages recoverable[98] include either value compensation for replacing the animal, or compensation for veterinarian costs for

94 NJA 1978, 207.

95 NJA 1959, 552 implying that lay-offs of personnel even may be part of that duty when production is stopped.

96 See further no. 47 ff. above.

97 See no. 52.

98 In accordance with the Tort Liability Act, chap. 5 sec. 7.

treating the animal where this is possible and justifiable in terms of costs. Furthermore, consequential losses and income loss resulting from the harm to the animal are also recoverable, e.g. lost milk production. However, this depends on how quickly the farmer could otherwise resume milk production, having regard to the duty to limit losses.

If an animal merely eats feed containing GM material, property damage may **100**
be at hand because its function has been altered.[99] Unless an explicit guarantee as to the quality of the feed has been made, it is unlikely that liability will be strict instead of based on fault. Moreover, the farmer may only direct a claim against his seller and not make a direct claim against the producer, as elaborated earlier.[100] The damages recoverable[101] will primarily include the income loss resulting from the fact that the milk or meat of such animal cannot be sold as GMO-free, such loss will derive from the difference between the potential GM-free vending value and the actual vending value.

(e) Costs of disposal

Costs incurred by disposing of the contaminated production or animals **101**
constitute loss consequential to property damage. Such a loss is recoverable[102] unless the duty to limit one's losses implies that the animals or products should have been put to use.

2. Non-compensatory damages

Punitive or exemplary damages are not recognised as such. In certain spe- **102**
cial instances, not relevant to GMO liability, the preventive effect of damages is taken into account when ascertaining the quantum of damages to be awarded, on top of the quantum for actual loss. Such considerations can primarily be found within labour law, in certain instances of defamation and especially in connection with violations of anti-discrimination rules.

---

99 See no. 10.
100 See no. 11.
101 In accordance with the Tort Liability Act, chap. 5 sec. 7.
102 In accordance with the Tort Liability Act, chap. 5 sec. 7.

3. Other remedies

103 Reparation in kind or restitution in natura are not considered part of Swedish tort law but arise more in connection with property law.[103] Damages are thus the primary remedy available.

4. Costs of pursuing a claim

(a) General cost rule

104 The general rule for recovering costs for a civil trial is the loser-pays principle.[104]

(b) Costs of establishing causation

105 Costs relating to the inquiry into the causation of damage are not recoverable in damages as a loss resulting from property damage, especially if such monitoring is part of the injured person's activity,[105] but may in some instances be recoverable as trial expenses with regard to technical evidence or expenses for calling an expert witness if the costs are deemed reasonable and necessary for the plaintiff to protect his rights.[106]

106 However, costs which the injured person has incurred consequential to the damage, which are affiliated with the duty to limit losses, such as assessing alternative uses of contaminated produce, are recoverable in damages.

5. Advance cover

107 Any public compensation fund or mandatory private redress scheme (mandatory liability insurance) has thus far been rejected regarding any GMO liability.[107]

---

103 *Karlgren* (fn. 26) 206.

104 Code of Judicial Procedure, chap. 18 sec. 1.

105 *Hellner/Radetzki* (fn. 7) 416 f.; NJA 1983, 209.

106 Code of Judicial Procedure, chap. 18 sec. 8.

107 That is at least the opinion of the commission that has analysed the need for any reforms with regard to GMO liability, in the Government Official Report SOU 2007:46 (245 ff.). Nonetheless, a proposal for an Act on a public compensation fund, financed by fees paid by GMO farmers, has been submitted for consideration by the Government. However, no legislative measures have thus far been taken.

IX. Cross-border issues – Conflict of laws

1. Conflicts rules applicable before (or instead of) Rome II

Unlike the Rome II regulation,[108] Swedish law has traditionally not had any explicit provision regarding the applicable law in tort cases with trans-boundary aspects. Guidance has instead been provided by case-law. The leading precedent by the Supreme Court dates from 1969. It established *lex loci delicti* as the general principle governing such cases. That typically applies also to cases when one or both parties have a strong link to some other country. However, with respect to cases where the harmful activity occurred in a country other than that in which the harm occurred, existing case-law does not provide a conclusive answer as to applicable law. Although the 1969 precedent may be interpreted as indicating a preference for the country in which the action was undertaken, the question remains largely open.[109] This should be compared to the general rule of Rome II, i.e. that the law of the country in which damage occurs, i.e. *lex loci damni*, shall apply irrespective of whether the event giving rise to such damage occurred in this country.[110] **108**

A further change in Swedish law is the right of the parties, if pursuing a commercial activity, to agree on the applicable law before any event giving rise to damage occurs.[111] In Sweden, parties have not traditionally been able to agree on applicable law prior to the occurrence of the damage.[112] With respect to environmental damage or damage sustained by persons or property as a result of such damage, Rome II prescribes the application of the general rule described above, but also provides persons seeking compensation for damage with the right to base their claim on *lex loci delicti*. Such opportunity of choice has not previously been generally available under Swedish law. However, in a Nordic context, the Nordic Convention on the Protection of the Environment[113] between Denmark, Finland, Norway and Sweden has long provided that any person who suffers damage caused by an environmentally harmful activity in another contracting state has the right to have the question of compensation judged by rules no less **109**

108 Regulation (EC) No 864/2007 of the European Parliament and of the Council of 11 July 2007 on the law applicable to non-contractual obligations (Rome II), OJ L 199, 31.7.2007, 40–49.

109 *M. Bogdan*, Svensk internationell privat- och processrätt (6th ed. 2004) 280.

110 Reg. (EC) No 864/2007, art. 4 (1).

111 Reg. (EC) No 864/2007, art. 14 (1).

112 *M. Bogdan*, Svensk internationell privat- och processrätt (7th ed. 2008) 285.

113 13 ILM (1974) 511.

favourable than those of the state in which the activity is being carried out. It is not certain that activities involving GMOs are covered by the Convention's definition of environmentally harmful activities and thus covered by its provisions. However, given the Convention's wide definition of such activities, it does seem within the bounds of reasonable interpretation.[114]

2. Special regime for cross-border claims

110 There is no such regime.

X. Cases

1. Due to the adventitious presence of GMOs in a field, maize which is normally sold as conventional contains GMOs beyond the legal labelling threshold. This is not discovered before the final stage of the food production chain by the producer of taco chips. The whole production is lost since the supermarket chains refuse to accept delivery from the producer.

(a) Who can sue along the chain of distribution?

111 Property damage[115] has been caused to the product which has been manufactured with the maize containing GM material.[116] A tort law claim based on negligence can be made for losses resulting from that damage.[117] However, a direct claim cannot be made against any other than the injured person's immediate contractual partner, and thus also not against anyone further down the contractual chain, on the basis of a general rule whereby such claims are barred.[118] Liability may be strict if the seller has made an explicit guarantee, but it is unlikely that the court would establish an implied guarantee in this connection.[119] Instead, a negligence assessment

114 See art. 1 which reads: "For the purpose of this Convention, environmentally harmful activities shall mean … and the use of land, the seabed, buildings or installations in any other way which entails or may entail environmental nuisance by water pollution or any other effect on water conditions, sand drift, air pollution, noise, vibration, changes in temperature, ionizing radiation, light etc.".

115 See no. 10.

116 See no. 47 ff.

117 See no. 50–53.

118 See no. 54 f.

119 See no. 56.

will be made. That assessment will not only take regard of fault in connection with the production but also the question of whether the supplier negligently omitted to examine the product before delivery or to supply important information about the product.[120] Recoverable losses[121] entail income loss, i.e. difference between the expected vending value of the GMO-free product and the possible vending value of the product containing GMO. Additional costs which the buyer incurs due to the disclosure of GMO contamination are also recoverable as consequential losses. If it is impossible for the buyer to sell the product to another supermarket, or any other person for that matter, value compensation for the vending value (whole sale price) of a GMO-free product will be awarded.

(b) Would the case be solved differently if the GMO content was below the labelling threshold?

While the above-described in principle applies, it is questionable whether such a contamination would amount to negligence attributable to the supplier. Nevertheless, if a party has assumed an obligation to deliver his product GMO-free to an extent which goes beyond the required standard, it may be considered that he has negligently contributed to the damage. 112

(c) Would the case be solved differently if the admixture was not adventitious, but occurred due to the disregard of segregation rules, for example?

No. Possibly more extensive losses would be incurred, which should be recoverable.[122] 113

(d) Would the case be solved differently if the GMO found was not admitted for production in your jurisdiction?

That would be a factor which would be considered under the assessment of negligence. Such GMO presence certainly implies that the duty of care had been disregarded, but that must be ascertained in casu.[123] 114

120 See no. 53.
121 See no. 91 ff.
122 See no. 11.
123 See no. 39.

(e) If the admixture had occurred on a non-GM field and it transpires that the GM seeds were blown from: (i) neighbouring fields; or (ii) a truck passing by, would the farmer of the affected field be liable for all or part of the loss caused further down the distribution chain?

115 In the first instance (i), the farmer of the non-GM field would be liable towards his contractual partner for losses incurred by that party, under the condition that the farmer was negligent in supplying his contractual partner with GM contaminated produce. He may indeed have a duty of care entailing the examination of his produce for GMOs having regard to the presence of surrounding GM-fields.[124]

116 In the second instance (ii), the farmer's duty of care would with reason be formulated more narrowly.

2. Twenty years after the sale of GM maize used for food products, it turns out that it has certain disadvantageous health effects for humans.

(a) Can the producers be held liable at this point for risks unknown at the time of growing the maize? Who would be liable?

117 Strict product liability for personal injuries is provided for under the Product Liability Act. The primary liability lies with the producer or person who has created the product. However, if a safety defect originates from a part of the end product, the producer of that part as well as the producer of the whole product, are liable.[125] Nonetheless, a defence for development risks is also provided.[126] Furthermore, a claim under this Act is prescribed after three years, from when the injuring conduct took place.[127] The claim would thus be prescribed. If the aforementioned defence is successfully raised by the defendant, the claimant may still base a claim under general tort law for fault liability.[128] Negligence can thereby be based on the duty of care to monitor defects and injuries or losses arising subsequent to the release of the product on the market, or based on the defendant's negli-

124 See no. 53 ff.
125 Product Liability Act, sec. 2 (2).
126 See no. 42 ff.
127 See no. 86.
128 See no. 43.

gence to observe the development of scientific or other knowledge regarding risks associated with the product and thereby to issue warnings or take other precautions. However, a general damages claim would also be prescribed, since the period of prescription is ten years, starting at the time of the injuring conduct.

(b) Can compensation already be claimed at a point when the negative health effects have not yet materialised, but are to expected to according to scientific expertise/mere rumours?

Personal injuries generally entail physical or psychological injuries to the **118**
body, which can be medically substantiated. Within that notion, one does also consider poisoning to constitute a personal injury.[129] If it can be scientifically ascertained that an ongoing injury to the body is at hand, of which the effects have not been realised yet, personal injury may be at hand.[130] However, the mere fear of personal injury does not constitute a personal injury. Non-pecuniary damages may only be awarded for psychological suffering caused by criminal conduct, which is unlikely in the present context.[131]

(c) Would it make any difference if the GM maize had only been in use for feed, causing harm to the animals, which may or may not cause harm to humans consuming the meat as well?

Strictly interpreted, the notion of product under the Product Liability **119**
Act would mean that the supplier of the GM maize could not be held liable towards anyone other than the farmer.[132] That is at least the case provided that one does not consider that the feed becomes a product part of the animal.

129 *Hellner/Radetzki* (fn. 7) 103.
130 Cf. NJA 1947, 82 concerning animals.
131 See no. 17 f.
132 See no. 118.

3. The driver of a food logistics company discovers that a farmer, from where he regularly picks up agricultural products fails to obey mandatory segregation rules or food or feed hygiene standards, which may lead to the admixture of GM and non-GM produce sold and packaged separately by that farmer. Does he or his employer have a duty to warn, i.e. the recipient of the allegedly "non-GM" produce?

120 The question is whether an extra-contractual duty of the driver or logistics company to warn the recipient of the delivery is at hand. That is primarily an issue of liability for omitting to act in order to prevent harm to another, which is an objective prerequisite and not an issue within the purview of negligence. A duty to act must be identified with regard to the driver or the logistics company.

121 In Swedish tort law, there is a general assumption that extra-contractual liability requires a positive action and pure omissions to act do not give rise to liability except for certain instances.[133] The scope for such exceptions is unclear. The duty to act may primarily be derived from statutes and contracts. It has been contended that a duty to act may arise in relation to a party other than the contracting person in cases such as a person taking care of a child for another when the child harms a third person. However, this is a highly ambiguous generalisation. Thus, a duty to act in relation to a third person facing the risk of harm, because of an underlying contract which entails obligations between two other persons, cannot normally be established.[134]

122 Hence, based on the present situation and facts, it is highly doubtful whether such a duty would be imposed upon the driver or the logistics company.

---

133 *Hellner/Radetzki* (fn. 7) 111.
134 Ibid., 113.

Non-European Reports

Damage Caused by GMOs under Australian Law

Mark Lunney

I. General overview

1. Regulatory background

1 The regulation of the release of genetically modified organisms into the environment is governed by legislation both from the federal government and the various state governments. Because the Commonwealth of Australia is a federation and the federal government does not have legislative competence over gene technology, the introduction of legislation to regulate gene technology required an agreement between the federal, and state and territory, governments. The agreement was made in 1990 and after negotiations a series of Gene Technology Acts were passed by the federal government and in all the states and territories.[1] This legislation is known as the *Gene Technology Act*.

The primary federal legislation is the *Gene Technology Act 2000* (Cth) which came into force on 21 June 2001. This legislation establishes a process by which genetically modified organisms ("GMOs"), which include genetically modified crops, may be released into the environment. The Act creates an officer, the Gene Technology Regulator, who, together with an administrative department, is responsible for approving releases of GMOs into the environment. There are a number of different categories of release but the most important are applications involving intentional releases of GMOs into the environment.[2] Broadly, the approval process involves the Gene

1 *Gene Technology Act 2000* (Cth); *Gene Technology Act 2003* (ACT); *Gene Technology Act (New South Wales) Act 2003* (NSW); *Gene Technology Act 2001* (Qld); *Gene Technology Act 2001* (SA); *Gene Technology Act 2001* (Tas); *Gene Technology Act 2001* (Vic); *Gene Technology (Northern Territory) Act 2004* (NT); *Gene Technology Act 2006* (WA).

2 The legislation also regulates other dealings with GMOs such as research conducted in laboratories. There are also categories for Notifiable Low Risk Dealings (*Gene Technology Act 2000* (Cth) Part 6); Emergency Dealings (*Gene Technology Act 2000* (Cth) Part 5A), Exempt

Technology Regulator conducting a risk assessment on the particular GMO and requires the Regulator to consider the risks posed to health or safety of humans and to the environment[3] and the long and short term potential of the GMO to be harmful to other organisms and its ability to transfer, spread or persist in the environment. Unless satisfied that the risks are insignificant or able to be managed, a licence to release the GMO cannot be granted. However, the Regulator has construed the factors that must be considered under the legislation not to include impacts on trade, social and cultural effects as well as benefits that may be derived from gene technology or food labelling.[4] There have been suggestions that such considerations should be included as part of the Regulator's risk assessment but this was rejected by the Statutory Review of the Gene Technology Act in 2005, the Review Panel considering that the transitory nature of marketing and trade impacts made them unsatisfactory components of what was meant to be a technical science-based review of risk to humans and the environment.[5]

If the Regulator decides that it is permissible for the GMO to be released, a licence will be issued which may contain a series of conditions which must be complied with for the release to be lawful. Criminal sanctions are imposed on those who deal with a GMO (which includes growing it) in the absence of a licence or who breach licence conditions imposed by the Regulator[6] although special licences may be granted to those inadver-

---

Dealings (*Gene Technlogy Act 2000* (Cth) s 32(3); *Gene Technology Regulations 2001* (Cth) Reg 6, Sch 2 Part 1) and Inadvertent Dealings (*Gene Technology Act 2000* (Cth) s 40A).

3 The definition of "environment" in the *Gene Technology Act 2000* (Cth) is different from that in the Commonwealth's primary environmental protection legislation, the *Environment Protection and Biodiversity Conservation Act 1999* (Cth) which has led some commentators to criticise the Regulator's assessment as paying insufficient attention to environmental concerns: *S Kiyork*, The Gene Technology Act 2000 (Cth) and the licencing of Australia's first genetically modified crop: A case study in ignoring risks to biodiversity, (2005) 22 Environmental and Planning Law Journal 174; *Ch. Lawson/R. Hindmarsh*, Releasing genetically modified canola into the environment – deconstructing a decision of the Gene Technology Regulator under the Gene Technology Act 2000 (Cth), (2006) 23 Environmental and Planning Law Journal 22, 40 f. However, the Statutory Review of the Gene Technology Act 2000 and the Gene Technology Agreement did not recommend a widening of the definition in the Act and argued that the Regulator had correctly interpreted "environment" for the purposes of the Act (Commonwealth of Australia, Statutory Review of the Gene Technology Act 2000 and the Gene Technology Agreement (2006) 33, 51 f.).

4 Australian Government Department of Health and Ageing, Office of the Gene Technology Regulator, Risk Analysis Framework (2009) 23.

5 Commonwealth of Australia, Statutory Review of the Gene Technology Act 2000 and the Gene Technology Agreement (2006) 30 f.

6 Both fault-based and strict liability offences are created: *Gene Technology Act 2000* (Cth) ss 32, 33, 34, 35.

tently dealing with a GMO.[7] However, the legislation does not create any private rights of action between persons (e.g. between a person growing a GMO and someone who alleges he or she has suffered damage or loss as a result of that action) with one possible exception. Section 147 of the *Gene Technology Act 2000* (Cth) provides that if a person has refused or failed, is refusing or failing, or is about to refuse or fail, to do a thing and the refusal or failure is, or would be, an offence under the Act, the Regulator or *any other aggrieved person* (my italics) can seek an injunction requiring the person to do the thing.[8] There is no definition of "aggrieved person" in the Act but it has been suggested that farmers in the proximity of a GM farm might have grounds to bring such an action[9] but in any event the remedy only lies for breaches or potential breaches of the legislation.

2 As mentioned above, the constitutional limitations on the legislative competence of the Commonwealth meant that national legislation needed to be passed by every Australian jurisdiction for a national scheme to be effective. This could only be achieved by agreement with each of the state and territory jurisdictions within Australia and, as part of the agreement to implement uniform legislation, the Gene Technology Ministerial Council was created, a joint federal-state body which oversees the operation of the Gene Technology Regulator. The legislation provides that the Ministerial Council may issue policy principles for, amongst other things, recognising areas, if any, designated under state law for the purpose of preserving the identity of one or both of GM crops or non-GM crops, for marketing purposes.[10] A principle was issued in 2003 dealing with this issue with the result that six of the eight Australian jurisdictions passed legislation dealing with the issue.[11] The result was an additional layer of regulation applying to the releases of GMOs into the environment.

---

7 *Gene Technology Act 2000* (Cth) ss 40A, 46A, 49. If the licence is for the purpose of disposing of the GMO and the applicant came into possession of the GMO inadvertently, the usual procedures for granting a licence need not be followed.

8 Cf. the moratorium legislation (discussed in detail in the text below) in New South Wales where injunctions to prevent contravention of the Act can only be sought by the relevant government minister: *Gene Technology (GM Crop Moratorium) Act 2003* (NSW) s 32. The other state moratorium legislation does not provide for injunctions to be sought to prevent actual or imminent contraventions of the legislation.

9 *K. Ludlow*, Gene Technology Regulation and the Environment Protection and Biodiversity Conservation Act 1999 (Cth), (2004) 30 Monash University Law Review 165, 178.

10 It will be remembered that the Gene Technology Regulator does take such matters into account in the Regulator's risk assessment.

11 *Gene Technology (GM Crop Moratorium) Act 2003* (NSW); *Control of Genetically Modified Crops Act 2004* (Vic); *Genetically Modified Crops Free Areas Act 2003* (WA); *Genetically Modified Crops Management Act 2004* (SA); *Genetically Modified Organisms Control Act 2004* (Tas); *Gene Technology (GM Crop Moratorium) Act 2004* (ACT). Queensland and the Northern Territory have not passed moratorium legislation.

Broadly, the state "moratorium" legislation, as it is known, prevents the growing of GM or some types of GM crops in designated areas of the state or territory (or even the whole of the state or territory). As in the federal legislation, the moratorium legislation provides for criminal sanctions to be imposed on those who grow GM crops in an area designated GM free. Most jurisdictions require knowledge that the GM crop is being grown in a GM free area or at least recklessness as to this possibility before imposing criminal sanctions. There are provisions for exemptions and permits to be granted which have a similar effect to licences granted under the *Gene Technology Act*. As of June 2009, moratoria on GM crops, in whole or in part, remain in force in South Australia, Tasmania, the Australian Capital Territory and Western Australia. As a result of reviews of the moratorium legislation in Victoria and New South Wales in 2007, the former allowed its moratorium on GM canola (the only GM crop the subject of the moratorium) to expire in February 2008. In New South Wales, the result of the review was that the moratorium formally remained in place until 2011 but a procedure was introduced to allow the relevant state government minister to approve the commercial cultivation of GM crops approved for release by the Gene Technology Regulator.[12]

In most jurisdictions the moratorium legislation follows the form of the Gene Technology Acts in not expressly providing for private rights of action although there are provisions for the destruction of GM crops in non-GM areas which have not been exempted and for compensation to be paid for the loss of value of the crop by the state.[13] However, legislation in two of the states contains provisions that may affect private rights. In South Australia, a person found to have contravened the moratorium legislation can be ordered to pay the relevant government minister the amount of reasonable costs and expenses incurred through the minister having to take action to destroy unauthorized GM crops or material.[14] Moreover, if a person is convicted of an offence against the Act, the *court* recording the conviction may order the person convicted to pay to any per-

---

12 *Gene Technology (GM Crop Moratorium) Act 2003* (NSW) s 7A. This applies in addition to provisions allowing for exemptions to be granted. The relationship between the s 7A approval and the exemption provision is not clear but presumably s 7A is intended to govern large scale commercial releases whilst the exemption provisions deal with smaller scale releases (such as field trials).

13 *Genetically Modified Crops Free Areas Act 2003* (WA) s 9(1); *Genetically Modified Crops Management Act 2004* (SA) s 18(4); in Tasmania compensation may be payable for damage caused by inspections to determine if there has been a contravention of the Act as well as for orders for the destruction of crops: *Genetically Modified Organisms Control Act 2004* (Tas) ss 25A(2), 28(2).

14 *Genetically Modified Crops Management Act 2004* (SA) s 18(4).

son who has suffered loss or damage as a result of the conduct on which the conviction is based, or who has incurred costs as a result or expenses as a result of that conduct, to pay compensation for that loss or damage or for the costs and expenses incurred.[15] A similar provision is contained in the Western Australian legislation.[16] It is important to note, however, that any compensation in the latter examples is ordered as part of the criminal proceedings for a contravention of the legislation and is thus more like compensation awards in criminal cases than private law claims. The legislation does not allow for private rights of actions nor does it provide a mechanism for those who might have suffered loss or damage as a result of the contravention to bring this to the attention of the court.[17] Thus, their utility as compensation mechanisms is highly questionable.

Although the moratorium legislation does not create private rights of action, two jurisdictions have legislated to provide a defence to a private law claim arising out of GM contamination. In South Australia, if genetically modified plant material is present on any land and is attributable to the spread, dissemination or persistence of the material, and the original introduction of the material was not knowingly undertaken by any person who is or has been an owner or occupier of the land, no action can be brought on account of the fact that the material is present on the land or that the person dealt with the material. An exception applies if the owner or occupier deliberately dealt with a crop knowing it was genetically modified in order to gain a commercial benefit and that, in the interests of justice, another person's rights with respect to that material should be recognised.[18] The legislative provision in New South Wales is in similar terms except that it applies only to the spread of GM food plants.[19] These provisions are aimed at dealing with the well-known Schmeiser case in Canada where Percy Schmeiser was sued successfully by Monsanto for infringing its intellectual property rights by dealing with GM crops that had naturally spread to Schmeiser's land[20] but the provisions extend more widely

15 *Genetically Modified Crops Management Act 2004* (SA) s 24(1)(c).

16 *Genetically Modified Crops Free Areas Act 2003* (WA) s 10(3).

17 Nor is the standard of proof, or who or how it is to be discharged, discussed, e.g. who is responsible for showing that the loss or damage was caused by the contravention, and to what standard (balance of probability?); cf. *Genetically Modified Crops Free Areas Act 2003* (WA) s 10(2) which deals with claims by the minister for reimbursement for costs paid to a person whose crop was contaminated by another person's contravention of the Act, where it must be shown that the contamination was proved on the balance of probability.

18 *Genetically Modified Crops Management Act 2004* (SA) s 27.

19 *Gene Technology (GM Crop Moratorium) Act 2003* (NSW) s 32A.

20 *Monsanto Canada Inc v Schmeiser* [2004] 1 Supreme Court Reports (SCR) 902, although on the facts no damages were awarded as it could not be established that any profits were

than that situation. Thus, the provisions would provide a defence in an action against a landowner by a neighbour alleging that her crops had been contaminated by GM material emanating from the landowner's land if the landowner can show that the presence of the GM material on the land was inadvertent.[21]

This summary of the regulatory background to the release of GMOs into the environment provides the backdrop against which discussions about liability must be construed.

2. Special liability or redress scheme for GMOs

The question of whether damage caused by the introduction of genetically modified organisms should be the subject of a special liability or compensation regime has been considered a number of times by governments at both the state and federal level. A report of a Federal Parliamentary Committee in 1992 recommended strict liability for the deliberate and unauthorised release of GMOs and that liability for authorised releases should be mitigated.[22] However, neither of these recommendations were contained in the Gene Technology Bill which came before the Federal Parliament in 2000, and a Select Committee of the Senate that considered the question of liability thought that the existing common law provided adequate safeguards for those who might be harmed by the release.[23] More recently, the report of the Review Panel which considered the operation of the *Gene Technology Act 2000* (Cth) five years after its enactment considered that the com- 3

attributable to the infringement. Note that the result of the *Schmeiser* case, in terms of liability, might well be the same under these provisions because Schmeiser's conduct, which suggested he had deliberately cultivated what GM plants might have spread naturally to his land, could well be construed as deliberately dealing with the GM crop in order to gain a commercial benefit.

21 This assumes that any action for the spread would be based on the *presence* of the GM material on the defendant's farm and the failure to stop it from spreading. In New South Wales, the provision was thought by legislators to cover this situation: New South Wales Parliament, Hansard, Legislative Assembly, 4 December 2007, 5022 ("Agreement in Principle" Speech for GM Technology (GM Crop Moratorium) Amendment Bill 2007).

22 *House of Representatives Standing Committee on Industry*, Science and Technology, Genetic Manipulation: The Threat or the Glory (1992), as discussed in *K. Ludlow*, Genetically modified organisms and private nuisance liability, (2005) 13 Tort Law Review 92.

23 *Commonwealth Select Committee on Community Affairs*, A Cautionary Tale: Fish Don't Lay Tomatoes: A Report on the Gene Technology Bill 2000 (Canberra, 2000) para. 6.30–6.32. See also *House of Representatives Standing Committee on Primary Industries and Regional Services*, Work in Progress: Proceed with Caution. Primary Producer Access to Gene Technology (Canberra, 2000) para. 7.108. For a discussion of these reports and their predecessors see *Ludlow*, (2005) 13 Tort Law Review 92.

mon law liability regime provided an adequate balance between the interests of those who introduced GMOs into the environment and those who might be damaged by such introduction.[24] Similarly, two recent reports commissioned by state governments to determine if moratoria on GM canola should be lifted (in New South Wales and Victoria) also saw no need to introduce any special liability or compensation rules specific to GMOs.[25]

4 The result is that any liability resulting from the release of GMOs is left to the existing law, both common law and statute. The main torts applicable are those of negligence and private nuisance. Negligence will be dealt with in more detail in the course of this report but something needs to be said here of private nuisance.

5 The tort of private nuisance deals with rights and obligations between neighbouring landowners and is thus relevant in the GM context. As in English law, liability in private nuisance depends upon the plaintiff suffering an unreasonable interference with the use and enjoyment of land owned or occupied by the plaintiff. The control device for determining when an interference with use and enjoyment of land becomes actionable is the notion of "reasonable user"; if the defendant's use of land amounts to a reasonable use of land then any interference with the land of another will be reasonable. In determining whether there has been reasonable use by the defendant, a number of factors are considered. These include the nature of the interference with the plaintiff's land, as a defendant's use of land that results in "material physical damage" to land is more likely to amount to a nuisance than one that causes an interference with the ability of the land to be enjoyed (known as amenity damage). The sensitivity of the plaintiff's use of land will also be considered as the plaintiff cannot increase the liability of his neighbour by putting land to an especially sensitive use.[26]

6 In the GMO context, important, unresolved issues concerning liability in private nuisance include whether GM contamination of a neighbour's land amounts to material physical damage to the land or an interference with amenity, and whether the use of land for a specific purpose, such as organic farming, might be considered an especially sensitive use of land. For example, it has been suggested that a GM contamination that resulted in a food crop being at least 1% GM would be material physical damage as

---

24 Statutory Review of the Gene Technology Act 2000 and The Gene Technology Agreement (Canberra, 2006) 38–41.

25 *Victorian Department of Primary Industries*, Review of the moratorium on genetically modified canola in Victoria (2007) 63–66; *NSW Department of Primary Industries*, Independent Review of the Gene Technology (GM Crop Moratorium) Act 2003 (2007) 16 f.

26 F. *Trindade*/P. *Cane*/M. *Lunney*, The Law of Torts in Australia (4th ed. 2007) 167–193.

the food regulator, Food Standards Australia and New Zealand, requires labelling where the GM percentage exceeds 0.9%.[27] This would make it more likely that the neighbour from whose farm the GM material had spread (assuming this could be established) would not be found to be engaging in a reasonable use of land. It has also been suggested that GM contamination that did not constitute material physical damage to land but resulted in a farmer not being able to grow organic or non-GM crops would amount to an interference with the amenity of the land and be actionable as a private nuisance.[28] Moreover, there is considerable debate over the effect of licences granted under the Gene Technology Act or the state moratorium legislation on liability in private nuisance; in particular, whether they will act as a defence or otherwise provide protection for those who release GMOs into the environment under licence.[29]

3. State liability

Australia has no separate system of liability rules that applies only to the State (whether at federal or state level). Although there are some differences, 7
Australian law broadly mirrors the approach to State liability that is adopted in the United Kingdom; that is, that the ordinary rules applicable to liability in tort apply, with some modifications, to the State. In the tort of negligence, these modifications reflect the fact that in some cases the allegation is that the State ought to have exercised its powers so as to prevent harm being caused. Given the competing calls on the resources of the State, Australian courts and legislatures have imposed additional hurdles over and above the ordinary requirements for a negligence action to succeed in this category of case.[30] As negligence law is a matter that is governed by

27 *K. Ludlow*, The economic impact of genetically modified organisms as actionable damage in torts, (2005) 13 Torts Law Journal 159, 166. The author argues (at 167) that the same should not apply to non-legislative standards, such as those applied by organic certification bodies.

28 Ibid., 170. Cf. *Mark Lunney/Robert Burrell*, A farmer's choice? Legal liability of farmers growing crops (Department of Agriculture, Forestry and Fisheries, 2006) chap. 4.

29 Compare *Ludlow*, (2005) 13 Tort Law Review 92, 113–115, with *Lunney/Burrell* (fn. 28) chap. 6.

30 *Graham Barclay Oysters Pty Ltd v Ryan* (2002) 211 Commonwealth Law Reports (CLR) 540. These include the nature of the control exercised by the particular emanation of the state by virtue of the statutory powers conferred on it, the vulnerability of the plaintiff to injury, the actual knowledge the state or its emanation had of harm to the plaintiff, whether imposing any duty in negligence would interfere with core policy or quasi-legislative functions, and whether there are any supervening policy reasons which would preclude liability in negligence, such as incompatibility with a statutory scheme or the nature of the loss, e.g. if it is pure economic loss, which is only recoverable in limited cir-

the law of the individual states of Australia, the precise details of the modifications vary amongst the different jurisdictions. The net result of the common law and statutory modifications is that it is very unlikely that the Commonwealth Government, the state governments, or the Gene Technology Regulator could be sued in negligence over decisions relating to the release of GMOs into the environment.[31]

II. Damage

1. Recoverable losses

8 The primary losses recoverable in a tort action in Australia are for personal injury (including psychiatric harm) and damage to property (assuming that a cause of action in tort has been made out). Under certain limited circumstances, purely economic losses are recoverable. An important issue that is yet to be resolved in Australia is how any damage caused by GMOs might be classified. For example, if a conventional crop is contaminated by some GM material with the result that it cannot be sold as non-GM, is this contamination damage to property or only pure economic loss?[32] Put simply, it is generally easier to recover for damage to property than it is for purely economic losses.

2. Pure economic loss

9 Pure economic loss is recoverable in the tort of negligence in certain limited circumstances. The most recent decision of the Australian High Court

---

cumstances. The following quote from Gleeson CJ (at [21]) in the case cited above (which dealt with injury to health caused by eating contaminated oysters) reflects an attitude that might well apply to attempts to make the state liable for GMO regulation:
"The assumption that a State government owes to individual citizens a legal duty to care for the health and safety of all citizens, or all consumers of food or all consumers of oysters, by exercising its regulatory powers to the extent judged reasonable by a court, has far-reaching implications".

31 Amongst other reasons, because the decisions are made as part of detailed statutory schemes granting considerable discretion to the decision makers (see, e.g., *Gene Technology Act 2000* (Cth) s 30), the decisions are reviewable in administrative (public) law so there are already some grounds for challenging a decision, and the decisions are of a quasi-legislative character in that they involve weighing up of competing policy arguments and are inherently unsuitable for review in a private law action for negligence.

32 For a discussion of these issues see *M. Lunney*, Damage and Cross-Pollination by a Deliberately Released Genetically Modified Organism, University of New England Centre for Agriculture and Law, Ag Law Papers Series 1, 2004; *Ludlow*, (2005) 13 Torts Law Journal 159; *Lunney/Burrell* (fn. 28) chap. 2.

on this issue is *Perre v Apand* in 1999 where the seven justices of the High Court of Australia each delivered separate judgments.[33] This makes it difficult to determine with certainty what is required before liability in negligence for pure economic loss may arise but a number of factors are common to the judgments. Thus, for a duty of care in the tort of negligence to be owed to another to prevent causing that other pure economic loss through carelessness, questions considered are whether the plaintiff's loss was reasonably foreseeable by the defendant, whether imposing a duty of care on the plaintiff would lead to indeterminate liability, whether imposing a duty of care would impose an unreasonable burden on the plaintiff's autonomy, whether the defendants were vulnerable to economic loss from the plaintiff's conduct and whether the defendant had actual knowledge that its conduct could cause harm to the plaintiff.

In private nuisance, once the interference with the use and enjoyment of land has been established, consequential economic loss can be recovered.[34] It should be noted, however, that compensation for an interference with an amenity interest in the land in certain cases may be functionally similar to a claim for pure economic loss. Hence, compensation for the inability to farm land organically might well be valued as the difference between the value of an organic crop grown on the land as opposed to a GM crop. In theory, however, the claim is for damage to an interest in land rather than for a pure economic loss.[35] **10**

Pure economic loss is clearly recoverable in contractual claims. **11**

3. Mere fear of a loss

There is no direct authority on the issue, presumably because the commercial growing of GM crops is limited in Australia and only in 2008 were the first commercial food GM crops authorised. Nor have there been any cases on injury caused by fear of adverse consequences of modern technology. In Australia, such action would have to overcome two obstacles. First, fear in itself is not a recognised category of damage for which recovery is possible. It is only recoverable if the fear constitutes a **12**

33 *Perre v Apand* (1999) 198 CLR 180.

34 *Ludlow*, (2005) 13 Tort Law Review 92, 117 f.

35 However, an award for loss of amenity is not always functionally similar to an award for pure economic losses as an award for loss of amenity can be made even though there may be no adverse economic consequences (for example, where the interference was not permanent).

recognised psychiatric illness. Second, even if this kind of injury results from the fear, there must be some cause of action which would allow damages to be awarded for the injury. In the context of mobile phone radiation, for example, it is not easy to see what that cause of action would be given that it is not accepted that mobile phone radiation is harmful to health and that the operation of the relevant technology has had to be approved by a regulatory procedure which considers, amongst other things, any risks to health associated with the technology. Similar reasoning would apply to GMOs as there is a detailed regulatory process for approval of releases of GMOs into the environment.[36]

4. Standard of proof

13 There are no peculiar rules that would apply to losses which are recoverable in law and are caused by GMOs. As in the United Kingdom, once a cause of action has been established by showing that at least some loss has been caused by the GMO on the balance of probability, potential future losses may be valued by reference to the chances of that loss accruing. For example, assume that A can prove that B's careless handling of a GMO has caused him some personal injury. Assume further that the evidence is that, as a result of the carelessness, there is a 30% chance that A might suffer epilepsy within five years. A court would be able to award compensation for that possibility by calculating the extra damage that A would suffer if epilepsy did occur within five years and then discounting, or reducing, the figure to represent the chance that it might not happen. Thus in the example given A would receive 30% of the additional damages which would accrue if she did in fact suffer epilepsy as a result of B's negligence.

5. Nominal losses

14 The position here is no different from that in the United Kingdom. There are some torts for which nominal damages may be awarded but in the context of losses caused by GMOs, the torts most likely to give rise to liability – negligence and private nuisance – are torts for which actual damage must be established for an action to be brought.

---

36 Fn. 34, 107. Ludlow argues that such a plaintiff might be considered hypersensitive with the result that any action for private nuisance would fail. For criticism of the belief in the objective truth of science in this context see *Lawson/Hindmarsh*, (2006) 23 Environmental and Planning Law Journal 22.

6. Mass losses

There are no special rules for mass losses that would be applicable to the GMO context. 15

III. Causation

1. Uncertainty of merely potential causes

Broadly, Australian law does not relax the requirement of proving a causal link, on the balance of probability, between the defendant's act or omission alleged to be tortious and the injury for which the plaintiff is claiming damages. The plaintiff must prove to this standard that the defendant's conduct has caused or materially contributed to the plaintiff's injury. This has recently been affirmed by the High Court of Australia.[37] However, the High Court has allowed damages for "loss of a chance", in the context of trade practices legislation,[38] where the chance related to financial gain but it has not ruled on the issue in the context of other types of damage nor has it considered the mesothelioma cases that led the House of Lords to relax the causation requirement in that category of case.[39] Legislation in many Australian states now governs the issue of causation in actions involving fault and provides that it be proved that the defendant's negligence was a necessary condition of the harm. However, provision is also made for dispensing with this requirement by considering "whether or not and why responsibility for the harm should be imposed on the negligent party".[40] This would be sufficient for courts to adopt the solution to causal uncertainty favoured by the House of Lords in the mesothelioma cases. It should be noted, however, that this legisla- 16

---

37 *Road Traffic Authority v Royal* (2008) 82 Australian Law Journal Reports 870 at [144] where Kiefel J noted that the defendant's creation of a possible risk of injury was not sufficient to establish causation; it must be proved that the risk eventuated.

38 See *Sellars v Adelaide Petroleum NL* (1994) 179 CLR 332; for comment see *M. Lunney*, What Price a Chance?, (1995) 15 Legal Studies 1.

39 Ibid, [94], where Kirby J noted that no argument for a reconsideration of the traditional approach to causation in light of *McGhee v National Coal Board* [1973] 1 Weekly Law Reports (WLR) 1, *Fairchild v Glenhaven Funeral Services Ltd* [2003] 1 Appeal Cases (AC) 32 or *Barker v Corus (UK) Ltd* [2006] 2 AC 572 had been made. More recently, the High Court avoided consideration of whether the English cases referred to above were applicable in Australian Law (*Amaca Pty Ltd v Ellis* (2010) 84 Australian Law Journal Reports 226). However, in the closely related field of damage for loss of chance, the High Court has refused to accept that loss of a chance is actionable damage in medical negligence cases: *Tabet v Gett* [2010] HCA 12 (21 April 2010).

40 See, e.g., *Civil Liability Act 2002* (NSW) s 5D.

tion prevents a reversal of the burden of proof in cases based on fault; the burden always rests with the plaintiff in such cases.[41]

17 Causation is also required to be proved in cases of private nuisance. There is authority that, where a nuisance is caused by the combined acts of a number of people, all who contributed are liable even though, individually, their conduct would not amount to a nuisance.[42] However, in these cases the defendants have been proved to have made some contribution to the nuisance thus the position is similar to the material contribution test for causation in negligence cases. It would not, it is submitted, cover cases where all that could be proved against a defendant was that there was a risk that his conduct had contributed to the nuisance, such as a risk that material from a GM farm had contaminated a neighbour's property.

2. Complex causation scenarios

18 There are no special rules that apply in the scenarios contemplated above. The possibility of modification of ordinary rules of causation as described in the previous paragraph would also apply in this category of case.

3. Force majeure

19 The expression "force majeure" refers in the Australian context to events that intervene between the defendant's tortious conduct and the plaintiff's damage, the defendant alleging that the intervening act prevents causal responsibility for the plaintiff's damage being attributed to the defendant. The intervening act might be the act of the plaintiff, a third party, or a natural occurrence.

20 It is extremely difficult to provide anything more than broad general rules on the effect of intervening acts as courts have repeatedly stressed that decisions are fact-specific.[43] With this in mind, it might be said that, at least since it has been possible to apportion liability between plaintiff

---

41 Ibid., s 5E.

42 *Pride of Derby & Derbyshire Angling Association Ltd v British Celanese Ltd* [1952] 1 All England Law Reports (All ER) 1326 (not considered on appeal: [1953] Ch 149); cited with approval in the Australian cases of *Barker v Permanent Seamless Floors Pty Ltd* [1983] 2 Queensland Reports (Qd R) 561, *Bonnici v Ku-ring-gai Municipal Council* [2001] Community Services Appeals Tribunal of New South Wales (NSWSC) 1124. More generally see *D. Nolan*, Nuisance, in: K. Oliphant (ed.), The Law of Tort (2nd ed. 2007) para. 22.74.

43 *Trindade/Cane/Lunney* (fn. 26) 560; *RP Balkin/JR Davis*, Law of Torts (4th ed. 2009) para. 9.13.

and defendant where the plaintiff has been found to have been contributorily negligent with respect to the injury, it is rare that an intervening act of the plaintiff will be found to have broken the causal connection between the defendant's tort and the plaintiff's injury. Similarly, since the introduction of contribution legislation allowing apportionment between concurrent tortfeasors, courts have generally preferred findings that attribute at least *some* causal responsibility to the first negligent party whilst attributing a greater degree of responsibility to the later intervening negligent act where the original negligence placed the plaintiff at risk of injury from a subsequent negligent act.[44] However, although Australian law does not formally recognise different degrees of culpability within the tort of negligence, the more extreme, or gross, the intervening act of negligence the more likely it is that it will be held to break the causal link between the original negligence and the plaintiff's injury.[45] Finally, if the intervening act is the deliberate and wrongful (i.e. criminal or illegal) act of a third party, it will normally be held to break the causal connection between the defendant's negligence and the plaintiff's injury. The exception is where the scope of the defendant's obligation to the plaintiff (referred to in Australian law as the scope of the duty of care) extends to taking reasonable steps to prevent the plaintiff suffering damage by the deliberate and wrongful conduct of a third party of a type that actually occurs.[46] The scope of the duty of care will extend this far only where there is a "special" relationship between the plaintiff and defendant (e.g. employee and employer) or, more rarely, between the defendant and the third party.

A brief mention must also be made of intervening acts of nature. If the **21**
intervening act of nature was foreseeable and the defendant failed to take reasonable precautions against the risks associated with such an intervening act, the defendant will remain liable if the act of nature materialises and causes damage.[47] These cases involve breach of duty (e.g. carelessness) rather than intervening causation. However, where the negligence does not relate to the failure to take precautions against acts of nature, but an act of nature intervenes independently of the negligence and results in damage to the plaintiff, it has been suggested that such cases should be con-

44 See, e.g., *Chapman v Hearse* (1961) 106 CLR 112, where the first defendant driver, responsible for a car accident, was held to have made some causal contribution to the death of the plaintiff's husband, a doctor who had stopped to render assistance and was hit by the driver of a car that did not see him, the second defendant, to whom a greater causal responsibility was allocated.

45 *Mahony v Kruschich (Demolitions) Pty Ltd* (1985) 156 CLR 522.

46 *Modbury Triangle Shopping Centre Pty Ltd v Anzil* (2000) 205 CLR 254.

47 *Balkin/Davis* (fn. 43) para. 9.13.

sidered as remoteness of damage cases (e.g. the question is whether damage caused in that way can be regarded as a type or kind of damage that was a reasonably foreseeable consequence of the original negligence).[48]

22 In the GM context, a number of points may be made. First, the deliberate, wrongful conduct of a third party that results in the unwanted spread of GM material, even if the conduct is only possible because the negligence of the defendant allowed the third party to act in that way, will not result in liability unless there is some special relationship between the plaintiff and defendant. The relationship between adjoining landowners is unlikely to meet this requirement. Second, although possible, it is unlikely that a GM farmer could argue that an intervening natural event, such as high winds or excessive rainfall, were not foreseeable. A defendant would need to take reasonable precautions against these acts of nature and a failure to do so which caused damage would make the defendant liable. However, if a GM farmer carelessly left GM material where it might accidentally be blown onto a neighbour's property, but, in fact, it spread to the neighbour's property through an unprecedented flash flood, it may be that the defendant would escape liability, either because the damage was too remote a consequence of the negligence or because the flood broke any causal connection between the negligence and the damage.

23 There is very little authority on the position of intervening acts in private nuisance. If a landlord lets premises to a tenant for a purpose which necessarily will result in a nuisance, then the landlord is liable along with the tenant. However, if the nuisance arises from the tenant's use of the premises, rather than the purpose for which it is leased, only the tenant is liable.[49]

4. Threshold to prove causation

24 Australian law adopts the "balance of probability" standard in civil cases.

5. Special rules on causation

25 N/A.

48 *Trindade/Cane/Lunney* (fn. 26) 560.

49 *Balkin/Davis* (fn. 43) para. 14.37. There is some authority that an owner is liable for the acts of licencees, as opposed to lessees, for acts of the licencees which amount to a nuisance and of which the occupier has knowledge: *Lippiat v South Gloucestershire Council* [2000] Queen's Bench (QB) 51.

IV. Types of liability

1. Fault liability

(a) Special rules governing fault

There is no suggestion that the fault requirement in cases involving GMOs should be varied from the usual requirements. As noted above, tort legislation in most Australian jurisdictions now prevents the burden of proof for causation from being shifted from the plaintiff. **26**

(b) Impact of specific rules of conduct

As noted above, the release of GMOs in Australia is governed by a two-tier regulatory process. If releases are approved by the granting of licences or exemptions, conditions may be imposed on these licences or exemptions. Moreover, conditions may be imposed by Food Standards Australia and New Zealand as part of their approval process for food products containing elements of GMOs. **27**

None of the legislation imposing this regulatory structure specifically deals with civil liability, it being thought existing common law rules were adequate. However, it is likely that any conditions imposed by the regulatory bodies will be important in determining civil liability. For example, if the Gene Technology Regulator imposes certain conditions on the commercial release of a GMO, and those conditions are not complied with, this would be good evidence of carelessness on the part of the grower, and perhaps even the distributor and manufacturer of the GM seed. Conversely, if the conditions are complied with, this would be good evidence that reasonable care had been exercised in growing the GM crop and it would be difficult to establish negligence. In the absence of express statutory provisions providing that compliance with regulatory requirements is a defence to a civil action (and such legislation would have to be passed at the state level as civil liability at common law is in general a state matter), compliance or lack of it can only be evidence of carelessness but given the extensive regulatory regime it is likely to be extremely influential evidence. The same reasoning would apply to any conditions attaching to exemptions under the state moratorium legislation. In private nuisance, it has been argued that the presence of licences and permits will be influential in determining both the locality in which the alleged nuisance takes place and the substantiality of the interference. For example, if a **28**

GMO release is licensed in a particular area, it is more likely that the carrying on of GM farming in that area will be regarded as a reasonable use of land.[50] It has also been argued that an interference may not be regarded as substantial if the release is pursuant to licences and exemptions under the Gene Technology Act and state moratorium legislation, at least where there is no material physical damage to land as a result of the release.[51]

29 Apart from compliance with the formal regulatory requirements for a release, considerable significance is likely to be attached to co-existence protocols and the best practice that is suggested in them. The Statutory Review of the Gene Technology Act 2000 recommended that the Commonwealth and states work together to develop a national framework for non-GM and GM crops to address market considerations and noted that the state of Queensland had already developed a model framework for co-existence.[52] In light of the differing rules amongst the states on allowing GM food crops, such a national scheme seems unlikely in the short term but industry has already developed protocols for marketing GM and non-GM canola.[53] Whilst compliance with any requirements set out in these protocols will assist a GM grower to argue that he or she has acted carefully, it has been suggested that questions relating to liability should not be determined by voluntary standards set by interested parties.[54]

2. Product liability

(a) Development risk defence

30 Interestingly, Australia adopted the text of the Product Liability Directive when it was introduced into the *Trade Practices Act 1974* (Cth) as Part VA of that legislation. In s 75AK(1)(c) it is a defence if "the state of scientific or technical knowledge at the time when they were supplied by their actual

50 *Ludlow*, (2005) 13 Tort Law Review 92, 109–111. Note, though, that locality is relevant only where there is no material physical damage to land: *St Helen's Smelting Company v Tipping* (1865) 11 Clark & Finnelly's House of Lords Reports New Series (HLC) 642, *Balkin/Davis* (fn. 43) para. 14.15.

51 *Ludlow*, (2005) 13 Tort Law Review 92, 113–115.

52 Statutory Review of the Gene Technology Act 2000 and the Gene Technology Agreement (2006) 97.

53 Principles for process management of grain within the Australian supply chain: A guide for industry in an environment where GM and non-GM grain is marketed, Industry Report prepared under the Single Vision Grains Australia Process (2007).

54 Hence, it has been argued that voluntary standards set by organic certification bodies should not determine whether the plaintiff has suffered damage to property: *Ludlow* (2005) 13 Tort Law Review 92, 106; *Ludlow*, (2005) 13 Torts Law Journal 159, 166 f.

manufacturer was not such as to enable that defect to be discovered". This language varies slightly from that in Article 7(e) of the Directive in that the relevant time is the time of supply by the actual manufacturer, not the time the producer put the product into circulation, but in substance it is the same. However, there are few cases discussing the operation of the section but those that do suggest a limited reading of the section. Thus, in a case involving contaminated oysters, it was held that the defence applied even though the defect – the presence of a hepatitis virus – was well known to exist. This was because it was impracticable to test for the virus without destroying the oyster. Such an interpretation focuses not on the knowledge of the defect but on the reasonableness of the steps available to detect it and introduces a fault-based component to product liability under Part VA.[55]

(b) Alternative routes

There are no specific alternative routes that are available for compensating losses caused by agricultural products. However, it is possible that no-fault workers' compensation schemes could cover such losses if they were attributable to the employment of the person suffering the injury. **31**

(c) Impact of compliance with rules and regulations

I have discussed this question in IV 1(b) above. **32**

3. Environmental liability

(a) Environmental liability regime

The primary Commonwealth statute dealing with environmental liability is the *Environment Protection and Biodiversity Conservation Act 1999* (Cth). It establishes a regime to protect areas of national environmental significance and the environments of Commonwealth land. Each of the states and territories also has their own environmental protection legislation.[56] **33**

---

55 *Trindade/Cane/Lunney* (fn. 26) 644.
56 *Lunney/Burrell* (fn. 28) chap. 7.

34 The Commonwealth Act requires that certain actions, known as controlled actions, have to be approved under the Act to be lawful. Controlled actions are actions that are likely to have a significant impact on an area of national environmental significance or Commonwealth land.[57]

35 It has been argued that the approval process under the Gene Technology Act has no effect on the *Environment Protection and Biodiversity Conservation Act 1999* with the result that, if the release of a GMO into the environment amounts to a controlled action for which an approval has not been obtained, the releaser will be committing an offence under the Act even if the release had been authorised under the Gene Technology Act.[58] This has now been accepted by the Gene Technology Regulator who advises that controlled actions under the *Environment Protection and Biodiversity Conservation Act 1999* require formal assessment and approval by the Australian Government Health Minister under that Act.[59] It is not clear whether the same applies to approvals that are necessary under state environmental protection legislation.[60]

(b) Claimants in cases of environmental harm

36 The Commonwealth legislation allows a person who suffers loss or damage arising from a contravention of the Act or the regulations passed under it to claim from the person responsible for the contravention an amount equal to the other person's loss or damage. Damage is defined to include the costs of taking action to prevent further damage to the environment as a result of the contravention.

37 Although the Act does provide for this private right of action, its practical use as a remedy is limited by the requirement that the conduct contravene the Act. As the types of conduct that contravene the Act are limited, there may be relatively few cases where the GM crops cause loss to an adjoining farmer in circumstances where the Act is breached. Similar limitations apply to most of the state environmental protection legislation.

---

57 See *Ludlow*, (2004) 30 Monash University Law Review 165, 185–194.

58 Ibid.

59 Australian Government Department of Health and Ageing, Office of the Gene Technology Regulator, Risk Analysis Framework (2009) 9.

60 It has been argued that the purpose of the licence under the Gene Technology Act would be defeated if it did not operate to free the licencee from actions under environmental protection legislation (*Lunney/Burrell* [fn. 28] para. 7.6) but this appears not to be the case in respect of the Commonwealth legislation.

However, there are two states that provide for a general environmental duty of care and allow private individuals to sue those responsible for the breach. Environmental harm is defined broadly and would include damage to crops and property and probably the resulting economic loss.[61] However, breach of this statutory duty is fault-based and it may be that these actions add little to the common law action for negligence.[62] **38**

(c) Special liability regime for losses sustained by individuals

There are no special liability regimes to cover losses of this kind other than those described above. Individuals must rely on the torts of negligence and private nuisance to claim for any losses suffered as a result of damage to the environment. These torts do not allow a plaintiff to claim for environmental damage per se but it may be that damage to crops (for example) might also amount to environmental damage and in that sense private law actions in negligence and private nuisance can be brought for environmental damage. **39**

(d) Cartagena Protocol

Australia has not signed the Cartagena Protocol. Treaty obligations do not form part of domestic law unless express legislation is passed incorporating them into that law. **40**

4. Other strict liability regimes

N/A. **41**

V. Vicarious liability

1. Scope of vicarious liability

The position in Australia broadly mirrors that in the United Kingdom. In general, an employer is not liable for the torts committed by independent contractors hired by the employer. There are a limited category of cases in which a non-delegable duty of care is imposed on the employer; these are **42**

61 *Lunney/Burrell* (fn. 28) chap. 7. The two states are South Australia and Tasmania.
62 Ibid.

situations where the employer owes a duty to be careful to the plaintiff and the duty cannot be discharged merely by delegating the performance of that duty to another, including an independent contractor. Thus, if the independent contractor discharges the duty carelessly, the employer is vicariously liable for the tort committed by the independent contractor.[63]

43 In the GM context, the most important of the exceptional categories is the "control of dangerous activities" category. If an independent contractor carries out a dangerous activity on the occupier's land for the benefit of the occupier, the occupier can be liable to persons outside the land for damage caused by the independent contractor. The notion of a "dangerous activity" is notoriously vague but the category was affirmed by the High Court of Australia in 1993.[64] Thus, if growing a GM crop can be considered a dangerous activity, this form of strict liability could attach to an employer who hires an independent contractor to discharge functions relating to the growing, distribution and marketing of GM crops. In light of the regulatory hurdles that must be overcome before GM crops may be released into the environment this seems a highly unlikely characterisation.

2. Liability for people further up the food or feed production chain

44 Apart from contractual warranties, there is no general principle by which liability can be passed from one party in the feed and food production chain to another as against the plaintiff. As between the parties, it is possible by contract to indemnify one party for any losses it suffers as a result of liabilities incurred by third parties but these provisions have no effect between the plaintiff and the tortfeasor. As in the European Directive, it is also possible under Australian product liability legislation that a supplier may become liable as the manufacturer if the manufacturer cannot be identified.[65]

3. Can someone further down the feed or food chain include someone further up it in a trial against him/herself?

45 The rules of civil procedure generally allow any defendant to seek permission to have another potential defendant joined to the action. Hence, a

63 *Trindade/Cane/Lunney* (fn. 26) 772–781.

64 *Burnie Port Authority v General Jones Pty Ltd* (1994) 179 CLR 520.

65 *Trade Practices Act 1974* (Cth) s 75AJ.

farmer being sued for GM contamination of adjoining crops could seek permission to join a seed retailer or manufacturer as a party to the action. However, this can only be done against a party who is also potentially liable to the plaintiff so it will be discussed in detail under multiple tortfeasors.

VI. Multiple tortfeasors

In general, multiple tortfeasors (an expression not used in Australia; the expressions are joint, several and concurrent tortfeasors) responsible for causing the same damage to the plaintiff are liable in solidum at common law; that is, each is liable in full to the plaintiff for the loss that the plaintiff has suffered. A plaintiff may sue only one of a number of multiple tortfeasors and is allowed to recover in full off that single defendant. Before the trial of any action, and subject to the appropriate procedural rules of court, it is open to either plaintiff or defendant to seek permission to join another party to the proceedings as an additional defendant. The benefit to the plaintiff is that, if the plaintiff wins against both defendants, judgment will be given against both defendants so that the plaintiff will be able to enforce the judgment against both defendants if necessary.[66] From the original defendant's perspective, the joining of an additional defendant allows for the possibility of apportionment between the defendants if both are found liable so that, as between the defendants, one may be ordered to pay 60% of the plaintiff's damages and the other 40% (for example). This form of apportionment, known as contribution, is also available in later proceedings; hence a defendant against whom a judgment has been ordered can originate later legal proceedings against a party that might also have been liable to the original plaintiff and seek a contribution towards the damages the original defendant has paid to the original plaintiff. 46

Whilst these rules will apply to potential actions involving the release of GMOs into the environment, there are no special rules that apply to this category of case. 47

Where the plaintiff suffers personal injury, the above rules apply. However, where the plaintiff suffers damage to property or pure economic loss, most jurisdictions in Australia have introduced proportionate liabil- 48

66 An important point if, for example, one of the defendants is insolvent. Even if judgment is obtained against more than one defendant, the plaintiff cannot, of course, obtain more than the full monetary amount of the judgment.

ity.[67] The rules are complicated and vary by jurisdiction but their general effect is that they limit the liability of any one defendant in proportion to their responsibility for the damage. Hence in the example above, under a system of proportionate liability, the first defendant would only be liable for 60% of the damage suffered by the plaintiff and the second defendant for only 40%. If one of the defendants cannot pay the plaintiff (for example, because of insolvency) this risk is borne by the plaintiff.

49 Again, the proportionate liability regime will apply to claims for pure economic loss and damage to property resulting from the release of a GMO.

VII. Defences

1. Licence/permission to grow GM material

50 Licences granted under the *Gene Technology Act* or exemptions or permits under the state moratorium legislation do not operate as a defence against common law civil liability triggered by the release of a GMO; they are not akin to the defence of statutory authority.[68] However, as argued above, compliance with licence conditions will be good evidence that a person responsible for the release of the GMO has exercised reasonable care in dealing with it.

2. Consent/assumption of risk

51 It is unlikely that consent/assumption of risk will play any role in liability arising out of the release of GMOs. This is because the assumption of risk defence in Australia requires both knowledge of the risk and voluntary acceptance of the adverse outcomes that may occur if that risk is run. Given the regulatory regime for the release of GMOs it can plausibly be argued that someone who consumed a GM food crop (for example) had no knowledge of any risk. If the GMO was approved for release and for food consumption, a consumer would be entitled to assume there was no risk. Even in situations where knowledge of the risk could be established (such as where a non-GM farm adjoins a GM farm and the risk is cross-pol-

---

67 For detail see *Trindade/Cane/Lunney* (fn. 26) 803–807. Broadly, proportionate liability applies to fault-based claims and to claims for breaches of consumer protection legislation in the states and Commonwealth where the damage suffered is pure economic loss or damage to property.

68 *Ludlow*, (2005) 13 Tort Law Review 92, 116; *Lunney/Burrel* (fn. 28) chap. 6.

lination) it would be very difficult, absent an express contractual agreement, to establish that the neighbouring farmer had consented to bear any damage that might occur as a result of contamination.

3. Third-party influence

This question would be answered by reference to existing common law principles. As regards sabotage or intervention by a third party, there is generally no obligation to take reasonable steps to prevent the deliberate and wrongful act of a third party from causing harm to the plaintiff. Some kind of relationship between the defendant and the third party would be required and this is very unlikely to be the case in sabotage. **52**

Conversely, if the conduct of the neighbouring farmer amounts in law to contributory negligence, legislation in all Australian jurisdictions allows for the plaintiff's damages to be reduced. However, it is very unlikely that continuing to plant a non-GM crop next to a neighbour who is growing a GM crop will be categorised as contributory negligence. It can hardly be unreasonable to use one's own agricultural land to grow the crops one wishes to grow. **53**

4. Prescription

Each jurisdiction in Australia has its own limitation statute. There is no suggestion that any special limitation rules will apply in the case of GMOs. **54**

5. Other defences

Related to 3. above, it is possible that the defence of illegality might be applicable in rare circumstances. For example, if the spread of GM material was the result of the unauthorised entry onto a neighbour's land and resulted in damage to the person who made the unauthorised entry, any action by that person alleging damage as a result of the spread of GM material could be met by the defence of illegality. This is especially likely as the Gene Technology Act creates the offence of interfering with dealings with GMOs.[69] **55**

---

69 *Gene Technology Act 2000* (Cth) s 192A.

VIII. Remedies

1. Pecuniary compensation

(a) Bodily harm

56 No special rules apply.

(b) Property losses

57 No special rules apply.

(c) Economic losses

58 There are no special rules that apply. Assuming that a cause of action had been established, whether these losses could be recovered would depend on the ordinary rules of remoteness of damage. It has been suggested that loss of profits in respect of contaminated crops and profits lost whilst the land was being restored are recoverable.[70] However, the ordinary rules as to mitigation of damage would apply so that, if the organic farmer, acting reasonably, was able to grow a non-organic crop and failed to do so, this failure to act would reduce the damages awarded.[71]

(d) Harm to animals

59 If the harm consists of an injury that can be remedied by treatment, the costs of that treatment are recoverable if the harm is caused by a tort. If the harm is the loss of the animal, the replacement cost of the animal would be recoverable together with any losses from contracts which cannot now be performed, provided the plaintiff has acted reasonably to mitigate those losses.[72] If the plaintiff does not wish to replace the animal, a damaged chattel is compensated on a "going concern" basis. The future earning capacity of the animal is reduced to net present value and then added to a figure that represents the capital value of the chattel.[73]

---

70 *Ludlow*, (2005) 13 Torts Law Journal 159, 183.

71 *Trindade/Cane/Lunney* (fn. 26) 714 f.

72 Ibid., 752 f.

73 But this figure is not the resale value at the date of the death since future profits can only be awarded on the basis that the animal was being retained: Ibid., 752.

There is little authority on what constitutes "damage" for the purpose of determining if a plaintiff has suffered property damage. On one view, what is required is some kind of physiological change to the animal that results in the value of the animal decreasing. On this view, and assuming the consumption of the GM contaminated feed resulted in a reduction in the sale price of the animal, the owner has suffered damage to property.[74] Conversely, it may be that the physiological change must pass a certain threshold before the damage can be categorised as damage to property.[75] It has also been suggested that different rules apply to determining whether there has been damage to property in a negligence action and "material physical damage" for the purpose of an action in private nuisance.[76] The result is that it cannot be said definitively whether an animal that eats GM contaminated seed is damaged but it is clear that the more significant the physiological effect of eating the GM contaminated feed the more likely it is that the physiological change will amount to damage to property. Of course, as noted above, whether the owner can recover for this damage will depend on establishing that the damage was caused tortiously. **60**

(e) Costs of disposal

Again, assuming a cause of action had been established, this would be governed by the rules relating to remoteness of damage. It has been suggested that the costs of removing and disposing of the invading GMO and any of the plaintiff's organisms contaminated by the invasion would be recoverable.[77] **61**

2. Non-compensatory damages

The primary tort actions applicable to the deliberate release of GMOs into the environment – negligence and private nuisance – require proof of damage to constitute the action. In principle, aggravated and exemplary (or punitive) damages can be awarded in appropriate cases; in particular the limits that apply to the latter awards in the United Kingdom do not apply in Australia.[78] However, a number of jurisdictions have legislated **62**

74 *Lunney* (fn. 32); *Lunney/Burrell* (fn. 28) chap. 2.
75 *Ludlow*, (2005) 13 Torts Law Journal 159, 179–181.
76 Ibid., 164–169, 179–181.
77 *Ludlow*, (2005) 13 Torts Law Journal 159, 183.
78 See *Uren v John Fairfax & Sons Ltd* (1966) 117 CLR 118.

to limit awards of exemplary damages in cases of personal injury and these would apply to injury caused by the release of GMOs.[79]

3. Other remedies

63 Injunctions are available in the tort of private nuisance if the conduct that constitutes the nuisance is continuing at the date of the trial. If a non-GM farmer could prove that the continued use of her neighbour's land for GM farming would amount to a nuisance, an injunction to prevent that land from being so used would be granted. As in England, there are statutory provisions allowing damages to be granted in lieu of an injunction but this jurisdiction is exercised sparingly and only where the damage to the plaintiff's rights as a result of the tortious conduct is small.[80] It is unlikely that GM contamination will be viewed in this way.

64 There is doubt as to whether an injunction can be granted to prevent negligent behaviour but, if this is possible, then it would apply as well in the GM context.

4. Costs of pursuing a claim

(a) General cost rule

65 The general rule is the "loser pays" principle.

(b) Costs of establishing causation

66 In principle, the recoverability of this kind of loss would be a question of remoteness of damage. If the spread of GM material amounted to a tort, it is likely that the costs would be recoverable as they would be damage of a type or kind that was a reasonably foreseeable consequence of the GM contamination. However, if the *only* damage claimed for was the monitoring costs, recoverability would depend on establishing a cause of action relating to this kind of pure economic loss and this may prove difficult given the limited circumstances in which tort claims may be brought for this kind of loss.

---

79 See, e.g., *Civil Liability Act 2002* (NSW) s 21; *Civil Liability Act 2003* (Qld) s 52.

80 *Trindade/Cane/Lunney* (fn. 26) 185 f.

5. Advance cover

There is no specific liability or redress scheme to which advance cover could attach. However, the Regulator has the authority to impose a requirement, as a condition of any licence, that the licence holder be adequately insured against any loss, damage or injury that may be caused to human health, property, or the environment by the licensed dealings with the GMO.[81] More generally, Part 4 of the Application for Licences for Dealings with a GMO involving both intentional and non-intentional releases of GMOs into the Environment, which deals with the suitability of the applicant, requires the applicant to disclose financial details,[82] and this may be something that the Regulator can take into account in determining if the applicant is a suitable person to be a licensee.[83] **67**

IX. Cross-border issues – Conflict of laws

Because of Australia's geographic location, most tort actions involving GMO's will involve domestic Australian law. In this regard two feautures of Australian conflict of law rules should be noted: first, that Australian state, territory and federal courts have significent freedom to transfer proceedings to other Australian courts[84] and second, that for any tort committed in Australia the governing law is the law of the state or territory where the tort was committed (the lex loci delicti)[85]. Whilst many areas of tort law will be common throughout the Australian jurisdictions, this will not be so where legislative amendments to the common law have been made in only one or a number of jurisdictons (eg. the adventitious presence defence discussed in I.1.3 above). **68**

81 *Gene Technology Act 2000* (Cth) s 62(3). However, a perusal of the licence conditions for licences granted by the Regulator suggest that no such condition has been imposed in any licence.

82 Australian Government, Department of Health and Ageing, Office of the Gene Technology Regulator, http://www.health.gov.au/internet/ogtr/publishing.nsf/Content/forms-1 (at 24 November 2009).

83 *Gene Technology Act 2000* (Cth) s 58(1), (2).

84 M Davies, AS Bell, PLG Brereton, *Nygh's Conflicts of Law in Australia*, (8th edn, 2010) ch 6.

85 Ibid, Ch 20. The same choice of law rule applies for international torts with a possible exceptian for maritime torts.

X. Cases

1. Due to the adventitious presence of GMOs in a field, maize which is normally sold as conventional contains GMOs beyond the legal labelling threshold. This is not discovered before the final stage of the food production chain by the producer of taco chips. The whole production is lost since the supermarket chains refuse to accept delivery from the producer.

(a) Who can sue along the chain of distribution?

69 The primary claims here would be contractual, through both the terms of the agreement between the parties and any implied terms under the *Sale of Goods Acts* and the *Trade Practices Act 1974* (Cth). It may also be possible to sue under the *Trade Practices Act 1974* (Cth) for misleading or deceptive conduct in trade or commerce, the misleading conduct being to represent the maize as GM free.[86]

70 On the facts presented, it is difficult to see that any claims in tort would arise in these circumstances. Note also that the statutory defences for the adventitious dealing with genetic material would also apply to the owner and occupier of the land from which the maize was produced.[87]

(b) Would the case be solved differently if the GMO content was below the labelling threshold?

71 Under Australian law it is difficult to see how any of the parties would suffer economic loss in those circumstances.

(c) Would the case be solved differently if the admixture was not adventitious, but occurred due to the disregard of segregation rules, for example?

72 The answer assumes that the person being sued is the GM producer who is responsible for the contamination. Assuming that the contamination was as a result of carelessness, members of the chain of distribution could

86 *Trade Practices Act 1974* (Cth) s 52.

87 See Section I above.

attempt to sue that party in negligence for the pure economic loss suffered as a result of the careless contamination of the crop. As in English law, Australian courts are reluctant to impose a duty of care in tort to take care to prevent causing pure economic loss to others. In particular, more is required than that it be reasonably foreseeable that pure economic loss may result from the defendant's conduct. Moreover, courts are reluctant to impose duties in tort where the parties have entered into a series of contracts that defines their rights and obligations. Primacy is given to contractual remedies in this situation. Although the "contractual structure" argument would not apply, these plaintiffs are unlikely to be considered "vulnerable" as they would have been able to protect themselves by virtue of their contracts with other parties in the chain of distribution. However, if the producer of the maize has suffered damage to property as a result of the contamination, it may be easier for that party to claim as it is generally easier to establish a duty of care in negligence, or an unreasonable user in private nuisance, where the plaintiff has suffered damage to property.

(d) Would the case be solved differently if the GMO found was not admitted for production in your jurisdiction?

73 If a GM farmer had deliberately used a GMO that was not licensed for release in Australia, this would be very strong evidence that the GM farmer had acted carelessly. Even in this case, however, it would still be necessary to establish that a duty of care in relation to pure economic loss was owed to the other parties.

(e) If the admixture had occurred on a non-GM field and it transpires that the GM seeds were blown from: (i) neighbouring fields; or (ii) a truck passing by, would the farmer of the affected field be liable for all or part of the loss caused further down the distribution chain?

74 It is extremely difficult to see that any tort claim would lie against the farmer of the non-GM field which was contaminated. If the non-GM farmer was neither aware of the contamination when it occurred nor had any reason to suspect it, Australian law would struggle to find that the farmer had been careless or was engaged in an unreasonable use of land. Moreover, in a number of jurisdictions the statutory defence relating to adventitious presence referred to in Section I above would apply.

2. Twenty years after the sale of GM maize used for food products, it turns out that it has certain disadvantageous health effects for humans.

(a) Can the producers be held liable at this point for risks unknown at the time of growing the maize? Who would be liable?

75 It is difficult to see that the producers would be liable. The conduct to which such claims would relate would be the original supply of the maize and at that date there was nothing to suggest that the original supply was tortious (i.e. the producers have not been at fault). In theory, the strict liability regime for defective products under the *Trade Practices Act 1974* (Cth) might apply but it is highly unlikely that the maize would be considered defective in light of the regulatory process for approval and it may also be that the development risks defence would also apply.

(b) Can compensation already be claimed at a point when the negative health effects have not yet materialised, but are to expected to according to scientific expertise/mere rumours?

76 It is unlikely that any claim could be made at this point as no loss has been suffered. Moreover, Australian law does not recognise an action to recover the costs associated with monitoring the plaintiff's medical condition to determine if the negative health effects have occurred, at least if the injury does not subsequently occur.[88] If the plaintiff had suffered a recognised psychiatric illness as a result of worrying about the possible adverse consequences it may be possible that this would give rise to a claim if the original supply was tortious.

(c) Would it make any difference if the GM maize had only been in use for feed, causing harm to the animals, which may or may not cause harm to humans consuming the meat as well?

77 It would make no difference to the answer to (b) above as regards claims for personal injury. If the harm to animals amounted to damage to property, a claim might lie if a cause of action could be established.

---

88 For discussion see *John Fleming*, Preventive Damages, in: N.J. Mullany (ed.), Torts in the Nineties, chap. 3.

3. The driver of a food logistics company discovers that a farmer, from where he regularly picks up agricultural products fails to obey mandatory segregation rules or food or feed hygiene standards, which may lead to the admixture of GM and non-GM produce sold and packaged separately by that farmer. Does he or his employer have a duty to warn, i.e. warn the recipient of the allegedly "non-GM" produce?

Australian tort law does not impose liabilities for "pure" omissions in the **78** law of tort. Assuming there is a contract between the food logistics company and the recipient, there is an implied term of the contract that requires the company to exercise due care and skill in the performance of its task and this may extend to providing warnings to the recipient where it has knowledge that the produce may be non-GM. Where there is no contract, some other basis for the duty to warn must be found. Traditionally, the duty to warn has been founded on occupation of land, the manufacture of goods or some special relationship between the parties giving rise to an obligation on the part of a party to act positively to provide a warning. None of these are present in the fact scenario outlined. Moreover, there is no statutory requirement (other than on the licence holder[89] to advise the Gene Technology Regulator that breaches of licence conditions are occurring.

89 *Gene Technology Act 2000* (Cth) S 65.

Damage Caused by GMOs under Brazilian Law

Clarissa Bueno Wandscheer, Ana Paula Myszczuk and Roseli Rocha dos Santos

I. General overview[1]

1. Special liability or redress scheme for GMOs

1 The Biosafety Act (Lei de biossegurança, law 11.105/2005) does not foresee any mechanisms for environmental recovery in cases of environmental damage, nor does the Executive Provisional Order no. 131 which has authorised the release of GM soy. In this case, in order to identify responsibility for environmental recovery, all legal dispositions for environmental protection must be analysed. For these purposes, it is necessary to start with the Brazilian Federal Constitution of 1988, which in its art. 225 establishes the possibility of triple responsibility – administrative, civil and penal – in matters concerning environmental damage. Furthermore, § 3 of the same article rules that "conduct and activities considered harmful to the environment will subject the offenders, persons or companies, to penal and administrative sanctions, independently of obligations regarding repair of damage". Therefore, the "system of environmental responsibility is multiple and must be articulated jointly and systematically. Thus, it should be remembered that breach of obligation or duty, even when arising from only one relevant act, may result in various types of penalisation and in cumulative responsibilities, by virtue of the different responsibilities corresponding to the different sanctions provided by our legal system".[2] Through analysis of the above-mentioned constitutional article, it

1 It is important to emphasise that many of the questions herein have no precedents in Brazilian law. Therefore, they have been answered according to current legislation and recent court decisions on environmental issues and civil responsibility in general.

2 *J.R. Morato Leite*, Dano ambiental: do individual ao coletivo extrapatrimonial (2nd ed. 2003) 114.

is also inferred that any person or corporation may be held responsible for damage inflicted upon the environment.

Reinforcing constitutional understanding, Law 6.938/1981, which institutes National Environmental Policy, establishes civil responsibility in environmental issues in an objective form, which means that the agent's fault is presumed and thus, only causation must be demonstrated in order to identify the responsible party. Accordingly, § 1 of art. 14 establishes that "notwithstanding the application of the penalties provided for in this article, the polluter is obliged, whether at fault or not, to indemnify or to repair the damage inflicted on the environment or on the third parties that have been affected by its activities." The State and Federal Public Attorney's Offices have competence to file lawsuits for civil and criminal responsibility for any damage to the environment. Similarly, art. 20 of the Biosafety Law rules that "notwithstanding the application of the penalties provided for by this Law, the parties responsible for the damage inflicted on the environment and on third parties will respond, jointly, for their indemnification or for their integral recovery, whether or not fault is proven". 2

In administrative matters, liability may accrue when the agent, whether person or corporation, acts in disobedience of official authorisations (issued in this case by the CTNBio[3]) coming from those responsible for agreements on the development of potentially polluting activities relating to GMOs. In penal matters, responsibility is subjective; thus, it becomes necessary to prove intention or fault. The agents that fulfil the requisites for particular penal offences may be sentenced to the penalties provided in the Environmental Crimes Law (Lei de crimes ambientais, act 9.605/1998), as well as those provided by the Biosafety Act (law 11.105/2005). In the environmental field, the objective of legal penal provisions is to protect the environment in all its aspects, inhibiting human actions that may damage it, and protecting society's most relevant interests.[4] 3

Civil liability in Brazil is established by the Civil Code (Novo Código Civil Brasileiro, 2002) in its art. 43 and 927, as follows: 4

> "Art. 43. Legal entities of domestic public law are liable for the acts of their agents who cause damage to third parties, notwithstanding the right to recourse against the parties who inflicted the damage, if there is, on their part, fault or intention.

3 Commissão Técnica Nacional de Biossegurança (National Biosecurity Technological Committee).

4 *Leite* (fn. 2) 115.

Art. 927. Those who through illicit acts cause damage to another, shall be obliged to repair it.

Sole Paragraph. There will be an obligation to repair damage, independently of fault, wherever it is thus specified in law, or when the activities engaged in by the author of the damage imply, by their nature, risks to the rights of third parties."

5 It is noteworthy that civil liability does not require the occurrence of an illicit act, as it would in classical cases of civil responsibility. On the contrary, activities may have received official authorisation, in other words, the agent, person or legal entity, may have authorisation to release the GMO in the environment; nonetheless, it is not exempt from civil responsibility for damage inflicted, that is, the obligation to restore the environment to the *status quo ante* (see also the above-mentioned Law 938/81). Another difference is that in classical tort law, damage arises from a relationship between individuals, whereas in environmental civil responsibility the damage affects diffuse and collective rights, as in the case of the right to an ecologically-balanced environment, provided by the *caput* of art. 225 of the Federal Constitution. According to Leite, "the concrete finality of this generic responsibility (classical civil), is to punish and to make the agent repair damage, as well as to avoid new damage that could occur".[5] On the other hand, liability in environmental matters evokes strict liability theory, which adopts the concept described as "the idea of civil liability arising from the activities of the agent of the damage, and in this way, establishing the obligation to indemnify for the exercise of certain undertakings or certain conduct developed by said agent".[6] Moreover, the "exercise of said activity of risk presupposes attainment of a certain profit, of economic nature in general, that arises as a consequence of the activity which is itself potentially damaging [risk-benefit]".[7]

6 Furthermore, it should also be stated that strict liability requires the establishment of causation, which means "that in light of the activity carried out by the agent, a potential environmental offender, it is necessary to indemnify any damage caused, as a way of protecting the social interests that were breached".[8]

---

5 *Leite* (fn. 2) 113.

6 *J.R. Alvarez Vianna,* Responsabilidade civil por danos ao meio ambiente (2006) 101.

7 *P. Stolze Gagliano,* A responsabilidade extracontratual no novo Código Civil e o surpreendente tratamento da atividade de risco. Jus navegandi, Teresina, ano 7, no. 64, abr. 2003. Available at: http://jus2.uol.com.br/doutrina/text.asp?id=4003 (accessed 25 June 2009).

8 *Vianna* (fn. 6) 107.

Certain jurists consider that this strict liability on environmental matters is excluded in situations involving *force majeure*, acts of God and third parties' acts,[9] according to interpretations of art. 37, § 6 of the Brazilian Federal Constitution of 1988, which construe the responsibility of Public Powers as more restricted than is provided for in art. 14 of the National Environmental Policy Law. According to Toshio Mukai: **7**

> "In the similitude of strict liability of the State within national law, strict liability for environmental damage follows the theory of created risks (admitting causes for exclusion of fault such as acts of the victim and third parties, force majeure and acts of God), and not those of integral liability for risks (which does not admit excluding causes), on the exact and express terms of § 1 of art. 14 of the Law no. 6.938/81, which, (...) only engages the liability of any party for environmental damage if there is proof of the effective action (activity) of that party, directly or indirectly, causing said damage".[10]

For Barros, "when degradation occurs through clandestine acts of the agent, there is no Government responsibility. In case of ecological accidents, the Government shall be held responsible only in cases of grave fault. In cases of acts of God, the Government will not be held responsible, but in force majeure cases, such must be deemed risks inherent to the services rendered".[11] **8**

On the other hand, another branch of opinion supports the idea that acts of God may not always exclude the liability of the agent – whether a person or a legal entity – as regards the obligation to repair damage to the **9**

---

9 A different position is taken by José Afonso da Silva who, based on Sérgio Ferraz's assertion, observes that: "the consequences of adopting strict liability in this matter are: a) the irrelevance of the intention to cause damage (the presence of the damage is enough); b) irrelevance of the measurement of subjective intention (what is important is that someone was party to the causation, and with that participation, has in any way been held objectively liable; c) the inversion of obligation to present proof; d) irrelevance of the illicitness of the activity; e) attenuation of causation – it is sufficient that the agent's activity has the potential to cause ecological damage, for immediate reversal of the obligation to present proof, immediately producing the presumption of liability and reserving for the respondent the obligation of trying to free himself of fault." In: *J.A. da Silva*, Direito Ambiental Constitucional (2004) 313.

10 *T. Mukai*, Responsabilidade civil objetiva por dano ambiental com base no risco criado. Available at: http://www.oab.org.br/comissoes/coda/files/artigos/%7B2A131C28-7CA1-4DF1-8A5D-5FD2A6E457B0%7D_Responsabilidade%20civil%20objetiva%20por%20-dano%20ambiental%20com%20base%20no%20risco%20criado.pdf (accessed 30 June 2009) 8.

11 *A.C. Ribeiro Barros*, Quebrando tabus em relação à responsabilidade civil causada pelos danos nucleares no Brasil. Available at: http://www.artigonal.com/direito-artigos/3quebrando-tabus-em-relacao-a-responsabilidade-civil-causado-pelos-danos-nucleares-no-brasil-345042.html. Published on 28 February 2008 (accessed 1 July 2009).

environment. In many cases, those phenomena are easily predicted and there are prevention mechanisms with technologies available on the market. For instance, there are systems of protection against atmospheric electrical discharges, containment lagoons for residue before emission into rivers, etc. Accordingly, for Vianna, "an act of God is not sufficient motive for exemption from the obligation to indemnify. With regard to these facts, when identifying causation, there must be evaluation as to whether the act of God, together with the activity carried out by man, was the real cause for such damage".[12] For instance: "lightning striking a silo is the cause of an explosion and its consequent damage. The existence of the silo was the condition of the occurrence. Had it been empty, there would have been no explosion, and therefore no damage would have occurred."[13]

10 In this same vein, a decision was proffered by the State Court of Parana, which ruled that there can be liability even in the case of *force majeure*.[14]

11 Furthermore, damage arising from the acts of third parties does not, according to any hypothesis, exclude liability for damage caused to the environment. Some damage is perfectly avoidable, and therefore the costs of implementing preventive mechanisms must be assumed by the agents. Some cases can be identified as internal and external acts of God. The former are "those episodes that relate to the risks inherent in the activities carried out". The latter in turn, do not have any relationship to said activity. Thus, they refer to an unusual occurrence which itself initiates the damaging event, without any input, direct or indirect, near or remote, from the agent of the activity. Only in this last hypothesis, where there is no trace of causation, could there be exemption from the obligation to indemnify.[15]

12 Other authors prefer the theory of created risks and their integral reparation, and theorise that:

> "...risks created are those produced by activities and assets of the agents which multiply, enhance or have the potential to create environmental damage. The risk is created when a person makes use of mechanisms, instruments or means that enhance the danger of damage. In these hypotheses, the persons who caused the damage must account for the

12 *Vianna* (fn. 6) 109.

13 *A. Pasqualotto*, Responsabilidade civil por dano ambiental: considerações de ordem material e processual, in: A.H.V. Benjamim (ed.), Dano ambiental: prevenção, reparação e repressão (1993) 444–470.

14 Tribunal de Justiça do Estado do Paraná – Embargos de Declaração Civil 0450460-9/01 – 9th Câmara Cível – Relator Desembargador Edvino Bochnia – Diario de Justica da Uniao (DJ) 30.01.2009.

15 *Vianna* (fn. 6) 110.

> losses which have been caused, since they involve created risks or damage, rather than fault. [...T]he integral reparation of environmental damage means that environmental damage must be wholly restored, and not merely partially, thus permitting more effective protection of environmental assets."[16]

It must be concluded, therefore, that the majority of theories adopt the theory of integral risks in environmental issues, since the basis for the establishment of liability is the risk of the activity, rather than fault. This is different from what occurs in cases of classical tort law. **13**

In Brazilian law, many rules can be identified which also establish strict liability. Among them are: **14**

- Law 7.092/1983, regarding damage arising from road transportation of dangerous products;
- Law 7.542/1986, on liability for damage to ships, third parties and the environment;
- Law 7.661/1988, on damage to natural and cultural resources of seaside zones;
- Law 7.802/1989, on damage caused to the health of persons and to the environment, arising from activities related to pesticides and similar products;
- Law 7.805/1989, on damage caused to the environment by mining activities;
- Law 8.171/1991, on damage caused to the environment arising from agricultural activities;
- Law 8.974/1995, on activities relating to bio-genetics; and
- Law 10.308/2001, on radioactive residue.

2. State liability

In legal cases to which the State[17] is a party, classical liability is not applicable, since the Brazilian Federal Constitution rules, in its art. 37, that: **15**

---

16 *Leite* (fn. 2) 129.
17 Apart from some public enterprises and capital hybrid companies that, whereas they were created by the State, do not apply the rules of public law but of private initiative.

> "The direct or indirect public administration of any of the powers of the Union, the States, the Federal District and the municipalities, as well as their foundations, shall obey the principles of lawfulness, impersonality, morality, publicity, efficiency and also the following: …
>
> § 6. Public legal entities and private legal entities rendering public services shall be liable for damages that any of their agents, acting as such, cause to third parties, assuring the right of recourse against the liable agent in cases of intention or fault."

16 This means that the State is strictly liable, thereby including the Union, the States, the Cities and the Federal District, regarding acts of their agents in the exercise of public functions, when damage to third parties arises. As a consequence, any federative unit may be held responsible, as well, for damage to the environment arising from active or passive activities (supervised or not, with or without granting of environmental licenses).

17 Furthermore, the State liability can be based on art. 3, IV of the Law 6.938/1981 (National Environmental Policy Law), which establishes: "Art. 3 – For the objectives of this Law, the following definition shall apply: (…) IV – polluter, the person or legal entity, public or private that is responsible, directly or indirectly, for any activities causing environmental degradation." Therefore, all "those that, for multiple reasons, become involved in conduct causing environmental damage, will be obliged, because of the joint liability between the parties, to respond for indemnification."[18]

18 In this same regard Machado states that

> "there is joint liability when, in the same obligation, there is more than one creditor, or more than one debtor, as a party with a right or obligation to the whole debt.(…) Art. 20 of the Biosafety Act (law 11.105/2005) when putting in the plural the term 'those responsible', indicates the possibility of searching for more than one responsible party. There may be serial co-participation in genetic manipulation, but joint liability excludes nobody, neither the liable parties – legal entities or institutions, private or public, nor the public servants who operate in the chain of decision – CTNBio, CNBS [National Council for Bio-Safety], Ministries nor public organs."[19]

19 Therefore, we see that federal entities are included among the persons that can be held liable. This same understanding is clear in several decisions issued by the Superior Court of Justice, which has recognised State

---

18 *Vianna* (fn. 6) 111.

19 *P.A. Leme Machado*, Direito Ambiental Brasileiro (2008) 1050 f.

liability in face of the obligation to supervise environmental preservation.[20] The same Court has stated that in case of environmental damage, the competence of the Port Administration does not exclude, but rather completes legitimacy to supervise and sanction State organs for environmental protection.[21]

As seen above, the Brazilian State can be held liable for damage caused by activities dependent on environmental licensing,[22] since art. 14 of the Law governing the National Politics for the Environment (Law 6.938/81) determines that those who contribute directly or indirectly to the damage or environmental pollution shall be held liable. It must be stressed that GMO release in the environment is allowed only with the approval of the CTNBio and with the authorisation or environmental license of the competent organ, IBAMA (Brazilian Institute for the Environment and Renewable Natural Resources), that is to say, after approval of the studies and presentation of environmental impact reports. In order to grant the license, IBAMA must take the environmental macro-zoning of the areas where the GMO will be released and relevant bio-geographical and socio-economical particularities into consideration. An environmental impact study must be carried out and a report elaborated, as well as a contingency plan for environmental damage caused by derivative GMOs. Resolution no. 305 issued by CONAMA (National Council for the Environment), determines the necessary procedures for the licensing of genetically modified organisms, for confined areas of research, for field research, and finally for commercial release. It is important to note that environmental licensing is approved only once for each product, which is to say, the party responsible for the manufacture of the product shall undertake all procedures and agriculture may make use of the modified seeds, as previously authorised by the government, as long as they are used within the zone that has been approved for the respective GMO. **20**

Fiorillo reports the four situations in which it is possible to analyse the co-participation of the State in the case of environmental damage from activities depending on authorisation or license from the authorities: **21**

---

20 *Special Appeal* – 604725/PR., Second Chamber, published in Diario de Justica da Uniao (DJU) 22/08/2005, 202, reported by Minister *Castro Meira.*

21 *Special Appeal* – 467212/RJ – First Chamber, published in DJU 15/12/2003, 193, reported by Minister *Luiz Fux.*

22 Environmental licensing is under the liability of federal, state, district and municipal environmental authorities, with their competences established by Law 6938/1981 and Resolution 237 of the Conselho Nacional do Meio Ambiente (Conama).

"1) If there was no EIA (Environmental Impact Study)/RIMA (Environmental Impact Report), and the authority was convinced by the RAIAS (Simplified Environmental Impact Report), the authority will be responsible to the extent to which there is a causal link between its action and the damage incurred, in other words, concurrence of action and damage.

2) If there was an EIA/RIMA, and it was (wholly) favourable to the granting of the license, there is no ensuing State responsibility, since license concession was an act arising from the law.

3) If there was an EIA/RIMA, and it was unfavourable, (wholly or in part), and the license was granted, there is joint responsibility including the State, since there is a causal nexus between the act of license concession and damage caused to the environment.

4) If there was an EIA/RIMA, and this was unfavourable and the license was not granted, as a rule there would be no State responsibility, except in cases in which the State had remained inert and, due to omission, allowed environmental damage to occur."[23]

22 The liability to which the author makes reference is joint, which means that the State (Union, States, Cities, and Federal District) may be held jointly liable with the author of the damage (person or legal entity); thus, they are mutually obliged to remedy environmental damage. If damage to the environment is a result of inertia on the part of the authorities, the latter share joint responsibility for adversities. This means that:

"the State, pursuant to the Principle of Powers and Duties, is liable when it overlooks or forgoes a legal obligation, or omits fulfilment of obligations that stem from its Policing Power, such as an accident that is caused by an animal left out on a road. In environmental matters, liability in cases of omission is great, because those who have the obligation to avoid the incurrence of damage, through vigilance or surveillance, for instance, and are negligent in this regard, are held liable. It is not enough that the State does not do what it should not do, it must also fulfil its obligations and comply with its duties."[24]

23 When not a matter of joint liability in which the Powers (Union, State, City or Federal District) respond jointly with the direct agent of the damage, there may still be subsidiary liability since, as established by art. 225 of the Brazilian Federal Constitution, the environment is a right of present and future generations and it is everyone's obligation to ensure

23 *C.A. Pacheco Fiorillo*, Curso de direito ambiental brasileiro (2009) 143 f.

24 *E. Séguin*, O direito ambiental: nossa casa planetária (2000) 300.

its existence. Therefore, if the direct agent of damage does not have the means to restore the degraded environment, the State shall be obliged to do so, thus assuring collective rights to an environment which is ecologically-balanced, an asset of common usage for the people of present and future generations.

II. Damage

1. Recoverable losses

In Brazil there are as yet no technically-established criteria for diagnosis of environmental damage. However, there are some alternatives for the recovery of environmental assets that have suffered damage or for avoiding damage to any environment which may be threatened. **24**

With respect to these considerations, it is possible to verify that, based on the Constitution of the Federative Republic of Brazil, (art. 225 § 2) and on the complementary environmental laws (Law 6.938/81, art. 4, VI), whenever there is environmental damage, the *status quo ante* shall be restored, i.e. recovered to its original condition. It is recognised that this modality of recovery is extremely difficult and complicated to achieve. Another possibility for recovery is environmental compensation, which means that the degradation of one area must be compensated by the recovery of another, in similar conditions and in terms of preference (localisation, diversity, climate, etc.). Environmental compensation may be of four different types, according to Leite: jurisdictional, extra-jurisdictional, pre-established and by autonomous means. **25**

> "Jurisdictional ecologic compensation consists of impositions established by irreversible court judgment which order the degrader to replace the injured asset with an equivalent one or to pay an amount in money. It is a compensation imposed by the judiciary, arising from an environmental lawsuit. The extra-jurisdictional compensation, in turn, is executed by means of a compromise for the change of conduct, establishing such change among the public organs and the potential polluting agents, which are obliged to obey legal exigencies. (...) The pre-established compensation mechanisms can be understood as those which are formulated by the legislator, independently of administrative and jurisdictional penalties (civil or penal), and have as their aim the compensation of negative impacts on the environment, arising from societal risks. The fourth and last type of ecological compensation refers to autonomous funds for ecological compensation, also called

alternative solutions for indemnification for environmental assets. In addition to civil liability, these funds are financed by potential polluting agents who invest in shares for the finance of reparation."[25]

26 Lastly, some legal opinion finds that the best way to recover is to replace the environmental assets which were damaged by an amount in money, which means pecuniary indemnification. Monies paid must be designated to funds for environmental protection.

27 In fact, with regard to the type of environmental recovery, it is necessary to carry out case by case evaluation, since in Brazil there are still no administrative or judiciary decisions ordering the recovery of areas after contamination by GMOs, while this contamination may have differential or varying impact on the environment. Experience in some cooperatives has shown that contamination may occur, more commonly, through the use of equipment that has not been properly cleaned and storage of conventional products together with GMOs, that is to say by non-compliance with procedures for separating seeds and products.

2. Pure economic loss

28 Economic losses are considered particular losses and may be indemnified, according to each case. However, it is necessary to observe that the damage to the environment, as a collective asset that belongs to the entire community, is hard to measure in merely economic – and therefore quantitative – terms.

29 In cases of private damage, the general rule of civil law, established in art. 942 of the Brazilian Civil Code, obliges the party who causes damage to proceed to its repair. To avoid possible and purely economic damage arising from GMO contamination, with respect to GMOs' commercial release, Brazilian law demands that studies on environmental impact be carried out and disseminated through reports presented to the IBAMA and that once this has been done, a formal request be forwarded to CTNBio, to decide whether or not the GMO in question can be released.

30 Therefore, in some cases, for instance in which conventional or organic seeds have been contaminated by GM grain in a defined area or plot, the party who is responsible for the damage will be obliged to provide indemnification. Furthermore, it is relevant to note that some companies, like

25 *Leite* (fn. 2) 212–214.

the IMCOPA, which exports conventional grains with preserved identity, have established previously, through contracts with rural producers, the amounts of indemnification to be provided in case of delivery of grain that does not comply with agreed upon specifications that may be considered to give rise to company liability, in the case that the latter cannot fulfil its contracts with European buyers. It is also worth mentioning the case of a certain type of certified export soy which, for some reason, has not been certified for a particular plot and becomes contaminated with GM grains and is sold as conventional grain to a European importer. In the event that contamination is verified, the certifying company may be held responsible for damage caused to the importer.

3. Mere fear of a loss

In Brazilian law, indemnification for merely suspected damage or threat of damage is not admissible. As a rule, damage must be proven in order to request indemnification. According to Antunes: **31**

> "The prevailing position of our Courts is that environmental damage must be real and present. This is to say that jurisdiction is activated only after damage has been caused. Simple infringement of legal rules, such as zoning norms, is not enough, according to the law, to characterise damage to the environment. (...) The risk itself, upon which environmental responsibility is founded, is not considered much; risk must materialise as an 'accident' in order to receive reparation."[26]

However, court decisions may concede protection in case of environmental threat in Brazil, even if no question has been raised regarding the threat of damage for release of or research on a GMO. Therefore, the simple threat of damage or environmental contamination which may be linked to a defined GMO is admissible only as an exception, generally related to missing requisites, such as the presentation of environmental impact reports, on the occasion of GMO release at the CTNBio. One example is the judgment that suspended the release authorisation given by CTNBio for sales of "round-up ready" soy seeds, without the inclusion of a previous environmental impact study into proceedings, all according to the judgment given by the Regional Federal Court of the 1st Region.[27] **32**

26 *P. de Bessa Antunes*, Direito ambiental (2008) 237.

27 Lawsuit 01000146611 – no. 2000.010.00.14661-1 – DF – 2nd Court – reported by Judge *Assusete Magalhães* – DJ 14.03 2001, 84.

4. Standard of proof

33 As regards the collective aspect, collective/diffuse environmental damage is immeasurable; therefore, as a rule, the values related to the administrative fines provided by law, including the limits, and those related to civil aspects will be ascertained according to the amount necessary to repair damage.

34 With regard to individual aspects, it is possible to claim indemnification for losses and ceasing of profits, that is, indemnification for the time in which the activity is rendered impossible due to the environmental damage, most specifically the contamination or pollution of the area. The greatest difficulty is the establishment of the amount and extent of the damage suffered. In a recent decision on the quantification of damage to private individuals, the courts imposed the amount of one minimum monthly wage, for six months, according to the strict liability theory.[28]

5. Nominal losses

35 Although, based on the Brazilian Civil Code, indemnification of any kind of loss may be understood as a possibility – there is no specific law or court decision expressly recognising indemnification for minimum damage to the environment, including GMO contamination. Therefore, in order to claim indemnification, losses must be significant enough to cause effective damage to the environment and to individuals.

6. Mass losses

36 There are no specific rules for damage caused by GMOs, for instance, in cases of release in the environment, of contamination or damage to human health. General rules for civil liability or consumer laws are applied in these cases. Thus, if losses occur, the general rules stated above must be applied, e.g. those of joint liability among all the parties who are directly or indirectly responsible for the damage. A comparison may be drawn with damage caused by nuclear accidents, following the reasoning expounded in a question discussed below. Nonetheless, there is no guaran-

28 Court of Justice of the State of Parana, Civil Lawsuit 0479369-9, 10th Civil Chamber, Relator Desembargador Marcos de Luca Fanchin, Justice Diary published 19.01.2009.

tee that Brazilian courts will apply the same theory with regard to liability for environmental damage.

III. Causation

1. Uncertainty of merely potential causes

Reference should be had to the observations pertaining to the section on damage, in which it is shown that the main positions taken by the Brazilian courts on issues of environmental damage have only related to damage which actually occurred. Regarding environmental damage caused by victims' own actions, it should be noted that that there are two different rulings for such cases, one pertaining to consumer legislation, the other to civil and corporate relationships. According to consumer relationships, which follow the rules of strict liability, if the victim has exclusive responsibility for what has occurred, the supplier cannot be held liable in any way. This is valid for cases in which the victim's actions are the only and exclusive cause of damage. In cases in which there is concurrent fault, which means that the causes of the damage are partly due to the victim and partly to the supplier, the latter must assume a portion of blame. When dealing with a civil or corporate relationship, the rule of classical liability is applicable, wherein the fault of the agent and the attitudes of the victims are taken into account in order to corroborate or to verify causation; the possible mutual fault and liability of both parties must be analysed at the time when evaluation of the liability for environmental damage is carried out. **37**

In scenarios in which it is impossible to discover exactly what has occurred, for example, in a case of contamination of a significant amount of conventional grain with GMO grain, there are also two hypotheses of civil liability. With regard to consumer relationships, joint liability may involve the whole production chain. Thus, regarding damage suffered by the consumer, it is of no importance where in the production chain the damage occurred. The only relevant factors are the occurrence itself and the absence of fault on the part of the victim. It is worth mentioning that, in cases in which one of the agents of the production chain has incurred damage caused by another agent, he may, by lawsuit *in rem verso,* claim the recovery of losses suffered. In this case, the rules of classical liability will be followed. Similarly, when analysing a corporate or civil relationship, the liability of the agents throughout the production chain will be determined by the rules of individual and classical liability. **38**

2. Complex causation scenarios

39 In Brazil there are no special rules for civil liability in the case of chain production of goods. In these cases, the agent's liability for damages is rather complicated, since the classic rule of liability is applied. What can change this scenario is the type of contract between the agents, which may establish core obligations. The case of a food-processing business which acquires conventional grain from various producers can be used as an example. In their contracts, this company may have specified that it will only accept the delivery of certified conventional grain from the contracting parties, with or without preserved identity. Therefore, the company will not receive any other kind of grain and the reason why the grain agreed upon was not delivered lie beyond the scope of the contract. Similarly, the food business may write up a contract with a given buyer for the sale of conventional products. In case of default on the delivery of said products, the company will be liable for the damage, even if the contamination may have occurred before it arrived in the factory. However, in cases in which the contract does not establish the delivery of a product with certain qualities, the rule of classical liability will be applicable and if it is impossible to determine who caused the damage, the proprietor of the thing will bear the damage, by the principle of *res perit dominus*.

3. Force majeure

40 The Brazilian Civil Code considers liability to be diminished when damage occurs due to *force majeure*, acts of God and fault of the victim. As art. 393 stipulates, "the debtor does not respond for the damage occurring due to acts of God or force majeure, unless he has expressly assumed responsibility for them". Besides that, it establishes that "force majeure or acts of God are inevitable occurrences whose effects have not been possible to avoid or to predict".

41 Thus, the author or agent of the damage would be released from liability for damage caused by acts of God (facts of nature) or *force majeure* (accident). Yet the situation is different when it comes to environmental issues, as has been explained above. In that sense, Lima states:

> "In civil liability for environmental damage, the principle of the polluter-pays or of liability rules (...) indicate that the polluter is obliged to correct or to recover the environment. He must bear the costs resulting therefrom and may not proceed with the polluting action. In addition to this, there is the liability of the agents, before third parties, for direct

> or indirect action concerning the natural resources. One of the most important consequences of this principle is the strict liability of the polluter. The national doctrine aims to embrace integral liability theory, which is inherent to the civil liability of the agent who caused the environmental damage."[29]

As has already been noted in reference to the former topic, the adoption of integral risk liability theory is not unanimous, so that it would remain up to the judiciary whether, in concrete cases, to decide on exclusion or not. There is no uniformity in the courts' decisions, as can be seen in conflicting rulings.[30] **42**

4. Threshold to prove causation

Since strict liability is admissible in Brazilian law, demonstration of causation is necessary. In some cases, courts have dispensed with proof of *nexus*, such as in some cases of damage caused to legal reserve areas, as can be seen in the judgment issued by the Superior Court of Justice,[31] in which we read that "the judgment taken as reference concludes that proof of causation can be *dispensed with* to establish the liability of the owner of the area inside the environmental reservation, with the restrictions imposed by law, on the issues caused by the environment" (emphasis in original). **43**

The same line is followed in a decision of the Superior Court of Justice,[32] ruling that: **44**

> "Article 225 *caput* rules that everybody has a right to an ecologically balanced environment, and imposes on Public Authorities and the collectivity the duty of defending and preserving it for present and future generations. The appealing State has the duty of preserving and of inspecting the preservation of the environment. In this case, the State, exercising its duty of inspection, should have required the realisation

29 R. *Medeiros Garcia de Lima*, Responsabilidade civil pelo dano ambiental. Available at: http://74.125.47.132/search?q=cache:TFqvvliMlpcJ:www.ejef.tjmg.jus.br/home/files/publicacoes/palestras/resp_dano_ambiental.pdf+dano+ambiental+e+caso+fortuito&cd=3&hl=pt-BR&ct=clnk&gl=br (accessed 30 June 2009).

30 Special Appeal no. 282.781/PR, Reported by Minister Eliana Calmon, published in the DJU of 27.05.2002; Special Appeal no. 218.781/PR, Reported by Minister Milton Luiz Pereira, published in DJU of 24.06.2002 and Special Appeal no. 218.120/PR, Reported by Minister Garcia Vieira, published in DJ of 11.10.1999.

31 Special Appeal 620.872. DF, lawsuit 2007/0160344-7, 1st S., Reported by Minister Castro Meira, published in DJ 17.11.2008.

32 Special Appeal 604725, PR, 2nd Court, reported by Minister *Castro Meira* and published in DJ 22.08.2005, 202.

of an Environmental Impact Study and its respective report, as well as the public hearings on the issue, or even the suspension of the works which have caused the environmental damage. The designation of money from the State of Paraná to the city of Foz do Iguaçu (action), the absence of the inspection, with reference to the licenses granted and those that should be provided by the state entity (omission), have concurred in producing environmental damage. All these circumstances, therefore, are apt to characterise the causation behind the occurrence and so legitimate the strict liability of the appealing party. Therefore, independently of the existence of fault, the polluter, however indirect (State – appealing) (according to Article 3 of the Law no. 6.938/81) is obliged to indemnify and to repair the damage caused to the environment (strict liability). Once the issue of the passive liability of the appealing party has been settled, the requisites for the civil liability are given (action or omission, causation and damage) and it must be observed also that such liability (strict) is joint, legitimating the inclusion of the three Powers of the nation as the responding parties of the lawsuit, as was requested by the Public Attorney's Office (facultative litisconsortium)."

5. Special rules on causation

45 There is no law defining criteria to determine causation in case of the release of GMOs into the environment. The applicable rules are general, such as the identification of the GMOs in areas where although they were not planted, the presence of the modified products in crops was verified. Nonetheless, there are hypotheses in which there is no agent liability: a) when no risk was created; b) when there is no environmental damage; and c) when there is no causal link between the damage and the subject who created the risk. However, these hypotheses must be proven in order to exempt from liability.

IV. Types of liability

1. Fault liability

(a) Special rules governing fault

46 The Consumer Protection Code (Código de Defesa do Consumidor, CDC, law 8.078/1990), lists among consumers' rights the effective prevention and repair of patrimonial, moral, individual, collective or diffuse damage (art. 6, VI). The Code also determines facilitation of the defence of rights, including the reversal of burden of proof in the consumer's favour during civil lawsuits, when the judge deems that allegations have the appearance of truth, or when the consumer is less-sufficient, according to the general rules of experience (art. 6, VIII). The objective of reversing the burden of proof is not only to treat consumer and supplier equally, but to ensure the former effective and adequate protection within the lawsuit. The possibility of reversing the burden of proof consists of attributing to the consumer an advantage within a lawsuit, relieving him/her of the burden of proving a determined fact which without the reversal would have to be demonstrated. Thus, the burden of proof that the occurrence did not actually take place is transferred to the supplier. In this way, reversal means absence of burden on the consumer and the creation of a burden of proof for the supplier. That reversal, however, does not take place automatically and can only be granted at the discretion of the judge, in specific procedural situations: when the allegations have the appearance of truth and the consumer is deemed to be less-sufficient.

(b) Impact of specific rules of conduct

With regard to civil liability, the customary and statutory rules are taken **47** into consideration, and in case of damage, a breach of legal, contractual or customary rules is examined by applying the civil liability rules on contracts or illicit acts in accordance with the Brazilian Civil Code.

However, in matters referring to environmental liability, the results of **48** carelessness, lack of expertise or negligence do not relieve the parties who are liable for the activity which caused the environmental damage of a duty to repair. But these facts will determine whether or not indemnification is possible by the agent who is directly liable for the acts which were found to be careless, negligent or lacking in expertise. The rules applicable in civil issues are still those of strict liability. In the administrative area, fines can be enhanced, and in the penal area such activities will constitute crime, since activities without proper authorisation from the public authorities are illicit.

2. Product Liability

(a) Development risk defence

49 In Brazil, the Consumer Protection Code (CDC, Law 8.078/1990) pursues the same goals as Directive 85/374/EC, in other words, consumer protection. This Code establishes that risk may emerge from any moment of the production chain of any product or service. Thus, any consumer who feels that s/he has suffered damage or threat of damage may demand due indemnification from any of the agents participating in this chain. Furthermore, rights that are established by the CDC do not exclude others that emerge from treaties or international conventions that Brazil is a party to, ordinary domestic legislation, regulations issued by the pertinent administrative authorities, as well as those that can be derived from general principles of law, analogy, customs and equity. Furthermore, the Code establishes that when an offense has more than one author, all authors must jointly provide reparation for damage as indicated by consumer norms. (art. 7, sole para.).

50 Attention should be given to the fact that according to the CDC, the consumer is any person or legal entity that acquires or uses a product or service as a final consumer. This includes all persons who, even if they have not yet been identified, have intervened in consumer relations and the third parties that, even without participation in such relations, have suffered damage caused by the service or product (art. 2). The supplier would include any person or legal entity, public or private, citizen or foreign national, as well as entities that participate in activities of production, assembly, creation, construction, transformation, import, export, distribution or commercialisation of products or service provision (art. 3).

51 Furthermore, the CDC adopts the strict liability theory (art. 12) which says that "the manufacturer, producer, builder, whether citizen or foreign national, and the importer are all responsible, *regardless of whether fault has been proven*, for reparation of damage consumers have suffered through faulty design, manufacture, formula, handling, presentation or packaging of products, as well as for the insufficient or inadequate provision of information on use and risks", (emphasis in original). As protection for the manufacturer, producer or importer, the law establishes the situations in which the former will not be held responsible if the opposite is proven, that is, it is shown that the product was not placed on the market, or that, having been placed on the market, no defects exist or finally, that the consumer or a third party are exclusively at fault (art. 12, § 3).

Regarding legal determination, court positions taken have shown concern not with consumer health but rather with products containing GMOs, since risks to health have not been proven. The decision of the Court of Justice of the State of Santa Catarina ruled to deny any injunction order in relation to withdrawal of food products containing genetically modified organisms from circulation.[33] **52**

The CDC ruling differs from Directive 85/374/EC with regard to the time limit for carrying out reparations for damage caused by a product or service, which is five years counting the period from the moment that their authorship has been recognised (art. 27 CDC). **53**

(b) Alternative routes

In addition to the CDC, there is also the Biosafety Act (law 11.105/2005) which establishes that activities and projects that involve GMOs and derivatives, related to teaching practices that include the use of live organisms, scientific research, technological development and industrial production remain limited to the State or private entities, which must assume responsibility for compliance with the precepts of this law and its regulations, as well as for the consequences or effects produced by non-compliance (art. 2). Furthermore, it stipulates that public or private, national, foreign or international, financing agencies or sponsors of the above-mentioned activities or projects must demand presentation of the Certificate of Quality in Bio-security, issued by the CTNBio, under penalty of assuming co-liability for any effects resulting from non-compliance with legislation. **54**

Within the Brazilian system, entities involved in the financing of activities or ventures that are potentially polluting are not held liable; therefore, alternatives for financial compensation are harder to come by. This leads to increased interest, in the Brazilian case, in the idea of creating tariffs to cover the costs of environmental liability, through the establishment of insurance for environmental damage. In this manner, insurance companies, on the basis of possible damage that could emerge from known causes, are able to calculate the amounts that entrepreneurs/businessmen/producers should disburse to back up their activities. There are particular laws in Brazil which establish indemnification limits in case of accidents, such as: Law 6.367/1976 and Law 6.195/1974, that deal with urban **55**

33 Tribunal de Justiça do Estado de Santa Catarina, Agravo Instrumento 2004.000086-3, 3rd Câmara de Direito Público, reported by Desembargador *Luiz Cézar Medeiros,* DJSC 15.02.2006.

and rural labour accidents; Law 2.681/1912, on railway accidents; Law-decree 32/1966, clean air codes and Law-decree 277/1967. Nonetheless, there is no specific legislation regarding the limits to be observed regarding damage caused by GMOs. This serves to favour the interests of the entrepreneur/businessman/producer to the detriment of the interests of the collectivity and an ecologically sound environment. Nonetheless, it is worth noting that,

> "The implementation of this mode (environmental insurance) can only be allowed if there is no exclusive and tariff-generating mechanism developed for the reparation of damage to the environment. This means that if environmental damage, concretely considered, extends beyond the limits that have been fixed by the insurance policy, the agent of degradation would not be exempted from indemnifying, as the case would require, since indemnification must be whole and never partial."[34]

56 Lastly, it is important to emphasise that in Brazil insurance companies do not cover risks arising from GMO-related activities, mainly due to the difficulties involved – when dealing with environmental damage of diffuse authorship – in establishing risk patterns and the possibility of losses that extend over a long latency, that is, a long period of time. The major obstacle lies in the risk factor, since activity that is supposedly safe is actually uncertain in terms of its effects. This in spite of the fact that there is agricultural insurance in Brazil for such things as drought. Nonetheless, an area of interest for coverage for GMO producers would be "to cover the costs of legal defence [rather than] coverage for civil liability as such".[35]

(c) Impact of compliance with rules and regulations

57 Rules regarding the functioning of production (of seeds, for example) that requires the necessary environmental licensing (previously-obtained licenses for installations and operations) exempt the producer/industry from environmental liability in penal and administrative areas. Nonetheless, they do not exclude civil responsibility on environmental matters (environmental recovery or compensation) that is based on strict liability, regardless of fault. Regarding the consumer, the producer/industry will be

34 *Vianna* (fn. 6) 178.

35 *W. Polido*, Humanização do atendimento à saúde – Aspectos Bioéticos e a positivação do Direito: O contrato de seguro e as novas ciências. Revista Brasileira de Direito do Seguro e da Responsabilidade Civil, MP. Ano I, 2009, 251–310.

held liable for product flaws and thus, respect of rules may not be enough to avoid attribution of liability, since product conditions will determine whether there may be blame. It should be emphasised that the fact that the person who is liable is able to prove that all regulations were duly followed may have an influence on decisions regarding indemnification costs.

3. Environmental liability

(a) Implementation of the Environmental Liability Directive

Obviously Brazil has not adopted European directives, but within Brazil- **58**
ian legislation there are numerous examples of obligations on the part of firms/industries that are known to pollute, to contribute to funds for environmental conservation/preservation or to contribute directly to such activities as a way of minimising damage inflicted upon the environment. Such is the case of the Fund for Diffuse Rights, established by Law 7.347/1985 which stipulates that, in public civil action, pecuniary penalties/fines must be contributed to this fund in case of damage to the well-being and interests of the Union; when damage affects the goods or interests of one of the states of the Federation, fines must be paid directly to the respective state funds. It is understood that this action is not subject to prescription since its goal is the protection of inalienable collective rights. For instance, a Minas Gerais State Law (law 12.503/1997) created a regional programme for water conservation that forces public and private firms that have a service concession for supplying water and electrical energy to invest 0.5% of their operational revenues in the protection and preservation of hydrographic basins.

(b) Environmental liability regime beyond the scope of the Directive

Brazilian legislation can be said to diverge from Directive 2004/35, insofar **59**
as there are special provisions for liability regarding nuclear material as established by Law 6.453/77 and the Vienna Convention, to which Brazil subscribes. In this regard, most Brazilian legal doctrine defends the theory of integral risk liability, according to which the Union is responsible whether or not fault can be adjudicated and with no possibility to argue for excluding causes, such as fortuitousness and *force majeure*. Fiorillo's position is as follows: "Regarding civil liability for damage caused by

nuclear activities, such will be ascertained according to the system of strict liability, as prescribed by art. 21, XXlll of the Federal Constitution. In this manner, the absence of any type of exclusion of liability (including fortuitousness and force majeure), the non-existence of limits regarding the amount of indemnification and the solidarity (joint character) of liability"[36] will apply. Other scholars such as Toshio Mukai and Celso Antonio Bandeira de Melo, claim that the theory of administrative risk liability is also applicable in cases of environmental damage caused by nuclear accidents. Regarding all else, commentaries on triple liability for environmental matters, as discussed above, remain relevant.

(c) Claimants in cases of environmental harm

60 Recovery of the environment after damage is the responsibility of the agent who has engaged in the activity that has caused harm. Civil liability is strict in this case and extends over the entire production chain, since the party who is interested in carrying out the specific activity assumes all risks it entails, including those that affect the environment. The above discussion (General Overview 1 and 2) in which this matter was dealt with should be referred to.

(d) Special liability regime for losses sustained by individuals

61 Art. 1.277 of the Brazilian Civil Code stipulates that "the owner or holder of a building is entitled to bring to a halt interventions that may be harmful for the security, tranquillity and health of its inhabitants as caused by usages given to neighbouring property". Furthermore, "interferences are prohibited, considering the nature of usage, location of building, zoning norms and the customary limits of tolerance on the part of those who reside in the area". According to Delton, "the discipline of neighbourhood relationships is sustained by the balance between the rights of owners or users to enjoy their goods without encroaching on the sphere of the rights of others, with their neighbours maintaining their own right to request ceasing of activities and indemnification in case of harm".[37] Thus established, it is not just any damage that can lead to possible indemnification.

---

36 *Fiorillo* (fn. 23) 282.

37 *D. Winter de Carvalho,* Dano ambiental futuro: da assimilação dos riscos ecológicos pelo direito à formação de vínculos jurídicos intergeracionais, Universidade do Vale dos Sinos (doctoral thesis) 2006, 169.

As Delton continues, "it is not just any bothersome activity that can be considered damage; rather, that which is considered excessive and intolerable, or as is commonly stated, abnormal. That which does not go beyond the realm of the normal may be considered to lie within the domain of ordinary aspects of neighbourly life that individuals must simply deal with under the imperative of life in society".

Therefore, existence and quantification of damage become contingent on the judge's decision, making reparation through administrative processes impossible. On the contrary, corresponding judicial appeal is necessary. For this reason, the individual environmental tutelage of a neighbourhood may be considered a mechanism of environmental protection, even if it is devoted initially to repairing harm done to individuals. It should be emphasised that neighbours' rights do not depend on proof of fault. Therefore, the Civil Code places neighbourhood conflicts under the auspices of "a directive which awards discretion to the judge, who will verify whether rights have been offended to the extent that goes beyond the ordinary limits of tolerance".[38] **62**

(e) Cartagena Protocol

Brazil signed the Cartagena Protocol on 22 February 2004, and as one of the undersigned is obliged to abide by the agreement, both in terms of risk prevention and transgenics, and regarding the initiative of member countries on the adoption of those measures that protocol implementation requires. The Cartagena Protocol was promulgated in Brazil by Decree 5.705, 16 February 2006. Upon ratifying the protocol, Brazil committed itself to adopting the necessary measures for safe GM trade as well as those necessary for consumer and environmental protection. In fact, these measures were implemented through the Biosafety Law and other regulations that deal with these issues within the country. It is worth noting that even before implementation of the treaty and the Biodiversity Convention itself, the 1988 Brazilian Constitution had already incorporated an evident concern with environmental protection and recognition of human dignity. Since the country belongs to the Mercosul, there are also some specific treaties which, although not directly referring to GMOs, do present norms on environmental protection and conservation, such as the Environment Agreement Frame (no. 02/01) and the creation of the Ad Hoc Group on Agricultural Biotechnology. There have, however, **63**

38 *Leite* (fn. 2) 143.

not yet been any judicial cases in which the above-cited protocol has been applied by the Brazilian government.

4. Other strict liability regimes

64 Within Brazilian law there is an explicit form of dealing with the crime of releasing GMOs into the environment *without a proper license.* The Biosafety Law establishes, through its arts. 27 and 29, that it is a crime to release or get rid of GMOs into the environment, a violation of the norms established by the CTNBio and by organs and licensing and inspection bodies, providing for sentences of one to four years in prison and fines. It is also a crime "to store, stock, transport, trade, import or export GMOs and their derivatives, without due authorisation or in violation of norms established by the CTNBio and organs and licensing and inspection bodies". In this case, prison sentences of one to two years and fines may be applied. Furthermore, the Law on Environmental Crime (9.605/1998) establishes, in its art. 61, that it is a crime to "spread disease or plagues or species that may cause damage to agriculture, animal husbandry, fauna, flora or the ecosystem."

V. Vicarious liability

1. Scope of vicarious liability

65 Arts. 12, 14, 16 and 18 of the Consumer Protection Code (CDC), in dealing with liability for flaws in products or services, deals with the above-mentioned vicarious liability; furthermore, § 1 of art. 25 establishes that if there is more than one party liable for damages, all are held liable.

66 However, when the CDC includes legal mechanisms that have explicit provisions for all agents who have participated in the production or supply chain as sharing liability for problems with the product, these means that it becomes vitally important to develop detailed knowledge of the entire process involving said product, until the moment that the consumer receives it. The manufacturer will be held liable, as will whoever supplied the ingredients used in manufacturing the product, and for this reason it becomes absolutely indispensable that information on all elements of the production chain is sufficiently brought to light, whether for purposes of damage compensation for the consumer or so that the diverse suppliers may share the liability for causal elements leading to possible damage.

2. Liability for people further up the food or feed production chain

According to the Brazilian system of civil liability, all actors may be held liable for the damage they cause. Thus, in the case of the seed producer who furnishes the farmer with GMO seeds, rather than conventional ones, the farmer who then plants and sells his produce as conventional to a food processor who makes a particular non-GMO product, when at the end of the process contamination is discovered, those who have been harmed may seek to receive compensation for their losses from any part of the whole (previous) agro-industrial food chain. This means that each agent may retrospectively seek recourse from those who are responsible for the damage caused, with the major controversy centred on establishing proof of precisely the moment in which contamination took place. **67**

3. Can someone further down the feed or food chain include someone further up it in a trial against him/herself?

The issue of the inclusion in the lawsuit of defendants other than those specified by the claimant follows a two-fold system in Brazil. In the face of a lawsuit for the reparation of damage incurred in a consumer relationship, there is no legal provision for the inclusion of any defendants who are not those indicated by the claimant. This is due to the fact that the CDC, in order to facilitate consumer protection, expressly denies the intervention of third parties (art. 88), in view of the vicarious liability that extends over the entire production change for harm done to the consumer. Thus, in the event that the consumer chooses to hold the retailer that sold him/her the conventional product contaminated with GM liable, the latter is unable to obtain the "call to procedure", for example, with respect to the wholesaler who sold him/her the contaminated product. The defendant may request the recovery of the harm the consumer has suffered and that he/she has been wrongly obliged to assume, not being the party who caused the damage, through a lawsuit that is regressive and independent of the consumer. **68**

On the other hand, when a lawsuit for the reparation of damage caused in a civil or business relationship where liability is personal or classical is underway, the defendant may summon those who s/he sees as liable for the damage, regardless of whether they are situated above or below him/her in the production chain. Thus, if another party is proven to be at fault for fraud or damage, the defendant may be exempted from liability. **69**

VI. Multiple tortfeasors

70 On this matter, the system is also two-fold. In situations of damage caused to the consumer and the environment, liability is strict and vicarious and therefore extends itself over the entire production chain, yet only one agent of those who have contributed to the occurrence may be held responsible for the entirety of the damages. Only in a regressive lawsuit can damage compensation be sought, through the proportional compensation by all those who had a part in producing the damage. On the other hand, if damage was caused in a civil or business relationship where liability follows from classical theory, each of those parties that had a role in the damage may be charged proportionally.

VII. Defences

1. Licence/permission to grow GM material

71 Environmental licensing releases parties involved from administrative liability, given its recognition that the activities undertaken are being carried out within the legally-defined parameters. In a similar sense, authorised activities cannot be considered criminal. Yet there can be civil liability, through risk theory (strict liability). Therefore, it becomes possible to demand environmental reparation. Yet at present there are no cases of environmental contamination through GMOs that have been analysed by the judiciary.

72 With regard to GMOs that have been released only for testing, the company that is liable for research development must have a plan for prevention or containment of damage to the environment and human and animal health. In these cases, liability for damage is strict, and there should be reparation, even by those that have followed the above-mentioned plan and have complied with all bio-security instructions.

2. Consent/assumption of risk

73 The final consumer in the production chain is protected by the CDC, as stated above. We should emphasise that the only way that the supplier may escape being held responsible is by demonstrating that the consumer or a third party is completely at fault. Furthermore, the CDC itself requires the correct labelling of products containing GMOs, following

the rules established by Decree no. 4.680/2003. It should be further emphasised that suppliers are prohibited from marketing products that entail risks for or harm to human health. Thus, even if the consumer is aware that s/he is consuming GM products that are later proven to cause damage to his/her health, the supplier can be held liable.

3. Third-party influence

In situations of strict liability, such as those of the supplier, blame may 74
only be avoided when the existence of a faulty third party or consumer is proven. Nonetheless, we must be certain about what constitutes a third party in Brazilian law. For example, if sabotage has been carried out by an employee of the supplier, this agent cannot be considered a third party, since the employer is responsible for employee's acts. Furthermore, if sabotage is carried out by another firm, one that the supplier has contracted, for example, for product transportation, this agent may similarly not be considered a third party, since the supplier is liable for the partners that s/he has chosen for his/her business activities. An example of a third party would be a neighbour who does not respect the rules established for separating cultivation of GM and non-GM grains, leading to damage on others' property.

4. Prescription

Within the civil sphere, for individual damage that goes beyond the con- 75
sumption sphere, there is a three-year statute of limitations. While there is no time limit for complaints to be filed for defence of diffuse and collective rights, that is for the protection of an ecologically-balanced environment, the main position taken by the Brazilian courts is to adopt a statute of limitations of twenty years, called "extraordinary limitation". For consumer defence, the CDC establishes a period of five years for the filing and prescription of claims, that is for the consumer to seek indemnification for damage suffered. Within the penal ambit, prescription varies according to the type of sentence. Four years is the longest sentence stipulated for environmental crimes linked to GMOs, as foreseen both by the Biosafety Law and within the Law on Environmental Crimes. Thus, the prescription for crimes related to GMOs is eight years. That is, the consumer has eight years to seek indemnification for damage.

5. Other defences

76 Legislation does not establish this expressly, but in court decisions and within legal theory some alternatives can be found: the first hypothesis regards the difficulty that exists in linking the damage that occurred and the polluting source, given technical verification which cannot always demonstrate unequivocally that damage has been caused. Furthermore, consequences of the damage may manifest themselves over time, thus representing a continuous type of harm; it is also possible that damage may come from an indeterminate source. Lastly, the distance between the source and the effects produced may imply movement across borders and boundaries. In view of these situations, Colombo has argued that, when examining cause and effect relationships in cases of environmental damage, the criteria of certainty should be substituted by that of verisimilitude.[39] Other possible allegations include denial of liability for the act or fact that is represented as damaging to the environment or submitting that the supposed event did not occur, that the occurrence was not harmful to the environment or that conduct was authorised through the necessary legal means and a license was provided by the corresponding and competent authority. These hypotheses were questioned in the decision regarding environmental damage caused by waste disposal in an improper location.[40]

VIII. Remedies

1. Pecuniary compensation

(a) Bodily harm

77 Regarding GMOs, art. 20 of the Biosafety Act (law 11.105/2005) applies; this determines that "without prejudice to application of the penalties provided by law, the persons liable for environmental damage and to third parties shall respond, jointly, for indemnification or integral recovery, independently of assessment of fault". This provision is similar to that of the Consumer Protection Code, insofar as the parties concerned here are comparable to the "consumers" dealt with in the latter. There are sen-

39 *S. Colombo*, Dano ambiental, available at http://www.boletimjuridico.com.br/doutrina/texto.asp?id=1256 (accessed 30 June 2009).

40 Justice Court of Santa Catarina, Ação Civil 97.000396-0, 5th Civil Chamber, Relator Desembargador João Martins, DJ 21.03.2002. Justice Court of Santa Catarina, Ação Civil 97.000396-0, 5th Civil Chamber, Relator Desembargador João Martins, DJ 21.03.2002.

tences granting indemnification to consumers in cases where they have been harmed by a lack of sufficient information provided on the inside of medications that are sold.[41] A similar decision may be pronounced regarding GMOs, since there are still no pronouncements regarding their usage in medications in Brazil today.

(b) Property losses

Compulsory rural insurance covers cases of agricultural losses arising 78
from damage caused by accidents, natural phenomena, plagues or diseases affecting livestock, crops and other activities relating to rural activities.[42] The beneficiary of insurance compensation is the rural producer, person or legal entity who has complied with all obligations to the country, as stipulated by law, and who has contracted the rural insurance in the modalities provided by law, as defined by the Agriculture, Cattle Raising and Supply Ministry.[43] There is no provision for cases of GMO contamination, whether directly through direct planting or by cross pollination. This is

41 Justice Court of Rio Grande do Sul, Apelação Civil 700086673, 10th Chamber Reported by Desembargador Luiz Ary Vessini de Lima and issued on 16.09.2004.

42 As provided in Art. 16 of Decree 61.867/1967, which regulates compulsory insurance.

43 Limits and percentages of subvention of premiums for 2009:

Modalities of insurance	*Groups of crops*	*Percentages of premium*	*Limitation in R$*
Agricultural	Beans, corn of the second crop, maize.	70%	96,000
	Plums, oats, canola, *caqui*, barley, rye, fig, kiwi, linen, apples, nectarines, pears, peaches, sorghum, triticale and grapes.	60%	96,000
	Cotton, rice, corn and soybeans.	50%	96,000
	Avocado, pineapple, pumpkin, squash, lettuce, garlic, peanuts, *atemoia,* bananas, potatoes, eggplant, beets, cocoa, coffee, cashew, sugar cane, onions, carrots, *cherimóia, chuchu*, cauliflower, peas, chicory, *fava*, sunflower, guava, *graviola, jiló*, orange, lychees, lime, lemon and other citric fruits, papaya, *mamona*, manioc, mango, passion fruit, water melon, melon, strawberry, bell pepper, *pinha*, okra, cabbage, sisal, tangerine, tomato, string beans and all other vegetables and legumes.	40%	96,000
Cattle		30%	32,000
Forests		30%	32,000
Fish farms		30%	32,000
Maximum amount covered by insurance			192,000

due, among other factors, to the lack of control on the part of authorities over GMO crops in national territory. The release of the "round-up ready" soybean, resistant to the Round Up herbicide, is one of those cases, since the government had authorised the commercialisation of the crops even before the release of the GMO in the environment.

79 Environmental Impact Studies have been required only from the date of the ruling that is to say, in 2005 with the advent of the new Biosafety Law. Therefore, in the case of losses, the general rule of civil, administrative and penal liability will be applied, as described above.

(c) Economic losses

80 There are no special protection mechanisms. There is a growing demand on the part of conventional farmers for indemnification of losses suffered due to contamination. However, these complaints have not yet arrived in the courts but are under analysis by the administrative authorities and non-governmental entities.

81 AS-PTA[44] has been publishing cases of contamination with GMOs in cases of organic farmers who have been forced to pay royalties to the Monsanto Company. In one article they stated that "farmers who never planted RR [round-up ready] soybeans were obliged to pay royalties to the Monsanto Company, as if they had purchased seeds from that company. Under threat of being subjected to the payment of fines, in addition to royalties, they were coerced into adopting transgenic seeds. Today, many of them are no longer able to buy non-transgenic seeds."[45]

82 The Government of the State of Paraná has "proceeded to a series of inspection operations to verify contamination in conventional soybean seeds for sale in the state. 283 tons of conventional seeds contaminated with transgenics were apprehended, involving 11 companies which sold seeds; the contamination amounted to 9% in some lots of seeds."[46]

83 It must be stressed that, according to the rules of civil liability in Brazil, both the ceasing of profits and future damage may be the object of com-

44 Assessoria e Serviços a Projetos em Agricultura Alternativa, Brazilian ONG founded in 1984.

45 Published in http://www.jornaldaciencia.org.br/Detalhe.jsp?id=42949 and http://www.aspta.org.br/.

46 Special edition, 21 June 2007. Available at: http://www.aspta.org.br/por-um-brasil-livre-de-transgenicos/contaminacao-x-coexistencia/coexistencia-impossivel-contaminacao-de-soja-convencional-em-medianeira-pr/.

pensation, which thus includes everything that may be lost due to contamination. Therefore, in theory, the complainant may ask not only for compensation for material losses in organic seeds, for instance, but also for the expenses necessary for restoration of the contaminated area and the farmer's loss in terms of revenues.

(d) Harm to animals

There are no specific rules for damage caused by GMOs, nor rulings providing for ways of calculating damages arising from animal consumption of GMOs. Therefore, liability is based on contract clauses, the Consumer Protection Code, the Brazilian Civil Code and the Biosafety Law. According to the aforementioned, all material damage suffered by proprietors may be recovered in the event that the feed affected animal health, or because the contract provided for feeding animals with GMO-free products, for certification purposes. **84**

(e) Costs of disposal

There are no specific provisions for the recovery of loss on sales of contaminated animals. Therefore, with the general rule of civil liability, emerging profits and losses may be recovered. Thus, it can be affirmed that the damage or losses incurred by producers who have to sell contaminated products may be indemnified. It is worth noting that if the producer suffered only a decrease in profits and not a total loss, only those losses effectively suffered will be eligible for indemnification. **85**

2. Non-compensatory damages

As we have explained above, Brazilian laws concerning collective rights provide, administratively, for fines, activity embargoes and building interdiction, among others things; in the civil area, there are penalties providing for the reparation and/or recovery of damage, and in the penal area sentences of imprisonment or restriction of rights may be applied. In the individual context, every case must be analysed with respect to the eventual possibility of indemnification for material and moral damages. **86**

3. Other remedies

87 There are no express provisions for the replacement of the so-called contaminated products with others not containing GMOs. However, the Brazilian Civil Procedure Code, in its art. 461, establishes a mechanism to specifically request the rendering of the obligation. Thus, the debtor may be obliged to render the obligation, to deliver the product or animal exactly as agreed upon, whether by substitution of the contaminated good or animal or product that does not comply with specifications, or by the acquisition of the specified product at own expense. Similarly, the Consumer Protection Code rules that the consumer, in case of product damage or defect, may ask for its substitution for another which complies with the specifications that have been agreed upon. Other forms of reparation, such as the rendering of community service, are reserved for the penal area.

4. Costs of pursuing a claim

(a) General cost rule

88 In Brazilian law, the rule is that the defeated party pays the costs of the lawsuit and the lawyers' fees upon its termination. Thus, it becomes necessary for the claimant to anticipate these costs when the lawsuit is initiated. Similarly, the claimant must pay for expert witnesses, inspection and all other technical proof that they require. The defendant is obliged to pay for the technical proof which he requests and, in the case of the reversal of burden of proof, must also bear the costs of the proof as required by the claimant.

(b) Costs of establishing causation

89 In Brazilian law, it is possible to ask for the recovery of the expenses which were incurred to establish the *causal nexus*. If the producer had to hire specialists to determine the existence of contamination, he may, in the reparatory claim, ask for indemnification of those expenses. During the lawsuit, the amounts spent by the victim will be part of the amounts which must be covered by the defeated party.

5. Advance cover

As a "general rule, insurance for environmental risks covers only accidental events, even if the origin, the generating cause, is of diffuse nature – cumulative, repetitive or gradual."[47] Cover for sudden or accidental pollution cases is governed by the general rule for Civil Responsibility, depending on contractual clauses generally called "particular provision for the additional covering for sudden pollution". In Brazil, re-insurance is common in order to minimise risks, preserve the stability of the insurance companies and warrant the payment of accident premiums to the insured. The Institute of Re-insurance of Brazil – (today, the IRB-Brasil RE)[48] – was created with a very well defined objective: to strengthen the development of the national insurance market through the creation of the re-insuring market in the country. The option of "environmental insurance must be detached from any compulsory measure, particularly those that require the contracting of insurance, since it is totally incompatible with the nature of the risks and the still precarious stage of development of this insurance segment in this country".[49] Therefore, we can conclude that at present, insurance contracting is not compulsory with respect to environmental damage by GMOs, but there is discussion in the National Congress about the possibility of establishing compulsory environmental licensing. **90**

IX. Cross-border issues – Conflict of laws

1. Conflicts rules applicable before (or instead of) Rome II

Brazil will not adopt the Rome rules, since they are applicable only to the European Community. In reference to the Mercosul, there are no provisions about jurisdiction or applicable law in cases of illicit acts that transcend its borders. The national legislation must be applied in those cases. **91**

In civil legislation, jurisdiction over cases arising from licit and illicit acts belongs to the States of the Federation, through the State Justice Courts, with the exception of acts involving domestic and foreign public legal entities, or indirect administration of the Union. Those acts will be judged by the Federal Courts. **92**

---

47 W. *Polido*, Seguros para riscos ambientais (2005) 332.
48 More information at http://www2.irb-brasilre.com.br/site/.
49 W. *Polido*, Seguros para riscos ambientais, Revista Brasileira de Risco e Seguro, V.1. No.0. December 2004, 142.

93 According to Brazilian law, crimes involving GMOs have already triggered decisions ruling that jurisdiction belongs to the Federal Courts. It must be observed that, regarding illicit acts relating to GMOs released in Brazil, there have been cases regarding the soy crops of 2003–2004, planted without authorisation by CTNBio, involving transportation, selling and release of forbidden products in the national territory. However, no legal action was taken against the farmers of southern Brazil for planting seeds which were forbidden at the time. Such actions are classified as crimes according to the Biosafety Law. Although there is a corresponding legal provision, no penal action was proposed against the farmers. Instead, the law which regulates crops obliged them to sign a term of conduct adjustment, establishing a term for the selling of said soybeans until 31 January 2004, as well as the impossibility of their usage as seeds, as well as the compulsory express indication that the grains contain GMOs. Non-compliance with such rules leads to fines established by the authorities, in a minimum amount of R$ 16,110 (sixteen thousand, one hundred and ten Reais), varying in proportion to the seriousness of the infraction.

2. Special regime for cross-border claims

94 In Mercosul there is no agreement providing for the liability of member countries in cases of environmental damage, and much less in cases of release of GMOs in the environment. Bilateral agreements do not hold any provision for this type of liability either. However, constitutional rules for international matters may be applied, constituting a list of positions that must be adopted by the country of Brazil. Art. 4 of the Federal Constitution of Brazil determines that this country will be ruled, in its international relationships, by the principles of national independence, respect for human rights, the self-determination of peoples, non-intervention, equality between nations, defence of peace, peaceful solutions to conflicts, rejection of terrorism and racism, cooperation between peoples for the progress of humanity and the granting of political refuge. It also states that the Federal Republic of Brazil seeks the economic, political, social and culture integration of Latin American peoples, aiming for the creation of a Latin American community of nations. The principles adopted by the Brazilian Federal Constitution state that Brazil must help in the prevention, reduction and restriction of environmental damage across borders. The international law treaties which Brazil has signed are applicable, together with the Cartagena Protocol and the Convention for the Biological Diversity.

Lastly, as explained before, there is no possibility in Brazil to contract insurance for a GM plantation, since there is no control on the part of authorities and there are no technical parameters to identify the risk potential of the activity. The only possibility would be insurance for damage against third parties in the modality of RC, but never against the environment. The illegal entry of GM soy into the country (2003–2004 crops) was not enough for the country to propose any liability lawsuit involving Argentina, the country of origin of the GMO planted in Brazil. **95**

X. Cases

1. Due to the adventitious presence of GMOs in a field, maize which is normally sold as conventional contains GMOs beyond the legal labeling threshold. This is not discovered before the final stage of the food production chain by the producer of taco chips. The whole production is lost since the supermarket chains refuse to accept delivery from the producer.

(a) Who can sue along the chain of distribution?

According to the civil liability system, those who have suffered damage due to GMOs may ask for due compensation. Thus, any of the agents may, by regressive lawsuit or by asking for inclusion as respondents in the indemnification lawsuit, claim recovery of the damage suffered. Some companies have included in their contracts a series of core obligations, establishing fines in case of delivery or tentative delivery of products outside specifications, and/or the duty to deliver the specified product, even if said product must be obtained from another producer or manufacturer. Furthermore, if, at any time in the production chain, a contract for the certification of corn was signed establishing the accepted limits for contamination and was not complied with, the certifying company will also account for the damage caused by the low quality of its services. This is also the case if accidental contamination occurs due to the fault of a neighbour who planted GM corn and did not take the necessary precautions. The country may also be held liable for the contamination if precautions for inspection did not comply with the specific regulations on planting GM corn. **96**

(b) Would the case be solved differently if the GMO content was below the labelling threshold?

97 If the product was within the specifications of the contamination limit, markets could not refuse it and there would be no grounds for damage recovery in case of accidental contamination. On the other hand, if the product were within the standards considered safe for human consumption, the solution would be a recall operation to identify it as "containing GMO" so that it could be returned to the consumer market.

(c) Would the case be solved differently if the admixture was not adventitious, but occurred due to the disregard of segregation rules, for example?

98 Results are independent of the way in which contamination occurred. The difference lies in the responsibilities regarding third parties, since civil liability is independent of fault and penal liability, in cases in which due authorisation has been given, would not be considered. As described above under (a), if at any moment in the production chain there was a contract for corn certification, within accepted contamination limits, and this certification did not take place or was done incorrectly, the certifying company shall account for the damage caused by improper rendering of services.

(d) Would the case be solved differently if the GMO found was not admitted for production in your jurisdiction?

99 This would be relevant in criminal law, since non-authorised GMO production is considered a crime. In civil law, the consequences would be the same as those described above, with possible responsibility of all parties who contributed to the damage.

(e) If the admixture had occurred on a non-GM field and it transpires that the GM seeds were blown from: (i) neighbouring fields; or (ii) a truck passing by, would the farmer of the affected field be liable for all or part of the loss caused further down the distribution chain?

Generally speaking, the producer may not be held liable in this case for contamination, since he did not act with fault or intention, and the damage is due to the exclusive fault of a third party. However, if it was established in the contract of purchase that the producer should deliver only non-GM corn, the company may refuse to receive said product. **100**

2. Twenty years after the sale of GM maize used for food products, it turns out that it has certain disadvantageous health effects for humans.

(a) Can the producers be held liable at this point for risks unknown at the time of growing the maize? Who would be liable – the seed producer/farmer/food producer/distributor/ etc.?

The understanding of court decisions is that lawsuits which do not have a term provided by law can be initiated within 20 years of the fact or occurrence. In the case described above, it would not be possible to ask for indemnification, except in cases in which this understanding of the court should change substantially. However, in theory, since it is a case involving a consumption chain, the whole production chain may be held liable for the damage. In theory, the consumer could establish the liability of the market where he bought the product, the distributor, manufacturer, farmer or producer of the seed. However, said liability would only be possible within a system where the product could be traced back to its origin, through a certification system with preserved identity. Furthermore, the company which produced the seeds could be held responsible for the risk of the activity of seed development that may cause damage in the future, whether to the consumer, producer, distributor, or any other business. **101**

(b) Can compensation already be claimed at a point when the negative health effects have not yet materialised, but are expected to according to scientific expertise/mere rumours?

102 The only damages which may be claimed are for losses which have effectively happened or, at least, can be proven as certain to occur in the future. The mere possibility that a particular illness may occur does not allow victim to claim indemnification. Therefore, the person may file claims for medical treatment of the illness, degeneration or acquired disease, as well as claims for preventive treatments that will lessen the impact on his/her health or allow him/her to avoid contracting the illness.

(c) Would it make any difference if the GM maize had only been in use for feed, causing harm to the animals, which may or may not cause harm to humans consuming the meat as well?

103 No, the liability would be just as described above.

3. The driver of a food logistics company discovers that a farmer, from where he regularly picks up agricultural products fails to obey mandatory segregation rules or food or feed hygiene standards, which may lead to the admixture of GM and non-GM produce sold and packaged separately by that farmer. Does he or his employer have a duty to warn, i.e. warn the recipient of the allegedly "non-GM" produce?

104 Keeping in mind that in Brazilian law liability for damage caused extends throughout the whole production chain and the employer is held liable for the acts of his employees, it is the duty of the employee to warn the producer of the difficulties that may occur in grain separation. In addition to this, it is possible to have an express contractual clause ruling liability on that matter.

Damage Caused by GMOs under Canadian Law

Jane Matthews Glenn*

I. General overview

1. Special liability or redress scheme for GMOs

Canada is one of the pioneers in GM agriculture. The main GM crop, canola (oilseed rape), was approved for unconfined release in 1995, and cultivation commenced in 1996, the first country in which this occurred.[1] GM agriculture now accounts for about 80% of canola and 65% each of corn and soybean produced in Canada. Labelling of GM foodstuffs is not mandatory, and hence is not done. There are no traceability requirements for GM products and no threshold on the amount of GM products a foodstuff may contain. Canadians have thus been consuming GM products in unknown quantities on a regular basis for over a decade. Some are aware of this, but many are not. The only food clearly free of GMOs is organic, and this places the burden of segregation, and the traceability and labelling that goes with it, on organic rather than GM or conventional producers. All of this makes responding to the Questionnaire somewhat speculative. 1

The Questionnaire focuses on issues of civil liability. Under the Canadian Constitution, private law is a matter of provincial jurisdiction.[2] Although the structure and functioning of the courts is basically the same in all 2

* I am very grateful to my colleague, Daniel Jutras, for his generous guidance in the early stages of my research for this report. Any errors or omissions are mine alone.

1 *T. Demeke/D.J. Perry/W.R. Scowcroft*, Adventitious Presence of GMOs: Scientific overview for Canadian grains, Canadian Journal of Plant Science 86 (2006) 1 ff. at 10.

2 Constitution Act, 1867, Revised Statutes of Canada (R.S.C.) 1985, App. II, No. 5, s. 92(13) (establishing provincial jurisdiction over "Property and Civil Rights in the Province"). "Civil Rights" refers in this context to private law rights rather than to modern notions of "human rights". Note that for Quebec, the original Civil Code of Lower Canada (C.C.L.C.), in force since 1866, has been replaced by a revised Civil Code of Quebec (C.C.Q.), adopted in 1991 and in force since 1994.

Canadian provinces and follows the English model, for historical reasons the substantive private law of Quebec is based on French civil law and that of other Canadian provinces on English common law. At the risk of over-simplification, the civil law's approach to delictual liability is unified, requiring proof of fault, injury and a causal link between the two,[3] whereas the common law's approach is disunified, recognising a number of nominate torts (e.g. trespass, nuisance), each with its own requirements, in addition to the fault-based tort of negligence based on duty, breach and damage. This report considers both approaches, although it treats Canadian common law more extensively than Quebec civil law.[4]

3 There are no specific liability or other compensation or redress regimes covering losses caused by GMOs in Canada at either the federal or provincial levels; nor are these losses addressed by a special regime with otherwise wider scope. A specific regime was not envisaged when Canada first authorised the unconfined release of GM canola, and none was recommended by the Canadian Biotechnology Advisory Committee in a major study released in 2002,[5] despite calls for one by such groups as the National Farmers Union (many members of which are organic farmers).[6] The CBAC was of the view that issues of liability and compensation for damages were adequately addressed under general civil liability principles, although it acknowledged that "practical difficulties" – such as proving a duty of care and a breach of that duty – might render this remedy "illusory".[7] This caution is borne out by the lack of success of Hoffman and Baudouin, two Saskatchewan organic grain farmers, in their class

3 Art. 1457 C.C.Q.; a fourth requirement is proof of mental capacity. See generally Book 5 ("Obligations"), Title 1 ("Obligations in General"), which deals with both contractual and delictual obligations.

4 The leading Canadian texts on civil liability are, for the common law, *L.N. Klar*, Tort Law in Canada (4th ed. 2008); *P. Osborne*, The Law of Torts (Essentials of Canadian Law Series) (3rd ed. 2007); *A.M. Linden/B. Feldthusen*, Canadian Tort Law (8th ed. 2006) and *G.H.L. Fridman*, The Law of Torts in Canada (2nd ed. 2002), and, for the civil law, *J.-L. Baudouin/P. Deslauriers*, La responsabilité civile (7th ed. 2007) and *J.E.C. Brierley/R.A. Macdonald* (eds.), Quebec Civil Law: An introduction to Quebec private law (1993) (especially Title 5: Obligations). Canadian statutes and court decisions, both federal and provincial, are readily available on the Canadian Legal Information Institute's open website, http//www.canlii.org.

5 Canadian Biotechnology Advisory Committee (C.B.A.C.), Patenting of Higher Life Forms and Related Issues (June 2002) http://strategis.ic.gc.ca/eic/site/cbac-cccb.nsf/eng/h_ah00113.html.

6 National Farmers Union (N.F.U.), Policy on Genetically Modified (GM) Foods, point 9 (requesting that the federal government "legislate efficient and accessible mechanisms to enable liability claims to be effectively pursued") http://www.nfu.ca/policy/GM_FOOD_POLICY.misc.pdf.

7 C.B.A.C. (fn. 5) 17 & 14. The C.B.A.C. Report does not provide reasons in support of these somewhat contradictory conclusions, nor do there appear to be any in-depth background studies supporting them.

action against Monsanto Canada and Bayer Cropscience (below, no. 10, 14, 30, 35, 39–41, 51, 58).

2. State liability

Like other Commonwealth jurisdictions, Canadian provinces and the federal government abandoned the historical position of sovereign (or "Crown") immunity by legislation adopted in the 1940s and 1950s.[8] The federal Crown Liability and Proceedings Act[9] is the most important in regard to GMOs, as the federal government is responsible under the Constitution for patenting of GMOs[10] and (generally speaking) for the resulting commercialisation of GM seeds and crops.[11] The federal Act provides that the Crown "is liable for the damages for which, *if it were a person*, it would be liable" in respect of torts committed by its servants.[12] Crown liability thus follows basically the same rules as standard civil liability (i.e. common or civil law rules, depending on the province in which the wrong occurs). This means that the federal Crown could be subject to two different sets of liability rules, civil and common, for a wrongful act that causes damage across Canada. A good illustration of this is a set of class actions resulting from the 2003 bovine spongiform encephalopathy (BSE) crisis in Canada. Although the "mad cow" at the heart of the crisis was located in Alberta, the economic consequences of the ban on Canadian cattle and beef products in international markets were felt across the country. Four class actions were commenced immediately, in Ontario, Saskatchewan, Alberta and Quebec, against the federal government for its alleged negligence in regulating the cattle industry (notably for permitting the sale of feed incorporating ruminant meat and bone meal (RMBM) well after this was recognised as a cause of BSE) as well as against the manufacturer of the feed in question. The actions from the three common law provinces have been consolidated, so that there are now two class actions being pursued against the federal government, the common law action of *Sauer v. Canada (Attorney General)* in 4

8 See generally *P.W. Hogg/P.J. Monahan*, Liability of the Crown (3rd ed. 2000) and *K. Horsman/G. Morley*, Government Liability: Law and Practice (2007); art. 1376 C.C.Q.

9 R.S.C. 1985, c. C-50 (adopted as S.C. 1952–53, c. 30).

10 Constitution Act (fn. 2) s. 91(22) ("Patents of Invention and Discovery").

11 S. 91 (2) ("Regulation of Trade and Commerce"). This is interpreted as including international and inter-provincial trade, with intra-provincial trade coming under the provincial head of jurisdiction of "Property and Civil Rights".

12 Crown Liability and Proceedings Act (fn. 9) s. 3(b)(i) [emphasis added]. Note that a successful action in judicial review is not a precondition to civil action: e.g. *River Valley Poultry Farm Ltd v. Canada (Attorney General)* (2009) 95 Ontario Reports (O.R.) (3d) (Ont. C.A.) par. 30 (action in negligence).

Ontario and the civil law action of *Bernèche v. Canada (Procureur Général)* in Quebec (below, no. 5, 11, 31, 51). The defendants applied to have the Quebec action stayed until final judgment in the Ontario action, but this application was denied by the Quebec courts both at trial and on appeal.[13]

5 A major concern with holding the State civilly responsible for its wrongful acts is to balance this responsibility owed to private individuals against its responsibility to govern effectively, owed to the general public. For this reason, the Quebec provision establishing State responsibility subjects it "to any other rules of law which may be applicable to them"; these rules can be either statutory or general principles derived from case law.[14] In the common law provinces, the balance is found in the way courts apply a two-stage stage test, applicable in negligence actions generally, to determine whether a duty of care exists. This test was set out by the House of Lords in 1978 in *Anns v. Merton London Borough Council* and refined subsequently by the Supreme Court of Canada.[15] The first stage looks at factors arising from the relationship between the parties to see if they show the requisite foreseeability and proximity to give rise to a prima facie duty of care; the second stage looks at factors outside this relationship to see whether there are any residual policy considerations that negate the imposition of a duty of care. Each stage poses problems in actions against public authorities.[16] At the first stage, a particular difficulty in proving proximity is the requirement that the duty of care be grounded in the legislative intent of the governing statute: the plaintiff must show that the statute provides, even implicitly, a private law duty of care to those in his position (although the statute might also include duties owed to the public).[17] In the case of *Sauer*, for example, the Saskatchewan Court of Appeal accepted

---

13 *Bernèche v. Canada (Procureur général)* 2006 Quebec Cour Supérieure (QCCS) 3046 (Que. S.C.), leave to appeal refused 2006 Quebec Cour d'Appel (QCCA) 984 (Que. C.A.) *(sub nom Ridley Inc. v Bernèche)*.

14 Art. 1376 C.C.Q.; *Finney v. Barreau du Québec* [2004] 2 Supreme Court Reports (S.C.R.) 17 (S.C.C.) par. 27–30.

15 [1978] Appeal Cases A.C. 728 (H.L.) as refined particularly in *Cooper v. Hobart* [2001] 3 S.C.R. 537 (S.C.C.) and *Edwards v. Law Society of Upper Canada* [2001] 3 S.C.R. 562 (S.C.C.). See also *J. Neyers*, Distilling Duty: The Supreme Court of Canada amends *Anns*, Law Quarterly Review (Law Q. Rev.) 118 (2002) 221; *S.G.A. Pitel*, Negligence: Canada remakes the *Anns* test, Cambridge Law Journal (Cambridge L.J.) 61 (2002) 252.

16 See *J. Matthews Glenn*, "Government Wrongs": Civil liability for GMO regulation in Canada, Journal of Environmental Law and Practice (J. Envtl L. & Prac.) 18 (2008) 169. See also *J. Matthews Glenn*, GM Agriculture in Canada: A courtroom drama, in: L. Bodiguel/M. Cardwell (eds.), The Regulation of Genetically Modified Organisms: Comparative approaches (Oxford University Press, forthcoming).

17 *Klar* (fn. 4) 359 argues that this "is a misguided approach, which in spirit runs counter to *Saskatchewan Wheat Pool*" (the leading case on "breach of statutory duty", discussed below, no. 27).

as arguable that proximity was demonstrated by "the many public representations by Canada that it regulates the content of cattle feed to protect commercial cattle farmers among others".[18] Existence of such a duty is denied by the courts more often than it is accepted, and one possible way of reconciling the cases is to suggest that those in which a private law duty is denied concern what might be called "risk reduction" measures (e.g., failure to control the spread of disease[19]) and those in which a duty is recognised involve "risk creation" measures (e.g., herbicide spraying[20]). It is very similar to the differing treatment the common law accords to nonfeasance and misfeasance.[21] Approving GM technology for commercial release would fall close to the "risk creation" end of the spectrum.

At the second stage of the *Anns* test, a residual policy consideration that **6**
might negate the imposition of a duty of care on a public authority is the general principle of maintaining separation between the activities of the legislature and the judiciary, coupled with the inherent difficulties of having judges decide questions involving the allocation of scarce resources. This has led the courts to distinguish between "policy" decisions, immune to judicial scrutiny, and "operational" decisions, subject to government liability. However, the boundary between the two categories has shifted back and forth over time,[22] and courts now appear to move consideration of it from the overtly policy-oriented second stage of *Anns* to the more fact-oriented first stage, and use it to reinforce decisions about proximity.[23] It

---

18 *Sauer v. Canada (Attorney General)* (2006) 79 O.R. (3d) 19 (Ont. S.C.) [*Sauer* trial], affirmed (2007) 225 Ontario Appeal Cases (O.A.C.) 145 (Ont. C.A.) [*Sauer* appeal] par. 62 leave to appeal refused 2008 CanLII 36470 (S.C.C.) (motions to strike pleadings denied). Compare *Attis v. Canada (Health)* (2008) 93 O.R. (3d) 35 (Ont. C.A.), leave to appeal refused 2009 CanLII 19874 (S.C.C.) (silicone gel breast implants) and *Drady v. Canada (Health)* (2008) 300 Dominion Law Reports (D.L.R.) (4th) 443 (Ont. C.A.), leave to appeal refused 2009 CanLII 19876 (S.C.C.) (jaw implants).

19 *Williams v. Canada (Attorney General)* (2005) 257 (D.L.R.) (4th) 704 (Ont. S.C.) (Severe Acute Respiratory Syndrome, or SARS) (stressing distinction between acting and failing to act in negligence law); *Eliopoulos v. Ontario* (2006) 276 D.L.R. (4th) 411 (Ont. C.A.), leave to appeal refused 2007 CanLII 19108 (S.C.C.) (West Nile Virus).

20 *Ring v. The Queen* (2007) 268 Newfoundland & Prince Edward Island Reports (Nfld. & P.E.I.R) 204 (Nfld & Lab. S.C., Trial Div.) (army base pesticide spraying; class action certified).

21 See *McMillan v. Canada Mortgage and Housing Corp.* (2007) 75 British Columbia Law Reports (B.C.L.R.) (4th) 359 (B.C.S.C.) par. 31–38, affirmed (2008) 86 B.C.L.R. (4th) 273 (B.C.C.A.), leave to appeal refused 2009 CanLII 21725 (S.C.C.), and *Horsman/Morley* (fn. 8) 5.10.

22 See *Linden/Feldthusen* (fn. 4) 684 ff.

23 See e.g. *Williams*, (2005) 257 D.L.R. (4th) 704, par. 72: "Statutory duties of Ministers to make policy decisions are quintessentially duties owed to the public and not to private individuals but, as the cases where liability is found in respect of operational decisions illustrate, an implementation of decisions made in the exercise of such powers or duties can create a relationship of proximity with persons affected".

is thus not easy to predict how the courts will apply this distinction to a given situation. In the case of the approval of GM technology for release, the most probable outcome at the present time is that the decision to have the Canadian Food Inspection Agency (CFIA) apply what might be styled (at the risk of over-simplification) an "American-style, product-based, substantial equivalence, agronomic-focused" approval model, rather than a "European-style, process-based, precautionary principle, holistic-focused" one[24] would be classified as a policy decision, but that certain aspects of it – a perceived conflict of interest between the CFIA's regulatory role and its Ministry's enthusiastic promotion of biotechnology, lack of transparency and possible undue industry influence with the attendant dangers of regulatory capture, very limited provision for public input, and inadequate post-release monitoring and control – would be classified as operational in nature and thus open to a negligence action.[25] The limited scope of Canada's regulatory process is of increasing concern in the face of a probable application by Monsanto for the unconfined release of GM wheat.[26]

7 The above discussion concerns the possibility of actions against the government for losses occasioned by GMOs. What of the converse, actions by the government against the biotechnology companies for whatever losses it incurs from GMOs? This possibility is suggested by a number of on-going actions by eight Canadian provincial governments, including Quebec, against tobacco companies to recover the cost of health care provided to persons suffering from lung cancer and other tobacco-induced illnesses.[27]

---

24 For critical analyses of the Canadian regulatory framework see *S. Benda*, Canadian Seed Regulation: Precautionary principle – not really! Precautionary approach – sort of? Precautionary and logical – within limits, European Food and Feed Law Review 4:1 (2009) 23 ff.; *M. Phillipson*, Are Genetically Modified Crops in Canada Under-Regulated? J. Envtl L. & Prac. 18 (2008) 195 ff.; Institute of the Environment, University of Ottawa, Practicing Precaution and Adaptive Management: Legal, institutional and procedural dimensions of scientific uncertainty (2005) http://www.ie.uottawa.ca/English/Reports/JBPP_Final_Report.pdf; and *Royal Society of Canada*, Elements of Precaution: Recommendations for the regulation of biotechnology in Canada (2001) http://www.rsc.ca/foodbiotechnology/GMreportEN.pdf..

25 See generally *J. Matthews Glenn*, Footloose: Civil responsibility for GMO gene wandering in Canada, Washburn Law Journal 43 (2004) 547, 566 ff.

26 See particularly *Phillipson*, J. Envtl. L. & Prac. 18 (2008) 195 ff.

27 E.g. *British Columbia v. Imperial Tobacco* [2005] 2 S.C.R. 473 (S.C.C.) (preliminary action, upholding constitutionality of B.C.'s Tobacco Damages and Health Care Cost Recovery Act, S.B.C. 2000, c. 30). See generally *J. Berryman*, Canadian Reflections on the Tobacco Wars: Some unintended consequences of mass tort litigation, International and Comparative Law Quarterly (I.C.L.Q.) 53 (2004) 579 ff.

II. Damage

1. Recoverable losses

Some intentional torts, such as trespass and defamation, are actionable at common law without proof of loss, but damage is a necessary component of negligence and most other tort actions, and of delictual responsibility generally in Quebec. Compensation for physical damage to persons or property is readily recoverable, but common law courts have been more reticent than Quebec courts about compensating for less tangible harms such as economic loss, psychiatric or psychological damage ("nervous shock") and fear of loss. 8

2. Pure economic loss

Canadian courts are more open than their English and American counterparts to claims for pure economic loss, although those in the common law provinces are more circumspect than Quebec courts about compensating for economic loss than for physical damage.[28] A common law negligence action for pure economic loss follows the same two-stage *Anns* analysis as for general negligence (above, no. 5–6), with the court first determining whether a prima facie duty of care exists – i.e. a duty of care to avoid causing economic loss -and then determining whether there is any overarching policy consideration that would serve to negate such a duty. A major hurdle at the first stage, here as well as in negligence actions generally, is to establish proximity. This is facilitated by judicial recognition of a number of established or analogous categories in which a duty is presumed. In economic loss cases, these are negligent supply of dangerous or defective goods or structures, negligent misrepresentation, negligent performance of a service, liability of public authorities, and relational economic loss – i.e. economic loss suffered by the plaintiff as a result of injury to another's person or property).[29] At the second stage, a key policy argument against 9

---

28 See generally *B. Feldthusen*, Economic Negligence: The recovery of pure economic loss (5th ed. 2008); *J. Stapleton*, Comparative Economic Loss: Lessons from case-law-focused "middle theory", University of California at Los Angeles (U.C.L.A.) Law Review 50 (2002) 531. For Quebec, see *Spar Aerospace v. Hughes* [2002] 4 S.C.R. 205 (S.C.C.) par. 46 ("in contrast to common law jurisdictions, there is no principle prohibiting the recovery of pure economic loss in Quebec").

29 First set out in *Canadian National Railway v. Norsk Pacific Steamships Co.* [1992] 1 S.C.R. 1021 (S.C.C.); list of categories recognised as open, not closed, in *Martel Building v. R.* [2000] 2 S.C.R. 860 (S.C.C.).

imposing liability in economic loss cases is a concern about indeterminacy of liability ("liability in an indeterminate amount for an indeterminate time to an indeterminate class"[30]). Quebec courts, on the other hand, address the concern of indeterminacy indirectly through a restrictive analysis of causation, rather than considering it directly as the common law courts do.

10 It remains possible that pure economic loss caused by GMOs is recoverable under Canadian common law notwithstanding the Saskatchewan courts' refusal to certify a class action for such loss in *Hoffman and Beaudoin v. Monsanto Canada and Bayer Cropscience Inc.*[31] This was a class action initiated by two Saskatchewan organic farmers against two biotechnology companies for compensation for the economic losses flowing from their inability to cultivate canola because of cross-fertilisation from surrounding GM crops. One of the requirements for certification is that the pleadings disclose a cause of action, and both the trial judge and the Court of Appeal held that this had not been done, as the plaintiffs failed at each stage of *Anns*. At the first stage, the plaintiffs had to show the requisite foreseeability and proximity. The trial judge described proximity as "a close and direct relationship of such a nature that the defendants may be said to have been under an obligation to be mindful of the plaintiffs' legitimate interests" and held that the plaintiffs in *Hoffman* had not alleged "the existence of any expectations, representations, or reliance", nor "physical harm to themselves or their property", nor a "special relationship between themselves and the defendants": "Indeed, they have not alleged any relationship at all ... that would give rise to an argument for sufficient relational proximity to support a *prima facie* duty of care."[32] At the second stage, the trial judge in *Hoffman* stressed the problem of indeterminacy, a concern that is compounded in class actions where the plaintiff must demonstrate the existence of an identifiable class whose claims raise common issues; moreover, the appeal court called the prior regulatory approval of com-

30 An oft-cited statement of Cardozo J. in *Ultramares Corporation v. Touche* (1931) 255 N.Y. 170, 179.

31 [2005] 7 Western Weekly Reports (W.W.R.) 665 (Sask. Q.B.) [*Hoffman* trial], affirmed (2007) 283 D.L.R. (4th) 190 (Sask. C.A.) [*Hoffman* appeal], leave to appeal refused 2007 CanLII 55334 (S.C.C.). See e.g. *K. Garforth*, When Worlds Collide: Biotechnology meets organic farming in *Hoffman v. Monsanto*, Journal of Environmental Law 18 (2006) 459; *H. McLeod-Kilmurray*, Hoffman v. Monsanto: Courts, class actions and perceptions of the problem of GM drift, Bulletin of Science, Technology and Society (Bull. Sci., Tech. & Soc'y) 27 (2007) 188; *J. de Beer*, The Rights and Responsibilities of Ag-biotech Patent Owners, University of British Columbia Law Review 40 (2007) 343.

32 *Hoffman* trial, [2005] 7 W.W.R. 665, par. 67.

mercial release "a powerful policy reason" for denying a duty of care[33] (below, no. 51).

However, the twinned cases of *Sauer* and *Bernèche* are more open to recovery for pure economic loss. They are worth watching, as the facts and issues echo those of *Hoffman*. Both are class actions brought by two cattle producers, one in Ontario and the other in Quebec, for economic loss resulting from the closure of international markets following the diagnosis of a BSE case in western Canada. A preliminary motion by the defendants in *Sauer* to strike the pleadings for failure to disclose a cause of action was refused both at trial and on appeal.[34] The Ontario Court of Appeal rejected the feed manufacturer's first-stage argument, based on *Hoffman*, that there was no proximity between it and the plaintiff because he had not purchased or used its feed. The Court distinguished *Hoffman* as turning on its own facts, and accepted as arguable that feed manufacturers and cattle producers were all linked as part of a single integrated industry, and by the effect a single infected cow had on all producers.[35] The trial judge stressed the danger of overstating the problem of indeterminacy at the second stage. "Indeed, there is no authority for the proposition that extensive liability is equivalent to indeterminate liability. ... In as much as there is a concern about making a defendant an insurer for the wrongs of the world at large, there should also be a concern that the policy limiter on the imposition of tort liability does not itself become a de facto indemnification for all tortfeasors faced with significant liability in terms of quantum."[36] Finally, both the trial and appeal courts held that as a matter of policy, the manufacturer's compliance with government regulations did not negate the existence of a duty of care. The Court of Appeal expressly disagreed with the decision in *Hoffman* on this point: "So far as *Hoffman* ... applies a similar government policy to negate a prima facie duty of care, I would respectfully decline to follow it."[37] Both actions against the feed manufacturer have now been settled out of court, and 11

33 *Hoffman* appeal, (2007) 283 D.L.R. (4th) 190, par. 60. The fact that the organic certification bodies had no policy about the presence of GMOs when GM canola was first released commercially was another policy reason to deny the existence of a duty: par. 61.

34 *Sauer* trial, (2006) O.R. (3d) 19, *Sauer* appeal (2007) 225 O.A.C. 145, leave to appeal refused 2008 CanLII 36470 (S.C.C.).

35 *Sauer* appeal, ibid., par. 39. It also stressed the extent of economic loss over physical damage in underscoring proximity: "This economic catastrophe dwarfs the impact of the loss of one cow on the cattle farmer who purchased the Ridley feed.".

36 *Sauer* trial, (2006) 79 O.R. (3d) 19, par. 84.

37 *Sauer* trial, ibid., par. 53; *Sauer* appeal, (2007) 225 O.A.C. 145, par. 48.

both actions against the Crown have now been certified and will be heard on the merits, one in Ontario and the other in Quebec.[38]

3. Mere fear of a loss

12 It is unlikely that mere fear of loss triggered by GMOs is recoverable under Canadian law as it presently stands. Both the common law and Quebec's civil law compensate for existing physical damage and mental harm, although Quebec law is more open than the common law to claims for mental harm. The principle of recovery for nervous shock (i.e. psychiatric or psychological damage) is now generally accepted in Canada, with recovery being controlled under both common and civil law by three requirements: firstly, that the damage be triggered by something of an immediate or traumatic nature (as the word "shock" suggests); secondly, that psychiatric or psychological damage to a person of normal susceptibility ("ordinary fortitude") be reasonably foreseeable; and thirdly, that the damage be serious enough to be evidenced in some recognisable way (e.g. heart attack, miscarriage, clinical depression, etc.).[39] Under these constraints, it is difficult to imagine that a GMO-related event could serve as the trigger for a claim in nervous shock in Canada, where people consume GM food on a regular basis.

13 There are signs that the common law's more restrictive approach may be further evolving, to include claims that come closer to a claim for a mere fear of loss. One line of cases recognises as actionable claims for the costs of medical monitoring or other preventive or diagnostic procedures where the plaintiff fears physical damage that has not yet materialised. Examples include fear of life-threatening illness from ingesting harmful weight loss pills,[40] fear of heart disease from being implanted with a faulty heart valve[41] and fear of cancer from herbicide spraying.[42] But

---

38 *Sauer*: (2009) 2460 O.A.C. 256 (Ont. S.C.), leave to appeal refused (2009) 246 O.A.C. 256 (Ont. S.C., Div. Ct.) (certification of class action); *Bernèche* 2007 QCCS 2945 (Que. S.C.) The causes of action in *Bernèche* are limited to gross negligence, faute lourde and bad faith, as those are what his lawyers alleged when introducing the action, and a motion to strike pleadings outside these areas was recently granted: *Bernèche v. Canada (Procureur général)* 2008 QCCS 2815 (Que. S.C.), aff'd 2008 QCCA 1581 (Que. C.A.).

39 See e.g. *Odhavji Estate v. Woodhouse* [2003] 3 S.C.R. 262 (S.C.C.) and *Mustapha v. Culligan of Canada Ltd.* (2008) 293 D.L.R. (4th) 29 (S.C.C.).

40 *Wilson v. Servier Canada* (2000) 50 O.R. (3d) 219 (Ont. S.C.), leave to appeal refused (2000) 52 O.R. (3d) 20 (Ont. Div. Ct.), leave to appeal refused (S.C.C. 2001), settlement approved 2005 CanLII 7128 (Ont. S.C.).

41 *Andersen v. St Jude Medical Inc.* (2003) 67 O.R. (3d) 136 (Ont. S.C.).

42 *Ring* (2007) 268 Nfld. & P.E.I.R 204. Non-medical examples include fear of motor damage from recalled gasoline additive: *Scarola v. Shell Canada Ltd.* 2003 CanLII 44004 (Que. S.C.) (class action certified), 2004 CanLII 10081 (Que. S.C.) (settlement approved).

although all are based on fear of future physical loss, the claims are really examples of claims for damages for present economic loss (i.e. the cost of medical monitoring); moreover, the fears are objectively reasonable, as they were triggered by a recall of the product in question or other concrete evidence. This approach recalls "loss of chance", which similarly recasts uncertain future damage as present loss; however, Canadian courts have tended to look at the notion as an aspect of causation, and rejected it in this context (below, no. 21). A second line of cases recognises the possibility of compensation for the emotional distress caused by fear of disease, even without proof of psychiatric or psychological symptoms.[43] Examples include fear of contacting hepatitis-B after receiving notice that they should be tested for it,[44] fear of developing cancer or other serious illness after being exposed to asbestos,[45] fear of contracting tuberculosis after learning of exposure to someone with it,[46] and fear of death after being implanted with a defective defibrillator.[47] However, these examples are all at very early stages in the procedure, at a point where the question before the courts is whether the pleadings raise a cause of action. Moreover, unlike GMOs, the feared outcome is one recognised as possible, even probable, by reliable scientific and medical opinion. As well, it is unlikely that fear of loss triggered by GMOs would meet the requirement of Quebec's Civil Code that a "future injury" must nevertheless be "certain and able to be assessed" in order to be taken into account in awarding damages.[48]

---

43 See *J. Cassels/C. Jones*, The Law of Large-Scale Claims: Product liability, mass torts and complex litigation in Canada (2005) 181 ff.

44 *Anderson v. Wilson* (1999) 175 D.L.R. (4th) 409 (Ont. C.A.) (application for certification of class action; "Given the uncertain state of the law on tort relief for nervous shock, it is not appropriate that the court should reach a conclusion at this early stage and without a complete factual foundation": par. 18), leave to appeal refused (S.C.C. 1999).

45 E.g. *Clifford Morris et al. v. Johnson Controls* (2002) 169 Manitoba Reports (Man. R.) (2d) 183 (Man. Q.B.) (motion to strike pleadings denied; American precedents a key factor: "Canada and the United States of America are now very closely connected through trade and the necessary treaties governing commerce between the two countries. The court is further prepared to take judicial notice of the fact that many employees on both sides of the border work for bi-national or multinational firms. In many instances they are covered by common insurers or at least connected through the established practice of reinsurance. To say that established law in the United States of America cannot even be considered as novel in Canada would deny the Plaintiffs in my opinion their entitled opportunity to have this area of the law, on their particular tried facts, re-examined": par. 30).

46 *Healey v. Lakeridge Health Corp. & Ross* 2006 CanLII 36247 (Ont. S.C.).

47 *LeFrançois v. Guidant Corp.* 2008 CanLII 15770 (Ont. S.C.) (class action certified; pleadings held adequate notwithstanding defendant's submission (at par. 40) that "unless and until death or a serious injury actually occurred, such a risk could not be considered to be compensable harm sufficient to support a cause of action in negligence"), leave to appeal refused (2009) 245 O.A.C. 213 (Ont. Div. Ct.).

48 Art. 1611 C.C.Q.

4. Standard of proof

14 At the stage of decision on the merits, the losses envisaged by this study, even future losses, must be established by the plaintiff under the usual standard of "balance of probabilities" or "preponderance of evidence" applied in private law actions generally.[49] However, such cases could possibly go through several preliminary stages, with differing standards of proof. In the common law provinces, the final stage of deciding on the merits could be preceded by a motion by the defendant to strike some or all of the pleadings for failure to disclose a cause of action and by an application by the plaintiff for certification of a class action (below, no. 17). Both of these apply a "plain and obvious" test,[50] a lower evidentiary threshold than on the merits, although the burden of proof falls differently in the two situations. In motions to strike, it is the defendant who must demonstrate that it is "plain and obvious" that the facts as pleaded do not disclose a cause of action (rather than the plaintiff being required to prove that they do), whereas in certification applications, the burden falls on the plaintiff to show that it is not "plain and obvious" that the pleadings do not disclose a cause of action. This requirement that the plaintiff prove what is in effect a double negative was criticised in *Hoffman*, and an arguably more stringent test was applied both at trial ("a reasonable prospect of success") and on appeal ("an apparently authentic or genuine cause of action on the basis of the facts as pleaded and the law that applies").[51] Although Saskatchewan cases now apply *Hoffman* at the certification stage, courts elsewhere continue to apply the "plain and obvious" test.[52] In Quebec, the plaintiff in a class action must demonstrate that "the facts as alleged *seem to justify* the conclusion sought", which the Supreme

49 E.g. art. 2804 C.C.Q. ("Evidence is sufficient if it renders the existence of a fact more probable than its non-existence, unless the law requires more convincing proof.").

50 *Hunt v. Carey Canada Inc.* [1990] 2 S.C.R. 956 (S.C.C.) (asbestos; motion to strike cause of action denied; stressed (at 990–1) that a claim should not be struck simply because it raises an arguable, difficult or important point of law: "Indeed, I would go so far as to suggest that where a statement of claim reveals a difficult and important point of law, it may well be critical that the action be allowed to proceed. Only in this way can we be sure that the common law in general, and the law of torts in particular, will continue to evolve to meet the legal challenges that arise in our modern industrial society.").

51 *Hoffman* trial, [2005] 7 W.W.R. 665, par. 36, *Hoffman* appeal (2007) 283 D.L.R. (4th) 190, par. 53.

52 Compare e.g. *Brooks et al v. Canadian Pacific Railway* (2007) 283 D.L.R. (4th) 540 (Sask. Q.B.) and *LeFrançois* (2009) 245 O.A.C. 213 (Ont. Div. Ct.) par. 8 ff.

Court interprets as requiring that the action raise "a serious colour of right" to be determined by the court.[53]

5. Nominal losses

In the common law provinces, nominal damages signalling an infringement of rights can be awarded for torts that are actionable without proof of damage (e.g. trespass and defamation, but not nuisance or negligence).[54] Awards of nominal damages are not possible in Quebec, however, as proof of prejudice resulting from a fault is an essential element of civil liability.[55] **15**

6. Mass losses

Litigation of mass losses is facilitated in Canada through recourse to class actions, which adapt the rules of procedure while leaving the substantive rules unchanged.[56] A leading case on class actions, *Western Canadian Shopping Centres Inc. v. Dutton*,[57] attributes the present-day interest in class actions to the rise of mass production, diversification of corporate ownership, advent of the mega-corporation, and recognition of environmental wrongs;[58] it identifies three important advantages for class actions over a multiplicity of individual suits: judicial economy, improved access to justice, and behaviour modification (by ensuring that wrongdoers do not ignore their obligations to the public).[59] More practical advantages include, for the plaintiff, the possibility of determining injury on an **16**

53 Code of Civil Procedure (C.C.P.) art. 1003(b) [emphasis added] and *Guimond v. Quebec (Attorney-General)* [1996] 3 S.C.R. 347 (S.C.C.).

54 *Osborne* (fn. 4) 283.

55 *Aubry v. Éditions Vice-Versa* [1998] 1 S.C.R. 591 (S.C.C.) par. 28.

56 See generally *Cassel/Jones* (fn. 43) 309 ff.

57 [2001] 2 S.C.R. 534 (S.C.C.). See also *Hollick v. Toronto (City)* [2001] 3 S.C.R. 158 (S.C.C.) and *Rumley v. British Columbia* [2001] 3 S.C.R. 184 (S.C.C.).

58 "A faulty product may be sold to numerous consumers. Corporate mismanagement may bring loss to a large number of shareholders. Discriminatory policies may affect entire categories of employees. Environmental pollution may have consequences for citizens all over the country. Conflicts like these pit a large group of complainants against the alleged wrongdoer. Sometimes, the complainants are identically situated *vis-à-vis* the defendants. In other cases, an important aspect of their claim is common to all complainants. The class action offers a means of efficiently resolving such disputes in a manner that is fair to all parties." (*Dutton*, [2001] 2 S.C.R. 534, par. 26).

59 *Dutton*, ibid., par. 27 ff. The Court in *Dutton* labelled the third advantage "efficiency and justice" but "behaviour modification" is now the usual term: e.g. *Pearson v. Inco Limited* (2006) 78 O.R. (3d) 641 (Ont. C.A.) (industrial pollution) par. 85 ff.

aggregated or statistical basis[60] and, for the defendant, the fact that settlements and decisions cut off their liability to those who have not opted out of the class.[61]

17 Class actions have come to the fore over the last several decades in Canada with the adoption of legislation modelled on American examples.[62] The first province to do so was Quebec, in 1978, with Ontario and British Columbia following suit in the 1990s and all but one of the remaining provinces in the 2000s.[63] Typical statutory requirements for certification of a class action are that: the pleadings disclose a cause of action; there is an identifiable class; the claims of the class members raise common issues; a class action is the preferable procedure; and the plaintiff would fairly and adequately represent the interests of the class.[64] Preferability is established by canvassing the three advantages set out in *Dutton*, and the requirement can be met even where there are substantial individual issues; in most provinces, the common issues need not predominate over the individual issues. Failure to satisfy the court of any one of the certification requirements means that the class action will not go forward. Appellate courts accord considerable deference to a trial judge's assessment of certifiability, although perhaps more so in regard to the more discretionary, procedural requirements (whether a class action is the preferable procedure and whether the plaintiff adequately represents the

---

60 E.g. *St. Lawrence Cement Inc. v. Barrette* [2008] 3 S.C.R. 392 (S.C.C.) (successful class action for compensation for property damage resulting from cement plant emissions). ("Given the trial judge's discretion and the difficulty of assessing environmental problems and annoyances, we consider [his] use of average amounts to have been reasonable and appropriate in the circumstances": par. 116).

61 E.g. *Walls v. Bayer Inc.* (2007) 217 Man. R. (2d) 66 (Man. Q.B.) (procedural motion about class action certification notice to class members).

62 Their precursor, "representative actions", had long been recognised by courts of equity, but they fell into disuse in the late 19th century, partially because of the stricter approach to them following the fusion of law and equity and partly because of the growth of the limited liability company (with legal personality): *Dutton* [2001] 2 S.C.R. 534, par. 19 ff.

63 See *Cassel/Jones* (fn. 43) 309 f. and S. *Penney*, Mass Torts, Mass Culture: Canadian mass tort law and Hollywood film, Queen's Law Journal 30 (2004–05) 205, 216 f. The most recent legislation is Nova Scotia's (Class Proceedings Act, S.N.S. 2007, c. 28) and now only the smallest province, Prince Edward Island, is without a separate act.

64 Drawn from Saskatchewan's Class Actions Act, S.S. 2001, c. C-12.01, s. 6. (*Hoffman* [2005] 7 W.W.R. 665) was the first case to be decided under this Act). In Quebec, the authorisation requirements are that: the claims raise "identical, similar or related questions of fact or law", the facts alleged "seem to justify the conclusions sought", the composition of the group makes the application of the usual rules for bringing an action "difficult or impracticable to apply", and the proposed representative "is in a position to represent the members adequately": art. 1003 C.C.P.

class) than to the more substantive requirements.[65] Many of the cases discussed in this report are class actions.

III. Causation

1. Uncertainty of merely potential causes

18 Canadian common and civil law both apply basically the same approach to proving causation in negligence or delict: the first hurdle, causation in fact, is assessed under the "but for" (or necessary condition) test and the second hurdle, causation in law, requires foreseeability.[66] The first asks whether, on the balance of probabilities, the injury would not have occurred but for the defendant's negligence.[67] This test is applied even where the defendant's actions are not the sole cause of the plaintiff's injury: if the other causes are not wrongful, the defendant will be solely liable for the entire loss; where the other causes are wrongful, there can be apportionment (i.e. claims over) between the defendants (below, no. 50); and where the plaintiff's own actions contribute to the loss, the rules of contributory fault apply (below, no. 20). The test is also applied, in principle, even when meeting it is difficult for the plaintiffs, such as cases involving long latency periods, multiple defendants and scientific uncertainty about possible causes. This is the subject of on-going controversy in Canada, and the Supreme Court of Canada has recognised on several occasions that it might be prepared to apply a less-demanding approach, based on "material contribution", when application of the usual test was

---

65 *Soldier v. Canada (Attorney General)* [2009] 4 W.W.R. 445 (Man. C.A.) (aboriginal treaty rights) par. 20 ff.; see also *Anderson* (1999) 175 D.L.R. (4th) 409 and *Bayer Inc. v. Pardy* (2005) 246 Nfld. & P.E.I.R. 157 (Nfld & Lab. C.A.). Note that under the Quebec legislation, appeal of certification is asymmetrical in the sense that the plaintiff may appeal a refusal to authorise but the defendant may not appeal an authorisation: art. 1010 C.C.P.

66 *Baudouin/Deslauriers* (fn. 4) par. 1–627 sums up the Quebec approach in the following question: "[Q]uels sont les faits qui rendaient objectivement possible la création de préjudice, et dont les conséquences étaient normalement prévisible pour l'agent?".

67 *Athey v. Leonati* [1996] 3 S.C.R. 458 (S.C.C.); *Brierley/Macdonald* (fn. 4) 440 ("showing the causal link between the damage and the wrongful act of the defendant . . . means convincing the court, on the balance of probabilities, that the delict or quasi-delict was a necessary condition for the realization of the damage"). For cogent criticism of continued recourse to the "but for" test, see *Cassel/Jones* (fn. 43) 203 ff. They discuss alternatives such as probabilistic discounting, enterprise theory and market share, burden-shifting, loss of chance and risk-based liability, and suggest (at 204) that "dealing with large-scale claims on a probabilistic basis may require more than procedural innovation and advanced methods of proof of causation. In fact, it may require that we re-examine the necessity of establishing precise causal connection between plaintiff and defendant *at all.*" [emphasis in original].

unworkable (not simply difficult).[68] In *Resurfice Corp. v. Hanke*,[69] the Court stressed that in order for a court to be entitled to apply the "material contribution" test, two requirements must be met: firstly, it must be impossible for the plaintiff to prove the defendant's negligence using the "but for" test, and this impossibility must be due to factors outside the plaintiff's control (such as current limits of scientific knowledge); and secondly, "it must be clear that the defendant breached a duty of care owed to the plaintiff, thereby exposing the plaintiff to an unreasonable risk of injury, and the plaintiff must have suffered that form of injury. In other words, the plaintiff's injury must fall within the ambit of the risk created by the defendant's breach." In these exceptional circumstances, "it would offend basic notions of fairness and justice to deny liability by applying a 'but for' approach."[70] The Court gave two examples of such exceptional circumstances: one is "where it was impossible to say which of two tortious sources caused the injury"[71] and the other is "where it is impossible to prove what a particular person in the causal chain would have done had the defendant not committed a negligent act or omission."[72] In such situations, both the common and civil law in Canada make all defendants equally liable for the damage caused (below, no. 50). However, the Supreme Court recently refused leave to appeal a Quebec Court of Appeal decision on the merits in a class action against a number of municipalities for their prolonged failure to control the presence of ragweed, as required by regulation.[73] The Court of Appeal agreed with the trial judge's conclusion that the long distances over which air-blown ragweed pollen travels made it impossible for the plaintiff to establish "on a preponderance of proof that the colonies of ragweed in flower discovered after August, in each of the Defendants' territories, contributed to the regional pollen count . . . to a degree that was *mate-*

68 The "material contribution" approach has been applied by the British House of Lords (whose decisions have strong persuasive authority in common law Canada) in *Fairchild v. Glenhaven Funeral Services Ltd.* [2002] 3 Weekly Law Reports (W.L.R.) 89 and *Barker v. Corus (UK) Ltd.* [2006] 2 A.C. 572.

69 (2007) 278 D.L.R. (4th) 643 (S.C.C.).

70 Par. 25.

71 Par. 27, citing *Cook v. Lewis* [1951] S.C.R. 830 (S.C.C.) (hunting accident; impossible to say which of two carelessly fired shots injured victim).

72 Par. 28, citing *Walker Estate v. York Finch General Hospital* [2001] 1 S.C.R. 647 (S.C.C.) (HIV-tainted blood; impossible to prove that donor would not have given blood if had been properly warned against doing so).

73 *Nadon v. Montreal (Ville de)* [2008] 10 Recueil de Jurisprudence du Québec (R.J.Q.) 2600 (Que. C.A.), leave to appeal refused 2009 CanLII 19885 (S.C.C.). Note, however, that in *Western Bank v. Alberta* [2007] 2 S.C.R. 3, par. 8, the Supreme Court recently clarified that "refusal of leave should not be taken to indicate agreement with the judgment sought to be appealed from anymore than the grant of leave can be taken to indicate disagreement. In the leave process, the Court does not hear or adjudicate a case on the merits.".

rial to the suffering of *most* members of the class anywhere within [the region], i.e. to a degree that if removed would prevent or appreciably reduce their suffering."[74] The plaintiff's difficulties in proving causation in the case are very similar to those that could arise in a case involving GMOs.

The second hurdle, foreseeability, is the means by which the liability of defendants for the consequences caused in fact by their wrongful actions is limited to those that are not too remote. At the risk of over-simplification, remoteness depends on foreseeability of the type of injury that occurred, and foreseeability depends on the state of knowledge at the time of injury. As Khoury and Smyth point out,[75] there is little uncertainty about the possibility of property and financial injury being caused by release of GM genes into the environment, as their propensity to wander is now accepted. However, there is considerable scientific uncertainty about the possible impacts of GMOs on human health and on the ecosystem, and this could be very real impediment to pursuing a tort action for these sorts of injuries, particularly because the doctrine of remoteness and foreseeability was introduced into the law of torts to limit an otherwise potentially unlimited liability of defendants. Khoury and Smyth therefore advocate importing the precautionary principle into the private law remoteness test as a means of redressing the evidentiary balance when dealing with uncertainty. **19**

Finally, as for contributory fault (i.e. causes lying within the victim's own sphere), all Canadian provinces now provide for apportionment of liability between victim and wrongdoer based on their degree of fault. In common law Canada, this required legislative reform reversing the traditional common law rule that a plaintiff's contributory negligence, no matter how minor, served as a complete defence to an action.[76] In Quebec, on the other hand, it was the courts which opted to follow the civilian model of shared responsibility over the common law approach, a choice which was codified **20**

74 *Nadon* [2008] 10 R.J.Q. 2600, par. 44, quoting par. 316 of trial judgment [emphasis in original].

75 *L. Khoury/S. Smith*, Reasonable Foreseeability and Liability in Relation to Genetically Modified Organisms, Bull. Sci., Tech. & Soc'y 27 (2007) 215, 224 ff.

76 Because courts did not want to be used by tortfeasors to obtain relief from the consequences of their own wrongdoing, the same reason that they did not permit actions for contributions between multiple tortfeasors (below, no. 50): *Klar* (fn. 4) 550. Most provinces have a single statute dealing both with contributory negligence and with apportionment between multiple tortfeasors, but some have separate statutes. See e.g. Contributory Negligence Act, Revised Statutes of Nova Scotia (R.S.N.S.) 1989, c. 471; Tortfeasors and Contributory Negligence Act, R.S.M. 1987, c. T90 (Manitoba); and Negligence Act, R.S.O. 1990, c. N.1 (Ontario).

in 1994 when the Code was revised.[77] Canadian courts thus now determine the degree to which the plaintiff is responsible for the harm suffered and reduce the damages accordingly; if this cannot be determined, both parties are deemed to be equally at fault. One author describes the apportionment process as being "conventional, arbitrary and lenient" – conventional because the courts apply similar rough-and-ready reductions (20%, 30% or 50%), arbitrary because these rough-and-ready reductions do not particularly reflect degrees of fault, and lenient both as to the degree of reduction of damages (rarely over 50% and often less than 30%) and the manner in which the standard of care is applied.[78]

2. Complex causation scenarios

21 Other than the "material contribution" approach (above, no. 18), there are presently no special rules applied in more complex cases to determine causation. This possibility was considered at length by the Supreme Court of Canada in *Snell v. Farrell*,[79] with the Court acknowledging a concern that the traditional approach to causation was inadequate in complex cases. "This concern is strongest in circumstances in which, on the basis of some percentage of statistical probability, the plaintiff is the likely victim of the combined tortious conduct of a number of defendants, but cannot prove causation against a specific defendant or defendants on the basis of particularised evidence in accordance with traditional principles".[80] The Court considered a number of possible alternative approaches, including apportionment of liability among defendant manufacturers on the basis of market share and reversing the burden of proof, but in the end held that the traditional approach, properly applied, was adequate to the task. However, it indicated openness to alternate approaches if need be: "If I were convinced that defendants who have a substantial connection to the injury were escaping liability because plaintiffs cannot prove causation under currently applied principles, I would not hesitate to adopt one of these alternatives".[81] "Loss of chance" (above, no. 13) is another possible causal tool in complex cases, but it was rejected as such by the Supreme

77 Art. 1478, par. 2. See *Baudouin/Deslauriers* (fn. 4) par. 1–654. On revision of the Code, see above (fn. 2).

78 *Osborne* (fn. 4) 106. The leniency towards plaintiffs is attributed to the effect of insurance, particularly the comparative rarity of first-party insurance.

79 [1990] 2 S.C.R. 311 (S.C.C.) (cataract operation medical liability case; rejecting House of Lord's decision in *McGhee v. National Coal Board* [1973] 1 W.L.R. 1).

80 *Snell* [1990] 2 S.C.R. 320.

81 Ibid., 326 f.

Court of Canada in *Laferrière v. Lawson*,[82] a Quebec medical liability case decided the year after *Snell*, in favour of the traditional causal analysis. On the other hand, "market share" has been accepted as an arguable alternative at a preliminary motion stage of a product liability class action, *Gariepy v. Shell Oil Co.*,[83] and adopted by statute in a number of provinces in the context of their on-going litigation against tobacco companies to recover the cost of health care (above, no. 7).

3. Force majeure

The position in both the common law provinces and Quebec is that proof of force majeure (or "act of God") would assist the defendant by negating a causal link between his actions and the plaintiff's injury, provided the intervening natural event was so "unforeseeable and irresistible"[84] that the defendant could not be expected to guard against it. 22

4. Threshold to prove causation

The requirement throughout Canada of proof "on the balance of probabilities" implies that the comparable threshold is simply "more likely than not" (i.e. a 51% threshold) rather than a more stringent requirement. This was stressed by the Supreme Court of Canada in *Snell v. Farrell*, in which the Court observed that causation "need not be determined by scientific precision" and is "essentially a practical question of fact which can best be answered by ordinary common sense rather than abstract metaphysical enquiry".[85] 23

---

82 [1991] 1 S.C.R. 541 (S.C.C.) (doctor negligently failed to report tumour as malignant; plaintiff therefore lost chance to undertake aggressive treatment and ultimately died of cancer). See *Klar* (fn. 4) 451 ff., *Baudouin/Deslauriers* (fn. 4) par. 1–351 ff., and *L. Khoury*, Causation and Health in Medical, Environmental and Product Liability, Windsor Yearbook of Access to Justice 25 (2007) 135 ff. [comparing French and Quebec law].

83 (2000) 51 O.R. (3d) 181 (Ont. S.C) (plumbing pipe resin), citing U.S. cases including *Sindell v. Abbott Laboratories*, 26 Cal. 3d 588 (1980). (For further proceedings, see *Gariepy 2* (2002) 23 Carswell's Practice Cases (C.P.C.) (5th) 360 (Ont. S.C.)). See also *Ragoonanan Estate v. Imperial Tobacco Canada Ltd* (2000) 51 O.R. (3d) 603 (Ont. S.C.), affirmed (2008) 236 O.A.C. 199 (Ont. Div. Ct.) (application for certification of class action denied, but for reasons other than lack of arguable cause of action; trial judge suggesting that, had particular manufacturer of cigarette causing damage not been known, plaintiff could have proceeded against all manufacturers on market-share basis for failure to market "fire safe" cigarette); see also below (fn. 223) (costs motion).

84 Art. 1470 C.C.Q.

85 *Snell* [1990] 2 S.C.R. 311, citing U.K case of *Alphacell Ltd. v. Woodward* [1972] 2 All England Reports (All E.R.) 475 (H.L.) 490.

5. Special rules on causation

24 None that are not mentioned in response to other questions.

IV. Types of liability

1. Fault liability

(a) Special rules governing fault

25 Fault liability rules are not treated in a special way in the type of cases envisaged here. The burden of proof is reversed on occasion in the Quebec Civil Code (e.g. responsibility for autonomous act of things, below no. 45) but in *Snell v. Farrell,*[86] the Supreme Court of Canada disapproved of the idea of reversing the burden to facilitate proof in difficult cases. Under the adversarial approach adopted by all Canadian courts, including those of Quebec, the general rule of "he who alleges must prove" applies, so that the burden is on the plaintiff to prove the claim and on the defendant to prove any defences to it. In the context of an *Anns* argument (above, no. 5–6, 9), for example, the plaintiff must prove the existence of a first-stage prima facie duty of care, and the defendant must demonstrate the existence of second-stage policy considerations rebutting the duty of care. This is sometimes described as shifting the evidentiary (or tactical) burden of proof, but the Court stressed in *Snell* that this description is not strictly accurate. It is simply a matter of weighing the evidence: the legal or ultimate burden remains with the plaintiff throughout; the strength of the plaintiff's evidence may be sufficient to give rise to an inference being drawn adverse to the defendant unless the defendant adduces evidence to the contrary; and where the defendant has superior knowledge of the facts (as in malpractice actions and perhaps also GMO cases), very little evidence need be adduced by the plaintiff to give rise to an inference of causation in the absence of evidence to the contrary.[87]

26 Nor has the standard of care been altered. It is the care that a reasonable person[88] would take when faced with a given risk. Who a reasonable per-

86 *Snell* [1990] 2 S.C.R. 311.

87 This common sense approach is very close to the common law maxim res ipsa loquitur, discarded by the Supreme Court of Canada in *Fontaine v. British Columbia (Official Administrator)* [1998] 1 S.C.R. 424.

88 Sometimes described, more vividly, in the civil law as "un bon père de famille" and at common law as "the man on the Clapham omnibus", with a suggested Canadian variant being "the person on the Yonge Street subway": *Linden/Felthusen* (fn. 4) 141.

son is, is in principle determined objectively, although the standard is relaxed for persons of limited knowledge and capacity (e.g. children, mentally ill) and tightened for those of superior knowledge and capacity (e.g. professionals, such as doctors and lawyers, and presumably also biotechnology companies). Assessment of risk involves balancing probability of loss (i.e. greater risk of injury) and extent of loss (i.e. risk of greater injury) against the object of the activity (i.e. its social utility) and cost (i.e. the ease or difficulty with which the risk can be avoided). Applying these principles to GMOs would probably involve looking closely at different types of GMOs. The present generation of pesticide-resistant GMOs would likely fare badly, as the probability and extent of loss to organic and conventional crops and to the environment is high, whereas the social utility (convenience to farmers) is low and the risk can be easily avoided (by not using GM technology). How the next generation of pharmaceutically and nutritionally enhanced or drought-resistant GMOs would fare is more open: little is known about the probability of loss, but should loss appear, it could be extensive; such GMOs would have a higher social utility than pesticide resistant ones, but the risk of harm to some could be avoided by improving people's nutrition and health in more traditional ways.

(b) Impact of specific rules of conduct

In the absence of a specific statutory or regulatory provision imposing (or **27** perhaps excusing) civil liability, breach of a government-imposed rule is simply one of the factors a court would weigh to determine the existence of fault-based liability; it is neither proof of negligence per se nor does it raise a presumption of negligence requiring the defendant to disprove its existence. This was decided by the Supreme Court of Canada in *Canada v. Saskatchewan Wheat Pool*,[89] which clarified the interrelationship between specific statutory rules and fault-based liability (i.e. "breach of statutory duty"), an area of the law that previously consisted of "arcane and obfuscative rules that few could articulate, let alone deal with on a basis of principle".[90] The notion of "statutory duty" is to be understood widely: it is not restricted to federal and provincial legislation, but also encompasses subsidiary regulations, municipal by-laws, sports federation by-laws, and the like.[91] Customary rules, such as good farming practices, were outside the

89 [1983] 1 S.C.R. 205 (S.C.C.). See also *Galaske v. O'Donnell* [1994] 1 S.C.R. 670 (S.C.C.).
90 *Cassels/Jones* (fn. 43) 105.
91 Ibid., 106.

scope of the old "breach of statutory duty" rules; they would always have been a relevant factor to be weighed as indicia of negligence, just as more formal government-imposed rules now are since *Saskatchewan Wheat Pool.*[92]

28 It should perhaps be noted that it is difficult to establish whether or not there are government-imposed set-back requirements for the cultivation of GM crops in Canada. If there are any, they are not posted on the government website in any readily accessible way. Perhaps they are imposed as conditions on the individual authorisation for unconfined release, as a recent newspaper article suggests.[93] There are also some industry-imposed standards.[94] However, the pleadings in *Hoffman* indicate that set-backs are not respected by GM farmers: "*Any* isolation distance will reduce the amount of cross-pollination occurring between fields. The Plaintiffs state that having *no* buffer strips greatly compounded the contamination of the environment."[95] The plaintiffs claimed that the biotechnology companies are under a duty to advise GM farmers of the need for a buffer strip.

2. Product liability

(a) Development risk defence

29 The Quebec civil law approach to product liability is more similar to Europe than to the common law provinces (below, no. 30). The general rules of fault-based delictual liability applied until the Code was revised in the early 1990s, at which time a stricter approach influenced by the Product Liability Directive was adopted. The Code now makes manufacturers (as well as distributors and suppliers: below, no. 48) liable for injury caused "by reason of a safety defect in the thing", with safety defects being defined to include manufacturing, design and labelling defects.[96] These provisions do not apply to agricultural produce unless it has been trans-

---

92 *Linden/Feldthusen* (fn. 4) 216 ff.

93 M. *Lalonde*, Feds okay genetically modified corn: Risk evaluation not done, The [Montreal] Gazette, 25 July 2009, A10 ("The CFIA has also conditionally authorized for SmartStax a reduction in the size of the buffer zone, or "refuge", normally required around genetically engineered corn.").

94 For example, the Canadian Seed Growers Association recommends a buffer strip of 600 metres for GM canola, and some organic standards require three times that distance: *Hoffman* trial, [2005] 7 W.W.R. 665, par. 49.

95 Ibid. [emphasis in original].

96 Art. 1468 and 1469 C.C.Q. ("where having regard to all the circumstances, it does not afford the safety which a person is normally entitled to expect, particularly by reason of a defect in the design or manufacture ..., poor preservation or presentation ..., or the lack of sufficient indications as to the risk and dangers it involves or as to safety precautions").

formed in some way.[97] The Code also incorporates the development risk defence by excusing liability if it can be proved that "according to the state of knowledge at the time that he manufactured, distributed or supplied the property, the existence of the defect could not have been known, and that he was not neglectful of his duty to provide information when he became aware of the defect".[98] As the common law adheres to a fault-based approach for product liability, the state of the defendant's knowledge at a given time has been held relevant to determining whether he acted reasonably.[99]

(b) Alternative routes

The product liability regime applied in common law Canada is not strict **30**
liability but rather fault-based.[100] Duty and damage are approached in the same way as in any negligence case; as under civil law, breach is defined to include defects in manufacturing (i.e. when the product departs dangerously from the manufacturer's specifications), design (i.e. when the manufacturer's specifications themselves create excessive risk of injury) and labelling (i.e. when the manufacturer fails to warn of hidden dangers inherent in intended or reasonably foreseeable uses of the product[101]). A GMO example of a manufacturing defect might be the unintended presence of GMOs in food or feed products sold as GM-free; circumstantial evidence of this presence would raise a presumption of negligence, but it could be open to the manufacturer to rebut the presumption by showing that it had taken all reasonable precautions to prevent admixture.[102] In

97 *C. Masse*, La responsibilité civile, in: Barreau du Québec et Chambre des notaires du Québec (eds.), La réforme du Code civil: Obligations, contrats nommés (1993) 235 ff., 301.

98 Art. 1473 C.C.Q., par. 2. (Par. 1 excuses the defendants on proof that "the victim knew or could have known of the defect, or could have foreseen the defect").

99 E.g. *Endean v. Canadian Red Cross Society* (1997) 148 D.L.R. (4th) 158 (B.C.S.C.) (tainted blood; class action certified), settlement approved [2000] 1 W.W.R. 688 (B.C.S.C.); see also (1998) 157 D.L.R. (4th) 465 (B.C.C.A.) (no tort of spoliation; destruction of documents a procedural wrong, not an independent cause of action). See also *Berryman*, I.C.L.Q. 53 (2004) 584 ("Reasonable care will be determined against the knowledge known at the time the product was manufactured. . ..This is an approach peculiar to negligence and marks the difference with strict liability where the state of a shifting level of knowledge would be irrelevant to determine liability.").

100 See generally *S.M. Waddams*, Products Liability (4th ed. 2002). See also *Cassels/Jones* (fn. 43) 95 ff.; *Berryman*, I.C.L.Q. 53 (2004) 582.

101 The leading case on duty to warn is *Lambert v. Lastoplex Chemicals Co.* [1972] 2 S.C.R. 569 (S.C.C.) (highly flammable sealant).

102 The question of a manufacturer's liability for defective components raises questions of vicarious liability (below, no. 46). See also *Osborne* (fn. 4) 134, who points out that common law Canadian reliance on negligence creates "a technical inconsistency in pro-

assessing design defects, the common law courts favour a "risk-utility" approach that asks, "Do the potential risks of the design significantly outweigh its utility?" Relevant considerations include the utility of the product to the public as a whole and to the consumer, the likelihood of harm, the availability of a safer design, the cost (both in terms of functionality and price) of the safer design, the ability of the consumer to avoid harm by careful use of the product, the ability of the consumer to become aware of the risks, and the manufacturer's ability to spread the costs related to improving the safety of the design.[103] Examples of design defects include asbestos insulation,[104] pesticides and herbicides,[105] silicone gel breast implants,[106] cigarettes,[107] and diet pills[108] – and there is no reason in principle why GMOs should not find their way onto this list, at least first-generation pesticide-resistant GMOs. Labelling defects was one of the issues in *Hoffman*, in which Monsanto was strongly criticised in the pleadings for failing to warn GM farmers of the need to retain adequate set-backs to protect the crops of non-GM farmers (above, no. 28). The propensity of GM genes to wander and cross-fertilise is a recognised hidden danger, but whether the mere presence of GMOs in food or feed constitutes a hidden danger, giving rise to a duty to warn, is an open point under the present state of scientific knowledge. The Canadian government obviously thinks GMOs are not a danger to health, as it has consistently rejected calls for mandatory labelling of GM products in favour of voluntary labelling (which in fact does not happen).[109] Mandatory label-

ducts liability law" between the strict liability under contract law of a vendor to a purchaser (an implied warranty that the goods are free from defect) and the reasonable care standard imposed on manufacturers. Presumptions of negligence based on circumstantial evidence (the former res ipsa loquitur [fn. 87]) attenuated but did not eliminate this inconsistency.

103 *Klar* (fn. 4) 369 f. These considerations and the "risk-utility" approach generally, recall the factors taken into account in assessing risk under standard of care (above, no. 26). For an example of the "risk-utility" approach, see *Ragoonanan Estate* (2000) 51 O.R. (3d) 603.

104 E.g. *Canadian Indemnity Co. v. Canadian Johns-Manville Co.* [1990] 2 S.C.R. 549 (S.C.C.).

105 E.g. *Van Oirschot v. Dow Chemical Canada Inc.* (1993) 31 Alberta Law Reports (Alta. L.R.) (3d) 212 (Alta. C.A.).

106 E.g. *Hollis v. Dow Corning Corp.* [1995] 4 S.C.R. 634 (S.C.C.).

107 E.g. *Spasic Estate v. Imperial Tobacco Ltd.* (2000) 188 D.L.R. (4th) 577 (Ont. C.A.).

108 E.g. *Wilson* (2000) 50 O.R. (3d) 219.

109 Canadian General Standards Board, Voluntary Labelling and Advertising of Foods that Are and Are Not Products of Genetic Engineering, Doc. CAN/CGSB-32.315-2004. See generally *M. Vreeman*, Labelling Policy for GM Foods: Pragmatism in action or policy failure? Current Agriculture, Food and Resource Issues 4 (2003) 107 ff.; *G.P. Gruère*, A Preliminary Comparison of the Retail Level Effects of Genetically Modified Food Labelling Policies in Canada and France, Food Policy 31 (2006) 148 ff. See also *Matthews Glenn*, Washburn Law Journal 43 (2004) 571 ff.

ling is supported overwhelmingly by the public, but is opposed by both the biotechnology companies, who fear the loss of market share, and the food industry, which does not want the responsibility of tracing GMOs into the finished product. This opposition might be used against them, depending on their state of knowledge at any given time, should the ingestion of GMOs prove to have adverse health consequences.[110]

(c) Impact of compliance with rules and regulations

Proof of compliance with all rules and regulations does not protect a defendant from liability any more than proof of breach requires a finding of wrongdoing (above, no. 27). Breach of statutory rules and compliance with them are thus, in a sense, two sides of the same coin, with breach or compliance being but one of the factors to be balanced in determining whether a defendant has acted wrongfully. As the Supreme Court of Canada put it in the leading case of *Ryan v. Victoria (City)*, "mere compliance with a statute does not, in and of itself, preclude a finding of civil liability. Statutory standards can, however, be highly relevant to the assessment of reasonable conduct in a particular case. ... This allows courts to consider the legislative framework in which people and companies must operate, while at the same time recognising that one cannot avoid the underlying obligation of reasonable care simply by discharging statutory duties."[111] In assessing reasonableness, courts will weigh such factors as whether the statutory standards were drafted with the circumstances of the case in mind, whether the standards are specific or general, and whether they allow for discretion in their manner of performance.[112] *Ryan* was applied in the "mad cow" case, *Sauer*,[113] to counter the feed manufacturer's argument that it had complied with the applicable regulations allowing inclu- **31**

110 As *Linden/Felthusen* (fn. 4) 647 put it, "The courts have rightly shown no sympathy at all for the manufacturers' complaint that 'sales would be prejudicially affected' by clear warnings." And *Berryman*, I.C.L.Q. 53 (2004) 583 notes that the duty to warn is subject to a higher degree of care for products that are ingested or otherwise consumed by humans.

111 [1999] 1 S.C.R. 201 (S.C.C.) par. 29 (injury from being thrown from motorcycle when wheel caught in rail tracks).

112 Ibid., par. 39 f. ("Where a statute authorizes certain activities and strictly defines the manner of performance and the precautions to be taken, it is more likely to be found that compliance with the statute constitutes reasonable care and that no additional measures are required. By contrast, where a statute is general or permits discretion as to the manner of performance, or where unusual circumstances exist which are not clearly within the scope of the statute, mere compliance is unlikely to exhaust the standard of care.").

113 *Sauer* appeal (2007) 225 O.A.C. 145.

sion of ruminant meat and bone meal in cattle feed. That compliance with existing rules and regulations does not negate a finding of wrongdoing was recently reaffirmed by the Supreme Court of Canada as the position under Quebec law in *St. Lawrence Cement v. Barrette.*[114] Indeed, the case seems to go even further, and require statutory immunity, not just statutory authority, to excuse a defendant (below, no. 51).

3. Environmental liability

(a) Implementation of the Environmental Liability Directive

32 N/A.

(b) Environmental liability regime beyond the scope of the Directive

33 N/A.

(c) Claimants in cases of environmental harm

34 There is nothing similar in Canada to the Environmental Liability Directive, dealing specifically with GMO-caused harm to the environment. In theory, no harm should arise where the GMOs have been approved for release and the approval procedure has adequately assessed their environmental impact.[115] The Canadian approval process compares GM plants to their conventional counterparts for: their potential to become an agricultural weed or be invasive of natural habitats; their potential for gene flow to wild relatives whose hybrid offspring may become more weedy or more invasive; their potential to become a plant pest; their potential impact on non-target species, including humans; and their potential impact on biodiversity. It typically concludes (from the example of canola) that the GM plant “does not present altered environmental interactions when compared to currently commercialized canola varieties”.[116] However, the approval procedure has

---

114 [2008] 3 S.C.R. 392, par. 34 ff., citing *Morin v. Blais* [1977] 1 S.C.R. 570 (S.C.C.) and contrasting Quebec and French law on this point.

115 See C. *Rodgers*, Biotechnology, Property Rights and Unknown Risk: What role for environmental liability? in: L. Bodiguel/M. Cardwell (eds.), The Regulation of Genetically Modified Organisms: Comparative approaches (Oxford University Press, forthcoming).

116 See e.g. *Hoffman* trial [2005] 7 W.W.R. 665, par. 13–18.

been criticised on a number of counts (above, no. 6) – including its narrow, agronomic approach to the environment – and doubts about the wider environmental effects of even approved GMOs remain.[117]

One possibility for recovery for environmental loss is actions by private individuals, and this is recognised either in specific statutes or more generally. The legislative examples provide discrete statutory causes of action independent of negligence, and the plaintiff is thus not required to prove fault but only breach of statute, causation and damages.[118] In lieu of a comprehensive overview of all provincial and federal environmental legislation, examples drawn from two provinces – Saskatchewan and Quebec – give a flavour of the sort of provisions that might be found.[119] Saskatchewan's example is the Environmental Management and Protection Act, discussed in *Hoffman*, which gives "any person" a right to compensation from "the person responsible for a discharge" for loss or damage incurred as a result of "the discharge of a substance" (e.g. GMOs) into the environment; "loss or damage" is defined widely to include "loss of use or enjoyment of property" as well as "pecuniary loss, including loss of income".[120] The trial judge accepted that this provision could have constituted a possible cause of action had the other requirements for certification of the class action been met, although the Court of Appeal disagreed in the circumstances.[121] Quebec's examples are drawn from two different statutes. The earlier provision, adopted in 1978 as an amendment to the Environment Quality Act, provides "any natural person" with the right to apply for an injunction to prohibit acts or operations interfering with their "right to a healthy environment and to its protection, and to the protection of the living species inhabiting it, to the extent provided for by this Act" and any regulations, etc., adopted under it; the person in question need not have suffered any personal loss or damage, but need only be a resident of the province "frequenting a place or the immediate vicinity of a place in 35

117 Such as the recent approval of Monsanto's "SmartStax" GM corn without an environmental assessment: *Lalonde*, The [Montreal] Gazette, 25 July 2009, A10. An assessment was not thought necessary because each of the eight different genetically engineered herbicide-and-pesticide-resistant genes combined in SmartStax had already been approved individually (albeit on the basis of substantial equivalence with conventional corn).

118 *Osborne* (fn. 4) 162.

119 See generally *E.L. Hughes/D. Iyalomhe*, Substantive Environmental Rights in Canada, Ottawa Law Review 30 (1998–1999) 229 ff.

120 S.S. 2002, c. E-10.21, s. 15. See generally *J. Kelly Brown*, Contaminated Site Liability in Saskatchewan: On the "right track" to remediation? J. Envtl L. & Prac. 12 (2003) 55 ff.

121 *Hoffman* trial, [2005] 7 W.W.R. 665, par. 168, *Hoffman* appeal (2007) 283 D.L.R. (4th) 190, par. 66 f.

respect of which a contravention is alleged".[122] The more recent provision, adopted in 2006 as an amendment to the economic and social rights chapter of the Quebec Charter of Human Rights and Freedoms, recognises that "every person has a right to live in a healthy environment in which biodiversity is preserved, to the extent and according to the standards provided by law".[123] The remedies available under this statute are wider than under the environmental statute, and can include awards of damages and mandatory injunctions as well as ordinary injunctions; standing to sue, on the other hand, is narrower, as the person must have a personal interest in the outcome of the litigation rather than being someone who simply frequents the place or its immediate vicinity. However, the environmental rights recognised under both Quebec statutes are not open and free-standing, but are rather limited to the right to enforce otherwise existing rights.[124] These provisions have been invoked, sometimes successfully, in a number of cases.[125] More generally, a non-statutory possibility open to private individuals is a claim in nuisance by a landowner for compensation for environmental harm caused to his land, as the Supreme Court of Canada recognised in *British Columbia v. Canadian Forest Products Ltd.*[126] (below, no. 36, 57).

36 Another possibility is action by a public authority, and Canadian law recognises that the Attorney-General of a province, as defender of the public interest (i.e. as parens patriae), can bring a suit in public nuisance for harm to the environment.[127] A public nuisance is an activity that constitutes an unreasonable interference with the public interest through either damage to public property or to the comfort and sensibilities of a suffi-

---

122 R.S.Q., c. Q-2, ss. 19.1–19.7. See generally *L. Giroux*, La Loi sur la qualité de l'environnement: Grands mécanismes et recours civils, in: Développements récents en droit de l'environnement (1996) 263 ff.

123 R.S.Q., c. C-12, s. 46.1. This amendment was added by s. 19 of the Sustainable Development Act, S.Q. 2006, c. 3. See generally P. *Halley*, L'Avant-projet de loi sur le développement durable du Québec, McGill International Journal of Sustainable Development Law and Policy 1 (2005) 59 ff.

124 The provision of the Charter is wider in this regard, as it requires only that the right be recognised "by law", whereas the Environmental Quality Act limits the protection to rights recognised in and under the Act (as well as under a provision in the Act respecting Land Use Planning and Development, R.S.Q., c. A-19.1, relating to agricultural odours).

125 E.g. *Nadon v. Anjou (Ville)* [1994] (R.J.Q.) 1823 (Que. C.A.); *Regroupement des citoyens du quartier St-Georges v. Alcoa* [2007] 6 R.J.Q. 1581 (Que. S.C.).

126 [2004] 2 S.C.R. 74 (S.C.C.).

127 The fact that it is provincial Attorneys-General who bring public nuisance actions is an advantage in the context of GMOs, as they are more likely to complain of environmental damage from GMOs than their federal counterpart, since it is a federal authority that approves them for release in the first place.

ciently large class of the public, and the usual list of protected interests includes health, safety, morality, comfort or convenience.[128] Including "the environment" on the list, as the Supreme Court accepted in *Canadian Forest Products*, widens the reach of public nuisance. *Canadian Forest Products* was an action brought by the province against a logging company for damage caused by a carelessly caused forest fire, in which the Crown claimed both as landowner and as parens patriae for the environmental loss of trees in protected areas as well as those in harvestable areas. "The result has been serious *physical* damage to 1,491 hectares of formerly green forests. One could reasonably anticipate that the environmental impact, apart from 'diminution of the value of the timber', was also significant. Erosion problems have likely been aggravated. Fish habitat likely threatened. Water supply to the local community to some extent degraded. Forest vistas replaced with the skeletons of blackened trees."[129] The Court accepted that the Crown could bring an action to enforce "the public interest in an unspoiled environment", for "[i]f the Crown cannot do so, who (if anyone) can?", and stressed the "deep roots in the common law" of the notion that there are public rights to the environment that reside in the Crown.[130] The Court also accepted that a public nuisance can be remedied by an award of damages, and not just by an injunction as is usually the case. It recognised that such actions raise "important and novel policy questions", including "the Crown's potential liability for *inactivity* in the face of threats to the environment, the existence or non-existence of enforceable fiduciary duties owed to the public by the Crown in that regard, the limits to the role and function and remedies available to governments taking action on account of activity harmful to public enjoyment of public resources, and the spectre of imposing on private interests an indeterminate liability for an indeterminate amount of money for ecological or environmental damage."[131] In the end, a majority of the Court held that this was not a proper case to consider such difficult issues, as damages had been pleaded in commercial, not environmental, terms (below, no. 57).

128 E.g. *Ryan* [1999] 1 S.C.R. 201, par. 52 (with the Supreme Court describing public nuisance as "a poorly understood area of the law").
129 *Canadian Forest Products* [2004] 2 S.C.R. 74, par. 57 [emphasis in original].
130 Ibid., par. 64, 74 ff. (majority) and 158 (minority).
131 Ibid., par. 81 [emphasis in original].

(d) Special liability regime for losses sustained by individuals

37 Public nuisances are actionable by private individuals (and not just the Attorney-General) where they have suffered damages that are unique to them, different from the harm suffered by members of the public. This "special damages" requirement is easily met where the plaintiff can show personal injury or property damage (even if many others show similar injuries). It is more difficult to show where the damage is less tangible, such as pure economic loss, personal discomfort or inconvenience: in this case, the courts look for damage that is different in kind or, perhaps, substantially different in degree from the damages suffered by others.[132]

38 The main environmental liability regimes available to private individuals are private nuisances in the common law provinces, or neighbourhood annoyances (troubles de voisinage) in Quebec. The two regimes are basically similar,[133] and attempt to balance one person's freedom to use his land as he pleases against another's right to protection from unreasonable interference with the use or enjoyment of his own land.[134] The interference can be localised, affecting one or two neighbours, or widespread, affecting all properties in an extended neighbourhood. This raises what Cassels and Jones describe as the "catch-22" of mass nuisance claims: "if the claims are so widespread as to warrant a class action, they may be too widespread to be actionable [as private nuisances] by individuals at all".[135] A nuisance action sanctions results (whether the plaintiff suffered excessive disturbance) and not conduct (whether the owner acted prudently), and is thus a matter of strict liability.[136] Where the result is physical damage, this is usually regarded as unreasonable per se, but where it is interference with comfort and convenience, courts balance the gravity of the plaintiff's harm with the utility of the defendant's activity, looking at such factors as the intensity and duration of the interference, the plain-

---

132 *Klar* (fn. 4) 721 ff; *Linden/Feldthusen* (fn. 4) 561 ff.

133 *Drysdale v. Dugas* (1896) 26 S.C.R. 20 (S.C.C.).

134 *Linden/Feldthusen* (fn. 4) 568 ff. Art. 976 C.C.Q. talks of "neighbourhood annoyances that are ... beyond the limit of tolerance they [i.e. neighbours] owe each other". Art. 976 is a codification of earlier case law, and was added to the Code as part of the 1994 revision.

135 *Cassels/Jones* (fn. 43) 117.

136 E.g. *St. Lawrence Cement* [2008] 3 S.C.R. 392, par. 37 ff., 86. This is the leading Canadian case on nuisance (neighbourhood annoyances) at the present time; other recent pollution emission cases are *Hollick* [2001] 3 S.C.R. 158 (municipal land fill; certification of class action denied) and *Pearson* (2006) 78 O.R. (3d) 641 (nickel factory; certification of class action approved).

tiff's sensitivity, the character of the neighbourhood, etc.[137] Unlike negligence, there is little discussion about whether mere fear of harm constitutes interference with comfort and convenience so as to be actionable in nuisance.[138] The economic importance of the defendant's activity to the community has been an important factor weighing in the defendant's favour in the past, but this could change as courts increasingly give more weight to environmental concerns.[139] Similarly, courts are increasingly reluctant to accept the defence of statutory authority.[140] Finally, Klar suggests a possible trend to use nuisance law "as a means of shifting the accident costs of activities from individual victims to those actors who are best equipped to distribute them among the activities' beneficiaries".[141]

Applying the principles of nuisance or neighbourhood disturbances to **39**
environmental damage caused by GMOs is difficult.[142] Who may sue? Will courts follow the traditional approach of limiting standing to those with a proprietary interest in land, or will they follow the suggestion of opening the action to mere occupants suffering personal injury?[143] Who may be sued? Will courts continue to require, as seems likely, that the action be brought against someone in occupation or control of neighbouring land (i.e. GM farmers), or will they accept nuisance actions against the biotechnology companies in appropriate cases? This last possibility was left slightly ajar by the trial judge in *Hoffman*, although the Court of Appeal,

---

137 *Osborne* (fn. 4) 362 ff. Under art. 976 C.C.Q., the "limit of tolerance" is assessed "according to the nature or location of their land or local custom". Note that the character of the neighbourhood is simply one factor to be considered; in other words, there is no defence of "coming to a nuisance".

138 E.g. *Linden/Feldthusen* (fn. 4) 570 cites a 1927 trial judgment, *Shuttleworth v. Vancouver General Hospital* [1927] 2 D.L.R. 573 (B.C.S.C.) (holding that an isolation hospital was not nuisance), and compares it with an older English case, *Metropolitan Asylum District v. Hill* (1881) 6 App. Cas. 193 (holding that a smallpox hospital was nuisance).

139 See *St. Lawrence Cement* [2008] 3 S.C.R. 392, par. 80; see also e.g. *114957 Canada Ltée (Spraytech, Société d'arrosage) v. Hudson (Town)* [2001] 2 S.C.R. 241 (S.C.C.) (upholding validity of municipal by-law prohibiting pesticide use).

140 In a nuisance action, the defence of statutory authority requires both that the activity was authorised by statute and the damage was the inevitable consequence of the activity. E.g. *Ryan* [1999] 1 S.C.R. 201, par. 54 ff., *St. Lawrence Cement* [2008] 3 S.C.R. 392, par. 97 ff. (where the Court seems to equate defence of statutory authority with provision of statutory immunity) (see below, no. 51).

141 *Klar* (fn. 4) 734 ff. (citing in support cases dealing with socially beneficial activities that are not themselves nuisances but have inadvertently misfired in some way, even just once, causing damage [e.g. burst water mains, backed-up sewers]).

142 See generally *C.P. Rodgers*, Liability for the Release of GMOs into the Environment: Exploring the boundaries of nuisance, Cambridge L.J. 119 (2003) 371 ff.

143 Canadian courts have applied the wider approach, but most decisions were reached before the restrictive decision of the House of Lords in *Hunter v. Canary Wharf Ltd.* [1997] A.C. 655.

in rejecting the cause of action, stressed that the GM genes emanated from lands that were not controlled by the biotechnology defendants.[144] As for actions against neighbouring farmers, what might be the impact of "right-to-farm" legislation protecting farmers from nuisance actions arising from carrying out agricultural operations using "normally accepted farm practices"?[145] Finally, what of damages? Will courts accept that the presence of GM genes constitutes physical damage, or will they hold that the affected plants, because still alive, remain intact? And if courts classify the damage as interference with comfort and enjoyment, what weight will they accord to the defences of character of the neighbourhood (is there extensive GM cultivation in the area?) and possible sensitivity of the plaintiff (e.g. organic producers, environmentally sensitive areas)?

40 Another recourse available to individuals in the common law provinces regarding recovery for losses of an environmental nature is an action under the principle of *Rylands v. Fletcher*.[146] It is a strict liability regime, the elements of which are a non-natural use of land, escape from one property to another of something liable to do mischief, and damage. These requirements were recently considered in *Brooks v. Canadian Pacific Railway*, and the court reaffirmed that the second element, escape, had to be from one property to another and that the third element, damage, could be personal injury as well as property damage, but not pure economic loss.[147] The most contested element is the first, non-natural use of land, and the Supreme Court of Canada has accepted the rather narrow definition that it must be "some special use bringing with it increased danger to others", not merely "the ordinary use of land or such use as is proper for the general benefit of the community."[148] Where the use of land is regarded as inherently dangerous (e.g. storage and use of explosives, dangerous chemicals or biological agents), it is readily classified as non-natural; but where the danger is less apparent, the courts tend to balance it against other factors such as the normalcy and utility of the use and the context in which it takes place. Examples of generally accepted agricultural prac-

144 *Hoffman* trial, [2005] 7 W.W.R. 665, par. 122 (discussing *In re StarLink Corn Products Liability Litigation, Marvin Kramer v. Aventis CropScience USA Holding Inc.* (2002) 212 F. Supp. 2d 828 [U.S. Dist. Ct, N.D. Illinois]); *Hoffman* appeal, (2007) 283 D.L.R. (4th) 190, par. 63.

145 See generally *J. Kalmakoff*, The Right to Farm: A survey of farm practices protection legislation in Canada, Saskatchewan Law Review 62 (1999) 226 ff.

146 (1868) L.R. 3 H.L. 330, affirming (1866) L.R. 1 Ex. 265.

147 (2007) 283 D.L.R. (4th) 540, par. 100 ff. (derailment of railway cars containing anhydrous ammonia; application for certification of class action dismissed).

148 *Tock v. St. John's Metropolitan Area Board* [1989] 2 S.C.R. 1181 (S.C.C.) (flooded basement from blocked sewer; held actionable under nuisance but not *Rylands v. Fletcher*) 1189, quoting the Privy Council in *Rickards v. Lothian* [1913] A.C. 263 (P.C.) 280.

tices to which *Rylands v. Fletcher* is applied include herbicide drift from aerial spraying and well contamination from manure spreading.[149] In regard to GMOs, the last two elements, escape and damage, raise issues similar to those canvassed under nuisance, that is, proof of the source of the escaped gene and proof of damage. As for the first element, it is an open question whether courts will regard GM agriculture as inherently dangerous, or only dangerous in some contexts and not others. *Rylands v. Fletcher* was rejected as a cause of action in *Hoffman* because the commercial release and sale of GM seed by Monsanto did not constitute an "escape" under the rule; accordingly, it was not necessary to decide whether GM canola was a dangerous substance or whether planting it for field trials was a non-natural use of land.[150]

A final possibility under the common law is trespass, which is direct physical interference with land. However, airborne pollutants such as GMOs are likely to be classed as indirect interferences, and thus analysed under private nuisance rules (above, no. 38). Arguments based on trespass were rejected in *Hoffman* for this reason, and also because the GM crops were not grown on Monsanto's land.[151] Incursions onto the land of another are treated at civil law under the general delict provisions.[152] **41**

(e) Cartagena Protocol

Canada is a party to the UN Convention on Biological Diversity (which it signed and ratified in 1992). However, like the other major producers of GM crops, it is not a party to the Treaty's Cartagena Protocol on Biosafety, even though it was adopted in Montreal; Canada signed the Protocol in 2001, but has not ratified it. On the other hand, it has both signed and ratified the International Treaty on Plant Genetic Resources for Food and Agriculture (in 2002) and is thus a party to it as well as the biodiversity convention. **42**

Canadian customary constitutional law, inherited from Britain, distinguishes between treaty-making (the prerogative of the federal executive) and treaty-implementation (a legislative function); the latter follows the **43**

---

149 *Fondrick v. Gross* [2004] 6 W.W.R. 367 (Sask. Q.B.) and *Metson v. R.W. DeWolfe Ltd.* (1980) 14 Canadian Cases on the Law of Torts (C.C.L.T.) 216 (N.S. S.C., Trial Div.).

150 *Hoffman* trial, [2005] 7 W.W.R. 665, par. 97.

151 *Hoffman* trial, ibid., par. 125 ff.; *Hoffman* appeal (2007) 283 D.L.R. (4th) 190, par. 58, 63 ff.

152 They are also punishable by fine under the Agricultural Abuses Act, R.S.Q., c. A-2, s. 2, but only when committed by a person (as opposed to a thing).

general division of legislative powers set out in the Constitution.[153] Applying this to GMOs, the result is that the federal executive (i.e. the Cabinet) has the authority to sign and ratify (or not) any treaty, which then binds Canada internationally. However, the treaty remains without effect domestically until it has been implemented by appropriate legislation – federal legislation for international agreements affecting the patenting of GM technology and the unconfined release (i.e. commercialization) of GM seeds as well as the inter-provincial and international sale of the resulting GM crops, and provincial legislation for international agreements affecting such matters as public lands, the use of private land, tort law and intra-provincial sale of goods. This asymmetrical approach complicates Canada's position with regard to international agreements.

4. Other strict liability regimes

44 Other than nuisance and the principle in *Rylands v. Fletcher* (above, no. 38, 40), the common law has few strict liability regimes,[154] and none that are relevant to harm caused by GMOs. Osborne attributes the paucity of strict liability torts to "the relentless expansion of the tort of negligence", but notes the existence of "a paradox in the Canadian law of torts in respect of strict liability. On the one hand, the courts have shown no willingness either to expand the existing heads of strict liability or to create new heads of strict liability. On the other hand, there is a willingness, in certain situations, to impose a covert strict liability under the guise of applying traditional negligence principles."[155] Another possible liability regime is the tort of conspiracy, which is an intentional rather than a strict liability tort. In common law Canada, actionable conspiracy arises either where two or more persons agree to act together for the predominant purpose of causing injury to the plaintiff, whether the means used are lawful or unlawful, or where they agree to act together unlawfully and the conduct is directed at the plaintiff in circumstances in which the defendants knew or should have known that injury to the plaintiff would result.[156] The injury is usually economic, but the Supreme Court of Canada accepted that this tort might be applicable to product liability and personal injury

153 Constitution Act 1867 (fn. 2). See generally *J.H. Currie*, Public International Law (Essentials of Canadian Law Series) (2nd ed. 2008); *Labour Conventions Reference* [1936] S.C.R. 461 (S.C.C.).

154 These are liability for fire, dangerous animals ("scienter" action) and cattle trespass.

155 *Osborne* (fn. 4) 324 f.

156 *Cement LaFarge v. B.C. Lightweight Aggregate Ltd.* [1983] 1 S.C.R. 452 (S.C.C.) 471 f.

cases in *Hunt v. Carey Canada Inc.*,[157] where the alleged conspiracy was to withhold information about the health risks associated with asbestos. Conspiracy was also accepted recently as a cause of action in *LeFrançois v. Guidant Corp.*,[158] a class action involving faulty defibrillators.

There are two additional civil law liability regimes that might be applicable. One is abuse of rights, a jurisprudential development which was included in the 1994 revision of the Code as art. 7: "No right may be exercised with the intent of injuring another or in an excessive and unreasonable manner which is contrary to the requirements of good faith." While most Quebec commentators, like their French counterparts, regard abuse of rights as fault-based (i.e. as constituting civil fault in the exercise of a right), the Supreme Court of Canada has recently presented a more nuanced view: "Where a right exists, ... the usual application of the concept of fault is qualified. The holder of the right has a sphere of autonomy in exercising that right. In such a context, it thus becomes crucial, when analysing civil liability, to consider the nature of the right in issue and the circumstances in which it is exercised, since ... an *abuse* of rights must be found in order to show fault. Once an abuse is found, the holder of the right loses the protection of the sphere of autonomy that flows from the right. Violation of a standard of conduct is therefore inextricably linked to the concept of abuse of rights."[159] The second regime is responsibility for injury resulting from the autonomous act of a thing. However, the Code provides that responsibility can be avoided by the person entrusted with custody if he "proves that he is not at fault",[160] a proviso which makes the regime fault- rather than risk-based. Quebec courts are more insistent than French ones on this aspect of the regime.[161] **45**

157 [1990] 2 S.C.R. 956 (S.C.C.).

158 2008 CanLII 15770 (Ont. S.C.). The Court also rejected (par. 44) the defendant's argument that conspiracy "was excluded by the plaintiffs' allegations that the defendants were operated as a single entity.".

159 *St. Lawrence Cement* [2008] 3 S.C.R. 392, par. 29 [emphasis in original], citing *J. Ghestin/ G. Goubeaux,* Traité de droit civil, vol. 1, Introduction générale (3rd ed. 1990) 694 as supportive of this approach.

160 Art. 1465 C.C.Q.

161 *Brierley/Macdonald* (fn. 4) 459.

V. Vicarious liability

1. Scope of vicarious liability

46 Vicarious liability is a form of strict liability which holds one person responsible for the misconduct of another because of the relationship between them. This responsibility supplements rather than replaces that of the actual wrongdoer, so that both are regarded as joint wrongdoers (below, no. 50). The categories of relationships attracting vicarious liability are not closed, but the main ones are the employer/employee relationship and the principal/agent one,[162] with liability limited to acts of employees "within the course of employment" and of agents "within the scope of their authority" (actual or apparent). Vicarious liability does not arise when the relationship is one of independent contractor and client, and a main difficulty facing the courts is thus to distinguish an employee from an independent contractor. The traditional test is the "control" test, but others (e.g. the "entrepreneur" test, the "enterprise" or "organisation" test) can also play a role.[163] For the Supreme Court of Canada, the central question is whether the person engaged to perform the services is performing them as someone in business on his own account, and this depends not just on the level of control the employer has over the worker's activities but also on such factors as whether the worker provides his or her own equipment or hires his or her own helpers, the degree of financial risk taken by the worker, the degree of investment and management responsibility held by the worker, and the worker's opportunity for profit in the performance of the tasks.[164] Certain duties are regarded as "non-delegable", for which the employer or principal remains personally liable notwithstanding a delegation, even to an independent contractor. Osborne suggests that non-delegable duties include a duty of care in regard to intrinsically dangerous or hazardous activities and a duty not to create a nuisance,[165] both of which might be relevant to GMOs.

162 Art. 1463 C.C.Q. provides, "The principal is liable to reparation for injury caused by the fault of his agents or employees in the performance of their duties; nevertheless, he retains his recourses against them.".

163 *Osborne* (fn. 4) 345 ff.

164 *671122 Ontario Ltd. v. Sagaz Industries Canada Inc.* [2001] 2 S.C.R. 983 (S.C.C.) par. 47 ff. (holding marketing company hired by supplier to be independent contractor; stressing that the various factors constitute "non-exhaustive list", with "no set formula" as to application and with relative weight depending on "the particular facts and circumstances of the case").

165 *Osborne* (fn. 4) 354. Others include strict statutory duties to do a positive act (e.g. road maintenance), a duty to provide lateral support, and the custodial duties of a bailee

Vicarious liability is also an issue in cases brought against interlocking companies. If it is not found, plaintiffs are required to prove negligence separately against each company, with all the difficulties this entails. The Ontario Court of Appeal set out a particularly stringent test for finding vicarious liability in this situation, which requires not only "complete control so that the subsidiary does not function independently" but also that "the subsidiary must have been incorporated for a fraudulent or improper purpose or be used by the parent as a shield for improper activity".[166] On the other hand, a class action has recently been certified, without reference to the test, against a US manufacturer and marketer of health products and services, a wholly owned US subsidiary manufacturing and marketing the product in question, and a wholly owned Canadian subsidiary of the latter distributing the product in Canada. The action was brought against all three "on the basis that their businesses were organised as if they consisted of a single entity. Each of the defendants was stated to be the agent of the others and therefore vicariously responsible for the acts and omissions set out in the ... statement of claim."[167] This is an issue for a country such as Canada where separate Canadian subsidiaries of foreign companies, including biotechnology companies, are the norm. **47**

2. Liability for people further up the food or feed production chain

The Quebec Civil Code specifically provides that both distributors ("a person who distributes the thing under his name or as his own") and suppliers ("any supplier of the thing, whether a wholesaler or a retailer and whether or not he imported the thing") are subject to the same responsibilities for injury of third persons as the manufacturer of a product (above, no. 29).[168] Although not quite the same thing, Canadian common law courts have held product manufacturers responsible for defects in component parts supplied by third parties, and Klar raises two different views **48**

---

for reward. The courts have yet to provide an underlying principle justifying the imposition of responsibility for the acts of independent contractors in these situations.

166 *Haskett v. Equifax Canada Inc. et al* (2003) 63 O.R. (3d) 577 (Ont. C.A.) par. 61, applied in *LeFrançois* 2008 CanLII 15770, par. 48.

167 *Serhan Estate v. Johnson & Johnson* (2006) 269 D.L.R. (4th) 279 (Ont. Div. Ct.) par. 11. This is somewhat similar to the "enterprise liability" approach which, in a sense, "pierces the corporate veil" of limited liability, and fixes liability on the major shareholder (the parent company) of the subsidiary: *Cassels/Jones* (fn. 43) 300 ff. (although this would not assist where the parent company is the wrongdoer).

168 Art. 1468 par. 2 C.C.Q.

about this.[169] One is that under the Canadian negligence-based approach to product liability, the manufacturer should not be held liable for defective components unless the manufacturer was itself at fault in using them (e.g. negligent choice of supplier, failure to inspect component, etc.); this view rejects arguments that manufacturers should be vicariously liable for the negligence of suppliers or that the duty on manufacturers to manufacture safe products is non-delegable. The other view is that to require plaintiffs to prove which component was defective, who supplied it and the negligence of the supplier is too onerous, and that it would be fairer for the manufacturer to be held liable, who can then claim over against suppliers, either in contract or in tort.

3. Can someone further down the feed or food chain include someone further up it in a trial against him/herself?

49 Under typical rules of civil procedure, a defendant in an action may, but is not obliged to, either bring a cross-claim against one or more of the other defendants or commence third party proceedings against someone who is not a party to the action.[170] Whether or not this will shift the loss to the other person is not a procedural question but rather depends on the court's assessment of the substantive law as applied to each.

VI. Multiple tortfeasors

50 In Canada, both the common and civil law approach the liability of multiple wrongdoers in much the same way, and hold multiple wrongdoers jointly and severally liable (solidarity) to the plaintiff where the damage results from a single wrongful act or from two or more wrongful acts causing the same damage. Each wrongdoer is thus fully responsible to the plaintiff for the damage caused, and payment by one relieves the others from responsibility to the plaintiff;[171] however, the defendants are entitled between themselves to contribution of an apportioned share of the damages paid. This is provided for in the Quebec Civil Code, and required special legislation in the common law provinces as contribution was not recognised at

---

169 *Klar* (fn. 4) 369.

170 E.g. Rules 28 and 29, Ontario Rules of Civil Procedure, Revised Regulations of Ontario (R.R.O.) 1990, Reg. 194 as amended.

171 Art. 1526 C.C.Q. (under "solidary obligations").

common law.[172] Apportionment of liability is done by the courts in a rough-and-ready way (e.g. equally, 2/3-1/3, etc.) in proportion to the seriousness of the fault of each. Both systems also recognise joint and several liability (solidarity) in certain situations where it is impossible to choose between possible wrongful sources of the damage (above, no. 18).[173] Apportionment also applies where one of the multiple wrongdoers is the plaintiff (above, no. 20).

VII. Defences

1. Licence/permission to grow GM material

It is unlikely that government permission to grow GM crops serves as a defence against liability in Canada. Admittedly, Canadian government approval for the unconfined release of the defendants' varieties of GM canola was described in *Hoffman* as "a powerful policy reason for not fastening them with the duty of care as pleaded",[174] but other cases are less categorical. In *Sauer* (the mad cow case), for example, the trial judge rejected the feed manufacturer's argument that a government regulation authorising the incorporation of ruminant meat and bone meal into cattle feed constituted a bar to the plaintiff's action in negligence against it; the Court of Appeal agreed, and expressly declined to follow *Hoffman* on this point: "So far as *Hoffman* ... applies a similar government policy to negate a *prima facie* duty of care, I would respectfully decline to follow it."[175] Other negligence examples in which the existence of regulatory approval was either rejected as a defence or not raised in argument include *Ring* (herbicide),[176] *Walls* (cholesterol drug)[177] and *Andersen* (heart valve).[178] Authorisation plays a stronger role in common and civil law nuisance actions (above, no. 38) and actions under *Rylands v. Fletcher* (above, 51

---

172 Art. 1478 ff. C.C.Q. (under "apportionment of liability"). For the common law legislation, see under contributory negligence (fn. 76); whether the legislation applies to all tort actions or is restricted to negligence varies from province to province, as the legislation is not uniform: see *Klar* (fn. 4) 551 ff.

173 Art. 1480 C.C.Q.; *Cook* [1951] S.C.R. 830.

174 *Hoffman* appeal, (2007), 283 D.L.R. (4th) 190, par. 60 (see above, no. 10).

175 *Sauer* trial, (2006), 79 O.R. (3d) 19, par. 53, *Sauer* appeal (2007) 225 O.A.C. 145, par. 48.

176 (2007) 268 Nfld. & P.E.I.R 204 (Nfld & Lab. S.C., Trial Div.) (holding that regulatory approval did not itself negate liability, as liability depended on method of use regardless of regulatory approval). See also also *Willis v. FMC Machinery & Chemical Ltd* (1976) 68 D.L.R. (3d) 127 (P.E.I. Trial Div.) (herbicide approved for use, but defendant still held liable for crop damage).

177 *Walls et al v. Bayer Inc.* (2005) 189 Man. R. (2d) 262 (Man. S.C.), affirmed (2005) 257 D.L.R. (4th) 435 (Man. C.A.) (class action certified).

178 (2003) 67 O.R. (3d) 136.

no. 40), where statutory authorisation is a defence if, but only if, the damage caused is the inevitable and unavoidable consequence of the authorised activity. Osborne criticises the acceptance of statutory authority as a defence, and suggests that only statutory immunity (not simply statutory authority) should be so accepted.[179] This position seems to be reflected in the Supreme Court of Canada's recent decision in *St. Lawrence Cement*, where the Court rejected the defendant's argument of statutory authority because the legislation in question did not exempt the company from the application of ordinary law: "When the legislature *excludes* the application of the ordinary law, it generally does so expressly."[180]

2. Consent/assumption of risk

52 Canadian common law courts still notionally apply the old common law rule that consent or voluntary assumption of risk (volenti non fit injuria) is a complete defence to an action, but have defined it so narrowly that it is now marginalised although not abolished. To succeed, the defendant must prove that the plaintiff has consented, either expressly or impliedly, to accept both the physical risk (i.e. the danger of being injured in fact) and the legal risk (i.e. abandonment of the right to sue the defendant). The former is fairly straight-forward to prove, but the latter is difficult. Possible plaintiffs may be required to sign an express waiver (when participating in dangerous sporting events, for example[181]) but such waivers are interpreted strictly against the defendant. This strict approach was developed by the Supreme Court in several "drunk driver/willing passenger" cases,[182] but applies generally in all cases of negligence and, seemingly, also *Rylands v. Fletcher* but perhaps not nuisance, at least where the plaintiff "comes to the nuisance".[183] Defining consent narrowly has the advantage of leaving more room for finding contributory negligence, and thereby apportioning responsibility more equitably between the parties. This thus comes close to Quebec's position that although assumption of risk (which is interpreted narrowly, as in the common law provinces) "may be considered imprudent having regard to the circumstances", it does not

179 *Osborne* (fn. 4) 335.

180 *St. Lawrence Cement* [2008] 3 S.C.R. 392, par. 97 ff. [emphasis added]. The polluting cement plant in question had been authorised by a special act adopted in 1952.

181 *Crocker v. Sundance Northwest Resorts Ltd.* [1988] 1 S.C.R. 1186 (S.C.C.) (intoxicated participant in inner tube race down mogul hill held not to have consented, although had signed waiver and refused to withdraw when asked by defendant's employee).

182 *Osborne* (fn. 4) 198.

183 *Linden/Feldthusen* (fn. 4) 592.

entail renunciation of an action against the defendant.[184] The narrow approach would make proof of assumption of risk of the dangers of GMO difficult in the Canadian context: assumption of risk implies knowledge of the risk and its extent, but GM products have been approved as risk-free and, in the absence of labelling, are consumed in ignorance.

3. Third-party influence

In both the common law provinces and Quebec, the courts recognise intervening third-party behaviour (novus actus interveniens) as a defence to an action only if the third party behaviour is unforeseeable.[185] Otherwise, courts tend to treat the two wrongdoers as multiple tortfeasors. An approach based on foreseeability is a change from the old common law view, which tended to treat the intervening event as breaking the chain of causation and excusing the defendant. A similar foreseeability approach is applied in nuisance actions and those under *Rylands v. Fletcher*, although "[i]n theory a 'no negligence' defence is incompatible with strict liability, the purpose of which is to allocate losses on the basis of cause alone. Its acceptance indicates the general ambivalence towards strict liability, and the modern tendency to blur the line between strict liability and fault liability."[186] 53

4. Prescription

The limitations period applying to the cases envisaged by this study are the same as in the standard tort cases.[187] In the common law provinces, this is typically a two-year period, running from the date "the cause of action arose". This was previously interpreted to mean the date "the 54

184 Art. 1477 C.C.Q.

185 Quebec regards unforeseeable acts of third parties as a form of force majeure (above, no. 22); see *Baudouin/Deslauriers* (fn. 4) 652 ff. *Klar* (fn. 4) 492 ff. stresses the importance of distinguishing this situation from one in which the defendant is responsible for another's misconduct. In this case, responsibility depends on duty (whether the defendant was under a duty to the plaintiff to take reasonable care to prevent the misconduct), and should not properly be regarded as a novus actus problem. Examples are where the defendant is responsible for the actions of those under the defendant's care and control (e.g. prisoners) or where the defendant has negligently provided a stranger with the opportunity to act wrongfully (e.g. ignition keys left in car, leading to joy ride and accident).

186 *Osborne* (fn. 4) 334.

187 See *Linden/Feldthusen* (fn. 4) 111 ff.

damage was incurred"; however, occasional harsh results (e.g. when claims were barred before the plaintiff knew of their existence) led the courts to interpret it as the date "the damage was discovered." This change has been applied by the Supreme Court of Canada on a number of occasions[188] and is now making its way expressly into provincial legislation.[189] In Quebec, the limitation period is three years, with the period running from "the day the damage appears for the first time" in cases where the right of actions arises from "moral, corporal or material damage appearing progressively or tardily."[190] The "discovery" approach goes some way to attenuate limitation period difficulties that might arise in so-called "long-tail" torts (i.e. those in which the damage takes a long time to appear or be recognised), and this would be of assistance in possible GM actions. For example, in *Endean v. Canadian Red Cross*, a tainted blood class action, the Court determined that "evidence of the evolving scientific knowledge of the existence of HCV [hepatitis C virus] and of the available tests for its presence in blood ... will establish the starting point for consideration of the postponement provisions of the *Limitations Act*."[191] Limitation periods are suspended during class action proceedings.[192]

5. Other defences

55 None that are not mentioned in response to other questions.

VIII. Remedies

1. Pecuniary compensation

(a) Bodily harm

56 There are no special rules on compensating bodily harm caused by GMOs, and the regular remedial rules would thus apply.[193] These provide for the payment of damages for pecuniary and non-pecuniary losses. Damages for

188 E.g. *Central Trust Co. v. Rafuse* [1986] 2 S.C.R. 147 (S.C.C.) par. 73 ff.
189 E.g. Limitation Act, S.O. 2002, c. 24, s. 4 ff.
190 Art. 2925, 2926 C.C.Q.
191 (1997) 148 D.L.R. (4th) 158, par. 61.
192 E.g. Class Proceedings Act, S.O. 1992, c. 6, s. 28; art. 2908 C.C.Q.
193 On remedies generally, see *J. Cassels/E. Adjin-Tettey*, Remedies: The law of damages (Essentials of Canadian Law Series) (2nd ed. 2008).

pecuniary losses include compensation for pre-trial losses (special damages), which are established precisely, and compensation for future losses (general damages), which are estimated and include the costs of future care and loss of future earning capacity. Damages for pecuniary losses are intended to be fully compensatory; they are not capped, and a defendant's capacity to pay is not taken into consideration in setting the amount. Damages for non-pecuniary losses include compensation for pain and suffering, loss of amenities, permanent disability and loss of expectation of life. They are not fully compensatory in the sense that they are based not on an evaluation of what the plaintiff has lost but rather on an assessment of the goods and services required to provide solace for the pleasure and enjoyment lost (described as a "functional approach"); they have been capped in personal injury cases since 1978.[194] Damages are usually awarded as a lump sum, to be paid immediately, although periodic payments through structured settlements are provided for under some provincial legislation or by court-approved consent.

(b) Property losses

Here again the regular rules apply, under which plaintiffs are entitled to **57**
payment of damages that will restore them to their position before the property was damaged. This amount is not capped, and is assessed to reflect either the amount by which the market value of the property has been diminished or the cost of reinstating the land to its pre-damaged position. In principle, diminution in value is the usual approach, with cost of reinstatement being appropriate if it is lower and sometimes acceptable even if higher. The appropriate test for the latter is "the reasonableness of the plaintiff's desire to reinstate the property ... judged in part by the advantages to him of reinstatement in relation to the extra cost to the defendant in having to pay damages."[195] In *British Columbia v. Canadian Forest Products Ltd.*,[196] the Supreme Court of Canada accepted that a landowner (i.e. the Crown claiming as the owner of public lands) could, in principle, claim for environmental damage caused to its lands by a carelessly set forest fire (above, no. 36), comprising compensation for loss of

194 Principles set by the Supreme Court of Canada in a 1978 "damages trilogy" (*Andrews v. Grand & Toy Alberta Ltd.* [1978] 2 S.C.R. 229, *Thornton v. Prince George School District No. 57* [1978] 2 S.C.R. 267 and *Arnold v. Teno* [1978] 2 S.C.R. 287), capping non-pecuniary losses at CAD 100,000 (now slightly over CAD 300,000 when adjusted for inflation).

195 *Fondrick* [2004] 6 W.W.R. 367, par. 98, quoting *H. McGregor*, McGregor on Damages (14th ed. 1980) [now 17th ed. 2003].

196 [2004] 2 S.C.R. 74.

use value, passive use or existence value, and inherent value.[197] It could be assessed on the basis of evidence relating to the nature of wildlife, plants and other organisms protected by the environmental resource in question, the uniqueness of the ecosystem from an environmental perspective, the environmental services provided by the resource, the recreational opportunities provided by it, and the subjective or emotional attachment of the public to the damaged or destroyed area. However, no evidence had been led on those points in the case, and the Court rejected the Crown's invitation to abandon a search for "more precise measures of loss and rational (though perhaps indirect) techniques of assessment of environmental losses" and to adopt a more broad-brush approach similar to that applied in regards to "solace" for non-pecuniary losses suffered in personal injury cases.[198] These principles could be applicable to environmental damage caused by GMOs.

(c) Economic losses

58 The regular remedies apply. Recovery is allowed for economic losses consequential to personal injury (e.g. loss of income) or property damage (e.g. necessary expenses, loss of profits) as well as in the limited "pure economic loss" situations (above, no. 9). In all cases, the principle is restitutio in integrum, restoration to the position the plaintiff would have been in but for the defendant's wrongful conduct.[199] Note, however, that the trial judge in *Hoffman* was of the view that the plaintiffs could not argue that GM canola was inherently harmful or dangerous to their crops as it had been declared safe by the approving body, and could not claim against the defendants for their economic loss since "*[b]y definition*, such losses are not the direct result of the defendant's action" (but followed rather from the stringent organic requirements of the organic certification associations).[200] Finally, plaintiffs are required to take all reasonable steps to mitigate the losses flowing from the wrong, with defendants being responsible for any additional costs reasonably incurred by a plaintiff in an attempt to mitigate; failure to mitigate is taken into consideration when assessing overall

---

197 *Canadian Forest Products*, par. 138 ff., citing *Ontario Law Reform Commission*, Report on Damages for Environmental Harm (1990) and *F.B. Cross*, Natural Resource Damage Valuation, Vanderbilt Law Review 42 (1989) 269 ff.

198 Par. 141, 151.

199 See generally *Rainbow Industrial Caterers v. Canadian National Railway* [1991] 3 S.C.R. 3 (S.C.C.).

200 *Hoffman* trial, [2005] 7 W.W.R. 665, par. 23, 73 [emphasis added].

damages payable by the defendant. Applying these principles to GMO contamination, an organic farmer could be under an obligation to grow conventional crops if this would not put his organic certification at risk and if it would not interfere with the restoration of the field.

(d) Harm to animals

If chattels (i.e. movables) are damaged, the owner is entitled to compensation reflecting the loss in value, usually measured by the cost of repair or replacement. In the case of an animal harmed by eating GM-contaminated feed, this would be the market value of the animal if the animal died or the difference between its value before and after the event (i.e. the organic premium) otherwise. Both measures would thus factor in a cow's revenue-producing potential.[201] In principle, one would think that the mere fact that an animal eats GM-contaminated feed would constitute harm for which the owner would be compensated by the person responsible for the contamination (assuming the contamination occurred negligently); but courts might take the same restrictive approach to animal damage as the trial judge in *Hoffman* did to crop damage (above, no. 58). 59

(e) Costs of disposal

Here again, one would assume that under the principle of restitutio in integrum, the costs of disposing of contaminated crops or animals is recoverable, and in this case, this assumption is borne out for GM production by both practice and judicial decision. As a matter of practice, Monsanto regularly pays the costs associated with removing unwanted GM canola from farmers' fields, with the farmers undertaking to remain silent about it in return.[202] And in 2008 Percy Schmeiser (who had lost a patent dispute with Monsanto[203]) was awarded CAD 660 (approx. € 440) in a Small Claims action against Monsanto for the clean-up costs of unwanted canola that had re-appeared in his fields in 2005, an award free of the usual non-disclosure restraint.[204] 60

---

201 See e.g. *Sauer* (motion to strike pleadings (2007) 225 O.A.C. 145, certification of class action (2009) 246 A.C. 256) which is in fact a claim for pure economic loss (loss of income), as his cattle were not themselves damaged.

202 The non-disclosure agreement makes the incidence of GM contamination difficult to estimate.

203 *Monsanto Canada Ltd v. Schmeiser* [2004] 1 S.C.R. 902 (S.C.C.).

204 *M. Hartley*, Grain Farmer Claims Moral Victory in Seed Battle, The [National] Globe and Mail, 20 March 2008, A3.

2. Non-compensatory damages

61 All Canadian provinces recognise punitive, or exemplary, damages. In Quebec this is governed by art. 1621 of the Civil Code, which requires that the awarding of punitive damages be "provided for by law".[205] No such restriction applies in the common law provinces, where punitive damages have long been recognised by the courts,[206] and the principles set out in the common law cases are often referred to in Quebec decisions. Punitive damages are awarded sparingly, to punish conduct that is "so malicious, oppressive and high-handed that it offends the court's sense of decency",[207] to denounce such conduct, and to deter others from similar conduct. Its availability is largely restricted to intentional torts or breaches of fiduciary duty, but can be awarded, as appropriate, in cases of breach of contract or of negligence or nuisance.[208] Punitive damages are regularly claimed (or so it seems) in class actions, particularly product liability cases (even though the cause of action is fault-based rather than intentional).[209] The Supreme Court of Canada accepted that the question of punitive damages is certifiable as a common issue where the plaintiff alleges "systemic negligence", that is, "negligence not specific to any one victim but rather to the class of victims as a group"[210] – which is usually the case in product liability actions. In Canada, the quantum of punitive damages is neither capped nor established as a fixed ratio of compensatory damages.[211] The Supreme Court recently stressed that the governing principle for quantum is "proportionality", and this is established by looking at such factors as the blameworthiness of the defendant's conduct, the degree of vulnerability of the plaintiffs, the harm or potential harm directed at the plaintiff (or class of plaintiffs), the need

---

205 E.g. Charter of Human Rights and Freedoms (fn. 122) art. 49 (providing for punitive damages where there is "unlawful and intentional" interference with protected rights or freedoms (in addition to compensatory damages available where interference is simply "unlawful")). See *Quebec (Public Curator) v. Syndicat national des employés de l'hôpital St-Ferdinand* [1996] 3 S.C.R. 211 (S.C.C.). As mentioned (above, no. 35), the *Charter* now protects a right "to live in a healthy environment" (art. 46.1).

206 See especially *Whiten v. Pilot Life Insurance Co.* [2002] 1 S.C.R. 595 (S.C.C.) par. 50 ff.; see also *Vorvis v. Insurance Corp. of British Columbia* [1989] 1 S.C.R. 1085 (S.C.C.).

207 *Hill v. Church of Scientology of Ontario* [1995] 2 S.C.R. 1130 (S.C.C.) 1208.

208 *Whiten* [2002] 1 S.C.R. 595, par. 67. Negligent conduct warranting punitive damages could be conduct which "goes beyond carelessness and reveals a high-handed and callous disregard of a plaintiff's rights": *Cassels/Adjin-Tettey* (fn. 193) 264.

209 Examples include *Ring* (2007) 268 Nfld. & P.E.I.R 204 (pesticide spraying), *Scarola* 2003 CanLII 44004 (gasoline additive), *Endean* (1997) 148 D.L.R. (4th) 158 (tainted blood) and *Walls* (2005) 257 D.C.R. (4th) 435 (cholesterol drug).

210 *Rumley* [2001] 3 S.C.R. 184, par. 34.

211 The amount awarded in Canada is generally low, partially because of the rarity of jury trials in civil cases.

for deterrence, the advantages wrongfully gained by the defendant, and other penalties imposed or likely to be imposed.[212]

3. Other remedies

Injunctions are a possible remedy, especially in nuisance actions where they **62**
are the norm.[213] These are usually negative (orders to cease), but can be positive (orders to do) although courts are generally hesitant to accept surveillance burdens that might be difficult to fulfil. Injunctions have been accepted as a possible remedy in class actions, and in *Regroupement des citoyens du quartier St-Georges Inc. v. Alcoa Canada Ltée*,[214] an industrial pollution class action, a range of positive (providing for reparation in kind) as well as negative injunctions were considered. Restitutionary remedies such as the imposition of a constructive trust or an accounting and disgorgement of profits are also conceivable, although rarely awarded in negligence and nuisance actions. Both were discussed and accepted as possible remedies in one class action, *Serhan Estate*,[215] and claimed in another, *LeFrançois*, because of the "alleged callous and intentional disregard of the rights of class members for the purpose of advancing their own economic interests".[216]

4. Costs of pursuing a claim

(a) General cost rule

In Canada, each party is responsible for his or her own costs of litigation **63**
(i.e. legal fees and disbursements, including for expert witnesses, etc.) although some (or very occasionally all) of the winning party's costs are

212 *Whiten* [2002] 1 S.C.R. 595, par. 111 ff. See also art. 1621 C.C.Q. (listing as factors "the gravity of the debtor's fault, his patrimonial situation, the extent of the reparation for which he is already liable to the creditor and, where such is the case, the fact that the payment of damages is wholly or partly assumed by a third party").

213 Abatement, a self-help remedy, is also possible, although is regarded with disfavour by the courts.

214 [2007] 6 R.J.Q. 1581.

215 *Serhan Estate* (2006) 269 D.L.R. (4th) 279 (diabetes monitoring system; class action certified). The discussion centred around, on one hand, the doctrine of "waiver of tort", described as "one of the lesser appreciated areas within the scope of the law of restitution", one which has "nothing whatever to do with waiver and really very little to do with tort" (par. 51, citing *P.D. Maddaugh/J.D. McCamus*, The Law of Restitution (2005) 24–1) and, on the other, the Canadian approach to the constructive trust, as set out in *Soulos v. Korkontzilas* [1997] 2 S.C.R. 217 (S.C.C.). Both are outside the scope of this paper.

216 *LeFrançois* 2008 CanLII 15770 (heart defibrillator) par. 18.

usually shifted to the losing party under the principle that "costs follow the event" (i.e. "loser pays").[217] In the common law provinces, awards of costs are at the discretion of the judge hearing the case, who takes into account a number of factors in determining what is "fair and reasonable" for a losing party to pay. Legal fees are assessed to reflect what is fair considering the lawyer's experience, hourly rate and number of hours spent on the file; disbursements are similarly assessed to reflect what a losing party could reasonably be expected to pay. In the end, the amount awarded must be generally proportional to the amount of damages claimed, the amount awarded by the court, the complexity of the issues and the importance of the dispute.[218] Usually, the amount thus awarded only partially indemnifies a successful party for his litigation costs (an award sometimes described as "ordinary costs" or "party-and-party costs", depending on the province), with the average award covering about 60% of full costs. Occasionally, courts may assess costs on a "substantial indemnity" (about 90%) basis (sometimes called "special costs" or "solicitor-and-client costs") to sanction an unsuccessful party for unnecessarily increasing the time and expense of the proceedings (e.g. through frivolous or vexations behaviour, bringing unwarranted motions, refusing a reasonable offer of settlement, etc.).[219] In Quebec, the approach is much less broad-brush than in the common law provinces, as costs are awarded according to set tariffs;[220] this results in costs awards much lower than in the common law provinces. Quebec courts may award legal fees higher than the tariff in "important" cases;[221] and like their common law counterparts, they may assess the successful party's legal fees at higher than the tariff rate (called "extra-judicial fees") in the event of procedural misconduct by the unsuccessful party.[222] As for class actions, costs awards in Quebec and some common law provinces (e.g. Ontario) follow the general cost-shifting approaches outlined above,[223] whereas other provinces (e.g.

217 See generally *M. Orkin*, The Law of Costs (2nd ed. 2003) and *Penney*, Queen's Law Journal 30 (2004-05) 205–259.

218 See e.g. Ontario Rules of Civil Procedure (fn. 170) rule 57.01.

219 "Full indemnity" (100%) costs awards are also possible, although rarely awarded, to sanction more egregious conduct; and an award of costs can be made against a losing party's lawyer personally, to sanction his conduct (e.g. engaging in sharp practices).

220 E.g. Tariff of Judicial Fees of Advocates, Revised Regulations of Quebec (R.R.Q.) 1981, c. B-1, r. 13; Regulation respecting Indemnities and Allowances Payable to Witnesses Summoned before Courts of Justice, R.R.Q. 1981, c. C-25, r. 2. See also art. 477 C.C.P.

221 Tariff, art. 15. This is not often done, even though the Tariff dates from 1978 and the rates have been described as "derisory": *Baudouin/Deslauriers* (fn. 4) par. 1–339.

222 E.g. *Royal Lepage Commercial Inc. v. 109650 Canada Ltd.* 2007 QCCA 915 (Que. C.A.).

223 Art. 1050.1 C.C.P.; Class Proceedings Act (fn. 192) s. 31 (providing that in exercising its usual discretion, court may consider "whether class proceeding was a test case, raised a

British Columbia, Saskatchewan) have adopted a "no-cost-recovery" model except in cases of procedural misconduct.[224] Finally, contingent fees are a form of lawyer-financing of the costs of litigation. These have been recognised in Canadian provinces starting in the 1960s, and their operation in cost-shifting situations presents interesting challenges.

(b) Costs of establishing causation

Under the adversarial system applied throughout Canada, including Quebec, each party must pay the costs of proving their claims, subject to recovering some from the losing party if successful (above, no. 63). **64**

5. Advance cover

There are no GMO-specific financial guarantees in Canada. The National Farmers Union suggested in 2000 that the federal government set up an industry-funded compensation scheme for affected farmers, but this has not been done.[225] **65**

novel point of law or involved a matter of public interest": discussed in *Gariepy 2* (2002) 23 C.P.C. (5th) 360; applied in *Ragoonanan Estate v. Imperial Tobacco Canada Limited* 2009 CanLII 4239 (Ont. Div. Ct.) (costs denied to successful defendant); not applied in *Kerr v. Danier Leather Inc.* [2007] 3 S.C.R. 331 (S.C.C.) (representative plaintiff held responsible for successful defendant's costs). Class action plaintiffs in both provinces can apply for public funding – in Quebec for assistance with both legal fees and disbursements, in Ontario for assistance with disbursements only.

224 Class Proceedings Act, R.S.B.C. 1996, c. 50, s. 37; Class Actions Act (fn. 64) s. 40. No-cost-recovery was recognised by the Court of Appeal in *Hoffman* [2005] 7 W.W.R. 665, par. 46 as being an incentive to plaintiffs, as the risk of an adverse award of costs is "a significant financial risk, one which serves as a disincentive in relation to individual action"; at the same time, it cautioned that removing this risk can serve "as an incentive to the unscrupulous to commence less than genuine class actions for the primary purpose of pressuring the defendant into a settlement, a settlement induced not by fear of being found to have engaged in wrongdoing but by concern over the enormous costs associated with class action litigation".

225 N.F.U. (fn. 6) point 9 (requesting the federal government to compel "companies which own patents on GM seeds or livestock to set up contingency funds to compensate for product liability"). The preamble describes the policy as an attempt "to introduce precaution and prudence into a process of GM food proliferation driven by profit", and point 8 of the policy stresses that it is "unreasonable to allow genetic modification companies to privately reap profits and not require that they also assume all costs.".

IX. Cross-border issues – Conflict of laws

1. Conflicts rules applicable before (or instead of) Rome II

66 The type of cases under survey presents difficult substantive and procedural cross-border issues. The substantive issue is choice of law. The basic rule in tort actions in Canada is that the law of the place of the wrong (lex loci delicti) applies, as the Supreme Court of Canada affirmed in *Tolofson v. Jensen*.[226] *Tolofson* was a motor vehicle accident case, and the place of the wrong was thus easily indentifiable. However, the Court recognised that identification can raise "thorny issues" where an act occurs in one place but the consequences are directly felt elsewhere: "In such a case, it may well be that the consequences would be held to constitute the wrong." The Court also noted that where the wrong "arises out of some transnational or interprovincial activity . . . territorial considerations may become muted; they may conflict and other considerations may play a determining role."[227] What these "other considerations" might be was suggested by the Court in an earlier case, *Moran v. Pyle National (Canada) Ltd.*, where the Court spoke of a foreign defendant who "carelessly manufactures a product in a foreign jurisdiction which enters into the normal channels of trade and he knows or ought to know both that as a result of his carelessness a consumer may well be injured and it is reasonably foreseeable that the product would be used or consumed where the plaintiff used or consumed it".[228] Although *Moran* was a jurisdictional case and the Court cautioned that "[t]he rules for determining *situs* for jurisdictional purposes may not be those which are used to identify the legal system under which the rights and liabilities of the parties fall to be determined",[229] the approach of the Court in the two cases is reflected in the Quebec Civil Code, which identifies the "obligation to make reparation for injury caused to another" as being governed by "the law of the country where the injurious act occurred" ("où le fait générateur du préjudice est survenu"), but notes that "if the injury appeared in another country" then the law of the other country applies if the author of the injurious act "should have foreseen that the damage would occur."[230] Choice of law is rarely discussed in the type of cases considered in this report. Choosing is

226 [1994] 3 S.C.R. 1022 (S.C.C.) (labelling its approach a "critique and reformulation" of the common law rule).

227 Ibid., par. 42.

228 [1975] 1 S.C.R. 393 (S.C.C.) 409.

229 Ibid., 397.

230 Art. 3126 C.C.Q.

unimportant where the action is in negligence and both the act of the defendant and the wrong to plaintiff occur in (different) common law provinces, as the law (the *Anns* analysis) is the same in each province. It becomes important where one of the provinces is Quebec, or another country, and the practice in such cases suggests that the courts follow *Moran* and treat the place where the injury occurs as the lex loci delicti.[231]

The main cross-border procedural issues relate to choice of forum (i.e. court jurisdiction), recognition of foreign judgments and class actions. **67**
The first two issues are correlative, as the Supreme Court of Canada recognised in its landmark decision in *Morguard Investments Ltd. v. De Savoye.*[232] This was a recognition-of-judgment case, and the Court reasoned that recognition and enforcement of foreign judgments should follow as a matter of course where the court had properly exercised its jurisdiction in the matter: this would be the case where there was a "real and substantial connection" between the place in which the court was located and subject matter of the action. *Morguard* was an interprovincial case, and the Court stressed constitutional reasons for a liberal approach to recognition, such as the intention to create a single country, the existence of common citizenship and personal mobility, economic integration and, in particular, a constitutional structure of the Canadian judicial system which constitutes an implied "full faith and credit" clause as in the United States and Australia.[233] In Quebec, the Civil Code is generally favourable to the recognition of foreign judgments, and provides that all are recognised and enforceable in Quebec except where they fall under enumerated exceptions (one of which is that the foreign court had not properly assumed jurisdiction);[234] foreign courts are recognised as having jurisdiction in the same circumstances in which Quebec courts have jurisdiction in conflicts cases (the "mirror effect") "to the extent that the dispute is

231 E.g. *Hunt* [1990] 2 S.C.R. 956 (asbestos manufactured in Quebec by Quebec companies; injury occurred in British Columbia; common law applied).

232 [1990] 3 S.C.R. 1077 (S.C.C.). See also *Hunt v. T&N plc* [1993] 4 S.C.R. 289 (S.C.C.) (Quebec statute blocking transfer of documents to foreign (i.e. B.C.) court held invalid).

233 "The Canadian judicial structure is so arranged that any concerns about differential quality of justice among the provinces can have no real foundation. All superior court judges – who also have superintending control over other provincial courts and tribunals – are appointed and paid by the federal authorities. And all are subject to final review by the Supreme Court of Canada, which can determine when the courts of one province have appropriately exercised jurisdiction in an action and the circumstances under which the courts of another province should recognise such judgments. Any danger resulting from unfair procedure is further avoided by sub-constitutional factors, such as for example the fact that Canadian lawyers adhere to the same code of ethics throughout Canada...": *Morguard* [1990] 3 S.C.R. 1077.

234 Art. 3155 C.C.Q.

substantially connected with the country whose authority is seised of the case."[235] As for class actions, finally, the main issue is the application of the above procedural rules in multi-jurisdictional class actions (pan-Canadian or even international). This is a large, open topic and much could be said.[236] There is no overarching legislation dealing with the issue, and the Supreme Court of Canada has expressed the view that resolution is a matter for the legislatures, not the courts.[237] Present practice varies, with some cases being dealt with at a national (or at least common law provinces) level,[238] but others being litigated in separate provincial class actions.[239]

2. Special regime for cross-border claims

68 There is no special cross-border regime in Canada applicable to GMOs.

235 Art. 3164 C.C.Q. (Quebec courts have jurisdiction where the defendant is domiciled in Quebec, or "if the dispute has a sufficient connection with Quebec" – or where actions outside Quebec are impossible or cannot reasonably be required: art. 3134 and 3136). Both the common law and civil law courts recognise that jurisdiction otherwise properly assumed (jurisdiction simpliciter) should be declined if the foreign courts are better placed to decide (forum non conveniens). On the interrelationship between recognition of foreign judgments and forum non conveniens under Quebec law, see *Canada Post Corporation v. Lépine* [2009] 1 S.C.R. 549 (S.C.C.).

236 See e.g. *T.J. Monestier*, Personal Jurisdiction over Non-Resident Plaintiffs in Multi-Jurisdictional Class Actions: Have we gone down the wrong road?, Queen's Faculty of Law Legal Studies Research Paper Series, Working Paper No. 08–05 (2008), http://ssrn.com/abstract=1291246; *C. Jones*, The Case for the National Class, Canadian Class Action Review 1 (2004) 29 ff.; *G. Saumier*, USA-Canada Class Actions: Trading in procedural fairness, Global Jurist Advances 5:2 (2005) art. 1; *J. Walker*, Crossborder Class Actions: A view from across the border, Michigan State Law Review (2004) 755 ff.

237 *Canada Post* [2009] 1 S.C.R. 549.

238 E.g. the BSE litigation (above, no. 11), where one action, *Sauer*, was brought on behalf of all Canadian cattle farmers except those of Quebec, and another, *Bernèche*, on behalf of those of Quebec. This division reflects the basic choice-of-law watershed in Canadian private law.

239 E.g. the cholesterol drug litigation, where at least six separate class actions have been brought, one in Quebec and at least five in common law provinces, one of which (Manitoba) is seeking certification for a national class made up of residents of the remaining provinces: see *Walls* (2007) 217 Man. R. (2d) 66.

Damage Caused by GMOs under US Law

A. Bryan Endres*

I. General overview

1. Special liability or redress scheme for GMOs

The United States does not have a stand alone GMO compensation/redress regime. Nor is there a single legislative statute governing safety assessment and approval for commercial use. Rather, the United Sates has separate regulatory and civil liability regimes comprised of a patchwork of federal regulatory legislation[1] to assess the health, safety and environmental impact of GMOs (The Coordinated Framework for the Regulation of Biotechnology[2]) and an independent, common law-based tort system to allocate civil liability for harms caused by GMOs.[3] As discussed later, liability 1

* Assistant Professor of Agricultural Law, University of Illinois.

1 Three federal statutes serve as the foundation for regulating the health, safety and environmental impact of GMOS: The Plant Protection Act, 7 United States Code (U.S.C.) §§ 7701–7786; the Federal Insecticide, Fungicide, and Rodenticide Act, 7 U.S.C. §§ 136–136y; and the Federal Food, Drug, and Cosmetic Act, 21 U.S.C. §§ 301–399a.

2 Coordinated Framework for the Regulation of Biotechnology, 51 Federal Register 23,302 (26 June 1986). For a brief description of the responsibilities of the United States Department of Agriculture (USDA), the Environmental Protection Agency (EPA) and the Food and Drug Administration (FDA) under the Coordinated Framework for the Regulation of Biotechnology, see *A.B. Endres*, Coexistence Strategies, the Common Law of Biotechnology and Economic Liability Risks, 13 Drake Journal of Agricultural Law 119–125 (2008).

3 When Congress elects to regulate activities with environmental implications, it generally includes specific liability provisions authorizing citizen-initiated enforcement actions (i.e., citizen suits). *See* Comprehensive Environmental Response, Compensation and Liability Act (CERCLA), 42 U.S.C. § 9659(a)(1); Clean Air Act, 42 U.S.C. § 7604; Clean Water Act, 33 U.S.C. § 1365. Because the Coordinated Framework used existing statutes to craft the regulatory regime for GMOs rather than an independent law, Congress did not provide a statutory-based claim for citizen-initiated civil liability. See *A.B. Endres*, GMO: Genetically Modified Organism or Gigantic Monetary Obligation? The Liability Schemes for GMO Damage in the United States and the European Union, 22 Loyola of Los Angeles International and Comparative Law Review (Loy.L.A.Int'l & Comp.L.Rev.) 453, 481 f. (2000). Moreover, the Federal Insecticide, Fungicide, and Rodenticide Act, under which the Environmental Protection Agency registers GMOs for pesticide tolerances, is some-

for harms caused by GMOs may, in some circumstances, include economic damages. Currently, there are no plans at the national level to introduce statutory liability or compensation schemes with respect to GMOs. Although the implementing agencies periodically revise their regulations, Congress has not discussed an omnibus regulatory statute to consolidate or significantly revise the current regulatory structure.

2. State liability

2 With respect to government civil liability, general rules of sovereign immunity, including the federal government's limited waiver of sovereign immunity under the Federal Tort Claims Act,[4] most likely would apply. In general, the limited waiver of sovereign immunity applies in actions to recover for injury or loss of property, or personal injury or death caused by the negligent or wrongful act or omission of any employee of the federal government while acting within the scope of the employee's office. Conceivably this could apply in a situation in which a government employee's tortious conduct in handling a GMO caused a physical injury to property or person. However, because there has not yet been a case asserting liability on the part of the state for GMO damage, the common law system has not addressed this issue specifically. As a practical matter, legal challenges to federal action in the GMO context have arisen not under the tort system, but rather within the context of administrative law (challenges to the adequacy of the administrative procedures followed by the federal government with respect to approving field trials and decisions to deregulate specific crop varieties).[5]

---

what unique among environmental statutes as it does not contain a citizen suit provision. *J.G. Miller/C.N. Johnston*, The Law of Hazardous Waste Disposal and Remediation (West Pub. Co., 1996) at 382.

4 Federal Tort Claims Act, 28 U.S.C. § 1346(b)(1).

5 Courts have resolved four recent challenges to the federal government's oversight of field trials and decisions to deregulate specific crop varieties. *Center for Food Safety v. Vilsack*, (N.D. Cal. No. 08-00484-JSW, 21 September 2009) (finding USDA failed to full assess the environmental impact of genetically engineered sugar beets prior to the agency's deregulation decision); *Geertson Seed Farm v. Johanns*, 541 Federal Reporter, Third Series (F.3d) 938 (9th Cir. 2008) (affirming permanent injunction against USDA's deregulation of genetically engineered alfalfa until the agency completes an environmental impact statement), reversed on appeal to the Supreme Court, *Monsanto Co. v. Geertson Seed Farms*, Case No. 09-475 (June 21, 2010) (overturning the trial court's permanent injuction prohibiting planting of GM alfalfa, but leaving in place the trial court's factual finding that the USDA failed to fully assess the environmental impact). *International Center for Technology Assessment v. Johanns*, 473 Federal Supplement, Second Series (F. Supp. 2d) 9 (D.D.C. 2007) (granting summary judgment on plaintiffs' National Environmental Policy Act claim that the agency failed to properly assess potential impact of field trial for geneti-

II. Damage

1. Recoverable losses

In the United States' common law-based system, recovery for losses caused by GMOs remains somewhat unsettled due to the scarcity of case law upon which to make solid conclusions. Despite this limitation, one important court decision to date has clarified that some losses caused by the commingling of GMOs unapproved for human consumption with the domestic food supply are recoverable under various common law theories.[6] In *In re StarLink Corn Products Liability Litigation*, corn farmers brought a class action lawsuit against the creator and manufacturer of the StarLink corn variety, alleging that defendants disseminated a product unapproved for food use that contaminated the corn supply, increasing farming costs and depressing corn prices. Plaintiff sought to recover damages based on state common law negligence, strict liability, nuisance, and conversion claims. The court held that the plaintiffs alleged viable claims for losses if (1) their crops were contaminated by pollen from StarLink corn on a neighbouring farm or (2) their harvest was contaminated by commingling with StarLink corn in a transport or storage facility. As discussed in no. 4 below, the court refused to allow the farmers' claims for damages arising from unknowingly purchasing seed contaminated with the StarLink DNA (i.e., the seed suppliers' inventory had been contaminated with the StarLink variety). Several scholars, both before and after the *StarLink* court decision, have defined further the potential losses resulting from GMO commingling. These scholars generally have focused on the liability scenarios arising from farmer versus farmer litigation, basing their respective analyses **3**

cally engineered grass); *Center for Food Safety v. Johanns*, 451 F. Supp. 2d 1165 (D. Hawaii 2006) (finding that the USDA's issuance of a permit for the open-air field testing of experimental, genetically engineered, pharmaceutical-producing plant varieties violated the Endangered Species Act and the National Environmental Policy Act).

6 *In re StarLink Corn Products Liability Litigation*, 212 F. Supp. 2d 828 (N.D. Ill. 2002). For a more thorough discussion of the *StarLink* case, see *D.L. Uchtmann*, StarLink^{TM} – A Case Study of Agricultural Biotechnology Regulation, 7 Drake Journal of Agricultural Law 159 (2002). In addition to the Starlink corn litigation, a federal Jury in St. Louis, Missouri, in the Case *In re Genetically Modified Rice Litigation*, Case No. 06 -md-01811 (Dec. 4, 2009), awarded two test plaintiffs approximately $ 2 million in compensatory damages for economic loss from the contamination of their rice crops with an experimental biotech rice variety-Liberty Link 601 rice. For a more thorough discussion of the case, see *T.P. Redick/ A.B. Endres*, Jury Verdict Against Bayer for Liberty Link rice breaks New Ground in Biotech Liability, 26 Agricultural Law update 2 (2009). Subsequent jury verdicts in other cases included $ 48 million for twelve rice farmers in Arkansas (April 15, 2010) and $ 1.5 million to a Mississippi farmer (Feb. 8, 2010), with hundreds of additional trials pending in various States in the southern United States. For information on recent events in the rice litigation see http://www.bayerricelitigation.com.

on established tort concepts within the agricultural context.[7] Some potentially viable claims include: negligence, nuisance, trespass and, in some circumstances, strict liability. Although the defendant in the *StarLink* case was the seed developer, there is no indication from the case that the successful claims would not apply in disputes among neighbouring farmers. As of this writing, no reported case in the United States has assessed liability in a farmer versus farmer litigation regarding GMO contamination via pollen drift, shared farm machinery or other sources of admixture.

2. Pure economic loss

4 In sum, the economic loss doctrine in the United States provides that physical injuries to persons or property are compensable; solely economic injuries are not.[8] Resolution thus turns on determining which GMO-caused injuries are "physical." The *StarLink* and *Sample v. Monsanto*[9] cases provide the best examples of the application of "physical injury" and economic loss doctrine with respect to agricultural use of GMOs in the United States. The general rule from both *StarLink* and *Sample* is that absent a physical injury, plaintiffs may not recover for drops in market prices. Nor may the plaintiffs recover for any additional costs, such as testing procedures, imposed by a marketplace seeking assurances of genetic purity.[10] But, if there was some physical harm to plaintiffs' crop, plaintiffs could avoid the economic loss doctrine's bar to recovery.[11] For example, in *Sample*, plaintiffs argued that Monsanto's domestic commercialisation of genetically engineered plant varieties, which were unapproved for export to the European Union, damaged the entire domestic corn and soybean industry (via lost revenue) when the European Union rejected biotech corn and soybean originating from the United Sates. Plaintiffs did not claim any physical contamination or injury to their specific property (i.e., their crops or harvest). Absent evidence of a physical injury, the court granted Monsanto's motion

7 *P.J. Heald/J.C. Smith*, The Problem of Social Cost in a Genetically Modified Age, 58 Hastings Law Journal 87 (2006); *D.L. Kershen*, Legal Liability Issues in Agricultural Biotechnology, 44 Crop Science 456 (2004); *M.R. Grossman*, Genetically Modified Crops in the United States: Federal Regulation and State Tort Liability, 5 Environmental Law Review 86 (2003); *T.P. Redick*, Biopharming, Biosafety, and Billion Dollar Debacles: Preventing Liability for Biotech Crops, 8 Drake Journal of Agricultural Law 115 (2003); *M.R. Grossman*, Biotechnology, Property Rights and the Environment, 50 American Journal of Comparative Law 215 (2002); *Endres*, 22 Loy.L.A.Int'l & Comp.L.Rev. 453 (2000).

8 *In re StarLink*, 212 F. Supp. 2d at 838.

9 *Sample v. Monsanto Co.*, 283 F. Supp. 2d 1088 (2003).

10 *In re StarLink*, 212 F. Supp. 2d at 842.

11 *In re StarLink*, 212 F. Supp. 2d at 843.

for summary judgment on the nuisance tort claim.[12] In contrast, in the *StarLink* case, the court held that non-StarLink corn crops sufferred a physical injury when pollinated by StarLink corn. The pollen caused the non-StarLink corn plants to develop the Cry9C (StarLink) protein, which rendered what would otherwise have been a valuable food crop unfit for human consumption. Similarly, non-StarLink corn was damaged when commingled with StarLink corn post-harvest. Although the commingling did not change the physical characteristics of the non-StarLink corn, once mixed, re-segregation of the combined product into edible and inedible parts was impossible – the court considered the entire batch tainted as it could only be used for the limited domestic and industrial purposes for which StarLink had government approval.[13] On the other hand, plaintiffs' claims for economic damages resulting from impurities in the seed itself would fall under the economic loss doctrine's bar to recovery. Plaintiff seed purchasers suffering injury from the genetically impure seed could, in theory, have negotiated at the time of sale an express warranty from the seller that the seed is 100% non-biotech. Of course, this express warranty would increase the price of the seed. In the alternative, the farmer could accept the commercial risk of operating without the seed warranty and the attendant potential loss in price for delivering a harvest with genetically engineered DNA.[14] A remedy in the former case would rest in contract (breach of warranty), rather than tort. In the later scenario (purchase of seed without a warranty), the United States' court system generally is reluctant to substitute a tort remedy when parties have the opportunity, but fail to negotiate a contractual remedy.[15] Plaintiffs' injury from the contaminated seed is a question of contracts and whether the buyers received what they had bargained for.[16] Regardless of the theoretical possibility for bargaining between farmer and seed supplier, most contracts for the sale of seed in the United States contain a variation of the International Seed Federation's Model Conditions of Sale Disclaimer in which the seed supplier expressly disclaims all guarantees that the seed is

12 *T.P. Redick/D.L. Uchtmann*, Coexistence Through Contracts: Export-Oriented Stewardship in Agricultural Biotechnology vs. California's Precautionary Containment, 13 Drake Journal of Agricultural Law 207, 217 f. (2008).

13 *In re StarLink*, 212 F. Supp. 2d at 841.

14 *Redick/Uchtmann*, 13 Drake Journal of Agricultural Law, 215 f. (2008).

15 *In re StarLink*, 212 F. Supp. 2d at 842 (citing *Hapka v. Paquin Farms*, 458 North Western Reporter, Second Series (N.W.2d) 683, 688 (Minn. 1990).

16 *In re StarLink*, 212 F. Supp. 2d at 842. See also *Redick/Uchtmann*, 13 Drake Journal of Agricultural Law, 216–218 (2008) (discussing the economic loss rule as applied in the *StarLink* case). For a discussion of seed purity laws and farmer liability concerns in the United States, see *A.B. Endres*, Revising Seed Purity Laws to Account for the Adventitious Presence of Genetically Modified Varieties: A First Step Towards Coexistence, 1 Journal of Food Law & Policy 131 (2005).

free of genetically engineered DNA.[17] As a result, most claims for damages caused by seed impurity in the United Sates would fail via either enforcement of the express disclaimer of warranty in a contract action or the economic loss doctrine in a tort action.

3. Mere fear of a loss

5 For the reasons articulated within the context of the economic loss doctrine discussed above, allegations of mere fear of losses, absent physical injury, would fail to state a common law claim in the United States. The *Sample v. Monsanto* case is instructive. In *Sample*, the court found for the defendant as a matter of law because plaintiffs offered no evidence of physical injury. One plaintiff did allege that he belonged to a "cooperative" and thus "may have experienced property damage" as a result of commingling grain with other members of the cooperative. This speculative statement, unsupported by any evidence, resulted in the granting of summary judgment for the defendant on the tort claims under the economic loss doctrine.[18]

4. Standard of proof

6 Under the United States system of party initiation and party presentation for civil claims, the party bringing the suit must provide the "facts" needed for the requested resolution of the dispute.[19] Failure by the plaintiff at the pleading stage of the case to allege sufficient facts so as to state a claim for relief (i.e., allege the facts necessary to later prove a cause of action for negligence) will result in dismissal.[20] After the pleading stage, the plaintiff bears a "burden of production," which requires the plaintiff to produce the evidence to establish his or her claim.[21] For trial purposes, yet another standard applies–the "burden of persuasion."[22] Again, the burden of persuasion falls upon the plaintiff (or the party asserting the pleading).[23] In the

---

17 *Endres*, 1 Journal of Food Law § Policy, at 154 f. (2005).

18 *Sample v. Monsanto*, 283 F. Supp. 2d at 1093 f. (2003).

19 *C.A. Wright & K.A. Graham, Jr.*, Federal Rules of Evidence, 21B Federal Practice and Procedure § 5122 (2009).

20 Federal Rules of Civil Procedure 12(b)(6), Motion for Failure to State a Claim upon which Relief can be Granted; Federal Rules of Civil Procedure 12(c), Motion for Judgment on the Pleadings.

21 *Wright & Graham* (fn. 19) § 5122.

22 Ibid., § 5122.

23 Ibid., § 5122.

civil context, the weight of this burden of persuasion is "the preponderance of the evidence."[24] Although courts have struggled to precisely define the preponderance of the evidence standard,[25] one relatively straightforward explanation of this standard by the United State Supreme Court is "that the existence of the fact is more probable than its nonexistence."[26]

5. Nominal losses

For some causes of action in tort, courts may award nominal damages where a tort exists, but the tort caused no harm or the amount of harm was not significant.[27] For causes of action in which actual damages is a required element of the tort (e.g. negligence), courts will not award nominal damages. In the context of GMOs, trespass and nuisance causes of actions could result in the award of nominal damages.[28] Nominal damages may also result when the plaintiff has sought significant damages, but failed to prove significant harm. Although an unsettled area of law, for intentional, wilful or malicious acts, punitive damages may be awarded in addition to nominal damages.[29] 7

6. Mass losses

With respect to mass losses, the United States does not cap recovery of actual damages. Punitive damages assessed for a mass loss or in response to an isolated claim, however, at some point, may be in a ratio so much in excess of compensatory damages as to render the punitive award in violation of the due process clauses of the Fifth and Fourteenth Amendments to the United States Constitution. The United States Supreme Court, in recent years, has struggled with establishing a precise test for determining when the ratio of compensatory to punitive damages is constitutionally improper.[30] A full discussion of this issue, however, is beyond the scope of this report. 8

24 Ibid., § 5122.
25 Ibid., § 5122.
26 *In re Winship*, 397 Supreme Court Reports (U.S.) 358, 371 f. (1970) (J. Harlan, concurring).
27 Restatement (Second) of Torts § 907, comment b.
28 Restatement (Second) of Torts § 907, comment b. See also *Pion v. Dwight*, 497 North Eastern Reporter, Second Series (N.E.2d) 20 (Mass. App. Ct. 1981) (finding for plaintiff, but awarding no damages for wrongful interference with easement).
29 *J.D. Lee/B. Lindahl*, 4 Modern Tort Law: Liability and Litigation (2nd ed.) § 35:18 (2009); *J.J. Kircher/C.M. Wiseman*, Punitive Damages: Law & Practice (2nd ed.) § 5:21 (2009).
30 See *BMW of North America, Inc. v. Gore*, 517 U.S. 559 (1996); *State Farm Auto Insurance v. Campbell*, 538 U.S. 408 (2003); *Philip Morris USA v. Williams*, 549 U.S. 346 (2007).

III. Causation

1. Uncertainty of merely potential causes

9 As noted in no. 6, above, the burden of persuasion rests on the party asserting the cause of action. This burden of persuasion attaches to all elements of the cause of action, including causation. Accordingly, factual uncertainty in the GMO liability context, which in the United States is a civil action, requires establishing facts proving causation by a preponderance of the evidence. Although variations exist among the state courts, in cases in which causation may lie partially within the victim's own sphere (e.g. the victim violated segregation rules/good farming practices), a determination must be made (by the jury) as to whether the victim's actions were the "sole proximate cause" of his or her injury.[31] If so, then the defendant's actions were not the proximate cause of the victim's injury and liability may not attach due to lack of causation.[32] So long as the victim's actions were not the sole proximate cause,[33] the comparative negligence doctrine would apply to apportion damages according to the respective parties' fault.[34] Under this doctrine, the fact finder would allocate negligence in terms of a percentage and reduce the compensatory damages due to the victim in proportion to the relative amount of negligence attributed to the victim.[35] Comparative negligence, therefore, does not squarely address causation, but rather is a method to apportion damages for harm caused by multiple individuals, one of whom is the victim.

2. Complex causation scenarios

10 To address complex liability scenarios involving multiple potential tortfeasors, courts have crafted at least four types of collective liability the-

31 57B American Jurisprudence 2d Negligence § 995 (2009).

32 57B American Jurisprudence 2d Negligence § 992 (2009). Some courts, however, disregard the issue of "sole proximate cause" and forward the issue to the jury for a determination under comparative negligence standards discussed below. 57B American Jurisprudence 2d Negligence § 994 (2009).

33 Various additional defenses may exist, such as the last clear chance, assumption of risk, or superseding cause doctrines; all of which are beyond the scope of this discussion.

34 States adopted comparative negligence statutes to replace the harsh effects of the common law rule of contributory negligence, which acted as a complete bar to plaintiff's recovery of damages no matter how slight the contributing negligence in relation to the primary tortfeasor. See *Kramer v. Petisi*, 940 Atlantic Reporter, Second Series (A.2d) 800 (Conn. 2008) (discussing the rationale for comparative negligence schemes).

35 57B American Jurisprudence 2d Negligence § 954 (2009).

ories: alternate liability, enterprise liability, concert of action and market share liability. Courts have not yet applied these theories in the GMO context. In the seminal case of *Summers v. Tice*,[36] the California Supreme Court adopted the first collective liability theory – alternate liability. In *Summers*, two hunters simultaneously fired their weapons in the same direction. The plaintiff suffered injuries to his head and was unable to prove which defendant's shot caused the injury. To solve this problem, the court shifted the burden of proving the identity of the actual tortfeasor to the defendants and imposed joint and several liability.[37] In other scenarios, courts have imposed enterprise liability when the wrongdoing of an entire industry is viewed as a single "enterprise." In *Hall v. E.I. Du Pont de Nemours*,[38] plaintiffs were unable to prove the manufacturer of a defective blasting cap because the product was destroyed in the blasting process. Accordingly, plaintiffs joined several manufacturers and the industry trade association as defendants. Plaintiffs presented evidence that the manufacturers had contributed to an industry-wide safety program. The court held that in order to sustain a claim of enterprise liability, plaintiffs must demonstrate defendants' joint awareness of the risks at issue and their joint capacity to reduce or affect those risks.[39] The third form of collective liability – concert of action – requires a plaintiff to prove that defendants engaged in a common plan to commit a tort. Although not applied in any GMO cases to date, courts have applied the concert of action theory in diethylstilbestrol (DES) litigation.[40] The rationale underlying concerted action liability rests upon the principle that "[a]ll those who, in pursuance of a common plan or design to commit a tortious act, actively take part in it, or further it by cooperation or request, or who lend aid or encouragement to the wrongdoer, or ratify and adopt his acts done for their benefit, are equally liable."[41] In the DES litigation context, the plaintiff in *Bichler v. Eli Lilly & Co.*, successfully sued one of the three manufacturers of DES supplying a particular pharmacy (out of a total of 147 suppliers nationwide). Finally, market share liability is a hybrid of alternate

36 *Summers v. Tice*, 199 Pacific Reporter, Second Series (P.2d) 1 (Cal. 1948). See also Restatement (Second) of Torts § 433B.

37 *Summers*, 199 P.2d at 5.

38 *Hall v. E.I. Du Pont de Nemours*, 345 F. Supp. 353 (D.C.N.Y. 1972).

39 *Hall*, 345 F. Supp. at 379. In *Hall*, the court emphasised the special applicability of this rule to industries composed of a small number of units. "What would be fair and feasible with regard to an industry of five or ten producers might be manifestly unreasonable if applied to a decentralized industry composed of thousands of small producers." Ibid.

40 *Bichler v. Eli Lilly & Co.*, 55 New York Reports, Second Series (N.Y.2d) 571 (1982). See also Restatement (Second) of Torts § 876.

41 *Bichler*, 55 N.Y.2d at 580 (quoting *Prossor*, Torts (4th ed.) § 46).

and enterprise liability in the United States. In accordance with this theory, courts apportion liability according to the defendant's particular share of the market. Often courts will require only joining enough defendants to represent a significant share of the market. In the DES context, some courts have approved use of market share liability, holding that it is reasonable to measure the probability that one of the joined defendants supplied the allegedly injury-causing product by the percentage of the product sold by each defendant.[42] Again, it is important to note that courts have not yet had a factual scenario appropriate to consider applying these liability theories to cases involving GMOs.

3. Force majeure

11 As a general rule, force majeure does not alter causation in analysis in the United States, especially in light of the adoption of comparative negligence standards. For a court to excuse liability under the doctrine of force majeure, human activities cannot have contributed to the loss in any degree.[43] To the extent the loss is attributed to human action, the respective human defendant would be liable under comparative negligence standards.

4. Threshold to prove causation

12 As discussed in no. 9 above, the burden of persuasion for both judge and jury in a civil action requires establishing facts to prove causation by a preponderance of the evidence.

5. Special rules on causation

13 In the relatively few GMO cases decided to date, courts have not applied special rules with respect to causation. GMO-based litigation is a civil action and follows the applicable jurisdiction's civil procedure and evidentiary rules, regardless of subject matter.

---

42 *Sindell v. Abbott Laboratories*, 26 California Reports, Third Series (Cal.3d) 588, 611 (1980).

43 *Moore v. Goltlieb*, 848 New York Supplement, Second Series (N.Y.S.2d) 328 (2007); *Cangialosi v. Hallen Const. Corp.*, 828 New York Supreme Court Appellate Division Reports, Second Series (A.D.2d) 565 (2001); *Michaels v. New York Central Railroad Co.*, 1894 WL 4160 (N.Y. 1894).

IV. Types of liability

1. Fault liability

(a) Special rules governing fault

Fault rules in cases involving alleged harm from GMOs would follow the standard rules applied in civil litigation. As discussed in no. 6, the burden of persuasion falls upon the plaintiff. The standard applied in all civil actions for damages is "the preponderance of the evidence" standard. 14

(b) Impact of specific rules of conduct

For fault-based liability (e.g. negligence,[44] nuisance[45]), custom is one approach to determine the reasonableness of a defendant's action. Widespread and longstanding practices typically, but not necessarily, are reasonable within a negligence analysis. For example, if the industry custom is not updated adequately, compliance with the outdated custom will not preclude liability for unreasonable behaviour. Moreover, custom is normally viewed as evidence of a reasonableness determination rather than a substitute for the appropriate standard of care.[46] To the extent specific statutory or regulatory rules exist, and the court has adopted those rules as establishing the standard of care,[47] the unexcused violation of the statutory/regulatory rule is negligence "per se."[48] The *StarLink* case is one such example. In *StarLink*, the Environmental Protection Agency issued, in accordance with the Federal Insecticide, Fungicide, and Rodenticide Act, a limited registration of the pesticide compound found in the transgenic 15

44 Negligence is either "(a) an act which the actor as a reasonable man should recognize as involving an unreasonable risk of causing an invasion of an interest of another, or (b) a failure to do an act which is necessary for the protection or assistance of another and which the actor is under a duty to do." Restatement (Second) of Torts § 284.

45 Nuisance is an activity or condition that interferes unreasonably with the plaintiff's use and enjoyment of land. *M.R. Grossman*, Anticipatory Nuisance and the Prevention of Environmental Harm and Economic Loss from GMOs in the United States, 18 Journal of Environmental Law and Practice 107, 109 (2008). The concept is further defined as a "nontrespassory invasion of another's interest in the private use and enjoyment of land." Restatement (Second) of Torts § 821A.

46 *Endres*, 13 Drake Journal of Agricultural Law, at 146 (2008); Restatement (Second) of Torts § 295A.

47 Restatement (Second) of Torts § 286 (describing criteria for a court's adoption of statutory or regulatory rules as the standard of care).

48 Restatement (Second) of Torts § 288B.

seed.[49] The limited registration allowed the commercialisation of the new variety, but specifically required 660 foot buffer zones to minimise commingling with non-StarLink varieties and segregation of the harvested crop with subsequent distribution to only domestic non-food uses.[50] Defendants allegedly failed to enforce these regulatory requirements and the court, accordingly, allowed the claim of negligence per se to proceed.[51] A final custom-based analysis in the nuisance context also warrants brief discussion. Possessors of land have a right to the reasonable use and enjoyment of their property. A nuisance is an action that interferes with this right. Accordingly, the movement of GMO pollen from one property to another could, in some circumstances, constitute an unreasonable interference with the pollen recipient's use and enjoyment of his or her property.[52] Plaintiffs in the *StarLink* case successfully alleged that cross-pollination with StarLink varieties from a neighbouring farm could be a nuisance.[53] Most courts, however, have limited this doctrine to instances in which the nuisance-generating activity is a "new" occurrence. Specifically, a plaintiff cannot "come to the nuisance" and thereafter claim harm from a pre-existing condition or use of neighbouring land.[54] A hypothetical example of a coming to the nuisance defence might be the new establishment of an organic farm in an area with a long history of GMO use. Although the pollen drift from the GM plants may interfere with the ability of the new organic farm to maintain desired genetic purity, the organic operation "came to the nuisance" and thus may not be able to enjoin the actions of the prior-established GM farming operations. A similar concept also applies within the context of strict liability. Courts, in determining whether to classify an activity as "abnormally dangerous," and thus subject to strict liability, will examine, along with five other factors, the "inappropriateness of the activity to the place where it is carried on."[55] Accordingly, a farming operation using GMOs in an area surrounded by organic or non-GM production methods might be subject to

---

49 Certain Companies, Approval of Pesticide Product Registrations, 63 Fed. Reg. 43,936 (17 August 1998).

50 *Uchtmann*, 7 Drake Journal of Agricultural Law, at 185 (2002).

51 *In re StarLink*, 212 F. Supp. 2d at 852.

52 *Grossman*, 50 American Journal of Comparative Law, at 230–35 (2002); *Endres*, 22 Loy.-L.A.Int'l & Comp.L.Rev., at 491–94 (2000).

53 *In re StarLink*, 212 F. Supp. 2d at 844–47.

54 For a discussion of this "coming to the nuisance" doctrine and its limited application with respect to negligent or other unlawful actions that generate the unreasonable interference in the use of the land, see *Grossman*, 50 American Journal of Comparative Law, at 234–35 (2002).

55 Restatement (Second) of Torts § 520(e); *Endres*, 22 Loy.L.A.Int'l & Comp.L.Rev., at 488 f. (2000).

strict liability for any harm caused by pollen drift, while an identical farming operation in a different area (e.g. an area dominated by GM farms) could be protected from nuisance actions under the coming to the nuisance defence.

2. Product liability

(a) Development risk defence

16 The United States does not have a federal statute equivalent to the EC Product Liability Directive. Rather, the United States has a common law-based products liability regime based on a complex combination of negligence and strict liability principles according to the type of claim: manufacturing defect, design defect or failure to warn.[56] Some states have enacted individual product liability statutes for their respective jurisdictions. This summary, however, focuses on the common law system. A manufacturing defect is a departure from the product's design that gives rise to liability without fault.[57] In contrast, a product alleged to have a design defect meets the defendant's specifications, but the specifications themselves pose an unreasonable risk of injury. Accordingly, there is a reasonableness analysis to this claim whereby the ability to reduce foreseeable risk of harm posed by the adopted design of the product is balanced with the utility and cost of an alternative design.[58] Courts may consider a number of factors in determining whether the alternative design is reasonable and its omission renders the complained of product unreasonably safe. Factors include the magnitude and probability of the foreseeable harm, accompanying instructions/warnings, the consumer expectations regarding the product, and the relative advantages/disadvantages of the product as designed.[59] In most jurisdictions, the burden is on the plaintiff to demonstrate that a reasonable alternative design would have reduced the foreseeable risks of harm.[60] The reasonableness analysis embedded within the design defect claim seems to incorporate an "unforeseeable defects" defence. Moreover, the law gener-

56 Restatement (Third) of Torts – Products Liability § 2 (rationale and general application of strict liability and negligence rules to defective products).

57 Restatement (Third) of Torts – Products Liability § 2(a).

58 Restatement (Third) of Torts – Products Liability § 2(b).

59 Restatement (Third) of Torts – Products Liability § 2, comment f.

60 Restatement (Third) of Torts – Products Liability § 2, comment f. Three notable exceptions to this requirement are for cases involving product malfunction, safety standard violation or egregiously dangerous design. Restatement (Third) of Torts – Products Liability § 2, Reporters' Notes, comment d.

ally does not hold a business liable for the failure to recall a product after sale or distribution unless directed pursuant to a statute or administrative regulation.[61] The rationale for this rule is that if every improvement in product safety were to trigger a common-law duty to recall, manufacturers would incur tremendous costs each time they improved the safety of their product line.[62] With respect to allegations of inadequate instructions or warnings, courts have adopted a "reasonableness" test to assess the adequacy of the instructions or explicit warnings.[63]

(b) Alternative routes

17 As noted above, the United States, with some state-level variation, resolves product liability claims based on the common law. Accordingly, claims regarding allegedly defective food and agricultural products follow the general common law rules.[64] However, with respect to a manufacturing defect for a food product, a variation in the traditional strict liability rules applies. Specifically, the harm-causing ingredient in the food that is an alleged manufacturing defect constitutes an actionable defect only if a reasonable consumer would not expect the food product to contain that ingredient.[65] For example, the presence of a one-centimetre bone in a fish filet is not an unanticipated adulteration and thus would not violate the reasonable consumer expectation test.[66] In at least one state, once the plaintiff has established a prima facie case that the food is not merchantable, the burden or persuasion shifts to the defendant to prove that the plaintiff should have reasonably expected the ingredient.[67]

(c) Impact of compliance with rules and regulations

18 Although a product's non-compliance with a product safety statute or administrative regulation renders a product defective, a product conform-

---

61 Restatement (Third) of Torts – Products Liability § 11.

62 Restatement (Third) of Torts – Products Liability § 11, comment a.

63 Restatement (Third) of Torts – Products Liability § 2(c).

64 Restatement (Third) of Torts – Products Liability § 7, comment a.

65 Restatement (Third) of Torts – Products Liability § 7. This rule has replaced the foreign-natural test for manufacturing defects in most jurisdictions. Restatement (Third) of Torts – Products Liability § 7, Reporters' Note 1.

66 *Morrison's Cafeteria of Montgomery, Inc. v. Haddox*, 431 Southern Reporter, Second Series (So. 2d) 975, 978 (Ala. 1983).

67 *O'Dell v. DeJean's Packing Co.*, 585 Pacific Reporter, Second Series (P.2d) 399 (Okla. Ct. App. 1978).

ing to applicable safety statutes or regulations does not preclude, as a matter of law, a finding of product defect.[68] This reflects the traditional view that government safety requirements (absent pre-emption) represent a minimum standard and a higher standard may apply in some cases.[69] However, many jurisdictions give compliance with government safety standards substantial weight when assessing an allegedly defective design.[70]

3. Environmental liability

(a) Implementation of the Environmental Liability Directive

Not applicable. 19

(b) Environmental liability regime beyond the scope of the Directive

The United States does not have a single environmental harm statute similar to the Environmental Liability Directive (2004/35/EC). At the federal level, a variety of statutes address environmental harm, usually with respect to a particular media or product (e.g. Clean Water Act, Clean Air Act, Federal Insecticide, Fungicide, Rodenticide Act (pesticides), Comprehensive Environmental Response, Compensation and Liability Act (hazardous waste), Endangered Species Act). In addition, the National Environmental Policy Act (NEPA)[71] provides an umbrella level of environmental protection by imposing a procedural requirement on the federal government to assess potential impacts on the environment prior to any action by an agency. Specifically, the act requires all federal agencies to prepare a detailed statement to accompany any decision significantly affecting the environment. The detailed finding, known as an "Environmental Impact Statement," must address the proposed action's unavoidable adverse environmental effects, alternatives to the proposed action, and any irreversible resource commit- 20

---

68 Restatement (Third) of Torts – Products Liability § 4. Because an alleged manufacturing defect falls under strict liability, compliance with a statute or regulation is not relevant to the analysis.

69 Restatement (Third) of Torts – Products Liability 4, comment e.

70 Restatement (Third) of Torts – Products Liability § 4, comment e (citing *Sims v. Washex Machine Corp.*, 932 S.W.2d 559 (Tex. App. 1995) ["Compliance with government regulations is strong evidence, although not conclusive, that a machine was not defectively designed."]).

71 42 U.S.C. §§ 4321–4370a.

ments.[72] Although a procedural rather than substantive statute, plaintiffs, in a variety of lawsuits have successfully used NEPA to challenge agency decisions regarding use of GMOs. Most recently, in *Center for Food Safety v. Vilsack* and *Geertson Seed Farm v. Johanns,* plaintiffs asserted that the USDA failed to adequately assess the environmental impact of the deregulation of genetically engineered sugar beets and alfalfa, respectively. In both cases, the court agreed, reversing the agency's deregulation decision and ordering it to prepare an Environmental Impact Statement addressing specific areas of concern.[73] In *International Center for Technology Assessment v. Johanns*, the court granted plaintiffs' NEPA claim asserting that the agency failed to properly assess the potential environmental impact of an open-air field trial of genetically engineered grass.[74] Similarly, in *Center for Food Safety v. Johanns*, the court found that USDA's issuance of a permit for the field testing of experimental, genetically engineered pharmaceutical-producing plants violated NEPA and the Endangered Species Act.[75] This emphasis on the role of citizen suits reflects the current state of the tension between the regulatory agencies and segments of the public opposed to further expansion of the use of GMOs. As Miller and Johnston note, "Congress included citizen suits in the environmental statutes, in part, because of suspicion that the executive branch would not perform its responsibilities and, in part, because ... even if the executive branch performed heroically, it could not address every violation."[76] Of course, the perception of who is the "hero" in what seems to be a series of legal challenges of every recent agency decision relating to a new genetically engineered plant variety depends on one's perspective. To the biotech industry, this raises substantially the cost of bringing to market a new crop that otherwise would have government approval.[77] For opponents, citizen suits serve as an important environmental check on the government's regulatory review process.

---

72 42 U.S.C. § 4332(C).

73 *Center for Food Safety v.Vilsack*, (N.D. Cal. No. 08-00484-JSW, 21 September 2009); *Geertson Seed Farm v. Johanns*, 541 F.3d 938 (9th Cir. 2008), reversed with respect to the permanent injunction remedy by *Monsanto Co v. Geertson Seed Farms*, U.S. Supreme Court, Case No. 09-475 (June 21, 2010).

74 *International Center for Technology Assessment v. Johanns*, 473 F. Supp. 2d 9 (D.D.C. 2007).

75 *Center for Food Safety v. Johanns*, 451 F. Supp. 2d 1165 (D. Hawaii 2006).

76 *J.G. Miller/C.N. Johnston*, The Law of Hazardous Waste Disposal and Remediation (West Pub.Co. 1996) 383.

77 *T.P. Redick/A.B. Endres*, Litigating the Economic Impacts of Biotech Crops, 22 Natural Resources & Environment 24, 24–26 (2008).

(c) Claimants in cases of environmental harm

With the exception of three federal statutes, the Clean Water Act,[78] the Oil Pollution Control Act[79] and the Comprehensive Environmental Response, Compensation and Liability Act (CERCLA or Superfund), the government is unable to recover losses for harm to biodiversity or natural resource damages. Section 107 of CERCLA provides one example. CERCLA imposes liability for "damages to, injury to, destruction of, or loss of natural resources, including the reasonable costs of assessing such injury, destruction, or loss."[80] However, the general citizen suit provisions of CERCLA, discussed in footnote 3, above, do not apply with respect to such "natural resource damages." Rather, only the United States, individual States and Indian tribes may bring claims for recovery of natural resource damages. Moreover, any sums recovered through these actions must be used solely to restore, replace or acquire the equivalent of the damaged resources.[81] Most environmental laws merely impose civil or criminal penalties for violations. For example, the Endangered Species Act authorises civil penalties up to $ 25,000 and criminal penalties up to $ 50,000 for each violation.[82] In private litigation (e.g. nuisance or trespass claims with an environmental component), at least one commentator notes that punitive damages, in excess of nominal compensatory damages awarded to the plaintiff, could substitute for explicit consideration of natural resources damages.[83] In the GMO context, the USDA, on occasion, has imposed penalties for regulatory violations that result in environmental harm. In addition to civil penalties, the agency generally enters into a settlement agreement in lieu of protracted litigation in which the responsible party agrees to restore the environment. For example, in 2002, ProdiGene, Inc. entered into a consent agreement with the USDA and FDA in which it paid a civil penalty of $ 250,000 and agreed to reimburse the government for its costs in securing approximately 500,000 bushels of soybeans commingled with corn geneti- 21

78 33 U.S.C. § 1321(f)(4) (authorizing recovery for costs incurred in the restoration or replacement of natural resources damaged or destroyed by oil or hazardous substances in water).

79 33 U.S.C. § 2706(d) (authorizing damages for the cost of restoring damaged natural resources, the diminution in value of natural resources pending restoration and the cost of assessing damages).

80 42 U.S.C. § 9607(a)(4)(C).

81 *Miller/Johnston* (fn. 77) 736 (citing *Artesian Water Co. v. Government of New Castle County*, 851 F.2d 643, 649 (3d Cir. 1988).

82 16 U.S.C. § 1540(a), (b).

83 *A.B. Klass*, Punitive Damages and Valuing Harm, 92 Minnesota Law Review 83 (2007).

cally engineered to produce pharmaceuticals.[84] Syngenta, the developer of the Bt10 corn variety that was not intended for general release into the environment, agreed to a $ 375,000 settlement with USDA[85] and a $ 1.5 million fine to EPA.[86] These fines, however, are at the discretion of the agency. For example, although responsible for the unauthorised release of genetically engineered rice that contaminated the long grain rice supply, shut down exports and prompted the filing of several class action lawsuits, Bayer Crop Science was not assessed civil penalties.[87] Whereas, in 2007, the Scotts Company LLC entered into a settlement agreement for failure to follow permit conditions for field trials of its genetically engineered grass that entered the environment. The company paid a $ 500,000 civil penalty and agreed to monitor and destroy uncontained GM grass.[88]

(d) Special liability regime for losses sustained by individuals

22 Common law torts in the United States, such as nuisance, trespass or negligence, award compensatory damages to restore the injured party to their pre-tort position, including damages for pain and humiliation.[89] In the environmental context, this could include cleanup costs or diminution in value of the impacted property. Harm that does not impact the plaintiff's property or person (e.g. loss of biodiversity), however, falls outside the realm of compensatory damages. Accordingly, a generalised harm to the environment may not have a legally recognised impact on an individual's person or property, and thus escape a compensatory damages calculation. Professor Klass, however, argues that punitive damages – damages designed to punish and deter future tortious conduct[90] – could be applied to account for generalised harm to the environment that is outside the realm of private property interests.[91] *Johansen v. Combustion Engineering, Inc*,[92] presents an example of Professor Klass' argument. In this case, a

84 *Endres*, 1 Journal of Food Law & Policy, at 132 no. 6 (2005).
85 *A.B. Endres*, Coexistence Strategies in a Biotech World, at 208, no. 9.
86 EPA Fines Syngenta for Distributing Unapproved Corn, 26 Biotechnology Law Report 20 (2007).
87 *A.B. Endres*, Coexistence Strategies, the Common Law of Biotechnology, at 133–35; USDA, Biotechnology, Noncompliance History, available at http://www.aphis.usda.gov/biotechnology/compliance_history.shtml.
88 USDA, Biotechnology, Noncompliance History, available at http://www.aphis.usda.gov/biotechnology/compliance_history.shtml.
89 Restatement (Second) of Torts § 903.
90 Restatement (Second) of Torts § 908(1).
91 *Klass*, 92 Minnesota Law Review, at 87 f (2007).
92 *Johansen v. Combustion Engineering, Inc.*, 170 F.3d 1320 (11th Cir. 1999).

jury awarded twenty-three landowners a total of $ 47,000 in compensatory damages, but $ 45 million in punitive damages, for defendant's operation of a mine that polluted the environment (and plaintiffs' property).[93] Public nuisance actions – claims for an unreasonable interference with a right common to the general public – are more likely to account for environmental damages.[94] These claims, however, are generally brought by a government entity, rather than a private individual. Private parties may only sue in public nuisance if they suffer an injury different in kind than that of the general public (a "special injury"),[95] thereby hampering the ability of private parties to sue for environmental harms.[96]

(e) Cartagena Protocol

The United States is not a party to the Cartagena Protocol. Representatives **23** from the United States, as well as the grain trading industry, however, actively participate in the negotiation process in an attempt to minimise potential negative impacts on trade in GMO commodities.[97]

4. Other strict liability regimes

Not at this time. **24**

V. Vicarious liability

1. Scope of vicarious liability

The general common law rule in the United States is that an employer of **25** an independent contractor is not liable for the harm caused by the act or

93 *Johansen v. Combustion Engineering, Inc.*, 170 F.3d 1320 (11th Cir. 1999). An appeals court later reduced the punitive damage award to $ 4.35 million, still an amount far in excess of the $ 47,000 in actual compensatory damages. Ibid.

94 Restatement (Second) of Torts § 821B. A public nuisance involves a significant interference with the public health, safety, comfort, or convenience; is illegal; or is of a continuing nature or has produced a long-lasting effect that the actor has reason to know will be significant. Ibid. For a discussion of public nuisance and GMOs, see *Grossman*, 18 Journal of Environmental Law and Practice, at 114 f (2008).

95 *Grossman*, 18 Journal of Environmental Law and Practice, at 114 (2008) (citing Restatement (Second) of Torts § 821B).

96 *M.S. Madden/G.W.Boston*, Law of Environmental and Toxic Torts (3rd ed. 2005) 66–68.

97 For an overview of the position of the domestic grain industry with respect to the Protocol, see *T.P. Redick*, The Cartagena Protocol on Biosafety: Precautionary Priority in Biotech Crop Approvals and Containment of Commodities Shipments, 18 Colorado Journal of International Environmental Law and Policy 51 (2007).

omission of the independent contractor, or the contractor's servants.[98] This basic rule, however, has been eroded by exceptions to the extent that the general rule only applies "where there is no good reason for departing from it."[99] Three broad categories for exceptions to the rule are: (1) negligence of the employer in selecting, instructing or supervising the contractor; (2) non-delegable duties arising out of some relation between the employer and the public (or the particular plaintiff); and (3) work that is specially, peculiarly or inherently dangerous.[100] Other than a potential determination that GMOs are specially, peculiarly or inherently dangerous (an unlikely proposition in the United States given the current permissive approach to GMO technology), one specific exception to the general rule has particular application in the GMO context–precautions required by statute or regulation. If a statute or regulation imposes a duty to provide particular safeguards for the safety of others, retention of an independent contractor will not shield the employer from liability for failure to provide such safeguards.[101] For example, in the genetically engineered rice commingling event, the seed developer used sub-contracts to carry out the field testing of the GMO rice. The developer, however, had a statutory and regulatory duty to prevent commingling with rice not approved for commercial release. This would be a classic situation in which the seed developer could shield itself from liability by employing an independent contractor – and almost certainly would result in vicarious liability for the seed developer.

2. Liability for people further up the food or feed production chain

26 To the extent the independent tortious conduct of multiple persons is a legal cause of an indivisible injury, the multiple tortfeasor law of the applicable jurisdiction will determine whether the defendants are subject to joint and several liability, several liability, or a hybrid type of liability.[102] The multi-tortfeasor liability rules allows for the judicial sorting-out of responsible parties joined to the original action, or a subsequent action. In other

98 Restatement (Second) of Torts § 409.

99 Restatement (Second) of Torts § 409, Comment b.

100 Restatement (Second) of Torts § 409, Comment b.

101 Restatement (Second) of Torts § 424.

102 Restatement (Third) of Torts: Apportionment of Liability § 17. Comment "a" to the Restatement notes that there is no majority rule currently with respect to the type of liability regime.

words, after the plaintiff is "made whole," the potentially responsible parties can resolve their respective contributions to the damages amount based on fault (taking into account the rules of the particular jurisdiction). One point with respect to timing bears mentioning here: in many states, even if the tortious actions occurred at different times (e.g. at different steps along the food supply chain), courts may impose joint and several liability so long as the injury cannot be apportioned based on the causal contribution of each tortfeasor.[103] For further discussion of comparative fault and joint and several liability, see no. 10 above. In a hypothetical situation of a dangerous GMO food product reaching the consumer, each potential defendant (retailer, crop wholesaler, farmer, seed producer, etc.) could be held jointly and severally liable for the indivisible harm in accordance with the product liability rules discussed in no. 16–18, above. This holds true even if one member in the supply chain (e.g. the retailer) represents itself as the manufacturer of the product (or that the product was manufactured on behalf of X, by Y).[104] In this circumstance, both X and Y would be subject to liability as "manufacturers."[105]

3. Can someone further down the feed or food chain include someone further up it in a trial against him/herself?

To the extent there is alleged fault by a person above or below the supply chain, the original defendant, as a third-party plaintiff, can join (implead) additional defendants (third-party defendants) in accordance with Rule 14 of the Federal Rules of Civil Procedure. Third-party defendants may subsequently implead a fourth-party defendant, and so forth under the same rule. However, a court in a products liability (or other) action may bifurcate the trial on the issues of liability and damage to reduce the complexity of the legal theories and minimise the risk of jury confusion.[106] It is through this process that an original defendant may escape liability by shifting the loss onto another segment of the supply chain. In addition, many state laws provide sellers with a statutory (non-contractual) right to seek indemnity by impleading a third-party defendant, even if the third-party defendant would not be liable until the plaintiff succeeds against the original defendant on the products liability claim. *Jeub v. B/G Foods,* 27

103 Restatement (Third) of Torts: Apportionment of Liability § A18, Reporters' Note, comment b.
104 Restatement (Third) of Torts: Products Liability § 14.
105 Restatement (Third) of Torts: Products Liability § 14.
106 *J.K. Levin/K. Oakes*, 7 Federal Procedure § 16:311.

Inc.[107] presents an early case on this issue. In *Jeub*, plaintiffs sought damages from B/G Foods for allegedly being served adulterated food in their restaurant. B/G Foods subsequently filed a third-party complaint alleging that the adulterated food was canned "Swift Premium Ham," a product of Swift and Company. As such, the adulterated condition of the canned ham allegedly was caused solely and entirely by the negligence and unlawful conduct of Swift and Company, which should indemnify B/G Foods. Accordingly, the court ordered the addition of Swift and Company to the original proceeding. Adulterated food as a result of GMO contamination could follow a similar pattern whereby the developer of the unsafe plant variety would be impled into an initial products liability claim against a manufacturer (or anyone else along the supply chain).

VI. Multiple tortfeasors

28 Joint and several liability, as well as comparative fault liability, applies in the United States. For additional discussion, see no. 10 and 25–26 above.

VII. Defences

1. Licence/permission to grow GM material

29 With respect to private liability actions, as discussed in no. 18, above, a product conforming to applicable safety standards (e.g. received USDA and EPA approval for commercialisation) does not preclude a finding of product defect under the products liability laws.[108] This reflects the traditional view that government safety requirements (absent pre-emption) represent a minimum standard, and a higher standard may apply in some cases.[109] Many jurisdictions, however, give compliance with government safety standards substantial weight when assessing an allegedly defective design.[110] Products approved for testing only, have attendant restrictions on containment/segregation from the food/feed supply and, to the extent those items migrate into an unauthorised supply chain, would present an example of negligence per se (as discussed in no. 15, above). Outside the products liability realm, private law actions of trespass, nuisance, negligence or strict lia-

107 *Jeub v. B/G Foods, Inc.*, 2 Federal Rules Decisions (F.R.D.) 238 (D. Minn. 1942).
108 Restatement (Third) of Torts – Products Liability § 4.
109 Restatement (Third) of Torts – Products Liability § 4, comment e.
110 Restatement (Third) of Torts – Products Liability § 4, comment e.

bility for non-economic losses may proceed, regardless of regulatory approval status.[111] Although there have been numerous disputes among farmers regarding use of GMOs, for various reasons, none have led to a civil legal liability lawsuit.[112] With respect to regulatory liability, the USDA has acknowledged that the low level presence of unapproved genetically engineered DNA in commercial commodities and seeds is an inevitable result of large-scale field tests of experimental crops. As a result, USDA has proposed a rule that such occurrences would be non-actionable by the agency.[113] This proposed rule, however, would not pre-empt private party actions for harm to the person or property in accordance with the rules discussed above (e.g. negligence, trespass, nuisance, strict liability). Regulatory liability, however, may arise from other agencies with responsibilities under the Coordinated Framework. The FDA has a unique regulatory role with respect to GMOs. Although the FDA does not issue formal "approval" of new GMOs entering the market, the agency encourages firms to engage in voluntary consultations prior to commercialisation. At the end of the consultation process, the FDA will issue a letter indicating that the agency has no further questions based on the data submitted.[114] This "no further questions" letter, however, does not pre-empt the baseline rule in the Federal Food, Drug, and Cosmetic Act that prohibits the introduction of adulterated food into interstate commerce.[115] To the extent the new GMO product is later found harmful, the agency has authority under the Act to seize the product and asses civil and/or criminal penalties.[116] Accordingly, even apparent regulatory approval by the agencies of the Coordinated Framework does not preclude subsequent private or regulatory liability claims.

---

111 See discussion in no. 3 and 4, and sources cited in footnote 7.

112 *D.L. Kershen*, Legal Liability and Agricultural Biotechnology: Ten Questions (April 2009), available at http://agribiotech.info/details/KershenFinal%2003%20layout.pdf. Professor Kershen outlines several explanations for the lack of litigation between farmers, including: agronomic compromises to obtain coexistence, adoption of tolerances or thresholds, and the general importance of neighborliness within the rural community. In addition, seed companies generally have been willing to quietly settle disputes resulting from cross-field contamination rather than engage in a costly, public trial.

113 USDA, Introduction of Genetically Engineered Organisms: Draft Programmatic Environmental Impact Statement – July 2007, at 155.

114 Guidance for Industry; Recommendations for the Early Food Safety Evaluation of New Non-Pesticidal Proteins Produced by New Plant Varieties Intended for Food Use, 71 Fed. Reg. 35,688 (21 June 2006).

115 21 U.S.C. § 331(a)-(d).

116 21 U.S.C. §§ 333, 334.

2. Consent/assumption of risk

30 As noted in the paragraph above, the Federal Food, Drug, and Cosmetic Act prohibits the sale or distribution of adulterated food. There is no statutory exclusion for waiving a right to safe food products. Accordingly, knowing consumption of GM products that are adulterated would not preclude regulatory liability. With respect to private tort liability, the assumption of the risk doctrine would bar a damage recovery only if the plaintiff voluntarily exposed himself to the hazard (in this case GMOs), with full knowledge of the condition (harmful effect of GMOs). Absent full knowledge of the danger, assumption of the risk defence would not apply. *Belinky v. F.& R. Lazarus Corp.*[117] is instructive. In *Belinky*, the plaintiff ate lunch at the defendant's restaurant. The sandwich did not taste "right," but the plaintiff nonetheless consumed the food, which caused a case of gastroenteritis. The court refused to apply the assumption of the risk defence because the plaintiff did not eat the food with the knowledge that it was contaminated or adulterated–he merely accepted food that, according to his knowledge, did not taste as good as one would expect for that type of food.

3. Third-party influence

31 Concepts of comparative fault would apply to apportion liability. See discussion in no. 10, 26 and 27. However, two common law rules operate in a scenario in which the unlawful acts of a third-party may impose liability on another actor. First, an individual may be negligent if he or she commits an act or omission which involves an unreasonable risk of harm to another through the foreseeable actions of a third person.[118] Moreover, if the likelihood that the third person may act in a particular manner is the hazard that makes the actor negligent, the third person's act, whether innocent, negligent, intentionally tortious or criminal, does not prevent the actor from being liable for the resultant harm.[119] Accordingly, if an actor engages in an activity relating to GMOs, and the sabotage or improper conduct by a neighbour is foreseeable, the actor may also be found liable for this independent negligence. This, of course, would not excuse the liability of third party. Although in a different context, contract law reinforces the common law's general reluctance to excuse liability completely due to the acts of a

117 *Belinky v. F.&R. Lazarus Corp.*, 1979 WL 208925 (Ohio App. 1 March 1979).
118 Restatement (Second) of Torts § 302.
119 Restatement (Second) of Torts § 449.

third party. For example, it is generally held that prevention of performance of a contract as a result of the interference of a third party will not excuse the breach.[120]

4. Prescription

Statutes of limitation for tort claims are a matter of state law. Where a duty, and therefore liability for its breach, would not exist but for a statute, actions for damages are governed by the statute of limitations for negligence or the liability imposed by the statute.[121] For example, in Illinois, an action for the recovery of lands has a twenty-year statute of limitations[122] and actions for personal injury have a two-year statute of limitations.[123] Other states have similar rules. **32**

5. Other defences

Based on the common law cases decided as of this writing there are no other defences that might be relevant in a GMO liability case. **33**

VIII. Remedies

1. Pecuniary compensation

(a) Bodily harm

Standard common law and statutory remedies would apply for compensating bodily harm caused by GMOs. The United States does not have a special liability/redress regime for GMOs and generally does not impose caps on damages. **34**

120 *Goldston Brothers v. Newkirk*, 64 South Eastern Reporter, Second Series (S.E.2d) 424, 432 (N.C. 1951).
121 Corpus Juris Secundum, Limitations of Actions § 96.
122 35 Illinois Compiled Statutes 5/13-101.
123 35 Illinois Compiled Statutes 5/13-202.

(b) Property losses

35 Standard common law and statutory remedies would apply for compensating property losses caused by GMOs. The United States does not have a special liability/redress regime for GMOs and generally does not impose caps on damages.

(c) Economic losses

36 Standard common law rules would apply for compensating economic losses caused by GMOs. For a discussion of the application of the pure economic loss rule in the United States, see no. 4 above. An organic farm in the United States contaminated with GMOs, however, would not necessarily lose its certification status. In fact, despite the ubiquitous adoption of GMOs in the United States, the author is not aware of a single farm that has lost certification status due to the low-level presence of transgenic material. The preamble to the National Organic Program's (NOP) implementing rule specifically addressed the issue of pollen drift and organic certification. Although the NOP specifically prohibits the use of GMOs, the rule provides that the presence of GMOs as a result of drift or other contamination does not preclude organic certification so long as the organic operation did not use GMOs and took reasonable steps to avoid contact with GMOs, as specified in the operation's organic system plan.[124] Although this rule interpretation seems to protect the organic farmer from the adverse impacts of GMOs, there may be a hypothetical situation whereby the organic system plan is unable to prevent GMO pollen drift and thus the farm would lose certification. The legal system has not dealt with an analogous hypothetical situation and thus predicting the result in a common law system is difficult. One common law theme, however, is applicable: the requirement of a plaintiff to mitigate damages in a nuisance action. The failure to mitigate damages may reduce the overall damages award otherwise due to the plaintiff.[125] Extrapolated to the issue of a farmer restoring a GMO contaminated field to organic standards, this rule would seem to require the farmer to mitigate the total damages, of which one method would be to produce non-certified crops during the transition back to organic status.

124 National Organic Program, Final Rule, 65 Fed. Reg. 80548, 80556 (21 December 2000).
125 1 Comparative Negligence Manual (3rd ed.) § 79.

(d) Harm to animals

Conceptually, the physical harm to property in animal form via consumption of GMO feed (and thus disqualification for the organic market) is no different than the physical harm to an organic plant that is cross pollinated or otherwise contaminated with GMO DNA. In either event, the property at issue no longer qualifies for its intended use. This is similar to the contamination of harvested crops with the StarLink corn. The commingled corn was no longer fit for its intended use, but relegated to lower-price domestic feed or industrial uses. Accordingly, the measurement of compensatory damages in the animal contamination scenario could include the lost potential for producing milk or meat for the organic market or other specialty market for the expected life of the animal. To the extent there are lost profits for a period of time (i.e., during the time to reconvert the animal to non-GM or organic production), rules established for mismating in the breeding of animals may be analogous. For example, the owner of a male animal that is entrusted with a third party's female for mating purposes, is liable to the female's owner for lost profits if the male's owner negligently allows the female to be mismated.[126] **37**

(e) Costs of disposal

To the extent the commingled crops or the GM-feed provided to the animals had government approval, the crops or animals could be sold in the commodity (GM) market. The general rules of compensatory damages would seek to compensate the owner in a manner intended to restore the victim to his or her pre-tort state of affairs. To the extent the victim incurred disposition costs in excess of those that would have been incurred for disposition in the desired non-GM market, the costs would fall under compensatory damages. **38**

2. Non-compensatory damages

For a discussion of punitive damages available in tort, see no. 8 and 22. **39**

126 *Barnard v. Sweet*, 1963 WL 6356 (Pa. Com. Pl., 3 September 1963).

3. Other remedies

40 In addition to nominal, compensatory and punitive damages, plaintiffs in tort proceedings may seek injunctive relief. Traditionally, courts have imposed injunctions in cases where there is an inadequacy of the remedy at law, inadequacy of monetary damages, and irreparable injury.[127] In considering imposition of an injunction against certain actions (e.g. enjoining cultivation of GMOs), courts must conduct a comparative analysis of several factors: "(a) the nature of the interest to be protected, (b) the relative adequacy to the plaintiff of injunction and of other remedies, (c) any unreasonable delay by the plaintiff in bringing suit, (d) any related misconduct on the part of the plaintiff, (e) the relative hardship likely to result to defendant if an injunction is granted and to plaintiff if it is denied, (f) the interests of third persons and of the public, and (g) the practicability of framing and enforcing the order or judgment."[128] Sometimes a single factor may be dominant and warrant issuance/denial of a plea for an injunction.

4. Costs of pursuing a claim

(a) General cost rule

41 Although states may alter these rules, the general standard in the United States legal system is that each party must bear its own costs of litigation, regardless of outcome.[129] Successful suits against the federal government for some causes of action may result in the recovery of costs and fees under the Equal Access to Justice Act.[130]

(b) Costs of establishing causation

42 As noted above, the general rule in the United States provides that each party must bear their own costs of litigation, regardless of outcome.

127 Restatement (Second) of Torts § 933, comment a.

128 Restatement (Second) of Torts § 936(1). See also *Monsanto Co. v. Geertson Seed Farms*, United States Supreme Court, Case No. 09-475 (June 21, 2010) (reversing imposition of a permanent injunction against planting genetically modified alfalfa pending USDA completion of the Environmental Impact Statement required under the National Environmental Policy Act).

129 For a discussion of this "American Rule," see *D.B. Dobbs*, Awarding Attorney Fees Against Adversaries: Introducing the Problem, 1986 Duke Law Journal 435 (1986).

130 5 U.S.C. § 504.

Accordingly, the costs of proving causation, unless also determined to be imbedded into a compensatory damage award, would not be recovered by the prevailing party.

5. Advance cover

The United States does not have a special liability/redress regime for GMOs. Accordingly, there are no requirements for financial guarantees such as mandatory insurance coverage for GMO producers. **43**

IX. Cross-border issues – Conflict of laws

1. Conflicts rules applicable before (or instead of) Rome II

The Rome II Regulation is not applicable in the United States. With respect to choice of law, most states have a statute directing the court, in limited situations, to apply a specified jurisdiction's law. Absent a statutory directive (which is the majority of the cases), courts must balance the following factors in selecting the choice of law: (a) the needs of the interstate and international systems, (b) the relevant policies of the forum, (c) the relevant policies of other interested states and the relative interests of those states in the determination of the particular issue, (d) the protection of justified expectations, (e) the basic policies underlying the particular field of law, (f) certainty, predictability and uniformity of result, and (g) ease in the determination and application of the law to be applied.[131] When considering these factors in the cases under survey, it is important to note that the difficulties and complexities involved with respect to tort claims have prevented the courts from developing a precise choice of law rule or set of rules.[132] There is a general principle, however, that the local law "of the state of the most significant relationship" should be applied.[133] Of course, this begs the question of how to analyse the most significant relationship. With respect to personal injury torts, the law of the location of the injury generally applies.[134] **44**

131 Restatement (Second) of Conflict of Laws § 6. For a more detailed discussion of choice of law principles in the United States, see the comments to the Restatement (Second) of Conflict of Laws § 6.

132 Restatement (Second) of Conflict of Laws § 6, comment c (discussing difficulty in crafting a precise set of rules for torts).

133 Restatement (Second) of Conflict of Laws § 6, comment c; Restatement (Second) of Conflict of Laws § 145(1).

134 Restatement (Second) of Conflict of Laws § 146.

Therefore, the choice of law for an action alleging personal injury from consuming GMO foods would be the location of consumption. Similarly, the choice of law for actions seeking recovery for harm to property (land or personality) would be the local law of the state where the land is located or, for personality, where the injury occurred.[135] Therefore, in a nuisance or trespass action alleging harm to property from GMO pollen drift, the choice of law would be the state in which the damaged property is located.

2. Special regime for cross-border claims

45 There is no special regime that applies to resolve cross-border claims for compensation under tort.

X. Cases

1. Due to the adventitious presence of GMOs in a field, maize which is normally sold as conventional contains GMOs beyond the legal labelling threshold. This is not discovered before the final stage of the food production chain by the producer of taco chips. The whole production is lost since the supermarket chains refuse to accept delivery from the producer.

(a) Who can sue along the chain of distribution?

46 This scenario in the United States is simplified by the absence of a labelling threshold set by statute or regulation. Absent a regulatory requirement for labelling, these claims would arise not as a matter of regulatory law, but rather breach of contract whereby the contract between the food processor specified delivery of non-GM taco chips to the retailer.[136] Presumably, the food processor's contracts with its suppliers also specified delivery of non-GM raw materials (e.g. non-GM maize). Therefore, any claims along the distribution chain would be for breach of contract. Although breach of contract would be the primary claim, an additional claim for negligence certainly is feasible. Parties along the distribution chain presumably knew of the intended use of the raw materials and the importance of maintaining

135 Restatement (Second) of Conflict of Laws § 147.

136 Of course the contracting parties would have had an opportunity to allocate the risk of adventitious presence in the contract.

genetic purity. Accordingly, they owed a duty of ordinary care outside of any contractual duty to preserve the non-GMO content at each subsequent stage in the supply chain. For example, the farmer delivering purported non-GMO grain to the local grain elevator had a duty to prevent the commingling of his GMO-tainted grain with the GMO-free grain already in storage regardless of his contract with the elevator for the delivery of non-GMO grain. The elevator operator, likewise, had a duty to prevent commingling when delivering his or her stored grain to the next point in the supply chain. This duty to prevent commingling continues along each step of the chain of distribution. The common practice in the United States in these situations demanding non-GMO content is to conduct genetic testing of the delivery to ensure purity. Accordingly, the standard of care owed by a reasonably prudent person to fulfil this duty would be to conduct some type of genetic testing to prevent commingling. Failure to test and thereby creating a commingled product is a breach of this duty, resulting in an injury to those products in the next stage of the supply chain. The *StarLink* case, discussed in no. 3 and 4, held that commingling of GMO and non-GMO grain in the supply chain is a physical injury thereby avoiding application of the pure economic loss rule.[137] Therefore, this scenario, in addition to a claim for breach of contract, could present a viable claim for negligence. Restricting this analysis to common law claims, however, would change if the GMO found in the product was not authorised for general production – a topic discussed in no. 48 and 49 below. In such cases, regulatory liability also would attach along the distribution chain.

(b) Would the case be solved differently if the GMO content was below the labelling threshold?

47 As noted in the previous paragraph, without a government-imposed labelling rule in the United States, this is an issue of contract and tort law, rather than regulatory law. Accordingly, analysis of the breach of contract or negligence would be measured by the threshold specified in the underlying contract, rather than a regulatory-based standard. Therefore, the analysis is the same as in no. 46.

137 *In re StarLink Corn Products Liability Litigation*, 212 F. Supp. 2d 828, 841 (N.D. Ill. 2002).

(c) Would the case be solved differently if the admixture was not adventitious, but occurred due to the disregard of segregation rules, for example?

48 There are no regulatory-based segregation rules for approved GMOs in the United States. However, because this is primarily a matter of contract law, the truly adventitious presence of GMOs, versus a wilful disregard for segregation, would not alter the analysis provided in the proceeding paragraphs. In either situation a party breached the terms of the contract – delivery of non-GMO product. This scenario, however, would present a stronger case of negligence than described in no. 46. Even if members of the supply chain did not have a duty of ordinary care to prevent contamination of the products in the next stage of production (i.e., assuming a court rejected the duty analysis outlined in no. 4), each member of the supply chain would have a duty to abide by segregation best practices, assuming knowledge that the intended use of the product was for a non-GMO product. The breach of the duty to follow segregation best practices resulted in a physical harm to the product, and thus a claim for negligence.

(d) Would the case be solved differently if the GMO found was not admitted for production in your jurisdiction?

49 Resolution of potential breach of contract and negligence causes of action is not dependent upon the regulatory approval of the GMO. In fact, the non-approval (and thus non-commercialisation) of the GMO could present a potential defence to negligence because standard genetic testing of the product for potential GMO contamination would not include testing of products not yet commercialised. Testing agencies must have the necessary technology (e.g. primers for PCR analysis) to test for the adventitious presence of genetically modified DNA. PCR testing will find and measure known contaminants, but not DNA for which a primer is unavailable. Accordingly, genetically modified material that has not received regulatory approval is unlikely to be detected because the DNA sequence data is probably not available to the testing lab.[138] To the extent ordinary care to prevent commingling would only mandate a standard DNA test, and the

138 See generally, *Endres*, 1 Journal of Food Law & Policy 131, 158 f. (2005) (discussing DNA testing and the need for an international database of DNA sequences and analytical procedures to detect unauthorized genetically engineered DNA).

defendant completed this test, there may not be a breach of duty.[139] Absent special circumstances, ultimate liability in this case would rest on the developer of the unauthorised GMO – most likely for negligence per se. As discussed in no. 15, the unexcused violation of a statutory or regulatory rule is negligence per se. In the *StarLink* case, the seed developer failed to follow the segregation conditions placed on the product unapproved for food use, resulting in a claim for negligence per se. In the scenario presented above, the product is unapproved for any use. Accordingly, any escape into the general food distribution channel would be a violation of the seed developer's regulatory restrictions on experimental use (e.g. containment of pollen during field trials; confirmation of destruction or other disposition of harvest from field trials). This violation would support a claim for negligence per se. With respect to regulatory liability, the Federal Food, Drug and Cosmetic Act prohibits the sale, distribution or delivery of an adulterated product.[140] Assuming the unapproved GMO was an "adulterated" product, the FDA has authority under the Act to seize the product and assess civil and/or criminal penalties against each party introducing the adulterated product into the stream of commerce.[141] For further analysis of the FDA role, see the discussion in no. 29.

(e) If the admixture had occurred on a non-GM field and it transpires that the GM seeds were blown from: (i) neighbouring fields; or (ii) a truck passing by, would the farmer of the affected field be liable for all or part of the loss caused further down the distribution chain?

For the reasons discussed in the immediately preceding paragraphs, common law interpretations of both contract and tort liability could impose all or part of the losses on the farmer of the affected field. The comparative fault of other parties in the supply chain (e.g. for failure to test the genetic purity of the product; delivering product that contaminates the supply chain) may mitigate the farmer's ultimate responsibility to bear all of the loss.[142] Moreover, the farmer may be able to transfer a substantial part of his or her liability to the original tortious act – pollen drift or escape from 50

139 Although in light of USDA's acknowledgment that the low level presence of unauthorized GMOs is inevitable, the presence in grain shipments could be a foreseeable occurrence, giving rise to a duty to test. See no. 29 and footnote 113.

140 21 U.S.C. § 331(a)–(d).

141 21 U.S.C. §§ 333, 334.

142 For a discussion of comparative fault rules, see no. 10 and 25.

the truck. In the case of pollen drift from a neighbouring field, claims of negligence, nuisance or trespass may apply.[143] With respect to the truck, negligence and trespass would be viable causes of action.[144]

2. Twenty years after the sale of GM maize used for food products, it turns out that it has certain disadvantageous health effects for humans.

(a) Can the producers be held liable at this point for risks unknown at the time of growing the maize? Who would be liable?

51 As discussed in no. 16 and 18, one who sells or distributes a defective food that causes harm is liable under the products liability rules regardless of when the harm arises (a latent defect).[145] Although outside the specific context of food, both the DES and lead paint cases provide common law support for a claim for delayed injury from a dangerous product whose risks were unknown by the manufacturer at the time of sale, such as the presented GMO scenario.[146]

(b) Can compensation already be claimed at a point when the negative health effects have not yet materialised, but are to expected to according to scientific expertise/mere rumours?

52 A plaintiff may not maintain a lawsuit for damages until a cause of action has arisen. For some torts, a cause of action exists irrespective of harm; in other torts, it exists only when harm has resulted.[147] A cause of action for negligence requires harm to a person or thing.[148] Accordingly, if the harm has not yet materialised, there is no cause of action for negligence. For products liability claims, the cause of action also accrues as of the time of the injury – the manifestation of the harm.[149] Except in very limited situations, there is no liability for negligent conduct that threatens bodily harm, but

---

143 See discussion in no. 3, especially articles cited in footnote 7.

144 See discussion in no. 3, especially articles cited in footnote 7.

145 Restatement (Third) of Torts: Products Liability § 7.

146 *Collins v. Eli Lilly Co.*, 342 N.W.2d 37, 57 (Wis. 1989) (DES); *Thomas v. Mallett*, 701 N.W.2d 526, 564 (Wisc. 2005) (lead paint).

147 Restatement (Second) of Torts § 910, comment a.

148 Restatement (Second) of Torts § 899, comment c.

149 Restatement (Second) of Torts § 899, comment c.

only results in emotional distress (e.g. fear of a negative health effect that has not yet materialised).[150] Only extreme and outrageous conduct that intentionally or recklessly causes severe emotional distress is subject to liability.[151] In the case presented, it is unlikely a plaintiff could successfully pursue a cause of action for severe emotional distress due to the lack of intent or reckless conduct (assuming previous government approval of the GMO in question). With respect to latent, but undiagnosed diseases, asbestos exposure cases provide some guidance for application in the GMO context. In *Wilber v. Owens-Corning Fiberglass Corp.*,[152] the plaintiff previously suffered from asbestosis, a condition caused by inhaling asbestos fibres, but did not seek compensation in court. Plaintiff subsequently developed mesothelioma, a separate disease also caused by exposure to asbestos. He filed a personal injury suit subject to a two year statute of limitation. Had plaintiff brought suit for the first injury – asbestosis – he could have sought recovery for the increased risk of developing mesothelioma, provided there is a reasonable medical certainty of developing the disease.[153] There must have been, however, an underlying physical injury (e.g. asbestosis) upon which to add this risk-based claim. Some courts, however, have refused to allow these "increased-risk" claims resulting from exposure to a harmful substance. These courts do state that if the plaintiff was to develop the exposure-related disease in the future, an action for damages could be brought at that time.[154] Accordingly, it is unlikely a plaintiff could maintain a successful lawsuit in the United States for negative health effects from GMOs that have not yet materialised into any physical injury.

(c) Would it make any difference if the GM maize had only been in use for feed, causing harm to the animals, which may or may not cause harm to humans consuming the meat as well?

The change in facts from scenario (b) to (c) does not change the analysis **53**
according to the common law in the United States stated in no. 52. Absent the unique facts required for a claim of emotional distress, a plaintiff may not maintain a personal injury action based on the fear of harm, absent some underlying physical injury.

150 Restatement (Second) of Torts § 924, comment a.
151 Restatement (Second) of Torts § 46(1).
152 *Wilber v. Owens-Corning Fiberglass Corp.*, 476 N.W. 2d 74 (Iowa 1991).
153 *Wilber*, 476 N.W. 2d at 77.
154 *Wilber*, 476 N.W. 2d at 77 (citing *Herber v. Johns-Manville Corp.*, 785 F.2d 79 [3d Cir. 1986]; *Jackson v. Johns-Manville Sales Corp.*, 727 F.2d 506 [5th Cir. 1984]).

3. The driver of a food logistics company discovers that a farmer, from where he regularly picks up agricultural products fails to obey mandatory segregation rules or food or feed hygiene standards, which may lead to the admixture of GM and non-GM produce sold and packaged separately by that farmer. Does he or his employer have a duty to warn, i.e. warn the recipient of the allegedly "non-GM" produce?

54 In the United States, there is no mandatory segregation rule. Therefore, any regulatory-based liability would have to arise from the food or feed hygiene standards. The standards, however, do not prohibit the admixture of GM and non-GM products. Assuming *for the sake of argument* that the food or feed hygiene standards prohibited the admixture of GM and non-GM produce under adulteration or misbranding standards, the driver of the food logistics company (and the company itself via the doctrine of respondeat superior) would face liability under the Federal Food, Drug, and Cosmetic Act. The Act prohibits not only the sale or introduction of an adulterated or misbranded food into interstate commerce, but the delivery for introduction, and even the receipt of such products for later delivery into interstate commerce.[155] Accordingly, regardless of any independent duty to warn, the driver would be subject to regulatory liability under the Act. There is an exception to liability, however, for persons who in good faith merely transport/deliver a misbranded or adulterated product, provided that they have a guarantee attesting that the article is not misbranded or adulterated.[156] This safe harbor clause would not apply in this scenario because the driver had knowledge of the regulatory violations and thus cannot assert a good faith defence. With respect to common law liability, absent special circumstances (e.g. parent, common carrier, etc.), an actor is under no duty to take actions to prevent the harm of another or come to another's aid.[157] The classic example is a blind person about to step into the street in front of a moving car. The fellow pedestrian is under no duty to warn the blind person.[158] This rather strict rule, however, does not apply if the active force causing the harm is under the control of the bystander.[159] For example, if a factory owner sees a blind person wander into his factory with moving machines, and the blind person is about to approach the machinery, the owner is negligent if he permits the machinery to continue moving when he

155 21 U.S.C. § 331(a)–(d).
156 21 U.S.C. § 334(c).
157 Restatement (Second) of Torts § 314.
158 Restatement (Second) of Torts § 314, illustration 1.
159 Restatement (Second) of Torts § 314, comment d.

could, by the exercise of reasonable care, stop the machinery before contact with the blind person.[160] In the delivery driver scenario presented above, the driver has the ability to control the harm – in this case the further commingling of the farmer's contaminated goods with products further up the distribution chain. This is analogous to the factory owner example and thus could result in breach of a duty to act for the protection of others.

160 Restatement (Second) of Torts § 314, illustration 2.

Special Reports

Damage Caused by GMOs under International Environmental Law

Vanessa Wilcox

I. Introduction

1 The advancement of modern biotechnology brought with it growing concerns over the possible adverse effects of such technology on human health and the environment. Although the initial reaction in some quarters was one of protectionism, with the EU imposing a de facto moratorium on the authorisation of GMOs for several years, increased regulations at international, regional and national level have since been adopted to reconcile the respective needs of trade and environmental protection and the growing global biotechnology industry. That said, given that relatively little can be ascertained of the full effects of genetic engineering in the long term, biotechnology, and agro-biotechnology in particular, is still received with a considerable degree of scepticism and resistance. This is especially so not only in the EU but also in developing countries which play host to the greater parts of the world's biodiversity. Indeed, it was the determination of these and other potential GMO recipient states to unite in constructive efforts that led to the conclusion of an international regulatory framework in the field of GMOs. In turn, this paved the way for a global debate on the need for an international liability framework for damage resulting from GMOs. The adoption of such a specific regime to supplement existing regulatory initiatives would bring GMO regulations in line with a series of other sectoral environmental liability regimes that have been adopted in the recent past.

2 The country reports which precede this paper review existing domestic instruments through which GMOs are regulated and through which any ensuing damage, be it to property, persons or common goods, can be redressed. Increasingly, national regimes are coloured by regional regulations and these by multilateral environmental agreements requiring global cooperation. In this paper, the author summarises the key provisions of the Cartagena Protocol on Biosafety – the only international legally

binding instrument to address the issue of biosafety: Part III. Though not covered in great detail here, the report should be read with an eye to existing World Trade Organization (WTO) agreements relevant to GMOs for a more comprehensive and contextual appreciation of the Protocol.

As a consensus element, delegations conceded to the establishment of a biotech-specific liability regime which was still in draft at the time of writing. In Part IV, the present paper details relevant elements of the Proposed Operational Texts on Liability and Redress in the Context of Art. 27 of the Biosafety Protocol. This is followed by a conclusion in Part V. Before then, however, Part II outlines the key instruments that marked the emergence of the global environmental agenda and in so doing sets the background to the current regime. **3**

II. The Regulation of Modern Biotechnology: Relevant International Environmental Instruments

Although GM engineering or recombinant-DNA technology was first applied in the 1970s, for much of the 1980s biotechnology was confined to the laboratory. As early on as the 1980s, however, concerns over the intrinsic nature of such techniques surfaced and in 1985, the United Nations Industrial Development Organisation (UNIDO) together with the United Nations Environmental Programme (UNEP) and World Health Organization (WHO) established an Informal Working Group on Biosafety which was later joined by the Food and Agriculture Organization (FAO). With UNIDO in lead, the inter-agency Group worked towards a Voluntary Code of Conduct for the Release of Organisms into the Environment which was finalised in July 1991. In addition, various agencies set up biosafety training programmes and issued guidelines in a concerted effort to steer the industry. **4**

In 1987, the United Nations World Commission on Environment and Development (WCED) had published a report, Our Common Future, which approached environmental and developmental challenges as a single issue to be solved by collective multilateral action.[1] It was through this report, the efforts of the Working Group and those of other affiliations, that the groundwork for the first international Earth Summit, the United Nations Conference on Environment and Development, held in Rio de Janeiro on 3–14 June 1992, was laid. It was Our Common Future that „crystallised and popularised“ the notion of sustainable development **5**

1 Report of the World Commission on Environment and Development, Our Common Future, 20 March 1987.

which, in addition to the environment, was the principal theme of the Summit. The Conference saw governments from 178 countries assemble to address urgent problems on environmental protection and socio-economic development. There were 5 major outcomes to the Rio Earth Summit. However, the three most relevant to this study are the endorsement of the Rio Declaration, the adoption of Agenda 21 and the signing of the Convention on Biological Diversity.

1. The Rio Declaration on Environment and Development, 1992

6 The Rio Declaration, which reaffirms and builds upon the Stockholm Declaration on the Human Environment 1972,[2] sets out 27 Principles which define the rights of people to development, and their responsibilities to safeguard the common environment. Although no mention was made of biotechnology in the Declaration, Principle 15 establishes a key element which has since been incorporated into the biotechnology regulatory framework as far as environmental protection and management is concerned:

> In order to protect the environment, the precautionary approach shall be widely applied by States according to their capabilities. Where there are threats of serious or irreversible damage, lack of full scientific certainty shall not be used as a reason for postponing cost-effective measures to prevent environmental degradation.

2. Agenda 21

7 Unlike the Rio Declaration, Agenda 21 specifically addressed the impact of the emerging field of biotechnology. In essence, the document is a comprehensive blueprint for sustainability in the 21st century. While acknowledging that biotechnology would not, by itself, resolve all the fundamental problems of the environment and development, the Agenda, under its Chapter 16, recognised the significant contribution which the field could have in enabling, inter alia, the development of better health care and enhanced food security through sustainable agricultural practices.

8 Five areas were defined in which action had to be taken: (a) increasing the availability of food, feed and renewable raw materials; (b) improving human health; (c) enhancing protection of the environment; (d) enhancing

2 The Declaration of the United Nations Conference on the Human Environment, adopted at Stockholm on 16 June 1972.

safety and developing international mechanisms for cooperation; and (e) establishing enabling mechanisms for the development and the environmentally sound application of biotechnology. By so doing, the Agenda, inter alia, set out a non-binding global action plan for the international harmonisation of biotechnological safety.

3. The Convention on Biological Diversity

With 190 Parties, the Convention on Biodiversity is one of the most broadly subscribed international environmental treaties in the world. Pursuant to its policy on the environment,[3] the European Community signed the Convention in 1992 and approved it in December 1993[4] as did its Member States.[5] The U.S. has signed, but not ratified the treaty. **9**

The Convention, which entered into force on 29 December 1993, was conceived as a practical tool for translating the Principles of Agenda 21 into reality. Its objectives are three-fold: **10**

- the conservation of biological diversity;
- the sustainable use of its components; and
- fair and equitable sharing of the benefits arising out of the utilisation of genetic resources.[6]

---

3 Art. 130r Treaty on the European Union of 1992.

4 93/626/EEC: Council Decision of 25 October 1993 concerning the conclusion of the Convention on Biological Diversity, Official Journal (OJ) L 309, 13.12.1993, 1–20.

5 Austria (signed 13.6.1992; ratified 18.8.1994); Belgium (signed 5.6.1992; ratified 22.11.1996); Bulgaria (signed 12.6.1992; ratified 17.4.1996); Cyprus (signed 12.6.1992; ratified 10.7.1996); Czech Republic (signed 4.6.1993; approved 3.12.1993); Denmark (signed 12.6.1992; ratified 21.12.1993); Estonia (signed 12.6.1992; ratified 1.7.1994); Finland (signed 5.6.1992; accepted 27.7.1994); France (signed 13.6.1992; ratified 8.18.1994); Germany (signed 12.6.1992; ratified 21.12.1993); Greece (signed 12.6.1992; ratified 4.8.1994); Hungary (signed 13.6.1992; ratified 24.2.1994); Ireland (signed 13.6.1992; ratified 22.3.1996); Italy (signed 5.6.1992; ratified 15.4.1994); Latvia (signed 11.6.1992; ratified 14.12.1995); Lithuania (signed 11.6.1992; ratified 1.2.1996); Luxembourg (signed 9.6.1992; ratified 9.5.1994); Malta (signed 12.6.1992; ratified 29.12.2000); Netherlands (signed 5.6.1992; accepted 12.7.1994); Poland (signed 5.6.1992; ratified 18.1.1996); Portugal (signed 13.6.1992; ratified 21.12.1993); Romania (signed 5.6.1992; ratified 17.8.1994); Slovakia (signed 19.5.1993; approved 25.8.1994); Slovenia (signed 13.6.1992; ratified 9.7.1996); Spain (signed 13.6.1992; ratified 21.12.1993); Sweden (signed 8.6.1992; ratified 16.12.1993); United Kingdom (signed 12.6.1992; ratified 3.6.1994). Australia is also a party to the Convention (signed 5.6.1992; ratified 18.6.1993) as is Brazil (signed 5.6.1992; ratified 28.2.1994) and Canada (signed 11.6.1992; ratified 4.12.1997).

6 Art. 1 Convention on Biological Diversity. All three recognised EU candidate States, i.e. Croatia, the Former Yugoslav Republic of Macedonia and Turkey, have ratified the Convention. All the EFTA States are parties.

11 During negotiations leading up to the draft of the Convention, a division emerged between the Like Minded Group (comprised of China and the Group of 77 developing countries, excluding Argentina, Chile and Uruguay), sided by the European Union on the one hand and developed nations such as the U.S. and Japan on the other. Anxious about the ongoing scientific uncertainty over the potential long-term effects of GMOs (or LMOs (Living Modified Organisms) as the Convention drafters preferred) and concerned not to be used as a giant experimental testing ground for U.S. biotech companies, the Like Minded Group sought the incorporation of comprehensive biosafety provisions into the Convention. Per contra, the U.S. and Japan opposed this, stating, inter alia, that lengthy biosafety rules would be trade-restrictive and open the industry to protectionist abuse.[7] As a compromise, Parties conceded to the inclusion of "soft-law" guidelines on biosafety and kept the prospects of a detailed biosafety instrument open: art. 8 addresses "In-situ Conservation" and art. 19 deals with the "Handling of Biotechnology and Distribution of its Benefits". In particular, art. 8(g) provides that each Contracting Party shall, as far as possible and as appropriate:

> "Establish or maintain means to regulate, manage or control the risks associated with the use and release of living modified organisms resulting from biotechnology which are likely to have adverse environmental impacts that could affect the conservation and sustainable use of biological diversity, taking also into account the risks to human health."

12 Art. 19(3) invited the Parties to:

> "...consider the need for and modalities of a protocol setting out appropriate procedures, including, in particular, advance informed agreement, in the field of the safe transfer, handling and use of any living modified organism resulting from biotechnology that may have adverse effect on the conservation and sustainable use of biological diversity."

13 While by art. 19(4), Parties were directed to:

> "...provide any available information about the use and safety regulations required by that Contracting Party in handling [living modified] organisms, as well as any available information on the potential adverse

7 R. *Falkner*, Negotiating the Biosafety Protocol, in: C. Bail/R. Falkner/H. Marquard, The Cartagena Protocol on Biosafety: Reconciling Trade in Biotechnology with Environment & Development? (2002) 6.

impact of the specific organisms concerned to the Contracting Party into which those organisms are to be introduced."

The First Meeting of the Conference of the Parties to the Convention on Biological Diversity (COP 1) took place in Nassau, Bahamas in 1994. The Conference of the Parties is the governing body of the Convention. Among other things, it reviews and advances the implementation of the Convention through the decisions it takes at its periodic meetings. In accordance with art. 19(3) of the Convention on Biodiversity, the Conference of the Parties established an Open-ended Ad Hoc Group of Experts on Biosafety to consider the need for and modalities of a protocol.[8] At COP 2 held in Jakarta, Indonesia in 1995, the Conference of the Parties, having considered the report and recommendations of the Open-ended Ad Hoc Group of Experts, established an Open-ended Ad Hoc Working Group on Biosafety. Their mandate was to develop a draft protocol on biosafety, specifically focusing on transboundary movement of any LMOs resulting from modern biotechnology that may have adverse effects on the conservation and sustainable use of biological diversity.[9] **14**

Considering the time required to elaborate modalities and draft the protocol, the Conference of the Parties stressed the importance of the urgent finalisation of the Technical Guidelines for Safety in Biotechnology. The Guidelines had been initiated by the UK and the Netherlands following the Rio Summit in June, 1992 and were in response to the call in Chapter 16 of Agenda 21. Negotiated under the auspices of the UNEP, they would serve as an interim measure until a biosafety protocol was agreed and complement the protocol thereafter. The Guidelines, which were adopted in Cairo in December 1995, are based on the premise that adequate mechanisms for risk assessment, risk management and capacity-building through, inter alia, the exchange of information at national, regional and international levels can contribute significantly to safety in biotechnology.[10] They thus encourage the exchange of such information and advice is given on risk assessment and management. **15**

The first meeting of the Open-ended Ad Hoc Working Group on Biosafety was convened in Aarhus, Denmark in July 1996. By this time, the EU had **16**

8 Decision I/9: Medium-term Programme of Work of the Conference of the Parties, Conference of the Parties to the Convention on Biological Diversity.

9 Decision II/5: Consideration of the Need for and Modalities of a Protocol for the Safe Transfer, Handling and Use of Living Modified Organisms, Conference of the Parties to the Convention on Biological Diversity.

10 Art. 10 UNEP International Technical Guidelines for Safety in Biotechnology.

long since developed legislation to regulate the use of GMOs in response to the mounting politicisation of their long-term effects.[11]

III. The Cartagena Protocol on Biosafety

17 From the long and arduous discussions on modalities emerged a multifaceted transboundary agreement regulating trade in LMOs. The Protocol was adopted on 29 January 2000 in Montreal, Canada and entered into force on 11 September 2003. Today, 59 Parties have ratified the Cartagena Protocol, including most of those reported in this study.[12] The European Community signed the Protocol in May 2000[13] and Regulation (EC) No

---

11 See Council Directive 90/219/EEC of 23 April 1990 on the contained use of genetically modified micro-organisms, OJ L 117, 8.5.1990, 1–14, as amended by Directive 98/81/EC of 26 October 1998, OJ L 330, 5.12.1998, 13–31 and Council Directive 90/220/EEC of 23 April 1990 on the deliberate release into the environment of genetically modified organisms, OJ L 117, 8.5.1990, 15–27, as last amended by Council Directive 97/35/EC of 18 June 1997, OJ L 169, 27.6.1997, 72–73.

12 Austria (signed 24.5.2000; ratified 27.8.2002). See also, *M. Weissenbacher*, Austria (contained in this publication) no. 34; Brazil (acceded 24.11.2004). See also, C.B. Wandscheer/A.P. Myszczuk/R. Rocha dos Santos Brazil (contained in this publication) no. 63; Czech Republic (signed 24.5.2000; ratified 8.10.2001). See also, *J. Hrádek* (contained in this publication) no. 77 f.; Denmark (signed 24.5.2000; ratified 27.8.2002). See also V. Ulfbeck Denmark (contained in this publication) no. 25; Estonia (signed 6.9.2000; ratified 24.3.2004). See also *I. Kull*, Estonia (contained in this publication) no. 38; Finland (signed 24.5.2000; ratified 9.7.2004). See also, *B. Sandvik*, Finland (contained in this publication) no. 58; France (signed 24.5.2000; approved 7.4.2004). See also, *S. Taylor*, France (contained in this publication) no. 58; Germany (signed 24.5.2000; ratified 20.11.2004). See also, *J. Fedtke*, Germany (contained in this publication) no. 52; Greece (signed 24.5.2000; ratified 21.5.2004). See also, *E. Dacoronia*, Greece (contained in this publication) no. 47; Hungary (signed 24.5.2000; ratified 13.1.2004). See also, *A. Menyhárd*, Hungary (contained in this publication) no. 52; Italy (signed 24.5.2000; ratified 24.3.2004). See also, *A. Monti/F. Fusco*, Italy (contained in this publication) no. 40; Luxembourg (signed 11.7.2000; ratified 28.8.2002). See also, *P. Goergen*, Luxembourg (contained in this publication) no. 80 f.; Malta (acceded 5.1.2007). See also, *E. Buttigieg*, Malta (contained in this publication) no. 37; Netherlands (signed 24.5.2000; accepted 8.1.2002). See also, *I. Greveling/W.H. van Boom*, The Netherlands (contained in this publication) no. 39; Norway (signed 24.5.2000; ratified 10.5.2001). See *B. Askeland*, Norway (contained in this publication) no. 26; Poland (signed 24.5.2000; ratified 10.12.2003). See also, *E. Bagińska*, Poland (contained in this publication) no. 1; Slovenia (signed 24.5.2000; ratified 20.11.2002). See also, *R. Lampe*, Slovenia (contained in this publication) no. 25; Spain (signed 24.5.2000; ratified 16.11.2002). See also, *M. Martín-Casals/A. Ruda*, Spain (contained in this publication) no. 75; Sweden (signed 24.5.2000; ratified 8.8.2002). See also, *D. Langlet/P. Mielnicki*, Sweden (contained in this publication) no. 66; United Kingdom (signed 24.5.2000; ratified 19.11.2003). See also, *K. Oliphant*, England and Wales (contained in this publication) no. 36. Of the EFTA States, Iceland and Liechtenstein are yet to ratify the Convention. All three recognised EU candidate States have ratified the Convention. For Liechtenstein, see *M. Geiger*, Liechtenstein (contained in this publication) no. 34.

13 Council Decision 2002/628/EC was taken on 25 June 2002 to conclude the Protocol, on behalf of the Community.

1946/2003 was passed to ensure the coherent implementation of the provisions of the Protocol on behalf of the Community.[14] The world's biggest exporter of LMOs, the U.S., is not a party to the Protocol.[15] Nor is Argentina, Australia, Canada, Chile or Uruguay.[16] The Protocol pushes Parties to encourage such non-parties to adhere to its provisions and to contribute appropriate information for the shared access of its Parties.[17]

The Protocol as it stands today is best examined against the backdrop of its complex negotiations, the negotiating groups that emerged and the underlying motivations behind their negotiation stances.[18] The five major groups were: (a) as already mentioned above, though in the context of the Convention, the Like Minded Group of developing countries; (b) the Miami Group, composed of Canada, the U.S., Australia, Argentina, Chile and Uruguay; (c) the European Union; (d) the Compromise Group, consisting of Japan, Korea, Norway, Switzerland, Singapore, New Zealand and Mexico; and (e) the Central and Eastern European Group composed, inter alia, of Albania, Armenia, Azerbaijan, Belarus, Bosnia and Herzegovina, Bulgaria, Croatia, Czech Republic, Georgia, Hungary, Latvia, Lithuania, Moldova, Poland, Romania, Russia, Slovakia, Slovenia, the Ukraine and Yugoslavia. **18**

For the most part, whereas the Like Minded Group genuinely had environmental interests (and the interests of environmental NGOs and its traditional agricultural industry) at heart, the Miami Group negotiated on behalf of the agro-biotech industry with the aim of facilitating trade and keeping Protocol-related costs low. The EU had just come out of a credibility crisis goaded by BSE[19] and CJD[20] outbreaks in Britain and a dioxin-related food scandal in Belgium. To top this, it had to respond to a fresh outbreak of salmonella and the avian flu pandemic.[21] In negotiating the terms of the Protocol, therefore, the chief endeavours of the Union were to regain consumer confidence, protect its domestic agricultural industry and to bolster its defences "in the event of a WTO challenge to its regula- **19**

14 Regulation (EC) No 1946/2003 of the European Parliament and of the Council of 15 July 2003 on transboundary movements of genetically modified organisms, OJ L 287, 5.11.2003, 1–10.

15 See *A.B. Endres*, U.S.A. (contained in this publication) no. 23.

16 See *M. Lunney*, Australia (contained in this publication) no. 40 and *J. Matthews Glenn*, Canada (contained in this publication) no. 42.

17 Art. 24 Cartagena Protocol on Biosafety.

18 See, The Cartagena Protocol on Biosafety: A Record Of The Negotiations (2003).

19 Bovine spongiform encephalopathy or mad cow disease.

20 Creutzfeldt-Jakob disease, a human spongiform encephalopathy.

21 *M. Lee*, EU Regulation of GMOs: Law and Decision Making for a New Technology (2008) 4 f.

tory framework for safety in biotechnology…".[22] For geographic and, in part, accessional reasons, the Central and Eastern European Group were strongly inclined to support EU interests. Compromise by name and compromise by action: the Group shared interests with all the other groups and stakeholders.

1. Key Provisions of the Protocol in Outline

20 Based on the Precautionary Principle, the Cartagena Protocol seeks to protect biological diversity from the potential risks posed by LMOs. It does this by establishing a comprehensive regulatory system for the safe transfer, handling and use of LMOs, specifically focusing on transboundary movements of the latter though also taking risks to human health into account.

(a) Living Modified Organisms v Genetically Modified Organisms

21 The term "LMO" is derived from the Convention on Biological Diversity. It was negotiated into the Convention, despite strong initial opposition during draft talks, as a compromise term on the insistence of the U.S. which had objected to the use of the term "GMOs". The Protocol begins with a circular definition of "LMOs" as any *living* organism that possesses a novel combination of genetic material obtained through the use of modern biotechnology.[23] It then defines "living organism" and "modern biotechnology". The former is a biological entity capable of transferring or replicating genetic material.[24] GMOs may also include entities which are *not* capable of growing, i.e. are dead. Thus LMOs, though broadly equivalent to GMOs, are more accurately seen as a subset of the same. Common examples of LMOs include tomatoes, cassava, corn, cotton and soybeans.

22 *C. Bail/J.P. Decaestecker/M. Jørgensen*, European Union, in: C. Bail/R. Falkner/H. Marquard, The Cartagena Protocol on Biosafety: Reconciling Trade in Biotechnology with Environment & Development? (2002) 167.

23 Ibid. art. 3(g) Cartagena Protocol on Biosafety. Regulation (EC) No 1946/2003 which implements the provisions of the Protocol maintains the use of the term "GMO" as defined in art. 2(2) of Directive 2001/18/EC: art. 3.

24 Art. 3(h) Cartagena Protocol on Biosafety.

(b) Precautionary Principle

Perhaps one of the key features that justify the Protocol's perception as a progressive piece of international law is its incorporation of the Precautionary Principle enshrined in Principle 15 of the Rio Declaration (supra no. 6). The effect is that LMO imports can be blocked if uncertainty exists as to their safety thus shifting the onus on those who advocate them to prove otherwise. The Miami Group, composed of the six major agricultural exporters, saw the Principle as a political axiom, open to abuse. Instead, it preferred a regulatory framework based on scientific certainty and risk management: a framework consistent with existing international environmental agreements. On the opposing side, the Like Minded Group strongly and consistently advocated the inclusion of the precautionary approach. The European Union, which often stood between these polarised negotiating groups, stressed the Principle's importance to GMO trade and so again allied itself with the Like Minded Group. **22**

The Protocol is peppered with language allowing importing countries to invoke "possible" or "potential" threats to justify their trade embargos. Moreover, several provisions of the Cartagena Protocol expressly incorporate the precautionary approach. In addition to reaffirming it in its preamble, reference is made to the Principle in art. 1, 10, 11 and Annex III. In particular, art. 10(6) and 11(8) in relevant part state that: **23**

> "Lack of scientific certainty due to insufficient relevant scientific information and knowledge regarding the extent of the potential adverse effects of a living modified organism on the conservation and sustainable use of biological diversity...taking also into account risks to human health, shall not prevent that Party from taking a decision, as appropriate, with regard to the import of the living modified organism...in order to avoid or minimize such potential adverse effects."

Par. 4 of Annex III on risk assessment reads: **24**

> "Lack of scientific knowledge or scientific consensus should not necessarily be interpreted as indicating a particular level of risk, an absence of risk, or an acceptable risk."

Thus, in deciding whether to allow the import of GMOs, be it for intentional introduction into the environment or for direct use as food or feed, or for processing, and in conducting the risk assessment to facilitate the decision, importing countries may rely on the precautionary approach as justification for trade inhibitions. However, the right to cite the Precautionary Principle as good reason for a ban or as good reason to impose **25** onerous conditions is provisional: where additional relevant scientific or technical evidence becomes available that would support a review of the original decision, an exporting Party is granted discretion to require an importing Party to re-examine its decision on import under art. 12.

(c) Scope of the Protocol

26 The disagreements as to the scope of the Convention would resurface later during negotiations of the scope of the Protocol. Developing countries, some of which had no domestic LMO regulatory structures, advocated a Protocol on biosafety in the field of "safe transfer, handling and use of LMOs" whilst developed countries preferred a tightly drawn Protocol on "transboundary transfer of any LMO". The latter thought that art. 8(g) of the Convention, which required Contracting Parties to introduce means to regulate risks associated with the use and release of LMOs (though *only* as far as possible and as appropriate), was already effectively an international agreement on use and release.[25] Other differences of opinion were apparent: to the discontent of Canada, the U.S. and Australia, the EU advanced the inclusion of risks to human health; and, to the frustration of the Like Minded Group of developing countries, delegations from countries with powerful pharmaceutical industries sought to exclude the transboundary movement of LMOs which are pharmaceuticals for humans. The Parties eventually settled on art. 4 which provides:

> This Protocol shall apply to the transboundary movement, transit, handling and use of all living modified organisms that may have adverse effects on the conservation and sustainable use of biological diversity, taking also into account risks to human health.

27 The scope is, however, qualified. Firstly, it is limited to LMOs that may have *adverse effects* on the conservation and sustainable use of biological diversity. In the main, the Protocol applies to the transboundary movement of LMOs and has less restrictive provisions for the transit, handling and contained use of LMOs.[26] Moreover, the Protocol does not apply to pharmaceuticals intended for human use[27] or LMOs declared safe by the

---

25 H. *Marquard*, Scope, in: C. Bail/R. Falkner/H. Marquard, The Cartagena Protocol on Biosafety: Reconciling Trade in Biotechnology with Environment & Development? (2002) 290.

26 Art. 6 Cartagena Protocol on Biosafety.

27 Ibid., art. 5.

meeting of the Parties for time to time.[28] As the Protocol establishes an internationally binding framework of *minimum standards*, Parties may decide to subject such LMOs to detailed assessments so long as these are consistent with the objectives and provisions of the Protocol and their obligations under international law.[29]

The Protocol primarily deals with two types of LMOs and these neatly reflect the areas of prime focus of this study: 28

- LMOs intended for deliberate release into the environment; and
- LMOs intended for food, feed or for processing (LMOs-FFP or commodities).

The significance of this classification is that the two types of LMOs are subject to separate sets of procedures. While both sets are designed to ensure that recipient countries are provided with the information they need for making informed decisions about whether or not to allow LMO imports, one contains more rigorous notification and consent procedures than the other. 29

(d) The Advance Informed Agreement (AIA) Procedure

Based on the Rotterdam Convention on the Prior Informed Consent Procedure for Certain Hazardous Chemicals and Pesticides in International Trade (1998), the purpose of the AIA procedure is to ensure that importing Parties have both the opportunity and the capacity to decide on LMO imports in an informed manner. It thus details the course which LMO exporting Parties are to take in order to obtain the consent of importing Parties in advance of the *first* intentional transboundary movement of LMOs:[30] i.e. notification, acknowledgment, decision. The Protocol also provides a review procedure. Unless otherwise internally legislated, no prior consent need be obtained for the subsequent shipment of a similar consignment of LMOs. 30

The AIA procedure only applies to LMOs intended for deliberate release into the environment.[31] As aforementioned, less restrictive provisions apply to LMOs in transit, LMOs destined for contained use and LMOs-FFP.[32] The EU and the Miami Group united behind a common negotiat- 31

28 Ibid., art. 7(4).
29 Ibid., art. 2.
30 Ibid., art. 7(1).
31 Art. 7(2) Cartagena Protocol on Biosafety.
32 Ibid., art. 6, 7(3) and 18.

ing position on the exclusion of LMOs in transit from the scope of the Protocol altogether and to their disquiet, the Like Minded Group won, though only partially. The argument went that the inclusion of LMOs in transit in the AIA procedure would require the prior informed consent of every country en-route to the importer. By excluding this from the AIA procedure, the allies thus came some way in impeding costly bureaucratic processes. LMOs intended for contained use are destined for research and so the EU with its vast R&D framework advocated their inclusion in the Protocol's scope but exclusion from the AIA procedure.

32 Under the said procedure, a Party wishing to export LMOs must notify the competent authority[33] in the country of import regarding the proposed consignment.[34] Annex I to the Protocol details the information that the Party of export must service the authority with. These include the contact details of the exporter and the recipient, details of the LMOs and their intended use in addition to suggested methods of safe handling, storage, transport and use. The importing Party should acknowledge the notification within 90 days of its receipt.[35] An important feature to note is that throughout the Protocol, absence of a decision does not imply consent or refusal, unless otherwise specified by the Party.[36] Such explicit consent is consistent with the sovereign rights of states to decide on the importation of LMOs.

33 The Party of import must then decide whether to import the LMOs. To facilitate this decision, the exporting Party is obliged to furnish the competent authority of the importer with a risk assessment report, enabling the latter to identify and evaluate the potential adverse effects of the LMOs intended for transboundary movement on the likely potential receiving environment and on human health.[37] Annex III to the Protocol sets out guidelines and methodologies on how to conduct a risk assessment, which ought to be executed in a scientifically sound and transparent manner. The risk assessment in Annex III involves, among other things: (a) identifying any characteristics associated with the LMO that may have adverse environmental/health effects; (b) evaluating the likelihood of these effects being realised, taking into account the level and kind of exposure in the receiving environment; and (c) an estimation of the overall risk posed by the LMO.

---

33 Art. 19 Cartagena Protocol on Biosafety requires parties to designate a competent national authority responsible for performing the administrative functions required by the Protocol.

34 Ibid., art. 8(1).

35 Ibid., art. 9(1).

36 Ibid., art. 9(4), 10(5) and 11(7).

37 Ibid., art. 10(1) and 15.

Having evaluated the risks, and within 270 days of receipt of notification, the competent importing authority can either approve the import (with or without conditions), prohibit the import, request further relevant information or extend the deadline for the decision by a defined period.[38] All decisions, except an unconditional acceptance, must be accompanied by reasons on which they are based.[39] For example, where the authority feels there is lack of scientific consensus or knowledge on the potential adverse effects of a LMO, it may prohibit importation, basing its finding on the Precautionary Principle. Decisions are to be communicated to the Party of export and relayed to the Biosafety Clearing House (BCH) which was specifically established under the Protocol to facilitate the exchange of scientific, technical, environmental and legal information on, and experience with, LMOs.[40] The Protocol also enables importing Parties to specify in advance to the BCH, LMOs they wish to exempt from the AIA procedure.[41] 34

Finally, and as aforesaid, the Protocol empowers the LMO exporting Party with the discretion to request the competent recipient authority to review its earlier decision if it considers that a change of circumstances has occurred or additional relevant scientific or technical information has become available that may influence the outcome of the risk assessment upon which the decision was based.[42] Conversely, the importing country can also reverse its decision to import on the basis of fresh facts.[43] The exporting country cannot appeal the importing authority's decision save where additional developments ensue. 35

(e) A Simplified Procedure for LMOs-FFP

Despite the insistence of developing countries, LMOs-FFP are included in the scope of the Protocol but exempt from the rigorous AIA procedures.[44] This exclusion was based on the averment that LMO-based commodities destined for food, feed and processing do not pose a threat to biodiversity as they are not intended for deliberate release. As was implicit in the Miami Group's position on the matter, the inclusion of LMOs-FFP within 36

38 Ibid., art. 10(3).
39 Ibid., art. 10(4).
40 Ibid., art. 20.
41 Ibid., art. 13.
42 Ibid., art. 12(2).
43 Ibid., art. 12(1).
44 Ibid., art. 7(2).

the AIA compass would entail cost-excessive approvals. Parties may, however, internally legislate for LMOs-FFP to undergo the AIA procedure.

37 Commodities are covered by art. 11. A Party that decides to import LMOs-FFP must inform other Parties through the BCH within 15 days of making that decision. Annex II enumerates the minimum information required. The details are similar to that of Annex I, i.e. the contact details of the importer and exporter, the characteristics of the LMOs-FFP to be exported and suggested methods for the safe packaging, labelling, documentation, etc, of the LMOs. In addition, a risk management report consistent with Annex III must be drawn up. The decision to import LMOs-FFP may be taken under the importing Party's domestic regulatory framework which must be consistent with the objective of the Protocol.[45] Further, the importing country must make copies of its national laws, regulations and guidelines applicable to the import of LMOs-FFP available to the BCH.[46] The EU, with its robust regulatory framework for commodities, supported this move.[47] Further, although excluded from the AIA procedure, the notification and information costs associated with LMOs-FFP would still unfavourably influence their fiscal competitiveness to the advantage of the EU agri-industry.

(f) Handling, Transport, Packing and Identification

38 More specific labelling requirements apply to LMOs-FFP than LMOs intended for contained use or release into the environment. Whereas doc-

45 Ibid., art. 11(4).

46 Ibid., art. 11(5).

47 Regulation (EC) No 1829/2003 of the European Parliament and of the Council of 22 September 2003 on genetically modified food and feed, OJ L 268, 18.10.2003, 1–23; Regulation (EC) No 1830/2003 of the European Parliament and of the Council of 22 September 2003 concerning the traceability and labelling of genetically modified organisms and the traceability of food and feed products produced from genetically modified organisms and amending Directive 2001/18/EC, OJ L 268, 18.10.2003 24–28; Commission Regulation (EC) No 65/2004 of 14 January 2004 establishing a system for the development and assignment of unique identifiers for genetically modified organisms, OJ L 10, 16.1.2004, 5–10; Commission Regulation (EC) 641/2004 of 6 April 2004 on detailed rules for the implementation of Regulation (EC) 1829/2003 of the European Parliament and of the Council as regards the application for the authorisation of new genetically modified food and feed, the notification of existing products and adventitious or technically unavoidable presence of genetically modified material which has benefited from a favourable risk evaluation, OJ L 102, 7.4.2004, 14–25; Directive 2001/18/EC of the European Parliament and of the Council of 12 March 2001 on the deliberate release into the environment of genetically modified organisms and repealing Council Directive 90/220/EEC, OJ L 106, 17.4.2001, 1–39. See also *M. Lee*, EU Regulation of GMOs, Law and Decision Making for a New Technology (2008).

umentation accompanying the latter only need identify them as such, LMO-FFPs should clearly identify that they „may contain“ LMOs.[48] This was strongly supported by the EU, the Like Minded Group and the Compromise Group which were pro-consumer choice. The task of carving out the detailed requirements for this purpose is still in progress.[49]

(g) **Socio-economic Considerations**

During Protocol negotiations, developing countries argued that as well as scientifically quantifiable harm, LMOs posed other societal risks worthy of consideration in the decision making process. They proposed an extensive list of negative socio-economic factors which could be triggered by the widespread use of LMOs including increased dependence on multinational agri-tech companies and the loss of traditional livelihoods. Art. 26 in its final form is narrowly formulated. It reads thus: **39**

> The Parties, in reaching a decision on import under this Protocol or under its domestic measures implementing the Protocol, may take into account, consistent with their international obligations, socio-economic considerations arising from the impact of living modified organisms on the conservation and sustainable use of biological diversity, especially with regard to the value of biological diversity to indigenous and local communities.

It is narrowly formulated insofar as: (a) it limits the scope of socio-economic considerations to those affecting biodiversity (though conceivably, prospective importers with anti-trade sentiments could rationalise any import ban on socio-economic grounds);[50] and (b) the considerations are consistent with Parties‘ international obligations. This greatly tapers the socio-economic issues that can legitimately apply and so guards against the adoption of trade measures which enhance national discretion beyond WTO requirements. **40**

---

48 Ibid. art. 18(2). See also art. 2(2), art. 23, Annex 1 and II.

49 In Decision BS-IV/10, the Conference of the Parties serving as the meeting of the Parties to the Cartagena Protocol on Biosafety requested the Executive Secretary to organize an online conference to: (a) identify the relevant standards with regard to handling, transport, packaging and identification of living modified organisms; (b) identify where gaps exist; and (c) suggest possible modalities to fill the gaps; and to prepare a summary of the outcome of the conference, reflecting the full range of views expressed, for the consideration of the Conference of the Parties serving as the meeting of the Parties to Protocol at its fifth meeting which will be held in October 2010 in Nagoya, Japan.

50 *K.L. Holtby/W.A. Kerr/J.E. Hobbs*, International Environmental Liability and Barriers to Trade: Market Access and Biodiversity in the Biosafety Protocol (2007) 36.

(h) Unintentional, Illegal Transboundary Movements

41 While legitimate concerns of the risks inherent in modern biotechnology might be mitigated through capacity building,[51] raising public awareness,[52] increased regulation and improved domestic, regional and international coordination among government agencies, the unpredictable and potentially irreparable, adverse consequences of LMOs cannot be ignored. The Protocol sets out the measures which Parties are to take in the event of unintentional and illegal transboundary movements of LMOs in art. 17 and 25 respectively. Under art. 17(4) Parties are under a general commitment to notify and immediately consult the affected or potentially affected states to enable them to determine the appropriate responses and initiate necessary action, including emergency measures.

(i) Compliance

42 In accordance with art. 34, a Compliance Committee composed of 15 members was established at the Conference of the Parties serving as the meeting of the Parties to the Protocol (COP-MOP). Broadly, the Committee is tasked to promote compliance, address cases of non-compliance and to provide advice or assistance.[53] The task of the Committee is evidently crucial, inter alia, in ensuring the ex ante mitigation of potential risks.

(j) Relationships with other International Agreements

43 By virtue of its application to the "transboundary movement of LMOs", the Cartagena Protocol is not only concerned with environmental issues but also extends to trade. What then is the Protocol's position in relation to other international agreements, in particular, those on trade? Established on 1 January 1995, the WTO is the successor to the General Agreement on Tariffs and Trade (GATT). It is a multilateral organisation charged with administering rules of international trade between its 153 Member States which include all the EU Member States and the six-member coalition of the main GMO producers. As an organisation for "liberalising trade", the WTO has often been thought insensitive to concerns regarding the impact of trade on the environment and social development. Several WTO agree-

51 Art. 22 Cartagena Protocol on Biosafety.

52 Ibid., art. 23.

53 See http://www.cbd.int/biosafety/issues/compliance.shtml.

ments have implications for trade in GMOs including the Agreement on the Application of Sanitary and Phytosanitary Measures (SPS), the Agreement on Technical Barriers to Trade (TBT) and the Agreement on Trade-Related Aspects of Intellectual Property Rights (TRIPs). Indeed, in the past, the U.S. has not been shy to threaten the EU and has in fact brought the latter before the WTO to challenge its GMO regulatory framework.

Given existing trade frictions, the Miami Group supported the insertion of a "savings clause" designed to expressly subordinate multilateral environmental agreements, including the Cartagena Protocol, to the SPS and TBT Agreements and the GATT. This would thus limit the extent to which importing states could hinder trade on purported health, food safety and/or environmental grounds. It would essentially undermine the signatories' ability to call on the Precautionary Principle, for example, or socio-economic considerations to restrict trade. Moreover, the Protocol's labelling requirements for commodities may potentially cause conflict. Unsurprisingly, the EU and the Like Minded Group fiercely contested the inclusion of the WTO supremacy clause, citing that it would undoubtedly weaken the efficacy of the Protocol. **44**

The Protocol's place vis-à-vis other agreements is stated in three paragraphs of preambular text: **45**

> *Recognizing* that trade and environment agreements should be mutually supportive with a view to achieving sustainable development,
>
> *Emphasizing* that this Protocol shall not be interpreted as implying a change in the rights and obligations of a Party under any existing international agreements,
>
> *Understanding* that the above recital is not intended to subordinate this Protocol to other international agreements,

The provisions clearly embody compromise and reflect the various negotiating positions adopted by the delegates. To that extent, the priority of the Protocol relative to WTO agreements remains ambiguous.[54] It is not uninteresting to note, however, that should a dispute arise, the WTO dispute settlement text, Understanding on Rules and Procedures Governing the Settlement of Disputes,[55] provides some guidance on the interpretation of potentially conflicting obligations in international environmental treaties. In accordance with the Rules, the WTO tribunal would have to **46**

54 See also art. 2 and 26 Cartagena Protocol on Biosafety.
55 See ibid., art. 3. See also *Holtby/Kerr/Hobbs* (fn. 50) 59.

take into account all relevant international law, including the provisions of the Cartagena Protocol.

(k) Liability and Redress

47 Notwithstanding appeals to states under Principle 22 of the Stockholm Declaration on the Human Environment, 1972[56] and Principle 13 of the Rio Declaration on Environment and Development, 1992[57] respectively, there is at present no international liability framework directly applicable to biotechnology. Significant concerns over the absence of a liability and redress regime, seen as crucial for the effective implementation of the Protocol, surfaced at the fifth meeting of the Working Group on Biosafety in 1998. Several delegations, noting the precedent for state responsibility in customary international law, proposed the "zero option" against the inclusion of a liability and redress provision.[58]

48 Developing countries expressed the opposite view and pointed to the growing trend of sectoral environmental liability regimes that had been adopted in the recent past, e.g. on oil pollution, nuclear damage, the carriage of hazardous goods and substances and damage caused by space objects. They submitted that a liability and redress regime would serve as a compliance incentive. In addition, the regime would contribute to the implementation of the Polluter Pays Principle by internalising damage costs: it would be inappropriate in this era of precaution not to pre-empt an LMO catastrophe by elaborating rules and procedures in the field of liability and redress for damage resulting from transfrontier movements of LMOs. This group of countries thus advocated the inclusion of a substantive article covering liability and compensation in the Protocol.[59]

---

56 "States shall cooperate to develop further the international law regarding liability and compensation for the victims of pollution and other environmental damage caused by activities within the jurisdiction or control of such States to areas beyond their jurisdiction".

57 "States shall develop national law regarding liability and compensation for the victims of pollution and other environmental damage. States shall also cooperate in an expeditious and more determined manner to develop further international law regarding liability and compensation for adverse effects of environmental damage caused by activities within their jurisdiction or control to areas beyond their jurisdiction".

58 See Principle 21 Stockholm Declaration which is essentially reproduced in identical terms in Principle 2 of the Rio Declaration.

59 *K. Cook*, Liability: "No Liability, No Protocol", in: C. Bail/R. Falkner/H. Marquard, The Cartagena Protocol on Biosafety: Reconciling Trade in Biotechnology with Environment & Development? (2002) 371 ff.

Of further relevance was art. 14(2) Convention on Biodiversity which provides that: 49

> The Conference of the Parties shall examine, on the basis of studies to be carried out, the issue of liability and redress, including restoration and compensation, for damage to biological diversity, except where such liability is a purely internal matter.

Opponents of an extensive liability regime in the Protocol felt that the 50 Convention already catered for the issue. However, it was later conceded that as part of the Convention, any liability considerations under art. 14(2) would not be purposely geared towards damage caused by the transboundary movement of LMOs. Moreover, art. 14(2) is directed at "damage to biological diversity". It does not cover negative effects on health as foreseen by the Protocol under its art. 4. Following further discussions, the delegations reached a compromise: rather than endorsing the zero option or hastily word a provision on the subject, they agreed that a text on liability should be drawn in the timeframe deserving of such a complex task. The delegates thus settled on an enabling clause to the effect that details of a liability regime under the Protocol would follow. Art. 27 on Liability and Redress reads:

> The Conference of the Parties serving as the meeting of the Parties to this Protocol shall, at its first meeting, adopt a process with respect to the appropriate elaboration of international rules and procedures in the field of liability and redress for damage resulting from transboundary movements of living modified organisms, analysing and taking due account of the ongoing processes in international law on these matters, and shall endeavour to complete this process within four years.

Pursuant to art. 27, and on the recommendation of the Intergovernmental 51 Committee for the Cartagena Protocol on Biosafety (IPCC),[60] COP-MOP established an Open-ended Ad Hoc Working Group of Legal and Technical Experts on Liability and Redress in the Context of the Protocol at MOP 1 in February 2004.[61] The Group was tasked to elaborate international rules and procedures regarding liability and redress for damage resulting from transboundary movements of LMOs and to endeavour to complete this process within four years. By the time of MOP 4 in May 2008, the Working Group had made significant progress under their terms of reference. How-

---

60 ICCP Recommendation 2/1, UNEP/CBD/ICCP/2/15; ICCP Recommendation 3/1, UNEP/CBD/ICCP/3/10.

61 Decision BS-I/8: Establishment of an Open-Ended Ad Hoc Working Group of legal and technical experts on liability and redress in the context of the Protocol.

ever, the Parties agreed to establish a Group of Friends of the Co-Chairs, which comprised a smaller number of representatives of delegates, to continue the process of further negotiating international rules and procedures in the field of liability and redress for LMO damage. In its decision BS-IV/12,[62] COP-MOP agreed that the Group of Friends would meet twice before MOP 5 in October 2010 where they would present the outcome of their negotiations. The first meeting of the Group took place in February 2009 in Mexico and the second in Malaysia in June 2010.

IV. Proposed Operational Texts on Liability and Redress in the Context of Art. 27 of the Biosafety Protocol

52 As well as establishing the Group of Friends, COP-MOP also adopted a negotiating text which is annexed to decision BS-IV/12 as the basis for further work for the Group of the Friends of the Co-Chairs. The first section lays down a binding international agreement on an administrative approach (including a provision on civil liability); the second sets out guidance on civil liability for traditional damage; and the third addresses other provisions, namely a supplementary compensation scheme, settlement of claims and complementary capacity building measures.

1. Working Towards Legally Binding Provisions: The Supplementary Protocol on Damage Resulting from Transboundary Movements of LMOs to the Cartagena Protocol on Biosafety

(a) Administrative Approach

53 The administrative approach fostered in the Supplementary Protocol sets out a liability framework thus affording national competent authorities broad discretion, in part, when implementing its provisions. It is this flexibility that also characterises the EU's Environmental Liability Directive as a framework directive.[63] Like the Directive, the focus is on damage to biodiversity as opposed to traditional damage (damage to property, economic

62 Decision BS-IV/12: Liability and redress under the Cartagena Protocol on Biosafety.

63 Directive 2004/35/CE of the European Parliament and of the Council of 21 April 2004 on environmental liability with regard to the prevention and remedying of environmental damage, OJ L 143, 30.4.2004, 56–75.

loss, personal injury) which is left to the jurisdiction of national civil law systems. The objective of the Supplementary Protocol is to:

> "...contribute to ensuring that prompt, adequate and effective response measures are taken in the event of damage or imminent threat of damage to the conservation and sustainable use of biological diversity resulting from living modified organisms that find their origin in transboundary movements."

Commensurate with the functional scope of the Protocol, the Supplementary Protocol applies to damage caused by LMOs-FFP, LMOs destined for contained use and those intended for intentional introduction into the environment. Such damage may arise out of the transport, transit, handling and use of the LMOs provided these activities find their origin in a transboundary movement.[64] As aforementioned, damage in relation to the administrative approach is specifically limited to "damage to the conservation and sustainable use of biological diversity".[65] This is clear from the objective, definition, scope and context of the Supplementary Protocol. In additional to ecological damage per se, it applies to damage to human health insofar as it *arises from damage to biodiversity*.[66] In line with recent trends on liability regimes, a de minimis threshold of damage must be reached. To qualify as damage, the Supplementary Protocol adopts standard non-determined qualitative adjectives – the harm must be measurable and significant – thus ultimately leaving the construal of the adjectives to legislatures/the judiciary.[67] **54**

Like most international legal instruments, and consistent with the Polluter Pays Principle, the Supplementary Protocol channels liability to the LMO "operator". The definition of "operator" is yet to be determined. At present, three options are available, the broadest of which assigns liability to actors beyond those envisaged in the Parent Protocol. It enumerates an exhaustive list of persons: the developer, producer, notifier, exporter, importer, carrier, or supplier. As is considered below, the regime also imposes residual state liability on Parties.[68] **55**

Under the Supplementary Protocol, a causal link needs to be established between the damage and the activity in question. A rigorous nexus may, **56**

64 Art. 3(2) Supplementary Protocol on Damage Resulting from Transboundary Movements of LMOs to the Cartagena Protocol.

65 Ibid., art. 3(1).

66 Ibid., art. 3(1).

67 Ibid., art. 2(2)(d).

68 Ibid., art. 2(2)(g).

however, prove difficult to establish where damage is caused by LMOs. Discretion is thus left to competent authorities to elaborate on the causation standard in accordance with domestic law and the results will no doubt reflect domestic policies on LMOs.[69]

57 In the advent of an LMO catastrophe, the Supplementary Protocol places an obligation on operators to inform the competent authority, evaluate the damage and take response and *restoration measures*.[70] Again, the meaning of the italicized phrase is yet to be determined. Under one option, this implies taking reasonable action in the event of damage. The pre-emption of damage where an imminent threat of harm exists is currently square bracketed. Under the most elaborate option, this expressly extends to mitigating damage and restoring biodiversity.[71] The operator has a right of recourse to third parties for indemnity in respect of any compensation which he has to pay for damage.[72] Depending on the preferred policies of Parties to the Protocol, the role played by competent authorities will vary: discretion is left to Parties to legislate that in the event of LMO damage, and should the operator fail to do so, competent authorities may implement remedial measures on a subsidiary basis themselves.[73] This is without prejudice to the right of recourse of the authority against the operator.[74] It is perhaps fitting to note at this juncture that the Supplementary Protocol does not affect the rights and obligations of states under the rules of general international law with respect to the responsibility of states for internationally wrongful acts.[75]

58 In furtherance of LMO trade, the Supplementary Protocol, as with most international agreements on environmental damage, includes an exhaustive list of exemptions which operators can invoke to reduce or extinguish their liability. The following defences are foreseen: act of God (force majeure), act of war or civil unrest and an exemption grounded on reasons of national security is included.[76] One of the draft options also foresees whole or partial exoneration for interventional acts of third parties and compliance with compulsory measures and includes a regulatory permit and state of the art defence.[77] Various stakeholders have complained, how-

69 Ibid., art. 4(2).
70 Ibid., art. 7(1).
71 Ibid., art. 2(2)(f).
72 Ibid., art. 9.
73 Ibid., art. 7(4).
74 Ibid., art. 7(6).
75 Ibid., art. 15.
76 Ibid., art. 8(1).
77 Ibid., art. 8(2).

ever, that the inclusion of the state of the art defence defeats the Precautionary Principle upon which the Cartagena Protocol rests.

For the purpose of legal and evidential certainty, the Supplementary Protocol directs Parties to lay down limitations in time. A relative and/or absolute time limit for the recovery of costs and expenses may be provided.[78] Consistent with liability regimes that impose strict liability on operators, the Supplementary Protocol leaves to domestic discretion the option of laying down a financial limit on the amount of compensation for damage caused by the transboundary movements of LMOs.[79] While caps serve to limit operator liability and so facilitate insurability, the drawback is that victims of the operator's activity may not be fully compensated and in the end, the state – society at large, that is – would be secondarily liable for the damage. **59**

Parties may legislate for the coverage of liability through financial securities e.g. insurance of LMOs upon import. Further, they are urged to take measures to encourage the development of suitable financial security instruments and markets.[80] The Supplementary Protocol provides for a review of the effectiveness of the system a few years after its entry into force.[81] **60**

(b) Provision on Civil Liability

Nothing in the Supplementary Protocol prejudices the right of Parties to have in place or to develop their domestic law or policy in the field of civil liability and redress resulting from the transboundary movement of LMOs. Parties are thus invited to ensure that their national civil liability rules and procedures adequately provide for redress to damage resulting from the transboundary movement of LMOs. They may either develop their existing domestic laws, draft a specific civil liability regime or both. As evinced by the country reports preceding this paper, several signatories to the Protocol already have existing special liability regimes for LMO damage. These include Austria,[82] Finland,[83] France,[84] Germany,[85] Hun- **61**

78 Ibid., art. 10.
79 Ibid., art. 11.
80 Ibid., art. 12.
81 Ibid., art. 14.
82 Gentechnikgesetz or the Gene Technology Act which entered into force in 1994. See also *Weissenbacher* (fn. 12) no. 1 ff.
83 The Gene Technology Act (377/1995). See also *B. Sandvik* (fn. 12) no. 1 ff.
84 Law no. 2008–595 of 25 June 2008. See also *Taylor* (fn. 12) no. 1 f.
85 Gentechnikgesetz or Act on Genetic Engineering enacted in 1990. See also *Fedtke* (fn. 12) no. 1 ff.

gary,[86] Italy,[87] Norway[88] and Poland.[89] Slovakia[90] and Switzerland[91] also. Per contra, other jurisdictions such as England,[92] the Netherlands, Slovenia and Sweden continue to rely on pre-existing general liability rules for now. In creating or reviewing their national rules and procedures on civil liability, Parties are guided under the Supplementary Protocol by a prescribed list of elements which the regime ideally ought to address:

- Damage;
- Standard of liability: that may include strict, fault or mitigated liability;
- Channelling of liability, where appropriate;
- Financial security, where feasible or redress/compensation;
- Right to bring claims.

62 The regime also lays down a provision on reciprocal recognition and enforcement of judgments in accordance with applicable rules of procedures of domestic courts in respect of matters within the scope of the Supplementary Protocol and annexed guidelines on civil liability.[93]

2. Working Towards Non-Legally Binding Provisions on Civil Liability

63 It is envisioned that the legally binding regime on the administrative approach which includes a provision on civil liability will be complemented by non-legally binding guidelines on civil liability. To negotiate a

86 Act No XXVII of 1998 on Genetic Technology Activity as amended. See also *Menyhárd* (fn. 12) no. 1 ff. See also *Monti/Fusco* (fn. 12) no. 1 ff.

87 Decree-law no. 279 of 22 November 2004 (Dl 279/04), as amended. See also *Monti/Fusco* (fn. 12) no. 1 ff.

88 Lov om framstilling og bruk av gen-modifiserte organismer or the Norwegian Act on Genetic Technology, 2 April 1993. See also *Askeland* (fn. 12) no. 1 ff.

89 Law of 22 June 2001 on Genetically Modified Organisms (consolidated text published in the Journal of Laws 2007, no. 36, item 233). See also *Bagińska* (fn. 12) no. 1 ff.

90 *A. Dulak*, Economic Loss Caused by GMOs in Slovakia, in: B.A. Koch (ed.), Economic Loss Caused by Genetically Modified Organisms: Liability and Redress for the Adventitious Presence of GMOs in Non-GM Crops (2008) no. 417 ff.

91 *M. Müller-Chen*, Economic Loss Caused by GMOs in Switzerland, in: B.A. Koch (ed.), Economic Loss Caused by Genetically Modified Organisms: Liability and Redress for the Adventitious Presence of GMOs in Non-GM Crops (2008) no. 489 ff.

92 See *K. Oliphant* (fn. 12) no. 1, who writes that consultation about the introduction of a statutory redress scheme in respect of economic damage resulting from GMO presence in non-GM crops is currently in progress.

93 Art. 13 Supplementary Protocol on Damage Resulting from Transboundary Movements of LMOs to the Cartagena Protocol.

binding international civil liability regime for damage caused by the transboundary movement of LMOs would not only be laborious and slow but would also result in the duplication of existing national provisions, if it ever entered into force. The complementary non-legally binding instrument on civil liability thus preserves national sovereignty and merely assists and encourages Parties by way of guidelines to fill existing lacunae in their domestic civil liability frameworks.

As will be seen below, whereas the administrative approach is oriented to the "prompt, adequate and effective" remediation of ecological damage caused by the transboundary movement of LMOs, the soft-law guidelines on civil liability ensure that robust liability regimes are in place to cover, inter alia, traditional forms of damage caused by the transboundary movement of LMOs. **64**

The first part of the operational texts contains sections on state responsibility, scope, damage and the primary compensation scheme. The first and second sections are substantially similar to that on the administrative approach. Damage encompasses any damage to the conservation and sustainable use of biodiversity not redressed through the administrative approach, damage to health (loss of life/personal injury), property and loss of income (and other economic loss). Loss of or damage to cultural, social and spiritual values, or other loss or damage to indigenous or local communities or loss of or reduction of food security as a head of damage remains square bracketed in its entirety. This definition is certainly broader than the parameters visualised by the art. 27 Biosafety Protocol drafters when read in context with the scope of the Protocol. Causation is also left to be dealt with in accordance with domestic laws. **65**

The last section on primary compensation schemes is further subdivided. It covers the standard of liability (fault-based/strict/mitigated strict liability) and channelling of liability essentially leaving these to be fleshed out by the domestic law-maker. Existing national liability regimes applicable to GMO damage such as those in Austria,[94] Germany,[95] Poland,[96] Switzerland[97] and Norway,[98] among other jurisdictions, have strict liability regimes in place. A few words on mitigated strict liability would be appropriate here. Under the latter regime, a fault-based standard of liability **66**

94 *Weissenbacher* (fn. 12) no. 1.
95 *Fedtke* (fn. 12) no. 2 ff.
96 *Bagińska* (fn. 12) no. 3.
97 See *M. Müller-Chen* (fn. 91) 495.
98 *Askeland* (fn. 12) no. 3.

applies except to the extent that a risk-assessment has identified LMOs as ultra-hazardous and/or the operator's acts or omissions violate national law and/or written conditions of any approval. In this case, a strict liability standard applies.

67 Interim relief in the form of an injunction or a declaration may also be legislated. The exhaustive options of exemptions duplicate those under the administrative approach except for an additional defence: "the operator could not have reasonably foreseen the damage". The guidelines provide for recourse action and contain detailed options on the apportionment of liability/joint and several liability in the case of multiple operators. Limitations of liability in time and amount are also included and the part ends with a section on coverage which reproduces the section on coverage under the administrative approach.

3. Other Provisions

68 It is envisaged that the above-outlined intergovernmental instruments on safety and the participation of parties in various efforts to share knowledge through information networks should lessen dangers posed by LMOs to health, property and environmental integrity. In the event of damage caused by the transboundary movement of LMOs, the administrative approach as supplemented by robust civil liability regimes (once developed), will conceivably step in to remedy the damage. On their own, however, these redress mechanisms are lopsided, encumbering importing countries for the most part with implementation, monitoring and other administrative obligations. The last part of the operational texts is divided into three sections: (a) a supplementary compensation scheme; (b) settlements of claims; and (c) compensatory capacity building measures.

69 Like the Lugano Convention on Civil Liability for Damage Resulting from Activities Dangerous to the Environment,[99] a supplementary compensation scheme is foreseen. The appeal of the scheme is that risks posed by LMOs can potentially be channelled to those who benefit most from LMO trade by requiring them to contribute towards it. In Portugal for example, the existing compensation scheme is funded by a green tax levied on seeds while in Denmark, GMO farmers pay a certain amount per hectare of GM cultivation. Voluntary schemes such as that in the Netherlands, industry led schemes such as that in Germany and pure state com-

99 21.VI.1993. See art. 12. The Council of Europe Convention is not yet in force.

pensation schemes such as that in Slovenia currently exist or are in draft.[100] The proposed provisions also foresee residual state liability where claims are not satisfied under the fund. The scheme's terms of reference will be decided either by COP-MOP or by the Parties themselves. In either case, compensation schemes on damage resulting from transboundary movements of LMOs can certainly draw from the strengths and experiences of existing domestic funds. The Protocol scheme could thus potentially fill a regulatory gap as none of the reported jurisdictions currently has GMO compensation funds for cross-border claims.

The section on the settlement of claims contains a subsection on a special **70**
tribunal to step in in the event of a dispute between persons claiming LMO damage pursuant to the texts. It also contains a section on standing/right to bring claims. "[Affected] natural or legal persons" have locus standi and can, in addition to non-governmental organisations promoting environmental protection, require a competent authority to act in accordance with its duties under the texts as implemented. The text is yet to be finalised.

V. Conclusion

Poised to reconcile the respective interests of trade in biotechnology and **71**
the environment, the Cartagena Protocol on Biosafety saw its negotiators' efforts realised in September 2003 on its entry into force. For the most part, the composite text achieves its mandate under art. 19(3) of the Cartagena Convention and reflects a careful and sustained endeavour to reach a balanced compromise. In so doing, however, it failed to satisfy the aspirations of the six LMO exporter countries, none of which have ratified the Protocol to date. Five years after its entry into force, and in accordance with its art. 35, COP-MOP undertook an evaluation of the effectiveness of the Protocol, including an assessment of its procedures and annexes. Predictably, in all the submissions, the Parties responded that it was challenging to give a full scale assessment of the Protocol's effectiveness in the absence of much practical experience in its implementation.[101] What is certain, however, is that the efficacy of the Protocol will largely depend on its transposition. In particular, contracting states have ample leverage

100 *B.A. Koch*, Comparative Report, in: B.A. Koch (ed.), Economic Loss Caused by Genetically Modified Organisms: Liability and Redress for the Adventitious Presence of GMOs in Non-GM Crops (2008) no. 174 ff.

101 See UNEP/CBD/BS/COP MOP/4/INF/10, UNEP/CBD/BS/COP-MOP/4/14 and MOP 4 Decision BS-IV/15: Assessment and review.

to reflect socio-cultural perceptions in their policy choices to either facilitate trade or hamper imports based on science, socio-economic considerations, health and environmental factors.

72 The extent to which countries feel compelled to masquerade trade prohibitions as environmental protection measures could, however, be mitigated by a liability and redress mechanism consistent with the Polluter Pays Principle. The regime, which has exceeded the allotted four-year timeframe set for its completion, is still in progress. This is perhaps testament to the challenges involved in trying to satisfy the expectations of multiple stakeholders in one of the most worrisome and controversial issues of our time.

73 In an attempt to carve out a sectoral liability regime for damage caused by the transfrontier movement of LMOs, the successive drafters of the subsidiary instrument borrowed extensively from existing international environmental liability regimes. It would be pointless to examine the intricacies of the proposed texts since they are still a way from completion. Certain general observations can, however, be made. Firstly, the drafters of the proposed texts have certainly tried to cover all bases, combining public and private law initiatives with a compensation fund. The choice of a binding administrative instrument is clearly in response to the need, not just for prompt, adequate and effective remediation of environmental harm but also, for a common, easily transposable, framework throughout participating states. Conscious efforts were taken not to overly prescribe provisions in an attempt to avoid duplicity and also any introduction of inconsistencies into domestic laws. This should avoid the plight of several international legally binding instruments, like the Lugano Convention, which have been adopted but are not yet in force. Much discretion has thus been left in the hands of national policy makers to frame implementing provisions as they see fit.

74 A binding civil liability regime was neither a legally nor scientifically justifiable option. Pragmatism thus dictated that the civil liability instrument take the form of soft law provisions which set out broad guidelines thus influencing the direction of domestic law makers in their review of existing civil liability provisions applicable to LMOs or in their drafting of the same. This would facilitate quicker drafting and adoption though at the expense of consistency in protecting victims and the environment. Given the nature of the Protocol, it has been submitted that a uniform liability regime would have been more effective in addressing damage

caused by transboundary movements.[102] It is yet to be seen just how far the supplementary compensation fund will be used to shift the burden of LMO damage to its advocators and beneficiaries.

In summation, the liability provisions, which were a key demand by devel- **75**
oping countries, will principally benefit those countries which have no or inadequate liability regimes in place. Notwithstanding so, scope to add to the regimes in developed countries exists and to that extent, the texts, in their final form, are greatly anticipated.

102 See UNEP/CBD/BS/WS-L&R/1/2.

Cross-Border Damage Caused by Genetically Modified Organisms: Jurisdiction and Applicable Law

Thomas Kadner Graziano and Matthias Erhardt

Case scenario 1

X farms genetically modified (GM) corn on his field in Austria. The field is adjacent to Y's organic farm in Hungary. When X sows his field, some GM corn is taken by the wind and mixes with the organic corn stored in Y's warehouse. The contamination of Y's corn is caused by a faulty fence on X's farm.

Y (the organic farmer) sells his crop to Z's organic food company in the Czech Republic where it gets mixed with corn from other organic providers and is processed into cornflakes. When the organic food company Z learns about the contamination, they have to destroy several thousand tons of their products resulting in damage of € 5,000,000.

Which courts have jurisdiction and what law would apply in the two following situations?

(a) Z claims damages from Y on a contractual basis. Y pays him out and brings a claim for damages against X on a non-contractual basis.

(b) Y is in financial difficulties. Z thus considers suing X for damages on a non-contractual basis.

(c) Z brings a claim against Y on a non-contractual basis.

(d) The cornflakes are sold in 15 countries. When the customers, having acquired the cornflakes in one of these countries, learn about the contamination they claim damages from Z, the Czech producer.

Case scenario 2

GM wheat originating from a farm in France contaminates a nearby field with rare species in Belgium.

(a) *The contaminated area is owned by a state-run research institute. The institute takes the necessary measures to decontaminate the area.*

(b) *The contaminated field belongs to a private association. The Belgian administrative authorities take the necessary measures to decontaminate the area in order to prevent imminent danger to health and safety of the environment.*

The State-run research institute and the Belgian administrative authorities subsequently claim damages, from the company that owns and farms the wheat field in France, for the cost of decontaminating the area.

Would the claims fall under the scope of the Brussels I and Rome II Regulations?

I. Introduction

The Amsterdam reform treaty of 1997 granted comprehensive legislative powers to the European Union (EU) in the field of private international law (PIL).[1] In the years following the conclusion of the treaty, international jurisdiction and conflict of law issues became the subject of important EU legislative activities. 1

The first major legislative step by the European Union in the area of private international law was Council Regulation (EC) No. 44/2001 on jurisdiction and the recognition and enforcement of judgments in civil and commercial matters (Brussels I).[2] The latter has, for the most part, maintained the provisions of the 1968 Brussels Convention on the same subject matter and transformed it into EU law. The other regulation that we will cover in this contribution is a genuine novelty, introducing, for the first time, common European rules on the law applicable to non-contractual obligations. This regulation is titled Regulation (EC) No. 864/2007 on the law applicable to non-contractual obligations (Rome II).[3] 2

This contribution will first deal with the jurisdictional issues that may arise in claims for damage caused by GMOs. We shall therefore examine the Brussels I Regulation and the rules on jurisdiction it establishes (infra, II.). We will then address the issue of the applicable law in cases of cross- 3

---

1 Cf. Art. 61 c and 65 EC. This section of the Treaty is not binding for Denmark; as such, measures in the field of judicial cooperation in civil matters concern the Member States of the EU, excluding Denmark. The United Kingdom, on the other hand, has always chosen to opt in on these regulations. Denmark has effectively done so with respect to the Brussels I Regulation, see fn. 6.

2 Official Journal (OJ) L 12, 16.1.2001, 1–23.

3 OJ L 199, 31.7.2007, 40–49.

border damage caused by GMOs (infra, III.). The above case scenarios may serve to illustrate these issues.

II. Jurisdiction under the Brussels I Regulation

1. Scope of application

4 The scope of application of the Brussels I Regulation is determined by Art. 1 to Art. 4 of the Regulation. An examination of the scope of application involves four different aspects: (a) territorial; (b) personal; (c) material; and (d) temporal requirements.[4]

(a) Territorial requirement

5 Due to the immediate binding character of regulations, as stated in Art. 288(2) of the Treaty on the functioning of the European Union (hereafter TFEU),[5] each and every court in a EU Member State[6] has to apply the Brussels I Regulation when dealing with issues of international jurisdiction.

(b) Personal requirement

6 The Brussels I Regulation is applicable if the defendant is domiciled in a Member State of the European Union. This flows from Art. 2 through Art. 4 of the Regulation.

7 Most of the rules of the Brussels I Regulation are – as part of the EU law – to be interpreted autonomously, i.e. independently of the national laws of the EU Member States.[7] In order to determine whether a party is domiciled in the Member State whose courts are seised, the court shall, according to Art. 59 of the Brussels I Regulation, however apply its internal law (i.e. the *lex fori*). For companies or other legal persons or associations,

---

4 See, e.g., *U. Magnus*, in: U. Magnus/P. Mankowski (eds.), Brussels I Regulation (2007) Introduction note 42 f.

5 OJ 2008 C 115.

6 While Denmark was originally excluded from the scope of application, it was extended to Denmark by way of a bilateral agreement with the European Community that entered into force on 1 July 2007: Agreement between the European Community and the Kingdom of Denmark on jurisdiction and the recognition and enforcement of judgments in civil and commercial matters, OJ L 299, 16.11.2005, 62.

7 ECJ C-464/01, *Johann Gruber v. Bay Wa AG* [2005] ECR I-439.

Art. 60(1) of the Brussels I Regulation provides, on the other hand, an autonomous legal definition of their domicile. According to this article, a company or other legal person or association is domiciled at the place where it has one of the following: (a) its statutory seat, (b) its central administration, or (c) its principal place of business.

(c) Material requirement

The Brussels I Regulation applies to "civil and commercial" matters, **8**
Art. 1(1). In cases of damage caused by GMOs, action to repair the damage is often taken by state authorities (see our second case scenario). If the state authorities subsequently bring a claim for compensation the question is whether the compensation claim qualifies as "civil or commercial matter".

The Brussels I Regulation excludes "administrative matters" from its scope **9**
of application. However, the mere fact that a claim is brought to court by a public authority does not, in itself, exclude it from the Regulation's scope of application.[8] The European Court of Justice (ECJ) has, on several occasions, ruled on the question of qualification as "civil and commercial matter", as it is stated in Art. 1(1) of the Brussels I Regulation. Under the principles of interpretation, developed in a well established line of precedents by the Court, the Regulation and its scope of application "must be regarded as an independent concept which must be construed with reference first to the objectives and scheme of the [Regulation] and secondly to the general principles which stem from the corpus of the national legal systems".[9] In its 1976 landmark decision in the case *LTU v. Eurocontrol*,[10] the Court held that litigations involving a public authority and a person governed by private law will fall within the scope of application of the Regulation unless the authority "acts in the exercise of its powers". Thus, the crucial question in each case is to know whether or not a public authority was acting within its public powers – that is, in the exercise of state authority. In several decisions, the ECJ provided guidelines as to when a public authority is to be regarded as acting in the exercise of its public powers.

---

8 See on the following with further references *Th. Kadner Graziano*, Law Applicable on Cross-Border Damage to the Environment, in: Yearbook of Private International Law (Yb PIL), Vol. IX (2007) 71.

9 ECJ Case 29/76, *LTU Lufttransportunternehmen GmbH & Co. KG v. Eurocontrol* [1976] ECR 1541, par. 3, 5; quote from: ECJ Case 814/79, *Netherlands State v. Reinhold Rüffer* [1980] ECR 3807, Summary at par. 1.

10 *LTU v. Eurocontrol* [1976] ECR 1541.

10 In the case *Netherlands v. Rüffer,*[11] a German boat collided with a Dutch motor vessel and sank in a public waterway. An international treaty between Germany and the Netherlands provided that the Netherlands were to be responsible for the "removal of wrecks" in the waterway. The Dutch authorities had the wreck removed by a Dutch company and claimed compensation for the removal costs from the German owner of the sunken boat. The Dutch Hoge Raad (Supreme Court) referred the issue to the ECJ in order to clarify whether the claim for redress was a "civil and commercial matter" under Art. 1 of the Brussels Convention (now: the Brussels I Regulation). The ECJ held that if a state removes "a wreck in a public waterway, administered by the State responsible in performance of an international obligation and on the basis of provisions of the national law which, in the administration of that waterway, confer [upon the State] the status of public authority in regard to private persons", the State claiming recovery of such costs acts "in the exercise of its public authority powers".[12] The fact that the state brought a claim for damages once the removal had been effected did not, according to the ECJ, change the character of the measure taken and did not have the effect that it was to be qualified as a civil matter.

11 In subsequent cases, the ECJ took a broader view of the notion of "civil and commercial" matters in the context of Art. 1(1) of the Brussels I Regulation. In 1993, the ECJ had to decide on a damage claim brought by the parents of a pupil against a German state-school teacher for having caused the death of their son on a school trip to Italy. The teacher, who was supposed to have been supervising the pupils, had breached this duty and, as a result, caused the death of the claimants' son. In this case, the ECJ held that the claim was a "civil matter" even though the teacher was acting in the capacity of a civil servant and even though the case was covered by a scheme of social insurance under public law that, according to German law, excluded a direct damage claim against a teacher.[13] The court argued that "a teacher in a State school assumes the same functions vis-à-vis his pupils [...] as those assumed by a teacher in a private school" and held that "the right to obtain compensation for injury suffered as a result of conduct regarded as culpable in criminal law is generally recognised as being a civil law right".[14]

---

11 [1980] ECR 3807.

12 Ibid., par. 8, 9; For a critical analysis of this case-law, see G. *Betlem/Ch. Bernasconi*, European Private International Law, the Environment and Obstacles for Public Authorities, Law Quarterly Review (LQR) 2006, 124 at 132 f.

13 ECJ C-172/91, *Volker Sonntag* [1993] ECR I-1963, par. 20–22.

14 Ibid., par. 19.

In recent judgments, the ECJ held that actions against individuals for the recovery of expenses incurred by public authorities may fall within the scope of the Brussels I Convention/Regulation.[15] In the *Baten* case, a community in the Netherlands had paid monthly contributions to a woman under Dutch social assistance laws. These laws provided a right to recovery for the amounts paid "from persons who do not, or do not fully, meet their maintenance obligations following a divorce" and, accordingly, the Dutch community claimed recovery from the woman's ex-husband, living in Belgium. The ECJ held that such an action for recovery of sums paid by public authorities, as long as exercised in accordance with the rules and principles governing actions for recovery between private parties, is as well within the scope of application of the Brussels I Regulation.[16] **12**

If this principle is applied to the field of environmental law and GMOs, an action for recovery of expenses incurred for cleaning up the environment or removing whatever damage may have been caused by GMOs, could be regarded as a "civil matter" as long as the recovery is in accordance with the principles governing the right of recovery between private parties. In cases in which the above-mentioned requirements are fulfilled, claims for damage caused by GMOs may thus be considered "civil and commercial matters" in the sense of Art. 1(1) of the Brussels I Regulation[17] and they may then fall into its material scope of application. **13**

(d) Temporal requirement

According to Art. 66(1) of the Regulation, it "shall apply only to legal proceedings instituted [. . .] after the entry into force thereof." Art. 76 stipulates that the Regulation enters into force on 1 March 2002. Any tort case that is brought to court after this date thus falls into the temporal scope of application of the Brussels I Regulation. **14**

---

15 ECJ C-271/00, *Gemeente Steenbergern v. Luc Baten* [2002] ECR I-10489.

16 Ibid., at par. 34–37. See also *Betlem/Bernasconi*, LQR 2006, 124 at 134: "The parallel with the Rüffer scenario where a right of recourse for cost recovery of removal of a wreck was at issue is striking".

17 This point of view is further supported by an analysis of Directive 2004/35/EC on environmental liability, *cf. Th. Kadner Graziano*, Yb PIL, Vol. IX (2007) 71 at 83 ff.

2. General rule on jurisdiction

15 Art. 2 of the Regulation not only determines the personal scope of application of the Brussels I Regulation but also establishes its general rule of jurisdiction. In general, a person domiciled in an EU Member State must be sued in the courts of that Member State. This maxim, i.e. the actor sequitur forum rei-rule, may be one of the few truly globally recognised general principles on jurisdiction.[18] Alleged tortfeasors thus have to be sued in the country where they are domiciled or where they have their statutory seat, their central administration or their principal place of business, cf. Art. 60(1). It is important to remember that Art. 2(1) only deals with international jurisdiction. The question regarding which court is locally competent is governed by the national law of the state designated by Art. 2(1).

3. Special heads of jurisdiction

16 The Brussels I Regulation contains numerous special heads of jurisdiction. These may either be exclusive or alternative. *Exclusive* jurisdictions exist in the areas listed in Art. 22 of the Brussels I Regulation. They concern cases regarding rights to immovable property, the validity of the constitution, the nullity or the dissolution of legal persons or the validity of entries in public registers, and do not apply to the areas covered by this contribution. Further, insurance, consumer and employment matters are also exclusively governed by sec. 3 through 5 respectively of the Brussels I Regulation.

17 In our context, the special rule on alternative jurisdictions concerning non-contractual liability scenarios is of particular interest. According to Art. 5(3) of the Regulation, a person domiciled in a Member State may be sued in another Member State "in matters relat[ed] to tort, delict or quasi-delict, in the court of the place where the harmful event occurred or may occur".

18 Art. 5(3) seems, at first sight, easy to grasp. And indeed, in most scenarios, the place where the harmful event occurred can be determined easily. If, for example, the truck of company A that is located in country X causes an accident while driving through country Y and releases some of its GM-corn there, contaminating farmer B's organic field in country Y, then the harmful event occurs in country Y.

19 The situation is more difficult if a person *acts* in one country (or several countries) and the damage *occurs* in another country (or in several other

18 *J. Kropholler*, Internationales Privatrecht (6th ed. 2006) 615.

countries). In such cases we speak of "multilocal torts", "double or multiple locality cases" or "complex torts". Imagine, for example, that in country X the load of the truck in our example is negligently secured causing the load to be released in country Y, contaminating a field located there.

The ECJ had to deal with the question of whether the "place where the harmful event" occurs is to be understood in the sense of the place where the harmful *act* is committed or, on the contrary, where its *result* manifests. In its major precedent dealing with this question,[19] the Court held that the special head of jurisdiction in Art. 5(3) of the Brussels I Regulation provides for an exception to the actor sequitur forum rei-rule, and that this exception is based on a "particularly close connecting factor between a dispute and the court which may be called upon to hear it".[20] The Court stated that this close connection exists in the context of both the place of the action and the place of the result.[21] **20**

This case-law offers the plaintiff up to three different courts for his claim: (1) those of the country in which the defendant is domiciled, Art. 2(1); (2) those of the country where the defendant's act took place (if located in another country), Art. 5(3); and (3) those of the country in which the harm was sustained by the victim (if located in yet another country), Art. 5(3). In the field of international jurisdiction, this interpretation of Art. 5(3) is widely recognised in legal writing.[22] **21**

According to this interpretation, under Art. 5(3) of the Brussels I Regulation, the courts in the country in which the harmful act was committed and those in the country in which the damage was sustained thus both have jurisdiction. **22**

4. Jurisdiction in case scenario 1(a)

Having examined the legislative framework, we may now turn to our case scenarios and analyse them under the Brussels I Regulation. In scenario 1(a), the organic farmer Y from Hungary considers suing the Austrian farmer of GMOs X on a non-contractual basis. **23**

19 ECJ 21/76, *Bier v. Mines de Potasse d'Alsace* [1976] ECR 1735.
20 Ibid. par. 11.
21 Ibid., par. 20.
22 See, e.g., *P. Mankowski*, in: U. Magnus/P. Mankowski, Brussels I Regulation (2007) Art. 5 note 204; *H. Gaudemet-Tallon*, Compétence et exécution des jugements en Europe (3rd ed. 2002) note 216; *J. Kropholler*, Europäisches Zivilprozessrecht (8th ed. 2005) Art. 5 note 82.

24 a) According to the general rule on jurisdiction in Art. 2(1) of the Brussels I Regulation, the courts of the country in which the defendant is domiciled have jurisdiction, i.e. the Austrian courts for a claim of the Hungarian farmer Y against the Austrian farmer X.

25 b) Alternatively, Y may choose to sue X under Art. 5(3) in the courts of the place where the harmful event occurred. However, Art. 5(3) applies only if this country is different from the one designated by Art. 2(1) of the Brussels I Regulation.

26 For a claim against X, the harmful event that caused Y's damage was the contamination of Y's crop by X. X's *act* (or equally important: his *omission* in violation of a duty to act) that led to the contamination of the crops happened in Austria where X failed to secure his field with appropriate fencing. The country of the defendant's domicile is thus also the country where the defendant acted; Art. 5(3) consequently does not apply in respect to the courts in Austria since, as we have seen, the Austrian courts already have jurisdiction under Art. 2(1) of the Brussels I Regulation.

27 The damaging *result* manifested in Hungary, where Y's crops (the crops that were later sold to Z) were contaminated. Under Art. 5(3) of the Brussels I Regulation, the courts of the place in Hungary where Y's field is located thus have jurisdiction for Y's claim against X.

28 The Hungarian farmer Y thus has the choice to bring a claim for non-contractual liability against the Austrian farmer of GMOs X either in Austria, Art. 2(1) of the Brussels I Regulation, or in Hungary, Art. 5(3) of Brussels I.

5. Jurisdiction in case scenario 1(b)

29 a) According to the general rule on jurisdiction in Art. 2(1) of the Brussels I Regulation, the courts of the country in which the defendant is domiciled have jurisdiction. The Austrian courts thus have jurisdiction for a claim of the Czech organic food company Z against farmer X, domiciled in Austria.

30 b) Just like Y, Z may alternatively choose to sue X under Art. 5(3) in the courts of the place where the harmful event occurred.

31 For a claim of Z against X, the harmful event that caused Z's damage was the contamination of Y's crop by X that eventually led to the destruction of Z's organic products. X's *act* (or *omission to act* in violation of a duty to act) that led to the contamination of the crops happened in Austria where X failed to secure his field with appropriate fencing. Since the Austrian courts already have jurisdiction under Art. 2(1) of the Brussels I Regulation, Art. 5(3) does not apply in respect to these courts.

The damaging *result* manifested in Hungary, where Y's crops (the crops that were later sold to Z) were contaminated. The case presents, however, a particularity compared to the standard situation under Art. 5(3). Whereas in the standard case for Art. 5(3), the damage manifests itself in the sphere of the victim's legally protected interests, in our scenario it needed the further step of the sale of the contaminated crops by Y in Hungary to Z in the Czech Republic in order to cause damage to Z's interests. The source of Z's injury, attributable to X, lies however in the contamination of Y's crops that took place in Hungary. Under Art. 5(3) of the Brussels I Regulation, the courts of the place in Hungary where Y's field is located thus, arguably, have jurisdiction for Z's claim against X. **32**

c) The question then is whether the Czech courts also have jurisdiction since it is the Czech Republic where Z mixed the contaminated crops with corn from his other organic providers and where these crops were processed into cornflakes that could consequently not be sold under the label "organic". **33**

The ECJ has not yet had to deal with such an issue under Art. 5(3) of the Brussels I Regulation. It could be argued that X's act (or omission to act) caused damage not only in Hungary (where Y's crops were contaminated) but also in the Czech Republic where Y's contaminated crops were mixed with Z's other crops, causing further contamination. On the other hand, in order for X's act (or omission to act) to cause such further damage in the Czech Republic, the crops needed to be sold by Y to Z, to be mixed by buyer Z with his other crops, further spreading the contamination to buyer Z's final products and causing him damage to these other crops and loss of earnings as the cornflakes could no longer be sold as organic flakes. As for an action against X, this further damage is, arguably, to be considered *indirect or consequential damage* which is generally not held sufficient to establish jurisdiction of the country in which the consequential damage was suffered.[23] **34**

In order to establish jurisdiction of the Czech courts, Z could, on the other hand, try to argue that the contamination of the products he had bought from other producers was, with regard to his own legally protected interests, not *consequential damage* but the *first damage he suffered* due to X's act (or **35**

23 Cf. *Kropholler* (fn. 22) Art. 5 note 87; ECJ 364/93; Bundesgerichtshof (German Federal Court, BGH) in Entscheidungen des Bundesgerichtshofes in Zivilsachen (BGHZ) 98, 263.

omission to act). He could further argue that this damage happened in the Czech Republic where his whole production was rendered unmerchantable. This would, however, make jurisdiction for X entirely unforeseeable, in particular when taking into consideration that X was not part of the distribution line of the crops.

36 If our above line of reasoning is followed, Z may, for his claim against X, choose between the courts in Austria, Art. 2(1), and the courts of the place in Hungary where Y's field is located, Art. 5(3); whereas, arguably, the courts in the Czech Republic do not have jurisdiction to hear a claim by Z against X.

6. Jurisdiction in case scenario 1(c)

37 a) For a claim of Z against Y, i.e. against the organic farmer in Hungary that sold the crop to Z's organic food company, Hungarian courts have jurisdiction since Y has his domicile or his statutory seat, central administration or principal place of business in Hungary, Art. 2(1) and 60(1) of the Brussels I Regulation.

38 b) For Z's non-contractual claim against Y, jurisdiction may also be based on Art. 5(3) of the Brussels I Regulation. Since the Hungarian courts already have jurisdiction under Art. 2(1) of the Brussels I Regulation, Art. 5(3) does not apply with respect to these courts.

39 According to Art. 5(3) of the Brussels I Regulation, Czech courts have jurisdiction if the harmful event occurred in the Czech Republic. In case of a claim against Y, the harmful event is the delivery of contaminated corn. Once again, we are dealing with a complex tort: the delivery is organised in Hungary and takes place as it were from Hungary to the Czech Republic, whereas the contamination of the whole production line and thus the damage to Z occurs while the crops delivered by Y are mixed with the other producers' crops in the Czech Republic. Further, since the harmful result occurred in the Czech Republic, for a claim against Y, an alternative jurisdiction is given in the courts of the place where the damage occurred, namely, Z's place of business in the Czech Republic.

7. Jurisdiction in case scenario 1(d)

40 In scenario 1(d) (i.e.: the cornflakes are sold in 15 countries where customers learn about the contamination and claim damages from the Czech organic food company Z on an extra-contractual basis), Czech courts have jurisdiction under Art. 2(1).

Under Art. 5(3) of the Brussels I Regulation, Z's customers may bring an extra-contractual claim against Z both at the place where the harmful *act* was committed (if this country is different from the country of Z's domicile) and the place where the damage was sustained. The harmful act committed by Z was the marketing of the contaminated cornflakes labelled "organic" in the consumers' respective countries of residence; the damage occurred in the countries where the consumers bought the products, i.e. most likely the countries of their residence. **41**

In scenario 1(b), the customers can thus choose to go to courts in the Czech Republic, these courts having jurisdiction under Art. 2(1), and in the countries where the product was marketed and bought, Art. 5(3) of the Brussels I Regulation, if of course these countries are Member States of the EU. **42**

8. Jurisdiction in case scenario 2

Case scenario 2 illustrates the difficulties involved in classifying a case as civil or commercial matter under Art. 1(1) of the Brussels I Regulation. The difficulty lies in the fact that, in this scenario like in many others, the claim is brought by public authorities. **43**

a) In scenario 2(a), the state-run research facility decontaminates a field owned by the facility. The measures taken are not any different from what a private association or research institute would have had to do in the same situation. The fact that the field belongs to a public institution is merely incidental. The claim for compensation of the clean-up costs is, consequently, a civil action and falls under Art. 1(1) of the Brussels I Regulation. **44**

For the claim against the French company, French courts would have jurisdiction flowing from Art. 2(1) of the Brussels I Regulation. The Belgian courts of the place where the damage occurred, i.e. the place where the contaminated field lies, would also have jurisdiction according to Art. 5(3) of the Brussels I Regulation. **45**

b) In case scenario 2(b), the public authority acts in order to prevent imminent danger to health and to safety of the environment. It is exercising its public duty and authority to protect these goods in the general interest. The measures taken by the state are based on administrative law. Furthermore, only the state authority is empowered to take action in the general **46** interest of preserving public goods such as public health and the safety of the environment. Having regard to the criteria used by the ECJ in the case *Netherlands v. Rüffer*,[24] these aspects speak in favour of the public authority acting "in the exercise of its public powers". In case scenario 2(b), the claim for damages brought by the Belgian government is, according to this reasoning, to be regarded as an "administrative matter" as opposed to a "civil matter" in the sense of Art. 1(1). The Brussels I Regulation does then not apply.

47 If, however, the public authorities claim compensation for expenses incurred on the basis, and according to rules and principles of private law,[25] the matter falls, according to the ECJ's reasoning in the *Baten* case,[26] within the scope of application of the Brussels I Regulation. Applied to the field of environmental law and GMOs, an action for recovery of expenses incurred for cleaning up the environment, such as the action for recovery in scenario 2(b), could thus well be regarded as a "civil matter" if the public authorities seek recovery according to rules and principles governing the right of recovery between private parties. Whether or not the subjacent action taken by the state authority is a public or civil matter is not crucial in the context of a subsequent claim for redress. According to this line of reasoning, only if the action for recovery itself is based on public law rules would the action be outside the scope of the Brussels I Regulation.[27]

III. The applicable law under the Rome II Regulation

1. Introduction

48 Once a court has determined that it has jurisdiction, the question in cross-border cases then is which law to apply. Since 11 January 2009, Regulation

24 Cf. supra no. 10, and the case [1980] ECR 3807.

25 See, e.g., the rule of private law suggested in the Draft Common Frame of Reference (DCFR), Art. VI-2:209: "Burdens incurred by the state upon environmental impairment": "Burdens incurred by the state or designated competent authorities in restoring substantially impaired natural elements constituting the environment, such as air, water, soil, flora and fauna, are legally relevant damage to the state or the authorities concerned" [for which compensation can be claimed according to the DCFR], text of the DCFR in: *Ch. von Bar/E. Clive/H. Schulte-Noelke/H. Beale et al.* (eds.), Principles, Definitions and Model Rules of European Private Law – Draft Common Frame of Reference (DCFR) (2009).

26 Cf. supra no. 12.

27 For further details and many more arguments, see *Th. Kadner Graziano*, Yb PIL, Vol. IX (2007) 71 at 82 ff.

(EC) No. 864/2007 of the European Parliament and of the Council of 11 July 2007 on the law applicable to non-contractual obligations (Rome II)[28] is applicable in twenty-six of the twenty-seven Member States of the European Union.[29] The Regulation applies to events that gave rise to damage occurring after its entry into force on 20 August 2007 (20 days after its publication in the Official Journal).[30]

2. Scope of application of the Regulation and general remarks

When adopting the Rome II Regulation, the European legislator intended to create an instrument that would work in parallel with the Brussels I Regulation. Recital 7 of the Rome II Regulation expressly states this purpose by emphasising that "the substantive scope and the provisions of this Regulation should be consistent with [the Brussels I Regulation]". Therefore, much of what was said for the scope of application of the Brussels I Regulation should, in principle, also apply for the Rome II Regulation. **49**

According to Art. 3 of the Rome II Regulation, any law designated by the Regulation "shall be applied whether or not it is the law of a Member State". The Regulation thus is a *loi uniforme* and does not require that the other state(s) involved are EU Member States. **50**

For damage caused by GMOs, several conflict of law rules may have to be taken into consideration. Rome II favours freedom of choice regarding the applicable law so that the first rule to be considered is the rule on party autonomy, Art. 14 of the Rome II Regulation. One will further have to consider the rule on product liability, Art. 5 of the Rome II Regulation, if the damage was caused by a product containing GMOs, and/or the rule on **51**

28 OJ L 199, 31.7.2007, 40–49 (hereafter: "Rome II" or "Rome II Regulation"). See on the Rome II-Regulation, e.g., *A. Dickinson*, The Rome II Regulation: A commentary (2009); *Th. Kadner Graziano*, Das auf außervertragliche Schuldverhältnisse anzuwendende Recht nach Inkrafttreten der Rom II-Verordnung, Rabels Zeitschrift für ausländisches und internationales Privatrecht (RabelsZ) 73 (2009) 1; *Th. Kadner Graziano*, Le nouveau droit international privé communautaire en matière de responsabilité extracontracuelle, Revue critique de droit international privé (Rev. crit. DIP) 97 (2008) 445. For an overview of the PIL in force in Europe before the Rome II Regulation, see *Th. Kadner Graziano*, La responsabilité délictuelle en DIP européen (2004); *Th. Kadner Graziano*, Europäisches Internationales Deliktsrecht (2003).

29 The United Kingdom and Ireland are taking part in the adoption and application of this Regulation (see Recital 39). In accordance with Art. 1 and 2 of the Protocol on the position of Denmark, annexed to the Treaty on European Union and to the Treaty establishing the European Union, Denmark is not taking part in the adoption of the Rome II Regulation (see Recital 40). For the date of application, see Art. 32.

30 Art. 31 of the Rome II Regulation.

environmental damage, Art. 7, if the conditions for its application are met. If none of these specific rules apply, the applicable law will be determined by the general rule of Art. 4.

52 As far as product liability is concerned, regard must also be had to the 1973 Hague Convention on the Law Applicable to Products Liability.[31] The 1973 Hague Convention is in force in 11 countries, including 6 EU Member States.[32] According to Art. 28(1) of the Rome II Regulation, Rome II "shall not prejudice the application of international conventions to which one or more Member States are parties at the time when this Regulation is adopted and which lay down conflict-of-law rules relating to non-contractual obligations". In the EU Member States in which the Hague Convention is in force (i.e. France, Finland, Luxembourg, the Netherlands, Slovenia and Spain), the applicable law in product liability cases will thus be determined by the 1973 Hague Convention and not by Art. 5 of the Rome II Regulation.[33] This following presentation will, however, focus on the rules of the new Rome II Regulation.[34]

3. Party autonomy

53 Since the second half of the last century, the place of party autonomy has become more and more important in modern conflict of law in tort and delict.[35] Art. 14 of the Rome II Regulation follows this trend towards the acceptance of party autonomy. Under Art. 14(1), the choice of law must be "expressed or demonstrated with reasonable certainty by the circumstances of the case". Furthermore, as soon as non-commercial parties are involved, the choice of law has to be by virtue of an agreement that was

---

31 Text and list of Contracting States available online at: www.hcch.net.

32 The Convention of 2 October 1973 on the Law Applicable to Products Liability is currently in force in France, Luxemburg, the Netherlands, Norway, Finland, Spain, Slovenia, Croatia, Macedonia, Serbia and Montenegro; www.hcch.net.

33 For more information, see *Th. Kadner Graziano*, The Rome II Regulation and the Hague Conventions on Tort Law – Interaction, Conflicts and Future Perspectives, Nederlands Internationaal Privaatrecht (NIPR) 2008, 425–429.

34 For the main differences between the 1973 Hague Convention and the Rome II Regulation, see *Th. Kadner Graziano*, NIPR 2008, 425 at 3.2.

35 Party autonomy, without any manifest limitations, was recognised in Art. 35(1) of the Austrian PIL Act, in Art. 39(1) of the Lichtenstein PIL Act and Art. 6 of the Dutch PIL Act. Moreover, *ex post* choice of law was permitted, see Art. 42 of the German EGBGB, Art. 101 of the Belgian PIL Act, Art. 132 of the Swiss PIL Act, Art. 1219(3) of the 3rd part of the Russian Civil Code, see for French PIL: Cour de cass. 19.4.1988 (*Roho v. Caron et autres*), Revue Critique 1989, 68, note *H. Batiffol*.

"entered into after the event giving rise to the damage occurred".[36] As a result, when applying Rome II, the first question to be asked will be whether the parties have agreed on the applicable law.

4. Complex torts: The general rule

If the parties have not made a choice of law, objective connecting factors are required to determine the applicable law. **54**

According to Art. 4(1) of the Rome II Regulation, complex torts are ordinarily governed by the law "of the country in which the damage occurs irrespective of the country in which the event giving rise to the damage occurred and irrespective of the country or countries in which the indirect consequences of the event occur". The Regulation thereby adopts the principle according to which the law of the place where the *injury* occurs is applied and, generally speaking, moves away from the application of the law of the place where the event giving rise to the damage (the damaging *act*) occurs. If the injury or damage occurs in several countries, the laws of these countries will be applied to the damage that occurred in each country respectively.[37] **55**

For cross-border damage caused by GMOs – just as for all other activities that potentially give rise to damage across borders – this implies that a person causing damage in another country will be subject to the rules on liability of the country in which his or her actions have their effects. The person acting must, consequently, take into consideration the potential victims' legitimate expectations to be protected according to the level of protection provided by the law of the state where the injury occurs and the victim suffers harm. Hence, the law of the place where the person causing the damage *acted* will, from now on, be applied only in exceptional circumstances, see Art. 4(1) to (3) of the Rome II Regulation. **56**

Consequently, for cross-border torts caused by omissions, it is not the place where the person claimed to be liable should have *acted* that determines the applicable law; rather, the determining factor is the place where the *damage* which the person ought to have prevented occurs. **57**

---

36 For further information and details, see *Th. Kadner Graziano,* Freedom to choose the applicable law in tort – Art. 14 and 4(3) of the Rome II Regulation, in *W. Binchy/J. Ahern* (eds.), The Rome II Regulation on the Law Applicable to Non-Contractual Obligations: A New Tort Litigation Regime (2009) 113 ff.

37 Commission's 2003 Proposal, note 11.

5. Product liability

58 The general rule of Art. 4(1) will, however, only apply if no specific conflict of laws rule is applicable. In the case of damage caused by GMOs, one special rule that may be of particular importance is the rule on product liability, Art. 5 of the Rome II Regulation.

(a) Conditions of application

59 In Art. 5, the Rome II Regulation establishes a specific regime for tort and delict caused by a product. In order to know whether this regime applies, we have to determine under what circumstances GMOs fall into the scope of that rule.

60 First, the Regulation concerns the liability for a *product*. This notion is not specified in the Regulation itself. Two points of interest must be distinguished:

61 i) The first point concerns *processed products* contaminated by GMOs, for example, the corn flakes produced by Z in our first case scenario. In the case of damage caused by processed products contaminated by GMOs, Art. 5 would be applied to determine the law applicable to claims of persons having sustained damage by the processed product.

62 It is important to note that Art. 5 of the Rome II Regulation only gives an answer to the question as to the law of which country is applicable in a product liability case. Once the applicable law is determined, the question of who is liable under what conditions will then have to be analysed under this law. Whether a specific person within the production chain, e.g. the actual producer, the supplier of components, the wholesaler, the importer, or the final seller, are eligible to be held liable, or whether, in our first case scenario, X or Y can actually be sued, is a matter of the applicable substantive law. The same is true for liability under the 1985 Directive on Product Liability[38] and the question as to whether the 1985 Directive on Product Liability has been transposed in the relevant country's legal order (and which exceptions to the Directive's liability regime apply under the applicable law).

38 Council Directive 85/374/EEC of 25 July 1985 on the approximation of the laws, regulations and administrative provisions of the Member States concerning liability for defective products, OJ L 210, 7.8.1985, 29–33.

ii) The second point of interest is the question as to whether *agricultural products* (the corn in our first scenario) can be considered products pursuant to Art. 5 of the Rome II Regulation. Based on a common understanding of the term "product", one might argue that only manmade objects should be considered "products". In order to answer this question, it is helpful to look at other regulations or directives that deal with the same subject-matter – in particular the 1985 Directive on Product Liability. The 1985 Directive establishes a system liability for products and may hence serve as a suitable point of reference to aid the interpretation of the term *product*. **63**

When the Directive was first enacted in 1985, Art. 2 defined a product as meaning, "all movables, with the exception of primary agricultural products". After the BSE crisis in the late 1990s, the European Union was eager to enforce and "restore consumer confidence in the safety of agricultural products"[39] and accordingly amended the 1985 Directive by including agricultural products in its scope of application. Based on this decision of the European legislator, there is absolutely no reason *not* to consider agricultural products, such as GM corn, "products" within the terms of Art. 5 of the Rome II Regulation. Art. 5 thus applies not only to manmade objects but also to agricultural products and, in particular, to GMOs. **64**

(b) Cascade of relevant connecting points

Art. 5 of the Rome II Regulation establishes a sophisticated system of connecting factors for product liability cases. Art. 5 combines different criteria that must be fulfilled in order to arrive at the applicable law. The different criteria are hierarchical; as such, if the first rule does not apply, then the second is applied and if this rule cannot be applied, then the third is applied (and so on). In any given scenario, the court will commence the process of reviewing these connecting factors until one of them specifies the applicable law. **65**

i) *Party autonomy:* As mentioned above, Art. 14(1) of the Rome II Regulation favours party autonomy in the field of torts. It also applies in product liability cases if the parties make an "express" choice of the applicable law or make a choice which is "demonstrated with reasonable certainty by the circumstances". **66**

39 Recital 5 of Directive 1999/34/EC amending Council Directive 85/374/EEC on the approximation of the laws, regulations and administrative provisions of the Member States concerning liability for defective products, OJ L 141, 4.6.1999, 20–21.

67 ii) *Rattachement accessoire*: If the parties have not chosen the applicable law pursuant to Art. 14 of the Regulation, Art. 5(2) provides for *rattachement accessoire*. That means that if there is "a pre-existing relationship between the parties, such as a contract, that is closely connected with the tort/delict in question", this relationship may establish a closer connection between the parties taking precedence over the other connecting factor specified by Art. 5(1). Art. 5 leads to the result that, if parties to a delivery chain are in a direct contractual relationship with each other, the same law applies to product liability claims in contracts and torts between them.

68 Where the contract is between a professional and a person that is not pursuing a commercial activity, the limitations on party autonomy provided for in Art. 14(1) of the Rome II Regulation and applying to a choice of law before the damaging event occurred may apply mutatis mutandis to the *rattachement accessoire*.[40]

69 iii) *Parties' common habitual residence*: If "the person claimed to be liable and the person sustaining damage both have their habitual residence in the same country at the time when the damage occurs", the law of this country applies to claims for product liability, Art. 5(1) in conjunction with Art. 4(2) of the Rome II Regulation.

70 iv) *Habitual residence of the injured party*: Where neither of the previous connecting factors applies, the law applicable to the non-contractual obligation shall be the law of the country in which the aggrieved person had his habitual residence at the time the damage occurred, Art. 5(1)(a) of the Rome II Regulation. This solution allows the injured party to enjoy the standards of compensation that he or she is accustomed to and expects. It also guarantees equal treatment of persons who sustained damage in the same country, but caused by persons acting in different countries.

71 The legislator, however, chose to introduce a further condition for the application of the *lex laesi*: It is only applicable if the product was marketed in the country in which the damage was sustained, be it that it was marketed there by the person that is sued or by someone else.

72 v) *Place where the product was acquired*: If the product that caused the harm was not marketed in the country of the aggrieved person's habitual residence, the claim shall be governed by the law of the country in which the product was acquired if the product was marketed there, Art. 5(1)(b).

---

40 For more information, see *Th. Kadner Graziano*, in: W. Binchy/J. Ahern (eds.), The Rome II Regulation, 113 at 124 ff.; *Th. Kadner Graziano*, RabelsZ 73 (2009) 1 at 16 ff.; *Th. Kadner Graziano*, Rev. crit. DIP 97 (2008) 445 at 464 ff.

Arguably, the application of this law of the marketplace is not appropriate for bystanders, i.e. persons who are affected by a defective product but who did not themselves acquire it. Art. 5(1)(b) should, therefore, not be used in these cases, and the following rule in Art. 5(2) (place of damage) should be applied. **73**

vi) *Place where the damage occurred*: Finally, if none of the rules presented above applies, the law applicable to product liability cases shall be the law of the country in which the damage occurred if the product was marketed in this country, Art. 5(2). **74**

6. Environmental damage

(a) Scope of application and general remarks

The law applicable to liability for cross-border damage caused by GMOs may also be determined by the rule Rome II provides for damage caused to the environment. In order for Art. 7 of the Rome II Regulation to apply, it is necessary that damage be caused to the environment. According to Recital 24 of the Rome II Regulation, "[e]nvironmental damage should be understood as meaning adverse change in a natural resource, such as water, land or air, impairment of a function performed by that resource for the benefit of another natural resource or the public, or impairment of the variability among living organisms." Art. 7 of the Rome II Regulation makes clear that the provision covers both environmental damage (i.e. damage to natural resources as such, so-called ecological damage) and damage sustained by persons or property as a result of damage to the environment. **75**

On the one hand, the notion of environmental damage used in the Rome II Regulation is significantly wider than the notion of such damage used in Directive 2004/35/EC on environmental liability with regard to the prevention and remedying of environmental damage. The 2004 Directive provides a definition of environmental damage in its Art. 1. According to Art. 1 (c), environmental damage includes "land damage, which is any land contamination that creates a significant risk of human health being adversely affected as a result of the direct or indirect introduction in, on or under land, of substances, preparations, organisms or micro-organisms". This category seems rather narrow, especially as a *significant risk* of human health being affected is part of the definition. It is, at this stage, not possible to know to what extent GMOs affect human health. Art. 3 of the 2004 Directive, however, refers to the "occupational activities" that are **76** considered to potentially cause environmental damage and that are listed in Annex III of the Directive. The Annex mentions under point 11 the "deliberate release into the environment, transport and placing on the market of genetically modified organisms...". This proves that the legislator considers the manipulation of GMOs a potentially dangerous activity with regard to the environment.

77 On the other hand, the notion of environmental damage in the Rome II Regulation requires an "adverse change in a natural resource, such as water, land or air, impairment of a function performed by that resource for the benefit of another natural resource or the public, or impairment of the variability among living organisms." It may be argued that, in the light of the definition of environmental damage given in Recital 24 of the Rome II Regulation, plants farmed for human consumption, e.g. the corn in our first case scenario, are – even if grown organically – not considered natural resources and therefore not part of the goods and interests protected by Art. 7. In this case, the law applicable to damage to humanly grown plants will be determined by Art. 4 or Art. 5 (instead of Art. 7).

(b) The ubiquity approach of Art. 7

78 In the legislative process leading to Rome II, the question of whether environmental damage required a special rule or should be governed by the general rule of Art. 4 was most controversial. In the end, the decision was to apply Art. 4 (in general) but to modify this approach at the discretion of the plaintiff. The final version of Art. 7 now stipulates that in the case of environmental damage, the applicable law "shall be the law determined pursuant to Art. 4(1), unless the person seeking compensation for damage chooses to base his or her claim on the law of the country in which the event giving rise to the damage occurred."

79 Art. 7 of the Rome II Regulation penalises cross-border polluters as compared to polluters who commit the same act inside the borders of the state where the result occurs. The European legislator was aware of this fact and deliberately chose to favour the protection of the environment over the equal treatment of foreign and domestic polluters. Any potential polluter has to respect the most stringent standards of safety and diligence in order to avoid liability resulting from his actions.

80 According to Art. 7, the applicable law on environmental damage is the law of the country in which the damage occurs or the law of the county in which the event giving rise to the damage occurs, at the choice of the claimant. According to Recital 25 of the Rome II Regulation, "[t]he question of when the person seeking compensation can make the choice of the law applicable should be determined in accordance with the law of the Member State in which the court is seised."

7. The effect of foreign authorisations

In some legal systems, civil liability is excluded altogether if the act that **81** caused the harm was permitted by a public authorisation. If the case is governed by the law of the country in which the *damage* occurred, but the action that led to the damage is covered by an authorisation under the legislation of the country where the harmful *act* was committed, the question will be if and in what manner the public authorisation is to be taken into account when dealing with a compensation claim.

Before the entry into force of the Rome II Regulation, it was argued that, **82** according to the principle of territoriality governing public law, foreign administrative authorisations need not be given any weight at all.[41] However, totally ignoring a foreign authorisation may lead to a judgment not being recognised or enforced in the country in which the authorisation was issued; as such, disregard for the authorisation may be seen as a violation of the country's public policy (see Art. 34(1) of the Brussels I Regulation which allows Member States to not recognise a foreign judgment when it is manifestly contrary to public policy).

According to another opinion, a foreign administrative authorisation held **83** by the person causing the damage should, under certain circumstances, be taken into account when dealing with liability claims for damage caused by the relevant activity.[42] According to this opinion, a foreign administrative authorisation should be taken into consideration, provided that the authorisation in question is legal under international law and the conditions for obtaining the authorisation are the equivalent of those provided for under domestic law.[43]

---

41 See the German decisions in BGH 10.3.1978, Die deutsche Rechtsprechung auf dem Gebiete des Internationalen Privatrechts (IPRspr.) 1978, no. 40; OLG Saarbrücken 22.10.1957, Neue Juristische Wochenschrift (NJW) 1958, 752 at 754.

42 Rechtbank Rotterdam 16.12.1983, Nederlandse Jurisprudentie (NJ) 1984, n° 341, ch. 8.7; OLG Linz 15.6.1987, Juristische Blätter (JBl) 1987, 577 and Oberster Gerichtshof (OGH) 20.12.1988, JBl 1989, 239.

43 OLG Linz, JBl 1987, 577 (579).

84 It seems that the Rome II Regulation itself provides some support for this second point of view. Recital 34 states that, "[i]n order to strike a reasonable balance between the parties, account must be taken, in so far as appropriate, of the rules of safety and conduct in operation in the country in which the harmful act was committed, even where the non-contractual obligation is governed by the law of another country. The term 'rules of safety and conduct' should be interpreted as referring to all regulations having any relation to safety and conduct, including, for example, road safety rules in the case of an accident."

85 However, such administrative authorisations cannot be recognised as proper justification for the harmful behaviour, excluding any liability. Rather, courts have to take them into account and consider them as a relevant fact in the process of applying the substantial law designated by Art. 7 (or Art. 4) of the Rome II Regulation.[44]

8. The applicable law in case scenario 1

86 Having discussed the relevant conflict rules of the Rome II Regulation, we can now proceed to apply them to our case scenarios.

(a) Claim of Y against X

87 i) Arguably, Art. 7 of the Rome II Regulation (the rule on environmental damage) does not apply to the claim of Y against X since, in light of the definition of environmental damage given in Recital 24 of the Rome II Regulation, plants farmed for human consumption, e.g. the corn in our scenario, are not part of the goods and interests protected by Art. 7.[45]

88 The applicable law is, therefore, to be determined either by Art. 5 (the rule on product liability) or by the general rule on extra-contractual liability in Art. 4 of the Rome II Regulation. This is true for a claim both before Austrian or Hungarian courts.

89 ii) The crops that caused the damage to Y were not marketed by X. He was, however, producer of these crops which may lead to the application of Art. 5 of the Rome II Regulation (the rule on product liability). As we

44 For details, see *Th. Kadner Graziano*, Yb PIL, Vol. IX (2007) 71 at 78 ff.
45 Cf. supra no. 75.

have seen, Art. 5 applies not only to processed products but also to *agricultural products* such as the GM corn in our scenario.[46]

If the parties do not choose the applicable law, objective connecting factors apply. According to Art. 5(1)(a) of the Rome II Regulation, the law applicable to the non-contractual obligation shall be the law of the country in which the aggrieved person had his habitual residence at the time the damage occurred if the product that caused the damage was marketed in the country in which the damage was sustained. The crops produced by X were never purposefully marketed but taken by the wind to Y's estate. This should exclude the application of Art. 5(1)(a) of the Rome II Regulation. **90**

The same aspect should exclude the application of Art. 5(1)(b) according to which the claim shall be governed by the law of the country in which the product was acquired if the product was marketed there; finally, if none of theses rules applies, the law applicable to product liability cases shall be the law of the country in which the damage occurred if the product was marketed in this country, Art. 5(2). Here again, the fact that X's GM crops were not marketed at all should exclude the application of this rule. **91**

These considerations show that, although *agricultural products* such as the GM corn in our scenario fall, in principle, within the scope of Art. 5 of the Rome II Regulation, Art. 5 requires that the product be deliberately put onto a market. This was not the case for X's crops which were taken by the wind to Y's estate; consequently, the case falls within the scope of the general rule of Art. 4(1) of the Rome II Regulation. **92**

iii) According to Art. 4(1) of the Rome II Regulation, complex torts are ordinarily to be governed by the law "of the country in which the damage occurs irrespective of the country in which the event giving rise to the damage occurred." In scenario 1(a) the damage occurred when X's crops contaminated Y's stock in Hungary. Art. 4 thus leads to the application of Hungarian law to Y's claim against X. **93**

(b) Claim of Z against X

As we have seen, neither Art. 7 of the Rome II Regulation (environmental damage) nor Art. 5 (product liability) apply to a claim against X. The law **94**

46 Cf. supra no. 63–64.

applicable to Z's claim against X is thus to be determined according to the general rule in Art. 4 of the Rome II Regulation.

95 In scenario 1(b) the damage occurred when X's crops contaminated Y's stock in Hungary. The contamination of Y's crops in Hungary was also the source of the further injury suffered by Z. Here again,[47] in order to lead to an injury of Z's legally protected interests, it needed the further step of the sale of the contaminated crops by Y to Z. The source of Z's injury, attributable to X, lies however in the contamination of Y's crops that took place in Hungary. Art. 4 should thus arguably lead to the application of Hungarian law to Z's claim against X.

(c) Claim of Z against Y

96 Organic farmer Y sold and delivered a product that caused damage to Z, the producer of the corn flakes. Primary agricultural products are to be considered products in the sense and for the purpose of Art. 5 of the Rome II Regulation.[48] Since neither Hungary nor the Czech Republic are Members of the 1973 Hague Convention on the Law Applicable to Products Liability, the applicable law will be determined, before the courts of both countries, according to Art. 5 of the Rome II Regulation.

97 i) The parties have not chosen the law applicable to Z's claim in tort against Y, an option they had under Art. 14 of the Rome II Regulation.

98 ii) Art. 5(2) then provides for *rattachement accessoire*, i.e. the application of the law governing "a pre-existing relationship between the parties, such as a contract, that is closely connected with the tort/delict in question".

99 In our case scenario, Y and Z have concluded a sales contract. Since both parties to the contract have their places of business in different Contracting States of the 1980 United Nations Convention on Contracts for the International Sale of Goods (CISG), the conditions for the application of CISG set by Art. 1(1)(a) of the CISG are met and the contract between Y and Z is, in principle, governed by the CISG. As the CISG does not contain any rules on tort liability, *rattachement accessoire* is, in this case, excluded. In this case, the tort will be governed by "the law of the country in which the person sustaining the damage had his or her habitual residence when the damage occurred, if the product was marketed in that country", Art. 5(1)(a) of the Rome II Regulation.

47 For the parallel problem in the context of jurisdiction cf. supra no. 32.
48 Cf. supra no. 63–64.

Should the parties have excluded the application of the CISG, the CISG would not apply (see Art. 6 of the CISG). In the absence of a choice of law-clause, their sales contract would then, according to Art. 4(1)(a) of Regulation (EC) No 593/2008 of the European Parliament and of the Council on the law applicable to contractual obligations (Rome I),[49] be governed "by the law of the country where the seller has his habitual residence". The seller Y has his habitual residence in Hungary so that, pursuant to Art. 4(1)(a) of the Rome I Regulation, Hungarian law governs their contractual relationship and, according to Art. 5(2) of the Rome II Regulation, the Hungarian law (by way of *rattachement accessoire*) also governs Z's claim against Y for damages in tort. **100**

(d) The customers' claims against Z

In scenario 1(d), Z marketed a product that does not meet the description he made of it. He may, therefore, be liable to the customers for the damage they suffered due to the difference between the description and the actual product quality. In such a situation, the applicable law is to be determined by the rule on product liability, i.e. Art. 5 of the Rome II Regulation.[50] **101**

i) If the customers claiming damages acquired the product directly from Z, according to Art. 5(2) of the Rome II Regulation the *rattachement accessoire* may again apply. The customers' claim in torts will then be governed by the law of the same state that governs their contractual relationship. **102**

In case scenario 1(d), this would lead to the application of Czech law to a claim against Z in tort: In the absence of a choice of law-clause in the contract between Z and his customers, their contracts are governed by the law of the country in which the seller has his habitual residence or central administration, Art. 4(1)(a) of the Rome I Regulation on the law applicable to contractual obligations. Since Z has his central administration in the Czech Republic, the contracts with his customers are governed by Czech law. By way of *rattachement accessoire,* according to Art. 5(2) of the Rome II **103**

49 OJ L 177, 4.7.2008, 6–16.

50 This is true if the claim against Z is brought before the Czech courts, according to Art. 2(1) of the Brussels I Regulation; it is also true if, under Art. 5(3) of the Brussels I Regulation, a consumer sues Z in another country which is not a Contracting State of the 1973 Hague Convention on the Law Applicable to Products Liability. If, on the other hand, a customer sues Z, on the basis of Art. 5(3) of the Brussels I Regulation, in a Contracting State of the 1973 Hague Convention, the applicable law will be determined according to the rules of the 1973 Hague Convention, see Art. 28 of the Rome II Regulation and supra no. 52.

Regulation, Czech law would then also govern Z's alleged tortious liability towards his customers.

104 ii) If, on the other hand, the customers acquired the corn via intermediaries, such as supermarkets or organic food stores, there is no direct contractual relationship between the customers and the *rattachement accessoire* has to be ruled out.

105 The applicable law in that case is to be determined by pursuing the cascade established by Art. 5(1) of the Rome II Regulation:

- if the parties did not agree on the applicable law (Art. 14 of Rome II), and
- since there is no case for *rattachement accessoire* (Art. 5(2) of Rome II), and
- if the parties do not have their habitual residence in the same country (Art. 5(1) in conjunction with Art. 4(2) of Rome II),
- the tort will be governed by "the law of the country in which the person sustaining the damage had his or her habitual residence when the damage occurred, if the product was marketed in that country", Art. 5(1)(a) of the Rome II Regulation. Thus, if the cornflakes in our case scenario 1(d) were available in the supermarkets of the country of the consumers' habitual residence, the law of this country will apply to their claims against Z (independently of whether or not the claimants actually bought the cornflakes in that country).
- if the product was not marketed in the country of the consumers' habitual residence, the applicable law is the law of the country in which the product was acquired, if the crops were marketed there, Art. 5(1)(b) of the Rome II Regulation.[51]

106 In most cases, the person having suffered the damage would have acquired the product in the country of his or her habitual residence, where the product would have also been marketed; in these (most frequent) cases, Art. 5(1)(a) of the Rome II Regulation leads to the application of the tort law of the country of the customers' habitual residence.

51 In the rare cases in which the cornflakes were not marketed there either, "the law of the country in which the damage occurred" applies, "if the product was marketed in that country", Art. 5(1)(c) of the Rome II Regulation.

9. The applicable law in case scenario 2

a) As mentioned above,[52] the material scope of application of the Rome II and Brussels I Regulations are identical. Just as for the purpose of the Brussels I Regulation, for the application of the Rome II Regulation the claim of the state in case scenario 2(a) is to be considered a "civil matter" within the meaning of Art. 1(1) of both Regulations. **107**

Case scenario 2(a) provides a perfect example for the application of Art. 7 of the Rome II Regulation. The damage in this scenario falls within the definition given in Recital 24 of the Rome II Regulation for "environmental damage" (i.e. "an adverse change in a natural resource, such as water, land or air" etc.). According to Art. 7 and Art. 4 of the Rome II Regulation, the claim of the Belgian state-run research institute is, in principle governed by the law of the country in which the damage occurs (i.e. by Belgian law); however, the research institute can, according to Art. 7 of the Rome II Regulation, opt for the application of the law of the country in which the French farm acted, i.e. for the application of French law. **108**

b) In scenario 2(b), according to the first line of arguments presented above,[53] the state authorities intervened in their function as guardian of the public health and safety. They exercised their public and administrative powers so that the case does not qualify as a "civil and commercial matter", according to Art. 1(1) of the Rome II Regulation. Under these circumstances, the Rome II Regulation is not applicable. **109**

If, however, the public authorities claim compensation for expenses incurred on the basis, and according to rules and principles of private law, the claim for recovery of expenses incurred for the restoration of environmental goods is to be regarded as a "civil and commercial matter" and the Rome II Regulation applies.[54] The claim then is, in principle governed by the law of the country in which the damage occurred (i.e. by Belgian law), the Belgian administrative authorities having the possibility to opt for the application of the law of the country in which the French farmer acted (i.e. for the application of French law), Art. 7 and Art. 4 of the Rome II Regulation. **110**

52 Supra no. 49.
53 Supra no. 46 in the context of the Brussels I Regulation.
54 Supra no. 47. For further details and more arguments, see *Th. Kadner Graziano*, Yb PIL, Vol. IX (2007) 71 at 82 ff.

IV. Concluding Remark

111 The case scenarios used in this contribution illustrate the interaction of the Brussels I and the Rome II Regulations and the variety and the complexity of legal issues to be considered in cases of cross-border damage caused by GMOs. Since the Rome II Regulation has been applicable since January 2009 only, many of the issues raised as well as some of the solutions suggested have not been submitted to the courts yet.

Did Biotechnology Regulation Come to a Conclusion? An Insurers' Perspective

Thomas K. Epprecht

The hot phase around biotechnology regulation seems over. It is time to ask where we stand, whether all the regulatory problems have been resolved or what is still lurking behind the scenes, to pop up anytime. The technological leap that was triggered by biotechnology innovation notably accentuated the question of regulatory balance between economic freedom and responsible action, between unlimited progress and anxious standstill. In this paper we hypothesise that the present balance is unstable because there is still a fundamental disagreement on the purpose of regulation and the necessary safety standards. We address the influence of the regulatory and societal environment on insurability, the role of the insurance industry in society, and how the availability or denial of insurance cover in fact makes an impact on regulation.

I. Total safety thanks to regulation?

Safety is an indispensable part of any technology and we can no longer imagine using a technology without strict safeguards. But, perhaps because there is so much emphasis put on safety, there is a widespread belief that technical failure could be avoided completely – if the masters of technology truly wished to do so, and if regulation were only tight enough. However, it will never be possible to patch every hole in the safety net, either by more research, or by more loss prevention measures or by tougher regulation. It is impossible to understand every possible hazard in advance and economic feasibility also puts a limit on raising the safety standard as high as we might like. 1

This regulatory dilemma calls for priority setting: When risks are life-threatening or prone to result in major destruction, maximum safety or even renouncement is a priority. As the case may be, the question "how much safety is enough?" deserves different answers. An example: 2

3 If we would like to protect ourselves against any possible infection, measures against the common cold should be as equally fail-safe as those in a laboratory dealing with highly pathogenic organisms. As a consequence, we would have to cut off all our live contacts with other people and life would come to a standstill.

4 Technological progress poses a real challenge to regulation and its purpose of achieving and maintaining an adequate level of safety. Therefore, some would prefer regulation that guarantees a risk-free technology. But the price would be inflexibility and paralysis. To make progress, economy and society require a regulatory framework which is adaptable. Only then can regulation fulfil its important gatekeeper role in a changing environment and help to ensure that risk and benefit are shared fairly.

1. A sense of entitlement to safety

5 Being surrounded by safety all the time leads to taking safety as a given and tempts one to behave increasingly carelessly and to take more risks. The illusion that societal institutions are there to take complete care of people's safety is illustrated by the following newspaper item. It happened in Florida: A family man arranged his wife and children 15 feet in front of an alligator to take a souvenir photo. When he was urgently called upon to evacuate, he said "But there is no warning notice!" And apparently soon afterwards: "Who can I sue?" A society constituted by such individuals is at greatest risk since it assumes that risk can and should be avoided altogether. The result is a latent worry about everything (and everyone). At the time, when steady fear tilts over into constant anger, the "enemy" lurks just around the corner, finally bringing the values of a pluralistic society into disarray. At the other pole end of control, extremely risky or reckless behaviour of individuals gains acceptance since the desire to escape from the straitjacket of safety is ineradicable. Hence, both types of individual behaviour show a lost capability for dealing with risk responsibly.

2. Safety and a sense of responsibility

6 Therefore, the question is not whether total safety can be produced by regulation or guaranteed by manufacturers, but which risk management measures are appropriate to provide the right level of safety. In this sense, safety is of course an indispensable part of the right to operate, building a crucial element of trust in an operator's ability to master risk appropri-

ately. A responsible attitude towards potential dangers requires convincing preventive measures. Meeting them is not merit, but a must. Nevertheless, despite every precaution, and despite careful compliance with the most exacting safety standards, losses can occur, as proven by the existence of insurance. Thus, whosoever admits frankly that losses are possible, creates more trust than those who deny danger, and the former gains some breathing space at the same time.

II. Biotechnology regulation in light of compensation schemes

The outcome of present regulation, in particular on green[1] biotechnology is 7
ambiguous: in general, the goal is strictly to separate GM-crops and products from their non-GMO counterparts. This approach aims to protect consumer preferences and the respective traditional producers and markets, and less to guarantee safety. After implementation, that kind of regulation helped to calm the controversial debate and slowly opened a way for co-existence. But is this way walkable all the way down the road? What if, for example, GM-wheat comes to the market, making the strict separation of GM-crops from traditionally grown crops increasingly difficult and costly? And what if insurance cover is not available because there is no attractive prospect of return for the provider of financial security?

1. Strict liability

Several countries, primarily in Europe, have imposed strict liability for 8
damages from biotechnology. "Strict" always sounds good and promises that safety is the primary concern in the political arena. But in legal terms it simply means expanding the legal basis for claims which affects the frequency and extent of claims for losses. With strict liability (or no-fault liability), unforeseeable grounds for liability claims grow, because the liable party, usually the provider of technology, may be obliged to pay damages even for tortious acts or omissions committed by another party using the technology. This concept, originally developed for activities like operating

1 We use the example here of "green", i.e. crop, biotechnology because there has been and still is most dissent in this field, although this article is written in a broader general context.

nuclear power plants which hold a particular hazard potential, gives grounds for exemption only in case of a missing proven link between the alleged cause of the loss (incident, event) and the effect (harm, material and non-material damage), or if an act of God intervened.

9 It goes without saying that this concept must result in endless dispute once non-safety related matters of fact are subjected to strict liability, for example, pure financial losses, because of two major irresolvable problems:

- What is considered a loss is often subject to interpretation.
- It is often unclear who within a long value chain is the tortfeasor.

10 Liability based on mere assumption comes into play the more it is based upon a scientific understanding of (probabilistic) causality, in particular when – in reality always dispersive – percentages are considered to trigger liability. Exposure assessment becomes completely uncertain if a reversal of the burden of proof, extremely extended limitation periods, or a retroactive application of newly discovered duties of care become established on top of existing liability regimes.

2. Effects on risk transfer

11 Introducing strict liability principles for biotechnology eases access to compensation. As a consequence, the casualty insurance business is confronted with an extended loss potential, in particular for matters of fact which are not safety related at all, like pure financial losses due to commingling with GM-crops. This type of loss turns the idea of insurance on its head: When legislation is driven by the desire to steer production methods and segregated value chains rather than to ensure safety, the purpose of "risk transfer" turns simply into "cost transfer". In the long run, providing insurance cover promises losses rather than profits, and it is even exposed to moral hazard.

A legal request for compulsory insurance or other financial guarantees is not suited for convincing hesitant insurers to put economic concerns on hold. Compulsory insurance would hardly result in additional offers of insurance cover, in particular not to the full extent of the stipulated liability. Why: Because the freedom of contract applies, "compulsory" just means that the liable party must buy insurance cover or provide for other financial guarantees; it does not mean that insurers must actually offer cover. Thus, if a liable party cannot get the required cover (or cannot afford the premium), the compulsory insurance mandate will work to restrict the liable party's activities, limiting its market access. Regulators should take this aspect into consideration carefully.

III. Principles of regulation

Against the background of the difficulties mentioned above, it makes sense to recall some basic principles of regulation. "Regulation" is a general term which encompasses public law and private law[2] as well. In public debates about regulation, however, this distinction is seldom made. In respect of the role of insurance, though, it is important to make the distinction because the subjects and goals of regulation in these two areas affect insurability in a fundamentally different manner. The following bullet points make the contextually necessary distinctions. Simplifications were made for the purpose of understanding, especially for laypersons in legal matters (to whom the author also reckons himself).[3] **12**

1. Public law

- Public law governs the relationship between the *res publica* (state) and persons (citizens or corporate bodies). It is mainly there to provide the juridical instruments and framework to ensure the societal actors' living together in peace, safety, and fairness. Public law is the codex that enables public authorities and governmental agencies to control, approve, impose, or punish. **13**
- With regard to regulating technology, this part of the legal framework establishes a set of duties, prerequisites and bans to be observed by manufacturers in order to ensure that products are manufactured safely, that only safe products come onto the market, and that economic activity is possible to the benefit of the different societal actors.

2 The use of the broad term "regulation" may cause additional confusion in international debates since it encompasses the areas of public law and private law (civil law) as well, while these areas are not identical from country to country. One cause is the fundamental difference between the private law traditions of continental Europe (Roman origin) and the Anglo-Saxon tradition (common law) – and this just in the western world. Many transatlantic misunderstandings and irritations in legal matters may go back to these different legal traditions which certainly shape the way of thinking, particularly in public matters. An example: Different patterns of legal thinking may be one of the major reasons why Anglo-Saxons often show difficulties in understanding the precautionary principle as part of public law.

3 Disclaimer: the author is a scientist and not a lawyer. The statements made in this chapter reflect his personal view and experience in public debates and business-related activities.

- Following technological progress, considerations about necessary regulatory changes focus on the role and the instruments of governmental agencies in order to ensure effective enforcement of the regulatory goals. Regularly under review come enforcement instruments, approval criteria and the embedding and control of self-regulation.
- Changes in public law concerning technologies influence the way public authorities act as a gatekeeper. As a consequence, residual risks which require insurance cover are changed too. Although insurance companies use similar selection criteria to public authorities to distinguish the acceptable from the unacceptable, the protection goals are different. Clearly, the insurance industry is primarily interested in a good risk (management) quality, while public authorities have a general protection goal.

2. Liability law

14
- In a broader context, rules of private law are necessary to keep the balance between societal actors and the protection of their legitimate interests, property and health. In a way, private law fills a gap since the regulating effect of public law cannot resolve conflicting interests between citizens or private entities which continue to exist, and because damage can occur anyhow.
- Liability law is that section of private law that interests most in the context of casualty insurance. In its traditional core, it defines the legal responsibilities of a person or entity versus a third party in the case that the latter suffers from actions or omissions of the tortfeasor. Casualty insurance helps to implement liability law by securing payment of the financial indemnification an insured party owes to the victim, within the limits of cover.
- Liability law is scrutinised regularly in the course of technological change. The questions normally dominating the debate all centre around the adequacy of existing rulings. If a technology is associated with particular dangers, technology-specific tightening of the liability rules are up for debate, as has happened with biotechnology.

3. Regulatory harmonisation

In light of the distinction made above, transnational regulatory harmonisation is most effective for issues related to approval, trade and marketing, i.e.: 15

- Approval requirements, procedures, and standards;
- Threshold, labelling, testing and risk evaluation criteria;
- Good farming and manufacturing practices, such as separation distances, use of machinery, transport and storage facilities.

Harmonisation, however, is disadvantageous for liability and redress, since there are too many definitions of harm, understandings of hazard manifestation and admissible causes of action. 16

IV. Liability framework and insurance

Insurers face a similar problem with novelty, uncertainty and lack of data as do public authorities and approval bodies. This parallelism makes it tempting to believe that insurance can play a quasi-regulatory role or act as a judge in case of conflicting interests. It is true that insurers have an interest in evaluating risk rigorously, and it is also true that they constantly review exposure. But, while the regulators' goal is to protect society and thus limit avoidable exposure to hazard, insurance companies evaluate (only) the financial exposure within the extent of cover they provide. 17

In fact insurance companies make business with calculated risk; playing judge is not their job. Thus when they choose to provide cover or not, the issue is not to deliver a verdict of "good" or "evil", but to decide whether accepting a risk is an attractive proposition. If the exposure cannot be calculated, the cover is then limited. Yet, the more limitations that are placed on a cover, the further it will fall short of the regulator's goal of protection and precaution. 18

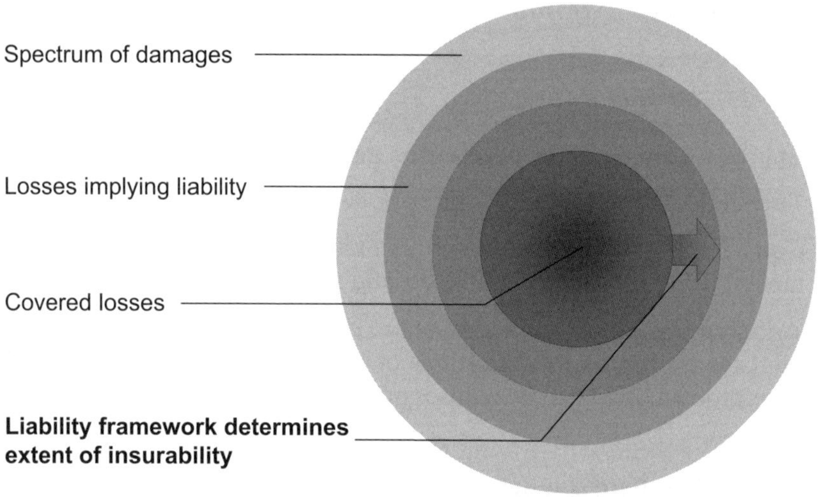

Fig. 1: A clear and practical liability framework promotes insurability

19 Therefore, if the insurance industry is to help society and regulators reach their protection goal, the statutory framework must also account for workable compensation schemes so that it will be possible to put a monetary value on losses subject to liability. To make such legal rule effective and workable, the availability of financial guarantees must be in line with the envisaged liability regime. In order to meet its obligations, a liable party can preferably transfer the financial burden to a risk carrier such as an insurer. However, this goal can only be achieved when risks are measurable, acceptable, affordable, etc., i.e. if the prerequisites of insurability are met. Therefore, the regulatory answers to some basic questions need to be clarified beforehand, for example:

- Is there consensus about what constitutes harm?
- Which events constitute personal injury, property damage, financial and environmental loss?
- Which rules apply for loss quantification, for example, in the case of commingling of GM-crops and non-GM-crops?
- What is an acceptable impact (including thresholds)?
- What constitutes exemption from liability?
- What is the limitation period?
- Who has the burden of proof?

V. The role of insurance

Insurance enables risk taking. Already in ancient times, societal communities knew the power of mutuality that made risk bearable for their individual members. Mankind made a crucial cultural accomplishment when it started to assist individuals to overcome bad luck, and as a whole it became able to take more risks and was no longer at the mercy of destiny. Later, modern mathematics, statistics and economics made risk calculable and economically acceptable: When a boat sunk in a storm or a breadwinner died early, life for those affected could continue almost normally if the burden of financial loss was shared, i.e. insurance was available. Today, the principle of mutuality is often discredited as, for example, discussions around life and health systems show. These discussions block out the purely economic reasons for mutuality: Risks can only be bearable if they are shared by many shoulders and over many years, also by those who believe themselves still to be far from risk. Considering payments on an average, insurance is not a means to save expenses for the insured individual, but to make losses affordable which may come big and fast, sooner or later and more often than expected. **20**

Basically, the role of insurance as described above has not changed. Insurance still enables individuals and corporations to cope better with risk, or more precisely to deal with the anticipated negative deviation from the expected as a consequence of disruptive events.[4] The insured in turn do not need to bear the possible negative financial consequences alone in case of a loss. Insurance assumes risk against premium, diversifies it and compensates losses within the scope of cover. Thus, insurance acts as a framework for organising a collective risk equilibrium, a "storage tank" that is fed with premiums and from which payments are taken for losses that occur despite every safety measure. As such, insurance contributes to economic growth and promotes innovation, turning threat into opportunity. **21**

Compared to cover for property losses as with the sunken boat mentioned above, exposure or liability insurance is a bit more difficult to estimate. There are two major differences: there is more uncertainty as regards extent of loss, because it is often subject to interpretation by a court, jury or settlement. Secondly, there is also uncertainty with regard to the point in time when an insurer is given notice of a loss, something that can hap- **22**

4 In this definition, risk not only encompasses the negative deviation from the expected (threat), but also opportunity, i.e. the positive deviation as a consequence or forecast of extraordinary success. Opportunity is reflected in expressions like "risk capital", "no risk – no fun" etc.

pen many years after the policy triggered was in force. That is why insurers are not keen on business with long tail exposure, for example, due to latent health risks of novel materials.

VI. Insurability

1. Basic principles ...

23 "Risk" is mathematically represented by the product of probability and average severity of a loss. For this formula to work, a statistically relevant number of losses must already have happened. In this classical model, statistical experience is the prerequisite for insurability. That means: Neither high frequency nor the possible extent of a loss hampers insurability in principle, as long as the required premium is calculable and affordable.

24 In short, the basic principles for underwriting include assessability (measurability, quantification); economic efficiency (profitability); randomness (fortuity); and mutuality (solidarity).

25 **Assessability** is a function of both the frequency and severity of risks that manifest themselves as losses. It is a key criterion since insurability requires transferred risks to be calculable and risk absorption to appear profitable, as well as affordable for the insured in terms of the necessary premium. In addition to actuarial and technical information on insurance markets, legal certainty is equally important.

26 **Economic efficiency**: In order to write insurance profitably over the long run, private insurers strive to charge premiums commensurate with the accepted risks. When selecting the risks he wants to cover, an insurer must take into consideration the expected variation, any trends, the legal certainty and the potential claimants' behaviour. The insurers' preference for allocating his capital in light of profitability is also a consequence of statutory solvency requirements which oblige insurers to calculate net premiums that are adequate to the risk exposure. Thus, the requirement of solvency, i.e. economic efficiency, is an element that also needs to be considered by legislators.

27 **Randomness:** An event is considered fortuitous or random if at least the point in time and occurrence of the damage are unpredictable: It must be impossible for the insured party's intention to influence the occurrence of a loss event. For example, where genetically modified seed is unlawfully commingled with conventional seed, it may remain unclear to what extent such an event stems from attempts to cut costs along the distribution

chain, and to what extent it happened fortuitously. Thus, as a rule, pure financial losses are excluded from insurance cover.

Mutuality: This basic principle of insurance expresses the joint effort and common will of a large number of insured parties to carry a hazard jointly. Particularly with compulsory insurance, a large number of policy holders are required – the large majority of which do not incur losses to an extent equivalent to the added premium they pay. Motor vehicle insurance or health insurance in Europe are good examples for a large number of policyholders that together make also extreme losses bearable. On the other hand, the more a liability regime targets a few industrial operators, the more unbalanced a risk portfolio becomes. This affects both insurability and premiums. 28

2. ... and additional criteria

Journalists often ask whether the object of an insurer's investigation, for example, biotechnology, is insurable. If the answer is "yes", they usually conclude that the technology must be safe, if it is "no", the verdict is "dangerous". However, such interpretation cuts too short, because – as mentioned above – measurability of potential losses, rather than safety, is the deciding factor. Technical progress and uncertain public acceptance may trigger legislative action to change existing rulings. This results in an increased exposure to formerly unexpected claims, in particular if the legal framework aspired to is inspired by the desire to avoid risk altogether. 29

Acceptance: Economically acceptable losses must be backed by a consensus on societal acceptability in order to be insurable. If there is mutual understanding about the losses a society is prepared to accept, society has consciously discounted insured risks: it knows what the price will be – as in traffic accidents, for example – and accepts the costs, both material and immaterial. Without public acceptance, however, the costs of a loss are in principle unlimited, because no compensation suffices. The affordable price can never bridge the gap; it is limited by the amount of money available. 30

Risk selection: As said before, in a first step the loss potential of insured activities is traditionally calculated from statistical loss experience. In a second step, knowledge about safety standards in a particular industry segment as well as of individual clients is considered. This allows the insurer to select preferred clients, steer capital allocation and adapt con- 31 tractual wordings and premiums. With new technologies, such as modern biotechnology, the insurance industry is well advised to put capital at risk with caution: Where knowledge is incomplete and acceptance shaky, the insurance industry is particularly in need of assessment criteria that help to identify preferred clients and calculable risk.

3. Availability of insurance cover

32 Existing casualty products of most insurers continue to offer cover for biotechnology related losses, in particular for bodily injury. Pure financial losses and ecological damage are mostly excluded from standard covers. Some insurers felt a need to render these exclusions more precisely and explicitly exclude losses resulting from commingling or cross-pollination (green biotechnology), or stipulate a GMO exclusion which often can be included again against extra premium and after individual risk assessment. A comprehensive overview – at least regarding green biotechnology – can be found in the previous study.[5]

VII. Conclusion

33 The bottom line as regards tightening regulation is whether legislative measures help to improve safety and compensate for loss adequately. This is the case for operations with an undoubtedly high risk potential, where the potential tortfeasor is unequivocal, and where causality is immediate and related to distinct production sites. But as with regulation of modern biotechnology, a main goal is compensation for pure financial losses as a consequence of market preferences, while the tortfeasor is often unclear because the origin of a loss may be just somewhere along the value chain.

34 The insurance industry is challenged by a technology which is considered scientifically safe but perceived nevertheless as a threat. Although so far there is no proof of a harmful causal relationship and hazard is subject to interpretation and debate, strict liability and reversal of the burden of proof or alike are a legal reality, difficult to cope with in an economically reasonable manner. Thus, a consistent approach to assessing exposures

5 *I. Ebert/Ch. Lahnstein*, GMO Liability: Options for Insurers, in: B.A. Koch (ed.), Economic Loss Caused by Genetically Modified Organisms. Liability and Redress for the Adventitious Presence of GMOs in Non-GM Crops (2008) 577–581.

also beyond technical criteria is indispensable, in addition to company-specific assessments. Indeed, cover should be limited to what is assessable. However, an uncompromisingly risk-averse approach to biotechnology would not be justified.

The promises of modern biotechnology present opportunities and threats; **35**
individuals and societal actors always have an option to decide whether the potential disadvantages outweigh the assumed benefits. It is a question of balance that is subject to negotiation, again and again. The minimum requirement in a pluralistic society is honesty in thinking and reasoning which calls for intellectual clarity and expertise, transparency, respect, tolerance and a sense of proportion.

The risk and regulatory issues associated with biotechnology are dynamic, **36**
global, and complex in scope. Efforts towards a global risk governance framework for biotechnology will continue to be codified in adapted law and regulation. Addressing the possible downside effects and the measures necessary to contain them accordingly clears the way towards the final aim of any regulation and risk governance system, of which insurance is a part, which is to enable progress and provide safety and opportunities both for society and the economy.

Liability to Third Parties for Damage Caused by GMOs: An Economic Perspective

Michael Faure and Andri Wibisana

I. Introduction

1 The problems sketched in this new GMO liability project are truly highly interesting from an economic perspective. In a first project we basically addressed the problems that arise as a result of so-called "admixture", being the presence of GMOs in non-GM crops.[1] The economic analysis of law has generally paid a lot of attention to the question of how legal instruments and legal rules in particular can promote social welfare in society. As we indicated in the earlier study much of the law and economics research has focused on uncertainties that arise as a result of the use of GMOs. In that respect for example questions arise whether GMOs should be used at all when uncertainties arise as to their consequences (also for third parties). This led to a lot of controversy, for example with respect to the application of the well-known precautionary principle.[2] When we wrote our first contribution (published in 2008) for the first GMO project (mainly dealing with problems of admixture) not that much literature (at least from an economic perspective) on liability for damage caused by GMOs had been published.[3] Since then a few more articles on this topic have been published or at least have come to our mind. We will therefore first of all update our previous study by incorporating this recent literature; moreover, whereas the

---

1 See our previous contribution *M. Faure/A. Wibisana*, Liability in Case of Damage Resulting from GMOs: an Economic Perspective, in: B.A. Koch (ed.), Economic Loss Caused by Genetically Modified Organisms. Liability and Redress for the Adventitious Presence of GMOs in Non-GM Crops (2008) 531 ff.

2 For a recent account see for example *H.T. Anker/M.R. Grossman*, Authorisation of Genetically Modified Organisms: Precaution in US and EC Law, European Food and Feed Law Review (EFFL Rev) 4 (1) (2009) 3–22.

3 For an exception see for example *D.A. Kershen*, Legal Liability Issues in Agricultural Biotechnology, Crop Science 44 (2004) 457.

previous study focused merely on the problems caused by "admixture", our current study is much broader and also focuses generally on third-party liability for damage caused by GMOs. Moreover, redress mechanisms (other than tort, such as compensation funds and insurance) will be addressed as well. Since this project focuses on third-party losses, the question can be asked to what extent GMOs can also cause damage to the environment as such and thus lead to environmental liability. The question equally arises to what extent the product liability regime can be applied.

It may be clear, however, that this is not a traditional ECTIL country report. 2
Hence we can only follow the Questionnaire to the extent that some of the items have indeed been discussed in the law and economics literature.

In the introduction it should be stated that for an economic analysis it is 3
always important to make a distinction between prevention and compensation. For an economist liability rules should primarily have a preventive effect and thus provide incentives to those dealing with GMOs to prevent damage from occurring to third parties. This preventive goal is in the economic literature distinguished from ex post compensation. Ex post compensation can be provided through a variety of mechanisms although tort law is in the economic literature usually not considered as compensation, but rather as a preventive mechanism. Hence it is rather insurance or compensation funds that could fulfil the particular compensation objective.

This contribution is structured as follows: first the types of liability applic- 4
able to GMO risks will be discussed and this will be the core of the paper since a distinction will be made between the general case where victims are third parties, the application of liability rules to environmental damage and the product liability case (II). An important and debated issue in GMO liability is the question how one should deal with causation. That will be dealt with in detail in section III. Related to the causation issue is the question how liability should be allocated if the damage is not caused by one tortfeasor, but if multiple tortfeasors are involved (IV). Defendants in a GMO related liability case may be able to call on several defences which are discussed in section V. Finally the question arises what remedies are available for the plaintiff if liability of the defendant is accepted. This involves inter alia discussion of damages, including the question whether from an economic perspective there should be liability for so-called pure economic loss, but also the question whether an injunction can be sought, and whether compulsory insurance may be indicated. These issues related to remedies are addressed in section VI. We briefly address cross-border issues in section VII and formulate a few concluding remarks in section VIII.

II. Types of liability

1. Economic criteria for strict liability

5 Economists use classic cost/benefit analysis to determine the level of care that will lead to a minimisation of the social costs of accidents. Not surprisingly, this can be found where the marginal costs of care-taking equal the marginal benefits in accident reduction. Indeed, since care-taking has its price as well, a legal rule should not give incentives to avoid every possible accident that could occur, but only accidents that could be avoided by investments in care, of which the marginal costs are lower than or equal to the marginal benefits in accident reduction. It might well be that extremely high care could well additionally contribute to a reduction of the accident risk but the marginal costs of care-taking in that case might well be much higher than the additional benefit in accident reduction. Investments in care would in that case be inefficient and scarce resources would be spoiled.[4] These levels of care where marginal costs of care-taking equal marginal benefits in accident reduction are referred to in the literature as optimal or efficient care levels.[5] We will now first address optimal liability rules in a unilateral case. With the unilateral accident situation we refer to the case where only one party (referred to as the injurer or tortfeasor) can influence the accident risk.

6 Looking at a unilateral accident situation, one can state that two legal rules, i.e. the negligence rule and strict liability, would give the injurer incentives for taking optimal care. If there were no liability at all, clearly the injurer would have no incentive for care-taking; therefore, in a no-liability situation, the externality will not be internalised and an inefficient outcome will follow. If a negligence rule is adopted, the injurer will take optimal care, provided the due care required in the legal system is equal to the optimal care as resulting from a marginal cost/marginal benefits weighing.[6] This can be easily understood. If the judicial system sets the due care standard correctly, the injurer can avoid liability by taking due care. Thus he will have to take care to avoid the accident, but if he does so he can avoid paying the expected damage. Of course, the injurer could take more care

---

4 This finding only holds in a risk neutral setting. In case of risk aversion, higher investments in care might well be efficient since a reduction of accident risk will in that case also remove the disutility of risk from a risk averse person.

5 See *W. Landes/R. Posner*, Georgia Law Review (GLR) 1981, 870 and *A.M. Polinsky*, Introduction to Law and Economics (1983) 38–39.

6 *S. Shavell*, Strict Liability versus Negligence, Journal of Legal Studies (JLS) 1980, 1–25 and *G. Calabresi*, Optimal Deterrence and Accidents, Yale Law Journal (YLJ) 1975, 656–671.

than the legal system requires him to do under a negligence rule, but he will have no incentive to do so since he can already escape liability by following the due care standard. The injurer could also spend less on care than the legal system requires him to. In that case he will have lower costs of care-taking, but he will have to pay damages in case an accident occurs. Since the optimal care standard was defined as exactly that level of care where the marginal costs of care equal the marginal benefits in accident reduction, taking less than the due care will not be interesting for the individual injurer since it will increase his total expected costs. Thus a negligence rule will lead to an efficient outcome as long as the legal system defines the due care as equal to the optimal care of the model.[7]

Also a strict liability rule will lead to the optimum in a case where only one party can influence the accident risk. The reason is quite easy. A strict liability rule basically states that the injurer has to compensate in any case no matter what care he took. It is sometimes argued that this will lead the injurer to take excessive precautions or to take no care at all since he is liable anyway. Neither of these statements seems true. By making the injurer strictly liable, the social decision is in fact shifted to the injurer. In a unilateral accident case it simply means that he has to bear all the social costs of accidents, i.e. his own costs of care-taking and the expected damage.[8] Therefore, he will take exactly the same decision, i.e. to minimise his total expected accident costs. This can be reached at the optimal care level. Therefore, the injurer will take optimal care since this is the way to minimise his total expected costs. Spending more on care would increase his costs of care-taking inefficiently and spending less on care would increase the expected damage inefficiently. **7**

How can the economic arguments in favour of strict liability be applied to the case of damage caused by GMOs? **8**

The first question to be answered would be whether handling of GMOs should be considered a unilateral accident, being an accident where only the injurer can influence the accident risk. If that were the case, we just mentioned that the economic model predicts that the advantage of the strict liability rule is that it will give the injurer optimal incentives for care.[9] If in this particular case the victim cannot influence the accident risk, strict liabi- **9**

7 For a recent discussion of the economics of strict liability versus negligence see *H.B. Schäfer/F. Müller-Langer*, Efficient Liability Rules, in: M. Faure, Tort Law and Economics (2009) 3–45.

8 *Polinsky* (fn. 5) 39; *S. Shavell*, JLS 1980, 11.

9 Also, it would provide optimal incentives to take an efficient activity level. See on the importance of the activity level also *Shavell*, JLS 1980, 1–25.

lity would be the first best solution to give the operator of the GMO optimal incentives to reduce the risk that GMO crops would cause damage to third parties.

10 The question, however, arises whether damage caused to third parties by GMO crops is always a truly unilateral accident. Depending upon the factual circumstances (which can of course significantly vary) some may argue that one could imagine circumstances where potential victims could also take measures to prevent the damage. If that were the case, one could argue that GMO damage becomes a bilateral risk in which also the potential victim could take efficient measures to prevent the risk. However, one could still argue that the influence of the operator of the GMO is probably still far more important than the influence of the victim. If that is the case, the outcome does not change and the strict liability rule remains warranted giving the operator of the GMO optimal incentives to take preventive measures. This clearly assumes that the operator who handles the GMO is in the best position to prevent the risk. However, as will be mentioned below, in this bilateral case, it remains important that a defence should be added to the strict liability rule to give victims incentives for prevention as well. However, if it appeared from the factual situation that it is as important to provide victims with incentives to prevent the risk as it would be to give similar incentives to the operator who handles the GMO, a negligence rule would be optimal.

11 Hence, GMO damage does not seem to be comparable with a classic environmental case. In the latter case, it is often argued that these are typical unilateral cases, where most of the influence of the accident risk comes from the potential polluter. Therefore, most argue in favour of a clear strict liability rule since the victim can usually do less than the potential polluter to avoid the risk. However, since potential victims in the GMO case may be professionals as well, the same line of reasoning does not apply. If in fact it appears that the influence of both the potential victim and the operator who handles the GMOs is equally important, a negligence rule might in fact be optimal.[10]

12 From the explanation above, it appears that whether GMO liability cases are of a unilateral or a bilateral nature depends upon the factual situation. One could certainly argue that cross-pollination from GM crops to non-GM crops constitutes a unilateral case because if organic or conventional farmers should also prevent the cross-pollination, they should change

10 The economic reason is that only negligence also provides incentives to the victim to adopt an optimal activity level.

their usual practices and, hence, incur high costs.[11] In this situation, if transaction costs between organic farmers and farmers planting GM crops are low, they may bargain as to determine who is in a better position to prevent the pollination (i.e. more cheaply). However, if transaction costs are high, a liability rule more suitable for the unilateral case should apply. In this regard, strict liability may be better than a negligence rule.[12] Transaction costs will certainly be high in the situation where damage, for example, is not related to admixture, but where it concerns damage to a third party who is not active in the field of agriculture at all.

oreover, another argument in favour of strict liability is related to the fact that the negligence rule works efficiently only if the judge is able to set the due care standard required in the legal system efficiently. However, in reality this may require very high information costs and judges may find it difficult to adequately balance the marginal costs of taking additional preventive measures against the marginal benefits in further reducing the accident risk. The advantage of strict liability is that, in that particular case, all information costs are shifted to the injurer. Under strict liability it is indeed the injurer who balances costs and benefits of prevention costs against the potential damage. Hence, in cases where one would assume that injurers are better able than judges to adequately perform the cost-benefit analysis required to set the due care level, this would be yet another important argument in favour of strict liability. Of course the latter argument may once more provide a strong case for strict liability when damage results from the use of GMOs. Indeed, for the judge to efficiently set a due care standard, high information costs would be involved. Given the highly technical nature of the risk it may be impossible for the judiciary to set care standards adequately.[13] Under strict liability it is the producer of GMOs who will bal- **13**

11 One commentator notes that if organic or conventional farmers are forced to prevent gene contamination, they may have to abandon their seed-saving practices and, given resistance of the hybrids, to use more toxic herbicides. See: *H. Preston*, Drift of Patented Genetically Engineered Crops: Rethinking Liability Theories, Texas Law Review (TLR) 81 (2003) 1159.

12 Another reason for applying strict liability is the non-reciprocal nature of damage possibly suffered by the organic or conventional farmers. In this case, a farmer who plants GM crops gains benefits from his crops, while at the same time exposing his neighbour to a risk to which he himself is not subjected, e.g. the risk of losing organic certification. See: *A.B. Endres*, "GMO": Genetically Modified Organism or Gigantic Monetary Obligation? The Liability Schemes for GMO Damage in the United States and the European Union, Loyola L.A. International and Comparative Law Review (ILR) 22 (2000), 491. A rationale behind this argument is probably related to the issue of the distribution of risk and benefit, in which those who gain benefits, while at the same time subjecting others to risks, should pay the damages if those risks materialise.

13 One has to balance this somewhat because the judge is of course aided in this process of weighing costs and benefits by the regulator. To some extent the regulator took over

ance costs and benefits and will subsequently choose the efficient care level. If producers are better informed than the judiciary on the optimal care necessary to reduce the risk (which is very likely in case of GMO risks), this constitutes yet another argument in favour of strict liability.

14 In addition, strict liability is more preferable than the negligence rule when one takes into account the activity level. Under the negligence rule, the injurer will escape liability as long as he takes the optimal level of care. As a result, as long as the injurer maintains the optimal level of care, he does not have to take into account the optimal level of his activity. This means that from society's point of view, the number of activities might be excessively inefficient. The situation will be different under strict liability. Since the injurer will be liable whenever the accident occurs, the injurer will pay adequate attention not only to the optimal level of precaution, but also to the optimal level of activity.[14]

15 Theoretically, the risk of an accident taking place depends not only on the level of precaution, but also on the number of activities.[15] Hence, since strict liability will induce the injurer to take into account both optimal precaution and activity level, we could argue that strict liability is superior to a negligence rule.

16 The above finding might also mean that the application of strict liability to GMO cases will reduce the number of activities involving GMOs. Whether or not this is a desirable result will mainly depend on the benefits of GMOs relative to possible costs resulting from the commercial release of GMOs.

2. Environmental liability

17 If one now turns away from the more traditional tort case (where a third party victim suffers loss resulting from the GMO) and assumes that damage is caused to the environment as such, the question arises how that affects

---

this process and set efficient regulatory standards on which the judge can rely in a negligence case. For further details see *M. Faure*, Economic Analysis of Tort and Regulatory Law, in: W.H. van Boom/M. Lukas/C. Kissling, Tort and Regulatory Law (2007) 399–404.

14 See *L. Kaplow/S. Shavell*, Economic Analysis of Law, NBER Working Paper Series No. 6960 (1999) 4 f.

15 Faure and Skogh have given an example about the risk of having a traffic accident, which depends not only on the care that the driver takes, but also on the number of kilo-metres driven over a certain period of time. Here the authors also conclude that, in terms of providing incentives to take optimal activity level, only strict liability will be efficient. See *M. Faure/G. Skogh*, The Economic Analysis of Environmental Policy and Law: and Introduction (2003) 252 f.

the choice between strict liability and negligence. An important difference between the classic tort case and the case where GMOs cause damage to the environment as such is that environmental harm usually makes the accident situation unilateral. There is indeed very little that the environment as such could do to prevent the accident. As we mentioned above, it is argued in the economic literature that in a unilateral accident model (where hence only the behaviour of the injurer influences the accident risk), strict liability will be efficient since it induces the injurer to adopt an efficient activity level and optimal care. Therefore it has been argued that there seems to be an economic rationale behind the tendency in case law and in many environmental statutes in legal systems to introduce strict liability for environmental damage: since the victim cannot influence the accident risk, strict liability will be first best to give the potential polluter optimal incentives for accident reduction.[16] A few important nuances have been formulated in the literature in this respect: strict liability (for environmental harm) assumes that the judge has accurate information on the extent of the damage. If courts err in assessing the damage, strict liability, so Cooter showed, will lead to under-deterrence.[17] Strict liability is also only efficient if the injurer is always held to fully pay for the consequences of the accident. If the injurer were insolvent or if the judge were to underestimate the amount of the damage, a negligence rule would be preferred if the judge could at least adequately fix the optimal level of care, even if there were uncertainty concerning the exact extent of the damage.[18] This hence leads to a balanced conclusion as far as the optimal liability rule is concerned for environmental damage caused by GMOs: given the unilateral nature of the damage, strict liability may be preferred. This, however, only holds if the injurers are not insolvent and if the courts can assess the damage correctly. Also, Trebilcock pointed at the fact that (in environmental, but also in other liability areas) strict liability may have undesirable effects if it is combined with other features such as shifting the burden of proof, joint and several liability and high (punitive) damages for (non-pecuniary) losses. The reason why strict liability regimes in the US (for example the regime in CERCLA, also referred to as superfund) is experienced as "crashing" is not so much because of the strict liability regime itself, but

16 For a detailed analysis see the contributions in *M. Faure*, Deterrence, Insurability, and Compensation in Environmental Liability. Future Developments in the European Union (2003).

17 See *R. Cooter*, Prices and Sanctions, Columbia Law Review (Colum.L.Rev.) 1984, 1343–1523.

18 Ibid.

because of the combination with joint and several liability and because of the retrospective nature of superfund liability.[19] Therefore the final judgment on the efficiency of strict liability for environmental harm may depend upon other issues such as causation and the magnitude of damages awarded.[20] Interestingly recent empirical evidence has shown that environmental liability regimes do indeed have a positive influence on the incentives of potential polluters to take additional preventive measures.[21]

3. Product liability

18 Although in this project the major focus is on damage caused by GMOs to third parties, it is of course possible, as the Questionnaire explicitly states, that there may be damage in contractual contacts as well. This is more particularly the case when damage is caused to parties who stand in a contractual relationship with one producing, manufacturing or delivering GMOs. In that particular case, product liability rules may play a role. Here again it is important to look at the contractual situation. The fact that a particular product (that by hypothesis has used GMOs) has first been delivered to a purchaser has important consequences for the structure of the liability rule, even if damage were not caused to the purchaser but to third parties. Economic analysis carefully distinguishes the various hypotheses that can be recognised in this respect.

(a) Contractual liability based on the Coase theorem

19 In, for instance, a product liability setting this means that if a purchaser of a product is fully informed of the possible defects and the product risk, he will always take into account the expected damage and add this to the market price to decide whether or not to purchase the product. The well-informed consumer will always take into account the full price of the pro-

---

19 For details see *J. Boyd*, A Market-Based Analysis of Financial Insurance Issues Associated with US Natural Resource Damage Liability, in: M. Faure, Deterrence, Insurability, and Compensation in Environmental Liability. Future Developments in the European Union (2003) 258–302.

20 See *M.J. Trebilcock*, The Social Insurance-Deterrence Dilemma of Modern North American Tort Law: A Canadian Perspective on the Liability Insurance Crisis, San Diego Law Review (San Diego L.Rev.) 24 (1987) 929–1002.

21 For a discussion of this literature see *M. Faure*, Environmental Liability, in: M. Faure, Tort Law and Economics (2009) 252 f.

duct, which includes the expected damage.[22] In that case the agreement on the distribution of risk might be reflected in the contract price. The price mechanism can have this signalling function to the consumer. If the market price reflects the expected damage, the consumer can know that the producer bears the accident risk. If, however, the market price only reflects the cost price of the product and not the expected damage, the well-informed consumer would know that he bears the accident risk himself.[23]

In this situation, where the consumer is fully informed of the accident risk, the optimal care and optimal production (activity) level will automatically be followed, irrespective of the legal rule. This result holds under both strict liability as well as under negligence and even in a no-liability setting. Indeed the fully informed consumer will, as is demonstrated, only purchase the product taking into account its full price. In a no-liability setting the consumer will add the expected loss to the (low) market price and still demand the product with the lowest full price. Applying the Coase theorem to the product liability case this means that an efficient result follows regardless of which legal regime applies. But this result depends upon the assumption that the consumer is fully informed. **20**

One has to understand that on paper this works perfectly, but these theories applying the Coase theorem to product liability of course rely on a very heavy assumption, being full information on the risk caused by the GMO products. It is more particularly this assumption of full information that has been subject to serious criticism. This can also be understood, more particularly in the case of GMOs. However, one should of course make a distinction between the types of consumers that purchase GMO related products. Again a distinction between commercial and non-commercial purchases may be indicated in this respect. Commercial purchases, aware that they purchase products where GMOs have been used, may be well aware of the risk. In that respect applying the Coase theorem is therefore not that strange. The story however changes in case of ordinary consumers who have no information whatsoever with respect to the risk related to the GMO use. Generally, it has been held that one solution to no information by consumers is to provide additional information on the risks **21**

22 *W.Y. Oi*, The Economics of Product Safety, Bell Journal of Economics (BJE) 1973, 3–28; *M. Adams*, Produkthaftung – Wohltat oder Plage – Eine ökonomische Analyse, Betriebs-Berater, Beilage 20/1987 zu Heft 31/1987, 1–24. And see *F. Silva/A. Cavaliere*, The Economic Impact of Product Liability: lessons for the US and the EU Experience in: G. Galli/J. Pelkmans, Regulatory Reform and Competitiveness in Europe, Horizontal Issues (2000) 294–298.

23 For simple examples see *S. Shavell*, Economic Analysis of Accident Law (1987) 78–80 and *Polinsky* (fn. 5) 95–97. See also *Silva/Cavaliere* (fn. 22) 295.

through legislative measures. If these were successful, the conditions for the Coase theorem could again be met. However, sometimes information may not help in providing better insights to the consumer. Recently behavioural law and economics have pointed to the fact that a consumer's choice is often subject to a variety of so-called heuristics and biases. As a result of this, consumers' ability to adapt their behaviour (even if information is provided) may be limited.[24] These biases may also play a role when consumers make decisions concerning the purchase of particular products. Hence the possibility to pure heuristics and biases by providing additional information may be limited.

22 Then the question how liability law should react to this underestimation of the risk by consumers arises. It is generally held that, if the consumer remains uninformed, non-liability will certainly not lead to an efficient result. The reason is that producers will in that case not have efficient incentives to invest in prevention of damage caused by GMOs. This is more particularly the case if the benefit of the investment (reduction of expected accident costs) is not recognised by the consumer.[25]

(b) A case for strict liability

23 It is generally held that a fault-rule can induce the producer in such a case to take efficient care, but only a strict liability rule will also lead to an optimal output of products. In this respect strict liability is considered to be a good remedy for risk underestimation by consumers.[26]

24 The latter conclusion is also true when harm is caused to third parties, e.g. bystanders, who have no connection at all with the producer. A Coasean solution is of course excluded under these circumstances since the transaction costs will be prohibitive. There is also no "contact" between a producer

---

24 See for an excellent summary of behavioural law and economics *C. Jolls/C.R. Sunstein/R.H. Thaler*, A Behavioural Approach to Law and Economics, Stanford Law Review (Stan. L.Rev.) 50 (1998) 1471–1550 and see also *R.B. Korobkin/T.S. Ulen*, Law and Behavioural Science: Removing the Rationality Assumption from Law and Economics, California Law Review (CLR) 88 (2000) 1051–1144.

25 This is especially stressed by Goldberg in a critique on the work by Oi: *V. Goldberg*, The Economics of Product Safety and Imperfect Information, BJE 1974, 683–688. For a reaction see *W. Oi*, The Economics of Product Safety: a Rejoinder, BJE 1974, 689–695.

26 See *Adams* (fn. 22) 12; *Polinsky* (fn. 5) 99; *Shavell*, JLS 1980, 4; formal proof of this statement is provided by *Shavell*, ibid., 14–17. See also *Silva/Cavaliere* (fn. 22) 295. A simple example is given by *Polinsky* (fn. 5) 98 f.

and third parties through the price mechanism.[27] Again, a fault rule would induce the producer to take efficient care, but consumers would purchase too many products. Therefore, the market price would be too low, since it would not reflect the expected accident costs and, given this demand, an excessive output of products will follow.[28] This inefficiency can again only be remedied by the introduction of a strict liability rule. Since under such a rule the market price will reflect the expected accident costs, efficient care will be taken by the producer and an efficient quantity will be produced.[29]

Up to this point we have shown that the economics theory does not jump **25** to hasty general conclusions with respect to the desirability of a fault or a strict liability rule. Such answer depends on whether the victim is a consumer or a third party and, if a consumer, a further distinction is made depending on whether or not the consumer underestimates the risk of a "product accident". However, if consumers do underestimate the risk or if the harm is caused to third parties, the literature clearly points to the advantages of strict liability from an efficiency viewpoint. These are mostly referred to as the full *internalisation* of the harm through the strict liability rule. However, the economic theory does not stop here. Until now it was assumed that the victim's behaviour had no influence on the accident risk, that both producers and consumers were risk-neutral and that the market was perfectly competitive. Other publications have shown that when these assumptions are relaxed, the advantage of strict liability, expressed earlier, no longer prevails.

It has been stressed that in a bilateral accident situation where the victim **26** also influences the accident risk, liability rules should also give victims an incentive to take appropriate care and not to engage too often in dangerous activities. If the victim has an important influence on the accident risk, most authors hold that a fault rule will be preferred to a strict liability rule, even if the latter is combined with a contributory negligence defence.[30] One of the reasons given is the unwillingness of courts to consider the contributory negligence of victims, which might easily affect their incentives. In addition, only a fault rule will lead to an efficient activity level by the victim. It is therefore held that if the victim has an important influence on the acci-

---

27 See *K. Hamada*, Liability Rules and Income Distribution in Product Liability, American Economic Review (AER) 1976, 230 and *C.G. Veljanovski*, The Economic Theory of Tort Liability-Toward a Corrective Justice Approach, in: P. Burrows/C.G. Veljanovski, The Economic Approach to Law (1981) 131.

28 See *Polinsky* (fn. 5) 103; *Shavell* (fn. 23) 49 f.

29 *Shavell*, JLS 1980, 3.

30 See on the debate whether a full contributory negligence rule is required or merely comparative negligence above at III A 2 10 and see 98–101.

dent risk and it is thus more important to control his activity level than it is to control the injurer's, a fault rule will be preferred.[31]

27 In sum, the economics theory advances strong arguments in favour of a strict liability rule when consumers underestimate the risk of a "product accident" or when the harm is caused to third parties. However, if the victim has a substantial influence on the accident risk, or if the producer has market power and consumers underestimate the accident risk or if one take into account the influence of insurance, strong arguments could be advanced in favour of a fault rule. Again, just as we stressed as far as the general choice between strict liability and negligence is concerned, everything depends on whether one considers the GMO risk as a unilateral or bilateral accident case. This choice may well depend on the type of victim: in the case of commercial users, accidents may be of a bilateral nature since the victim should also be given incentives to reduce the accident risk; if the victim is however a consumer without specialised information, the measures the victim could take to reduce the risk may be very limited. That would make the accident a unilateral type accident where a strict liability may have a preference.

28 It may be interesting to apply these insights from the economic analysis of product liability to Product Liability Directive of 25 July 1985. Other contributors will address the implications of the Product Liability Directive for GMO related risks. In this contribution, focusing on economic analysis, it may be worthwhile to point at some economic effects of the Product Liability Directive, which at the same time also provides a critique of the Directive.

(c) Economic analysis of the Product Liability Directive[32]

29 Apparently the drafters of the EC Directive judged that the best way to realise the "protection of the consumer", of which they spoke so much, was to implement a generalised strict liability rule for harm caused by defective products. The first thing that is striking when one compares this rule

31 *Shavell*, JLS 1980, 7 and 20; *Shavell* (fn. 23) 48–51. This will often be dependent upon the nature of the activity and the potential harm.

32 In the scope of this paper only a few brief remarks concerning the economic effects of the Directive can be made. For a more thorough analysis see *J. Finsinger/J. Simon*, An Economic Assessment of the EC Product Liability Directive and the Product Liability Law of the Federal Republic of Germany, in: M. Faure/R. van den Bergh, Essays in Law and Economics (1989) 185–214 and see for an economic analysis of the Product Liability Directive, *P. Burrows*, Products Liability and the Control of Product Risk in the European Community, Oxford Review of Economic Policy (OREP) 1994, 68–83 and *Silva/Cavaliere* (fn. 22) 292–323.

with the economic theory of product liability discussed above is that the Directive advances one single rule to deal with all product accidents, whereas the economic theory is much more balanced and detailed. Depending upon whether or not the victim is a consumer, whether the latter has information on the product risk, what the influence of insurance is, etc. economic theory would only hold that a strict liability rule is efficient in some case, but certainly not in all.

At the policy level it is, using all the relevant criteria, almost impossible to advise the legislator to use one single rule for all product accidents. There is even a question as to whether the legislator would have to define such a rule. An obvious alternative would be to let the judge decide what type of liability rule should govern the product accident. Thus case law could take into account all the relevant criteria and the application of the strict liability rule could be limited to certain product accidents. The Directive, which introduces one strict liability rule for all product accidents, disregards all the various factual situations which make different liability rules efficient. Thus, by using one single rule inefficiencies will unavoidably be created since the strict liability rule will also be applied where this would be inefficient taking into account the above mentioned criteria. It is also doubtful whether the savings in administrative costs by using one legal rule would outweigh the inefficiencies. The costs of applying the economic criteria of strict liability should not be that high either. Especially if they can easily be recognised (e.g. whether the victim was a consumer or a third party) an individualised product liability system, where strict liability will only be applied in some cases, could be used by the judge at relatively low costs. **30**

One of the criteria was also applied for a long time in legal practice. Indeed in most legal systems, product accidents whereby the victim stood in a contractual relationship with the producer were treated differently from those where the victim was a complete stranger to the producer. One of the reasons for introducing generalised strict liability in the Directive was that this different treatment was considered "unjust". However, this distinction is quite sound from an economic point of view. As was mentioned above, if the victim stands in a contractual relationship with the producer, a Coasean solution is in principle possible if the consumer is well informed of the accident risk. Such a solution is of course always excluded in the event that the victim is a third party because of the prohibitive transaction costs. So there is a good economic reason for treating both situations differently. **31**

32 Of course the EC Directive does not introduce a general strict liability rule for all damage caused by a product. The producer will only be liable when the damage was caused by a *defect* in the product. It looks like this is more a fault liability since a product is often defective because the producer did not take efficient care. However, the liability is indeed strict since the producer will always incur liability when harm was caused by a defect in his product even if he can prove that he took due care. The producer only escapes liability when the harm was not caused by a defect of the product. Given the broad notion of defect in the Directive, a product is almost already considered defective by the mere fact of having caused the harm.[33] Therefore, there is not, from an economic point of view, a substantial difference between strict liability for all damage caused by a product and strict liability for damage caused by a defect in the product as in the Directive.[34]

33 It was mentioned above that the general strict liability rule for all activities which is used in the Directive might create inefficiencies since it will also be applied in cases where a fault rule would be preferable from an efficiency viewpoint. It should not be forgotten, however, that the purpose of the Directive was the harmonisation of product liability law in Europe. If this should succeed, it could create benefits which could easily outweigh the disadvantages of the generalised strict liability rule discussed above. Indeed, if a single product liability rule were to be created in all Member States, this could contribute to the creation of equal marketing conditions and thus to the realisation of the internal market. This could bring substantial savings[35] which could easily outweigh the inefficiency of the use of a single rule. This, however, will not be the case in practice since the Directive can never bring a harmonisation of product liability law in Europe. So the inefficiencies remain without any compensating benefit for the realisation of the internal market.

34 One could go one step further and argue that the Directive not only creates several inefficiencies, but that it is also problematic from a distributive point of view. Many lawyers favour the introduction of strict liability since it would protect the consumer. It is even argued that strict liability would

---

33 One could argue that as soon as a product causes serious injuries it does not provide the safety which a person is entitled to expect and is therefore defective according to Article 6 of the Directive.

34 For an economic analysis of liability for defective products see *Shavell* (fn. 23) 58–60. For an analysis of the notion of defect in the Directive see *Finsinger/Simon* (fn. 32).

35 Compare R. *van den Bergh*, Subsidiarity as an Economic Demarcation Principle and the Emergence of European Private Law, Maastricht Journal of European and Comparative Law (MJECL) 1998, 129–152.

be necessary "to restore the broken balance" between producers and consumers. The Coase theorem teaches that also from a distributive point of view there will be no difference between a fault rule and a strict liability rule. Fully informed consumers will only take into account the full price of the product (cost price + expected accident costs) and will base their purchase decision upon this full price. Even if the legislator would like to protect the consumer by introducing a strict liability rule, the producer will still add the expected accident costs to the cost price. The consumer will again pay the full price since the expected accident costs are passed on to him, which is reflected in a higher market price. Since producers and consumers are bound through the price mechanism, every shift of liability to the producer will be passed on to the consumer. In this setting a legal intervention to redistribute wealth to the consumer by introducing a strict liability rule seems therefore useless.[36]

When consumer groups are heterogeneous, the introduction of strict liability will even have adverse distributional effects. Indeed, the greatest part of damages is lost income. The expected damages are therefore of course higher for high-income consumers than for low-income consumers. The producer will, however, take into account an average expected damage and will add this to his cost price to get one single market price for all consumers. The effect is that when low-income consumers still buy the product, they "pay" for the expected damages of the high-income consumers. Therefore strict liability in a product liability setting creates a redistribution from poor to rich consumers.[37] For the same reason Priest argued that the product liability explosion in the US hurts low-income groups in particular.[38] **35**

This effect will be stronger still when the strict liability rule not only applies to harm suffered by consumers, but is extended to damage caused to third parties, as the EC Directive does. In that case the market price will again be higher since the producer will also have to pay for damage caused to third parties. The increased expected damages will again be passed on to the consumer who pays a higher market price for the damage a product he bought can cause to third parties. The consumer has no possibility to pass on this increased price and therefore he, in effect, pays for the protection of third parties. **36**

So the expansion of strict liability to damage caused to third parties redistributes wealth from consumers (of the product) to third parties. Consu- **37**

36 See *Hamada*, AER 1976, 228–234.
37 See *Adams* (fn. 22) 6 f.
38 *G. Priest*, The Current Insurance Crisis and Modern Tort Law, YLJ 1987, 1521–1590.

mers will indeed have to pay for the protection of third parties. This again increases the adverse redistributive effect of poor consumers paying higher prices for the expected damages of third parties. Since lost income is an important part of compensation awards, again the high-income groups do indeed benefit from the redistribution.

38 This shows that the generalised strict liability which has been introduced by the EC Directive will not only create inefficiencies, but is also based on a rather narrow perception concerning the protection of consumers.

4. Influence of regulation

39 So far we presented liability rules from an economics perspective as instruments to provide incentives to prevent damage caused by GMOs. In reality of course, GMOs are subject to a great deal of safety regulation. The goal of this safety regulation is precisely the prevention of damage. Thus a much more important role will in practice probably be played by safety regulation than by liability rules, at least as far as prevention is concerned. This, by the way, also corresponds with the economic criteria for safety regulation as they have inter alia been developed by Shavell.[39] Indeed, information on the optimal ways to prevent damage caused by the use of GMOs is probably more readily available to a regulator than to the potentially liable operator. Thus the informational advantage is a *first* important criterion in favour of safety regulation. *Second*, there may be a serious insolvency risk. That plays from the moment that the damage that could result from the use of the GMO is higher than the wealth of the particular operator. The damage can virtually be catastrophically high, especially if one imagines damage along the food chain with far-reaching consequences to many consumers or at least leading to large economic losses. There is always the likelihood that operators are organised as legal entities. Legal entities enjoy limited liability and thus there is always the danger that they will externalise harm to third parties.[40] *Third*, there may be a risk that there is a long time lapse between the moment that the source of GMO damage takes place and the moment that the damage occurs. In

39 See *Shavell*, JLS 1984, 357–374; *S. Shavell*, A Model of the Optimal use of Liability and Safety Regulation, Rand Journal of Economics (Rand J Econ) 1984, 271–280. For a summary of this literature see supra no. 21.

40 See *H. Hansman/R.H. Kraakman*, Toward Unlimited Shareholder Liability for Corporate Torts, YLJ 1999, 1879–1939. Because of this danger of using the corporate structure for externalising harm to involuntary creditors, Hansman and Kraakman pleaded in favour of unlimited shareholder liability for corporate torts.

addition, it may be difficult for the victim to prove a causal relationship between his damage and the acts of a particular operator. These latency and causation problems may lead to situations whereby tort law is not used, even though the conditions for liability are fulfilled. When thus the threat of a liability suit will not provide a sufficient deterrent effect, this provides another argument in favour of regulation.

Although Shavell's criteria thus provide a strong argument to control GMO **40**
risks *ex ante* through regulation, in individual cases there can still be damage. Then again, liability under tort comes into the picture and the question of course arises how regulation influences the liability system and *vice versa*.

The first question that arises is whether a violation of a regulatory stan- **41**
dard concerning GMOs should automatically be considered a fault under tort law and thus lead to liability. Most legal systems consider a breach of a regulatory duty evidence of negligence per se.[41] One of the reasons for introducing safety regulation to control GMO risks is, as was mentioned above, that the regulator will usually possess better information to evaluate the efficient standard of care than the parties involved. Hence, regulation passes on information to the parties on the efficient standard of care, but equally to the judge. The judge may lack the necessary information to find out what the particular care is that could be required from the person handling the GMO. Therefore, the statutory standard can guide the judge in a liability case.

A more difficult question may arise as to whether compliance with a regu- **42**
latory standard could release an injurer from liability. Several authors argue that since a GM crop has undergone a pre-market test, in which a risk assessment has been carried out prior to the commercialisation of the crop, as long as the injurer has followed the requirements for planting the GM crop, such as establishing a buffer zone, he should not be liable for the damage of cross-pollination. In this case, the injurer may argue that the damage is in fact unforeseeable.[42] However, as Smyth and others have sug-

---

41 See for instance *K.S. Abraham*, The Relation between Civil Liability and Environmental Regulation: An Analytical Overview, Washburn Law Journal (Washburn LJ) 41 (2002) 379–398.

42 *C.P. Rodgers*, Liability for the Release of GMOs into the Environment: Exploring the Boundaries of Nuisance, Cambridge Law Journal (Camb LJ) 62(2) (2002) 390 and 400. However, the author argues that the authorisation itself does not constitute a defence. A firmer rejection of the idea of holding an injurer liable for unforeseeable damage is given by Bergkamp. The author argues that the deliberate release of GMOs undertaken in compliance with regulations and conditions prior to authorisation is quite unlikely to create foreseeable and significant risks. Accordingly, only "non-compliant GMOs or

gested, although regulatory standards have been followed, many species could still possibly wander to their wild relatives, which potentially would create environmental problems.[43] This means that a regulatory standard does not necessarily remove the risks of damage, since some species may wander beyond the buffer zone and pollinate with other plants and potentially cause damage as well. The question will be whether the impact of such pollination is significant. In this case, the injurer may still be found liable although he has complied with the regulatory standards.[44]

43 Consequently, although the non-compliance with a regulatory standard is a sufficient reason for holding an injurer liable, the reverse is not true: following regulation should, from an economic perspective not necessarily free the GMO operator from liability. The reason to reject this so-called regulatory compliance defence from an economic perspective is that the regulatory standard is in some cases merely a minimum.[45] The efficient

---

activities involving GMOs conducted in an irresponsible way" that could pose the risks should hence be subjected to liability. *L. Bergkamp*, Allocating Unknown Risk: Liability for Environmental Damages Caused by Deliberately Released Genetically Modified Organisms (2000), available at SSRN: http://ssrn.com/abstract=223068 25. The author goes on to refuse the proposal of singling out biotechnology in a specific liability system. He argues that concerns about the environmental impact of GMOs have been triggered by the fear of the unknown, unforeseeable, and even non-existent risks of GMOs, which cannot be adequately dealt with by a specific liability system. Ibid., 28 f.

43 *S. Smyth/G.C. Khachatourians/P.W.B. Phillips*, Liabilities and Economics of Transgenic Crops, Nature Biotechnology (Nat Biotechnol.) 20 (2002) 537 f. As quoted by Endres, a study conducted in the UK found that pollen from GM crops had been carried by bees 4.5 kilometres away from the test field. *Endres*, ILR 22 (2000) 456.

44 Khoury and Smyth argue that although a risk assessment prior to an authorisation of GM crops revealed the remoteness of risks, these risks could still be considered foreseeable based on public concerns. This is because, as the authors argue, the absence of knowledge does not mean the absence of public concerns about possible risks. As a result, the injurer will still be held liable if these risks materialise in the future. To support this argument, the author resorts to the precautionary principle, by which the injurer is liable when the uncertain risks of serious magnitude materialise in the future. *L. Khoury/S. Smyth*, Reasonable Foreseeability and Liability in Relation to Genetically Modified Organisms (2005), The 9th ICABR International Conference on Agricultural Biotechnology: Ten Years Later, Ravello, Italy, 20 f.

45 Some countries may not even have a minimum, but also sub-optimal regulatory standards for GMOs. See for example the critics of Bratspies concerning the US regulation on the commercialisation of Bt crops. The author argues that the agencies responsible for the release of Bt crops have abandoned the precautionary principle, and instead used the most optimistic estimates as the basis of their decision. In addition, there is no clear mechanism to ensure the growers' compliance with the requirement set by the seed companies, as it could be assumed that it is not in the companies' interest to enforce their requirement. See: *R. Bratspies*, The Illusion of Care: Regulation, Uncertainty, and Genetically Modified Food Crops, New York University Environmental Law Journal (New York Univ Environ Law J) 10 (2002) 346. Assuming that this allegation is true, releasing an injurer just because he has followed such non-optimal regulatory standards, may create too much damage in society, in which case the price of GM products does not represent the true social costs.

standard can be higher and thus liability should supplement regulation in this case to provide the GMO operator incentives to take efficient care to prevent the damage.[46] Exposure to liability provides the GMO operator with incentives to take all efficient precautions, even if this requires more than merely following the regulation. Holding operators liable beyond the mere compliance with regulatory standards is moreover an important remedy for the unavoidable capturing of administrative agencies which may lead to inefficiently low regulatory standards. Exposing the GMO operator to liability even though the operator followed regulation or the conditions of a license is thus, from an economic perspective, an important tool to guarantee that the operator will take efficient care.

In addition, it should also be noted here that the discussion of the regulatory standard should also include how the release of GMOs will be overseen. **44** In this case, it is not unusual that the seed company is responsible for the release of the GMOs.[47] Clapp argues that the tendency to apply lenient compliance monitoring, in which the compliance with regulatory standard depends largely on the self-reporting and self-monitoring conducted by the seed companies, have contributed to some cases of illegal release of GMOs in the US.[48] This point confirms that following a lenient regulatory standard should not be a defence, since such a standard will certainly fail to induce potential injurers to take the optimal preventive measures.

---

46 Compare *P. Burrows*, Combining Regulation and Liability for the Control of External Costs, International Review of Law and Economics (IRLE) 19 (1999) 227–242.

47 The commercial release of GMOs by individual farmers is usually based on contractual agreements, by which the seed company specifies certain procedures in planting its GM seed. For example, the company usually requires the GM growers to meet the permissible use of the GM seed and prohibits seed saving or seed transfer to third parties. Based on these agreements, the company will then monitor the farmers' compliance with the agreements. See *J. De Beer*, Biotrespass, Bulletin of Science, Technology and Society (Bull. Sci. Tech. & Soc.) 2007, 293.

48 Clapp finds out that in the three cases of illegal release of GMOs, i.e. the StarLink case (GM corn for animal feed is found in food supply), the Bt10 maize case (unapproved GM maize was found in the seed and food supplies), and the Liberty Link RICE 601 case (unapproved GM rice was found in rice supplies), the seed companies failed to detect the illegal release and inform the authority about the release, since in all three cases the illegal release was initially detected by other parties. In addition, Clapp also finds that rather than improving their monitoring systems, the companies either blamed the farmers for the illegal release, or asked the regulatory agency to issue a retroactive approval for the previously unapproved seeds. Accordingly, Clapp concludes that heavy reliance on voluntary corporate measures, such as self-monitoring and self-reporting, has failed to induce compliance with the regulatory standards. *J. Clapp*, Illegal GMO Releases and Corporate Responsibility: Questioning the Effectiveness of Voluntary Measures, Ecological Economics (Eco. Econ.), 66 (2008) 352–355.

III. Causation

1. General

45 Problems can of course arise as regards the requirement that a causal link should be established between the alleged damage and the presence of the particular GMO concerned. The economic analysis of law has paid a lot of attention to the requirements that should in general be attached to causation. It would lead us too far to discuss these in any detail at this moment.[49] Shavell explains in simple words that there is a good economic reason to limit the liability of an injurer to cases he has really caused. If the requirement of a causal link did not have this limiting effect on liability, the result would be that many potentially beneficial activities in society would no longer take place since in effect an operator would then also be held liable for damage which did not result from his acts. A liability for damage which is not the result of the individual activity of the operator would thus be considered as crushing, so Shavell holds.[50] Thus it makes sense to limit the liability of the operator who handled GMOs to the damage actually caused by the GM crop.[51]

2. Burden of proof

46 A first question that arises in this respect is on whom the burden of proof should lie in case there is uncertainty about causation. Uncertainty can arise for instance when there may be many different sources and it is not clear what precisely caused the damage to a non-GM crop. Also, there may be multiple causes. To all of these issues of causal uncertainty, there is both a procedural aspect (who should bear the burden of proof?) and an aspect of contents (how should the law deal with uncertainty over causation?).

47 Traditionally, the plaintiff, i.e. the victim, should bear the burden of proof regarding the elements of the liability rule upon which he bases his claim for damages.[52] He should, for example, prove that he has suffered damage

---

49 See G. *Calabresi*, Concerning Cause and the Law of Torts, University of Chicago Law Review (U.Chi.L.Rev.) 1975, 69–108; *Shavell*, JLS 1980, 463–516 and *Landes/Posner*, JLS 1983, 109–134.

50 See *Shavell*, JLE 1985, 587–609 and *Shavell* (fn. 23) 108.

51 For a recent overview of the economic literature related to causation see O. *Ben-Shahar*, Causation and Foreseeability, in: M. Faure, Tort Law and Economics (2009) 83–108.

52 In the absence of a specific liability system for GMOs, the victim should resort to one of several liability rules, namely trespass, private nuisance, negligence, strict liability, or product liability. Each of these rules has its own elements that should be proven by the

and that this damage was the result of exposure to a GM crop. Almost inevitably, the victim needs to rely on expert opinions to support his claim.[53]

48 However, one may argue that the court may place the burden of proof on those who can acquire information more cheaply. In this regard, shifting the burden of proof to the GM operators could be justified for several reasons. First, the shifting of the burden of proof may release the victims from the difficult task of proving their claims. Second, it could also be argued that GM operators have the control over GMOs, and thus they are in a better position to control any resulting damage from GMOs and to acquire information regarding the impact of GMOs. Third, as Clapp has argued, in cases when uncertainty is prevalent, such as in the development and release of GMOs, there is asymmetric information in favour of GMO operators.[54] In this case, the reversal of the burden of proof may function as a counter against this asymmetric information.[55]

49 Based on such an efficiency argument, the more important question will no longer concern the burden of proof, but rather the standard of proof. The issue of standard of proof is particularly important if we are faced with uncertainty concerning causality.

3. Causal uncertainty

50 There is a real likelihood that, as we mentioned above, many issues of causal uncertainty could arise in case non-GM crops cause damage to third

---

victim. For a brief summary about the elements of nuisance, negligence, and strict liability, readers may consult: *T.N. Vollendorf*, Genetically Modified Organisms: Someone is in the Kitchen with DNA, Who is Responsible when Someone Gets Burned? Mississippi College Law Review (Miss.C.L.Rev.) 21 (2001) 48–53.

53 The burden of proof borne by the victim might be reduced if GM products are required to be labelled with their genetic markers, as has been proposed in Europe. See *Endres*, ILR 22 (2000) 487.

54 *Clapp*, Eco. Econ. 66 (2008) 355 f.

55 Gollier argues that self-interested firms may exploit the favourable situation resulting from information asymmetry for their own benefit by, for example, introducing their inventions as soon as possible in order to pre-empt the market of these products. *C. Gollier*, Should We Beware of the Precautionary Principle? Economic Policy (Ec. Pol.) 16 (33) 2001, 313 f. See also: *C. Gollier/N. Treich*, Decision-Making under Scientific Uncertainty: The Economics of the Precautionary Principle, The Journal of Risk and Uncertainty (JRU), 27 (1) 2003, 98 (arguing that under the situation of uncertainty, competitive firms may introduce their products before fully assessing the possible impact of the products). With respect to GMOs, this situation might be indicated by poor pre-market testing of GMOs or inefficient voluntary control of the release as illustrated in three cases studied by Clapp.

parties. Uncertainty could, for example, arise concerning the question whether GM crops have indeed caused the damage, but uncertainty could equally arise concerning the question which of potentially many GM crops caused the loss of the third party. There can indeed be different sources of presence of GMOs, whereby the question arises how the law should deal with uncertainty when it cannot be established with certainty who the precise cause of the problem was. This is especially true if liability is channelled to the farmers, whereby the plaintiff has to prove which of the neighbouring GM farmers caused the damage. The situation will be less difficult for the plaintiff if liability is channelled to the seed company, since the damage in question is relatively less difficult to identify and link to the producing company.[56]

51 Potentially, the law could provide a variety of solutions to the causal certainty problem. One could, on the one hand, judge that as soon as there is any statistical chance that a certain activity (or product) may cause a certain damage, all victims receive 100% compensation of their damage. The second possibility is to refuse the claim of the victim unless there is 100% certainty that the tort caused the damage. The third possibility is to award compensation only when the probability that the damage was caused by the tort passes a certain threshold of, say, 50%. This threshold rule is a kind of "all or nothing" approach: if the probability is lower than the threshold, the victim receives no compensation at all; if the probability is higher than the threshold, the victim receives full compensation. This threshold rule is known in the American literature as the "more probable than not" solution, referring to the fact that the plaintiff must convince the judge that it is "more probable than not" that its damage was caused by the tort. The final solution is to take into account the probability that the tort caused a certain damage and to award compensation, taking into account this probability. This would mean that if the scientific expertise indicates that the likelihood of damage is, say, 40%, the victim can then receive compensation for 40% of his damage.

---

56 To illustrate this situation, one could refer the StarLink case, in which StarLink corn developed for animal feed was found in the food supply. In this case, it was not difficult to determine which was the responsible company, since this patented GM corn was produced by Aventis. However, it was not possible to determine which particular farmers caused the co-mingling between the GM feed corn with non-GM food corn. For detailed information of this case, see *Bratspies*, New York Univ Environ Law J 2002, 352 f. See also: *Smyth/Khachatourians/Phillips*, Nat Biotechnol. 20 (2002) 537 f.

The way the law should deal with causal uncertainty has been addressed extensively in the economic literature, for instance by Rosenberg,[57] Kaye[58] and Shavell.[59] 52

Let us address more closely the various options addressed above. The *first* option would be to award a victim total compensation of damage, even if the probability that his loss was actually caused by the injurer's activity was relatively low, say 30%. In such a case, this means that we also know that there is a 70% probability that the damage (e.g. a certain illness) was caused by another event. If an injurer is held liable for the full amount even if there was only a 30% probability that his activity caused a loss, this will lead to too few incentives to invest in socially desirable activity, such as e.g. the development of genetically modified organisms. 53

This shows that the first solution, simply arguing that in case of causal uncertainty the victims can claim full compensation, is inefficient and unjust. The same is obviously true for the *second* solution in which it would be required that the victim proves with 100% certainty that his damage was caused by the tort. That requirement would mean that in many cases injurers would escape the clutches of the law whereas their activities have effectively created an additional risk. That solution would therefore amount to under-deterrence. 54

This therefore leads us to the two other solutions, often seen in tort law, being either the requirement that a certain threshold should be passed or proportionate liability. 55

The threshold liability leads to a situation whereby the victim's claim is totally accepted if the probability passes the threshold of, say, 50%. If the probability passes the threshold, compensation is in full, but if the probability is lower than the threshold, the victim receives no compensation at all. The disadvantages of this hard and fast solution are obvious. One problem, both from a victim compensation as well as from a deterrence perspective, is that the probability of causation could systematically be lower than the threshold. Assume that the probability were systematically 40% that a certain cancer would have been caused as a result of a certain activity. If the threshold were 50%, this would mean that the enterprise 56

57 *D. Rosenberg*, The Causal Connection in Mass Exposure Cases: 'Public Law' Vision of the Tort System, Harvard Law Review (HLR) 1984, 851–929.

58 *D. Kaye*, The Limits of the Preponderance of the Evidence Standard: Justifiably Naked Statistical Evidence and Multiple Causation, American Bar Foundation Research Journal (Am Bar Found Res J) 1982, 487–516 and see. *S. Gold*, Causation in Toxic Torts: Burdens of Proof, Standards of Persuasion and Statistical Evidence, YLJ 1986, 376–402.

59 *Shavell*, JLE 1985, 587–609.

exposing persons to this 40% risk would systematically escape the clutches of the law. Victims would not be compensated and the incentives for accident reduction would be too low.[60] This seems inefficient and probably also unjust, since the enterprise has after all, in a number of cases, at least statistically, created certain losses. Assume that 100 cancer victims all file a lawsuit; in this particular example one would assume that 40 out of these 100 cancer cases would have been caused by the emissions emanating from the particular enterprise. However, for every individual, the probability of causation would always be below the 50% threshold, so that the enterprise would not be held to compensate the victims in any of these cases. That is a clear disadvantage of this "all or nothing" approach which is inherent in the threshold liability.

57 A more fine-tuned alternative can be found by translating the probability of causation by awarding the victim a proportionate amount of his damage. In practice, this would mean that if the probability that the victim's damage was caused by the injurer's activity was 40%, the victim would be compensated for 40% of his damage. From an economic perspective, the advantage of this proportionate liability is that it exposes the injurer precisely to the excess risk (in this case the additional number of cancer cases) that was caused by the (assumed wrongful) activity of the injurer. The enterprise will then, returning to the previous example, have to compensate 40% of all the damage of every particular victim, which amounts at the aggregate level to the same as compensating 40 out of 100 victims whose illness would have been caused by the enterprise.[61]

58 The result of this proportionate liability is that the injurer will receive optimal incentives for prevention, since he is exposed to liability for the risk which was caused by his activity.[62] A proportionate liability rule therefore provides optimal incentives for accident reduction, so it is generally held in the economic literature.[63]

60 Ibid.

61 So *Shavell*, JLE 1985, 587–609.

62 So *L. Bergkamp*, Liability and Environment (2001) 290 f.

63 See on this proportionate liability *J. Makdisi*, Proportional Liability: A Comprehensive Rule to Apportion Tort Damages Based on Probability, North Carolina Law Review (N.C.L.Rev.) 1989, 1063; *Landes/Posner*, JLS 1984, 417–34 and *G. Robinson*, Probabilistic Causation and Compensation for Tortuous Risk, JLS 1985, 797 f. For a discussion of the possible legal foundation of a proportionate liability rule see *A.J. Akkermans*, Grondslagen voor proportionele aansprakelijkheid bij onzeker causaal verband, in: W.H. van Boom/C.E.C. Jansen/J.G.A. Linssen (eds.), Tussen 'Alles' en 'Niets'. Van toedeling naar verdeling van nadeel (1997) 105–115.

This proportionate liability rule has been defended by several American scholars and is also defended in the economic analysis of law. The negative consequences of causal uncertainty could then be limited. A proportionate liability rule is less rigorous than the all or nothing approach of the reversal of the burden of proof.[64] The proportionate liability rule would indeed mean that all victims can claim a proportion of their damage equal to the amount by which the injurers contributed to the loss. Thus the exposure to liability of the enterprise corresponds precisely with the amount to which the injurers contributed to the risk.[65] This proportionate liability rule could, more particularly in cases of product liability, take the form of market share liability.[66] **59**

IV. Multiple tortfeasors

A related problem, also concerning causal uncertainty, is how one should handle a situation when multiple actors are involved. This can again have different sources. It could either be the case that there are potentially many GM crops that could have affected the non-GM crop (if the damage is the result of cross pollination). The other possibility is that there are various liable actors in the vertical production chain. But similar questions also arise when damage is caused to third parties. Again, as with the general issue of causal uncertainty, the law has basically a variety of options to solve this issue, the most realistic (and thus applied in the legal system) are, on the one hand, joint and several liability and, on the other hand, proportional liability. A so-called market share liability whereby the liability is apportioned according to the market share of the operator is an example of such a proportionate solution to multi-actor causation.[67] **60**

At first sight, a joint and several liability rule appears as a regime whereby the legal system deviates from the principle that a tortfeasor should only be held liable for the damage which was caused by its own behaviour. Under joint and several liability, the tortfeasor is held liable in full also for damage which was not caused by its own behaviour. **61**

64 See *G. Brüggemeier*, Liability for Water Pollution under German Law: Fault or Strict Liability, in: J. van Dunné (ed.), Transboundary Pollution and Liability: The Case of the River Rhine (1991) 88–91.

65 *Robinson*, JLS 1985, 798.

66 See also *P. Widmer*, Causation under Swiss Law, in: J. Spier, Unification of Tort Law: Causation (2000) 112 f.

67 For a recent overview of the economic literature see *L.A. Kornhauser/R.L. Revesz*, Joint and Several Liability, in: M. Faure, Tort Law and Economics (2009) 109–133.

62 One could therefore at first blush argue that joint and several liability seems inefficient since it leads to over-deterrence: the injurer's liability is not limited to the risk created by its own activity. However, such a simple conclusion is (as usual) indeed too simple. One may argue that a distinction should be made between the situation of full solvency of all the contributing tortfeasors on the one hand and the situation in which either one or more of them are insolvent. In case of full solvency of all the actors, one can argue that there is no efficiency loss caused by joint and several liability.[68] In that case, the injurer who has to compensate the victim can in turn exercise a redress against the other parties who contributed to the loss in proportion to their contribution. Assuming that the other tortfeasors are fully solvent, the one who first paid only prefinances the compensation of the victim and will be able to recover a part of the damage paid. Thus, in the end, also under joint and several liability, the extent to which every contributor has to pay should be proportionate to their contribution to the risk. In that sense, a joint and several liability rule, combined with a right of recourse and solvent actors amounts to a proportionate solution. The exposure to liability of every tortfeasor in this model is limited to its own contribution to the loss and thus optimal incentives would follow.

63 Of course one could wonder what the additional benefit is of a joint and several liability rule compared to the situation whereby the victim would have to sue every individual tortfeasor separately. One could make a victim protection argument, simply on the basis of the fact that for the victim it is often more difficult to prove a causal link with the action of one particular actor. Thus it certainly makes the life of the victim easier if the victim can claim full compensation from one injurer who then has to exercise a right of redress against the other parties who contributed to the loss. However, in addition to this distributional argument, there are undoubtedly efficiency arguments in this particular case as well. They are probably not linked to a benefit in administrative costs. Indeed, whether either the victim has to sue e.g. five different tortfeasors or the victim just sues one tortfeasor and the latter exercises a right of redress probably does not make much difference as far as administrative costs are concerned. However, one could make the argument that the joint and several liability rule may give *ex ante* excellent incentives for mutual monitoring between

68 For a detailed analysis of joint and several liability when all defendants are fully solvent see *L.A. Kornhauser/R.L. Revesz*, Sharing Damages Among Multiple Tortfeasors, YLJ 1989, 831–884 and for the analysis in case of limited solvency see *Kornhauser/Revesz*, Apportioning Damages Among Potentially Insolvent Actors, JLS 1990, 617–651.

potential joint tortfeasors.[69] Indeed, victims may well encounter difficulties in proving a causal link between the action of every particular tortfeasor and the loss he suffered. That may result in too few claims and hence in under-deterrence. Shifting the risk to the injurers would mean that they *ex ante* have an excellent incentive to mutually monitor their activities. Joint and several liability in fact shifts the risks of uncertainty concerning the proof of the causal link to the injurers. The victim has to sue just one of the many potentially liable injurers and claim full compensation. If the one injurer who is sued does not succeed in proving that others contributed to the loss, the damage will ultimately fall on him.

However, these arguments may no longer be valid under insolvency.[70] **64**
Indeed, the picture changes if the tortfeasors are no longer solvent. In that case, the risk of insolvency is shifted to the injurer who will be sued by the victim. If in that particular case one would assume that e.g. only the solvent injurer is sued by the victim and he has no right of recourse (given the insolvency of the others), the effect would be that one (solvent) injurer would be held to compensate for losses which he did not cause.[71] In case of insolvency, a joint and several liability rule may thus violate the principle that the injurer should only be held liable to compensate in the proportion to which he contributed to the loss.

V. Defences

1. Force majeure

A traditional defence accepted in almost every liability regime is force **65**
majeure (although it may have different interpretations). From an economic point of view, one can easily argue that, in case of force majeure, there should be no liability. Force majeure is generally a condition, not only for fault or strict liability, but for every liability in tort. It is related to the blame-

---

69 An argument in that direction is made by *T.H. Tietenberg*, Innovation in Environmental Policy, Economic and Legal Aspects of Recent Developments in Environmental Enforcement of Liability (1992) 139.

70 For an excellent analysis of the effects of various systems of extended liability, see Boyd and Ingberman who argue that under certain conditions extended liability may promote cost internalisation, but that there are serious drawbacks as well. Hence, they argue that other solutions should be examined to cure the problem of undercapitalisation, J. *Boyd/D. Ingbergman*, The Vertical Extension of Environmental Liability through Claims of Ownership, Contact and Supply, in: A. Heyes (ed.), The Law and Economics of the Environment (2001) 44–70.

71 Then joint and several liability would lead to overdeterrence, so Bergkamp rightly argues (*Bergkamp* (fn. 62) 301).

worthiness requirement, which requires that the injurer should have capacity for tortious liability. A tort will indeed, according to most legal systems, only make an injurer liable if the wrongful act can be imputed to him.

66 This condition of blameworthiness relates to the free will and the capacity of discretion of the tortfeasor.[72] This blameworthiness requirement also has a clear economic rationale. When the injurer did not act out of free will, liability cannot influence his incentives to take care and has, therefore, no economic meaning. A finding of liability which does not influence the incentives of the tortfeasor will only create administrative costs (caused by the transfer of the loss) without any compensating benefits in providing additional incentives to take care.

67 We refer here to the blameworthiness requirement simply as meaning that the injurer contributed in some way to the loss. The requirement of "blame" traditionally fits into a fault or negligence concept. In fact, in the context of strict liability, mere causation suffices. But if the injurer did not "cause" the accident, he should not be held (strictly) liable. Force majeure therefore should remain a defence, even under strict liability, since a finding of liability makes no sense if the injurer could not have influenced the risk.

2. Development risk

68 An important question, also with respect to GMO damage, is of course whether the operator handling GMOs should be allowed to call on the development risk defence. This would mean that the operator would not be liable if the damage could, according to the state of the technology at the time when the act took place, not be foreseen.[73] One would thus assume that an operator is handling GMOs and that certain negative consequences of GMOs for third parties could at that particular moment not have been foreseen by the operator. How should, from an economic perspective, the law deal with the fact that there may be situations where either the risks change or technology changes and, as a result, also the standard of care increases?

---

72 *H. Vandenberghe/M. Van Quickenborne/P. Hamelink*, Overzicht van rechtspraak. Aansprakelijkheid uit onrechtmatige daad (1964–1978), Tijdschrift voor Privaatrecht (TPR) 1980, 1170 f.

73 This is also the formulation chosen in the European Directive on Product Liability of 25 July 1985: liability is excluded if the producer can prove that, having regard to the circumstances, it is probable that the defect did not exist at the time when the product was put into circulation.

One could argue that holding a person liable for an unforeseen damage will not give an injurer an incentive to take more care, because unforeseeability means that the injurer's subjective probability of the occurrence of the damage is zero. In this case, although the injurer has to pay infinite damages, his expected damage remains zero because the subjective probability of the damage is zero and hence his optimal care is also zero. In this regard, holding an injurer liable for the unforeseen damage could actually reduce social welfare. **69**

However, one may argue that exposing an injurer to liability, regardless of the unforeseeability of the damage, is efficient as it will induce the injurer to acquire as much information as possible in order to prevent the damage. In addition, with regard to the GMOs case, one could also argue that although the exact magnitude of damage might be uncertain, the risk of damage caused by GMOs itself is a real threat.[74] In this regard, Repp argues that the fact that the planting of GM crops is usually undertaken with a contractual obligation to establish a buffer zone implicitly shows recognition of the possibility of cross-pollination and other possible impacts of GM crops. Hence, the unforeseeability of GMO damage is relatively hard to argue.[75] **70**

Consequently, it would be too easy simply to state that the tortfeasor will only be held to comply with the "old" standard of care and will never be liable for risks which he could not have foreseen. Indeed, it has equally been stated in the literature that the knowledge that there may be liability **71**

---

74 Ellstrand provides several conclusions with regard to gene flow from GM to non-GM crops. First, mating between crops and their wild relatives is common. In this case, mating between GM crops and non-GM relatives is also possible. Second, gene flow does not necessarily lead to serious impact. Third, natural hybridizations may lead to the problems of increased persistence and invasiveness of some unwanted plants. Fourth, natural hybridizations may also lead to the risk of extinction of wild relatives. Fifth, gene flow varies both between species and within species. Sixth, gene flow may occur at high rates and over great distances. See for more detailed information in: *N.C. Ellstrand*, Current Knowledge of Gene Flow in Plants: Implications for Transgene Flow, Philosophical Transactions: Biological Sciences (Phil Trans Biol Sci) 358 (1434) (2003) 1166–1168.

75 *R.A. Repp*, Biotech Pollution: Assessing Liability for Genetically Modified Crop Production and Genetic Drift, Idaho Law Review (Idaho L.Rev.) 36 (2000) 615. A similar opinion has been advanced by Endres, who argues that the possibility of cross-pollination is supported by some studies showing that transgenic pollen may be carried by vectors to a great distance even beyond the buffer zone. See *Endres*, ILR 22 (2000) 487. Lewis also shares this opinion by arguing that when released into the environment, GM crops may cross-pollinate with other plants due to wind or animal pollinators; therefore, so the author argues, the risk of cross-pollination "is almost guaranteed". *S.K. Lewis*, Attack of the Killer Tomatoes? Corporate Liability for the International Propagation of Genetically Altered Agricultural Products, Transnational Lawyer (Transnat Lawyer) 10 (1997) 186.

ex post will obviously give industrial operators incentives to obtain information about possible risks.[76]

72 The fact that there may be *ex post* liability even if technology changes is one of the powerful arguments made in law and economics in favour of liability for the so-called development risk. This should give an operator appropriate incentives to invest in research to acquire more information about risks and optimal technologies to prevent risks.

73 The question, however, arises whether this reasoning can also be used to justify a retrospective change of a liability rule or changes in the standard of care itself. The argument is hence a totally different one if not only the nature of the risk changes but the liability rule itself. The economics of tort law assumes that future incentives for prevention will be affected, given the legal regime in force. Hence, it is hard to defend that an *ex post* change in the liability rule will positively affect the incentives for proper behaviour which was not considered wrongful at all at the time when the act was committed by the industrial operator. One can expect an operator to assume that new risks may emerge, but hardly that the contents of the law will change. Requiring this would lead to an inefficiently high demand for preventive measures and thus to over-deterrence. Hence, retrospective liability indeed seems problematic, taking into account the deterrent function of tort law.

74 From this it follows that there is apparently a dilemma: on the one hand, it is obviously useful that the standard setting process in civil law is seen as a process of learning whereby the standard of care is not static, but dynamically changes over time.[77] It would obviously be wrong to state that due care standards should never change. There may be many reasons, for instance new technological insights, leading judges to the efficient decision that a more stringent standard of care can be applied. This new case law can, moreover, have an important signalling function for other parties in the market who can again adapt their future behaviour. But the question obviously arises what should be done with the individual defendant in the particular case in which a new standard of care is set? Should we sacrifice him for the benefit of a more efficient standard in the future and make him retroactively liable although his behaviour was not considered

76 This point is discussed in *Shavell*, JLS 1992, 259–270; see also L.T. *Visscher*/H.O. *Kerkmeester*, Kenbaarheidsvereiste en gewoonte als verweren tegen een aansprakelijkheidsactie: een rechtseconomische benadering, Tijdschrift voor Milieuschade en Aansprakelijkheidsrecht (TMA) 1996, 48–57.

77 This argument has been powerfully stressed by C. *Ott*/H.B. *Schäfer*, Negligence as Untaken Precaution, IRLE 1997, 15–29.

wrongful at the time when it was committed? There is a possible way out of this dilemma presented by – inter alia – the German Supreme Court.[78] The Supreme Court held in this particular case that an operator violated a general duty of care given the fact that technology had changed. However, at the same time, the Court also held that the operator was not to blame for the violation of the duty of care, since this was not foreseeable. This approach is known in the American literature as the "prospective overruling", meaning that a court follows an old duty of care in the particular case (with the result that there is no finding of liability), but announces that it will arrive at a different decision in the future.[79] This seems to be both an efficient and a just solution: on the one hand, a preventive effect is achieved for the future since future potential tortfeasors know that a new and more stringent due care standard will apply. On the other hand, it seems fair not to apply this new standard with respect to the particular defendant, who could indeed not have known that new rules would apply.

In sum, the discussion above makes clear that in fact a distinction has to be made (although the issues seem to be confounded sometimes) between on the one hand a retrospective application of a new liability regime and on the other hand the liability for development risks. A liability regime for risks which are not yet known today is not necessarily inefficient, precisely since, if this is known in advance, it will give incentives to require information on these new risks and on the optimal techniques to prevent them. Thus strict liability, also for development risks, might provide appropriate incentives for a dynamic investment in optimal preventive techniques. This however does not justify a retrospective application of new standards or new legislation, which could never positively affect future incentives for prevention. In other words: liability for development risks is not inefficient as long as it positively influences incentives for prevention and as long as the development risk liability is not a disguised retroactive liability.[80] **75**

The – justified – fear of retroactivity probably explains why legal systems are often reluctant to introduce liability for development risks. For **76**

78 See Bundesgerichtshof, 23 October 1984, [1985] Neue Juristische Wochenschrift (NJW) 16–20 and Bundesgerichtshof, 14 March 1995, [1985] NJW, 26–31.

79 This has been defended in the Dutch legal literature by *J. Drion*, Stare Decisis. Het gezag van precedenten (1950) and by *O. Haazen*, De temporele werking van een rechterlijke uitspraak, in: H.G. Schermers/Th.L. Bellekom/P.T.C. Van Kampen (eds.), De rol van de rechter in de moderne Westerse samenleving (1993) 171–207.

80 A simular – balanced – conclusion concerning the efficiency of a development risk defense is reached by *G. Wagner*, Haftung und Versicherung als Instrumente der Techniksteuerung, Versicherungsrecht (VR) 1999, 1450.

instance, in the context of the Product Liability Directive we can point to Art. 7 (b) which explicitly excludes liability if the producer can prove that, having regard to the circumstances, it is probable that the defect did not exist at the time when the product was put into circulation. Moreover, the real "state-of-the-art defence" is included in Art. 7 (e) which states that the producer shall not be liable if he can prove that the state of scientific and technical knowledge at the time when he put the product into circulation was not such as to enable the existence of the defect to be discovered.[81] However, Art. 15.1.b provides for an option for Member States to nevertheless introduce liability for development risks. This option has only been used by Luxembourg and Finland.[82]

77 As explained earlier, the foreseeability requirement could be relaxed as long as it does not constitute a disguised retroactive liability. It should also be noted here that the question of foreseeability of damage resulting from GMOs requires an analysis of the possible impact of GMOs. Some of which will certainly occur with a high level of certainty, while other may not. Such different levels of certainty inevitably merit special attention in the discussion of foreseeability.

3. Foreseeability requirement and possible impacts of GMOs

78 To begin the discussion of foreseeability, let us consider several possible impacts of GMOs as explained below.

79 In addition to human health problems,[83] GMOs may also create adverse environmental impacts, some of which are feared to be catastrophic. Typical concerns about environmental impacts of GMOs can be summarised as follows:

---

81 The state of the art defense has also been addressed in the American context by *Boyd/Ingberman*, JLS 1997, 433–473. They show that the "customary practice test" tends to induce inadequate safety, whereas the "technological advancement test" tends to induce excessive safety.

82 And by Spain for food and medical products as well as by France for products derived from the human body. See the overview of the transposition in domestic law, provided in the Green paper on the liability for defective products (COM (1999) 396 final of 28.7.1999) 35 f.

83 Concerns over the impact on human health include the issue of allergen and the use of antibiotic as gene marker. For discussion about possible impact on human health, see: *A. Bakshi*, Potential Adverse Health Effects of Genetically Modified Crops, Journal of Toxicology and Environmental Health, Part B: Critical Reviews (Jur Tox and Env Health) 6 (2003) 211–225.

(a) Gene transfer

There are, at least, three separate issues related to gene transfer from GM crops.[84] The first issue arises when a GM crop from one piece of land cross-pollinates with a non-GM crop of the same species on surrounding land. Where GM crops are grown in the same region as non-GM cultivars, opportunities for cross-pollination exist. One of the most cited studies on this issue is the work of Quist and Chapela published in *Nature* in 2001. In this study, the authors observed the transfer of genes, referred to as "introgression", from GM corn to non-GM corn fields in Mexico.[85] Economic damage from gene transfer ranges from the possibility of losing certification as an organic farmer,[86] or losing markets because the products have been "contaminated" by GM crops.[87] In addition, there is also a real possibility that **80**

84 Once released into the environment, genes from GM crops may be transferred, by several possible ways, into the genes of other plants. According to the Royal Society of Canada, GM crops can be categorised in the possibility for genes transfer into: 1) No possibility, where wild relatives are absent from the region where the crop is grown; 2) Low possibility, where crops species are either predominantly self-propagated (e.g. many cereals) or are infrequently propagated by sexual reproduction and flower (e.g. sweet potato, sugar cane); and 3) Moderate to high possibility, where the crops are grown in an area where their sexually compatible wild relatives are present (e.g. canola in Europe and North America, and rice in South East Asia). See: *Royal Society of Canada (RSC)*, Elements of Precaution: Recommendations for the Regulation of Food Biotechnology in Canada: An Expert Report on the Future of Biotechnology, 2001. Available at: http://www.rsc.ca/foodbiotechnology/GMreportEN.pdf, 124 f.
Others have characterised the risks of gene flow as follows: a) Very low and low risks; crops in this group show a low level, or even do not show evidence at all, of cross-pollination with wild relatives. Corn, soybean, tomato, and potato are considered to belong to this group; b) Moderate risk; crops, such as alfalfa, sugar beet, and sunflower, could be included into this moderate group. These species have the same genus and similar chromosome numbers with wild relatives, and there is some evidence about hybridisation of these species with their wild relatives; c) High risk; crops which belong to this group are considered to hybridise easily with their wild relatives. Examples of these crops include canola and sorghum. This characterisation is based on: *A.A. Snow/P.M. Palma*, Commercialization of Transgenic Plants: Potential Ecological Risks, Bioscience, vol. 47(2) (1997) 91; and *C.N. Stewart/Jr.M.D. Halfhill/S.I. Warwick*, Transgene Introgression from Genetically Modified Crops to their Wild Relatives, Nature Reviews Genetics (Nat Rev Gen) vol. 4, Oct. (2003) 810–812.

85 For a complete result, see *D. Quist/I.H. Chapela*, Transgenic DNA introgressed into Traditional Maize Landraces in Oaxaca, Mexico, Nature, vol. 414 (2001) 541–543.

86 Concerns of losing the entire organic market due to the mixture of crops has motivated Canadian groups of organic farmers to sue two giant GMO producers, Monsanto and Aventis. The farmers argue that the entire organic market for wheat, worth as much as $ 17.5 million, is threatened due to the commercialisation of GM wheat in Canada. See *A. Bouchie*, Organic Farmers Sue GMO Producers, Nat Biotechnol. 20 (2002) 210.

87 The famous case for this type of damage is the StarLink case, triggered by the finding of Bt genes, i.e. the so-called Cry9C genes, in corn-based products for human consumption. The genes supposedly originated from Aventis's StarLink, a GM corn specifically designated for animal feed. Due to its potential health impact, GM crops containing the Cry9C were rejected for human consumption by the US authority. This finding induced

a farmer whose land has been polluted by neighbouring GM crops could even be found guilty of patent infringement.

81 Concerns have also been given to the invasiveness of new hybrids caused by the transfer of genes from GM crops to their wild relatives.[88] GM crops, such as herbicide-tolerant and insect-resistant crops, have phenotypes that are designated to increase the fitness of these crops. Such fitness-related enhancement traits could be transferred into wild relatives.[89] The invasiveness of new hybrids could replace the wild species and, hence, reduce the genetic diversity and uniqueness of wild-native populations.[90] In this regard, as Johnson has argued, the development of resistance in wild plants may change plant population dynamics, increasing the risks of their invading agricultural land and natural ecosystems.[91]

82 The greatest concern about gene transfer is perhaps the possibility that the transfer from GM crops to their sexually compatible wild relatives may result in unwanted hybrids that are more persistent. According to Ervin et al., there is little doubt among scientists that genes will wander from crops into the wild. The concern is not whether the genes will move, but rather whether they will survive in the wild and, thus, significantly increase the persistence and invasiveness of particular wild plants, by conferring a certain trait on such undesirable plants which would make them more difficult to control.[92] Some studies in Western Canada show that

---

US corn farmers and producers to sue Aventis for the impact of admixture between Star-Link with corn for human food. No court decision was, however, made on this case, since the parties settled the case outside court, by which Aventis had to pay up to US$ 110 million. *Khoury/Smyth* (fn. 44) 12. See also: *Smyth/Khachatourians/Phillips*, Nat Biotechnol. 20 (2002) 537 f.

88 Although crop-to-wild hybridisation has been a common phenomenon in agriculture for years, the introduction of GM crops could increase the likelihood of such a hybridisation occurring.

89 N.C. *Ellstrand*, When Transgenes Wander, Should We Worry? in: B. Bailey/M. Lappé (eds.), Engineering the Farm: Ethical and Social Aspects of Agricultural Biotechnology (Washington, DC: Island Press, 2002) 65.

90 *L.L. Wolfenbarger/P.R. Phifer*, The Ecological Risks and Benefits of Genetically Engineered Plants, Science 290 (2000) 2088. If, however, the hybrids show lower fitness than their parents, the wild population may shrink. Hence, hybridisation seems to create a problematic situation between a population reduction and the invasiveness of the hybrids. See: *D.A. Andow/C. Zwahlen*, Assessing Environmental Risks of Transgenic Plants, Ecology Letters (Ecol Let), 9 (2006) 200.

91 *B. Johnson*, Genetically Modified Crops and Other Organisms: Implications for Agricultural Sustainability and Biodiversity, in: G.J. Persley/M.M. Lantin (eds.), Agricultural Biotechnology and the Poor: an International Conference on Biotechnology, (Washington, D.C.: Consultative Group on International Agricultural Research, 2000) 133.

92 *D.E. Ervin* et al., Towards an Ecological Systems Approach in Public Research for Environmental Regulation of Transgenic Crops, Agriculture Ecosystems and Environment (Agr. Ecos. & Env.) 99 (2003) 5.

three different herbicide-resistant canola (rape) varieties have cross-pollinated to create canola plants that are resistant to multiple types of herbicide.[93] The presence of such a "superweed" could force farmers to use older and less environmentally friendly herbicides.[94] In this regard, the large-scale introduction of GM crops might, thus, change wild relatives into new and worse weeds.[95] Along with the creation of invasive hybrids, the presence of persistent hybrids, which require a more toxic application of herbicides, could also be considered a threat to the biological diversity of weeds in natural ecosystems.[96]

(b) The development of resistance

The introduction of novel crops with fitness-enhanced genes could lead to **83**
an undesirable effect of resistance. Some biologists believe that sooner or later weeds will begin to become resistant, and, hence, more herbicides, which could be far more toxic than before, will be required.[97] Krimsky and Wrubel conclude that the increased use of "herbicide resistant crops" may increase the possibility of weeds developing resistance to herbicides. In addition, a single herbicide could also put pressures on weeds to evolve and adapt to this herbicide, which will eventually enable them to develop resistance to this herbicide. Hence, herbicide tolerant crops developed for a single application of certain herbicides could increase the likelihood of resistant development in weeds. To control these stronger weeds, farmers might respond by increasing the amount or toxicity of herbicides they use.[98]

The possibility of resistance development has also been shown in insects. **84**
Some research, at least in laboratory experiments, has observed the exis-

93 One of the study shows that spontaneous hybridisation occurred among three varieties of canola, two of which were transgenic canola, when they were planted close to one another. The hybridisation of these varieties, each of them was resistant to glufosinate, imidazolinone, and glyphosate, resulted in hybrid volunteers that were resistant to more than one of these herbicides. In addition, the study also found that the resistant alleles were able to move rapidly. See: *Ellstrand*, Phil Trans Biol Sci (2003) 1167.

94 *RSC* (fn. 84) 122 f.

95 *M.A. Altieri*, The Ecological Impacts of Transgenic Crops on Agroecosystem Health, Ecos Health, vol. 6, no. 1 (2000) 16.

96 *P.J. Dale/B. Clarke/E.M.G. Fontes*, Potential for the Environmental Impact of Transgenic Crops, Nat Biotechnol. 20 (2002) 571.

97 *D. Pimentel*, Overview of the Use of Genetically Modified Organisms and Pesticides in Agriculture, Indiana Journal of Global Legal Studies (Ind. Jour. Glo. Leg. Stu.) 9 (2001) 57.

98 *S. Krimsky/R. Wrubel*, Agricultural Technology and the Environment: Science, Policy, and Social Issue (1996) 47 f.

tence of insects that can develop resistance to the Bt toxin.[99] With this evidence, one could argue that the release of GMOs into the environment could increase the possibility of resistance development in some target insects. In this regard, Anderson argues that, in contrast to the occasional use of Bt in organic farming, crops that are genetically engineered to contain Bt (hereinafter called Bt crops) produce toxins all the time they are growing. This means that insects are continually exposed to the toxins, and therefore are under constant pressure to develop resistance.[100] The development of resistance to the Bt toxin might, thus, lead to serious environmental problems. *Bacillus thuringiensis*, once considered as a miracle for agriculture, will rapidly become useless, as pests become resistant to the bacteria that are massively and continually produced by Bt crops.[101] Some experts have even predicted that the development of pest resistance to Bt crops is inevitable; the question is how quickly this will occur.[102] The report of the US Environmental Protection Agency (EPA) in 1998 predicted that most target insects could be resistant to the Bt toxin within three to five years.[103]

85 According to the report of the Royal Society of Canada (RSC), the presence of insects that are resistant to Bt toxins, politically often referred to as "superpests", will result in two undesirable effects. *First*, since Bt is the most effective biological insecticide available to organic farmers, a diminishing of its effectiveness means that organic farming loses its control over pests, which could seriously affect a more ecologically friendly form of agricultural practice. *Second*, if conventional farmers resort to an increased use of chemical insecticides to control populations when GM plants no longer offer sufficient levels of protection against pest species, there is a possibility of a serious environmental impact resulting from the use of more toxic chemicals in agriculture.[104]

---

99 A laboratory research conducted by Tabashnik has reported an increase of resistance to Bt toxin in Lepidoptera (butterflies and moths), Coleoptera (beetles) and Diptera (mosquitoes and flies). For technical detailed, see: *B.E. Tabashnik*, Evolution of Resistance to *Bacillus thuringiensis*, Annual Review of Entomology (Ann. Rev. Ento.) 39 (1994) 47–79.

100 *L. Anderson*, Genetic Engineering, Food, and Our Environment: A Brief Guide (1999) 28.

101 *M.A. Altieri/P. Rosset*, Ten Reasons Why Biotechnology will not Help the Developing World, AgBioForum, vol. 2, no. 3 & 4 (1999a) 157.

102 *M.A. Altieri/P. Rosset*, Strengthening the Case Why Biotechnology Will not Help the Developing World: A Response to McGloughin, AgBioForum, vol. 2, no. 3 & 4 (1999b) 229.

103 *L. Anderson* (1999) 28.

104 RSC concludes that this case will result in two undesirable effects: i) Bt is the most effective biological insecticide available to organic farmers; the loss of it means the loss of control that would seriously jeopardise their livelihood and an expansion of

In addition, the report of the RSC indicates a potential impact of the "insect resistant crops" on secondary target pests. The reports states that although the use of Bt crops has decreased the number of sprays used against the target pest, it has also increased problems with secondary pests. In the past, these pests were controlled by the repeated use of insecticides against the primary pests. However, in the absence of the sprays, these pests are able to develop, since they are unaffected by the toxin.[105] **86**

(c) Effects on non-targets

Some authors have also expressed concerns about the impacts of Bt crops on non-target herbivores and insects, such as lacewings, ladybirds beetles, monarch butterfly larvae and soil biota.[106] Some studies have, indeed, indicated the possibility of such impact. For example, a laboratory study of Losey et al. observes 44% mortality rate in monarch butterfly larvae fed on milkweed leaves dusted with Bt corn pollen. No mortality occurred in monarchs fed on leaves with non-Bt corn pollen.[107] Another study on the impact of GM crops on non-targets is also given by Hilbeck, et al. In this case, Hilbeck, et al. observe that lacewings, an important and beneficial insect, experience higher mortality rate when fed with pests reared on Bt corn compared to when fed with the pests reared on non-Bt corn. The authors also detect no significant difference in mortality when the lacewings are fed with various preys exposed to corns with the same treatment, i.e. Bt corn or non-Bt corn. Hence, so the authors conclude, the difference in the mortality of lacewings is primarily due to Bt corn.[108] **87**

It is worth mentioning here that the prediction of laboratory studies about the effects on non-target species might be different with the situation in the natural environment.[109] For example, if in reality non-targets are not **88**

---

this more ecologically friendly form of agricultural practice; ii) the possibility of a serious environmental impact if conventional farmers resorted to increased use of chemical insecticides to control populations when GM plants no longer offer sufficient levels of protection against pest species. *RSC* (fn. 84) 139.

105 Ibid., 141.

106 *Ervin* et al., Agr. Ecos. & Env. 99 (2003) 6.

107 *J.E. Losey/L.S. Raynor/M.E. Carter*, Transgenic Pollen Harms Monarch Larvae, Nature 399 (1999) 214.

108 For the complete report, see: *A. Hilbeck* et al., Effects of Transgenic Bacillus thuringiensis Corn-Fed Prey on Mortality and Development Time of Immature Chrysoperla carnea (Neuroptera: Hrysopidae), Biological Control (Bio. Cont.), vol. 27, no. 2 (1998) 480–487.

109 *Wolfenbarger/Phifer*, Science 290 (2000) 2089.

exposed to Bt toxins, one certainly could argue that the risk posed by Bt crops on non-target species is actually low. However, a caveat given by the Ecological Society of America (ESA) might be worth considering: "if a Bt toxin kills pest insects, it also has the potential to kill other insects".[110] This potential should be seriously taken into account before the release of GMOs is allowed.

(d) Effects on soil ecology

89 GM crops are also believed to have a significant impact on soil ecology. The impact, for example, may also occur from the massive use of Bt crops. A study of Saxena, et al, for example has observed that toxins in Bt crops remain active in the soil, allowing the toxins to retain their insecticidal characteristics and, thus, prevents them from being degraded by soil microbes. The result is that the toxins persist in the soil for 234 days.[111] The accumulation of toxins, which could be released into the soil as farmers incorporate plant material into the ground after harvest, has the potential to create serious environmental problems in the future.

90 The various studies summarised above may serve as early warnings regarding the possible impact of GMOs. In addition, these findings might also serve to criticise the safety claims about GMOs. However, despite such findings, some argue that the impact of GMOs is still uncertain. In this case, it should be noted here that any concern or finding regarding the impacts of GMOs seem to have counter-arguments. Such counters may take several forms, for example, that there is a low probability of the impact occurring[112]

---

110 *A.A. Snow* et al., Genetically Engineered Organisms and the Environment: Current Status and Recommendations, Ecological Applications (Ecol. App.), vol. 15, no. 2 (2005) 393.

111 *D. Saxena/S. Flores/G. Stotzky*, Insecticidal Toxin in Root Exudates from Bt Corn, Nature, 402 (1999) 480. The authors, however, have conducted another research study which, as quoted in Andow and Zwahlen, observes that the toxins can persist in the soil for at least 365 days. See: *Andow/Zwahlen*, Ecol Let 9 (2006) 199.

112 The study of Chèvre et al., for example, observes that under natural conditions, gene flow from GM canola to its wild relatives is very rare and occurs only at a slow rate. The findings indicate that the probability of gene transfer from GM crop to its relatives is actually low. See: *A.M. Chèvre* et al., Gene Flow from Transgenic Crops, Nature, vol. 389 (Oct. 1997) 924. Scientists also argue against the study of Quist and Chapela concerning the introgression of Bt corn in Mexico. As cited by Mann, they conclude that the introgression of GM corn with its relatives has never occurred. Instead, they pointed to some methodological flaws in Quist and Chapela's study and argued that the findings have "no credible evidence ...to justify any of [their] conclusions". *C.C. Mann*, Has GM Corn 'Invaded' Mexico?, Science, vol. 295 (March 2002) 1617.

or that the impact will be rather limited or isolated.[113] Others may simply point out that the risks posed by GMO crops are similar to those posed by non-GM crops.[114] Taking into account these opposing views, one might argue that scientific evidence about the possible impact of GMOs is not conclusive. When damage does occur in the future, such argument will certainly serve as a reason to show that some damage was still scientifically unforeseen at the time of the release of the GMOs.

Whether the level of scientific information available today constitutes evidence to indicate the foreseeability of some damage will depend on how the courts interpret the foreseeability standard. This will eventually depend on the courts' interpretation of the level of scientific certainty regarding the impact of GMOs. **91**

In this regard, Khoury and Smyth argue that based on the current approach applied by the courts in considering the foreseeability requirement, the seed companies are likely to succeed in exercising the unforeseeability defence by resorting to the current uncertainties concerning the impact of GMOs. The more uncertain the impact is, the more likely that the company will escape liability.[115] To avoid this situation, Khoury and Smyth have proposed the incorporation of the precautionary principle into liability. They argue that by applying this principle, the courts will **92**

---

113 Some studies, for example, have also found some results contrary to the study of Losey, et al. concerning the impact of Bt crop on the Monarch butterfly. These studies point to some methodological flaws in the Losey, et al.'s study to indicate that the study was conducted in conditions that are not similar to natural conditions. In addition, they also show that GM corn pollen actually contains much less Bt toxin compared to that used by Losey, et al. Based on these grounds, these studies conclude that the impact of GM crops on the Monarch butterfly are negligible. See: *J.P. Wisniewski* et al., Between Myth and Reality: Genetically Modified Maize, an Example of a Sizeable Scientific Controversy, Biochimie 84 (2002) 1099–1100; *J.K. Kaplan*, Bt Corn not a Threat to Monarchs, Agricultural Research (Agr. R), Feb. 2002, 16–18; *M.K. Sears* et al, Impact of Bt Corn Pollen on Monarch Butterfly Populations: A Risk Assessment, Proceeding of National Academy of Science (PNAS), vol. 98, no. 21 (Oct. 2001) 11937–11942; and *J. Hodgson*, Monarch Bt-Corn Paper Questioned, Nat Biotechnol. 17 (July 1999) 627.

114 Corner and colleagues for example argue that the problems of crop-to-wild hybridisations and their impact are not new in agriculture, and cannot be associated only with GMOs. Because the hybridizations are equally likely to occur both via GM or non-GM crops, the problem of hybridizations should be resolved based on an agricultural strategy equally applied both to GM and non-GM crops. The authors hence argue that the impact will not be more severe and catastrophic than those resulting from non-GM crops. Similarly, they also argue that some possible impact such as the effect on non-target organisms, are familiar and inherent not only to GMOs, but also to all existing agricultural practices *A.J. Corner/T.R. Glare/J.P. Nap*, The Release of Genetically Modified Crops into the Environment. Part II: Overview of Ecological Risk Assessment, The Plant Journal 33 (2003) 28–31.

115 *L. Khoury/S. Smyth*, Reasonable Foreseeability and Liability in Relation to Genetically Modified Organisms, Bull. Sci.Tech. & Soc., vol. 27, no. 3 (2007) 226.

be induced to take into account the current concerns of the impact of GMOs. This means that the foreseeability requirement will be somehow relaxed to allow the courts to appreciate public concerns regarding the impacts of GMOs. They argue that when the principle is incorporated into the liability system, "acting in accordance with the prevailing levels of knowledge would no longer exonerate an individual who could be held liable for omitting to foresee and prevent risks that although unconfirmed may bring about injury in the future".[116]

4. Liability based on the precautionary principle for GMOs?

93 Khoury's and Stuart's proposal of incorporating the precautionary principle into liability is similar to Treich's and Gollier's opinions. They argue that in face of uncertainty, the threat of liability will prevent strategic-behaviour of competitive firms that otherwise will exploit the uncertainty of the damage of their products in order to gain market control for the products. In this case, firms will market their innovations as soon as possible without considering the possible impact of the innovations. Incorporating the precautionary principle into liability thus means that despite current uncertainties concerning some impact of GMOs, GM operators will be held liable if this impact materialises and their GM products turn out to be toxic.[117]

94 This section will not discuss whether the proposal of incorporating the precautionary principle into tort is justifiable.[118] Instead, this section attempts to discuss possible costs to the GM operator when such a proposal is implemented.

95 The application of the precautionary principle in tort law may take several forms. First, it may apply a negligence standard in which the failure to

---

116 See *Khoury/Smyth*, Bull. Sci.Tech. & Soc (2007) 228.

117 *Gollier/Treich*, JRU 27 (1) (2003) 98; N. *Treich*, What is the Economic Meaning of the Precautionary Principle? The Geneva Papers on Risk and Insurance, vol. 26, no. 3 (2001) 342.

118 It should be mentioned here that the proposal may run against the purpose of the precautionary principle. Craik and colleagues for example argue that the use of the principle in tort law may not be in line with the democratisation of the decision-making process, which constitutes an important message of the principle. They argue that unlike the policy makers, the courts may lack the capacity to address the issue of uncertainty. Accordingly, they often have to rely on conflicting testimonies. More importantly, the precautionary principle requires the democratisation of the decision-making process, such as board public consultation and participation. And it is not possible that this process is done by the courts. N. *Craik*/K. *Culver*/N. *Siebrasse*, Genetically Modified Crops and Nuisance: Exploring the Role of Precaution in Private Law, Bull. Sci. Tech. & Soc., vol. 27, no. 3 (2007) 211 f.

comply with the precautionary regulatory standard will result in liability. In this case, the principle is indirectly incorporated into liability in which the regulatory standard follows the precautionary approach and non-compliance with the standard will result in liability. Second, the defendant will be held liable for any damage because the damage per se is evidence that the defendant has failed to take precautionary measures to prevent it.

To answer the question of which of these possible applications will impose **96** the highest cost on individual GM operators, one certainly needs some empirical research. However, one could expect that the cost of compliance for the first option is lower than that of the second option. This is because under the first option, the operator will follow the regulatory standard in order to avoid liability. Since such a standard has been implemented, the precautionary principle, in the sense that it is established to prevent some possible impact although this impact is not yet conclusive, the cost incurred by the operator is the cost of complying with such a precautionary standard.

The second option, in which the operator will be held liable for any **97** damage simply because the damage is considered proof indicating the operator's failure to take precautionary measure in order to prevent the damage, will result in a rather high compliance cost. Under this option, the operator will be forced to consider all damage, whether or not it is foreseeable. In this situation, the operator is at least induced to continue increasing its investment in safety in order to avoid the damage.[119] Assuming that every increase in the level of care will also increase the compliance cost, the second option will eventually result in higher compliance costs compared to the first option.

5. Contributory negligence

We indicated above that both a strict liability rule and a negligence rule **98** will lead to the optimum in cases where the victim's care does not influence

119 To predict whether or not the operator will stop investing in safety, one could refer to the study of Farrow and Hayakawa. In this study, the authors interpret the precautionary principle according to the real option approach, and conclude that the application of the precautionary principle into the decision making process of whether one needs to continue investing in safety will increase the level of care and its cost. The presence of this multiplier functions to increase the threshold benefits for not investing in safety, in which the benefits of not investing in safety should be much higher than the costs of investing in safety. For a mathematical proof on the application of the precautionary principle as the real option approach, see *S. Farrow/H. Hayakawa*, Investing in Safety: An Analytical Precautionary Principle, Journal of Safety Research 33 (2002) 168–172.

the probability of an accident and where only care (and not the activity level) can influence the risk. Most accident situations are, however, "joint care" cases.[120] In this situation the risk is also influenced by the behaviour of the victim. A simple strict liability rule would not lead to the efficient result, since the victim has no incentive to spend on care. To remedy this problem, the victim might be considered "contributory negligent" if he does not take due care. A contributory negligence rule, as known under common law, excludes a right to compensation for the victim who did not take due care.[121]

99 Assuming that the legally required level of care for the victim is equal to efficient care, the victim will have the incentive to take optimal care. If he did not take due care, he would be found negligent and would receive no compensation. An efficient result will also follow both under a negligence rule and under a strict liability rule with a contributory negligence defence. In both cases the injurer will take efficient care and the victim will (being fully exposed to the risk), in order to avoid bearing the loss himself, take efficient care as well. Discussing the economic model of tort law, we therefore indicated that both a strict liability rule in combination with a defence of contributory negligence and a negligence rule (with or without contributory negligence) will give appropriate incentives to the victim to take efficient care.[122]

100 A comparative negligence rule has the effect of proportionally dividing the loss between the injurer and the victim if both committed a fault. Under this rule the right to compensation will be proportionally reduced if the victim was negligent. The injurer will still take efficient care to avoid liability, while the victim still takes care to minimise his own loss.[123] The efficiency of this rule is debated in the literature. Haddock and Curran point to difficulties in analysing the comparative benefits of comparative negligence versus a contributory negligence defence.[124] It is well known that Posner is an opponent of this rule.[125] According to him, the rule causes considerable administrative costs, without any compensating bene-

---

120 Although we already argued above that environmental pollution is probably a good example of a truly unilateral case.

121 For a recent overview of the economic literature in this respect see *M. Artigot/I. Golobardes/F. Gomez-Pomar*, Contributory and Comparative Negligence in the Law and Economics Literature, in: M. Faure, Tort Law and Economics (2009) 46–82.

122 *Brown*, JLS 1973, 340–342; *Calabresi*, YLJ 1975, 663; *Landes/Posner*, GLR 1981, 880–882.

123 *D. Haddock/C. Curran*, An Economic Theory of Comparative Negligence, JLS 1985, 59–63.

124 *Haddock/Curran*, JLS 1985, 59–63.

125 *R. Posner*, Economic Analysis of Law (1998) 187–189.

fits for the incentives to take care. Not only is an intervention of the legal system necessary to shift a part of the loss from the victim to the injurer, but judges will also have to examine the faults of both parties and the proportion to which they contributed to the loss. Posner argues that comparative negligence makes economic sense only when society wants to use the tort system to provide insurance to accident victims.

In sum, if a strict liability rule is proposed for GMO damage, some defence **101** should be added to take account of the behaviour of the victim, but this can either be a contributory or a comparative negligence rule. To be clear: a strict contributory negligence rule, meaning that the victim loses the claim on compensation entirely in case of his negligence, is practically no longer applied. Most legal systems have turned to a proportionate reduction of the compensation due to the victim. If on the other hand a negligence rule is applied to GMO damage, it is not strictly necessary to add a contributory negligence defence.[126]

VI. Remedies

1. Damage in tort

Law and economics scholars usually hold that the amount of damages the **102** injurer should pay should be at least equal to the victim's loss in order to provide optimal compensation to the injurer.[127] These so-called compensatory damages must be paid to the victim in order to give the victim an incentive to sue, which is essential to allow the tort system provide an effective and credible deterrent. The duty to pay compensatory damages to the victim will moreover prevent victims from taking inefficiently

126 A contributory negligence defence may arise, for example, in the form of the infringement of patent right of a GM crops producer. One could refer to the *Monsanto v. Schmeiser* case, in which the defendant was found guilty of the infringement of Monsanto's patent rights for herbicide-resistant canola. See: *J.L. Fox*, Nat Biotechnol 2001, 396 f. The ruling of the court of this case has, however, been severely criticised as the court ignored the fact that the defendant had not used glyphosate, a herbicide to which the patented GM canola is supposed to resist. Some authors argue that if the possession of hybrids containing the patented gene is already a sufficient ground for defendant liability for the infringement of a patent right, then the question of the defendant's intention should be seriously considered by the court. Otherwise, a farmer whose land has been contaminated by GM crops and, hence, unwillingly grows the hybrids, will be found guilty of the patent infringement. See: *M. Lee/R. Burrell*, Liability for the Escape of GM Seeds: Pursuing the 'Victim'? Modern Law Review (Mod Law Rev) 65 (2002) 523–525; also *Preston*, TLR 81 (2003) 1167–1169.

127 *R.A. Posner*, Economic Analysis of Law (6th ed. 2003) 192.

high precautions.[128] If the damages to be paid by the injurer would fall short of the harm so that the expected payments would be below the expected harm, the incentives to reduce the risk would be inadequate.[129]

103 Therefore the starting principle should be that the liable party should pay for the actual level of losses of the victim.[130] There is in addition extensive economic literature for instance on the question on how life should be evaluated in a tort case and more particularly on the evaluation of non-pecuniary losses. In addition economists hold that in some cases damages should outweigh a low probability of detection and should therefore be "punitive". The punitive damages are thus meant to provide appropriate incentives to injurers when, for instance, through his malicious acts, the probability of the tort being detected would be lower than one.[131]

104 A highly debated issue in the law and economic literature and an important issue for damage caused by GMOs is to what extent damages should be awarded for so-called pure economic loss.[132] Economic analysis has provided explanations for the traditional denial of compensation for pure economic loss in many legal systems. An important explanation is that an economic loss is merely a private loss for the victim, but not necessarily a loss for society (a so-called social loss). Compensation of an economic loss which would not at the same time constitute a social loss would thus, from an economic perspective, lead to a waste of administrative costs.[133] a problem is, however, that today certain differences still exist between legal systems as far as the recoverability of pure economic losses is concerned. Hence, more recent economic literature still holds that as long as a damage does not lead to a personal but only to a social loss the traditional denial of compensation can be justified. However, the large differences between the legal systems as far as compensation of economic loss is concerned cannot be fully explained on economic grounds.[134]

---

128 Ibid., 192.

129 So *S. Shavell*, Foundations of Economic Analysis of Law (2004) 236. For a recent overview see *L.E. Visscher*, Tort Damages, in: M. Faure, Tort Law and Economics (2009) 153–200.

130 *Shavell* (fn. 129) 237.

131 For an overview of the economics of punitive damages see *A.M. Polinsky/S. Shavell*, Punitive Damages, in: M. Faure, Tort Law and Economics (2009) 228–244.

132 For an excellent economic account see *G. Dari-Mattiacci*, The Economics of Pure Economic Loss and the Internalization of Multiple Externalities, in: W. van Boom/H. Koziol/C.H.R. Witting (eds.), Pure Economic Loss, vol. 9. Tort and Insurance Law (2004) 167–190.

133 See in this respect also the contributions in *M. Bussani/V. Palmer* (eds.), Pure Economic Loss in Europe (2003).

134 So *J. De Mot*, Pure Economic Loss, in: M. Faure, Tort Law and Economics (2009) 201–214.

2. Injunction

There is still another question that could be asked in relation to the remedies. What if the potential victim knows that the harm will occur or if the case concerns ongoing harm? Can injunctive relief be sought so that the judge can order the injurer to refrain from the damaging behaviour? In these particular cases it would, for instance, mean that the judge orders the injurer to undertake specific measures to avoid further damage resulting from the GMO. **105**

As far as injunctive relief is concerned, the economic literature makes a distinction between the way property rights are protected and the way in which other rights are protected. Economists point at the fact that the typical remedy in case of a violation of a property right is an injunction. Damages are the usual remedy for torts whereas an injunction is the usual remedy in case of a nuisance, hence a violation of a property right.[135] For the case of GMO mixture, this would hence mean that when a neighbour's property right (enjoying a non-GM crop) were endangered by the presence of a neighbour using GMOs, economists would thus predict that the remedy would be injunctive relief. However, the fact that a property right is granted and that the victim could theoretically use injunctive relief does of course not mean that this will necessarily be the result. The Coase theorem[136] predicts that parties may engage in bargaining and when transaction costs are low, this is precisely what will happen. Hence, the injurer may "buy" his right to pollute by paying damages to the victim. This would of course depend on what the efficient outcome is. But the Coase theorem holds that if transaction costs are equal to zero, successful bargaining can cure inefficient laws. Hence economists consider damages and injunctions as equally efficient remedies when transaction costs equal zero. Differences in efficiency thus depend on transaction costs.[137] If transaction costs are high, bargaining may be impossible. In this case the more efficient remedy is damages and no longer the injunction. The injunction could lead to an inefficient solution, whereas damages could be adjusted to the harm incurred. Precisely because in a nuisance context, where a property right protection is enforced, are transaction costs relatively low and the typical remedy will be the injunction. The injunction is more particularly more efficient than damages when the parties can bargain with each other. The reverse is thus true in a high transaction costs setting, which is typically **106**

135 See *R. Cooter/Th. Ulen*, Law and Economics (4th ed. 2003) 100.
136 *R. Coase*, The Problem of Social Cost, JLE 1960, 11–44.
137 *Cooter/Ulen* (fn. 135) 104.

the tort case. Then economists would predict that the efficient remedy should be damages and not injunctive relief. This is a finding in a well-known paper by Calabresi and Melamed, which is often quoted in the law and economics literature.[138] They argue as follows:

- When there are obstacles to cooperation (high transaction costs), the more efficient remedy is the award of compensatory money damages;
- When there are few obstacles to cooperation (low transactions costs), the more efficient remedy is the award of an injunction against the defendant's interference with the plaintiff's property.[139]

107 They therefore hold that when the nuisance is private and thus few parties are affected by it, the costs of bargaining will be low and the injunction may be the preferred remedy. As a result, the court will be spared the difficult task of assessing damages. From this law and economics perspective, the injunction is, however, not viewed as a remedy which would prohibit the offensive activity forever, but rather serve as a signal to the parties to resolve their dispute through bargaining. If the harmful externality is of the "public bad" type bargaining is impossible because of high transaction costs and an award of damages will be the more efficient remedy. Cooter and Ulen therefore hold that, in choosing between injunctions and damages, the court will have to examine the number of people affected by the externality. Only when the number of affected parties is low (as is often the case with GMO damage) may injunctive relief be warranted.[140]

108 If the court, however, tends to apply a permanent injunction and damages, the results might be different. In this case, the court should consider the social value of GM crops compared to the harm suffered by non-GM crops farmers. In this case, the court might look at the benefits of GM crops in general, ranging from an increasing productivity to serving as a solution to provide cheap and nutrition-rich food for the world. These benefits should, of course, be compared with the perils of GM crops and with the need to provide non-GM product as an alternative for society. In particular, the benefits of individual GM crops farmers might be compared with the damage suffered by individual non-GM crops farmers. If the value of GM crops exceeds the harm suffered by non-GM crops farmers, then periodical damages is a preferable remedy. This is because, as the potential Pareto criterion suggests, efficiency means that the winner

138 *G. Calabresi/D. Melamed*, Property Rules, Liability Rules and Inalienability: One View of the Cathedral, HLR 85 (1972) 1089.
139 See for a summary of Calabresi/Melamed, *Cooter/Ulen* (fn. 135) 104–107.
140 *Cooter/Ulen* (fn. 135) 168 f.

still gains after compensating the loser, and because GM crops, which are highly beneficial to society, are too important to be permanently stopped. An injunction to stop the use of GM crops might in that case be inefficient.

3. Financial caps?

A further question that could be asked as far as damages and remedies are concerned is whether there is any argument to put a financial cap or limit on the amount of damages due to the victim. To answer this question again a distinction has to be made between the contracts case (where *ex ante* bargaining was possible) and the tort case (where the victim is a third party and hence bargaining was impossible). In case of a contract, parties could of course *ex ante* agree to limit damages due to a specific amount which can be less than the actual loss suffered by the victim. If that were the case, it is an explicit agreement concerning allocation of risk which will undoubtedly also have an effect on the price agreed between the parties. In that particular case there is no objection against a limit. In fact it amounts to liquidated damages, an amount of damages *ex ante* agreed by the parties in case of breach of contract.[141] **109**

A limitation of liability is far more complicated in the tort case. In the literature it has been indicated that there may be good reasons to favour a strict liability rule for major industrial accidents,[142] the main reason being that only a strict liability rule would lead to a full internalisation of those highly risky activities.[143] This strict liability rule is in particular put forward in a so-called unilateral accident situation; this is where only one party influences the accident risk. Only with strict liability would the potential injurer also have an incentive to adopt an optimal activity level. This full internalisation is obviously only possible if the injurer is effectively exposed to the full costs of the activity he engages in and is therefore in principle held to provide full compensation to a victim. An obvious disadvantage of a system of financial caps is that this will seriously impair the victim's rights to full compensation. But if the cap is indeed set at a much lower amount than the expected damage, this would not only violate the victim's right to com- **110**

141 See supra fn 127.

142 Above we argued that it will depend upon the specific circumstances of the case whether there is an argument in favour of strict liability for GMO damage. A crucial factor in that respect is what the respective contribution of both injurer and victim to the risk of GMO mixture is. We therefore assume here that the influence of the injurer is more important and that therefore the legal system has adopted a strict liability rule.

143 *Shavell*, JLS 1980, 11 and *Shavell*, JLE 1985, 8.

pensation, but the above-mentioned full internalisation of the externality would not take place either. From an economic point of view, a limitation of compensation therefore poses a serious problem since there will be no internalisation of the risky activity. Indeed, if one believes that the exposure to liability has a deterrent effect, a limitation of the amount of compensation due to victims poses another problem. There is a direct linear relationship between the magnitude of the accident risk and the amount spent on care by the potential polluter. If the liability therefore is limited to a certain amount, the potential injurer will consider the accident as one with a magnitude of the limited amount. Hence, he will spend on taking care to avoid an accident with a magnitude equal to the limited amount occurring and he will not spend on the care necessary to reduce the total accident costs. Obviously, the amount of care spent by the potential injurer will be lower and a problem of under-deterrence arises. The amount of optimal care, reflected in the optimal standard, being the care necessary to reduce the total accident costs efficiently, will be higher than the amount the potential injurer will spend to avoid an accident equal to the limited amount.[144] Thus, as a result of the cap, too little care is taken.[145]

111 The conclusion is, however, different in case of bilateral accidents, where also the victim's behaviour may affect the accident risk. The standard argument against providing full compensation to victims (also of non-pecuniary losses) in case of bilateral accidents is that victims can take precautionary measures which are not always observable for judges and which can therefore not be fully accounted for in contributory or comparative negligence defences.[146] A limit on compensation in case of bilateral accidents may therefore be useful in cases where victims should be given additional incentives to reduce the accident risk. Whether caps are efficient in specific bilateral accident cases will depend on the circumstances. The question arises – inter alia – whether exposing the victim to risk is indeed necessary to provide these additional incentives or whether the victim's incentives can be

---

144 See *M. Faure*, Economic Models of Compensation for Damage Caused by Nuclear Accidents: Some Lessons for the Revision of the Paris and Vienna Conventions, European Journal of Law and Economics (EJLE) 1995, 21–43.

145 The reason for the under-deterrence is obviously the same as for the under-deterrence which results from the insolvency of the injurer. Under-deterrence arises because the injurer is not exposed to full liability, either as a result of his insolvency or as a result of a cap.

146 This point has been made by *Rea*, JLS 1982, 50–52, but also by *M. Adams*, Warum kein Ersatz von Nichtvermögensschäden, in: C. Ott/H.B. Schäfer (eds.), Allokationseffizienz in der Rechtsordnung, 214 and by *C. Ott/H.B. Schäfer*, Schmerzensgeld bei Körperverletzungen. Eine ökonomische Analyse, Juristen Zeitung (JZ) 1990, 564 f. See also *M. Faure*, Compensation of Non-pecuniary Loss: An Economic Perspective, in: U. Magnus/ J. Spier (eds.), European Tort Law, Liber Amicorum for Helmut Koziol (2000) 143–159.

optimally controlled via the contributory negligence defence. Also the amount of the cap remains important. If the cap were set too low, this would give incentives to the victim but it could equally lead to serious under-deterrence of the injurer.

4. Compulsory insurance?

Another question is of course whether there is an economic argument to force a potentially liable GMO producer to seek insurance cover. This again is an issue that has received a lot of attention in the law and economics literature. We will of course not summarise all of this literature within the scope of this study, but merely state the most important argument in favour of compulsory insurance from an economic perspective.[147] The most important argument to introduce compulsory liability insurance relates to the insolvency problem. According to this argument, the magnitude of the harm will often exceed the individual wealth of an injurer, whereby a problem of undercompensation of victims will arise. Lawyers would, hence, push forward compulsory insurance as an argument to guarantee effective compensation to the victim. **112**

It is, however, also possible to make an economic argument that insolvency will lead to under-deterrence problems which might be remedied through liability insurance. Indeed, this so-called "judgment-proof" problem has been extensively dealt with in the economic literature.[148] Insolvency may pose a problem of under-deterrence. If the expected damage largely exceeds the injurer's assets, the injurer will only have incentives to purchase liability insurance up to the amount of his own assets. He is indeed only exposed to the risk of losing his own assets in a liability suit. The judgment-proof problem may therefore lead to underinsurance and thus to under-deterrence. Jost has rightly pointed to the fact that in these circumstances of insolvency, compulsory insurance might provide a better outcome.[149] By introducing a duty to purchase insurance coverage for the amount of the expected loss, better results will be obtained than with insolvency whereby the magnitude **113**

147 For a more detailed discussion see M. *Faure*, Economic Criteria for Compulsory Insurance, Geneva Papers on Risk and Insurance (Geneva Pap Risk Ins) 31 (2006) 149–168.

148 More particularly by *Shavell*, IRLE 1986, 43–58.

149 *P.J. Jost*, Limited liability and the requirement to purchase insurance, IRLE 1996, 259–276. A similar argument has been formulated by M. *Polborn*, Mandatory Insurance and the Judgement-Proof Problem, IRLE 1998, 141–146 and by G. *Skogh*, Mandatory Insurance: Transaction Costs Analysis of Insurance, in: B. Bouckaert/G. De Geest (eds.), Encyclopedia of Law and Economics (2000) 521–537. Skogh has also pointed out that compulsory insurance may save on transaction costs.

of the loss exceeds the injurer's assets.[150] In the latter case the injurer will indeed only consider the risk as one where he could at most lose his own assets and will set his standard of care accordingly. When he is, under a duty to insure, exposed to full liability, the insurer will obviously have incentives to control the behaviour of the insured. Via the traditional instruments for the control of moral hazard the insurer can make sure that the injurer will take the necessary care to avoid an accident with the real magnitude of the loss. Thus Jost and Skogh argue that compulsory insurance can, provided that the moral hazard problem can be solved adequately, provide better results than under the judgment-proof problem.

114 Indeed, this economic argument shows that insolvency may cause potentially responsible parties to externalise harm: they may be engaged in activities which may cause harm which can largely exceed their assets. Without financial provisions, these costs would be thrown on society and would hence be externalised instead of internalised. Such an internalisation can be reached if the insurer is able to control the behaviour of the insured. This shows that if the moral hazard problem can be solved adequately, insurance even leads to a higher deterrence than a situation without liability insurance and insolvency.

115 Notwithstanding this advantage of liability insurance, the literature has equally pointed to many dangers of compulsory insurance and has thus formulated several warnings. They can be summarised as follows:

116 Compulsory insurance should only be introduced when there is a sufficient amount of supply of differentiated insurance policies on the market. This supposes that sufficient competition on insurance markets exists and that operators have the possibility to actually seek coverage.

117 Therefore, compulsory insurance should only be introduced when insurers have sufficient information on the particular risk that will be covered. If too little information on the risk is available, the risk might be uninsurable or the risk premium (to account for insurers' ambiguity) may be so high that insureds are not willing to pay it.

118 That insurers have information about the risk is also crucial since they need to be able to control the moral hazard problem through an appropriate risk differentiation.

150 See also *H. Kunreuther/P. Freeman*, Insurability, Environmental Risks and the Law, in: A. Heyes (ed.), The Law and Economics of the Environment (2001) 316.

Compulsory insurance should never be accompanied with a duty to accept for insurers. The possibility for an insurer to refuse cover to high risk individuals can be a socially desirable control of moral hazard. **119**

If at all, a legislator should merely impose a duty to seek financial coverage, but should at the same time provide a great degree of freedom to the market to choose the type of financial coverage which is desired. Hence, this can but should not necessarily be limited to insurance. Other alternatives could meet the financial security requirement as well. **120**

VII. Cross-border issues – Conflict of laws

One can briefly address how from an economic perspective conflict of law rules should address this type of transboundary damage. Although traditional handbooks in law and economics do not address conflict of law rules in any detail, the potential solution does not seem to be that complicated. Again, a distinction should be made between on the one hand the contract case and on the other hand the tort case. In a contract case parties can again negotiate *ex ante* on the applicable law and hence a choice of law regime will usually be agreed between the parties and next, enforced by the judge. **121**

As far as the tort case is concerned, from an economic perspective there should not necessarily be a preference for the application of the law of the state of the injurer rather than applying the law of the state of the victim. The most important issue is, however, that also in a transboundary context externalities may arise. Hence, the function of tort law in the transboundary context should again be the internalisation of externalities. The bottom-line should therefore be that the GMO producer in the injurer's state should be forced to take into account the damage suffered by the victim even if it took place in another state. This result can be achieved as long as the victim has the possibility to bring a lawsuit against the injurer for the damage suffered, so that this internalisation can take place. Depending upon the legal system, in some cases victims will be forced to bring the suit in the state of the injurer. Injurers are thus also liable for harm caused in a transboundary context. In other situations victims may have the possibility to file the suit in their own state whereby the judgment that is obtained can afterwards be executed in the other country. **122**

Thus, if the victims can sue the injurers according to the place that is most suitable to provide full compensation, the difference in liability system among countries would not create serious problems for the victims. The same conclusion may also be drawn if we look at the incentive to interna- **123** lise the damage. Consider for example two neighbouring countries: country A has a sub-optimal standard and country B has an optimal one. Suppose that the victim is a citizen of B, and the injurer is a citizen of A. If the victim can file the case either in A or B, the injurer will take an optimal level of care according to country B, otherwise he would be held liable.[151]

VIII. Concluding remarks

124 As we indicated in the Introduction, there is unavoidably some parallels between the economic analysis as applicable to damage resulting from admixture (which was the focus of the first project) and GMO damage to third parties (being the focus of the current project). The main difference between both situations is that in the first case, there may be a contractual relationship between injurer and victim which may, from an economic perspective, call for the application of the so-called Coase theorem. When damage is, however, suffered by third parties, transaction costs are prohibitive and hence the legal system will have to intervene to determine how to internalise the externalities resulting from the GMO.

125 There is, however, one case where the Coase theorem and contractual liability may still play a role, being product liability. As we indicated, the economic analysis of product liability relies on assumptions concerning the possibility for a victim to obtain accurate information about the product risk. The victim would, taking into account the potential damage and hidden defects, only base his decision on the full price of a product. However, given the fact that many victims may be uninformed of potential product defects, this analysis may not be very useful. It was more particularly for that reason that economic theory also showed that, in case of underestimation of risks by consumers, strict liability may be an appropriate liability rule. Only strict liability provides incentives to manufacturers to adopt both an optimal care level as well as an optimal activity (production) level.

126 Most of our analysis concerning the types of liability focused on the situation where third parties are victims. This is typically also the case where damage is caused to the environment as such and where hence the "victim" may be the government who sues on behalf of the environment. Given the fact that in those cases it is not very likely that the victim itself

151 This is not the case if the victim can only sue the injurer in the injurer's country. In this case, the injurer will only take a level of care that is sufficient for him to avoid liability, namely the less-than-optimal standard of his country (A). As a result, there will be too many activities in country A that may create externality in country B.

may have a great influence on the accident risk, economic analysis predicts that these are cases where also a strict liability rule may be warranted. We indicated that this may be different only in the situation where the victim may not be an individual but rather a corporate entity. In those cases, corporations can sometimes also take measures to reduce risks which would make the situation bilateral. In that case, liability rules should also provide incentives to the potential victim to adopt optimal care and activity levels. The strict liability rule may hence probably be most indicated when individuals are victims, whereas negligence may be indicated in bilateral situations where both injurer and victim are corporate entities who could take adequate measures to reduce the risk.

We also indicated that to a large extent, regulation will have an important **127**
influence on the scope of liability. However, economic analysis indicates that whereas a breach of regulation may give rise to a finding of liability, the reverse is not necessarily true: following a regulatory standard (e.g. concerning the way in which one should deal with GMOs) does not necessarily reduce or exclude liability. Indeed, it is possible that the efficient care standard is higher than the level required under regulation. For that reason, it is important to still hold the potential injurer liable even if he followed the regulatory standard. Thus supplementary incentives could be provided through this complementary use of liability rules.

An important problem in any environmental liability case and also in case **128**
of damage suffered as a result of exposure to GMOs is the issue of causation. The few cases that have already dealt with GMO liability also show that causation is an important issue to be dealt with in case law. We indicated that economic analysis proposed the use of a proportionate liability rule to deal with uncertainty over causation. The advantage is that thus the inefficiency and unfairness of so-called "all or nothing" approaches are avoided and that the injurer is precisely exposed to the damage to the extent that he contributed to the risk. Moreover, a proportionate liability rule has the advantage that the judge can simply follow the indications provided by experts on the likelihood that a specific damage was caused by an exposure to GMOs.

As far as the defences are concerned, we indicated that undoubtedly con- **129**
tributory negligence may play an important role, especially in those cases where victims can really have an influence on the accident risk. If a strict liability rule were introduced, it is important to add a contributory negligence defence.

In addition, with respect to the unforeseeability defence, we observe that the **130**
foreseeability requirement could take into account some possible damage of

GMOs that have been discussed. Since the prevailing uncertainties on such impact may serve as evidence for the unforeseeability of the impact, applying the precautionary principle might induce the courts to lower their foreseeability requirement. In this case, the currently available findings on such impact function as early warnings to indicate that such impact is foreseeable according to the current state of knowledge. However, if the application of the precautionary principle is interpreted as holding the GM operator liable for any damage simply because the damage is itself the evidence indicating the failure to take the precautionary measure, the cost incurred by the GM operator will be high because the operator is induced into continuing to take a higher level of care. The operator will stop doing so only if the benefits of not taking a higher level of care are much higher than the cost of taking the higher level of care.

Comparative Report

Damage Caused by GMOs: Comparative Analysis

Bernhard A. Koch

I. Introduction[1]

1. Casum sentit dominus

1 The most basic rule of tort law, as explained in the first report,[2] is "casum sentit dominus", or "the loss lies where it falls". This often forgotten principle makes clear that the law of delict is not an automatic route to compensation, awarded by way of reflex as sometimes suggested by the media, but insists on a set of more or less far-reaching requirements which need to be fulfilled before a duty to indemnify the victim is greenlighted. Therefore, if any of these requirements are not met in a single case, the loss stays with the victim, who may still recover elsewhere, though, e.g. from social or private insurance.

2 While it may be bitter for the victim, law needs to reconcile all interests involved, including those of the addressee of the victim's claim, but ultimately also of society at large. In the interplay of individual interests, compromises are sometimes necessary. For example, while someone may be disturbed by the noise of aircraft flying over her piece of land, she will

---

1 Please note that the TLD country codes will be used in the following footnotes to refer to the country reports in this volume, added by the marginal number of the relevant paragraph(s). AT therefore refers to the Austrian country report, CZ to the Czech Republic, DE to Germany, DK to Denmark, EE to Estonia, ES to Spain, FI to Finland, FR to France, GR to Greece, HU to Hungary, IT to Italy, LI to Liechtenstein, LU to Luxembourg, MT to Malta, NL to the Netherlands, NO to Norway, PL to Poland, SE to Sweden, SI to Slovenia, UK to England and Wales (not to the UK as a whole), AU to Australia, BR to Brazil, CA to Canada, US to the United States. In order to stay with two-letter abbreviations, the report on jurisdiction and conflict of laws is referred to as JC, the presentation of international environmental law as IE, the insurers' perspective as IN, and the economic analysis as EA. As it will be cited throughout the text, the previous study published in B.A. Koch (ed.), Economic Loss Caused by Genetically Modified Organisms (2008) will be referred to as "Economic Loss" only.

2 *B.A. Koch*, Comparative Report, in Economic Loss 585 (no. 12).

not be able to collect compensation for the alleged deterioration of the value of the property or for the impact upon her own enjoyment of the land unless certain limits are exceeded (the latter being more likely the closer to the airport the land lies, of course).[3] These limits are laid down in order to achieve a balance of interests, and maintaining an airport, being in the public interest, necessarily entails that those living in the area will have to live with the sound of approaching and ascending aircraft, just like house owners have to bear noise from through-traffic on nearby streets, or with fumes from their neighbours' occasional barbecue parties. In the case of GM farming, pollen may drift over very long distances, but its concentration obviously decreases the further it is carried, and at some point away from its origin the likelihood and/or impact of contamination is so low that a legal system generally favourable to GM farming may not link consequences to such remote admixture, so that the owner of the target field will also not be able to find compensation in tort law.

In this project, we wanted to identify what tort law requires in order to indemnify "classic" losses such as bodily injury and property damage if caused by GMOs in the food or feed supply chain, or if GMOs turn out to have a harmful impact upon the environment. This comparative report tries to highlight the key differences between the jurisdictions under survey, but also show where they are in accord. The prime focus is on the law of delict, since contractual liability will be addressed elsewhere.[4] At the end of the comparison, more general questions will be asked such as whether the existing differences should be levelled out by harmonisation (at least within the EU), and/or whether there is a need to promote alternative compensation models. **3**

2. Imaginable losses

In our previous project on liability for GMOs, we focused primarily on the pure economic loss sustained by a conventional or organic farmer whose produce had been contaminated with GMOs, so that a supposed premium on the market price for GM-free products could no longer be gained. Imaginable loss scenarios involving GMOs go much further though: **4**

3 But see DK no. 6.
4 This is subject of a separate study conducted within the framework of the CoExtra project.

- To begin with, such economic losses may equally be incurred by others up or down the chain of distribution such as wholesalers or food or feed producers. However, most of these issues will be governed by contractual liability, which is not within the ambit of this study.[5] Still, some jurisdictions expand contract law concepts into tort law, and this is where such losses may play a role also in tort practice.
- Anti-GMO activists in particular point at potential health risks of GM products to humans. If such a risk should indeed materialise, the injuries caused would affect interests enjoying the highest degree of protection in all legal systems.[6] However, consequences in tort law depend upon the degree of impact: mere concern, for example, about potential health risks, that does not amount to a condition which can be measured medically will typically be disregarded by all jurisdictions (*de minimis non curat praetor*).[7] As soon as the impact upon the human body can be diagnosed by medical science and triggers the need for treatment, however, the laws of delict will invariably be prepared to offer compensation to the victim subject to their standard conditions.
- Animals themselves may be harmed upon consumption of GM feed, but even if they remain unaffected, their produce may turn out to be harmful, or it may not be as equally marketable as before.
- The impact upon the environment may at first be just that – potentially detrimental effects of GM farming upon biodiversity is another theme often heard in public debates.[8] Harm to the environment, should it ever occur, may also trigger secondary losses to individuals, however, starting from property losses, e.g. of those owning land or animals affected by such effects, to – again – impacts upon the health of human individuals.[9]

5 The causes of such imaginable losses may lie along the entire supply chain – from the very first steps of the seed production to the ultimate distribution of (primary or secondary) agricultural products to consumers. Admixture may occur at all stages of the supply chain, and the further down damage occurs, the more difficult it will be to establish who caused the loss.

5 The responses to Case 1 show that the core cases that come to one's mind first would fall under contractual rather than delictual liability, thereby reducing the scope of loss scenarios falling under tort law proper.

6 See infra II.1.

7 See infra II.3.

8 This will be addressed infra VII.

9 See infra VII.2(b) and (c).

II. Damage

1. Losses to persons or things

When it comes to the "classic" losses for which tort law foresees remedies, jurisdictions worldwide are united in claiming to adhere to the principle of full compensation.[10] However, this only means that they are willing to remedy all those losses in full that they recognise as compensable, so there is a first limit to the indemnification of losses in a comparative overview already at the stage of defining damage. For example, while stepping on someone's toes clearly counts as an attack on that person's physical integrity, no jurisdiction will probably be prepared to offer her compensation for the pain she undoubtedly may suffer, arguing that such a minor interference is not damage recognised by tort law. "No law based on rational principles can impose damages on each and every act of carelessness. ... The law of delict would ruin itself ... if, say in a case of negligent damage to the environment, it not only compensated the owners of the contaminated land for their loss, but also awarded the inhabitants of the area compensation for loss of enjoyment due to the damage caused to the wildlife of the area."[11] 6

Apart from such *de minimis*-rules and related limitations, there are more substantial filters that may apply, and so compensation for that reason alone may be fuller in some jurisdictions than in others.[12] The objective of full compensation has further boundaries at the remedies stage, where there may be caps or other limitations that have an immediate impact upon the amount offered in compensation (not to speak of the limits of liability along the way towards remedies). These will be addressed further below.[13] 7

While there are at least some differences concerning the recognition of material losses as compensable in the various jurisdictions, there is clearly a marked difference between them as far as non-pecuniary losses are concerned. To begin with, not all legal systems recognise the same types of 8

10 *B.A. Koch/H. Koziol*, Comparative Analysis, in B.A. Koch/H. Koziol (eds.), Compensation for Personal Injury in a Comparative Perspective (2003) 407 (no. 40); *Ch. von Bar*, The Common European Law of Torts II (2000) no. 136 ff. See, e.g., CZ no. 17.

11 *Von Bar* (fn. 10) no. 1.

12 Cf. *B.A. Koch/H. Koziol*, Comparative Analysis (fn. 10) no. 41 ff. on the differences of the legal systems covered there with respect to compensation for personal injury (as far as, for example, visitors' costs, loss of earning capacity without loss of actual income, or the indemnification of the inability to run the household are concerned).

13 See infra IX.

immaterial harm, which comprises pain and suffering in its core, extends to losses of amenities, and can go further beyond infringements of personality rights[14] to what in Italy is recognised as *danno extrapatrimoniale* such as *danno biologico* or *danno esistenziale.*[15] A separate category concerns third party losses following fatal or non-fatal, but very serious bodily injuries of the primary victim, even though most jurisdictions have in the meantime at least acknowledged the equivalent of damages for bereavement in recognition of the loss of a loved one.[16]

9 Property losses also on their face seem to be recognised as compensable in the same way under all jurisdictions, but at least subtle differences remain, if only when it comes to the type of assessment (objective or subjective), the (lack of) recognition of sentimental or other personal affection, the compensability of repair costs even if not yet incurred, or similar facets.

10 Even though legal systems evidently differ not only in minor points of detail when defining what losses are recognised by the various tort systems in the first place,[17] one should note that these variations are nothing specific to the topic of GMO-related losses, but apply equally in other tort cases.

11 What is specific to the kinds of cases under survey here, however, is the question whether or not contamination with GMOs counts as "damage" to the land or to the crops of the non-GM farmer. This issue does not seem to have appeared before European courts yet and is therefore open to debate.[18] The solution will depend upon the courts' view of whether a mere physical change such as a GM seed taking root already counts as "damage", or whether it is the (monetary or other) consequences of such a change instead that are crucial, such as a loss in value on the markets or the like, but also the development of certain resistances[19] or a measurable

14 Cf., e.g., HU no. 17; NO no. 4.

15 See, e.g., *G. Christandl*, Eine kurze Darstellung der neuesten Entwicklungen im italienischen Nichtvermögensschadensrecht unter besonderer Berücksichtigung des *danno esistenziale*, http://www.personaedanno.it/cms/data/articoli/files/000180_resource1_orig.doc.

16 Germany may end up being the only jurisdiction left which does not accept mere grief as a loss to be indemnified by tort law now that the Netherlands are considering to introduce some type of compensation by statute: See e.g. *M. Faure/T. Hartlief*, The Netherlands, in H. Koziol/B. Steininger (eds.), European Tort Law 2006 (2008) 338 (no. 6).

17 An in-depth comparative analysis of the recognition of the various kinds of losses in 28 European jurisdictions will be given by *B. Winiger/H. Koziol/B.A. Koch/R. Zimmermann* (eds.), Digest of European Tort Law II: Essential Cases on Damage (forthcoming 2011).

18 See, e.g., UK no. 6.

19 *Munich Re*, Genetic Engineering – A Challenge for the Insurance Industry (2003) 55, 73; EA nos. 83 ff.

negative impact upon soil ecology.[20] GM-pro activists will of course argue that admixture with GMOs not only leaves (previously) GM-free land unharmed, but on the contrary improves its value due to the specific benefits of transgenic seeds.

2. Pure economic loss in particular

Pure economic loss has been dealt with extensively in the first volume that focused primarily on farmer-to-farmer problems. As in this previous work, pure economic loss is defined as damage suffered irrespective of injuries to the human body or to someone's property, but which instead directly affects the assets of the victim. It is typically distinguished from mere consequential loss, which is a secondary loss incurred after primary damage to persons or property was sustained, and which is indemnified together with the latter.[21] **12**

In order to continue the example of the previous sub-section, if one considers the drift of transgenic pollen onto a conventional or organic field as harm to the latter itself, the loss of the premia on GM-free products is but consequential to the damage to the land. If, on the other hand, the admixture as such is not yet perceived as a detrimental change to the property, so that the loss of bonus payments is the primary harmful event in the sphere of the victim, she has thereby suffered a pure economic loss. This already shows that there are no clear borderlines between the two concepts.[22] Still, the question is not of purely academic relevance – as stated for Australia (but equally applicable to all jurisdictions), "it is generally easier to recover for damage to property than it is for purely economic losses."[23] **13**

This is true because some jurisdictions are more reluctant to offer compensation for merely economic losses which are not just a secondary consequence to some primary injury to life, health, or property. A prime argument for such limitation is the floodgates argument:[24] the range of **14**

20 EA nos. 89 ff.

21 *M. Bussani/V. Palmer*, The notion of pure economic loss and its setting, in M. Bussani/V. Palmer (eds.), Pure Economic Loss in Europe (2003) 3 ff.; *W. van Boom*, Pure Economic Loss: A Comparative Perspective, in W. van Boom/H. Koziol/Ch. Witting (eds.), Pure Economic Loss (2004) 1 (nos. 5–8).

22 *Von Bar* (fn. 10) no. 25 ff. and the examples given in no. 32 (in particular the German case at fn. 175 where fish were fed with antibiotics, rendering them unmarketable, even though they were not harmed from a veterinarian point of view). Cf. the *Hoffmann* class action cited by CA no. 10 and the comparison of the *Starlink* and *Sample* cases in US no. 4.

23 AU no. 8.

24 *Bussani/Palmer* (fn. 21) 16 ff. See also EA no. 104.

claimants may explode if a tortfeasor had to compensate all those who incur a purely financial detriment without further harm – just think of the employer of an injured employee, or someone stuck in a traffic jam after an accident between two other cars. The more foreseeable the range of such third-party claimants gets, in particular by way of a special pre-existing relationship between them and the tortfeasor, or the more blameworthy the latter's conduct is, the less protection the tortfeasor deserves against their demands. Therefore, jurisdictions which distinguish pure economic loss as a separate category tend to recognise it as a compensable detriment to one's assets only if incurred within a contractual relationship,[25] or in cases of qualified fault (in particular if the tortfeasor acted with intent).[26]

15 Other jurisdictions do not distinguish pure economic losses from further types of damage and are therefore not biased against recognizing them as compensable.[27] However, there may be a *de facto* difference when it comes to applying the standard tests of liability, in particular when establishing causation.[28]

3. Mere fear of a loss in particular

16 As long as there is no scientific evidence of specific dangers of a certain technology, there obviously have also not been any occurrences where it was proven to have caused harm. Apart from lawmaking where legislators are called to foresee liabilities for potential future incidents of losses, such technologies are considered in tort law practice from two different angles: one is based upon assumptions of causation when the technology is merely suspected of triggering some actual loss that has already occurred. This will be dealt with in the subsection on causation below.[29] The second angle focuses on the assumption itself and examines whether the fear of a future loss caused by some technology (and therefore the fear of the technology itself) is compensable as such.

17 In the area of green biotechnology, these fears may arise in various settings, depending upon the role of the players involved:

---

25 E.g. AT no. 9; EE no. 10; LI no. 9; AU no. 11.

26 AT no. 9; DE no. 11 (both e.g. when acting *contra bonos mores*); FI no. 15; SE no. 13 (criminal act, especially weighty reasons).

27 DK no. 5; FR no. 6; GR no. 10; IT no. 10; LU no. 23; NL no. 5; NO no. 6; PL no. 18; SI no. 4; ES no. 14.

28 See, e.g., ES no. 14. Cf. also HU no. 19 and NL no. 6.

29 Infra III.2(b).

(1) Whether substantiated or not, consumers may fear of developing allergies or suffering from other health problems if the agricultural products they buy contain GMOs. This may, in turn,

(a) remain a mere emotional state, without having any measurable medical impact, or

(b) develop into an illness triggered by the fear, which then constitutes an actual injury to the person and as such falls under the commonly recognised types of harm in tort law.

(2) Farmers, on the other hand, may fear that their fields will be contaminated with GMOs, which, if true, might trigger economic losses if their conventional or organic products can otherwise be sold with a premium on the market for being GM-free.[30]

(3) A combination of the two may lead to a real loss in income: If consumers refuse to buy from a farmer whose produce they suspect contains GMOs, which in fact it does not, the latter will suffer an actual economic damage, even though it is just a secondary consequence of the fear of technology (that the victim does not even apply herself).

The economic losses of varieties (2) and (3) have already been addressed by our previous study.[31] With most jurisdictions requiring actual loss in order for tort law to step in,[32] there are still some where at least variant (3) may not be excluded from recovery just because of the type of loss at stake.[33] **18**

In the first scenario, all jurisdictions are in accord that mere fear as in alternative (a) is not compensable.[34] However, if the anxieties trigger psychosomatic manifestations or are otherwise medically measurable, victims in some countries may pursue tort claims for recovery subject to further requirements.[35] **19**

---

30 This is currently not the case in those European markets where both conventional and GM-maize compete. See, e.g., *M. Gómez-Barbero/J. Berbel/E. Rodríguez-Cerezo*, Adoption and performance of the first GM crop introduced in EU agriculture: Bt maize in Spain (2008, available at http://ftp.jrc.es/EURdoc/JRC37046.pdf) Table 11 (pp. 29).There are, however, organic premia which are the same as compared to both conventional and GM maize.

31 *Koch* (fn. 2) no. 35 with further references.

32 See, e.g., AT no. 10; BR no. 31; CZ no. 23; GR no. 13; MT 17.

33 *V. Ulfbeck*, Denmark, in Economic Loss 145 (no. 46); FI no. 24; HU no. 25; LI no. 10.

34 CZ no. 25–26; UK no. 9, FR no. 8 (with some lower courts obviously not entirely sure); cf. LU no. 27 (influence of the precautionary principle at the example of mobile phone radiation); ES no. 16 (probably more generous approach of the Spanish courts).

35 AU no. 12; CA no. 12 (but "difficult to imagine" in light of requirements); DK no. 6; FI no. 21 ff.; GR no. 13 (fn. 12); HU no. 25; NL no. 7; PL no. 21; ES no. 15 (mental trauma

20 Variant (1) may be treated differently by some jurisdictions otherwise denying compensation if the harm that the consumers fear is certain to occur in the future and just a matter of time unless prevented in the meantime.[36]

4. Nominal losses in particular

21 The congestion of court dockets with petty lawsuits has not only triggered procedural limits to access to courts, but also raised concerns how substantive law could cut back small tort claims. During the debate around the German tort law reform of 2002, there were (ultimately unsuccessful) plans to declare nominal losses incompensable.[37]

22 While the monetary benefit to claimants in such cases may be minimal, they nevertheless invest time and efforts into pursuing such trivial claims in order to be heard, and in order to achieve the recognition of the claim as such. The question therefore is whether jurisdictions hear such calls and accept nominal losses, e.g. by awarding symbolic "compensation". This may be an issue particularly when it comes to recognizing fear scenarios as outlined above, or if there is no economic loss because conventional crops achieve the same market price as their GM counterparts.

23 The question is answered in the affirmative in common law jurisdictions, where certain torts do not require proof of damage.[38] In most other jurisdictions, nominal losses are either expressly or at least in practice disregarded.[39] Only a few civil-law countries recognise symbolic losses or at any rate do not rule them out as such.[40]

---

even if no illness); SE no. 18 (only if criminal conduct). See also the Seveso case example given by the Italian report, IT no. 13.

36 See, e.g., CA no. 13; DE no. 12; FR no. 7; GR no. 12; MT no. 17.

37 *J. Fedtke*, Germany, in H. Koziol/B.C. Steininger (eds.), European Tort Law 2001 (2002) 229 (no. 6). Such plans are also in the current Austrian tort law draft; see *B.C. Steininger*, Austria, in H. Koziol/B.C. Steininger (eds.), European Tort Law 2007 (2008) 134 (no. 8).

38 AU no. 14 (though doubtful for GMO cases); CA no. 15; UK no. 12; US no. 7.

39 AT no. 12; CZ no. 30; DE no. 14; DK no. 8; ES no. 18; GR no. 17; HU no. 29; IT no. 15; LI no. 12; LU no. 30; MT no. 20; NL no. 12; NO no. 9; PL no. 27; SE no. 21 (but no general exclusion); SI no. 7; BR no. 35; CA no. 15 (Québec).

40 EE no. 15 (not per se excluded); FI no. 28 ("possible"); FR no. 13 ("recognised in French law"; see in particular the examples in fn. 29).

5. Special rules on mass losses

Losses caused by GMOs in the food supply chain in particular may exceed the scope of individual incidents by far and affect a multitude of victims. If we depart from the assumption for the time being that food products containing transgenic ingredients indeed trigger allergies or cause other bodily harm, as opponents predict, cases filed on behalf of all those affected may go far beyond what we know from recent pharmaceutical cases or the like. **24**

Only few jurisdictions in Europe currently foresee special rules on handling mass tort claims,[41] but this is just a momentary snapshot, as many are indeed considering introducing (or already have introduced[42]) procedural tools such as class actions modelled after or at least inspired by the U.S. correspondent.[43] This development is being pushed by efforts on the EU level to develop such mass claim models for consumers. **25**

Procedural alleviations as indicated have a considerable impact on substantive tort law as well: Some claims (particularly smaller ones) would never be brought into court otherwise, since fees and costs effectively serve as deterrent. Also, the scientific proof of a causal link may be too expensive a hurdle for individual claimants, even if they can ultimately recover such costs should they win their case. **26**

Another feature of mass tort claims from experience is that they tend to draw more media attention, which is particularly true for sensitive issues like GMOs, and media presence puts pressure on politicians to interfere ad hoc, whether within the limits of existing substantive and/or procedural law or not. **27**

6. Proving damage

Jurisdictions adhere to different standards of proof, which can have a crucial impact on the outcome of the case if filed in country A rather than country B.[44] While this is generally true for all elements of a tort claim, **28**

41 Most jurisdictions do not offer special provisions governing mass losses: AT no. 13; CZ no. 31; DE no. 15; DK no. 9; EE no. 16; FR no. 14 (but action pursuing recovery for collective losses, e.g. of an association); GR no. 18; HU no. 30; IT no. 16; LI no. 13; LU no. 31; MT no. 21; NO no. 10; SE no. 22; UK no. 13; AU no. 15; BR no. 36.

42 FI no. 29; NL no. 13; PL 29 (planned); ES no. 19–20; CA no. 16–17; US no. 9.

43 See FR no. 15–16 (collective loss claims, representative action); PL no. 29.

44 See, e.g., *K. Clermont/E. Sherwin*, A Comparative View of Standards of Proof, American Journal of Comparative Law (Am. J. Comp. L.) 50 (2002) 243. The impact of these differences on the proof of causation is addressed below at III.2(a).

some legal systems apply further special rules when it comes to proving the fact or the extent of a loss.[45]

29 In particular, courts seem to be more generous with respect to the exact value of the damage, although all require from the plaintiff proof that she has suffered some loss, even if the burden of proving further elements may be shifted onto the defendant.[46] There is a more or less express range of judicial discretion in valuing the loss, particularly if it appears difficult to quantify it in detail.[47]

30 More troublesome are claims where it remains uncertain (though not impossible) whether there will be any harm at all in the future. French courts, for example, are in practice willing to compensate such a mere chance of future damage,[48] whereas other jurisdictions at best allow declaratory judgments which freeze the question of liability (and stop the period of limitation from running), but leave the determination of remedies for the future to decide once the risk of a loss materialises.[49]

III. Causation

1. Linking damage to a cause

31 As a general rule recognised by all legal systems (though not necessarily expressly),[50] no-one is to compensate damage that he did not cause.[51] This simple rule is linked to an equally simple test, the *conditio sine qua non* (or but-for) test: if the victim had still incurred her loss had the defendant not intervened in the course of events, the latter's conduct cannot have caused the damage, and therefore he should not be liable for his

45 See, e.g., FI no. 26–27.

46 E.g. AT no. 11; FR no. 9; NL no. 9; PL no. 23; ES no. 17. But see GR no. 16 (claims may be rejected "as vague and unfounded" if exact quantum not proven).

47 AT no. 11; FI no. 25; FR no. 10; HU no. 27–28; NL no. 10–11; SE no. 20; UK no. 11.

48 FR no. 11; see also AU no. 13.

49 E.g. DE no. 13; PL no. 25.

50 *J. Spier/O. Haazen*, Comparative Conclusions on Causation, in J. Spier (ed.), Unification of Tort Law: Causation (2000) 127 (127 ff.). A comprehensive overview of the jurisprudence in 26 European jurisdictions is given by *B. Winiger/H. Koziol/B.A. Koch/R. Zimmermann* (eds.), Digest of European Tort Law I: Essential Cases on Natural Causation (2008, in the following: Digest I).

51 If someone is to be held liable for another, such as an employer for his employees, it is the latter that have set the actual cause and not the person ultimately liable, but a cause within the latter's sphere is still necessary to trigger his liability.

behaviour or source of risk within his sphere.[52] If it is clear, for example, that the GMOs that caused damage did not originate from the defendant's field, he will not be liable for such harm. If the only applicable rule of liability requires some culpable wrongdoing of the tortfeasor, the fact that, say, the physical condition of the victim would equally have deteriorated if the defendant's conduct had not taken place, the latter will not be liable, even if his behaviour as such was in violation of some rule.

There are exceptions to this general rule, which come into play when there is more than one possible cause of the loss at hand, and it remains unclear which of the several options really had a significant impact on the events ultimately triggering the damage. These exceptions will be addressed further below.[53] **32**

The but-for test is too rough, though, in order to identify only those tortfeasors who should be considered when trying to establish liability. With no further qualification, the parents of a tortfeasor would also pass the test because if they had not had that child, the latter could never have committed a wrong. There are, of course, less ridiculous constellations which all legal systems sort out in order to focus on those causes which are worthwhile pursuing in tort law. The filters they use are referred to as adequacy test, the theory of proximate causation or remoteness, and similar methods to reduce the list of causes to those which are close in space, time and other relevant factors.[54] **33**

Whatever gets caught by those filters (if it gets there through the but-for test in the first place) cannot trigger liability, no matter how negligent or dangerous the activity was. In light of obvious question marks as to the risks involved with GMOs, this hurdle will undoubtedly prove crucial in claims practice. **34**

2. Burden of proving causation

All actual or potential causes have in common; however, that they need to be proven as such before tort law can decide whether they may ultimately lead to liability of the person(s) in whose sphere the cause lies. **35**

---

52 The mirror image is equally true in case of omissions: if the defendant had failed to act, he in principle should not be liable for such omission if the damage would have occurred even if he had acted as expected.

53 Infra III.3.

54 *Spier/Haazen* (fn. 50) 130 ff.

(a) Standard of proof

36 As has already been explained, legal systems apply different standards of proof.[55] Some legal systems insist on a very high degree of likelihood that the alleged cause had indeed led to the harm at stake – even though one can rarely prove facts with certainty, that is in essence what these jurisdictions strive for, at least in theory.[56] At the other end of the spectrum lie those countries where it suffices to prove that some conduct or activity was more likely than not the cause of some harm, so instead of a probability above 90%, anything higher than 50% sharp – "a preponderance of the evidence" – is already satisfactory.[57] Some jurisdictions are in between, requiring a distinct degree of likelihood, but not necessarily close to certainty.[58]

37 These differences clearly affect the way claims regarding novel technologies make it through the courts in the various legal systems.[59] The higher the degree of uncertainty, the less likely will the claimant succeed in those jurisdictions which aim at certainty, and consequently allegations based upon mere fear of negative consequences will hardly serve as a basis for compensation claims or other remedies.

(b) Easing the burden of proving causation

38 The standard of proof affects the one who is charged with the burden of establishing causation, which traditionally is the claimant, who has to prove all requirements to grant her claim.[60] This burden may be shifted or entirely reversed, however, which imposes a more or less stringent duty upon the defendant to weaken the allegations by the claimant.

---

55 Supra II.6. These are also the key drivers in cases of causal uncertainty from an economic perspective: EA no. 49.

56 AT no. 18; CZ no. 47; DE no. 23 (but lower standard in pharmaceutical product liability); ES nos. 23, 31 (though some courts appear to be more generous); GR no. 24; HU no. 37; LI no. 18 (but lower degree of likelihood in draft GMO Act, no. 19); LU nos. 49–50; MT no. 25; SI no. 15.

57 EE no. 21; IT no. 20; NL no. 20; NO no. 15; AU no. 24; CA no. 23; UK no. 20; US no. 12. See also SE no. 37 (if strict liability of the Environmental Code applies).

58 DK no. 7; FR no. 28; PL no. 39 (generally "probability bordering on certainty", but sometimes only "sufficient degree of probability"); SE no. 36.

59 See also EA nos. 50 ff. providing an economic assessment of the various options outlined above.

60 See, e.g., AT no. 11; GR no. 14; UK no. 11.

The most far-reaching variation thereof is a presumption in favour of the 39
claimant which cannot be rebutted at all. As it seems, no legal system is going to such extremes in the case of damage caused by GMOs, even though French law comes close: if produce that was intended to be GM-free exceeds the labelling threshold because of involuntary admixture, any ensuing loss can be recovered upon mere proof that it was grown in the vicinity of a field where transgenic crops were cultivated.[61]

A less radical variation of the burden of proof is to presume a chain of cau- 40
sation from certain facts (that the claimant has to prove as a starting point), but to allow the defendant to overturn this supposition by proving that the impact of the conduct or activity for which he is charged did not in fact fit into this presumed course of events. This may be expressed by law, such as the Luxembourg Coexistence Law[62] or the Norwegian Pollution Act,[63] or develop from court practice, such as in the Netherlands.[64] In Greece, a reversal of the burden of proof may result from the analogous application of a civil code provision which manifests the principle that he who controls a source of danger has to account for the consequences should such risk materialise.[65]

A lesser form of modifying the standard of proof is to lower it by requiring 41
the claimant to convince the court that the defendant's activities were at least likely to have triggered the loss instead of asking for full conviction. This is the case, for example, in Austria, where the Gene Technology Act does just that by reducing the scope of what claimants have to prove to a mere likelihood (without setting a minimum degree of probability) that the GMO controlled by the defendant was prone to cause the damage.[66] The defendant can only rebut this if he shows that the claimants' loss was in fact not caused by the genetic modification, but by some other cause.[67] Also the Finnish Environmental Damage Compensation Act – as the draft

---

61 FR no. 29.

62 LU no. 51.

63 *B. Askeland*, Economic Loss Caused by GMOs in Norway, in Economic Loss, 361 (nos. 9–10). See also ES no. 24–25.

64 On the *omkeringsregel* developed by the Dutch Supreme Court, see NL no. 21. Cf. NO no. 11.

65 *E. Dacoronia*, Economic Loss Caused by GMOs in Greece, in Economic Loss, 246 (nos. 52–53).

66 A similar rule is included in the proposed Liechtenstein Act on the Use of Organisms. Art. 61 para. 2 of the draft provides: "If this evidence cannot be reliably provided or if providing such evidence would be an undue burden on the person required to provide proof, the court may accept the preponderance of the evidence." See also Sweden for a comparable approach developed by court practice.

67 AT no. 21.

law on coexistence – does not require full proof of causation from the claimant, who only needs to show the probability of a causal link between defendant's activity and her loss; however, such likelihood must exceed 50% significantly.[68] The German Act on Genetic Engineering, though otherwise similarly strict as its Austrian counterpart, does not go that far: the claimant still has to prove that her loss was caused by GMOs within the defendant's control, the only alleviation offered by the statute is a presumption that it were the modified qualities of the genes which triggered the loss. This can be rebutted by the defendant, however, if the latter shows that it could also have been the unmodified genes.[69]

42 Another way to reduce the burden of proving causation is by way of recognizing prima facie evidence, which is slightly more covert than the variety just mentioned, but in most cases has the same effect on the outcome of the case.[70] Here, the evidence evaluated by the judge upon her own assessment (and not required by statute) shows that the alleged cause according to experience tends to show exactly the kind of consequences at stake in the absence of proof that there are other possible explanations for such outcome. Let us for example assume that one day there may be scientific proof that a certain gene transferred into a crop triggers a specific allergic reaction upon consumption.[71] The claimants all show exactly those symptoms, and the defendant has cultivated, processed or distributed a product containing that very gene which made its way into the food consumed by claimants. In such a case, the judge may prima facie assume that the defendant caused their condition unless the latter can prove that there are other allergens within the sphere of the claimants which also might have triggered those reactions.

43 Finally, yet another approach is to conclude from the facts as evidenced that there is no other logical explanation but for the one presented by the claimant. This practice of "elimination" is expressly recognised in France, for example.[72]

68 FI no. 31.

69 DE no. 24.

70 AT no. 19 (particularly if protective law was violated, such as the Austrian GMO legislation); PL no. 33. See also *Dacoronia* (fn. 65) nos. 54–55.

71 Cf. the Brazil nut allergen case described in *Munich Re* (fn. 19) 50.

72 FR no. 19 ff. (see, e.g., the bee case cited at fn. 40); cf. NO no. 11.

3. Special problems of causation

44 Despite alleviations of the burden of proof it may still remain unclear what actually triggered the loss because there are other equally imaginable causes. If none of them exceeds the threshold of the standard of proof, no cause in the legal sense can be identified, and consequently no-one will be held liable. If, however, more than one constitutes a *conditio sine qua non*, the first question is whether the two or more causes have jointly triggered the loss, so that all of them pass the but-for test. Such a joint causation typically leads to joint and several liability of all those interacting, unless their individual share can be identified.[73] There may, however, also be cases where the multitude of possible causes does not pass the but-for test because the loss may still have occurred if any one of them is disregarded. Only a few of these scenarios shall be briefly considered by way of examples.

(a) Alternative causation

45 The classic scenario of such kind is often explained with the so-called hunters' case – two hunters fire a shot, a passer-by is hit by one stray bullet, but it cannot be proven from which of the two guns it originated. Several legal systems currently foresee joint and several liability of both hunters, if at all, so irrespective of who is sued by the victim, he has to indemnify the latter in full and subsequently seek recourse from the other hunter.[74] A more modern approach suggests to instead hold them liable proportionally, i.e. each of them only for half of the loss, arguing that it would be unfair to place the risk of insolvency of the other hunter solely upon the one who happens to be sued first – and who may not have caused the loss at all.[75] A radically different solution is followed by some other legal systems, essentially based on the argument that neither defendant can be proven to have

73 E.g. AT no. 52.

74 See *H. Koziol*, Comparative Report, in Digest I (fn. 50) 6a/29 (pp. 387 ff.); AT no. 14; DE no. 16–20; EE no. 18; ES nos. 21, 27; FR no. 25; *Dacoronia* (fn. 65) nos. 56 ff.; LI no. 14; NL no. 16 (but see no. 17 for an exceptional case of proportionate liability); PL no. 32 (if concurrent action of a group, not random individuals); SE no. 26 (but see no. 27); BR no. 38; CA no. 18; US no. 10.

75 See Art. 3:103 (1) of the Principles of European Tort Law (http://www.egtl.org/Principles/) and the commentary thereto in *European Group on Tort Law*, Principles of European Tort Law (2005) 47 ff. The proposed solution was adopted, e.g., in England (UK no. 16–17) and the Netherlands (NL no. 17, so far only in special legal – here: employment – relationship). Cf. US no. 9. It is also favoured by the economic analysis: EA nos. 51 ff., 60 ff.

caused the loss, so that the victim cannot recover in tort law at all.[76] So if, for example, conventional maize was stored in three different silos consecutively, and the admixture with a transgenic variant could have occurred in either of the three with the same degree of likelihood, none of those in charge of the silos can be held liable in the latter jurisdictions.[77] If all three had disobeyed cleaning requirements after storing GM maize and thereby are proven to have created a risk of admixture, which may have materialized in only one of the silos, each of the three will be held liable for the full loss in the first jurisdictions cited and then have to seek recourse from the other two internally.[78]

(b) Cumulative causation

46 Another multi-cause scenario where the but-for test fails is the case of so-called cumulative causation, where each of the several simultaneous activities alone would have triggered the whole loss, so the damage would have occurred even in the absence of all but one of these events, irrespective of which is to be disregarded hypothetically. In contrast to the previous scenario, there is no doubt that each activity constituted a cause if assessed individually, and so probably at least a majority of jurisdictions would opt for joint and several liability of all those involved.[79] This may also be a more frequent scenario in the GMO context, at least if there is widespread cultivation of a certain approved GMO which ends up being discovered in conventional or organic food or feed.

(c) Supervening causation

47 A related problem differs from cumulative causation only when looking at the timeline: If there are two consecutive rather than simultaneous causes, where the first one already triggers the full negative consequences that the second event would have equally caused had it happened earlier, one

76 CZ no. 41; DK no. 11; HU no. 35; LU no. 39 (but see nos. 40–42); SI nos. 9–12; cf. SE no. 27.

77 Cf. the Canadian ragweed example (CA no. 18).

78 Cf. the US theory of enterprise liability, where an entire industry producing the same risky product can jointly be held liable if the producer of the actual defective item cannot be identified; US no. 10.

79 E.g. *Dacoronia* (fn. 65) nos. 56 ff; UK no. 15 (if material contribution to the risk proven). See also *B.A. Koch,* Comparative Report, in Digest I (fn. 50) 7/29 (pp. 476 f.). Cf. PL no. 34–35 (two defective products from different producers as possible causes – Supreme Court overturning Court of Appeal's decision in favour of joint and several liability).

could disregard the second event unless it has led to further or aggravated harm. However, not all legal systems draw these consequences, as some would treat both events as causes that may trigger liability. Different solutions are not only offered in an international perspective, but even intranationally, as these problems seem to be disputed even within legal systems.[80]

Such problems may well arise in the GMO context if admixture occurred at two or more steps of the supply chain. In such cases, it is not always the first one in time that will be considered relevant in a tort law perspective. **48**

(d) Risks within the victim's own sphere

(i) Conduct of the victim or people within her sphere

If one of the actual or suspected causes for the victim's loss lies in her own sphere, this may reduce liability of those in charge of the other causes or exclude it altogether.[81] The operator of a storage site will therefore not be liable for the admixture occurring there if caused by the carelessness of one of the victim's own drivers. **49**

Generally speaking, any activity set by the victim which also meets the but-for requirement (including adjustments) will reduce her claim for compensation according to the respective percentages of likelihood that the victim's own activity was a cause of the loss and would fulfil the other requirements of liability if she were a third person.[82] **50**

A specific facet of a contributing risk within the victim's own sphere is predispositions or infections independent from the GMO activity. A tricky question arises, for example, if it remains unclear whether certain allergies from which the victim suffers were triggered by the consumption of GM food, or whether there were other external reasons or, say, the genetic condition of the victim, which are all at least equally probable causes. Most jurisdictions would probably stick to their "all-or-nothing" approach and consequently deny liability because of this "non liquet" **51**

---

80 *B.A. Koch*, Comparative Report, in Digest I (fn. 50) 8a/29 (pp. 501 ff.): "Academic debate in several countries has apparently not yet found a universally accepted answer to this question.".

81 E.g. AT no. 15; CZ no. 96 ff.; FR no. 24; GR no. 20 (at discretion of the court); LI no. 15; MT no. 44; SE no. 77.

82 On the mirror image rule in case of contributory conduct, see, e.g., *M. Martín-Casals*, Commentary on Chapter 8, in *European Group on Tort Law*, Principles of European Tort Law (2005) 130; *von Bar* (fn. 10) no. 518 (p. 546).

situation. Only a few so far are prepared to divide the loss between the victim and those responsible for potential external causes according to the respective likelihood.[83]

(ii) Risks outside both the tortfeasor's and the victim's sphere

52 If the origin of the loss lies beyond both parties' sphere, its consequences still remain with one of them, i.e. the victim. As explained at the beginning, the purpose of tort law is to highlight possibilities for the victim to shift her loss onto someone else. If that should fail, the loss lies where it falls.[84] However, if such extraneous cause should coincide with another potential event for which someone else could be held liable, the question is how legal systems balance the impact of these two (or more) influences upon the chain of causation.

53 At least one aspect triggers the same legal consequences in all jurisdictions covered, though not necessarily based upon the same dogmatic reasoning: Force majeure, i.e. the impact of an extraordinary natural phenomenon that is unavoidable and unpredictable, serves as a full defence against liability and leads to a complete exoneration of the defendant if its effects overshadow any other causal influence.[85] This does not mean, however, that any act of nature will serve to free the alleged tortfeasor from liability – a GM farmer cannot argue that the wind which blew pollen to neighbouring fields was force majeure simply because it was an act of nature. After all, winds are predictable, at least within limits of long-term experience. A once-in-a-century storm, on the other hand, may fall within the ambit of the defence, also because one is not required to take precautionary measures against such outrageous and rare events (unless foreseeable).[86]

54 Another external cause for which the alleged tortfeasor cannot be held accountable lies with third parties who interfere with the chain of causation. To the extent they do, the but-for test and its variants will decide.[87]

---

83 Cf. *B.A. Koch*, Comparative Report, in Digest I (fn. 50) 6b/29 (pp. 436 ff.).

84 Supra no. 1.

85 AT no. 17; CZ no. 43–46; DE no. 22; EE no. 20; ES nos. 28 ff.; FI no. 35 (but narrow interpretation of force majeure); FR no. 26 (but not if mere contribution to causing the loss); GR no. 22; HU no. 36; IT no. 19; LI no. 17; LU no. 43 ff.; MT no. 24; NL no. 19; NO no. 14; PL no. 36–38; SI no. 14; SE no. 33; UK no. 19K; AU 21 (remoteness; defence only if unforeseeable); BR nos. 40 ff.; CA no. 22; but see US no. 11. Cf. EA nos. 65 ff.

86 See AU no. 22.

87 However, in practice courts may tend to apply the test more favourably to the victim.

If they pass the test, the victim can go after them to seek compensation. The remaining question – whether she could recover from the original defendant what was actually caused by someone else – has already been answered: only if it remains unclear who of the two actually did cause her loss, legal systems may opt for joint and several liability of both of them, and they will then have to seek recourse internally, but not at the expense of the victim.

IV. Fault

The "most traditional, most widespread and – apparently – most important"[88] route to compensation in tort law leads through the test whether there was any culpable wrongdoing on the side of the defendant. Over time, the perception of blameworthiness has moved away from judging the individual individually to applying a more objective notion of fault, using a yardstick that has some imaginary third person in mind who in the position of the actual wrongdoer may have behaved differently – the "reasonable man" (or woman) who in practice is presumed to have superpowers beyond any cartoon hero, if all the standards of comparison developed in court were compiled to a catalogue of responsibilities: after all, what should have been done by the defendant is always assessed ex post, when the chain of causation has been analysed thoroughly, tempting judges to retroactively impose duties upon the defendant that were unbeknownst to the latter at the time of acting. However, after the event even the fool is wise, so duties of care identified in the aftermath of the loss are not necessarily obvious to the tortfeasor in advance. The concept of fault has thereby "become a mere phantom, an empty shell which serves as camouflage for a criterion of imputation which has almost nothing to do anymore with an individual blame towards the tortfeasor".[89] Nevertheless, it continues to be (or at least to be claimed) the backbone of tort law.[90] Unless there is a special liability regime in place which eases access to compensation for the victims (otherwise they will disregard it),[91] fault liability will always be in place to step in as at least one option to pursue one's claims. This also means that damage caused by the negligence of the actor 55

88 *P. Widmer*, Commentary on Chapter 4, in *European Group on Tort Law*, Principles of European Tort Law (2005) 64 (no. 1).
89 *P. Widmer*, Comparative Report on Fault as a Basis of Liability and Criterion of Imputation (Attribution), in P. Widmer (ed.), Unification of Tort Law: Fault (2005) 331, 357 (no. 68).
90 *Widmer* (fn. 89) 332 (no. 2).
91 See infra V.

are run-of-the-mill cases in tort law for which all legal systems foresee liability – things going wrong through human error deviating from common standards at any given point along the supply chain will easily be remedied this way once causation is established.

56 Apart from this move towards objective standards of care, there is also a noticeable tendency at least in Europe towards stricter varieties of liability, e.g. by shifting the burden of proving fault under certain circumstances[92] or even generally[93] (which in practice means that the defendant will hardly escape liability because proving that he was not at fault is often tricky).

57 If the duty of care is not determined by way of imagining what a reasonable person would have done under the circumstances, but by law which prescribes or forbids a certain behaviour in order to prevent the infliction of harm, any violation of such express duties will help to establish fault or even be considered faulty per se by most jurisdictions unless the defendant can prove that no-one could have adhered to that legislative standard in light of the facts of the case.[94] This may also apply to non-legislative standards which are common usage in a particular industry or trade, including protocols or guidelines promoted by interest groups, but also instructions by a GMO producer.[95] Therefore, if statutes or soft law foresee certain farming practices and coexistence measures in order to avoid unwanted admixture, any violation of such rules may be presumed to be faulty if the negative consequences that should have been avoided materialise due to such conduct. The reverse is not true, however – abiding by such standards does not per se exempt the defendant from liability.[96]

58 The case is simpler with unapproved GMOs – whoever made it possible to let them flow into the food or feed supply chain will easily be held liable for any negative consequences thereof that are deemed compensable, in case of qualified fault probably most (if not all) legal systems would even indemnify pure economic loss.[97]

---

92 E.g., ES no. 34; DK no. 17; *Dacoronia* (fn. 65) no. 62; IT no. 22 ff.; PL no. 47; BR no. 46.

93 E.g. EE no. 23; HU nos. 5–6; SI no. 17.

94 See, e.g., CZ nos. 52–53; DE nos. 4, 6, 39; DK no. 17; ES no. 36; FR no. 32 ("Any violation of regulations governing GM farming will automatically be considered a fault leading to liability . . ."); IT no. 26; LI no. 21, 23; LU no. 55; MT no. 28; NL no. 30; NO no. 18; PL no. 50; SE no. 41; UK no. 23; AU no. 28; US no. 15; but see HU no. 41; CA no. 27.

95 Cf. AU no. 29.

96 See infra IX.2. Support for this impact of compliance with or disregard of rules and regulations comes from an economic analysis: EA nos. 39 ff.

97 Cf. FR no. 34.

V. GMO-specific liability regimes

Several legal systems have decided to introduce a liability or other redress scheme tailor-made to losses caused by GMOs. To the extent these rules fall into the respective jurisdiction's environmental liability legislation, therefore dealing only with GMO-related environmental harm, they will be addressed in the section on environmental liability below.[98] **59**

Some of these special regimes provide for compensation to farmers who have suffered an economic loss due to the adventitious presence of GMOs on their fields. These systems have already been addressed by our previous study, which is why they will not be covered specifically again.[99] Furthermore, if special provisions merely refer to the general law of torts, they will also be disregarded in the following.[100] **60**

None of the legal systems represented in this study have introduced any alternative compensation scheme designed to absorb the losses under survey here, even though some of them have done so for the economic losses addressed in the earlier project. **61**

Of the jurisdictions foreseeing special liability rules for the types of harm that are the main focus here, one has to differentiate between those that have introduced rather general strict liability for damage caused by GMOs and others which draw a line between risks during research and development on the one hand and commercial cultivation after approval on the other hand. **62**

Finland, Hungary, Liechtenstein, Norway and Poland belong to the first group. **63**

Finland refers losses other than harm to the environment (which is covered by a separate regime) or those falling under its Product Liability Act to its general Tort Liability Act, but cuts out the requirement of fault in so doing. Therefore, damage to persons or property of the latter category is subject to strict liability.[101] Furthermore, in order to fill the gap left by the consumer losses limitation of the product liability regime, there is a separate strict liability rule for feed producers, thereby extending the fun- **64**

98 Infra VII.

99 See *Koch* (fn. 2) nos. 153 ff. Of the jurisdictions covered in both studies, Austria, Denmark, Finland, France, Germany, Hungary, Italy, Norway and Poland already provide for such special solutions, furthermore Luxembourg (where it was introduced after the manuscript to the previous book was finished, see LU nos 2–4).

100 E.g. EE no. 1.

101 FI no. 4.

damental ideas underlying product liability to losses to property in the commercial setting of feed buyers.[102]

65 *Hungary* chose a different legislative technique and by statute defined all loss scenarios involving GMOs – including approved ones – as "abnormally dangerous activities" within the meaning of the Hungarian Civil Code's general rule on strict liability. In order to prevent non-GM farmers from being held liable for losses caused by GMOs adventitiously present on their fields, this referral does not extend to risks inherent in GMOs themselves, but only to "genetic technology activities".[103]

66 Current Art. 24 of the *Liechtenstein* Act on Handling Genetically Modified or Pathogenic Organisms (GPOG) foresees strict liability for all damage originating from an enterprise or installation handling GMOs, unless caused by force majeure or grave fault of the victim or some third party.[104] The regulatory compliance defence is unavailable. Those potentially liable have to take out liability insurance. If the current draft bill replacing the GPOG becomes law,[105] the current liability regime will be more refined, providing inter alia for a lowering of the standard of proving causation and an explicit exclusion of the state of the art defence. If harm is caused to consumers or to agricultural or forestry businesses, liability will be channelled to the person holding the authorisation to use or release GMOs.

67 The *Norwegian* Act on Genetic Technology provides for strict liability for any activity that places or emits GMOs in the environment[106] (NO no. 2). The definition of "deliberate release" in § 9 of this Act includes release for commercial purposes and is not limited to contained use or research-related release.[107]

68 Art. 57 of the current Law on GMOs in *Poland* holds any "user" of GMOs (whether in containment or by deliberate release, including commercial cultivation or distribution) strictly liable for damage to persons, property, or to the environment.[108] As in the afore-mentioned jurisdictions, force majeure excludes liability, but the impact of the victim or of a third party

---

102 FI no. 5.

103 HU no. 14.

104 LI no. 1.

105 LI nos. 3–4. An English translation of the draft can be found on TRIS at http://ec.europa.eu/enterprise/tris/pisa/cfcontent.cfm?vFile=120099006EN.DOC.

106 NO no. 2.

107 See also the English translation of the whole Act at http://bch.cbd.int/database/attachedfile.aspx?id=602.

108 PL no. 1 ff.

must be the exclusive cause of the loss in order to exonerate the defendant. A pending draft bill to replace the existing legislation will essentially maintain the current liability rules.[109]

The second group is marked by a strict liability solution for the risks associated with the research and development stage, whereas potential losses occurring during commercial cultivation are merely referred to each country's product liability regime (which is modelled after the Product Liability Directive).[110] *Austria* and *Germany* appertain to this second group.[111] The special statutory regime addresses damage caused in laboratories, in field trials, etc., but does not extend to approved GMOs as soon as they enter the market. This duality of approaches leaves some gaps: not all imaginable risks fall under the concept of product liability (or environmental liability, which essentially absorbs risks that GMOs pose to the environment). In particular, damage to property which is not in private use such as harm to farm animals will fall under those jurisdictions' general tort law since none of the special regimes applies.[112] **69**

VI. Product liability

The most likely cause of action in delict arising out of the case scenarios under survey here will presumably be product liability, though possibly coupled with other causes of action: Apart from contractual liability, damage caused by agricultural products (whether genetically modified or not) can trigger liability of the producer towards users, consumers, as well as towards third parties (often referred to as innocent bystanders). The core of the claim is a product defect which causes harm to the human body or to property, and liability is typically channelled towards the producer. **70**

In the EU and to some extent even beyond,[113] this area of liability is harmonized by way of the Product Liability Directive (PLD).[114] This instrument was not only introduced to protect consumers, as repeatedly con- **71**

109 PL no. 9; an English translation of the draft can be found on TRIS at http://ec.europa.eu/enterprise/tris/pisa/cfcontent.cfm?vFile=120080581EN.DOC.

110 See infra VI.

111 AT nos. 1–2; DE no. 2.

112 However, those losses may be absorbed by alternative claims which cannot be found in statutory language: The German producer's liability even started with a case that may be of relevance in the GMO context (referred to as "Hühnerpest" or fowl pest case; see the translation of this German Supreme Court case BGH 26.11.1968, VI ZR 212/66, at http://www.utexas.edu/law/academics/centers/transnational/work_new/german/case.-php?id=760).

firmed throughout the recitals, but also "because the existing divergences may distort competition and affect the movement of goods within the common market", as stressed in the very first recital of the Directive. The latter argument was reaffirmed by the ECJ and put on at least equal footing as the former argument.[115]

72 The starting point of product liability as envisaged by the Directive is a product defect. Initially, "primary agricultural products and game" were not covered by the Directive, but this exception was given up by way of amendment in 1999[116] after such crises as the mad cow disease. A "defect" is given if the product "does not provide the safety which a person is entitled to expect", thereby taking into account, inter alia, the presentation of the product, its use according to reasonable expectation, and the time when it was put into circulation (Art. 6 PLD). The latter aspect also means that improvements of the product over time do not alter the standard according to which the safety of a product is being assessed that was put onto the market before the better product was developed. The focus on reasonable expectations of use would for example prevent liability of a feed producer whose products are used by another in the food chain, but most likely only for harm that would not have been caused had the feed been used as such, whereas problems of admixture may be seen independent of the use aspect, which aims at the ultimate consumer. One commonly distinguishes between three different types of product defects: design defects (if, for example, an allergy-prone protein is introduced by genetic engineering), manufacturing defects (if a food producer erroneously processes a bag of GM feed maize due to a mixup instead of the usual food product), and warning defects (if inadequate labelling did not alert the buyer of certain risks of use or limits to edibility).

---

113 Several non-EU Member States such as Switzerland have adopted the Directive as a model for their own legislation in this field. From the jurisdictions under survey here, this is not only true for the EEA Member States Liechtenstein and Norway (LI no. 24; NO no. 19), but also for Australia (!, no. 30), and Québec seems to have been influenced by the PLD as well (CA no. 29, but excluding primary agricultural products).

114 Council Directive 85/374/EEC of 25 July 1985 on the approximation of the laws, regulations and administrative provisions of the Member States concerning liability for defective products, OJ L 210, 7.8.1985, 29, as amended (see infra fn. 116). Criticism from an economic perspective is raised by EA nos. 18 ff., in particular 29 ff.

115 ECJ C-183/00, *María Victoria González Sánchez v. Medicina Asturiana SA*, [2002] ECR I-3901, no. 26.

116 Directive 1999/34/EC of the European Parliament and of the Council of 10 May 1999 amending Council Directive 85/374/EEC on the approximation of the laws, regulations and administrative provisions of the Member States concerning liability for defective products, OJ L 141, 4.6.1999, 20.

If an agricultural product is put into circulation without indication of GM content even though it would need to be labelled according to the legal standards, one could well argue that the product is therefore defective not only according to contractual standards defined, e.g., by the buyer's expectations, but also from a delictual point of view, since at least one of the reasons for introducing the labelling threshold was to give consumers an informed choice in order to cope with their concerns about possible health implications of GMOs.[117] The lack of proper information on the transgenic contents might at least lead to a presumption of potentially negative effects on the health of the consumer.[118] Since the whole issue will only fall under the product liability regime if damage was actually caused, the impact of the presumption may be decisive also for assessing causation.[119] **73**

The PLD regime is limited to damage caused to the human body or to property, the latter subject to two limitations: First of all, only property loss worth more than € 500 will be compensated under the Directive's regime.[120] Secondly, only damage to property is covered that is "intended for private use or consumption" and indeed "used by the injured person mainly for his own private use or consumption" (Art. 9 lit. b PLD). Therefore, harm caused by GM feed to farm animals, for example, will not fall under the scope of the implementing statutes.[121] **74**

The Directive imposes strict liability, i.e. liability irrespective of fault (and not just with a presumption thereof), primarily upon the producers. These **75**

---

117 Cf. Recital 22 of Regulation (EC) No 1829/2003 of the European Parliament and of the Council of 22 September 2003 on genetically modified food and feed, OJ L 268, 18.10.2003, 1: "[T]he labelling should give information about any characteristic or property which renders a food or feed different from its conventional counterpart with respect to composition, nutritional value or nutritional effects, intended use of the food or feed and health implications for certain sections of the population, as well as any characteristic or property which gives rise to ethical or religious concerns.".

118 Cf. CA no. 30. But see the Finnish draft bill on coexistence, which, if coming into force, would specifically rule out that a product is defective due to GMO admixture (FI no. 40).

119 Cf. supra III.2(b).

120 This lower threshold is unique to the tort laws of all Member States, which has triggered quite substantial resistance from countries such as France or Greece, who fought the matter with the Commission: ECJ C-52/00, *Commission v. France*, [2002] ECR I-3827; ECJ C-154/00, *Commission v. Greece*, [2002] ECR I-3879. The Court, however, ruled against them, probably unaware of the tort law situation in the Member States. See *B.A. Koch*, European Union, in H. Koziol/B.C. Steininger (eds.), European Tort Law 2002 (2003) 432 (nos. 62 ff., in particular no. 75).

121 However, that case may be resolved by a comparable strict liability rule specifically applicable to feed, e.g. in Finland (FI no. 5). See also FI no. 6 (special regime for seeds with equivalent provision absorbing cases that are filtered out by the PLD).

are defined as manufacturers of a finished product, of raw material or of any component parts, therefore anyone involved in the manufacturing process from the beginning to the end, when it leaves the production process and is circulated on the market. In case more than one "producer" within that meaning can be identified, they are all jointly and severally liable (Art. 5 PLD). In addition, the term is expanded to include those who present themselves as producers on the product without actually being involved in the manufacturing process (e.g. a supermarket's "own brand" food line). Furthermore, if the defective products came into the internal market from outside, the importer is also responsible like the actual producer abroad (Art. 3 PLD). If the producer cannot be identified, the supplier will be liable instead unless he reveals the identity within reasonable time. This applies *mutatis mutandis* to goods imported into the European market, so that the distributors have to compensate those injured unless they name the importer (while the name of the extra-EU producer is irrelevant in this case). Therefore, the local supermarket can be strictly liable for harm caused by GM food produced in, say, the U.S., if the company that imported these products into Europe is not made known to the claimants in due time. Otherwise, if for example tortilla chips contain GMOs that are harmful, the seed producer, the farmer who grew the maize, the producer of the masa flour, the company that processed the flour to chips, and the owner of the brand under whose name they are sold on the market (if separate legal entities) are all jointly and severally liable for losses caused by this defect, and the victims can choose who to go after first.

76 There are, however, limits to such at first sight far-reaching liabilities of all those involved in the manufacturing process:

77 To begin with, a producer as defined above is exempt from liability if he did not put the defective product into circulation, i.e. if it left his premises involuntarily, be it by accident or – even more so – by way of sabotage (Art. 7 lit. a and c PLD). Therefore, a GM farmer from whose fields pollen are blown to a neighbouring conventional field will not be considered a producer of whatever produce is derived from the latter field, since he did not intend to participate in the cultivation of that field. The same is true for testing labs which do not produce for the commercial market. This does not mean that they are completely off the hook – there may be other grounds of action against them, e.g. fault or some special liability standard, but not through product liability as designed by the Directive.

78 The next defence in Art. 7 lit. b PLD relates to the timing of the defect: if the defendant can prove that GMO content did not flow into the manufacturing process until after the part product that he produced left his premises, he will not fall into the group of potentially liable producers.

Two more defences expressly foreseen in the PLD are important: First, the question is to what extent "compliance of the product with mandatory regulations issued by the public authorities" (Art. 7 lit. d PLD) can really exonerate the producer, or in other words, whether coexistence legislation and permits[122] granted on its basis constitute "mandatory regulations" within that meaning. This may rarely ever – if at all – be the case, as it is meant to designate specific rules of conduct. Also, the producer would have to prove "that the defect leading to the damage is directly due to his observance of these rules and regulations with which he was obliged to comply".[123] **79**

Second, but of higher relevance in our context, Art. 7 lit. e PLD contains the so-called development risk defence, which Member States did not necessarily have to include into their implementing legislation, so that a producer cannot rely upon it in these countries (Art. 15 para. 1 lit. b PLD).[124] In the remaining countries, he is not liable if he can prove "that the state of scientific and technical knowledge at the time when he put the product into circulation[125] was not such as to enable the existence of the defect to be discovered".[126] If it therefore turns out that products containing certain GMOs are defective within the meaning of the Directive (i.e. bear certain health risks for consumers, for example), but this is not discovered until after these products have been put into circulation, the producer will not be liable for said defect. The defence does not extend to risks which were known as such but could not be detected according to state of the art technology.[127] There may be a grey zone where there is already a suspicion of harmfulness at the time the products are brought **80**

122 See infra IX.2.

123 MT no. 32; see also FR no. 42.

124 Finland and Luxembourg have opted against the development risk defence entirely (FI no. 39; LU no. 57), as did Norway (no. 19). It does not apply to food or food products for human consumption as well as to pharmaceuticals in Spain (ES no. 37); the latter also being true in Germany, where the application of the defence on products containing GMOs which were based on the statuory licensing requirement was also abolished (DE no. 40).

125 The ECJ has made clear that "Article 7(e) is not specifically directed at the practices and safety standards in use in the industrial sector in which the producer is operating, but, unreservedly, at the state of scientific and technical knowledge, including the most advanced level of such knowledge, at the time when the product in question was put into circulation."ECJ C-300/95, *Commission v. United Kingdom*, [1997] ECR I-2649.

126 See EA nos. 68 ff. on economic arguments in favour of and against the development risk defence.

127 ES no. 37, 44 (on the question of what constitutes knowledge).

onto the market, but no scientific proof yet. It is hard to imagine, though, how the latter could apply to GMOs approved for cultivation in the EU, which have undergone high-level scrutiny by the authorities and are only admitted to the market if greenlighted after thorough testing and analysis.[128]

81 The producer may still be liable despite the applicability of the development risk defence: while the latter excludes (strict) product liability, the producer may nevertheless have to account for his negligence in monitoring the product safety, for issuing adequate warnings or for failing to recall products.[129]

82 Another escape from product liability is along the timeline: while there are remarkable differences within Europe alone when it comes to the time periods applicable to delictual claims in general,[130] the PLD foresees certain clear-cut limits that have to apply irrespective of the legal system. The last chance to launch a claim against a producer under the Directive's regime is ten years after the latter has put the actual harmful product (not the first of its type) onto the market (Art. 11 PLD).[131] However, in addition to this absolute limit which extinguishes all claims, there is also a relative time limit of three years, which starts to run "from the day on which the plaintiff became aware, or should reasonably have become aware, of the damage, the defect and the identity of the producer" (Art. 10 PLD). If that day was later than seven years after the harmful product was circulated, the claimant has only the remaining time left of the long-stop period.

83 Finally there may also be a monetary limit to recovery: Member States may (but do not have to) cap liability for death or personal injury at € 70 million or any amount higher than that. Of the countries under survey here, Germany and Spain have opted in favour of such limits.[132]

---

128 See also IT no. 31.

129 See in particular NL no. 33, arguing why tortious liability for failure to recall can live alongside the PLD regimes and will therefore stand the scrutiny of the ECJ concerning the exclusivitiy of the PLD (infra no. 84). See also SE no. 43.

130 See infra IX.5.

131 It suffices if the producer passes the product on to a distributor, even if the latter is the former's fully-owned subsidiary, see ECJ 9.2.2006 C-127/04 *Declan O'Byrne v. Sanofi Pasteur MSD Ltd et al.* [2006] ECR I-1313.

132 The limit is € 85 million in Germany (§ 10 Product Liability Statute) and slightly above € 63 million in Spain (Art. 141 lit. b of the 2007 Consumer Protection Act; the latter therefore being in obvious violation of the Directive, but the consequence of a mere conversion from ECU to pesetas in the preceding legislation and from there to Euros, without reconsidering the original text of the PLD).

While the PLD led a somewhat dormant life in the first two decades after its initiation,[133] a series of ECJ decisions on the substance of the Directive seems to have reanimated the concept in the Member States,[134] above all probably due to the Court's radical view on the possibility (or mostly: impossibility) of other liability regimes to coexist with the Directive's regime, as was the practice in the Member States before, which was also the reason why the PLD was ignored in practice. In *González Sanchez*,[135] the ECJ made clear that the Directive not only sets a minimum standard of liability, but an exclusive one, thereby ruling out even national systems which are more favourable to consumers. The Court opined that the Directive "cannot be interpreted as giving the Member States the possibility of maintaining a general system of product liability different from that provided for in the Directive".[136] The Court conceded, however, that the PLD "does not preclude the application of other systems of contractual or non-contractual liability based on other grounds, such as fault or a warranty in respect of latent defects".[137] This seems to rule out other systems of strict liability for product defects, but does not give any guidance as to what falls under the Court's notion of "strict liability", if it has any. After all, liability without fault is not a closed concept in European tort law,[138] which was even acknowledged by the Commission in the product liability context.[139] **84**

This means for the time being that it is somewhat unclear which competing causes of action in the Member States will survive the test that the ECJ has thereby sketched out. While general fault liability may prevail, particularly if the focus is on some misbehaviour of the producer rather than his **85**

133 See the first two Commission reports on the application of the Directive: COM(95) 617 final and COM(2000) 893 final, the latter admitting expressly that "there is still limited experience", which it traces back to two factors: "the Directive was lately transposed in some Member States and, according to the possibility given to Member States under Article 13 of the Directive, national contractual or extra-contractual law or a specific liability regime is applied in parallel" (at p. 28).

134 This is maybe why the third report on the application of the PLD, COM(2006) 496 final, boldly states "that the Directive works by and large in a satisfactory way and that there is no need for amendments at present" (at p. 4).

135 ECJ C-183/00, *González Sánchez v. Medicina Asturiana SA* [2002] ECR I-03901.

136 *González Sanchez*, para. 30.

137 *González Sanchez*, para. 31.

138 Cf. *B.A. Koch/H. Koziol*, Comparative Conclusions, in: B.A. Koch/H. Koziol (eds.), Unification of Tort Law: Strict Liability (2002) 395 (in particular no. 2).

139 Report on the Application of Directive 85/374 on Liability for Defective Products, COM(2000) 893 final, 30: "[C]ase-law in several Member States tends to interpret the producer's liability under fault-based liability systems in an extensive way with the result that in practice the difference between fault-based and strict liability systems is getting blurred. In this situation and given that fault-based liability systems generally provide for a larger scope of consumer protection parallel applications are a practical consequence.".

quality as such,[140] victims may probably no longer choose strict causes of action alternatively, even if these should be more attractive for them, e.g. due to more far-reaching remedy options, longer periods of limitation or the like.[141]

VII. Environmental liability

1. The Environmental Liability Directive

86 Another act of harmonisation on the European level that has an impact on the loss scenarios under survey in this study is the Environmental Liability Directive (ELD).[142] By its name, it seems to deal with tort law like its namesake for products. However, the use of the word "liability" in the title is a mere leftover from earlier, more ambitious plans, which failed during the drafting process. The result of the extensive bargaining with interest groups is a mere administrative law regime, with the state as the sole actor on the victims' side (via its designated "competent authority"),[143] whereas individual claims are excluded altogether. Compensation for personal injury or damage to property therefore continues to be untouched by the unification process and therefore still falls under each national regime with all its differences.[144]

87 This shift to administrative law is not problematic; however, apart from the fact that language was retained throughout the draft that seems to point to private law without actually doing so: after all, it is virtually impossible to identify any individual who could sue in tort law on behalf of the environment as such.[145] At best, one could imagine certain interest

---

140 E.g. AT no. 23; DE no. 42; FR no. 39, 41; GR no. 31; HU no. 43; NL no. 32; SE no. 53.

141 E.g. FI no. 41; IT nos. 32–34.

142 Directive 2004/35/CE of the European Parliament and of the Council of 21 April 2004 on environmental liability with regard to the prevention and remedying of environmental damage, OJ L 143, 30.4.2004, 56; as amended by Directive 2006/21/EC of the European Parliament and of the Council of 15 March 2006 on the management of waste from extractive industries and amending Directive 2004/35/EC, OJ L 102, 11.4.2006, 15, and by Directive 2009/31/EC of the European Parliament and of the Council of 23 April 2009 on the geological storage of carbon dioxide and amending Council Directive 85/337/EEC, European Parliament and Council Directives 2000/60/EC, 2001/80/EC, 2004/35/EC, 2006/12/EC, 2008/1/EC and Regulation (EC) No 1013/2006, OJ L 140, 5.6.2009, 114.

143 Art. 12 ELD allows natural or legal persons affected by the threat or actual contamination, but also NGOs pursuing environmental interests, to request action from the competent authority, but the decision whether and how to act on such request stays with the latter.

144 See infra VII.2.

groups or others who seek compensation for work they have spent in restoring damage to the environment.[146] The best representative of the interests of the environment seems to be the state, however, where all these activities can be bundled, and who already is in charge of a series of measures to protect the environment. What the Directive does, therefore, is to try to shift expenditures for remedying specific losses from society at large to those who are responsible for the damage, though not necessarily in a sense of blame, but rather in order to let those who derive benefits for the activities that cause the losses pay. This is the essence of the "polluter-pays" principle, even though it has a negative connotation. It also serves "to induce operators to adopt measures and develop practices to minimise the risks of environmental damage so that their exposure to financial liabilities is reduced", as the second recital of the Directive points out.

As far as the protection of the environment is concerned, further expectations from the title of the Directive have to be dashed, however: it does not deal with all kinds of harm to the environment, but only addresses specific varieties thereof as defined by its Art. 2 (1), which lists damage to biodiversity (as confined by existing protection under EU or national law[147]), water damage (as regulated by the Water Framework Directive[148]), as well as land contamination that creates significant health risks for humans. Furthermore, the Directive is predominantly focused upon certain risky activities as laid down in Annex III to the Directive. Other conduct not listed explicitly only triggers liability if carried out at least negligently.[149] However, Annex III includes inter alia: **88**

145 Cf. UK no. 34: "n general tort law, harm to biodiversity and harm to the environment as such – as distinct from harm to the proprietary rights of particular persons – cannot give rise to liability.".

146 Cf. ES no. 72; FR nos. 48 (standing of associations for the defence of the environment), 52–54 (e.g compensation for "moral harm" granted to an association protecting birds); PL no. 65.

147 Apart from habitats or species designated by the Member State, the protection extends to all species and habitats protected under the Habitats Directive (Council Directive 92/43/EEC) as well as most threatened species and birds protected under the Birds Directive (Council Directive 79/409/EEC).

148 Directive 2000/60/EC of the European Parliament and of the Council of 23 October 2000 establishing a framework for Community action in the field of water policy, OJ L 327, 22. 12. 2000, 1.

149 The wording of Art. 3 (1) (b) ELD speaks of activities "whenever the operator has been at fault or negligent", which uses tort terminology in a rather awkward way. Some translators have tried to make sense out of this wording, so the German version speaks of "intentionally or negligently" ("vorsätzlich oder fahrlässig") instead, similarly the Italian translation ("doloso o colposo").

"10. Any contained use, including transport, involving genetically modified micro-organisms as defined by Council Directive 90/219/EEC of 23 April 1990 on the contained use of genetically modified micro-organisms.

11. Any deliberate release into the environment, transport and placing on the market of genetically modified organisms as defined by Directive 2001/18/EC of the European Parliament and of the Council."

89 The fact that the Directive is specifically concerned about environmental damage caused by GMOs is further evidenced by its Art. 18 (3) (b), which requires the Commission to specifically report to Parliament and to Council after some time[150] a review of

"the application of this Directive to environmental damage caused by genetically modified organisms (GMOs), particularly in the light of experience gained within relevant international fora and Conventions, such as the Convention on Biological Diversity and the Cartagena Protocol on Biosafety, as well as the results of any incidents of environmental damage caused by GMOs; ..."

90 While the main thrust of the Directive is aimed at the prevention of environmental harm, it also considers that it might fail, in which case remedial action[151] comes into play, which is marked by the afore-mentioned "polluter-pays" principle. If there is an immediate threat of harm to the environment, the "operator"[152] has to take necessary preventive measures, and, if the risk persists, inform the competent authority, who then can either order him or someone else at his expense to take further preventive action. In case environmental damage has already occurred, the operator (again) has to notify the competent authority, and also invest all efforts into preventing a deterioration of the situation as well as remedying the harm already sus-

---

Note that from an economic perspective, a strict liability approach would have been preferable: EA no. 17.

150 The Directive speaks of 30 April 2014, which may be delayed due to the late implementation of the Directive in the Member States, see infra no. 95.

151 Art. 2 (11) ELD defines remedial measures as "any action, or combination of actions, including mitigating or interim measures to restore, rehabilitate or replace damaged natural resources and/or impaired services, or to provide an equivalent alternative to those resources or services as foreseen in Annex II".

152 Art. 2 (6) ELD defines "operator" as "any natural or legal, private or public person who operates or controls the occupational activity or, where this is provided for in national legislation, to whom decisive economic power over the technical functioning of such an activity has been delegated, including the holder of a permit or authorisation for such an activity or the person registering or notifying such an activity".

tained, i.e. to restore the damaged natural resources either in kind or by way of recreating similar resources (Annex II ELD).

One of the problems unresolved by the Directive is the causation issue: environmental damage is one of the casebook examples for difficulties in establishing the causal link between the conduct of the defendant and the harm, even more so in the almost typical case where more than one possible polluter may have contributed to the loss. The list of Annex III is one way where the European legislator tried to indicate typical activities prone to cause environmental harm. However, this list cannot establish presumptions for individual cases. The Directive therefore has to admit in its Art. 4 (5) that it can "only apply to environmental damage or to an imminent threat of such damage caused by pollution of a diffuse character, where it is possible to establish a causal link between the damage and the activities of individual operators". Recital 13 further admits that the Directive does not present "a suitable instrument for dealing with pollution of a widespread, diffuse character, where it is impossible to link the negative environmental effects with acts or failure to act of certain individual actors". The Directive also leaves it to the national laws of the Member States to decide who, in a case of multiple proven causes, has to bear which share of the loss (Art. 9 ELD). **91**

The duty of the operator to at least finance preventive and/or remedial measures is limited by exceptions due to external causes (force majeure,[153] armed conflict or the like, third parties) or compliance with mandatory orders.[154] Furthermore, for cases where the operator is not at fault, Member States may optionally recognise the regulatory permit defence[155] or the equivalent of the development risk defence of the PLD, paying tribute to the state of the art.[156] If they do, they will not recover, for example, if the **92**

153 Art. 4 (1) (b) ELD defines this as "a natural phenomenon of exceptional, inevitable and irresistible character".

154 Art. 8 (3) (b) ELD. Cf. supra no. 79.

155 Cf. supra no. 57. Art 8 (4) (a) ELD allows member states to refrain from seeking recourse from the operator if damage was caused by an activity "expressly authorised by, and fully in accordance with the conditions of, an authorisation conferred by or given under applicable national laws and regulations". Of the countries under survey here, the defence was adopted by the Czech Republic, Denmark, England, Estonia, Greece, Italy, Malta and Spain (ES no. 39); with a lesser degree also by the Netherlands and Sweden (discretion). See *H. Lopatta*, The Environmental Liability Directive – Overview and State of Play (Slides for ELD Workshop 10.7.2009, available at http://www.biohost.org/eld/workshop09/docs/ELDWS2009_Lopatta_EC.pdf), Slide 14.

156 According to Art. 8 (4) (b) ELD, Member States can also waive the duty to compensate the costs of remedial actions if the operator can prove that the environmental damage was caused by an activity which "was not considered likely to cause environmental damage according to the state of scientific and technical knowledge at the time when

release of a GMO that later turns out to be harmful to the environment was specifically authorised or if it was not possible to anticipate the damaging effect on the basis of the state of scientific and technical knowledge at the time of release.[157]

93 The limitation period is yet different from the PLD and national regimes – the competent authority has to seek recourse for the costs spent on preventive or remedial actions "within five years from the date on which those measures have been completed or the liable operator, or third party, has been identified, whichever is the later" (Art. 10 ELD).

94 Though it has been vividly demanded by the EESC,[158] the Directive fails to foresee duties to take out insurance or to provide for any other financial cover in advance, claiming that this were not yet available on the market. Nevertheless, some Member States have included such duties into their implementing legislation.[159]

95 In contrast to the PLD, the ELD is meant to set only a minimum, not an exclusive standard: Art. 16 expressly allows "more stringent provisions" on the national level.[160] Seen together with the afore-mentioned optional provisions of the Directive, the legal situation in Europe even after its implementation is not really homogeneous as one would expect from such a legislative project.[161] Compliance with the Directive has not been better than with the PLD – at the end of the period set by the Directive

---

the emission was released or the activity took place". Of the countries under survey here, the defence was adopted by the Czech Republic, England, Estonia, France, Greece, Italy, Luxembourg, Malta and Spain; with a lesser degree also by the Netherlands and Sweden (discretion); *Lopatta* (fn. 155) Slide 14.

157 http://europa.eu/rapid/pressReleasesAction.do?reference=MEMO/07/157.

158 Opinion of the Economic and Social Committee on the Commission's proposal, OJ C 241, 7.10.2002, 162 (para. 4.4).

159 E.g. CZ no. 64–65; ES nos. 52 ff.; GR nos. 35–37; HU no. 47; SE no. 61. See also UK no. 32 (though no statutory requirement, but authorisations and permits may include financial guarantees within the discretion of the authority); similarly PL no. 58 (financial security optional). A "Study on the Implementation Efficiency of the Environmental Liability Directive (ELD) and related Financial Security issues" published in November 2009 by the European Commission in preparation for its report due under Art 14 para. 2 ELD (available at http://ec.europa.eu/environment/enveco/others/pdf/implementation_efficiency.pdf, subsequently referred to as "ELD Implementation Efficiency Study") inter alia gives an overview of currently available insurance products covering ELD risks.

160 Cf., e.g., the Estonian implementing law, which includes a more stringent definition of "operator" irrespective of the nature of his activity (EE no. 29); or the Italian Single Act which introduces a parallel civil liability regime (IT no. 37). On the impact of this flexibility upon the availability of insurance, see infra fn. 268.

161 On the variations, see the overview by *CEA*, Navigating the Environmental Liability Directive. A practical guide for insurance underwriters and claims handlers (2009, available at http://www.cea.eu/uploads/DocumentsLibrary/documents/1240585425-eld-best-practice-guide-update.pdf) 10 ff. and 56 ff.

itself, only three Member States had passed implementing legislation (Italy, Latvia and Lithuania), and the ECJ just recently ruled on cases filed by the Commission for failure to transpose,[162] with several applications withdrawn in the meantime due to (late) compliance.

2. Environmental liability beyond the scope of the Directive

(a) Environmental harm

Some Member States had already a more or less far-reaching system in force which attributed the costs of remedying environmental harm, thereby overlapping at least in part with the new regime proposed by the Directive. In contrast to the Product Liability Directive,[163] such a duplication of legal measures is admissible.[164] Therefore, in those jurisdictions there may be more than one way for the state to recover the costs of repairing the environment.[165] **96**

(b) Damage to individual claimants

Apart from dealing with harm to the environment as such, several jurisdictions have also special rules in force which foresee the indemnification of losses that are secondary to environmental harm and caused by the latter to individuals. If pollution causes personal injuries or damage to property, the victims do not have to resort to general rules of tort law (even though they could), but have an easier way to receive compensation, typically via strict liability. Many of these systems seem to be limited to specific parts of the environment and/or to certain types of activities that may be harmful,[166] whereas others are not subject to such limitations and respond to secondary environmental harm irrespective of the kind of activity causing **97**

162 ECJ C-368-08, *Commission v. Hellenic Republic*; C-417/08, *Commission v. United Kingdom*; C-422/08, *Commission v. Austria*.
163 Supra no. 84.
164 Supra at fn. 160.
165 E.g. CZ nos. 66 ff.; DE no. 3; EE nos. 31, 34; ES nos. 65 ff.; FI nos. 2, 11, 45 ff.; IT no. 37; UK no. 33 (several specific liability regimes).
166 E.g. DE no. 49 (not covering facilities for genetic engineering); DK no. 24. Cf. US no. 21.

it.[167] Yet others have the option to pursue claims based on general strict liability provisions.[168]

(c) Nuisance/neighbourhood laws in particular

98 Even if a country has not yet foreseen specific liability rules for environmental harm caused to individuals, at least property losses may already be recoverable under that jurisdiction's special rules for losses incurred within a neighbourhood, notwithstanding the general rules of tort law, of course. These provisions aim at resolving disputes arising out of a conflict of interests between owners of adjacent or nearby land. They do not so much focus on the conduct of those interfering with the neighbouring interest, but rather on that outcome as such. Therefore, these rules do not primarily deal with an assessment of behaviour and are thus typically irrespective of fault. Since they are result-oriented, they tend to compare the effect of emissions on neighbouring land with other influences on it from the surrounding area: the more common practice it is and/or the more likely comparable sources in the neighbourhood will lead to similar outcome, the less likely the claimant will prevail under these rules.[169] Conduct may still come into play if the impact on the land was aimed at it specifically by the neighbour, in which case more stringent standards apply to remedy the outcome.[170]

99 Apart from injunctions and similar titles to stop the neighbour from continuing to exert the influence upon the adjacent parcel of land, reparation may be another remedy foreseen under these circumstances.[171] If these claims are not already considered as part of tort law such as in common law jurisdictions,[172] they are thereby at least closely related to the law of delict inasmuch as they also aim at repairing (either in kind or by way of monetary compensation) any impermissible deviation from the *status quo ante*.

---

167 E.g. EE no. 31; FI no. 57; NO no. 23. Cf. CA no. 33.

168 HU no. 51; IT no. 39; NL nos. 36, 40 ff.; PL no. 62.

169 See *Koch* (fn. 2) nos. 67 ff. and the country reports cited therein. Another comparative survey is given by *B. Pozzo* (ed.), Property and Environment (2007). See also *Ch. von Bar*, The Common European Law of Torts I (1998) nos. 532 ff.

170 *Koch* (fn. 2) no. 70.

171 This does not seem to be the case in Hungary, where the only claim is an *actio negatoria* (HU no. 51).

172 On the tort of private nuisance, see UK no. 35. A historical overview is given by *B. Pozzo*, Property Rights in the Defense of Nature. From the Historical Evolution to the Contemporary Challenges: A Comparative Law Analysis, in Pozzo (fn. 169) 3, 18 ff. Also the Netherlands place these claims into tort rather than property law.

Most jurisdictions have such special rules in force.[173] While they share the same fundamental approach,[174] there are quite considerable differences which may be decisive in case of actual disputes.[175] For example, the standard applied to weighing what is still common to the area or already unreasonable varies from country to country. The same is true for the relevance of an authorisation for determining the availability of remedies, from a mere piece of fact considered for assessing the reasonableness of the neighbouring activity[176] to an exclusion of the claim altogether or to cutting off the possibility to enjoin it while retaining a claim for reparation.[177] The latter is an expression of the jurisdiction's efforts to preserve the balance of interests between the neighbours – both can pursue activities on their land as they wish, but if one thereby causes losses to the other, the equilibrium is restored by way of compensation.[178] **100**

Claims arising out of neighbourhood conflicts are less relevant to the cases under survey here as compared to the problems addressed in our previous study: while it can be crucial for determining the consequences of GMO admixture between two neighbouring farmers, it will not be helpful for damage caused along the food or feed supply chain which is not confined to a specific limited area in the vicinity of the defendant.[179] Nevertheless, as long as the claimants and the defendants fall within the definition of "neighbours" applied to these special rules, the latter may offer an alternative basis for resolving such disputes. **101**

173 E.g. AT nos. 31 ff.; CZ nos. 71 ff.; DE no. 51; DK no. 24; ES nos. 75–76 (apart from some regional rules no national statutory provision, but based on general doctrines); FR nos. 55–57 (see in particular the cases cited in no. 57: hormone treatment of crops deteriorating neighbouring lettuce, cement dust on claimant's crops preventing photosynthesis); GR nos. 43, 46; *P. Goergen*, Luxembourg, in Economic Loss 325 (nos. 34–35); HU no, 51; NO no. 25; PL no. 69; UK no. 35; cf. BR 61; CA nos. 35 ff.

174 Cf. *B. Pozzo*, Conclusions, in Pozzo (fn. 169) 355, 357 ("an inherent harmony, more than a gradual convergence").

175 But see *von Bar* (fn. 169) no. 545: "There are differences between the individual legal systems, but these only concern marginal issues, such as which types of emissions are covered, how the quantum of compensation is calculated, and whether the corresponding claim is in tort or property law.".

176 UK no. 35: "Regulatory consent ... may serve to 'crystallise' what is a reasonable land-use in the area in question.".

177 See *Koch* (fn. 2) no. 72.

178 Cf. *Pozzo* (fn. 172) 6 at the example of Germany; but see *von Bar* (fn. 169) nos. 536–537, who is critical of such claims "akin to that of a land-owner against the state which compulsorily purchases his land" as a matter of principle.

179 See also *von Bar* (fn. 169) no. 532: "The law governing relationships between neighbours does not cover personal injury or rights to personality."

(d) International liability regimes

102 The only project developing public international law rules on liability for changes to biodiversity through GMOs[180] is based upon the Cartagena Protocol on Biosafety, which is described in depth by the report on International Environmental Law.[181] It is still a work in progress, though it seems at present[182] that the path – if anywhere – is leading towards a regime modelled after the Environmental Liability Directive in principle, therefore rather an administrative regime, leaving civil liability matters to domestic law.

VIII. Other bases of liability

1. Vicarious liability

103 The classic term "vicarious liability" typically refers to a principal-agent relationship, out of which the former is liable for harmful conduct of the latter. Some jurisdictions perceive this as a case of strict liability, since fault of the liable person is not of the essence, others see this as at least a mixed regime between fault and strict, since the principal is only liable if the conduct of his agent did not meet the required standard of care, which is defined by duties of the principal, however.[183]

104 Despite these theoretical differences, there is a solid common core throughout Europe that the principal should be accountable for damage caused by

---

180 The so-called Lugano Convention or – in full – the Convention on civil liability for damage resulting from activities dangerous to the environment (text at http://conventions.coe.int/treaty/en/treaties/html/150.htm) has never reached any viable stage of signatures (9), let alone ratifications (0): http://conventions.coe.int/Treaty/Commun/ChercheSig.asp?NT=150&CL=ENG. An overview over this stillborn regime as it would have applied to GMOs is given, e.g., by *L. Bergkamp*, European Community Law for the New Economy (2003) 245 ff.

181 Supra 661 ff.

182 The latest available document at the time of drafting this comparative report was the Report of the Group of the Friends of the Co-Chairs on Liability and Redress in the Context of the Cartagena Protocol on Biosafety on the Work of its Second Meeting (http://www.cbd.int/doc/meetings/bs/bsgflr-02/official/bsgflr-02-03-en.pdf), which took place in February 2010.

183 "Liability has in effect become 'indirect': the master is responsible for the servant's fault." *Von Bar* (fn. 169) no. 181. Cf. *H. Koziol*, Commentary on Chapter 1, in *European Group on Tort Law*, Principles of European Tort Law (2005) 27 (Art. 1:101 no. 17).

his servants[184] while acting within their duties.[185] The agreement already stops, though, when it comes to the question whether those retaining independent contractors, i.e. persons acting for the former without his immediate control or supervision, but still in his interest,[186] should also be liable for the latter despite the lack of subordination as commonly associated with an employment relationship. "In most of the countries, the general principle is that there is no vicarious liability in tort for harm caused by an independent contractor."[187]

For our purposes, it is not so much interesting whether a farmer should compensate losses caused by his farm hands, or the operator of a storage facility for his workers, or a shipowner for his sailors – those questions would invariably be answered in the affirmative in most jurisdictions.[188] What really matters, though, is whether someone along the supply chain has to account for wrongs further up or down the chain, where independent operators have set the cause. This is deemed at least doubtful by all reporters,[189] though not entirely unimaginable by some.[190] **105**

184 A relationship of employment as such is not needed, it suffices if the servant/agent/employee has to follow the instructions of the principal as to the conditions of his tasks (typically what, when, where and how to act). See *von Bar* (fn. 169) nos. 191 ff.

185 See, e.g., the comparative overview by *S. Galand-Carval*, Comparative Report on Liability for Damage Caused by Others, Part I – General Questions, in J. Spier (ed.), Unification of Tort Law: Liability for Damage Caused by Others (2003) 289 (nos. 41 ff.). An exception is Austria, where bystanders and third parties who are not in a contractual or other special relationship of the employer can only sue the latter for wrongdoings of his employees if these were unfit or dangerous (AT no. 38). However, Austrian courts try to circumvent this outdated statutory approach by expanding the notion of contractual duties towards third parties more broadly (see also AT no. 38 on statutory exceptions to this narrow rule, in particular in the GMO legislation: no. 49). Cf. FR no. 62. See also *von Bar* (fn. 169) nos. 179 ff.

186 See *Galand-Carval* (fn. 185) no. 64 (fn. 86) on the problems of defining "independent contractors". Cf. GR no. 51.

187 *Galand-Carval* (fn. 185) no. 64. See also *von Bar* (fn. 169) 201: "[T]he basic rule that others are not liable for damage by independent contractors remains." E.g. DK no. 27; ES no. 80; FI no. 61 (but possible under special circumstances); FR no. 61; HU nos. 57–58; LU nos. 86–87, 90; NO no. 28; UK no. 40 (with exceptions); AU no. 42. But see AT nos. 36–39: principal in contractual relationship with victim liable for independent contractors; equally EE no. 41; PL no. 72; SE no. 68. Further exception: IT no. 42. Cf. NL no. 47.

188 With the aforementioned exception of Austria, see fn. 185. Cf., e.g., CZ nos. 79–80. However, in all these cases the duties of the employee have to be established first; cf. the responses to Case 3.

189 As the Austrian reporter summarises, "the scope of vicarious liability within the feed and food production chain, which typically consists of independent contractors, is limited" (AT no. 42), which is true for all jurisdictions covered.

190 See, e.g., LI no. 37, quoting a provision of Liechtenstein's GMO legislation which holds "the person responsible for the first placing on the market of a GMO . . . liable for those losses as well which are caused to third parties even if all rules of conduct have been respected", which was explained by the legislator's intention to protect "all those

106 One way to hold someone liable for another further *down* the supply chain is if he made the wrong choice in selecting that person, e.g. a producer selecting an unfit wholesale distributor. This may amount to *culpa in eligendo*, but is not a case of vicarious liability, as the producer will be liable (if at all) for his own fault.[191]

107 Another scenario where it merely seems that someone has to account for another *upstream* is in product liability, with e.g. a processor liable for shortcomings of raw materials supplied to him by another.[192] In that case, the processor will not be liable *for* the producer of the raw materials, but *himself* for distributing his own products that are defective,[193] even if only due to the flawed ingredient, which may allow him to sue the supplier in recourse.[194] As clarified by the ECJ,[195] victims of product defects cannot sue anyone else but a producer within the meaning of the Directive;[196] therefore they cannot base their claims against a seller or other distributor along the supply chain upon the product liability regime.

2. General strict liability

108 Some jurisdictions have general strict liability rules which may apply at least in some cases under survey here, provided that the use of GMOs falls under the definitions used, which can only be estimated for the time being as there have not been any actual court cases yet. Such rules address, for example, "hazardous substances",[197] "extremely dangerous activities"[198] or are even broader.[199] Under this condition, these provisions may serve as additional bases of claims to indemnify losses to persons or property. To the extent they compete with general tort law claims, victims are probably tempted to pursue this alternative path which does not require them to prove the tortfeasor's fault. However, it seems rather doubtful that

---

actors producing GMO-free products which suffer from a contamination further up the production chain". Cf. further NL no. 48.

191 AT no. 48; DE no. 57; EE nos. 44–45; ES nos. 87 ff.; FI nos. 67 ff.; FR no. 65; GR no. 54; HU no. 61; IT no. 49; LI nos. 39–40; LU nos. 94 ff.; NL nos. 50 ff.; NO no. 31; PL no. 80; SE no. 73; SI no. 29; UK nos. 43–44; AU nos. 46 ff.; CA no. 50.

192 Cf. supra no. 75. E.g. ES no. 85.

193 UK no. 41.

194 AT no. 42.

195 ECJ C-402/03, *Skov Æg v. Bilka Lavprisvarehus A/S,* [2006] ECR I-199. But see CA no. 48 (opposite solution in Québec).

196 Supra no. 75.

197 NL nos. 41 ff. (but "unlikely" that GMOs will be deemed as dangerous).

198 E.g. EE no. 3; HU no. 48; IT nos. 32–34.

199 Cf. the French general strict liability of the *gardien* for inanimate objects (FR no. 59).

courts are willing to subsume GMOs under any notion of ultra-hazardous objects, particularly if they are approved, though the circumstances of the case may ultimately be decisive (e.g. if large-scale bodily injury is at stake).[200]

IX. Defences

The notion of "defence" itself is not entirely identical in all legal systems. **109** In essence, it denotes counter-arguments that the defendant has to raise and prove in court in order to avoid or at least reduce liability.[201] This would apply to all responses of the defendant in reaction to the claimant's allegations or to presumptions held against him, including efforts to rebut evidence submitted by the claimant in order to support her case. These are reactions to moves by the claimant, however, whereas defences typically identify topics addressed at the defendant's initiative. However, as has been said elsewhere, defences are quite diverse by nature,[202] and only a few have been selected for the following comparative overview.

1. Regulatory compliance/permit defence

The so-called regulatory compliance defence is a rather weak tool – the fact **110** that one abided by rules prescribing (or prohibiting) certain conduct does not per se equalise the verdict that the defendant's behaviour was at fault.[203] It will at least be considered as one element, however, in the assess-

200 But see Hungary, where the legislator has stepped in and by statute defined such losses as falling under the general strict liability rule: supra no. 65 and HU no. 1.

201 On the problems of defining defences and bringing them together under one heading, see *B.A. Koch*, Commentary to Chapter 7, in *European Group on Tort Law*, Principles of European Tort Law (2005) 144.

202 *W. van Gerven*, Torts (The Common Law of Europe Casebook, 2000) 356: "It is practically impossible to bring these defences within a conceptual framework, which is valid for the … systems under examination.".

203 E.g. FR no. 33; *Dacoronia* (fn. 65) no. 64; but see DK no. 20; AU no. 28. An economic explanation for this weakness is given by EA nos. 42 f. A comparative view of this defence is given by *W. van Boom*, On the Intersection Between Tort Law and Regulatory Law – A Comparative Analysis, in W. van Boom/M. Lukas/Ch. Kissling (eds.), Tort and Regulatory Law (2007) 419 (nos. 43 ff.). See also *M. Lukas*, The Function of Regulatory Law in the Context of Tort Law – Conclusions, in the same volume, 449 (nos. 24 ff.), arguing that "the more detailed the regulatory law provisions determine the required conduct in a certain situation, the less room naturally is left for an autonomous evaluation of such conduct according to the general standards of tort law".

ment of the latter's conduct.[204] So even if a farmer can establish that he complied with all rules regulating the cultivation of his crops, he may still be found to have acted negligently under further conditions. Equally, a car driver cannot excuse himself by proving that he drove below the speed limit and on the proper lane if he otherwise failed to meet the standard of care required under all circumstances, which is not only defined by written rules.

111 The same is true for permits granted by the authorities: as these are never aimed at allowing the infliction of harm upon others (even though they may take this into account as a side-effect[205]), such permits will at best be considered when assessing in tort law the due standard of care, but the latter can certainly exceed the scope of the authorisation.[206] This is even spelled out in Art. 7 (7) of Regulation (EC) No 1829/2003 of the European Parliament and of the Council of 22 September 2003 on genetically modified food and feed:[207] "The granting of authorisation shall not lessen the general civil and criminal liability of any food operator in respect of the food concerned."[208] The regulatory permit defence may be specifically excluded in some liability regimes or even generally unavailable.[209]

2. Consent/assumption of risk

112 "As long as GMOs and GM products are considered to have no adverse effects on human health, the knowing consumption of these products by the victim will have no impact on the tortfeasor's liability."[210] Even though most jurisdictions recognise the voluntary assumption of risk as a defence in general,[211] this citation from the Austrian report will prob-

204 E.g. FI no. 70 ("may be relevant as one factor among others"); GR no. 56; NO no. 32; UK no. 45 ("may be relevant to the determination of fault or, in respect of private nuisance, unreasonableness"); AU no. 28.

205 This may be an unavoidable consequence of deciding upon the balance of interests as mentioned earlier (supra no. 100 at fn. 178). Cf. ES no. 91: "Licences or permits are issued as a rule even though the authorised activity may cause damage to third parties. In fact, authorisations are usually issued including a clause which leaves the rights of third parties in respect of damage caused to them unaffected …".

206 On the regulatory permit defence, see *Lukas* (fn. 203) nos. 27 ff.

207 OJ L 268, 18.10.2003, 1.

208 Similarly Art. 19(7) with respect to feed.

209 E.g. EE no. 46; FR no. 66; HU nos. 44, 63; IT no. 50; *B. Askeland*, Tort and Regulatory Law in Norway, in W. van Boom/M. Lukas/Ch. Kissling (eds.), Tort and Regulatory Law (2007) 205 (nos. 37–38: "[A] 'regulatory permit defence' is an unknown concept under Norwegian law."); LI no. 41; LU no. 98; MT no. 43; NL no. 53; PL no. 82; ES no. 91; SE nos. 74–75; AU nos. 6, 50; CA no. 49; US no. 29; cf. also supra at fn. 155.

210 AT no. 47.

ably apply universally, if only because awareness of a vague risk does not necessarily mean that the victim expected it to materialise.[212] Obviously, it would not apply to non-labelled products, or to GMOs not identified by the later victim, unless she deliberately and specifically took this into consideration. After all, consent requires knowledge of what one agrees to.[213]

On the other hand, a conventional farmer moving into an area with widespread GM cultivation will probably not succeed in seeking compensation for the consequences of admixture.[214] **113**

Furthermore, it is also possible to expressly waive the right to compensation in advance, as conventional farmers may do if they consent upfront by way of contract with their neighbour to the latter's GM cultivation.[215] **114**

3. Third-party influence

If damage was caused (in the legal sense) by a third party rather than the defendant, this will generally not be considered as a defence, but exclude liability of the defendant in the first place for lack of causation attributable to the latter.[216] If it was just one of several causal influences upon the bodily integrity of the victim or on other protected interests, the recognition of third-party influence as a defence depends upon the legal system and the models used therein. **115**

Apart from those that consider the impact of an activity upon the course of events according to the (neutral) degree of causation,[217] others recognise qualified fault by the third party as a ground for complete exoneration,[218] again others only if it was the sole cause (and not just one of several)[219] – and all that depending upon the basis of the claim.[220] As a rule of thumb, **116**

211 Cf. *von Bar* (fn. 10) nos. 512–513 (pp. 534–542).

212 CZ no. 96; DK no. 32; ES no. 93 (assumption of risk "blurry concept under Spanish law"); FI no. 74; FR no. 68; GR no. 58; LI no. 42; LU no. 100 ff. (defence only if abnormal risk); NO no. 33; PL no. 84 (public policy exception); SE no. 76 (consent no defence if harm "atypically more severe than could be expected"); UK no. 46; AU no. 51; CA no. 50. Cf. IT 50, arguing that knowingly distributing harmful products will outweigh the victim's assumption of risk.

213 Cf. FR no. 68.

214 See, e.g., US no. 15. An exception is France, where "coming to a risk" is no defence in neighbourhood cases: FR no. 69.

215 E.g. DE no. 59.

216 Cf. IT no. 51; UK no. 47.

217 E.g. FR no. 70; IT no. 51; LU no. 106; NL no. 55; SE nos. 81 ff.

218 E.g. GR no. 59; LI nos. 1, 43; NL nos. 44, 55.

the stricter liability gets, the more carefully will the effect of the third party's behaviour on the liability of the operator be scrutinized.

4. Prescription

117 The defence of prescription allows the defendant to escape liability despite all substantive requirements fulfilled, just because of the lapse of time.[221] A survey of the prescription periods in tort law and their scope of application throughout the jurisdictions covered in this study shows remarkable differences, however, when it comes to defining how much time must have passed, and what marks the start of the relevant period.[222] To the extent substantive tort law was harmonised including rules on prescription, there is no diversity, however, at least within the EU. This is true in particular for the respective provisions of the PLD.[223]

118 Apart from such singular unified solutions, most legal systems apply a combination of a shorter and a longer period, with different triggers. Typically, the short period starts with subjective elements such as the victim's (actual or imputed) knowledge of damage and/or tortfeasor, whereas the long period is set off with objective events such as the harmful conduct. The short period ranges from one[224] to ten years,[225] with a clear

---

219 E.g. AT no. 48; ES no. 94 (if third party intervention amounts to force majeure); FR no. 71 (like Spain – if equivalent of force majeure); HU no. 66; NO no. 34 (if defendant's influence was "unsubstantial" or "unimportant" in comparison); PL no. 86.

220 The line will be drawn more narrowly in cases of strict liability, where the impact of third-party conduct may be tested against stricter standards, e.g. FI no. 77. See also Art. 8 (1) PLD: "Without prejudice to the provisions of national law concerning the right of contribution or recourse, the liability of the producer shall not be reduced when the damage is caused both by a defect in product and by the act or omission of a third party.".

221 *Von Bar* (fn. 10) no. 545, therefore calls it "the morally weakest defence".

222 See generally *von Bar* (fn. 10) nos. 545 ff. and *R. Zimmermann/J. Kleinschmidt*, Prescription: General Framework and Special Problems Concerning Damages Claims, in H. Koziol/B. Steininger (eds.), European Tort Law 2007 (2008) 26.

223 Art. 10 (1) PLD provides: "Member States shall provide in their legislation that a limitation period of three years shall apply to proceedings for the recovery of damages as provided for in this Directive. The limitation period shall begin to run from the day on which the plaintiff became aware, or should reasonably have become aware, of the damage, the defect and the identity of the producer." See also Art. 11 PLD (claim extinguished 10 years after the actual product was put into circulation); supra at no. 82.

224 ES no. 97.

225 FR no. 73 (personal injury claims); SE no. 86.

majority opting for three years.[226] The long period varies from 3 to 30 years.[227]

The idea of the double time span is to strike a compromise between the interests of the victim on the one hand in preserving her claim, which she cannot pursue if she is unaware of it (yet), and the interests of the alleged tortfeasor on the other hand to have the file closed, also in light of the fading availability of evidence. "After all, prescription constitutes an essential means to defeat claims that may be unfounded."[228] Furthermore, there is "public's need for legal certainty as well as the courts' need for relief from overwork".[229] In light of the gravity of the harm, however, several jurisdictions extend their long-stop period[230] or do not apply it at all[231] in cases of personal injury or death, thereby shifting the focus exclusively onto the victim. Some jurisdictions have no long-stop period at all also in cases other than bodily harm.[232] **119**

The applicable rules on prescription are crucial for the cases examined here: Particularly in light of the uncertainties regarding potential hazards originating from GMOs, which may materialise far in the future, the key question is how such risks are being dealt with, and whether those applying and implementing agri-biotechnology today have to fear being sued in decades to come, for dangers they themselves do not even know yet.[233] A quick escape would be by raising the defence of prescription, but it may not be available, particularly not in future cases of bodily harm. Experi- **120**

---

226 AT no. 49; DE no. 61; DK no. 34; EE no. 49; FI no. 80; HU no. 67 (strict liability); LI no. 44; NO no. 35; PL no. 88; UK no. 48 (personal injury claims); BR no. 75; CA no. 52 (Québec).
Two years: CZ no. 100; MT no. 48; CA no. 52 (common law provinces).
Four years: CZ no. 104 (in business relationships); BR no. 75 (environmental crimes).
Five years: FR no. 73 (property damage or financial loss); HU no. 67 (fault); IT no. 52; NL no. 56; BR no. 75 (consumer claims).
Six years: UK no. 48 (other than personal injury).
Eight years: BR no. 75 (crimes related to GMOs).

227 3 years: CZ no. 100.
10 years: CZ no. 100 (if damage caused intentionally or in a business relationship); DE no. 61 (after injury other than bodily harm); EE no. 49; FI no. 80 (not bodily injury or environmental harm); PL no. 88 (standard); UK no. 80.
20 years: DK (no. 34); NL no. 36; NO. no. 36; PL no. 88 (crime or misdemeanour).
30 years: AT no. 49; DE no. 61 (after personal injury; in other cases after the harmful act); LI no. 44; NL no. 56 (contamination).

228 *Zimmermann/Kleinschmidt* (fn. 222) no. 51.

229 *Von Bar* (fn. 10) no. 545.

230 E.g. DE no. 61.

231 CZ no. 102; FI no. 80; NL no. 56; PL no. 89.

232 E.g. CA no. 52; ES no. 97; IT no. 52; SE no. 86.

233 One substantive counter-argument could be a state-of-the-art defence or the equivalent, see supra nos. 80 and 92.

ence with asbestos[234] and similar long-tail risks may serve as examples. So if, say, 31 years from now an approved GM crop turns out to be the overriding cause for a certain disease of the human body, the case will not proceed in jurisdictions such as Austria or England, but does not stop for being time-barred in the Netherlands, for example, where the victims have five years to pursue their claims.[235]

X. Remedies

1. Damages

121 All country reports are in accord that GMO loss scenarios are not treated differently to other tort cases when it comes to determining type and amount of monetary compensation. The only exception is Germany, where liability under the Act on Genetic Engineering is capped at € 85 million per event (as in the Product Liability Act and other German strict liability statutes).[236] The existing differences between the tort laws of the various legal systems as far as damages is concerned, in particular for non-pecuniary losses, are therefore equally true for the specific cases covered here.[237]

122 Non-compensatory awards such as punitive damages are not recognised at all in continental legal systems, while possible in common law jurisdictions subject to specific conditions.[238]

---

234 See e.g. the RAND report by *St. Carroll et al.*, Asbestos Litigation (2005), in particular 25 f., available for download at http://www.rand.org/pubs/monographs/2005/RAND_MG162.pdf, and the recent Munich Re publication Asbestos. Anatomy of a mass tort (2009), available at http://www.munichre.com/publications/302-06142_en.pdf.

235 Unless the basis of the claim is product liability, in which case the 10 year prescription period applies. See also the responses to Case 2, which are in accord that no liability attaches to such long-tail risks, even though the justifications may slightly differ.

236 DE no. 63.

237 See, e.g., *B.A. Koch/H. Koziol*, Compensation for Personal Injury in a Comparative Perspective (2002).

238 UK no. 55; AU no. 62; CA no. 59; US nos. 8, 22. See also BR no. 86; EA no. 103. A recent comparative survey is provided by H.Koziol/V. Wilcox (eds.), Punitive Damages: Common Law and Civil Law Perspectives (2009).

2. Non-pecuniary remedies

In many legal systems, reparation in kind is the prime response to losses in tort law.[239] In some of them, it is also the first option, while monetary compensation is (at least in theory) only available subject to further conditions, e.g. if reparation is unreasonable or too expensive.[240] In other systems, claimants can choose between both types of remedies, while allowing the defendant to avoid excessive efforts by payment of the money equivalent.[241] Again other jurisdictions start with monetary compensation as the prime remedy, with reparation in kind as the exception.[242] **123**

Particularly neighbourhood-related claims are not immediately linked to compensation, but first aim at enjoining the neighbour from intruding upon one's land. Such injunctions are the standard remedy in the abovementioned special regimes designed for conflicts concerning neighbouring land.[243] **124**

3. Advance cover

While there is no general requirement to take out liability insurance or seek any other form of advance cover against potential future losses, some of the special liability regimes mentioned above provide for such a requirement, linked to the conditions of the scheme.[244] **125**

XI. Specific aspects of cross-border claims

As far as tort claims are concerned, a special report in this volume[245] deals with questions of cross-border litigation and presents the regimes of jurisdiction and choice of law as harmonized within the EU.[246] It is therefore not necessary to analyse these matters in detail here. However, it may be useful to repeat the key findings of that report with an eye to risks associated with GMOs. **126**

239 But see NO no. 44; SE no. 103; UK no. 56.
240 AT no. 60; DE no. 70; HU no. 77; LI no. 53; LU no. 122.
241 NL no. 65.
242 EE no. 59; GR no. 72; IT no. 63.
243 Supra VII.2(c). See, e.g., AT nos. 3, 31 ff.; DE no. 30; FI no. 96; UK nos. 35, 56; cf. MT no. 62; AU no. 63; CA no. 60; US no. 40. Cf. EA nos. 105 ff.
244 AT no. 63; FR no. 89; DE no. 73; ES nos. 52 ff.; LI no. 56; LU no.129.
245 Supra 687 ff.
246 In case of Rome II with the exception of Denmark, JC no. 48 fn. 29.

127 In a cross-border tort claim, the victim can choose to sue in one of up to three different jurisdictions – at the defendant's domicile or principal place of business, in the country where the damaging event took place, or where the harmful result manifests.[247] Once the proper court under the Brussels I Regulation[248] is identified, the Rome II Regulation[249] will then determine which law to apply to the case at hand. In a standard tort case, the prime focus when it comes to resolving a conflict of laws is on the place where the (first) damage occurred, so on only the last of the three afore-mentioned aspects that are relevant for deciding upon jurisdiction.[250] The place of the damaging conduct comes into play again as an optional alternative at the choice of the claimant if she builds her case upon environmental liability.[251] If the basis of the claim is product liability, however, a rather complex multi-level checklist has to be worked through,[252] unless the Hague Product Liability Convention applies instead.[253]

128 No matter how complex the solutions offered by these regulations may be, they have certainly achieved their prime goal – to let the same rules decide upon these apparent "formalities", which have a crucial (and previously often underestimated) impact upon the solution of a case, throughout the EU. Therefore, concerns that cross-border contamination with GMOs may effectively lead to a collapse of the legal order due to the complexity of the matter and the diversity of solutions are completely unfounded – Community law already provides for a stable framework to place such claims and also identifies which laws apply, irrespective of where the cases are filed. This leads to a predictable regime, even if the target jurisdiction whose laws govern may foresee a solution that is not identical to expectations of the claimant in his own country. Cross-border lawsuits by definition leave the realm and shelter of one single jurisdiction, and there is no foundation whatsoever for a claim to be protected by one's own (domicile's) laws wherever one acts or causes effects, or wherever one is harmed. It is naïve to

247 JC nos. 15 ff. Obviously the choice is less if two or more of these places are within the same jurisdiction.

248 Council Regulation (EC) No 44/2001 of 22 December 2000 on jurisdiction and the recognition and enforcement of judgments in civil and commercial matters, OJ L 12, 16.1.2001, 1; JC nos. 4 ff.

249 Regulation (EC) No. 864/2007 on the law applicable to non-contractual obligations, OJ L 199, 31.7.2007, 40; JC nos. 48 ff.

250 JC nos. 55–57.

251 JC nos. 75 ff. This amelioration of the claimant's position is somewhat odd, considering that someone suffering personal injury or property damage only indirectly via some environmental harm has a choice which another person injured directly has not.

252 JC nos. 58 ff.

253 JC no. 52.

assume that one's own laws will apply if one causes harm to another in a different legal system, as it is to expect that exactly the same degree of protection is awarded by each jurisdiction worldwide. As an Englishman cannot expect to be allowed to drive on the left while on holidays in Spain, he can equally not expect to receive the same kind and amount of compensation like at home if injured on the road there. Equally, a Spanish GM farmer cannot grow GM maize on a field right on the French border[254] and expect to be completely immune against claims e.g. by French farmers whose conventional maize ends up being contaminated.

Other cross-border aspects which cannot be elaborated here are, for example, questions of claims against or recourse by or between social security carriers in cases of personal injury caused in one country and having financial implications in another; or international environmental damage falling under the scope of the ELD, which foresees international cooperation between the Member States for such cases.[255] **129**

XII. Alternatives to tort law

1. Insurance regimes

(a) Social insurance

While its cross-border dimensions cannot be analysed here, as just indicated, it is nevertheless important to at least highlight in principle the role of social security, which cushions the immediate financial consequences of harm to the human body irrespective of the cause, at least in all European systems covered. The question whether and under which conditions social security carriers are prepared to seek recourse for the amounts they paid out, at least in their own jurisdiction, obviously has an impact of the frequency and type of claims brought in tort law.[256] At any rate, one should bear in mind that, say, illnesses and other negative health conditions triggered by GMOs, whether directly or indirectly, will first be an issue for social security benefits in Europe, so victims will not suffer an immediate financial loss to the full extent when seeking treatment. This is clearly one prime difference to the economic losses examined **130**

254 Assuming for the sake of the argument that this is agronomically feasible anywhere along the borderline.

255 Art. 15 ELD.

256 An extensive comparative survey is provided by U. Magnus (ed.), The Impact of Social Security Law on Tort Law (2003).

in our first study, which also reduces the need for immediate external action such as state intervention.

(b) Private insurance

(i) General aspects

131 As already outlined in the comparative report to the previous study,[257] private insurance allows the pooling of risks among a larger group of risk-prone individuals, who thereby reduce the full impact of a loss on their assets by spreading the risk among all those who are equally exposed. "Insurance enables risk taking."[258] There are two approaches which need to be distinguished: first-party liability brings together those who are potential victims of a risk and who want to protect themselves against their own losses, whereas those who cause losses to others and therefore may be bound to indemnify these third parties will join in sharing this risk of being secondary payors of losses in what is referred to as third-party liability.

132 Either way, "the basic principles for underwriting include assessability (measurability, quantification); economic efficiency (profitability); randomness (fortuity); and mutuality (solidarity)."[259] As elaborated further by the special report on insurance aspects,[260] insurers need to have sufficient information to estimate how often a risk will materialise, and how much damage it will cause, while at the same time being confident that the loss is not certain to occur. They need to be able to calculate how high the premiums need to be priced in order to accumulate sufficient funds to pay the aggregate losses and to cover administrative costs with a certain margin of profit for the insurer, while at the same time taking into account that the higher the premium, the less demand there will be for the product. This in turn would weaken another important aspect of insurability, the size of the risk pool: the more insured participate, the better the risk can be spread, unless the additional customers are more risk-prone than average, so if only people living near a river buy flood insurance, the mere number of clients does not improve the mutuality aspect.

---

257 *Koch* (fn. 2) nos. 115 ff.
258 IN no. 20.
259 IN no. 24.
260 IN nos. 25 ff.

While the immediate prospects of having a range of suitable insurance products for the risks of cross-pollination and ensuing economic losses offered on the market are rather slim, as examined in our previous study,[261] the risks covered by this follow-up project may more likely be underwritten, even if at present mostly excluded. 133

(ii) First-party insurance

If it is not a purely economic interest, but rather one's life or health one wishes to provide for, existing private insurance products, which cover 22% of the population,[262] are seemingly capable to meet this demand and are readily available. After all, these risks are buffered by social insurance irrespective of the cause, which leaves the buffer overflow that still needs to be absorbed individually relatively well calculable, so that even risks posed by terrorism remain covered.[263] 134

While this is not equally true for property insurance, it still seems to be easier to underwrite in comparison to lost profits due to GMO admixture because the extent of the latter is much less foreseeable than the affected value of land or cattle, for example, even though type and frequency of harm are still not as predictable as in case of other risks. Furthermore, particularly in the farming industry, there are already existing insurance bundles to which these risks could be attached, which include insurances against natural hazards or diseases (e.g. multi-peril crop insurance). These do not yet cover such losses even if they lack an explicit exclusion of GMO risks, because traditional agricultural insurance limits coverage to certain named hazards.[264] 135

(iii) Third-party insurance

More recent product and environmental liability policies for farmers "usually have an explicit GMO exclusion".[265] However, this rather seems 136

---

261 *Koch* (fn. 2) nos. 115 ff.

262 CEA, The European Health Insurance Market in 2006 (CEA Statistics N° 35, 2008, available at http://www.cea.eu/uploads/DocumentsLibrary/documents/1222087447-cea-statistics-nr-35-european-health-insurance-2006.pdf) 35.

263 *Ch. Lahnstein*, Liability Insurance for Acts of Terrorism? in B.A. Koch (ed.), Terrorism, Tort Law and Insurance (2003) 252 (no. 3).

264 *I. Ebert/Ch. Lahnstein*, GMO Liability: Options for Insurers, in Economic Loss 577 (no. 5).

265 *Ebert/Lahnstein* (fn. 264) no. 7. Cf. *CEA* (fn. 161) 12: "Some policies exclude … genetically modified organisms." See also the ELD Implementation Efficiency Study (fn.

to be an overcautious reaction to the lack of experience with existing liability rules. After all, liability insurance is dependent upon tort law, which defines its insured events, and tort law in turn lives with the experience accumulated in practice, which is presently low or entirely missing with respect to commercially cultivated GM plants.

137 If other product defects or environmental risks of novel technology can be covered,[266] however, if only with a financial limit in the policy, one wonders why this should not equally be possible for GMOs.[267] One of the prime arguments raised against covering the financial losses of non-GM farmers due to adventitious admixture – uncertainty with respect to applicable liability rules – does not equally count for the cases covered here, as both product and environmental liability are already established concepts in all jurisdictions under survey.[268]

138 In order to (re-)enable the insurance industry to offer cover for liability risks of GMO operators, at least one step forward would be a clarification of the duties of care, in particular by legislating on good farming practice. Otherwise "insurers will have to impose well-defined rules of good professional practice in cultivating GM plants as a prerequisite for covering cross-polli-

---

159) 59 ("general unavailability of insurance for environmental damage caused by GMOs"). A few years ago, one of the world's largest reinsurers did not yet expect such a large-scale exclusion apparently: *Munich Re* (fn. 19) 99: "Although there are exceptions, … Munich Re does not see any general trend to exclude losses attributable to genetically modified organisms from public liability, product liability and environmental impairment liability covers in general, nor to impose any limits on them." They recommended "to exclude genetic engineering risks from the liability section" of property insurance policies, however, if these cover laboratories or other R&D installations with dangerous GMOs that could contaminate the environment or pose unknown health risks (103).

266 See the ELD Implementation Efficiency Study (fn. 159) 52: "Although the ELD-related insurance market remains small, it is growing and most insurers state that there is good cover available. In many MS national insurers compete with international groups and competition is described as healthy, at least across Western Europe.".

267 See also the ELD Implementation Efficiency Study (fn. 159) 59: "The concern regarding the lack of insurance cover for any harm caused by GMOs is partially offset by the limited number of companies that carry out activities that involve GMOs in the EU and the large size of a substantial proportion of these companies. The companies, therefore, should be able to obtain other evidence of financial security such as letters of credit and trust funds.".

268 Due to the flexibility of the ELD with respect to national deviations, however, the insurance industry insists that the alleged uncertainties of the times before the Directive continue to exist, cf. *CEA* (fn. 161) 9: "The manner in which the ELD has been transposed means that there is no harmonised liability system. This means that there is a strong possibility that there will be variations in enforcement. These issues pose quite significant challenges for the insurance industry for both underwriting and claims. At European level there is now an absence of one of the most important prerequisites for insurability, ie legal clarity and certainty.".

nation losses, at least where adequate state regulations are missing".[269] Either way, uncertainties about what farmers, transporters, warehouse operators, food or feed producers, etc. have to do in order to avoid liability could at least be substantially reduced.[270]

Another question peculiar to third-party liability is whether the legislator should channel liability towards one player along the supply chain in order to improve insurability. If all claims are aimed at one addressee, the risks are bundled, clarifying where a loss will be placed ultimately, thereby reducing uncertainties who along a chain of distribution will have to provide for cover. At the same time, the addressee of all claims will more easily be in a position to redistribute the costs of insurance. This has been acknowledged by the ECJ in support of channelling product liability onto the producer, thereby releasing interim suppliers from parallel litigation;[271] it is equally true for environmental liability and other scenarios.[272] Preselecting one out of several potential defendants by way of tort law steering also has an impact on cross-border scenarios if these potential addressees of claims are domiciled in different jurisdictions. There is a further, purely pragmatic reason speaking in favour of channelling liability in cases of damage caused by GMOs: if there are, for example, several farmers using the same GM seeds, it may be impossible to identify one out of them as the tortfeasor, but the transgenic traces will always lead back to the producer of these seeds, who could spread the risk (the insurance premium) via the price.[273] In the case of product liability, the PLD did not exhaust all possible advantages inasmuch as it fails to focus claims onto one out of several producers in a multi-step manufacturing process, **139**

269 *Ebert/Lahnstein* (fn. 264) no. 14.

270 Cf. supra no. 57.

271 "While acknowledging that the possibility of holding the supplier of a defective product liable in accordance with the provisions of the Directive would make it simpler for an injured person to bring proceedings, there would – it was observed – be a high price to pay for that simplicity, inasmuch as, by obliging all suppliers to insure against such liability, it would result in products becoming significantly more expensive. Moreover, it would lead to a multiplicity of actions, with the supplier seeking recourse in turn against his own supplier, back up the chain as far as the producer. Since, in the great majority of cases, the supplier does no more than sell the product in the state in which he bought it and only the producer is able to influence its quality, it was thought appropriate to concentrate liability for defective products on the producer." *Skov Æg* (fn. 195) no. 26.

272 But see *M. Faure/D. Grimeaud*, Financial Assurance of Environmental Liability, in M. Faure (ed.), Deterrence, Insurability, and Compensation in Environmental Liability (2003), 7, 165 f.: "Channelling … creates a greater risk exposure for the operator and therefore creates higher uncertainty for the insurer. If the channelling has any effect on insurability it is more likely that it decreases insurability …".

273 EA no. 50.

by allowing claims against both the producer of a defective component as well as against the manufacturer of the final assembled product where the defect persists.[274]

140 A further argument occasionally raised when discussing options to improve insurability is a call for mandatory insurance[275] – by forcing a certain risk group to provide for advance cover of potential losses, the legislator can create demand, which should be ample incentive for insurers to offer suitable products. However, as elaborated by the insurance report, a "legal request for compulsory insurance or other financial guarantees is not suited for convincing hesitant insurers to put economic concerns on hold".[276] They will continue to calculate their options as before, these may be too costly for their involuntary clients, so that the latter are prevented from proceeding with the activity to which the insurance requirement is linked. Ultimately, this is yet another way to steer market behaviour.[277] Apart from that, requiring insurance by law does not produce the data missing for a proper assessment of the insurability criteria.

2. Compensation funds

141 Some countries have set up compensation fund regimes in order to assist farmers who have suffered an economic loss due to the involuntary admixture of GM crops with their own.[278] Neither these nor any other separate funds are currently established that shall offer compensation for bodily injuries or property damage caused by GMOs to third parties. This seems strange at first sight in light of the fact that the by far higher ranking legally protected interests[279] are not considered at all to fall within the scope of the fund. The latter invariably offers easier access to compensation than the tort system, whereas purely economic interests are detoured from the classic route and assigned the fast track through the fund system.

---

274 Supra no. 75.

275 EA nos. 112 ff. See also *Faure/Grimeaud* (fn. 272) 180 ff.

276 IN no. 11. See also IN no. 28: "Particularly with compulsory insurance, a large number of insurance takers is required – the large majority of which do not incur losses. … [T]he more a liability regime targets a few industrial operators, the more unbalanced a risk portfolio becomes. This affects both insurability and premium".

277 An economic analysis of this question is provided by *M. Faure*, The View from Law and Economics, in G. Wagner (ed.), Tort Law and Liability Insurance (2005) 239 (nos. 5 ff., in particular no. 17, warning that insurers thereby "become the licensor of the industry").

278 See *Koch* (fn. 2) nos. 174 ff. with further references; cf. DK no. 2; IT no. 7; LU no. 129.

279 *Koch* (fn. 2) no. 32.

This curiosity is not as odd as it appears, however, if one considers that losses compensated by said funds are more or less clearly defined, with an equally narrow scenario in which they may arise. This makes a compensation regime substantially easier to devise as one that should cover a much broader range of legally protected interests. There are simply too many constellations imaginable out of which damage to persons or property may arise, no matter how speculative they are. Furthermore, with product liability well established, the core areas of realistically conceivable third party losses caused by GMOs seem to be covered, with further potential claims along the food or feed supply chain primarily falling under a contractual regime instead. Environmental harm is addressed by the administrative regime of the ELD in Europe. Also, the political interest to construct an alternative redress scheme is much higher if it needs to fill an apparent gap in the traditional compensation systems in order to stimulate a desirable economic activity. This is not equally true for losses that may affect individuals much more directly: by installing a special compensation scheme for bodily harm caused by GMOs, for example, the state seems to signal to its citizens that it wants to push a technology regardless of the expense – not necessarily a wise message sent out by a politician, even if misunderstood. Furthermore, any other industry would claim equal treatment, and ultimately one might ask why not all attacks on the human body, including, e.g., those in road traffic, are absorbed by some alternative compensation scheme rather than sent to the tort system. Ultimately, one may also wonder why the state should get involved at all – concerns about undesirable state aid also play a significant role in this context.[280] **142**

There are a few funds established in the countries under survey for the protection (and ultimately restoration) of the environment.[281] However, these funds rather serve as an interface between the state, who is in charge of taking measures to restore the environment, and private individuals who are called to contribute financially to these measures, typically ex post. **143**

280 *Koch* (fn. 2) nos. 210 ff.

281 E.g. ES no. 39; HU no. 48; MT nos. 12, 35, 51, 60; PL no. 9. See *Faure/Grimeaud* (fn. 272) 235 ff. on the pros and cons of compensation funds in the context of environmental liability.

XIII. Harmonisation needed?

144 As in the previous study,[282] let us briefly consider whether the current state of the laws in the EU calls for harmonisation of at least certain segments of tort law.

145 An assessment of the existing situation undeniably shows quite a colourful picture – there are ample differences between the tort laws of the Member States which seem to indicate that cross-border cases in particular will meet considerable difficulties.

146 However, unlike the scenarios examined in the first project, two core areas of liability that are relevant for the cases under survey here are already harmonised anyhow. This is particularly true for product liability, where the ECJ never tires of insisting that the PLD provides for an exclusive regime with no permissible national deviations.[283] But also the consequences of environmental harm caused by identifiable individuals have been subjected to a harmonised regime, even though it was shifted from private to administrative law, at the same time disregarding secondary losses caused to third parties. The solutions foreseen by the ELD are much more flexible than the PLD regime, which has led to quite some diversity between the Member States that have already implemented it.[284] However, one could at least argue that the EU legislator has crossed off this item on its agenda, after exhaustive and exhausting political wrangling. Seen through rose-coloured spectacles, one could also claim that this is what the legislator actually wanted to achieve. In any case, there is no indication that this discussion will be resumed in Brussels in the nearer future.

147 As to the remaining areas, the question marks indicated in the previous study are also valid here: To begin with, one would have to ask whether harmonisation is feasible at all.[285] If the solution strived for is anything less than a full unification of tort law as a whole, which is entirely unlikely for the time being in light of the recent failure of contract law harmonisation, any singular scheme only applicable to GMO-related losses will necessarily be an alien element to most jurisdictions, as it will have to disregard certain peculiarities of the local regimes which are crucial to the national claims and compensation culture.[286] This inevitably lays the

282 *Koch* (fn. 2) nos. 222 ff.
283 Supra no. 84.
284 Supra no. 95.
285 *Koch* (fn. 2) nos. 233 ff.
286 "A failure to thoroughly examine existing legal standards can lead to internal conflicts in the law, unjustified discrimination against particular activities, and general confu-

groundwork for a differing application of any supposedly unified European regime. The use of unavoidable terminology alone is a predetermined breaking point[287] – not to mention procedural differences which can have a decisive influence on the outcome of a case despite substantive law parallels.

Furthermore, also in this context it is at least questionable whether the internal market is truly negatively affected by the existing diversity.[288] While there admittedly is more international activity along the food and feed supply chains as compared to the rather local problem of economic losses caused to neighbouring farmers, these transnational legal problems for the most part seem to lie in the domain of product liability, and if not, the remainder primarily seems to be a question of contractual liability which is being disregarded here. Since the general set-up for cross-border tort claims is dominated by EU regulations in the meantime, one can no longer argue that, say, the applicable law in a given lawsuit is unpredictable – in contrast, it is clearly foreseeable which law under Rome II will govern the non-contractual liability matters addressed in this study, and even though the target jurisdictions may not provide for entirely uniform solutions, they are at least identifiable upfront, which allows for sufficient planning for all participants in cross-border trading. There is no apparent reason why the food or feed supply chain differs so dramatically from other industries (or why its participants deserve more protection) that it should call for special treatment. **148**

sion." *R. Lattimore*, Guide for Countries Considering Liability and Redress for LMOs, in CropLife International (ed.), Compilation of Expert Papers concerning Liability and Redress and Living Modified Organisms: A Contribution to the Article 27 Process under the Cartagena Protocol on Biosafety (2004, available at http://www.croplife.org/library/attachments/1da4b9ac-a8ec-48d3-a896-686c375b8b0a/7/2004 1 26-DOC-Independant Handbook-draft-e.pdf) 43.

287 See e.g. fn. 149. Cf. IN no. 16: "Harmonisation, however, is disadvantageous for liability and redress, since there are too many definitions of harm, understandings of hazard manifestation and admissible causes of action.".

288 *Koch* (fn. 2) nos. 242 ff. See also the outcome of an economic study submitted in preparation for the Environmental Liability Directive, which concluded that "[i]t seems unlikely ... that existing liability systems in EU Member States are currently creating any significant distortion of trade": *ERM Economics*, Economic Aspects of Liability and Joint Compensation Systems for Remedying Environmental Damage (Summary Report), Annex 2 to the Commission's White Paper on Environmental Liability, COM(2000) 66 final, 9.2.2000 (available at http://ec.europa.eu/environment/legal/liability/pdf/el_full.pdf) 37, 39.

XIV. Conclusions

149 "A survey of all EU Member States shows considerable differences between the various ways how non-GM farmers may be compensated for their economic losses resulting from the admixture of their crops with GMOs stemming from an adjoining field."[289] The same holds true for other losses that GMOs may cause along the food or feed supply chain.

150 However, as in the previous study, the combined country reports show that there are no lacunae in the laws regarding losses caused by GMOs. All provide ample solutions for such problems; they are simply not equally favourable for the victims. Success for the latter, however, is not an indispensable requirement of tort law after all. The question remaining therefore is whether there is a true need for an enhanced protection of potential victims. This is above all a political question, even though there are objective arguments that can be raised in favour of shifting losses, starting with unfavourable conditions for proving causation and the special qualities of GMOs in agriculture.

151 In core areas for the purpose of this study, the European legislator has already answered this question by introducing the Product Liability Directive, and by providing for a special administrative regime handling losses to the environment. Further uniform rules deviating from the respective national liability standards do not seem to be imperative in order to provide for the protection of third parties that may be harmed by GMOs in the future, particularly not for GMOs that have already been approved for food or feed production. After all, this would have to extend to all food- or feedstuffs, and ultimately to all commercial activities and their output, which all may some day turn out to be the source of a risk that presently seems to be speculative at best.

152 A few countries currently foresee strict liability specifically for the commercial use of GMOs,[290] although surprisingly not the countries which probably have some of the strictest liability regimes for potential economic losses of neighbouring farmers due to admixture.[291]

153 There are indeed several arguments that would speak in favour of doing without the requirement of fault when establishing liability for GMOs, not just economic ones.[292] Strict liability is the "classic" response by tort

289 *B.A. Koch*, Conclusions and Recommendations, in Economic Loss, 653 (no. 1).

290 Supra nos. 62 ff.

291 Austria and Germany only foresee strict liability for the R&D application of green biotechnology, but not for the commercial cultivation and distribution; see supra no. 69.

law to novel technologies, ever since the Industrial Revolution. In exchange for permitting activities whose consequences are not yet entirely foreseeable and which might entail risks to third parties, those who profit from the activity in turn have to compensate losses should they materialise[293] even if they did all that could be expected from them in light of the state of knowledge at the time of their conduct. A related argument is that also others who benefit at least indirectly from the advancement of science and technology should contribute to losses that individuals sustain, which is achieved by way of insurance cover that the user of the technology buys and whose costs are spread evenly by way of the price mechanism.[294] Furthermore, those who are allowed to conduct an activity whose risks are not entirely foreseeable yet are at the same time presumed to have the best knowledge and control of the technology; they are under a duty to monitor it in practice and are the first who could intervene if a specific danger becomes evident.[295] All this would only be relevant if the victims can prove that their losses indeed originated from the activity pursued by the defendant, so causation is an indispensable element of the claim even if fault is not of the essence.[296]

The more common the use of modern technology gets and the more information is gathered on potential hazards it may bring about, the more one can do to prevent the materialization of those risks, which in turn makes the technology safer. In those cases, one might reconsider applying strict liability, and the fault requirement may be reintroduced over time. This is actually the experience in the U.S., where, for example, airplanes were originally subjected to a strict liability regime, whereas meanwhile, "as aviation became a safer activity, the trend toward strict liability is reversed, and in most states liability for aviation ground damage is now subject to a negligence standard".[297] **154**

Irrespective of which liability model a jurisdiction selects for the use of GMOs in agriculture, including the choice not to deviate from the classic **155**

292 See EA nos. 5 ff.

293 "The old tort law principle of *cuius commodum, eius damnum* can certainly be regarded as an equally traditional rule underlying strict liability". *Koch/Koziol* (fn. 138) no. 71.

294 *Koch/Koziol* (fn. 138) nos. 72, 76.

295 *Koch/Koziol* (fn. 138) nos. 58 ff.

296 Depending on the circumstances, however, courts or legislators might be willing to soften the requirement of proving causation as described supra nos. 38 ff.

297 *G. Schwartz*, United States, in B.A. Koch/H. Koziol (eds.), Unification of Tort Law: Strict Liability (2002) 351 (no. 17). This is also one of the explanations for the "common usage" requirement of strict liability under the Restatements on Torts: Restatement (Third) of Torts: Liability for Physical Harm (Basic Principles), § 20 cmt. j.

fault regime, the call for legislative action is louder when it comes to technical or administrative regulations that define standards of care for operators along the supply chain. Such rules undoubtedly have a much more significant impact upon promoting coexistence than liability rules or alternatives thereto.[298] Even though these only set minimum standards from the perspective of tort law,[299] they are still important criteria for a judge who may have to retroactively define duties of care when called upon in the future to determine who shall bear the risk of GM farming in the past. For the time being, such rules are also useful to clarify the boundaries of permissible conduct at present, thereby also helpful to avert pleas for injunctions.

156 While both first- and third-party insurance options may seem desirable, the lack of experience with GMOs, but also the lack of clarity with respect to the application of liability standards at the moment still seems to prevent insurers from offering a range of suitable products that might satisfy the needs (and concerns) of those who are willing to produce, distribute or process transgenic crops.

157 Compensation funds do not seem to be useful to fill this evident gap on the insurance market, as the same uncertainties apply to their setup. In contrast to existing funds covering economic losses, the range of imaginable damage scenarios and the types of potential losses are too broad to allow devising a similarly tailor-made alternative redress scheme.

---

298 *Koch* (fn. 289) no. 12. There are also economic arguments in support of the finding that "a much more important role will in practice be played by safety regulations than probably by liability rules ..." (EA no. 39).

299 Cf. supra nos. 57 and 110.

Index

The two-letter codes refer to the reports as follows: AT stands for Austria, AU for Australia, BR for Brazil, CA for Canada, CR for the Comparative Report, CZ for the Czech Republic, DE for Germany, DK for Denmark, EA for the Economic Analysis, EE for Estonia, ES for Spain, FI for Finland, FR for France, GR for Greece, HU for Hungary, JC for the report on jurisdiction and applicable law, IE for the International Environmental Law Report, IN for the Insurance Report, IT for Italy, LI for Lichtenstein, LU for Luxembourg, MT for Malta, NL for the Netherlands, NO for Norway, PL for Poland, SE for Sweden, SI for Slovenia, UK for England & Wales, US for the United States of America. The numbers following these codes point to the marginal numbers in these reports.